ADVANCES IN
X-RAY ANALYSIS

Volume 9

ADVANCES IN
X-RAY ANALYSIS
Volume 9

Edited by

Gavin R. Mallett, Marie Fay, and William M. Mueller

**Proceedings of the Fourteenth Annual Conference on
Applications of X-Ray Analysis
Held August 25-27, 1965**

Sponsored by
University of Denver
Denver Research Institute

ℙ

PLENUM PRESS
NEW YORK

Plenum Press
A Division of Plenum Publishing Corporation
227 West 17 Street, New York, N. Y. 10011

ISBN 978-1-4684-7635-4   ISBN 978-1-4684-7633-0 (eBook)
DOI 10.1007/978-1-4684-7633-0

PREFACE

The papers presented in this volume of *Advances in X-Ray Analysis* were chosen from those presented at the Fourteenth Annual Conference on the Applications of X-Ray Analysis. This conference, sponsored by the Metallurgy Division of the Denver Research Institute, University of Denver, was held on August 24, 25, and 26, 1965, at the Albany Hotel in Denver, Colorado. Of the 56 papers presented at the conference, 46 are included in this volume; also included is an open discussion held on the effects of chemical combination on X-ray spectra.

The subjects presented represent a broad scope of applications of X-rays to a variety of fields and disciplines. These included such fields as electron-probe microanalysis, the effect of chemical combination on X-ray spectra, and the uses of soft and ultrasoft X-rays in emission analysis. Also included were sessions on X-ray diffraction and fluorescence analysis. There were several papers on special topics, including X-ray topography and X-ray absorption fine-structure analysis.

William L. Baun contributed considerable effort toward the conference by organizing the session on the effect of chemical combination on X-ray spectra fine structure. A special session was established through the excellent efforts of S. P. Ong on the uses and applications of soft X-rays in fluorescent analysis. We offer our sincere thanks to these men, for these two special sessions contributed greatly to the success of the conference.

We also extend our thanks to those who chaired the sessions. Through their diligent efforts the conference ran on a smooth and efficient schedule. The chairmen were Sheldon H. Moll, Advanced Metals Research Corporation, Burlington, Massachusetts; William L. Baun, Air Force Materials Laboratories, Wright-Patterson Air Force Base, Ohio; S. P. Ong, Philips Electronics, Mount Vernon, New York; A. F. Berndt, Argonne National Laboratory, Argonne, Illinois; W. J. Wittig, Union Carbide Stellite Company, Kokomo, Indiana; Frank Bernstein, General Electric Company, Milwaukee, Wisconsin; and W. V. Cummings, General Electric Company, Pleasanton, California.

Sincere thanks and appreciation go also to Mr. Frank Rivera, who took charge of the meeting aids and audio-visual equipment during the conference, and to Mrs. Mildred Cain, who so effectively transcribed the discussion sessions which then required only a small amount of editing to make them readable.

G. R. Mallett
M. J. Fay
W. M. Mueller

CONTENTS

A CAMERA FOR BORRMANN STEREO
X-RAY TOPOGRAPHS

F. W. Young, Jr., T. O. Baldwin, A. E. Merlini,* and F. A. Sherrill

Oak Ridge National Laboratory
Oak Ridge, Tennessee

ABSTRACT

A relatively inexpensive camera which was designed for taking Borrmann topographs with standard X-ray diffraction equipment is described. This camera has been used to take stereo pairs of Borrmann topographs by rotating the crystal around an axis normal to the diffraction planes. Topographs have been taken of nearly perfect copper crystals up to 0.2 cm thick using silver, molybdenum, copper, and chromium radiation, and comparisons have been made with topographs obtained with crystal monochromated radiation. Geometrical factors affecting the resolving power of the technique are briefly reviewed. In addition, the resolution inherent in the diffraction phenomenon is analyzed on the basis of the theory of anomalous transmission. Comparisons are made between calculated and observed image widths of a few dislocations.

INTRODUCTION

A number of X-ray diffraction techniques are available for the study of individual dislocations in crystals; these techniques and the advantages of X-ray diffraction topography were recently reviewed.[1,2] Topography in anomalous transmission (Borrmann topography) is the one which has been least applied and studied, probably because of its limited application to the few perfect crystals until recently available such as silicon and germanium. In the past few years anomalous transmission was found in crystals of zinc[3] and in nearly perfect crystals of copper,[4] and the interest in Borrmann topography for the study of dislocations in metal crystals was increased. It was believed that anomalous transmission topographs could show with a sufficient degree of sharpness only those imperfections lying very close to the crystal surface from which the diffracted X-rays exited. Experimental evidence and theoretical estimates of the projected size of the dislocation image have shown that this is not the case if the proper choice of the characteristic radiation is made (for example, Ag K_α and Mo K_α for copper crystals). The possibility of seeing dislocations lying at a depth of 0.06 to 0.08 cm from the exit surface[5] is a definite advantage of the technique for the study of thick crystals. Finally, it was believed impossible to obtain satisfactory stereo pairs if a strong Borrmann effect were present. In Lang topography stereo pairs can be obtained from the hkl and $\bar{h}\bar{k}\bar{l}$ reflections and have proved to be very useful for the study of the spatial arrangements of dislocations.[6] Even this objection has been eliminated since Noggle et al.[7] showed that Borrmann stereo topographs could be taken by rotating the crystal a few degrees around the normal to the diffraction planes. This method, which can also be used with thin crystals,[8] has the advantage of versatility with respect to the stereo angle.

* Guest scientist from Euratom, Ispra, Italy.
[1] References are at the end of the paper.

1

A camera for Borrmann topography was briefly described and some examples of topographs were presented in an earlier paper.[5] A camera of improved vertical resolution was subsequently designed. Good quality topographs can be taken with standard X-ray diffraction equipment and a relatively inexpensive camera. The purpose of this paper is first the description of such a camera and of some of the results obtained, and then the presentation of simple rules for the evaluation of the diffraction resolution in Borrmann topography.

PRINCIPLES OF BORRMANN TOPOGRAPHY

The conclusions of the theory of anomalous transmission[9,10] as they apply to X-ray topography can best be discussed with the aid of Figure 1. Suppose an X-ray beam is incident at the Bragg angle with respect to crystallographic planes which are normal to the crystal surface (symmetrical case, Laue geometry). Provided the crystal is nearly perfect, two standing wave fields for each polarization direction of the incident radiation are generated inside the crystal. One wave field has nodes at the atomic planes and is transmitted almost undiminished in intensity, while the other wave field has antinodes at the atomic planes and is highly absorbed. If the crystal is thick [μt (absorption coefficient × crystal thickness) ~ 10 or greater], the net energy flow is very nearly parallel to the diffraction planes, and only the wave field with polarization perpendicular to the plane of incidence and with nodes at the atomic planes is transmitted with appreciable intensity. Only this latter wave field will be considered in the remainder of the paper. At the exit surface the wave field splits into two beams of nearly equal intensity, which we will denote as the anomalously transmitted beam T and the anomalously reflected beam H. The anomalously transmitted beam has the same direction as the incident or direct beam D, but is displaced toward the reflected beam a distance which is proportional to the thickness of the crystal. Whenever the lattice is disturbed from its perfect periodicity, such as is the case for dislocations and other imperfections, the intensity of both beams in Borrmann transmission is normally reduced, and regions of reduced intensity or white images of the imperfections appear on a photographic plate placed near the exit surface of the crystal.

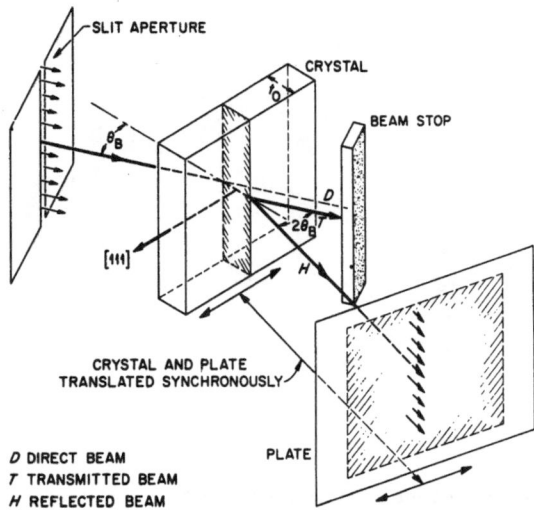

D DIRECT BEAM
T TRANSMITTED BEAM
H REFLECTED BEAM

Figure 1. Principles of the Borrmann technique for X-ray topographs. Stereo topographs are taken by rotating the crystal around the [111] axis.

In order to take topographs of large crystal areas it is necessary to block the anomalously transmitted beam and the direct beam. However, if the crystal is thick, $\mu t \gtrsim 10$, and remains stationary, operating the X-ray tube at relatively low kilovolt values allows one to record both beams emerging from the sample.

EXPERIMENTAL METHOD

The basic components of the camera consist of a translating table, rotary machine table, goniometer head, and a support for the counter and film shield (Figure 2). The translating table used is an Automatic Gages, Inc., Hi-Precision Ball Slide assembly, which according to the supplier has deviations of less than 0.0001 in. per inch of travel. The translating table is mounted on a Universal Vise and Tool Company rotary machine table through a shaft which is fastened rigidly to the center of the bottom slide of the translating table, and which fits into a bushing mounted in the center of the rotary machine table. When the rotary table wheel is turned, the translating table and counter support

1. BEAM SLIT
2. BEAM LIMITING SCREWS
3. FILM SHIELD
4. FILM CASSETTE
5. GEIGER COUNTER
6. TRANSLATING MOTOR
7. TRANSLATING TABLE
8. CRYSTAL
9. GONIOMETER HEAD
10. ROTARY TABLE
11. CONNECTING ARM
12. SHIELD TRANSLATING SCREW
13. GONIOMETER HEAD SHAFT
14. ROTARY TABLE WHEEL
15. DIAL INDICATOR GAGE

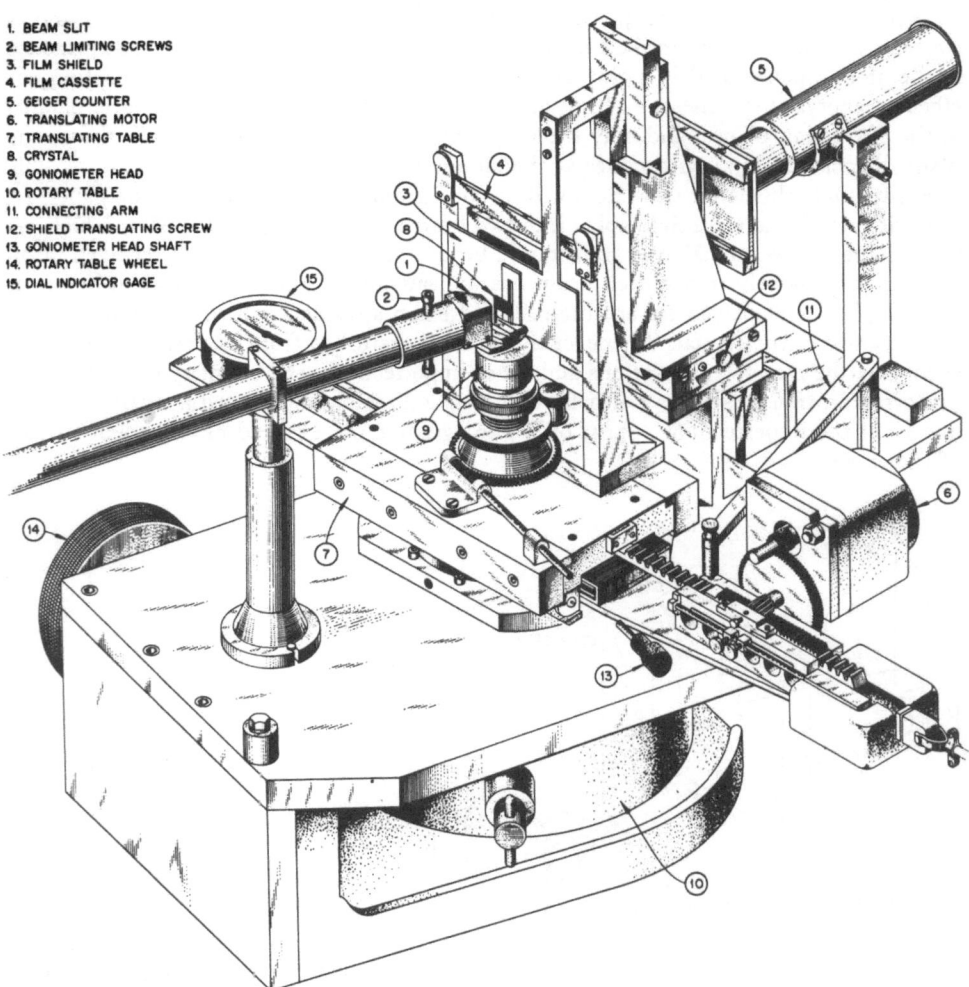

Figure 2. Detailed sketch of a camera for Borrmann topographs.

normally rotate with the same angular velocity; however, by disengaging the connecting arm the translating table can be rotated independently of the counter. The film or nuclear plate holder is fastened to the translating table in such a way that the crystal and photographic plate translate together to record the projection of the image of the dislocations in the crystal at unit magnification. A stationary film shield which serves to stop the direct and anomalously transmitted beam has a vertical lead slit with a height of about 3 cm and a width of 0.1 cm and can be moved both perpendicular and parallel to the diffracted beam.

In order to take topographs of large sections, the crystal is translated by means of a Bodine motor which is connected by a worm gear to the translating table. A microswitch, which serves to reverse the motor, is connected to translation stops which are set to allow the desired crystal area to be traversed. A translation speed of approximately $\frac{3}{16}$ in./min is used. Faster and slower translation speeds were also tested, but they did not affect the overall quality of the topographs. However, frame topographs obtained with a stationary crystal are usually better than the translation topographs, especially if the diffracted intensity is weak. With frame topographs a dial indicator gauge fastened to the bottom side of the translating table is very useful to indicate the investigated portion of the crystal. To find the desired Bragg reflection, first the detector is set at the proper 2θ angle by rotation of the rotary table wheel. The crystal reflection is found successively by rotation of the goniometer head shaft.

Strain-free lamellas of copper were acid-cut from nearly perfect bars of single crystals grown from the melt.[11] Previous to the acid cutting, most of these bars were irradiated in a nuclear reactor to a fast neutron dose of about 10^{17}nvt. This treatment hardened the crystals and it was then easier to handle the lamellas without deforming them.[5] All the copper crystals studied were cut with either the (111) or the (110) planes parallel to the large surfaces. The lamellas were fixed on a goniometer head and their final orientation with respect to the holder was adjusted to a precision of about 0.5° by means of the back-reflection Laue technique. After finding the proper reflection with the crystal in the Laue geometry, the beam stop was inserted and adjusted to allow the reflected beam to go through the 0.1-cm lead slit but to stop the transmitted and direct beams. The translation stops were then set so that the desired portion of the crystal was recorded on the photographic plate. The width of the incident beam was 0.06 cm at the crystal, the best compromise for intensity and resolution requirements. The vertical size of the incident beam was limited by two screws at the end of the beam tube.

For all detailed work, 2 × 2 in. Ilford L4 nuclear plates with 100 micron emulsion thicknesses are used. The plates are loaded in a film cassette covered with black paper or coated Mylar. The Mylar was found to be preferable for chromium radiation, since the texture of black paper produced an irregular background on the plate. For both silver and molybdenum radiation, an Ilford G film is placed behind the plate to give an indication of the plate exposure and to assist in determining the developing time. The 100 micron L4 emulsion absorbs about 60% of the diffracted beam for silver radiation and about 80% for molybdenum. The exposure time necessary to record a topograph is usually 16 hr or greater, depending upon the crystal thickness, the reflection studied, and the area of the crystal exposed. For instance, to record a 111 topograph of a 1 × 1 × 0.035 cm nearly perfect (< 10^3 dislocations/cm²) copper crystal using silver, molybdenum, or copper radiation requires a 24-hr exposure. Figure 3 is an example of such a topograph; most of the topographs which follow are enlarged sections of a topograph similar to the one in Figure 3. All of the topographs in this paper are positives; i.e., dark regions on the print correspond to regions of less X-ray intensity on the plate.

The photographic plates are processed in a special developer by the Eastman Kodak

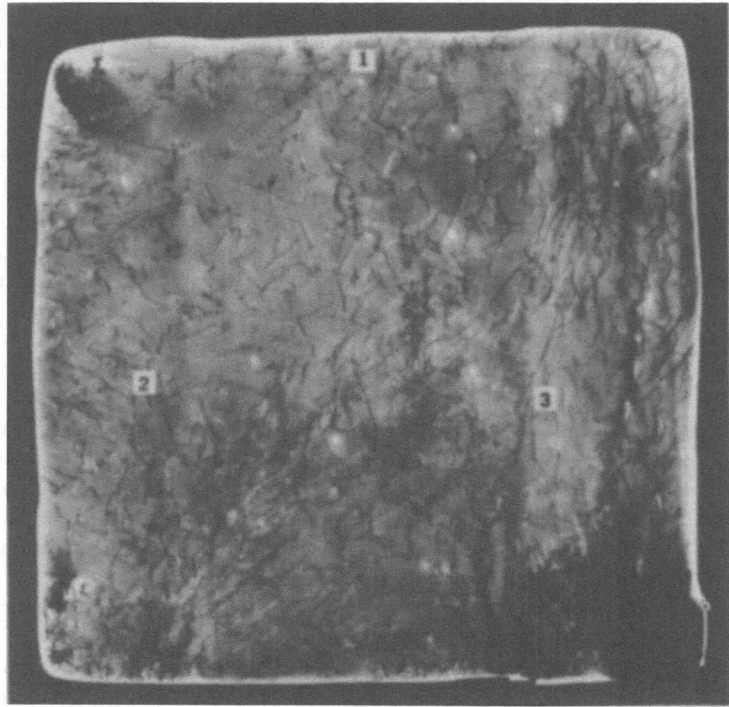

Figure 3. A 111 topograph of a 1 ×1 ×0.035 cm copper crystal using molybdenum radiation. The numbers 1, 2, 3 in the photograph refer to dislocations immediately to the right of the respective numbers.

Company designated as M-AA-1 (Metol-ascorbic acid). The formula for this developer* is Metol (Elon), 2.5 g; ascorbic acid, 10.0 g; potassium bromide, 1.0 g; Kodalk-balanced alkali, 35.0 g; and water to make 1.0 liter. (Either L-ascorbic acid or d-arascorbic acid can be used.)

The plates are developed with stirring for about 20 min in a 5°C developer solution, washed in a 1% acetic acid solution for about 10 min, fixed for $1\frac{1}{2}$ hr in a plain, non-hardening fixing bath (hypo and water), and then washed for about 2 hr in water. The resulting grain size is such that a topograph can generally be usefully enlarged 100 times, although some topographs have been enlarged as much as 600 times. As Lang[12] has pointed out, the useful enlargement will also depend upon the wavelength of the radiation used for the exposure, since the track lengths of the photoelectrons in the emulsion will increase considerably with decreasing wavelength.

RESOLUTION

The only collimation requirements for high-resolution anomalous transmission topographs are those necessary to limit the vertical divergence and to resolve the K_α doublet. With a standard X-ray diffraction tube, both of these requirements can be met

---

* The authors are indebted to Dr. T. H. James of the Eastman Kodak Research Laboratories for making this formula available to us.

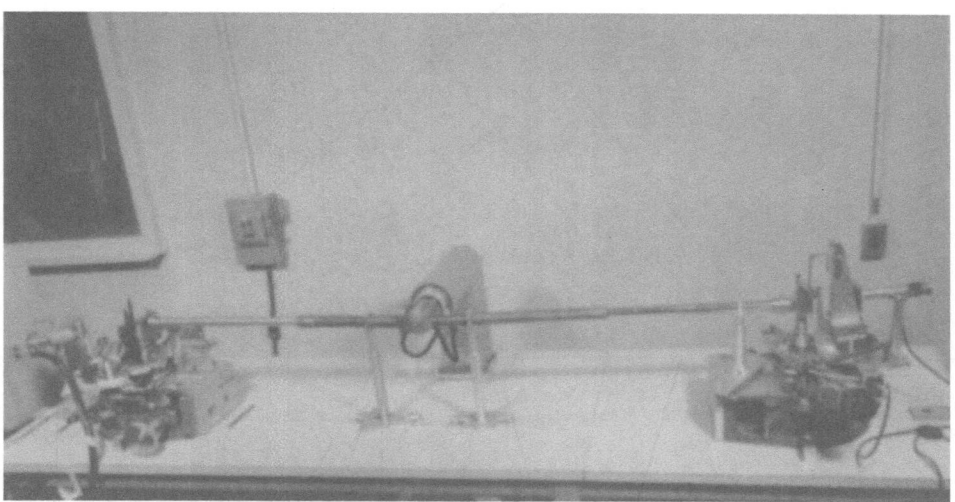

Figure 4. A photograph of the two Borrmann cameras used with a single GE-CA7 X-ray tube.

using a large source-to-sample distance. Figure 4 shows two of these cameras placed around a General Electric CA-7 X-ray tube. To limit the vertical divergence, the spot source (vertical dimension is 0.08 cm) is used, the source-to-crystal distance is 100 cm, and the crystal-to-nuclear-plate distance is 0.5 cm. The resulting vertical resolution W' is about 4 microns [$W' = (B/D)D'$, where B is source size, D is source-to-crystal distance, and D' is crystal-to-plate distance]. The apparent width of the focal spot (0.06 cm, with a takeoff angle of about 3°) and the simple slit of width 0.06 cm in front of the crystal permit the separation of the K_α doublet. For low-order reflections the chromatic width of the K_{α_1} line has a negligible effect on resolution.

The horizontal resolution is now determined by the phenomenon of anomalous transmission. An incident beam of horizontal extent L is broadened in anomalous transmission through a perfect crystal by the Borrmann delta (see Figure 5). This effect arises because the divergence of the incident beam is greater than the half-width $\Delta\theta$ of the anomalously reflected beam, thereby producing wave fields which are propagated in all directions within the 2θ span. Since the effective absorption coefficient decreases from μ at the 2θ extremes to $\mu(1 - \epsilon)$ along the diffraction planes, Borrmann delta Δ_B is less than 2θ and is given for the symmetrical case by[9,10]

$$\Delta_B \simeq 2 \text{ arc tan} \left[\left(\frac{2 \ln 2}{\mu t \epsilon + 2 \ln 2} \right)^{\frac{1}{2}} \cdot \tan\theta \right]$$

where μ is the absorption coefficient, t is crystal thickness, ϵ is the ratio of the imaginary parts of the scattering factors, and θ is Bragg theta. (This formula is based on the supposition that only intensities greater than those at half-maximum need be considered.) Assuming that a defect is a volume of diameter S in the crystal which scatters the wave field, the defect will cast a shadow (on the exit surface of the crystal and hence on the plate), and its horizontal width W will be determined by

$$W = 2Z \left(\frac{2 \ln 2}{\mu t \epsilon + 2 \ln 2} \right)^{\frac{1}{2}} \cdot \tan\theta + S$$

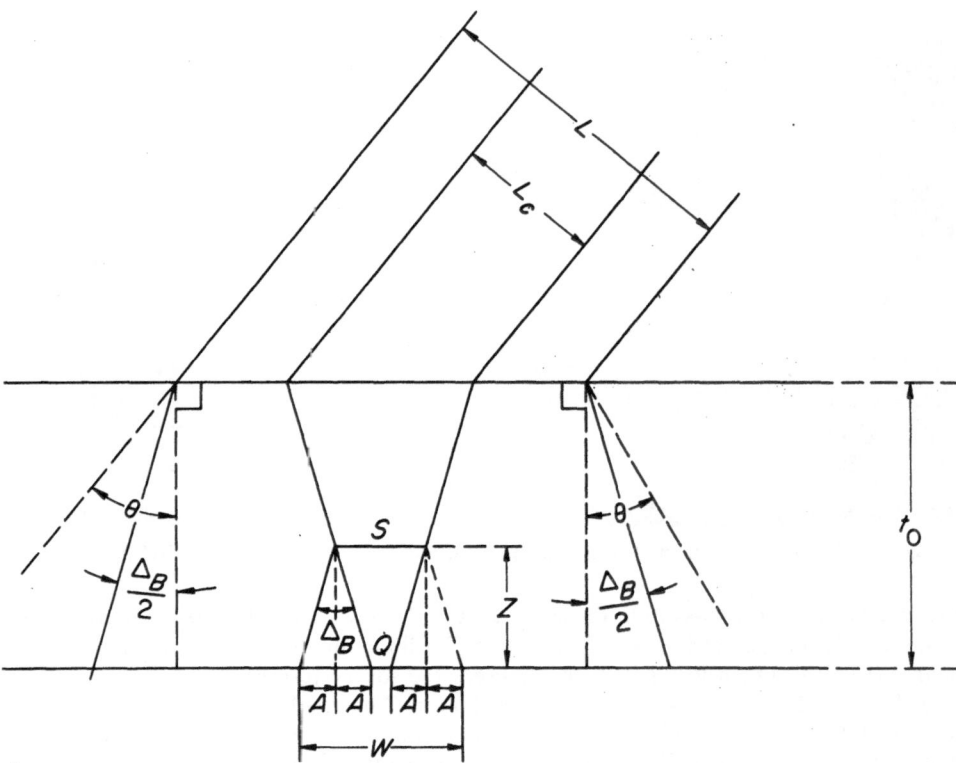

Figure 5. Effect of Borrmann delta upon image widths.

where Z is the distance from the exit surface. From Figure 5 it can be seen that the defect image W should consist of an intense part $Q \lesssim S$ and a diffuse part $W - Q$. The value Q is a function of Z and Δ_B. It should be remembered that this is horizontal resolution only, pertaining, for example, to a dislocation lying parallel to the vertical direction. It is apparent from an inspection of Figure 5 that for $L > L_c$ the extent of the incident beam does not affect the divergence, while if $L < L_c$ the resolution depends on L, and the effect of the Borrmann delta on horizontal resolution can be considerably decreased if a beam size $L \sim S$ is used. The angles in Figure 5 are exaggerated; for our slit width of 0.06 cm, $L \gg L_c$. However, additional slits, limiting either the vertical or horizontal width of the beam, have been used in special studies of resolution.

A simple rule for the visibility of a dislocation on a topograph can also be drawn from the above formula of the dislocation image size. If the dislocation lies at a distance Z' from the exit surface of the X-rays such that

$$S \ll 2Z' \left(\frac{2 \ln 2}{\mu t \epsilon + 2 \ln 2} \right)^{\frac{1}{2}}$$

the dislocation image is smeared out over the background intensity of the topograph and is no longer recognizable as an image.

STEREO TOPOGRAPHY

Lang[6] has described in some detail how stereo pairs of topographs of thin crystals taken with hkl and $\bar{h}\bar{k}\bar{l}$ reflections can be used to observe the arrangement of dislocations and defects in crystals. Since in anomalous transmission, the direction of the energy flow is along the plane of reflection, topographs of hkl and $\overline{hkl}$ reflections should give identical projections. This conclusion was verified experimentally. However, Borrmann stereo pairs can be taken by rotating the crystal around the normal to the diffraction planes (such as the 111 planes in Figure 1). Thus, in the stereo pair shown in Figure 6, the topograph on the left was made with the crystal rotated counterclockwise 5° about the normal to the diffraction planes, while the topograph on the right was made with the crystal rotated clockwise 5° about the normal. This stereo pair can best be viewed with the aid of two converging lenses with a planar separation of about 6 cm.

As is evident from examination of Figure 6, stereo topographs are also very useful in the study of relative positions of defects in thick crystals. In fact, they are nearly essential for crystals with relatively high dislocation densities (greater than 10^4 dislocations/cm^2). Stereo topographs provide the only clear way to establish the difference between the intersection of two dislocations and the change in direction of a single dislocation; this feature can be seen in Figure 6 by comparing a single topograph with the stereo pair. Furthermore, stereo topographs give clear evidence of the effect of Borrmann delta on W. Images of dislocations that are inclined steeply to the exit surface fan out quickly and become diffuse by the time they reach the entrance surface.

Figure 7 is a stereo topograph of a copper crystal containing a relatively small dislocation density (100 dislocations/cm^2), but a large number of black spots (4×10^5/cm^2). These black spots are formed during crystal preparation and are seen by stereo to be more or less uniformly distributed throughout the crystal. Although these defects have not been positively identified, some are believed to be vacancy loops.

The images of defects which lie near or intersect a surface of the crystal generally show "anomalous contrast." This is defined as contrast which reverses between the transmitted and diffracted beams, a definition analogous to that used in electron microscopy. (The images of defects well within the crystal show the same decrease in intensity in both beams.) The stereo pairs have shown conclusively the correlation between anomalous contrast images and the position of the defects relative to the crystal surfaces.

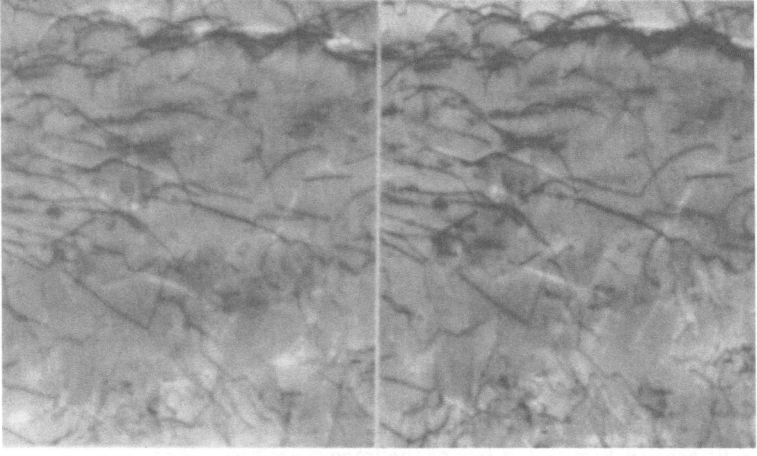

Figure 6. A 111 stereo pair of the crystal in Figure 4. The left photograph is rotated 10° about the [111] axis with respect to the right photograph. Dislocations that are sharp and appear nearer the viewer lie near the exit surface.

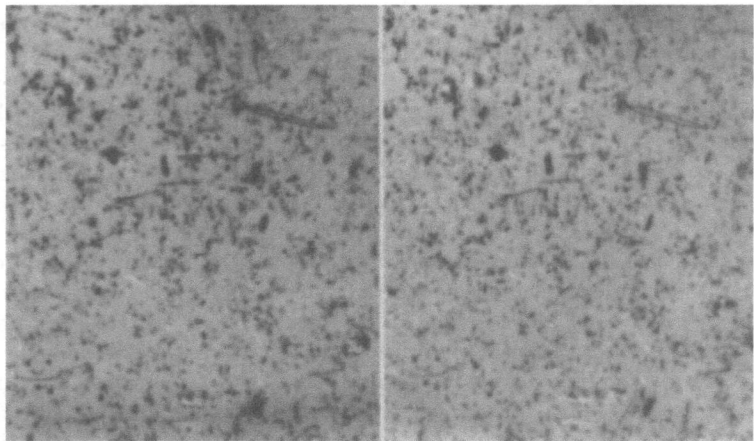

Figure 7. A 111 stereo pair (taken as in Figure 8) emphasizing the black spots and anomalous contrast.

If dislocations show no anomalous contrast they probably lie a distance $> \mu^{-1}$ from the surface.

CONCLUSIONS

In order to compare the resolution and sensitivity to strain of crystal collimated radiation with that obtained with the camera described here, a few double crystal topographs were taken using apparatus described previously.[13] Shown in Figure 8 are double crystal topographs of 111 parallel, 111 antiparallel, and 222 parallel positions along with a topograph taken with the camera described here, all taken with Mo K_{α_1} radiation of a crystal 0.035 cm in thickness. It is apparent that the resolution and sensitivity to strain is comparable between the double-crystal and single-crystal methods, as expected. The image widths were very nearly the same for the 111 topographs in contrast to those in topographs made using Bragg reflections, for which crystal-collimated radiation produces a greater sensitivity to strain than that obtained with uncollimated radiation.[14] In general the 222 topograph, Figure 8, shows both increased image widths and more contrast than the 111 topographs. This was generally found to be true also with the camera described here. The increased image widths probably result from the Borrmann delta, but may also indicate an increased sensitivity to strain for higher order reflections.

The image widths W of dislocation lines in the crystal shown in Figure 3 were measured and data for three lying parallel to the vertical direction are given in Table I; measurements were made on the nuclear plates with the aid of an optical microscope. Dislocations lying perpendicular to the vertical direction have widths smaller than vertical dislocations and they are not affected by Δ_B. The width increases and the image contrast decreases with increasing vertical divergence. Since there is no accurate method for measuring Z, only dislocations lying approximately in the middle of the crystal (as determined by taking topographs for which the entrance and exit surfaces are reversed), such as no. 1, and dislocations lying near the exit surface, nos. 2 and 3, are considered here. As can be seen in Figure 3, a few of the images are black (nos. 2 and 3 are examples) as opposed to the gray color for the remainder. These two dislocations are also black in topographs taken with Mo K_{α_1} and Cu K_{α_1}.

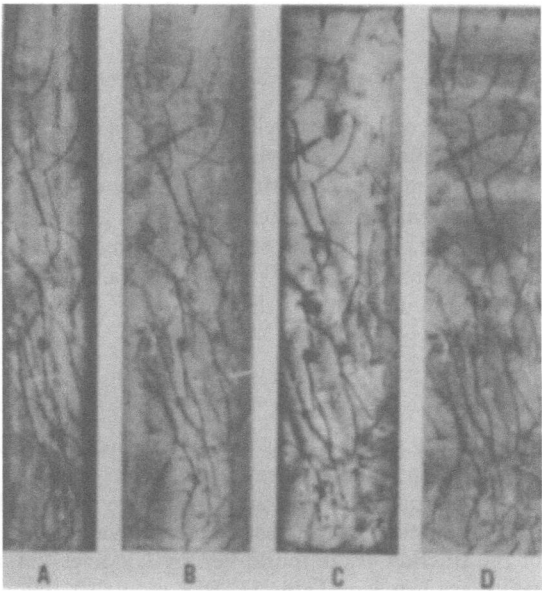

Figure 8. A comparison of topographs obtained using different techniques. (a) 111 Borrmann topograph of the crystal shown in Figure 3 obtained using a double crystal spectrometer aligned in the antiparallel (1, 1) position, (b) same as (a) except with a (1, −1) parallel arrangement, (c) same as (a) except in the (2, −2) position, (d) a topograph of the same section of the crystal obtained with the Borrmann camera. In all of these section topographs the X-rays entered the side of the crystal opposite to that used for the topograph shown in Figure 3.

Table I. Experimental Width in Microns of the Dislocation Images for Three Characteristic Wavelengths*

Dislocation	g·b	g × n	W Ag K_{α_1}	W Mo K_{α_1}	W Cu K_{α_1}
1	0.82	1	26 ± 5	27 ± 5	50 ± 10
2	0.82	0.98	6 ± 1	6	6
3	0.82	1	6	6	6

* The values **g**, **b**, and **n** are vectors parallel to the normal to the diffracting planes, to the Burgers vector, and to the dislocation direction, respectively.

Making use of the model indicated in Figure 5 and assuming that, since dislocations 2 and 3 show no anomalous contrast, $Z \approx \mu^{-1}$ (for Ag K_α $\mu^{-1} = 45$ microns), the image widths listed in Table I can be explained. From the observation that the images are black, the obvious supposition is that they correspond to the Q part (Figure 5). Inspection of Figure 5 shows that

$$S = 2A + Q$$

where

$$A = Z\left(\frac{2\ln 2}{\mu t\epsilon + 2\ln 2}\right)^{\frac{1}{2}}\cdot\tan\theta$$

hence the values for S for the different wavelengths can be determined and are shown in Table II ($\mu t\epsilon \simeq 8$ for silver and 16 for molybdenum and copper). The values of Δ_B and $\Delta\theta$ can be calculated from the dynamical diffraction theory and are included in Table II. It might be noted in passing that these values are in good agreement with ones obtained experimentally on this crystal. Now if the strain associated with an edge dislocation is estimated at a distance $S/2$ from the core[15] and this strain is equated to the angle of lattice curvature $\delta\theta$ at this point, it follows that

$$\delta\theta \simeq \frac{|b|}{12S}$$

where b is the Burgers vector, and the $\delta\theta$ at the extremes of S are listed in Table II.

The propagation directions of the wave fields have a span Δ_B corresponding to a beam divergence $\Delta\theta$. We suppose that the wave fields are bent with the lattice planes, with a curvature of $\delta\theta$ in the lattice planes producing a curvature of Δ_B in the wave fields. Then the product of the fraction $\delta\theta/\Delta\theta$ multiplied by Δ_B gives the angle of bend of the wave field. These values are listed in Table II. It is seen that this angle of deviation of the wave field is essentially the same for all the three radiations, thus accounting for Q, and hence W, having the same value in all three topographs. For $Z = 45$ microns, the lateral displacement at the exit surface of the image of a point at the extreme of S is about 0.5 micron, a very plausible figure.

The image for dislocation no. 1 is diffuse, the contribution from Δ_B being the greatest part. Since this dislocation is approximately in the middle of this crystal, $Z = 0.0175$ cm. The value W was calculated using the formula above and the values for S listed in Table II. The calculated values, listed in Table II, compare favorably with the experimental ones (the large experimental error limits indicate the diffuseness of the images). While the numerical agreement between the image widths obtained from the model and the experimental ones may be somewhat fortuitous, it appears that this model provides a basis for interpreting image widths in Borrmann topographs.

The increase in image width with distance of the defect from the exit surface, resulting from Δ_B, limits the crystal thickness which can usefully be examined by Borrmann topography. Since there is no change in the Borrmann delta between Ag K_α and Mo K_α radiation, the topographs for these two radiations are similar. But between Mo K_α and Cu K_α, there is a large increase in Δ_B, and the correspondingly large effect on the image width is apparent in Figure 9. It is difficult to trace the dislocations all the

Table II. Computations Concerning Image Widths of Dislocations in Table I

Radiation	$\Delta_B(°)$	$2A$ (microns)	S (microns)	$\delta\theta$ (sec)	$\Delta\theta$ (sec)	$\left(\frac{\delta\theta}{\Delta\theta}\right)\Delta_B$	Dislocation 1 W calculated (microns)
Ag	6	4.5	10.5	0.4	3	0.80	27
Mo	5.5	4.3	10.3	0.4	3	0.70	28
Cu	13	10.0	16	0.25	6.6	0.50	55

Table header spanning "Dislocations 2 and 3" over columns $\Delta_B(°)$ through $\left(\frac{\delta\theta}{\Delta\theta}\right)\Delta_B$.

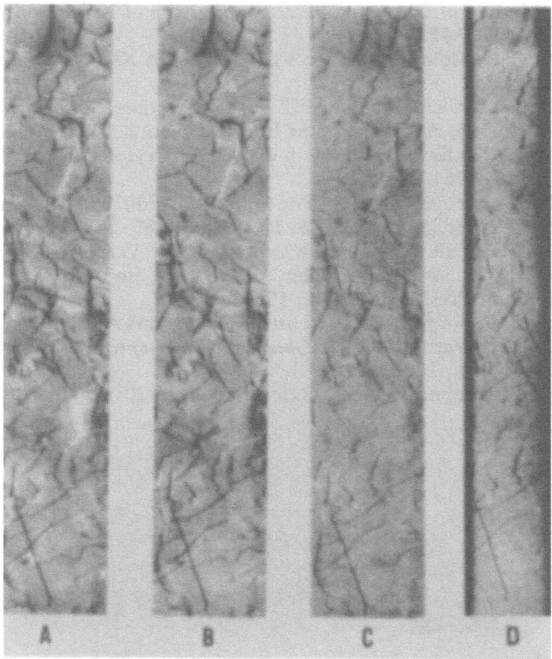

Figure 9. A comparison of topographs obtained using different characteristic radiations. (a) Ag K_α, (b) Mo K_α, (c) Cu K_α, (d) Cr K_α.

way through this crystal ($t = 0.035$ cm) in the topographs taken with copper radiation, while the corresponding thickness for silver radiation is about 0.08 cm.

ACKNOWLEDGMENTS

The performance of the Borrmann camera was made possible by the excellent workmanship of A. C. Kimbrough. Also, we are indebted to B. F. Day for great care in preparing prints of the topographs. This research was sponsored by the U.S. Atomic Energy Commission under contract with the Union Carbide Corporation.

REFERENCES

1. W. W. Webb, "X-Ray Diffraction Topography," in: J. B. Newkirk and J. H. Wernick, *Direct Observation of Imperfections in Crystals*, Interscience Publishers, Inc., New York, 1962, p. 29.
2. A. R. Lang, "X-Ray Diffraction Topography," in: G. L. Clark, *Encyclopedia of X-Rays and Gamma Rays*, Reinhold Publishing Corp., New York, 1963, p. 1053.
3. A. Merlini and S. Pace, "Anomalous Transmission in Zinc Crystals," *Nuovo Cimento, Suppl.* **1**: 531, 1963.
4. M. C. Wittels, F. A. Sherrill, and F. W. Young, Jr., "Anomalous Transmission of X-Rays in Copper Crystals," *Appl. Phys. Letters* **2**: 127, 1963.
5. F. W. Young, Jr., F. A. Sherrill, and M. C. Wittels, "Observation of Dislocations in Copper Using Borrmann Transmission Topographs," *J. Appl. Phys.* (in press).
6. A. R. Lang, "Studies of Individual Dislocations in Crystals by X-Ray Diffraction Microradiography," *J. Appl. Phys.* **30**: 1748, 1959.
7. T. S. Noggle, B. F. Day, F. A. Sherrill, and F. W. Young, Jr., "Stereo Images from Borrmann X-Ray Topographs of Copper Crystals," *Bull. Am. Phys. Soc.* **10**: 324, 1965.

8. K. Haruta, "New Method of Obtaining Stereoscopic Pairs of X-Ray Diffraction Topographs," *J. Appl. Phys.* **36**: 1789, 1965.
9. B. W. Batterman and H. Cole, "Dynamical Diffraction of X-Rays by Perfect Crystals," *Rev. Mod. Phys.* **36**: 681, 1964.
10. R. W. James, "The Dynamical Theory of X-Ray Diffraction," in: F. Seitz and D. Turnbull, *Solid State Physics*, Vol. 15, Academic Press Inc., New York, 1963, p. 55.
11. F. W. Young, Jr. and J. R. Savage, "Growth of Copper Crystals of Low Dislocation Density," *J. Appl. Phys.* **35**: 1917, 1964.
12. A. R. Lang and M. Polcarova, "X-Ray Topographic Studies of Dislocations in Iron Silicon Alloy Single Crystals," *Proc. Roy. Soc. (London) Ser. A* **285**: 297, 1965.
13. M. C. Wittels, F. A. Sherrill, and A. C. Kimbrough, "A Vertically Rotating Double-Crystal X-Ray Spectrometer," in: W. M. Mueller, G. Mallett, and M. Fay, *Advances in X-Ray Analysis*, Vol. 7, Plenum Press, New York, 1964, p. 265.
14. U. Bonse, "X-Ray Picture of the Field of Lattice Distortions Around Single Dislocations," in: J. B. Newkirk and J. H. Wernick, *Direct Observation of Imperfections in Crystals*, Interscience Publishers, Inc., New York, 1962, p. 431.
15. J. Friedel, *Dislocations*, Pergamon Press, Oxford, 1964, p. 20.

A MODIFICATION OF THE SCANNING X-RAY TOPOGRAPHIC CAMERA (LANG'S METHOD)

Mitsuru Yoshimatsu and Atsushi Shibata

Rigaku Denki Company Ltd.
Haijima, Akishima, Tokyo, Japan

and

Kazutake Kohra

University of Tokyo
Bunkyo-ku, Tokyo, Japan

ABSTRACT

A modification of the scanning X-ray topographic camera is reported. The specimen and photoplate are traversed back and forth independently in two directions rather than in the same direction as in the case of Lang's camera. The distortions of the photographs caused by geometrical arrangement can be eliminated through this construction so as to have a one-to-one correspondence. Examples of reflection photographs as well as transmission photographs are shown. Some of them are compared with those taken using Lang's camera. The dislocation images in the reflection photographs show a good one-to-one correspondence to those in the transmission photographs. The broadening of the dislocation images in the traversing direction is discussed. The present camera is especially useful for the studies of lattice defects in the thick specimens because the reflection photographs can be easily taken.

INTRODUCTION

X-ray diffraction topography has been used recently by many workers[1-3] for the study of imperfections in crystals. Among various methods, the scanning method developed by Lang[4] has been found to be one of the most convenient. In Lang's method the plate and specimen are usually traversed in the direction parallel to the surface of the specimen, and the plate is usually placed perpendicular to the diffracted beam. With this setting, distortion of the photograph or the different magnification between horizontal and vertical direction is in most cases unavoidable. The rate of distortion is given by $\cos\phi$, where ϕ is the angle between the specimen surface and the plate. Such image distortion is experienced especially with reflection photographs from specimen surfaces, where ϕ becomes unusually large. In the transmission case, which the Lang method used exclusively, heavy image distortion is often experienced, also.

It is very inconvenient to compare the photographs taken using various net planes of a specimen with one another because the magnitude of the distortion varies with the

[1] References at the end of the paper.

14

net plane used and one-to-one correspondence is not obtainable. Often such an in-convenience is experienced when the reflection photographs are taken from the surface of the specimen. This is one of the reasons why reflection photographs have seldom been taken with Lang's camera. If the photoplate and the specimen are traversed independently along their own surfaces, the distortion of the image will be avoided and a one-to-one correspondence of the image will be obtained. We have constructed a camera satisfying this condition, and as preliminary studies have taken some diffraction topographs of transmission as well as reflection cases.

INSTRUMENTATION

In Figure 1 the principles of the usual Lang method and of the present method are compared. In the present method, (b) and (c), the photoplate (P) and the specimen (C) are traversed back and forth in two independent directions with the same velocity and parallel to their own surfaces, while in the Lang method (a), they move together in one direction parallel to the specimen surface. It is possible with the present construction to take transmission photographs as in the case of the Lang method, and further to take reflection photographs from the specimen surface without distortion of the defect images. The reflection case corresponds to the Berg–Barrett method, but in the present method the separation of K_{α_1} from K_{α_2} is made so that higher resolution is obtained.

Figure 2 is a view of the modified camera. A goniometer head (center) is mounted on the specimen traversing mechanism. A photoplate cassette and its traversing mechanism (right) can be rotated around the specimen axis. A pulse motor is used for each traversing mechanism. One pulse makes a 1 mμ increment in this system, and accordingly the relative position between the specimen and the plate can be maintained with an accuracy of the order of 1 mμ. The maximum traversing distances are 25 mm for both the specimen and the plate. The camera is used in conjunction with the X-ray tube (left in Figure 2) having a fine focus of 40 × 40 mμ in effective size. The photoplate used is Ilford Nuclear Research Plate, emulsion type L_4.

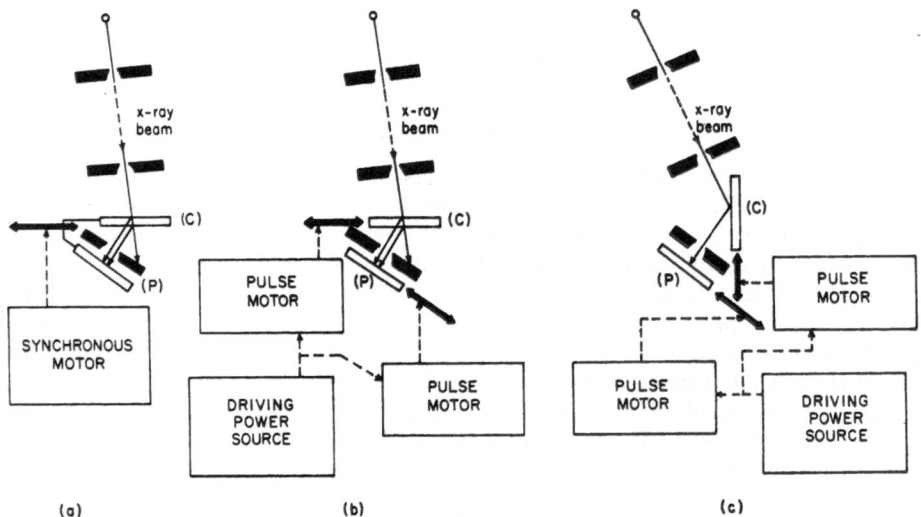

Figure 1. Comparison of the principles of the Lang camera and the present modified camera. (a) Lang camera, (b) modified camera, transmission photography, (c) modified camera, surface reflection photography; (C) specimen, (P) photoplate. Solid arrows indicate traversing directions.

Figure 2. A view of the modified type camera.

EXAMPLES OF PHOTOGRAPHS

In Figures 3 and 4, transmission diffraction photographs taken with the Lang method (left) and the present method (right) are compared. The specimen is a silicon single crystal of about 0.65 mm in thickness. The (111) (Figure 3) and (220) (Figure 4) reflections with Ag K_{α_1} radiation were used. While the magnification in the vertical direction is the same for both methods, the differences in magnification in the horizontal direction is about 90% and 40% in Figure 3 and 4 respectively. Figures 5 and 6 are enlarged photographs of a part of Figure 3 and of Figure 4, respectively, indicating the shortening of the dislocation images (left) in the traversing or horizontal direction. It is to be noted that the light and dark structure of the image is emphasized in the photographs on the right. The dislocation images in Figure 5 have almost the same sharpness in both left and right photographs. This shows that any serious vibration of the apparatus which would cause blurred images does not exist, even with the present use of pulse motors. The dislocation images in the right photograph of Figure 6 are broader in the traverse or horizontal direction than those in the left photograph. This broadening is due to the traversing system of the present modified camera and it is possible to reduce it to some extent with the use of a smaller incident beam aperture, as will be mentioned later.

Next, some photographs taken from the surface of the specimen (reflection case) are shown. For reference, a section pattern is shown in Figure 7 with a schematic diagram. The symmetrical (111) reflection of a silicon crystal with Ag K_{α_1} is used. The straight line on the left edge of the image corresponds to the reflection from the surface. The extension of the image toward the right is due to the reflection from the inner part below the surface. The image is wedge-shaped at top and bottom due to the decrease of the effective thickness, which occurs because the beam in the incident direction is transmitted through the side surface before reaching the lower surface parallel to the incident surface. It is to be noted that the images which are probably due to the dislocations lying inside the crystal are shown by the negative contrast against the background in the middle part of the section photograph. In Figure 8 transmission and reflection photographs from the same area of a silicon single crystal are shown. Both photographs were taken with the modified camera. A transmission photograph [Figure 8(a)] was taken of a (111) reflection using Ag K_{α_1} radiation, and a surface reflection photograph [Figure 8(b)] was taken of a (422) plane

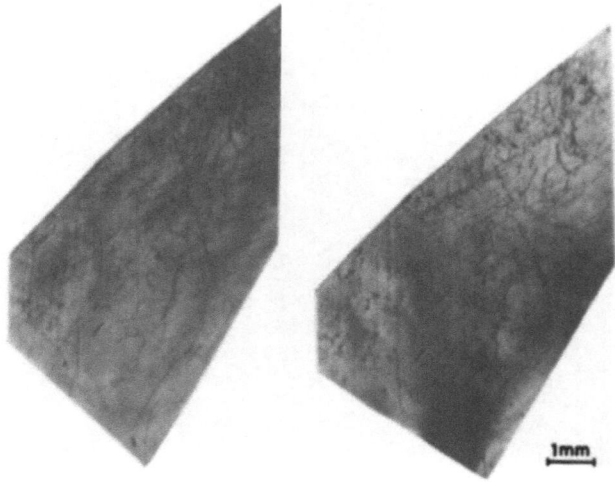

Figure 3. Transmission photographs of a silicon crystal, by the Lang method (left) and modified method (right); (111) reflection, Ag K_α .

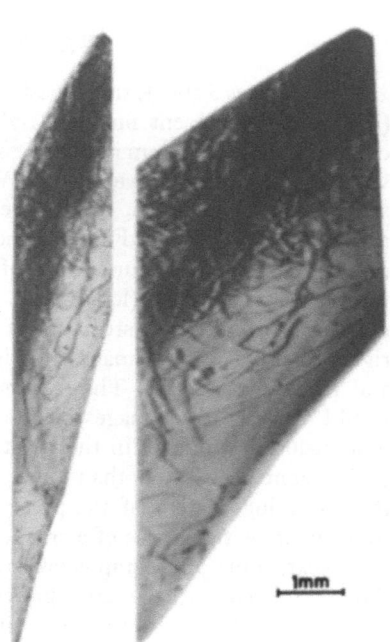

Figure 4. Photographs of a silicon crystal showing image shortening. Transmission cases by the Lang method (left) and modified method (right); (220) reflection, Ag K_{α_1}.

reflection, which is almost parallel to the crystal surface, using Mo K_{α_1} radiation. The images in the reflection photograph which correspond to the outcrops of the dislocation coincide well with the end of corresponding dislocations in the transmission photographs as indicated by arrows.

Consideration should be given to the sharpness of images in the traversing direction, as was pointed out in Figure 6. Figure 9 shows the schematic diagram of the surface reflection. The X-ray beam passing through a slit having aperture S encounters the crystal surface and is diffracted towards the photoplate placed perpendicular to the diffraction beam. The dislocation itself, found at P_0 on the surface, at time t_0 moves to

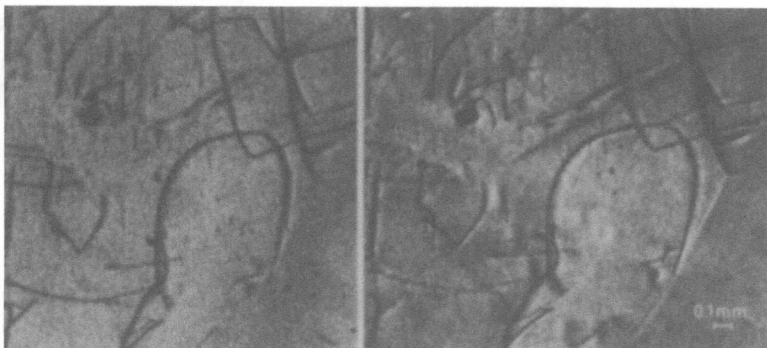

Figure 5. Enlarged photographs of a part of Figure 3. Lang method (left) and modified method (right).

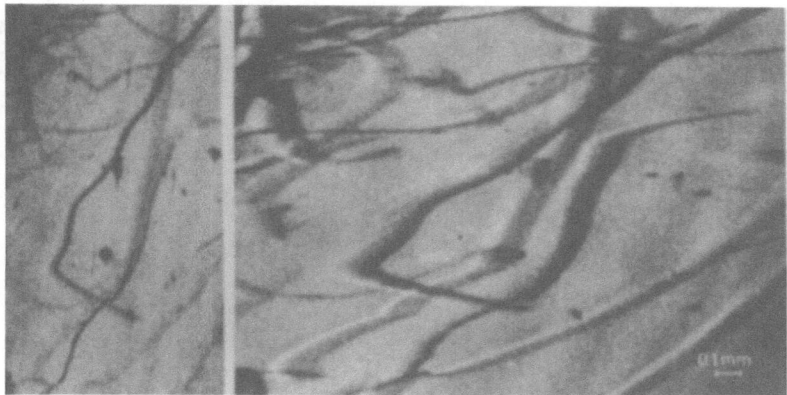

Figure 6. Enlarged photographs of a part of Figure 4. Lang method (left) and modified method (right).

P_1, the left end of the incident beam, after Δt according to the displacement of the specimen. On the other hand, the image of the dislocation moves a distance d' during the same time interval. However, the recorded image on the plate at time t_0 is moved a distance d during the same time interval, the same distance as the displacement of the specimen, since the crystal and the plate are moved with the same traverse velocity. Thus the broadening of the image is given by

$$L = d - d' = S \operatorname{cosec} \theta_1 (1 - \sin \theta_2)$$

where θ_1 and θ_2 are the incident and reflection angles respectively. From the above relation, it is seen that in order to obtain a sharp image, S should be small, and large values of θ_1 and θ_2 are preferable.

In Figure 10 three photographs are reproduced which were taken with different aperture values, S, using the same specimen of a silicon crystal. These show the effect of

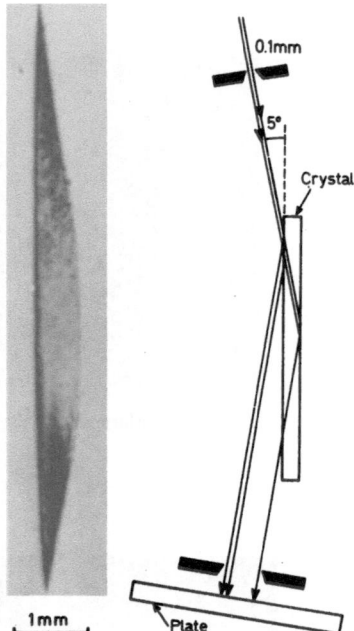

Figure 7. Section photograph by symmetrical (111) reflection, from the surface of silicon; Ag K_{α_1}. Schematic diagram shows the experimental arrangement.

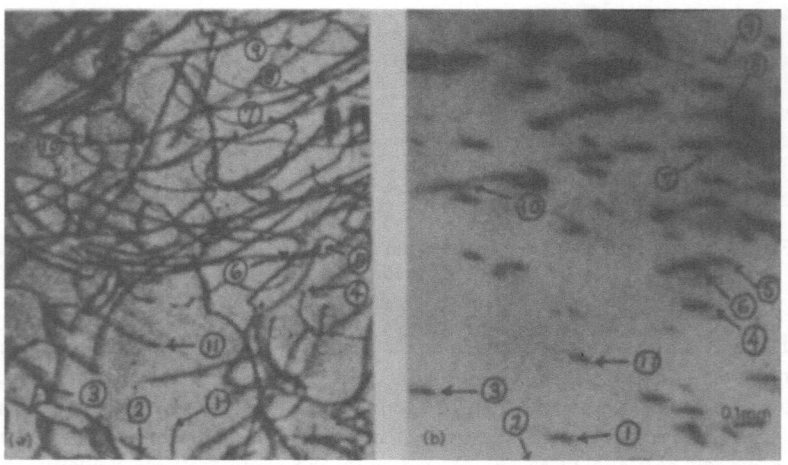

Figure 8. Diffraction photographs in the transmission case and the reflection case of silicon crystal. (a) Transmission case, (111), Ag K_{α_1}; (b) symmetrical reflection case, (422), Mo K_{α_1}. Arrows with numbers indicate identical positions in both photographs.

the aperture on the broadening or the sharpness of the images. Schematic experimental diagrams corresponding to the photographs of Figure 10 are shown in Figure 11. The values of S for the incident beam were 0.1 mm, 0.01 mm, and 0.01 mm for Figures 10(a), (b), and (c), respectively, and in Figure 10(c) another aperture of 0.3 mm was

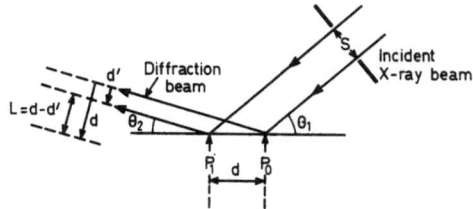

Figure 9. Schematic diagram of surface reflection. The value L is the broadening of the image. Incident X-ray beams are taken to be parallel.

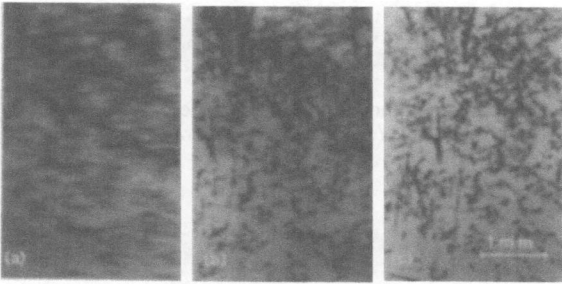

Figure 10. Effect of aperture S on image sharpness. Three photographs in the reflection case (a), (b) and (c) show the same area of a silicon crystal (see text).

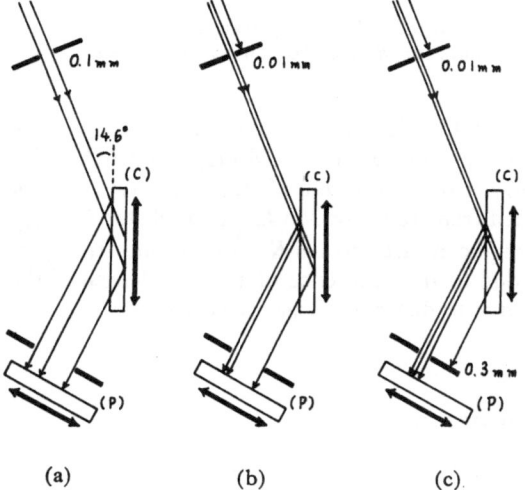

Figure 11. Experimental arrangements for surface reflection of Figure 10. (a), (b) and (c) correspond to (a), (b), and (c) in Figure 10. Dimensions are not given in exact proportion.

placed in front of the plate. The radiation was Ag K_{α_1} and a symmetrical reflection of a (422) plane with 14.5° glancing and reflection angles was used. Figures 12(a) and (b) are the enlarged photographs from the corresponding parts of Figure 10(b) and (c), respectively. The images in Figure 12(b) are from the surface area because of the additional use of a 0.3 mm aperture in front of the plate, as illustrated in Figure 11(c), while the images in Figure 12(a) represent the images of dislocation not only at the surface but inside the crystal.

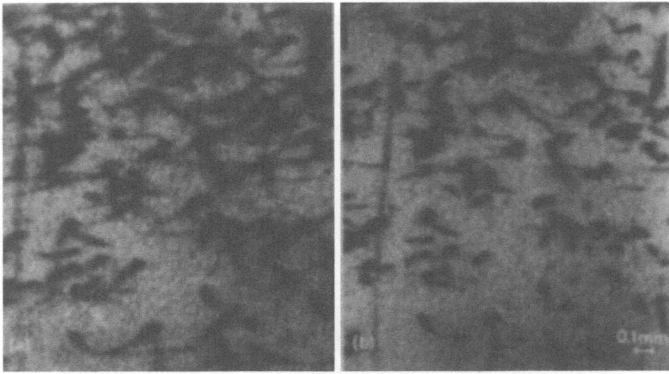

Figure 12. Diffraction photographs in the reflection case enlarged from the part of Figure 10. (a) and (b) correspond to (b) and (c), respectively, of Figure 10.

Figure 13. The difference in the image contrast with X-ray wavelength in the reflection case. Silicon crystal, (422) reflection. (a) Mo K_{α_1}, (b) Cu K_{α_1}, (c) Co K_{α_1}.

The difference in contrast among several radiations was observed in the case of reflection. Figure 13 gives an example. The radiations used were Mo K_{α_1} (a), Cu K_{α_1} (b), and Co K_{α_1} (c) respectively. The symmetrical reflection from a(422) plane was used for all radiations, and the reflection angle varied from 18.6° for Mo K_{α_1} to 53.8° for Co K_{α_1}. Evidently, Mo K_{α_1} gives the strongest image contrast and Cu K_{α_1} the second strongest. The difference in contrast is essentially due to the difference of the wavelength of the radiation used, although it may be partly due to distortion near the surface.

CONCLUSIONS

The experimental results shown here are only a few examples of the preliminary studies which have been made. The quality of the photographs will be improved. This type of camera is useful both for reflection and transmission in obtaining photographs without distortion. The photographs taken by various lattice planes can be compared directly not only with each other, but also with some other types of photographs, such as optical photographs of the etched surface, because of the good one-to-one correspondence of the images. In the reflection case, it is convenient to take photographs from different depths below the crystal surface with the correct use of aperture in front of the photoplate. By the use of the pulse motors, the satisfactory accuracy of the coincidence and of the movement of the specimen and photoplate is obtained and at the same time the setting of the specimen is possible to within an accuracy of 1 mμ over the wide range of 25 mm.

ACKNOWLEDGMENT

The authors thank Dr. Y. Shimura for his encouragement and Y. Kobayashi for his assistance in taking photographs, and also I. Kawai and M. Arimoto for invaluable technical assistances.

REFERENCES

1. Wolfgang Berg, "An X-Ray Method for Study of Lattice Disturbances of Crystals," *Naturwiss.* **19**: 391–396, 1931.
2. C. S. Barrett, "A New Microscopy and its Potentialities," *Trans. AIME* **161**: 15–65, 1945.
3. J. B. Newkirk, "Observations of Dislocations and Other Imperfections by X-Ray Extinction Contrast," *Trans. AIME* **215**: 483–497, 1959.
4. A. R. Lang, "The Projection Topograph: A New Method in X-Ray Diffraction Microradiography," *Acta Cryst.* **12**: 249–250, 1959.

LATTICE DEFECT RESEARCH BY KOSSEL TECHNIQUE AND DEFORMATION ANALYSIS

Masataka Umeno, Hideaki Kawabe, and Gunji Shinoda

Osaka University
Miyakojima, Osaka, Japan

ABSTRACT

An electron probe microanalyzer (EPMA) was applied for the deformation analysis of aluminum single crystals. The lattice distortions caused by tensile stresses were observed by Kossel patterns, which are sensitive in their change of shape to lattice distortion. The effects of lattice distortion would appear as splitting, tearing, bending, broadening, disappearance, and shift of Kossel lines. This distortion behavior can be analyzed successfully. The inhomogeneities and anisotropy appearing on every line were explained by the crystallographic consideration of slip mechanisms. The lattice distortions and corresponding changes in Kossel patterns depend on the direction of elongation; the deformation modes of those crystals which show typical fcc behavior in stress–strain curves can be reasonably explained by a fragmentation model. It was also found that there are some portions in Kossel patterns where some specific Kossel lines, i.e., {200} and {111}, are very sensitive to lattice deformation.

INTRODUCTION

Kossel techniques or pseudo-Kossel techniques have recently been given attention again by many researchers since the electron probe micro-analyzer (EPMA) was invented by Castaing.[1] The Kossel method was usually employed using a capillary X-ray tube invented by Fujiwara[2] and developed by Lonsdale[3] and many other researchers.[4–7] Since Lonsdale applied this method to the precise determination of lattice constants of diamond, it was recognized as an excellent method for precise lattice constant measurement. But capillary X-ray tube devices found their main roles as back-reflection type apparatuses and only a few examples of the transmission type have been reported in the literature, because high resolving power and high contrast of patterns can be obtained in back-reflection photographs. Since the invention of EPMA, this new instrument has been thought to have some advantages over capillary X-ray tube devices if suitable accessories are attached. Using an EPMA, the spot size of an electron beam for an X-ray source can be reduced to about $1\,\mu$ in diameter, and moreover, transmission Kossel patterns can be easily obtained without experimental difficulties. Since the appearance of EPMA some reports have been published concerning new methods of precise measurement of lattice constants from Kossel patterns and their applications to the metallurgical field.[8–10]

It was well known that Kossel patterns are very sensitive to the lattice distortions of specimen crystals. Many years ago one of the present authors investigated lattice defects in synthetic rubies, cold-worked macrocrystals of aluminum, and α-brass with

[1] References are at the end of the paper.

pseudo-Kossel technique using a capillary X-ray tube.[11-12] Imura applied this method for the deformation analysis of aluminum single crystals and obtained some interesting results from back-reflection photographs.[13,14] But with a back-reflection type apparatus, the distance between the X-ray source and the surface of the specimen is long and the investigated area on the crystal surface becomes large, so the patterns obtained by this method give information for macroscopic areas. Therefore, our intent was to make microscopic observations of deformation mechanisms in aluminum single crystals using transmission pseudo-Kossel patterns with an EPMA. For the purpose of microscopic observation a molybdenum foil was used as an X-ray target; the diameter of the examined area of the specimen then becomes several tens of microns.

EXPERIMENTAL PROCEDURE

A JEOL-JXA-3 EPMA was used to obtain pseudo-Kossel patterns and, by this electron optic system, an electron beam several microns in diameter was focused on the specimen surface. The present apparatus is not very different in principle from those reported elsewhere,[8-10] except for the sample holder, which was especially developed for this experiment and can hold a specimen of a maximum size of 40×70 mm. The specimens were electropolished 99.99% aluminum single crystals, $1.5 \times 12 \times 50$ mm, obtained by the strain-annealing method. Because the wavelength of Al K_α radiation is too long to satisfy Bragg's reflection condition, a pseudo-Kossel method with Mo K_α radiation was used. At the center of each specimen a molybdenum foil nearly 1 mm^2 in area and 50 μ in thickness was fixed on the surface by an electroconductive paint and a cross mark was scratched on the foil to distinguish the observation point at every stage of deformation. The specimen crystal was stretched along the longer axis with the target foil attached on the surface, and at various stages of deformation pseudo-Kossel patterns were taken and the corresponding slip status appearing on the crystal surfaces was observed.

It is important in analyzing the deformation mechanism to index every Kossel line correctly. A convenient method of indexing proposed by Lonsdale is the stereographic projection method. According to this method, the stereographic projection of any Kossel lines from crystals of a cubic system becomes circles or arcs in a unit circle represented by $x^2 + y^2 = 1$, and when the films are set parallel to (001) crystal planes, the position of the center (x, y) and the radius R of each circle or arc can be represented by

$$x = \frac{h}{l + \sin\theta\sqrt{h^2 + k^2 + l^2}}$$

$$y = \frac{k}{l + \sin\theta\sqrt{h^2 + k^2 + l^2}}$$

$$R = \frac{\cos\theta}{\sin\theta + l/\sqrt{h^2 + k^2 + l^2}}$$

where

$$\sin\theta = \frac{\lambda\sqrt{h^2 + k^2 + l^2}}{2a}$$

a is the lattice constant of the crystal, λ is the wavelength of the X-rays employed, and h, k, and l are the indices of the crystal plane from which the Kossel line was produced.

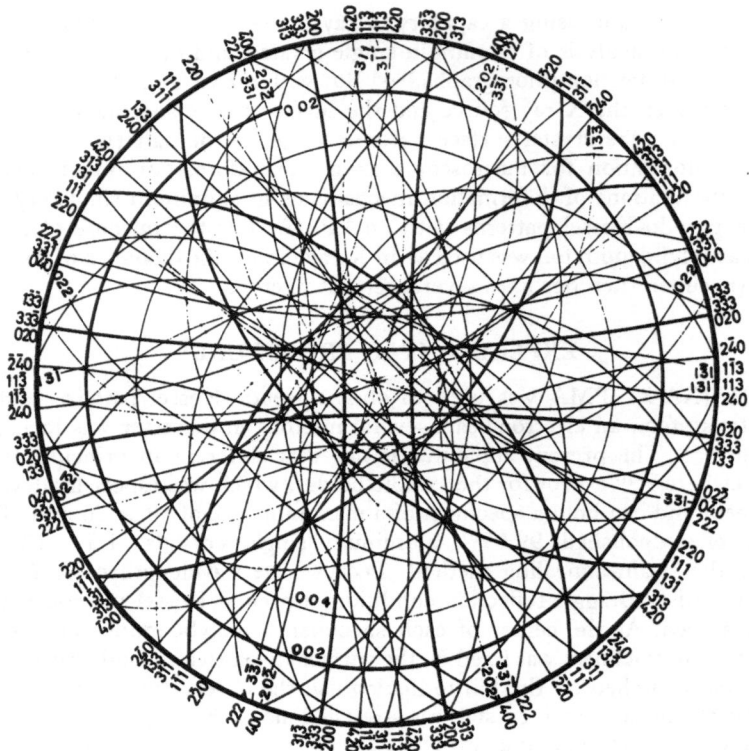

Figure 1. Standard (001) projection of pseudo-Kossel patterns of aluminum for
Mo K_α radiation.

Figure 1 shows the standard (001) stereographic projection of the Kossel pattern for aluminum with Mo K_{α_1} radiation. Patterns obtained from each specimen were compared with Figure 1 and the indices of each line could then be easily determined as shown in the figures which follow.

EXPERIMENTAL RESULTS

Tensile deformation was applied at room temperature for four specimens having different orientations. The directions of tensile axes and the orientations of crystal surfaces are shown in Figure 2, and Figures 3–6 are typical pseudo-Kossel patterns corresponding to specimen nos. 1, 2, 3, and 4, respectively. These figures were obtained at an accelerating voltage of 35 kV and by an exposure time of 7 min on fine-grain X-ray films placed parallel to the surfaces of the specimens. In these figures black lines represent absorption lines and white lines represent diffraction lines, for the photographed patterns are positive.

Figure 3 shows the pseudo-Kossel patterns of specimen no. 1 at various stages of deformation. Through crystallographic studies it was found that the ($\bar{1}\bar{1}1$) line and the ($1\bar{1}1$) line in Figure 3 correspond to the primary slip plane and critical slip plane of the specimen, respectively. It is conceived that in the deformation stages shown in Figure 3 only the primary slip plane is active in the specimen at the earlier stage of deformation. From Figure 3 the effect of slip on the {111} Kossel lines can be seen. At every stage of deformation the ($\bar{1}\bar{1}1$) line and the ($1\bar{1}1$) line change their shapes in different ways; that

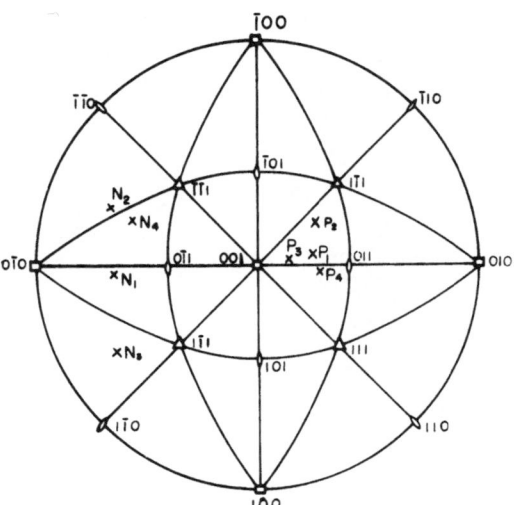

Figure 2. Stereographic projection of specimen orientations and elongation direction. P_1, P_2, P_3, and P_4 denote the orientations of specimen surface of specimens nos. 1, 2, 3, and 4, and N_1, N_2, N_3 and N_4 denote the elongation directions.

is, while the ($\bar{1}\bar{1}1$) line shows only splitting into several branches, the ($1\bar{1}1$) line shows both splitting and tearing of the line, whereas the (200) line remains sharp even at a higher deformation of 55% of extension for this specimen. Every picture in this figure, however, was not taken at exactly the same sample position.

Figure 4 shows a complicated deformation occurring in specimen no. 2. The ($\bar{1}\bar{1}1$) and (111) lines in this figure correspond to cross and critical slip planes, respectively. As the conjugate and primary slip planes are considered to be active at the earlier stages of deformation in this specimen, the effects of both primary and conjugate slips should appear on the {111} lines of this figure.

Figure 5 gives the patterns from specimen no. 3. In this figure a nearly central portion of the stereograph appears, and the behavior of lattice distortion on the Kossel pattern is quite different from the case of Figure 3, in which a rather off-center portion of the stereograph appeared. In Figure 5, the ($\bar{1}\bar{1}1$) line and the ($1\bar{1}\bar{1}$) line represent the primary slip plane and the conjugate slip plane, respectively, and the {220} line is the slip direction of primary slip. But in this figure the effects of slip are not observable on the {111} lines but appear on the {200} lines, which were stable both in Figure 3 and Figure 4. The {311} lines are also sensitive to the lattice deformation though the change of line shapes is not so informative as for those lines mentioned above.

In Figure 6 the line representing the active slip plane does not appear, but the changes of the Kossel patterns are very similar to those of Figure 5 except for some local differences observed in the {111} lines, which have peculiar distortions around the intersections to the {200} lines.

As for the differences in the patterns due to the different observation points, there are sometimes slight changes in the Kossel patterns of the same stage of deformation taken at two points separated by nearly 150 μ. But so far reproducibility of each observation point is satisfactory, and the differences in Kossel patterns by the observation points are given little attention in the present cases.

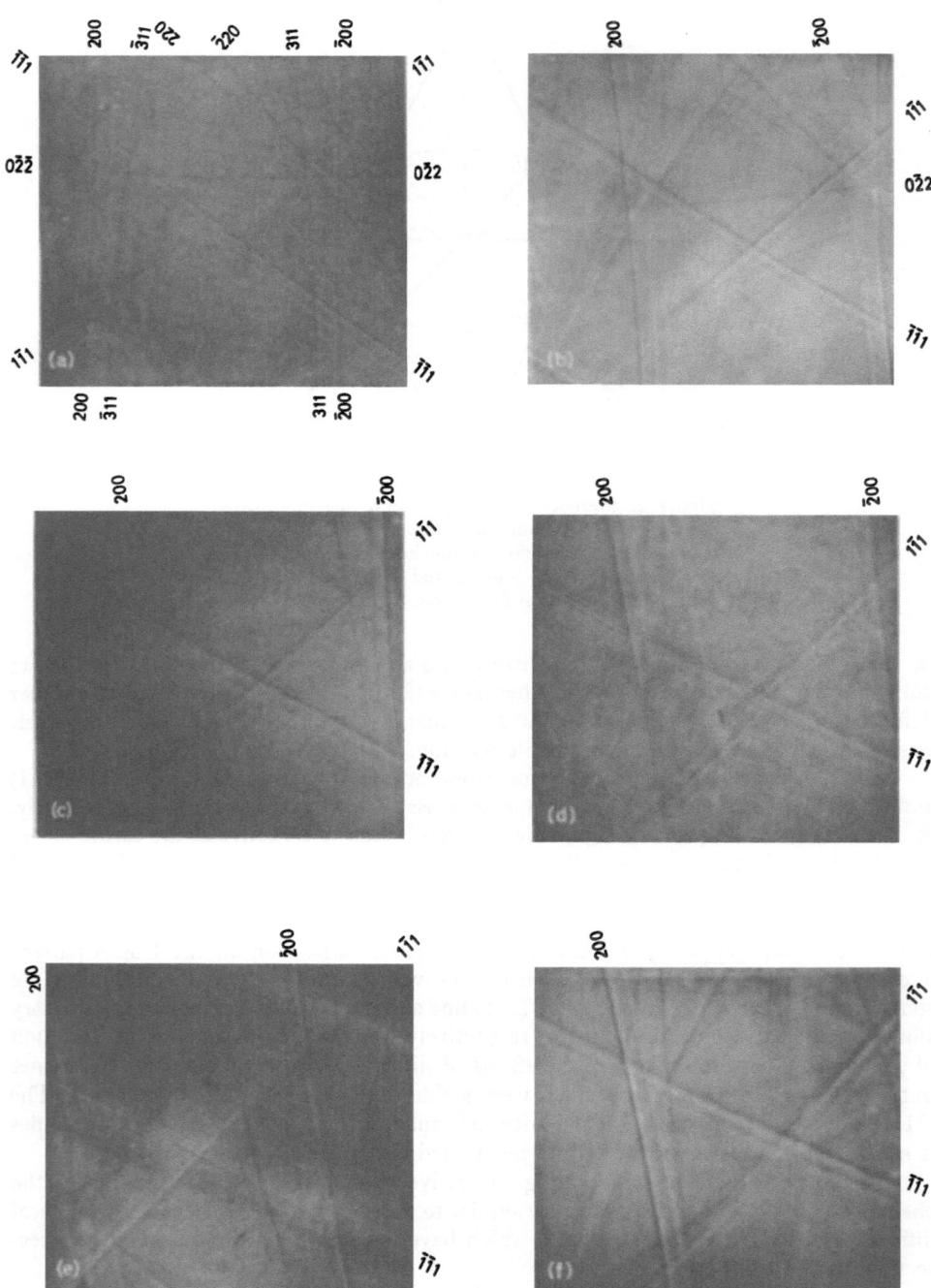

Figure 3. Pseudo-Kossel patterns from specimen no. 1 obtained at the tensile strains of (a) 0%, (b) 1.0%, (c) 2.5%, (d) 5.0%, (e) 6.6% and (f) 15.7%, with Mo K_α radiation, at 35 kV, for 15 min of exposure.

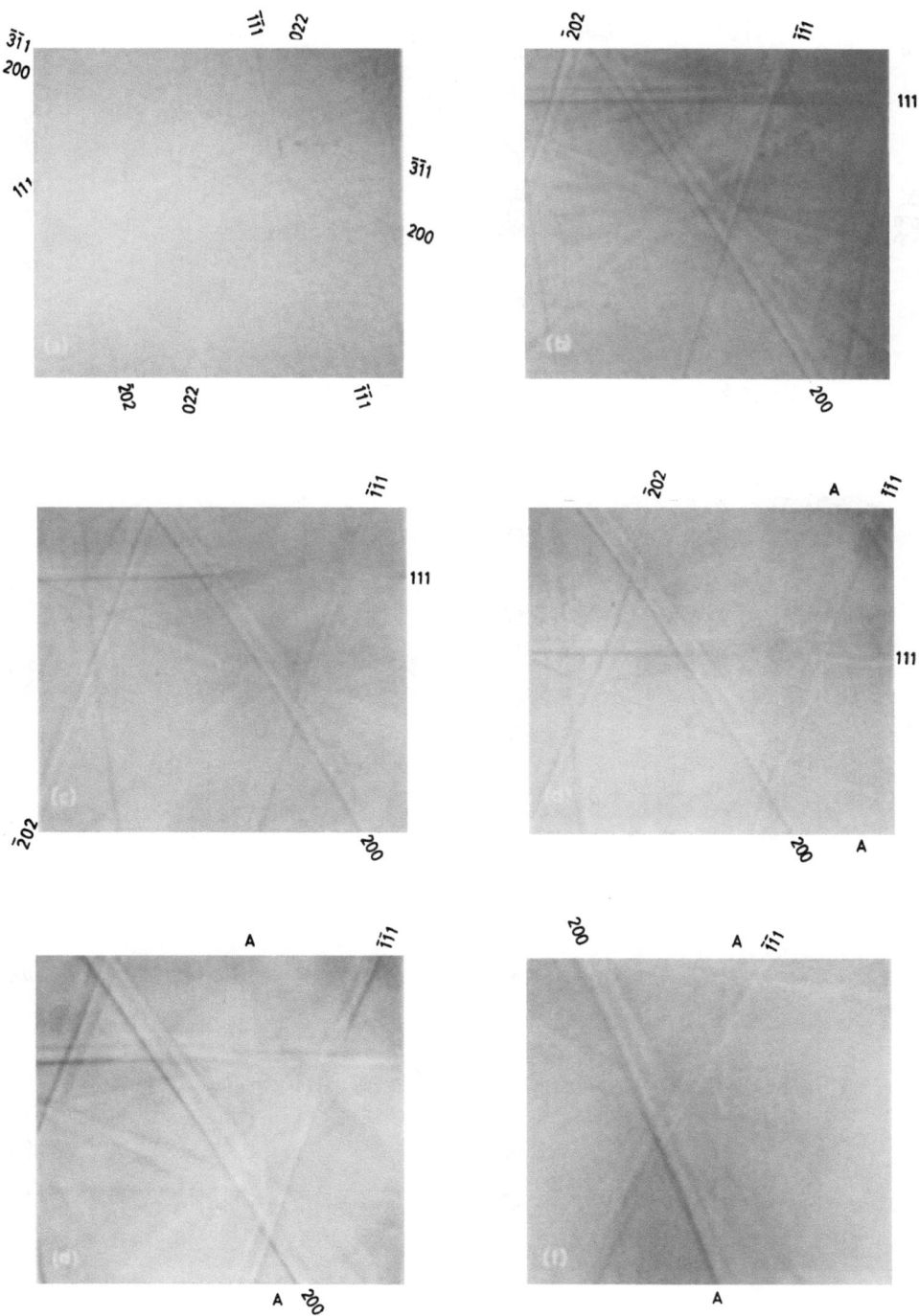

Figure 4. Pseudo-Kossel patterns from specimen no. 2 obtained at tensile strains of (a) 0%, (b) 1.8%, (c) 3.6%, (d) 5.4%, (e) 7.1%, and (f) 12.3%, with Mo K$_\alpha$ radiation, at 35 kV, for 15 min of exposure.

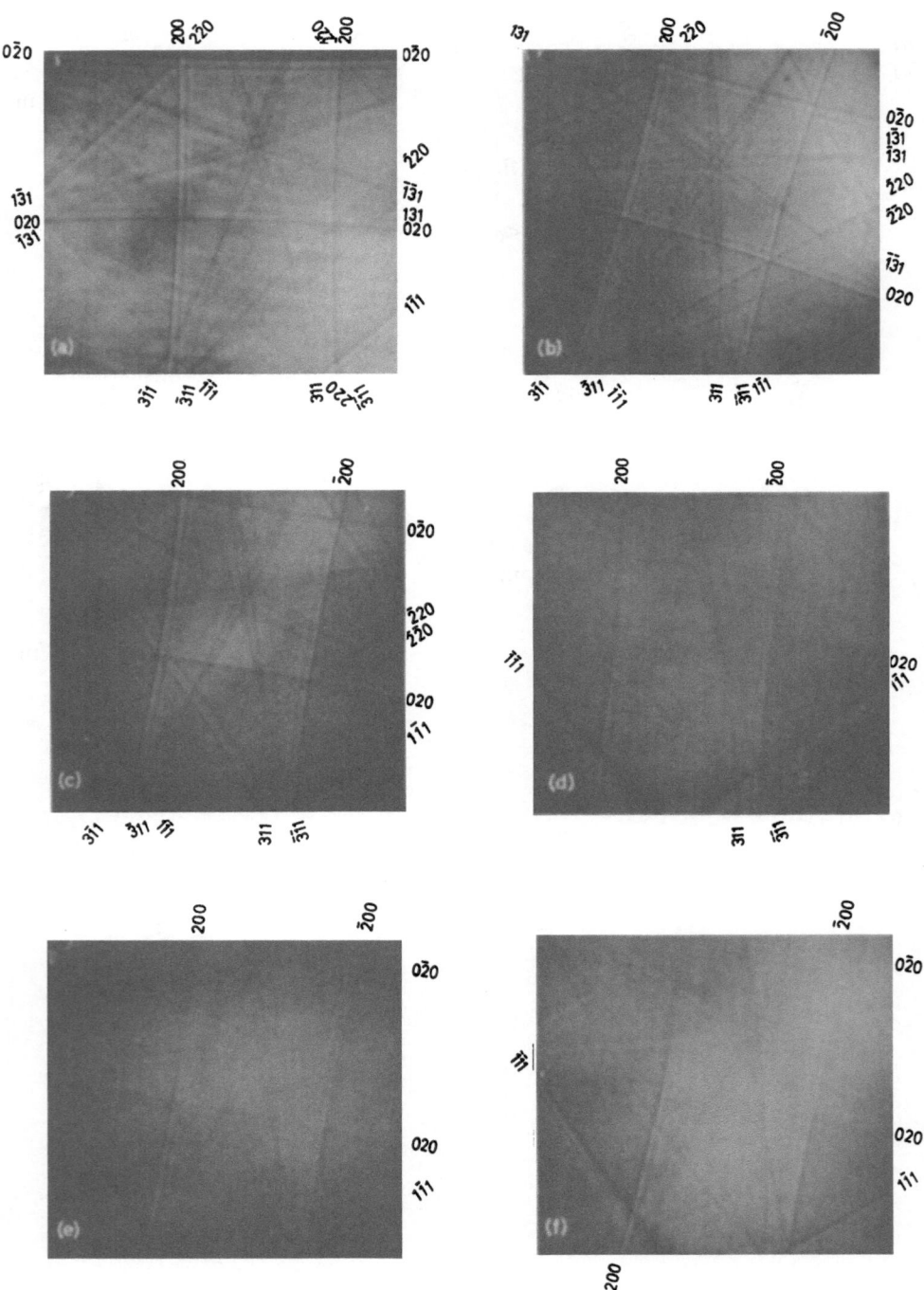

Figure 5. Pseudo-Kossel patterns for specimen no. 3 obtained at tensile strains of (a) 0%, (b) 0.5%, (c) 2.5%, (d) 5.3%, (e) 7.0%, and (f) 9.1%, with Mo K_α radiation, at 35 kV for 15 min of exposure.

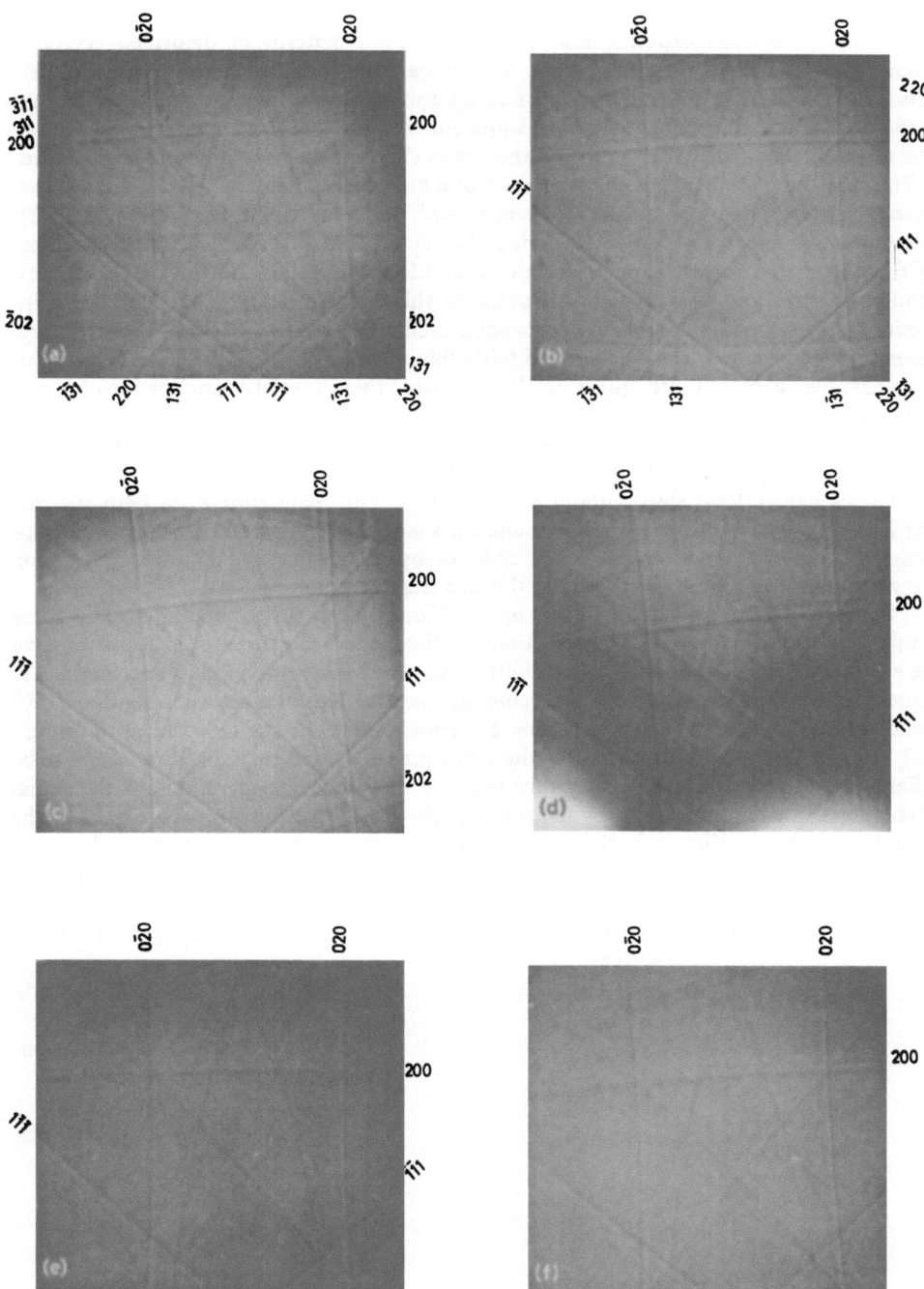

Figure 6. Pseudo-Kossel patterns from specimen no. 4 obtained at tensile strains of (a) 1.4%, (b) 2.3%, (c) 3.5%, (d) 4.7%, (e) 5.5%, and (f) 6.5%, with Mo K_α radiation, at 35 kV, for 15 min of exposure. K_{α_1} and K_{α_1} lines are clearly resolved on {311} and {222} planes in (a) and (b).

CONCLUSION

From the present experimental results on Kossel patterns of deformed crystals, several observations can be made. First, the lattice distortions in the crystal cause distortions of the Kossel lines in such ways as splitting, tearing, bending, and broadening of the lines; these phenomena are more apparent with the {111} lines than the other lines. The fact that the line corresponding to the active slip plane shows only splitting, as seen in Figure 3, indicates the possible existence of a kind of fragmentation mechanism in the crystal. That is, when a single crystal is stretched, slip is initiated on a specified {111} plane and the slip is considered to divide the crystal into several blocks of crystallites by the slip plane. A certain inclination of each block causes line splitting on the corresponding Kossel line. A schematic diagram of this fragmentation model is shown in Figure 7. Because the anomalous absorption coefficients are much larger than ordinary absorption coefficients, the existence of fairly thin inclined blocks of the crystallite might be detectable. From Figure 3(c) it can be estimated that the relative inclination of every block is distributed within three degrees of arc. Imura[14] also considered a similar fragmentation model in his experiments with a capillary X-ray tube to explain the line distortions appearing on the (022) line in back-reflection patterns. He observed that the (022) lines have a local shift without any splitting. The main difference between the phenomena observed in his experiment and ours is caused by the fact that his observations were done rather macroscopically and not by a transmission method. But there are no significant differences between the two fragmentation models.

Second, it is observed that even in the Kossel lines corresponding to the same family of planes, there are clear distinctions in distortion behavior. As {111} planes have the same angular distance with every {100} plane, the effects of slip on a specified {111} plane to {100} planes should be the same and so the Kossel lines representing {100} planes are considered to show the same behavior. But from the {200} lines in Figure 5(c), (d), (e), and (f) the distortion of the {100} planes caused by elongation seems to be different for each plane. The discrepancy between the above consideration and the actual phenomena can be understood by considering the locality of a distorted domain in the crystal as shown in Figure 8. If the thickness of a slipped domain is thin as shown in Figure 8, the intersections of this domain and the {100} absorption conics will occur at some local positions, shown, for example, by A and B in Figure 8, and so Kossel lines from the {100} planes should be distorted at the positions corresponding to A and B.

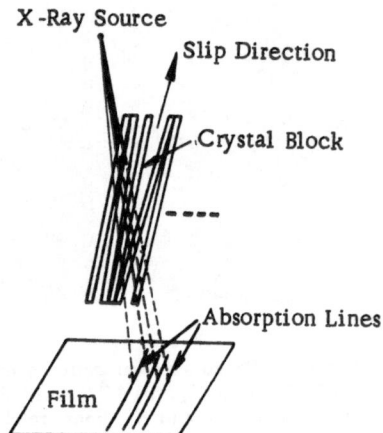

Figure 7. Fragmentation model to explain the line splitting appearing on the {111} lines in Figures 3 and 4.

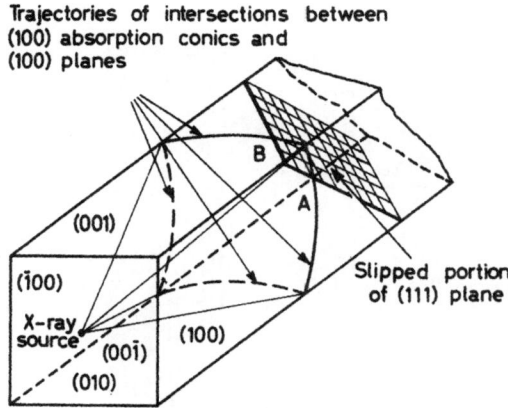

Trajectories of intersections between (100) absorption conics and (100) planes

Figure 8. The relation between absorption conics for {100} planes and the slip plane. The intersections of the conics and the slip plane cause inhomogeneous distortions in the {200} Kossel lines.

The heterogeneity or anisotropy of the lattice distortions in the crystal can also be seen on the {220} and the {311} Kossel lines, for example, see Figures 5 and 6. The distortion behavior of these lines can be understood in terms of the previous fragmentation model. If the lattice distortions of the {110} planes are mainly caused by a single slip on the primary slip plane, the lines from some of the {220} planes, which are perpendicular to the slip plane, should be more stable than the other {220} lines. The line corresponding to the plane whose normal is the slip direction should have slight bending or broadening, which was also observed as local line shifts in back-reflection pseudo-Kossel lines by Imura.[14] In Figure 5 there is a pair of stable {311} lines, namely the (311) and the ($\bar{3}\bar{1}1$), and the indices of these persistent lines correspond to the planes which are nearly perpendicular to the slip plane and parallel to the slip direction. This can also be explained reasonably by a fragmentation model.

A third observation was that, as shown in Figure 2, specimen no. 2 was stretched along a direction near the (100)–(111) boundary of the standard triangle. Many papers about work-hardening reported that in a single crystal of an fcc system, rapid hardening occurs by multiple slip when tensile stress is applied along such a direction. The complex distortion of the patterns with line splitting, tearing, and bending observed in Figure 4 seems to indicate a complicated deformation mechanism. In Figure 4 there is a peculiar diffraction line, indicated by A—A. The relative position of the line changes in every pattern, but its sharpness and shape never change, while other lines are distorted. These facts, and also the fact that the indices of this line could not be determined, might indicate that there is a small crystallite block in the specimen, small enough, possibly a few microns in diameter, not to be distorted by external stresses even though it may be rotated by an applied stress. Still, at the present stage it could not be concluded from Figure 4 whether the small crystallite already existed in the as-received state of the specimen or was created by external tensile stresses.

Fourth, it was observed that there are some regions in the stereograph in which the effect of lattice distortion is observable only on the Kossel lines of a specific family of planes. So for the purpose of deformation analysis, the orientations of specimen surfaces must be of certain orientations where the line corresponding to the primary plane will appear in the Kossel patterns.

APPENDIX

The pseudo-Kossel method described in this paper was applied to investigate the perfectness of various single crystals. Figure 9 was obtained from a laser ruby rod

Figure 9. A pseudo-Kossel pattern of ruby single crystal obtained from a laser rod of 5 mm in diameter with Ag K_α radiation at 50 kV and for 50 min of exposure.

Figure 10. A back-reflection pseudo-Kossel pattern of a synthetic ruby of nearly 5 mm in thickness, obtained by a capillary X-ray tube with Cu K_α radiation at 35 kV for 8 hr of exposure. This picture was taken in 1947 in our laboratory.

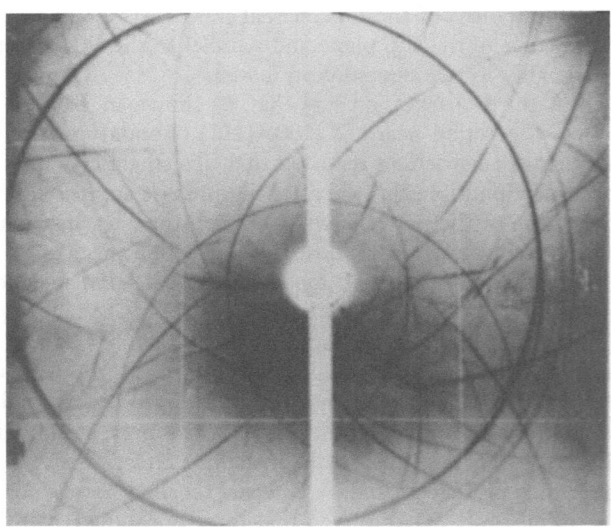

5 mm in diameter (manufactured by the Linde Company) with Ag K_α radiation. Usually synthetic rubies for laser experiments are made very carefully because high homogeniety is required, but this figure shows that even in the laser ruby some lattice distortions are present. As a comparison, a pseudo-Kossel pattern of an ordinary synthetic ruby, obtained in our laboratory with Cu K_α radiation with a back-reflection capillary X-ray tube, is shown in Figure 10. This figure also indicates the existence of macroscopic complicated lattice distortions which originated during the crystal growth.

ACKNOWLEDGMENT

The authors wish to express their thanks to Dr. Y. Amano for his kind discussion.

REFERENCES

1. R. Castaing, Thesis, University of Paris, 1951.
2. T. Fujiwara, *J. Sci. Hiroshima Univ. Ser. C* 7: 179, 1937.
3. K. Lonsdale, "Divergent-Beam X-Ray Photography of Crystals," *Phil. Trans. Roy. Soc. London* **240**: 219, 1947.
4. A. H. Geisler, J. K. Hill, and J. B. Newkirk, "Divergent Beam X-Ray Photography with Standard Diffraction Equipment," *J. Appl. Phys.* **19**: 1041, 1948.
5. T. Imura, S. Weissmann, and J. J. Slade, Jr., "A Study of Age-Hardening of Al–3.85% Cu by the Divergent X-Ray Beam Method," *Acta Cryst.* **15**: 786, 1962.
6. S. Weissmann and K. Nakajima, "Defect Structure and Density Decrease in Neutron-Irradiated Quartz," *J. Appl. Phys.* **34**: 611, 1963.
7. T. Ellis, L. F. Nanni, A. Shrier, S. Weissmann, G. E. Padawer, and N. Hosokawa, "Strain and Precision Lattice Parameter Measurement by the X-Ray Divergent Beam Method. I," *J. Appl. Phys.* **35**: 3364, 1964.
8. R. E. Hanneman, R. E. Ogilvie, and A. Modrzejewski, "Kossel Line Studies of Irradiated Nickel Crystals," *J. Appl. Phys.* **33**: 1429, 1962.
9. B. H. Heise, "Precision Determination of the Lattice Constant by the Kossel Line Technique," *J. Appl. Phys.* **33**: 938, 1962.
10. P. Gielen, H. Yakowitz, D. Ganow, and R. E. Ogilvie, "Evaluation of Kossel Microdiffraction Procedures: The Cubic Case," *J. Appl. Phys.* **36**: 773, 1965.
11. G. Shinoda and Y. Amano, "X-Ray Investigation on Artificially Prepared Jewels. Part I," *X-Sen* **6**: 7–13, 1949 (in Japanese).
12. Y. Amano, Thesis, Osaka University, 1960.
13. Tohoru Imura, "Study of the Deformation of Single Crystals by Divergent X-Ray Beams," *Bull. Naniwa University, Ser.* 2*A*: 51–70, 1954.
14. Tohoru Imura, "A Study on the Deformation of Single Crystals by Divergent X-Ray Beams (Part III), *Bull. Univ. Osaka Prefect. Ser.* 5*A*: 99–120, 1957.

DISCUSSION

H. Yakowitz (National Bureau of Standards): What was the distance from source to film?

G. Shinoda: About 8 cm.

H. Yakowitz: What was the thickness of the unstrained aluminum crystals?

G. Shinoda: 1.3 mm.

H. Yakowitz: I have one comment. I think you will find that the contrast will improve if you change the thickness of the crystals to the optimum thickness for molybdenum radiation. It was found by Lonsdale with a similar linear absorption coefficient to be of the order of half a millimeter. I think you will get a general improvement in contrast.

G. Shinoda: I agree. In a later experiment I used an 8-mm-diameter ruby rod sample. Using a suitable target we obtained photographs.

T. Reichard (Monsanto Company): Were your crystals stressed *in situ* while the patterns were made as opposed to being stretched? Was the force released at the time the patterns were made? Were they elongated and then released or were they being stressed *in situ*?

G. Shinoda: Stress was released when the specimen was set in the camera.

T. Reichard: They were not under tension?

G. Shinoda: Not under tension.

L. S. Birks (Naval Research Laboratory): How accurately can you measure the angle of tilt between these fragments from the lines on the patterns?

G. Shinoda: It depends on the widths of Kossel lines and on the wavelength of the radiation employed. In the present case it is about 20 minutes of arc.

THE CRYSTALLINE PERFECTION OF MELT-GROWN GaAs SUBSTRATES AND Ga(As, P) EPITAXIAL DEPOSITS

J. K. Howard and R. H. Cox

Texas Instruments Incorporated
Dallas, Texas

ABSTRACT

The preparation of semiconductor materials by epitaxial deposition has rapidly gained prominence in recent years. $GaAs_{(1-x)}P_x$ alloys of variable composition, deposited on GaAs substrates, have been prepared in an open-tube reactor using H_2, $AsCl_3$, and PCl_3 with GaAs source material. The deposits ranged in composition from GaAs to $GaAs_{0.5}P_{0.5}$ and were 10–30 microns thick.

X-ray diffraction topography was employed to evaluate the structural perfection of the monocrystalline GaAs substrates and the perfection of the deposited Ga(As, P) films. A scanning-reflection technique was developed to study the concentration, type, and spatial distribution of imperfections over large areas (~ 8 cm^2) to depths of 25 microns. Some areas of investigation included (1) the polycrystallinity of some deposits, (2) the effect of substrate perfection on the film quality, (3) the depth of mechanically induced damage, and (4) the structural perfection of magnesium-diffused GaAs and Ga(As, P).

INTRODUCTION

The structural perfection, homogeneity, and single crystallinity of semiconductor materials are of major concern in the development of crystal growth processes, because these parameters directly affect the performance of electronic devices fabricated from these materials. It is of particular importance to be able to identify the growth flaws and relate them to either the substrate perfection, preparation, or the deposition process.

A scanning-reflection X-ray topographic method[1] was selected to determine the degree of perfection of the GaAs substrates and the epitaxial Ga(As, P) deposits. This method has the particular advantages of being simple, fast, and nondestructive.

The gross features of substrate perfection were investigated and their effect on the film quality was established. Polycrystalline regions were identified in early deposits. Inhomogeneity and compositional gradients across the epitaxial film were investigated and attributed to the deposition process. Preliminary results of magnesium diffusion into GaAs and Ga(As, P) are also presented.

X-RAY TOPOGRAPHY AND THE SCANNING-REFLECTION X-RAY TOPOGRAPHIC METHOD

X-ray topography is an effective method of examining the defect character of various monocrystalline materials.[2] This method yields a photographic display of the lattice

---

[1] References are at the end of the paper.

35

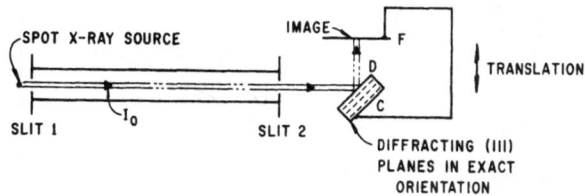

Figure 1. The experimental arrangement for the scanning-reflection
X-ray topographic method.

defects in the crystal. The imperfection is detected by placing a photographic plate near
the primary beam entrance surface (reflection geometry); the transmitted beam exit
surface is utilized in the transmission arrangement. The perturbed lattice diffracts with
more intensity than the unperturbed material enclosing the discontinuity. The point-to-
point variation in intensity revealed in a topographic image of the crystal interior corresponds to fluctuations in primary extinction.[3]

The transmission methods developed by Lang[4] and Bonse[5] were considered for the
investigation of the perfection of the Ga(As, P) epitaxial film (10–30 microns) which had
been deposited on a GaAs substrate (0.025 in.). These techniques demand long exposure
times (10–60 hr) and introduce the problem of separating substrate and film defects
since the defect array in the substrate and the film are superimposed in the transmission
topograph. The reflection method developed by Newkirk[6] was rejected because in some
instances the entire slice could not be examined.

The scanning-reflection method[1] was developed to study imperfections in the active
growth surface of the GaAs substrate; after deposition, this method was employed to
evaluate the film defects. Large area slices (≈ 8 cm^2) have been inspected with this
system.

The experimental arrangement is depicted in Figure 1. The source-to-sample
distance was 28 cm; the sample-to-film distance varied from 15 to 20 mm (as compared
to 0.1 mm in the Newkirk method). The incident beam I_0, restricted by slits S_1 and S_2,
impinges on the crystal C in proper diffracting position. Slit S_2 was adjusted until the
$K_{\alpha_1 \alpha_2}$ doublet was separated and only the K_{α_1} component diffracted. When the ribbon-
shaped beam strikes the crystal, the irradiated volume diffracted coherently to form a
topographic image of the flaw density in that volume element. The image was recorded
on a photographic plate which was placed perpendicular to the diffracted beam D. The
distribution of defects over the entire slice surface was obtained by translating the coupled
film-sample unit perpendicular to the incident beam I_0 or parallel to the slice.

The specific intent of this diffraction geometry was to separate the perfection of the
substrate from that of the epitaxial film. A topograph of the substrate surface reveals
the presence of grown-in defects and mechanically induced damage. After deposition,
the epitaxial film can be examined independently by the selection of a radiation and
reflecting plane in which the penetration depth of the incident beam is less than the
deposit thickness. If the incident beam is attenuated by absorption (mosaic case), then
the maximum penetration depth X (microns) is given by

$$X = \frac{- \ln A \cdot 10^4}{\mu/\rho \cdot \rho [\csc(\theta - \phi) + \csc(\theta - \phi)]} \tag{1}$$

where μ/ρ (cm^2/g) is the mass absorption coefficient of the material, ρ (g/cm^3) is the
density, θ is the diffraction angle, ϕ is the angle between the slice surface and the diffraction planes, and A is the fractional attenuation of the incident beam. A plot of X as a
function of the angle between the incident beam and the slice surface ($\theta - \phi$, related to
the operating reflection) is shown in Figure 2 for GaAs. This plot is a good approximation

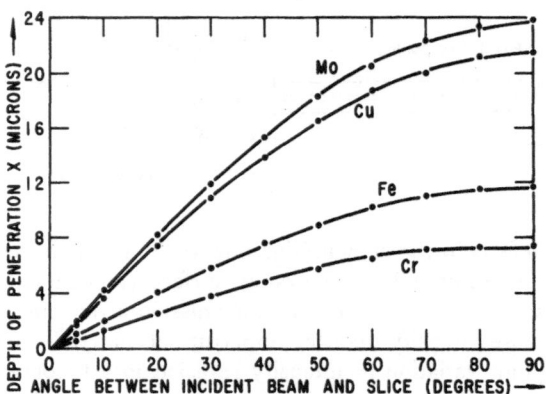

Figure 2. The depth of penetration of an X-ray beam in GaAs.

for Ga(As, P) if the phosphorus concentration is less than 30%. Epitaxial deposition of Ga(As, P) on the (111)-B face of GaAs (which is the polished face as opposed to the (111)-A face, which is the marked face) can be explored to a depth of 25 microns* by the employment of copper radiation and the (333) reflection ($\theta - \phi \approx 45°$). A set of planes can be used to form a topographic image if $\theta > \phi$ and appreciable intensity is realized from the reflection. The {440} and {422} planes in GaAs satisfy these conditions and are frequently employed.

EXPERIMENTAL

Substrate Preparation

The substrates were prepared from GaAs single crystals grown by the Czochralski method. Both tellurium-doped and chromium-doped semi-insulating crystals were used. The crystals were oriented on the (111)-B, 5° off toward the $\langle 100 \rangle$ using X-ray diffraction techniques, and 30-mil-thick slices were cut using an I.D. diamond saw. The substrates were chemically polished on pellon pan W paper employing a $NaOCl-H_2O$ solution to remove the saw damage. Just before the substrates were put into the epitaxial deposition reactor, they were etched for 5 min in cold $5H_2SO_4-H_2O_2-H_2O$ followed by 1–5 min in hot $5H_2SO_4-H_2O_2-8H_2O$.

Epitaxial Deposition

An open-tube, vapor-transport system was used to deposit epitaxial Ga(As P) on (111)-B GaAs substrates. The method employed was similar to that of Finch and Mehal.[8] Other epitaxial techniques for preparing Ga(As, P) alloys have also been reported in the literature.[9–13]

Vapor-phase, epitaxial deposition has particular advantages over other techniques of alloy preparation because high temperatures, high pressures, and long periods of time are not necessary to form homogeneous solid solutions. The technique utilizes the transport of halides in a hydrogen gas stream flowing through a decreasing temperature gradient. Halides are preferred because they do not act as donors or acceptors in Ga(As, P).[11] $AsCl_3$ and PCl_3 were selected because they can be prepared with a high degree of purity by laboratory distillation, and their vapor pressure is high enough so they can be transported as gases using gas saturation bubblers.

---

* Using the method employed by Dobrott *et al.*,[7] the Ga(As, P) film thickness was determined which completely attenuates the (333) GaAs reflection (copper radiation employed). The film thickness can be determined from an interferometer measurement.

The epitaxial deposits are made in the following way. $AsCl_3$ and PCl_3 are brought into the quartz reactor with a hydrogen gas stream and react with GaAs source material at 950°C. The reaction products (GaCl, arsenic, and phosphorus) are transported in the vapor phase to the cooler part of the reactor. The GaCl disproportionates in the decreasing temperature gradient to form $GaCl_3$ and elemental gallium. The gallium, arsenic, and phosphorus react to deposit single crystal Ga(As, P) epitaxially onto the substrate surface. Substrate temperatures are around 750–770°C, and the deposition rate is from 5 to 10 microns/hr. The various alloy compositions are produced by controlling the $AsCl_3$-to-PCl_3 ratio of the entering gas mixture.

X-Ray Procedure

Characteristic copper radiation from a spot-focus X-ray tube (Jarrel–Ash microfocus generator) was employed in this investigation. The effective size of the spot viewed at 6° was 100 microns. Nickel foil (≈ 0.002 in.) was placed over the film holder in order to filter the radiation diffracted by the sample. Kodak type A plates were used; the average exposure time was 2 hr.

The sample [Ga(As, P)/GaAs] was mounted on a goniometer head such that the (111)-B face was bathed in the incident beam. Angular rotations were used to position the diffracting planes in exact alignment. The diffracted beam intensity was monitored with a scintillation detector–ratemeter circuit. After orientation was completed, the scanning mechanism was engaged and the topograph was recorded.

RESULTS

Substrate Perfection

The dislocation density of the GaAs substrate material was determined directly from the X-ray topographs. The material examined yielded a defect concentration of 2×10^3 to 8×10^4 cm^{-2}. The dislocation density D (dislocations/cm^2) was determined from the relation

$$D = \frac{N \cdot 10^5}{X} \tag{2}$$

The value N is the number of dislocation images crossing a 1 mm line superimposed on the photographic plate, and X is the depth of penetration (microns) of the incident X-ray beam.

The ($\overline{4}40$) topograph of a polished (111)-B GaAs slice is shown in Figure 3. Numerous dislocations ($\approx 5 \times 10^4$ cm^{-2}) are revealed as curvilinear lines of enhanced contrast. Since the majority of these imperfections glide in the plane of the slice, an etch-pit analysis[14] of this slice would yield a gross error in relative perfection.

Mechanically induced damage in GaAs substrate material is also characteristic of substrate perfection. When the substrate is processed (sawed, lapped, polished), an expedient method to mark individual slices for identification consists of scratching a number on the (111)-A side with a diamond scribe. A ($\overline{3}\overline{3}\overline{3}$) topograph (Figure 4) of a polished (111)-B face clearly reveals the number "9" which had been *lightly* scribed on the opposite side. The slice thickness was approximately 0.025 in. This effect has been termed "kinking" by Renniger[15] and results in angular distortions of the atomic planes near the disturbance. Intense straining in the vicinity of the "9" image deforms the atomic planes to such an extent that diffraction does not occur in this region (arrow).

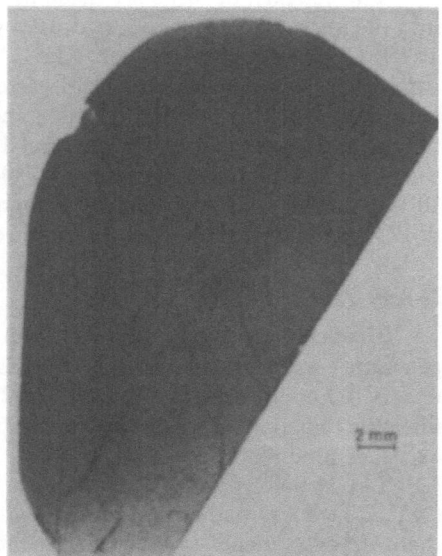

Figure 3. The ($\overline{4}\overline{4}0$) topograph of a dislocation array in GaAs substrate material.

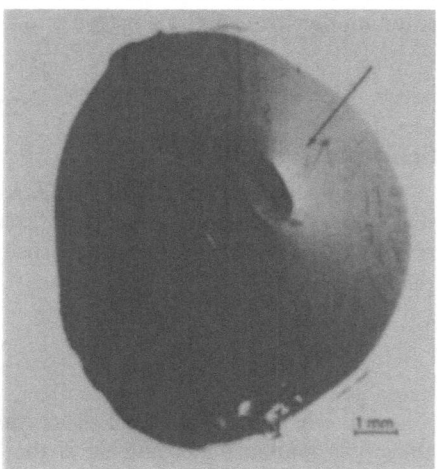

Figure 4. Evidence of scribe damage which propagated through 25 mils of GaAs. The diffracting planes are strained in the vicinity of the violation (arrow).

Epitaxial Film Perfection

The homogeneity and single crystallinity of the Ga(As, P) deposit are of optimum concern in the evaluation of film perfection. The dislocation density of the deposit is of lesser importance to device parameters. However, the nature and concentration of any imperfections in the substrate or deposit reflect the growth history of the epitaxial film. A close correlation between substrate perfection, reactor parameters, and deposit flaws may suggest a solution to perfection problems encountered in the growth process.

The Effect of Substrate Perfection on Deposit Quality

The (111)-A face of a sawed GaAs slice (≈ 0.030 in.) was mounted on a polishing block and the B face was chemically polished to remove saw damage.[16] After polishing, the slice may contain a residual strain field produced by the resulting slice bow.[17] This strain field was observed with X-ray topography.

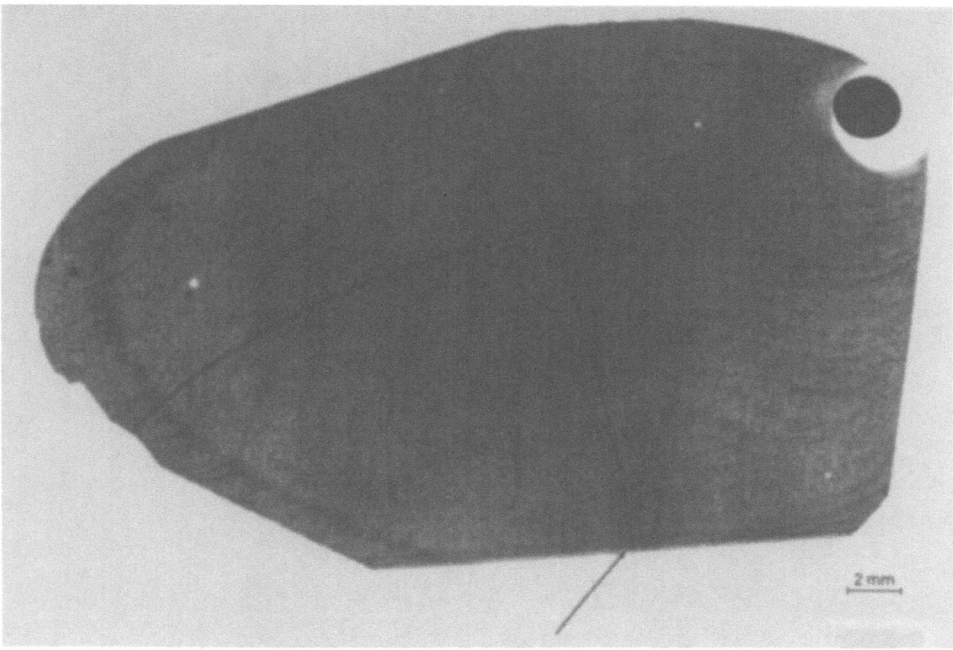

Figure 5. A ($\overline{3}\overline{3}\overline{3}$) topograph of a long-range strain field in GaAs. The lines traversing the image are scratches produced in the polishing operation (arrow).

A ($\overline{3}\overline{3}\overline{3}$) topograph of the (111)-B face (Figure 5) reveals regions of enhanced contrast corresponding to lines of constant lattice displacement. The lines are almost periodic in nature and propagate across the entire slice, terminating at a highly deformed edge. The direction of the strain field was [111] since maximum image contrast was observed for the ($\overline{3}\overline{3}\overline{3}$) topograph.[6] The lines (arrow) traversing the image are scratches produced in the polishing operation.

The ($\overline{3}\overline{3}\overline{3}$) topograph of the epitaxial deposit (Figure 6) yielded no evidence of the strain field in the substrate. Annealing of the substrate at the deposition temperature probably relieved the strain. The substrate slice is now lightly etched *before* polishing to reduce the substrate "bowing" and eliminates the "frozen-in" strain field.

The damage induced in the polishing operation has been shown to drastically affect the structural quality of the GaAs epitaxial film.[1] The existence of a deformed layer at the substrate surface produced a dense array of localized surface protrusions in the (111)-B GaAs deposit. A (100) GaAs substrate was processed in a similar manner and gross polishing damage was observed upon topographic inspection. These slices yielded no evidence of damage under optical examination. The (400) topograph (Figure 7) exhibited polishing scratches traversing the length of the slice in the [1$\overline{1}$0] direction. A series of {115} topographs depicted the same image; the contrast invariance of the image defines the lines as scratches rather than slip. The Ga(As, P) epitaxial film was deposited in a single-slice reactor; the thermal gradients across the seed were known to be small.[18] After deposition, extreme care was taken to prevent thermal shock by cooling the reactor very slowly to room temperature. The (115) topograph (Figure 8) reveals intense slip in ⟨110⟩ directions at the periphery of the image. Another damaged (110) substrate was subjected to the same deposition conditions; the epitaxial film also displayed slip. These

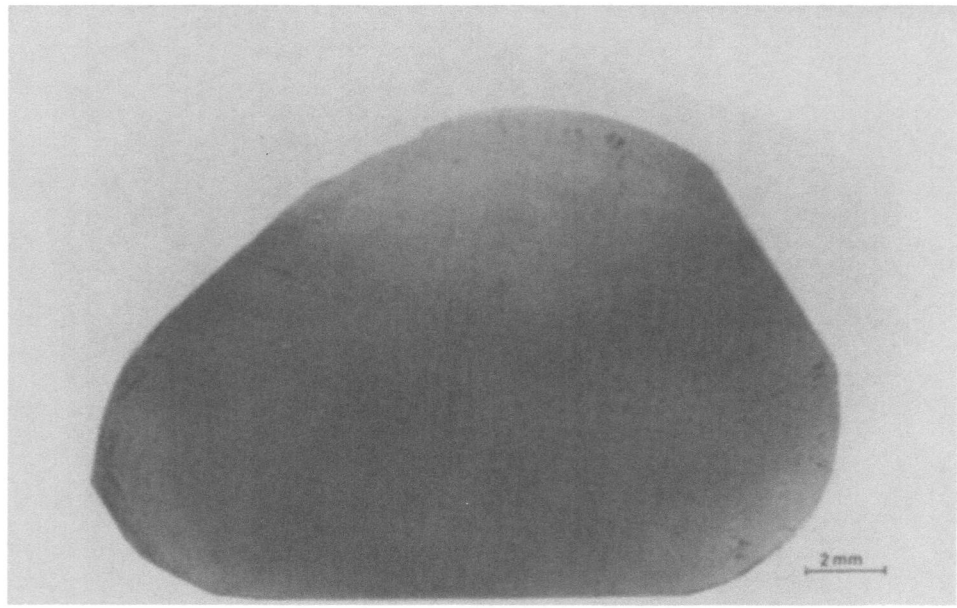

Figure 6. The ($\bar{3}\bar{3}\bar{3}$) topograph of the epitaxial film deposited on the strained substrate.

Figure 7. A (400) topograph of polishing scratches on a (100) GaAs substrate.

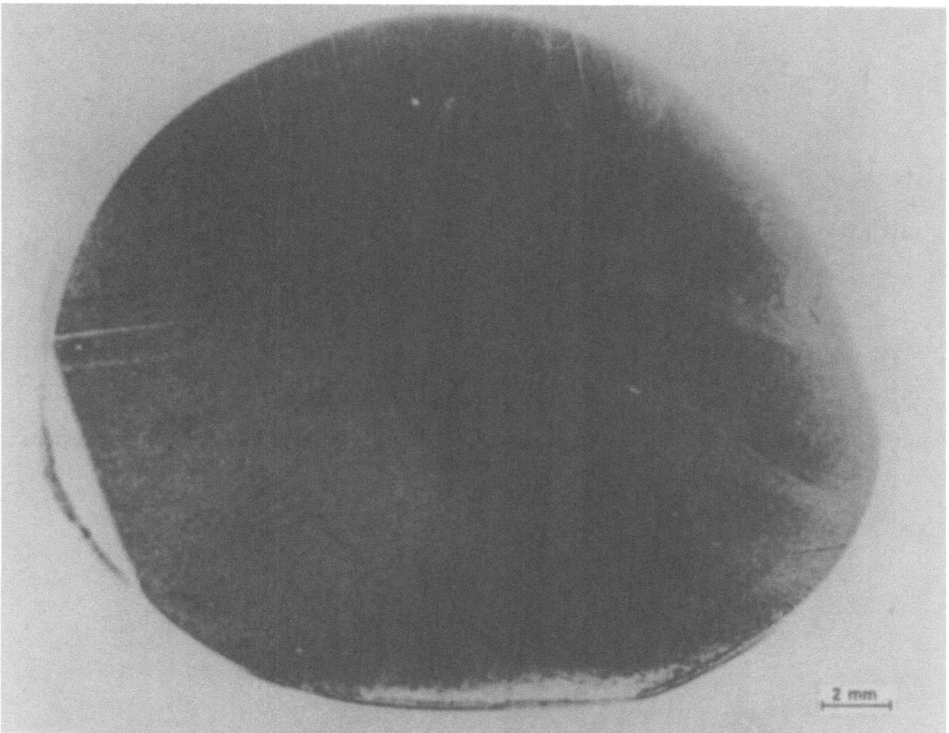

Figure 8. The (115) topograph of the Ga(As, P) film deposited on the damaged (100) substrate in Figure 7. Intense slip is observed in $\langle 110 \rangle$ directions.

results indicate that the damaged layer or the high growth rates, or both, produced strains in the film which were relieved by slip.

The Effect of the Deposition Process on the Deposit Quality

In the early deposits the primary goal was to grow single crystal films of Ga(As, P). Polycrystalline deposits can be detected by the employment of back-reflection Laue methods. The area of sampling in the Laue method was 1.0 mm², whereas the single crystallinity of the entire deposit was of interest. X-ray topography was used to detect deviations in single crystallinity over large-area deposits.

The ($\bar{1}\bar{1}\bar{1}$) topograph (chromium radiation) of a Ga(As, P) deposit (Figure 9) reveals numerous spots superimposed on an uniform intensity image. These spots were produced by diffraction from misoriented regions in the matrix. The local tilt in the diffracting planes associated with the misoriented region may have been produced by a non-uniform distribution of phosphorus or may actually be polycrystalline inclusions. The close separation of the matrix and spot images indicates that the angle of mismatch is small. The structure observed in this deposit was only detected in the first stages of the experimental work.

An inhomogeneous Ga(As, P) deposit can contain local variations in phosphorus concentration or inclusions of foreign impurities. The variations in composition may be gross or microscopic in nature. The variations in composition are easily detected by X-ray topographic examination. Compositional gradients across the Ga(As, P) deposit

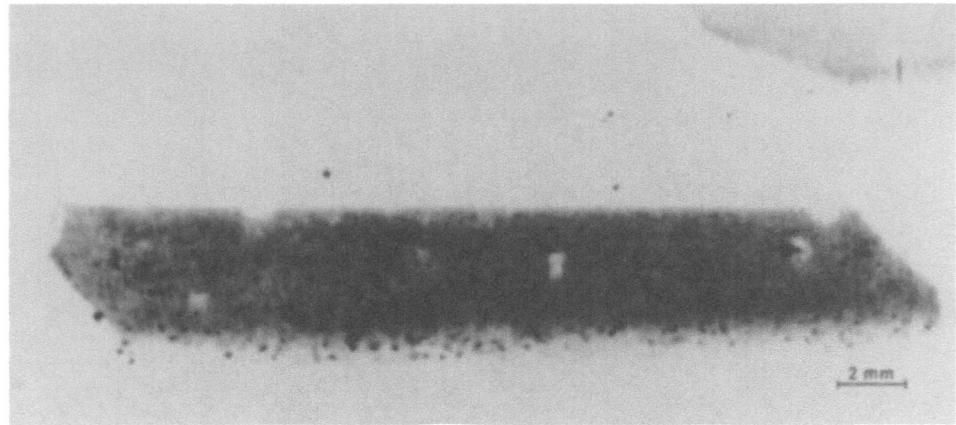

Figure 9. A ($\bar{1}\bar{1}\bar{1}$) topograph of a Ga(As, P) deposit revealing diffraction from misoriented regions in the matrix.

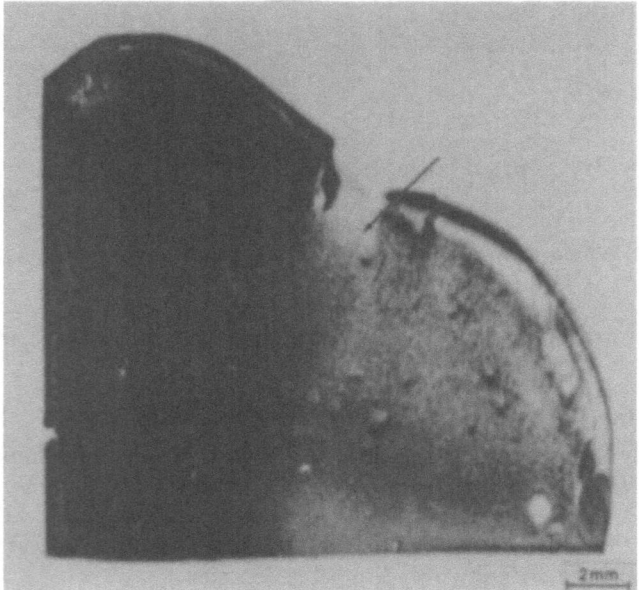

Figure 10. A planar gradient in phosphorus composition is displayed in the ($\bar{3}\bar{3}\bar{3}$) topograph. A linear function (arrow) separates the two regions of different composition (diffraction angle).

are undesirable for device applications. The discovery of a compositional gradient across the slice would appear to be impossible if a non-destructive evaluation was demanded. A (111)-B Ga(As, P) deposit which exhibited a broad line-profile (indicative of inhomogeneity) during an X-ray measurement of phosphorus concentration[7] was examined for inhomogeneity. The ($\bar{3}\bar{3}\bar{3}$) topograph of this film (Figure 10) exposes regions of dark and light contrast separated by a linear junction. An X-ray composition measurement on each side of the junction yielded a 3% variation in phosphorus. The phosphorus-rich matrix corresponds to the region exhibiting the light contrast.

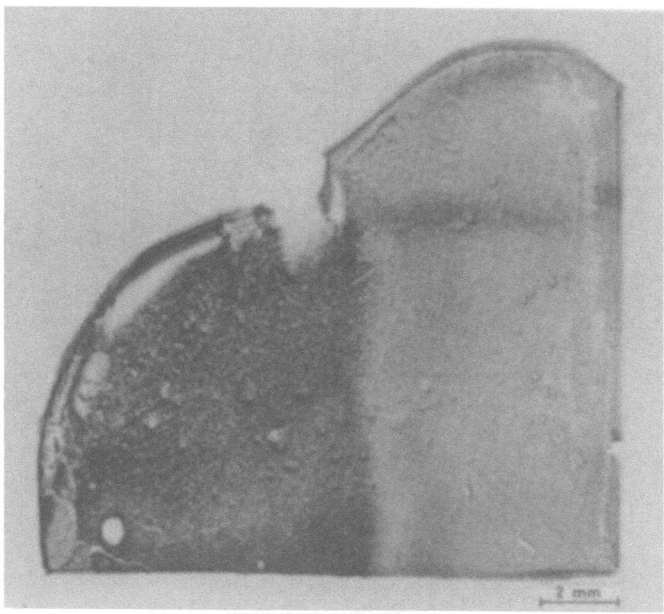

Figure 11. A reversal in contrast of the compositional gradient is obtained by the employment of the diffraction angle corresponding to the phosphorus-rich matrix.

The diffraction angle θ is related to the lattice parameter a_0 for the (hkl) reflection in cubic materials by the expression

$$\theta = \sin^{-1}\left[\frac{\lambda(h^2 + k^2 + l^2)^{\frac{1}{2}}}{2a}\right] \qquad (3)$$

where λ is the wavelength of the incident X-ray beam. A change in lattice parameter or strain produces a change in the diffraction angle. The contrast was reversed by using the diffraction angle which corresponds to the phosphorus-rich matrix (Figure 11). The reactor conditions under which this deposit was grown revealed that the gas stream was directed parallel to the junction. The thermal gradient in this direction was monitored by thermocouples placed near the seed holder. The radial gradient perpendicular to the gas stream was unknown. A difference of 3°C (across the slice) was measured parallel to the gas flow. Since the deposition of Ga(As, P) is temperature-dependent, a radial temperature gradient probably produced the compositional gradient. Other deposits have shown similar compositional gradients; the line-profile of these deposits also indicate the inhomogeneity.[7] The point-to-point variations in intensity observed in Figure 8 also indicate local variations in phosphorus concentration.

Two Ga(As, P) films were deposited simultaneously; inhomogeneity in both films was observed with X-ray topography. The films appear to be of high quality under optical inspection. Large circular regions of null contrast are incorporated into the matrix of one deposit (Figure 12). Figure 13 also displays gross variations in intensity across the image. Inspection of the reactor walls revealed that severe etching of the quartz had taken place. Emission spectrography performed on polycrystalline Ga(As, P) from the reactor walls (a standard method to monitor the impurities present in the deposit) yielded an abnormally high silicon content. It is strongly suspected that the circular regions in

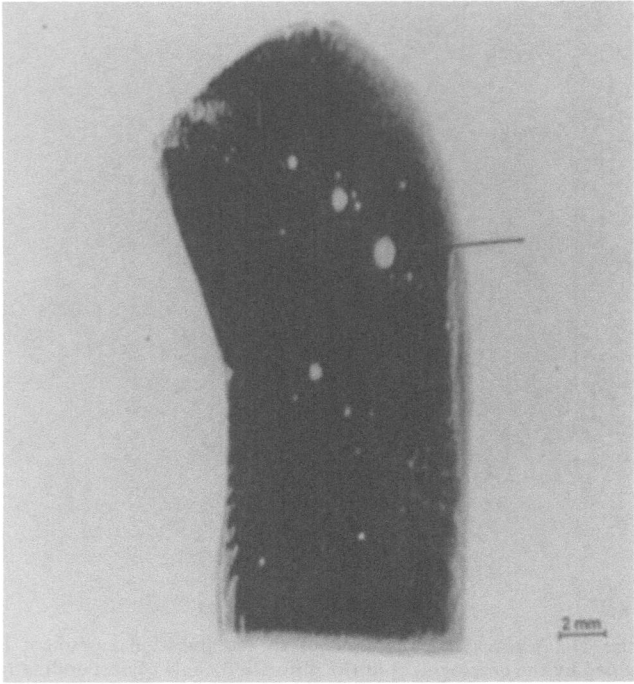

Figure 12. The (3̄3̄3̄) topograph of a Ga(As, P) deposit displays inhomogeneity. The circular regions (arrow) are thought to be silicon (or silica) inclusions which were incorporated into the deposit during epitaxial growth.

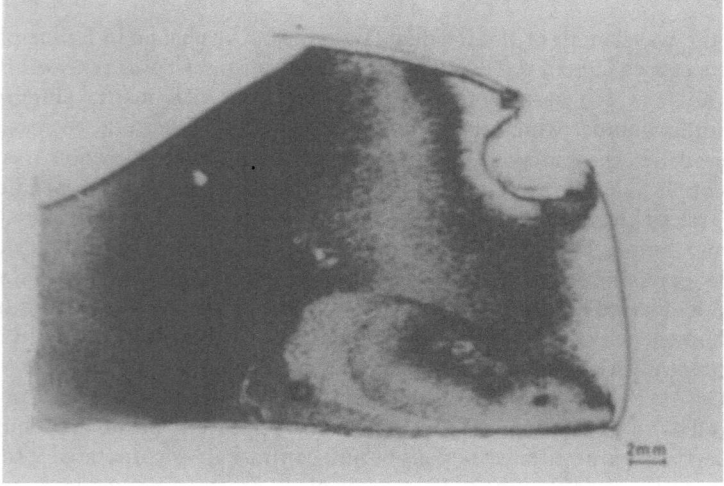

Figure 13. Another silicon-rich Ga(As, P) deposit.

Figure 12 reveal the incorporation of silica (or silicon) into the deposit. The variations in intensity observed in Figure 13 are also attributed to this same effect.

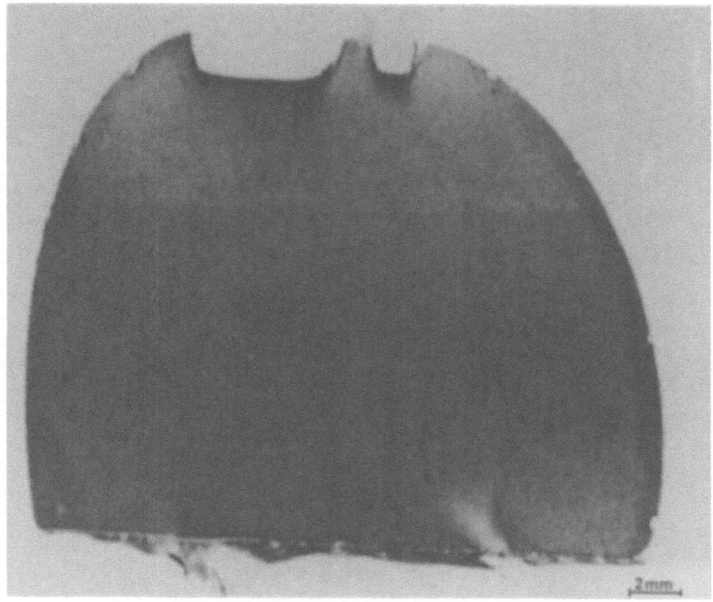

Figure 14. Magnesium-diffused GaAs.

Figure 15. Ga(As, P) deposit before diffusion.

Magnesium Diffusion into GaAs and Ga(As, P)

Magnesium-diffused GaAs and Ga(As, P) were examined to determine if any defects were introduced in the diffusion process. Diffusion was carried out in a sealed

Figure 16. ($\overline{3}\overline{3}\overline{3}$) topograph of the Ga(As, P) deposit after magnesium diffusion; generation of slip was observed (arrow).

quartz ampule at 1100°C; the resulting surface concentration was approximately
4×10^{18} cm^{-3}. The diffusion depth varied from 4–7 microns in the slices evaluated.

The ($\overline{3}\overline{3}\overline{3}$) topograph (Figure 14) of a (111)-B diffused GaAs slice revealed the absence
of any diffusion-induced dislocations or precipitates. Other diffused slices yielded a similar
result. A Ga(As, P) deposit (32% phosphorus) was inspected prior to diffusion (Figure
15) and a structurally uniform image was observed. After diffusion the ($\overline{3}\overline{3}\overline{3}$) topograph
(Figure 16) yielded an image with strong contrast variations; dislocation slip was also
observed (arrow). A phosphorus concentration measurement[7] (before and after diffusion)
indicated a loss of 3%. After removing 1 micron by etching (8H$_2$SO–H$_2$O$_2$–H$_2$O), a
topograph (Figure 17) revealed that the disturbed layer had been nearly removed. However, the diffusion-induced slip was still present. A topograph of another diffused
Ga(As, P) film (24% phosphorus) also showed irregular intensity variations, a loss of 4%
phosphorus (after diffusion) was also determined.

The structural uniformity of the magnesium-diffused GaAs slice suggests that the
heat treatment has a negligible effect on the diffused Ga(As, P). The anomalous surface
image is probably the result of an intensive strain field; dislocation slip relieved the
nonuniformly distributed stress. The loss in phosphorus content may result from a
magnesium–phosphorus reaction at the surface or from the high vapor pressure of
phosphorus at 1100°C. A complete analysis of this effect will be reported at a later date.

Figure 17. ($\bar{3}\bar{3}\bar{3}$) topograph of the same Ga(As, P) after removing
1 micron by etching.

SUMMARY

X-ray topography was employed to investigate some of the defects encountered in epitaxial films of Ga(As, P) which had been deposited onto single crystals of GaAs. An attempt was made to relate the origin of the defect to the substrate perfection or to the deposition process. The following conclusions can be drawn from this study: (1) Scribe marks on GaAs can produce extensive damage; (2) strains in the substrate do not seem to influence the deposit perfection; (3) compositional gradients can be identified with X-ray topography; and (4) magnesium diffusion into Ga(As, P) can produce dislocation slip. Other areas of interest which were suggested by this investigation are: (1) the effect of a mechanical damaged substrate on the film perfection for various substrate orientations, (2) the structural perfection of Ga(As, P) deposits grown in a reactor with a nonstoichiometric source, and (3) further investigations into inhomogeneity and its causes. The problems which were discovered during the topographic examination had not been detected by other methods.

ACKNOWLEDGMENTS

The authors gratefully acknowledge the cooperation and interest in this work shown by Mr. R. W. Haisty and Dr. R. D. Dobrott. Dr. L. G. Bailey is to be thanked for reading the manuscript. Valuable discussions on diffusion were supplied by Dr. H. A. Strack.

DISCUSSION

P. Lublin (General Telephone and Electronics): You mentioned a study of line profiles to detect these compositional variations. How was this done?

J. K. Howard: The sample (film/substrate) was mounted on a goniometer head at the end of a shaft; the shaft was inserted into a standard Norelco diffractometer. Characteristic copper radiation was employed and the diffracted radiation was rendered monochromatic with a focusing monochromator. The GaAs substrate was aligned to diffract the symmetric reflection. The sample was then scanned between the angular limits of the alloy; i.e., 2θ (GaP) to 2θ (GaAs). The Ga(As, P) reflection occurs between these angles. The peak shape of the alloy reflection (whether or not K_{α_1} and K_{α_2} are resolved) indicates qualitatively the degree of inhomogeneity.

P. Lublin: I would suggest that the electron probe would be an excellent instrument to give you composition profiles on the surface, as well as cross sectioning and then determining your concentration gradients across the thickness of your deposits, because you have no idea how the composition varies through the thickness of your deposit.

J. K. Howard: The scanning or counting time necessary to examine variations in composition over 1 in. diameter deposits would seem impractical. The gross inhomogeneity over large area deposits is easily identified in the X-ray topographs. The axial variations in composition can also be investigated with X-ray topography. The slice is cleaved perpendicular to the slice surface to expose a {110} plane; an X-ray topograph of the surface will display the compositional variations.

T. E. Reichard (Monsanto Company): I have a related question I was going to ask. Do you know quantitatively what change in arsenide phosphide ratio will give you a detectable contrast in your topograph?

J. K. Howard: We have experimentally detected a planar variation of 3% GaP (Figures 10 and 11). There have been no attempts to calculate the minimum variation that might be resolved.

T. E. Reichard: I was suspecting that the change in lattice parameter would be a more sensitive detection of the change in this ratio than direct electron probe microanalysis. Is this the impression that you have? That you could see a change in contrast that would not be picked up by electron probe microanalysis?

J. K. Howard: Yes.

T. E. Reichard: Do you find that you have to mount your wafers very carefully to avoid physical strains? Is it very easy to strain the wafer enough to affect your topograph?

J. K. Howard: We have rather thick slices and we usually mount them on clay. So there is no heat treatment or anything like this which could introduce a strain field. If you use very thin samples in silicon, 8 or 10 mils, then you can see the effect of the mounting operation, if you are not careful. However, we are talking about 30 to 40 mil slices and we have not seen any strain introduced by the mounting process.

N. Spielberg (Philips Laboratories): It is not clear to me how you would distinguish in your setup between a change in diffraction angle due to change in lattice constant and that due to tilt or twist of your deposit.

J. K. Howard: A change in lattice parameter produced by a local strain field could yield the same effect on the topographic image. The effect would, of course, depend on the magnitude of the strain. The strain and composition effects were separated in Figures 10 and 11; a 3% GaP variation was determined across the contrast gradient.

N. Spielberg: I believe the technique of rocking the crystals while scanning might help you in this regard.

J. K. Howard: True, but this eliminates the effect from the topograph where we want to see this effect.

REFERENCES

1. J. K. Howard and R. D. Dobrott, to be published.
2. W. W. Webb, "X-Ray Diffraction Topography," in: J. B. Newkirk and J. H. Wernick (eds.) *Direct Observation of Imperfections in Crystals*, Interscience Publishers, Inc., New York, 1962, p. 29.
3. R. W. James, *The Optical Principles of the Diffraction of X-Rays, Vol. II*, Bell, London, 1958, p. 294.

4. A. R. Lang, "Studies of Individual Dislocations in Crystals by X-Ray Diffraction Micro-radiography," *J. Appl. Phys.* **30**: 1748, 1959.

5. U. Bonse, "Zur rontgenographischen Bestimmung des Typs einzelner Versetzungen in Emkristallen," *Z. Physik* **153**: 278–96, 1958.

6. J. B. Newkirk, "The Observation of Dislocations and Other Imperfections by X-Ray Extinction Contrast," *Trans. AIME* **215**: 483, 1959.

7. E. W. Williams, R. H. Cox, and R. D. Dobrott, to be published.

8. W. F. Finch and E. W. Mehal, "Preparation of $GaAs_xP_{1-x}$ by Vapor Phase Reaction," *J. Electrochem. Soc.* **111**: 815, 1964.

9. G. E. Gottlieb, "Vapor Phase Transport and Epitaxial Growth of $GaAs_{1-x}P_x$ Using Water Vapor," *J. Electrochem. Soc.* **112**: 192–6, 1965.

10. M. Rubenstein, "The Preparation of Homogeneous and Reproducible Solid Solutions of GaP–GaAs," *J. Electrochem. Soc.* **112**: 426, 1965.

11. G. R. Antell, "Chlorine and Iodine as Impurities in InAs and GaP," *J. Appl. Phys.* **31**: 1686, 1960.

12. San-Mei Ku, "The Preparation and Properties of Vapor-Grown GaAs–GaP Alloys," *J. Electrochem. Soc.* **110**: 991–5, 1963.

13. E. M. Hull, "Epitaxial Growth of Homogeneous Solid Solutions of GaAs–GaP," *J. Electrochem. Soc.* **111**: 1295–6, 1964.

14. M. S. Abrahams, "Dislocation Etch Pits in GaAs," *J. Appl. Phys.* **35**: 3626, 1964.

15. M. Renniger, "Net Plane 'Interferometry' and Applications," in: G. N. Ramachandran, *Crystallography and Crystal Perfection*, Academic Press, New York, 1963, p. 145.

16. A. Reisman and R. Rohr, "Room Temperature Chemical Polishing of Ge and GaAs," *J. Electrochem. Soc.* **111**: 1425–8, 1964.

17. W. C. Dash, "Distorted Layers in Silicon Produced by Grinding and Polishing," *J. Appl. Phys.* **29**: 228–9, 1958.

18. G. L. Cheney, private communication.

THE APPLICATION OF CYLINDRICAL GEOMETRY FOR THE DETERMINATION OF CRYSTAL ORIENTATION

Robert D. Forest

Dow Chemical Company
Denver, Colorado

Richard J. Barton

University of Saskatchewan
Regina, Canada

and

N. C. Schieltz

Colorado School of Mines
Golden, Colorado

ABSTRACT

A procedure for conveniently determining the orientation of noncubic crystals is presented. The methods usually employed for cubic crystals, flat-film back-reflection Laue patterns interpreted with the aid of a table of interplanar angles, is not readily adaptable to noncubics. In general, tables of interplanar angles for noncubics do not exist and if the effort is expended to generate them, such a vast array of angles result that they are virtually impossible to use. It is thus necessary to use the symmetry of the pattern to identify the low-index planes. However, on flat-film geometry, due to the small angular range of the data, insufficient low-index points are present to permit orientation.

In order to alleviate this problem we have developed the necessary techniques for the interpretation of back-reflection Laue patterns employing cylindrical-film geometry. The necessary overlays and their use are presented along with some of the results obtained.

INTRODUCTION

The conventional flat-film back-reflection Laue cameras record an angular range of plane normals of approximately $\pm$ 30°. It has been our experience that often insufficient data is recorded on the film to permit a ready interpretation. This is especially true if the quality of the pattern is rather poor due to strained crystals which have resulted from experiments such as cleavage, crystal growing, and deformation. If one attempts to increase the angular range by placing the film nearer the sample, the resolution decreases, or if one attempts to increase the resolution, it can be done only at the expense of the amount of data collected; thus the technique is always an unhappy compromise.

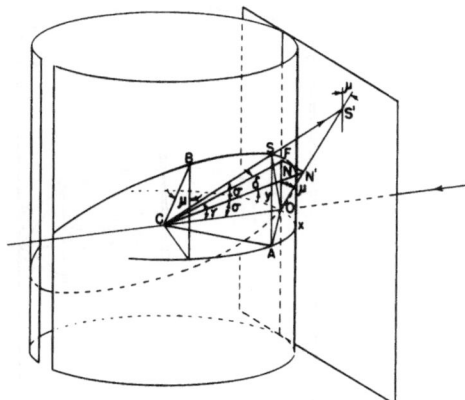

Figure 1. Geometric figure containing both flat film and cylindrical film.

Consequently, we have developed a cylindrical-geometry back-reflection Laue procedure that provides data over a range of ± 90° in azimuth and ± 30° in elevation. This records the diffraction spots of most planes of a crystal that have suitable d-spacing and scattering power.

In this paper the geometry is derived and a few illustrative examples cited. The mathematical equations used in the derivation of the required net, equivalent to the Greninger net for the flat-film geometry, do not necessarily represent the simplest derivation, but the one that could best be utilized on the computer available.

GEOMETRICAL CONSIDERATIONS

The information needed to determine the orientation of a crystal plane is the vertical angle γ and the δ angle of its normal, although the information available is the x and y coordinates of the diffraction spot on the film. Thus one needs a convenient relation between these sets of quantities.

The incident beam, the plane normal, and the diffracted beam define a plane whose intersection with the cylinder of the film gives an ellipse (Figure 1) whose semimajor axis $\overline{CB}$ is oriented at an angle μ to the cylinder axis and has as semiminor axis $\overline{CO}$, the camera radius. The distance $\overline{CB}$ can be obtained from the angles γ and δ, employing the relations in Figure 1.

$$\tan \mu = \frac{\overline{FN'}}{\overline{FO}}$$

where

$$\overline{FN'} = \overline{FC} \tan \sigma = \frac{R \tan \sigma}{\cos \gamma}$$

$$\overline{FO} = \overline{CO} \tan \gamma = R \tan \gamma$$

$$\tan \mu = \frac{R \tan \delta / \cos \gamma}{R \tan \gamma} = \frac{\tan \delta}{\sin \gamma}$$

From Figure 1 it is seen that

$$\overline{CB} = \frac{R}{\sin \mu}$$

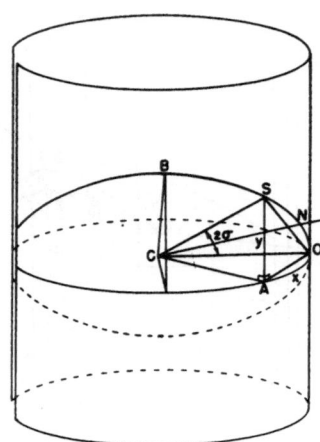

Figure 2. Geometric figure containing cylindrical film only.

Substituting for $\sin \mu$,

$$\overline{CB} = \frac{R \tan \delta}{(\sin^2 \gamma + \tan^2 \delta)^{\frac{1}{2}}}$$

The angle between the incident and the diffracted beam, 2σ (Figure 2), can be obtained from the relation

$$\cos \sigma = \frac{\overline{CO}}{\overline{CN'}} = \cos \gamma \cos \delta$$

where

$$\overline{CO} = \overline{CF} \cos \gamma$$

and

$$\overline{CN'} = \frac{\overline{CF}}{\cos \delta}$$

The distance $\overline{CS}$, which is needed to determine x and y, can be obtained from a consideration of the plane of the ellipse (Figure 3). Taking the polar equation for an ellipse with the pole at the origin,* or using the relationship $\tan 2\sigma = u/v$ and the equation $\overline{CS}^2 = u^2 + v^2$, we have

$$\overline{CS}^2 = \frac{\overline{CB}^2 + \overline{CO}^2}{\overline{CO}^2 \sin^2 2\sigma + \overline{CB}^2 \cos^2 2\sigma}$$

We can now evaluate y (Figure 2) from the relation

$$y^2 = \overline{CS}^2 - \overline{CA}^2 = \frac{R^2}{\sin^2 \mu \sin^2 2\sigma + \cos^2 2\sigma} - R^2 \qquad (1)$$

* Mathematical Tables, *Handbook of Physics and Chemistry*, 36th ed., Chemical Rubber Publishing Company, p. 326.

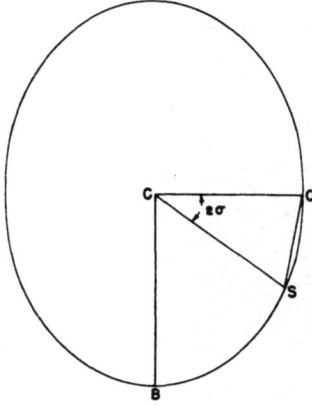

Figure 3. Ellipse formed by intersection of plane containing the incident and diffracted X-ray beams with the film cylinder.

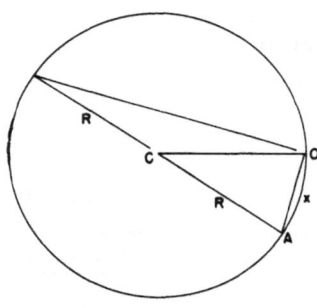

Figure 4. Circle formed by intersection of a plane normal to the film cylinder axis with the film cylinder.

where $\overline{CA} = R$. To determine x, consider the triangle CSO (Figure 3), which yields

$$\overline{OS}^2 = \overline{CS}^2 + \overline{CO}^2 - 2 \cdot \overline{CO} \cdot \overline{CS} \cos 2\sigma$$

and the relation

$$\overline{OA}^2 = \overline{OS}^2 - Y^2$$

In the plane perpendicular to the cylinder axis CAO (Figure 4),

$$\tfrac{1}{2} < ACO = \sin^{-1}\!\left(\frac{\overline{AO}/2}{R}\right)$$

and

$$\tfrac{1}{2} < ACO = \frac{X}{R}$$

thus

$$X = 2 \cdot R \cdot \sin^{-1}\!\left(\frac{\overline{AO}}{2 \cdot R}\right) \tag{2}$$

Equations (1) and (2) were evaluated on a computer, a table of x and y for a given set of values of γ and δ was determined, and the resulting net drawn. A 5° interval net is illustrated in Figure 5; we also have data for 1° and 2° nets.

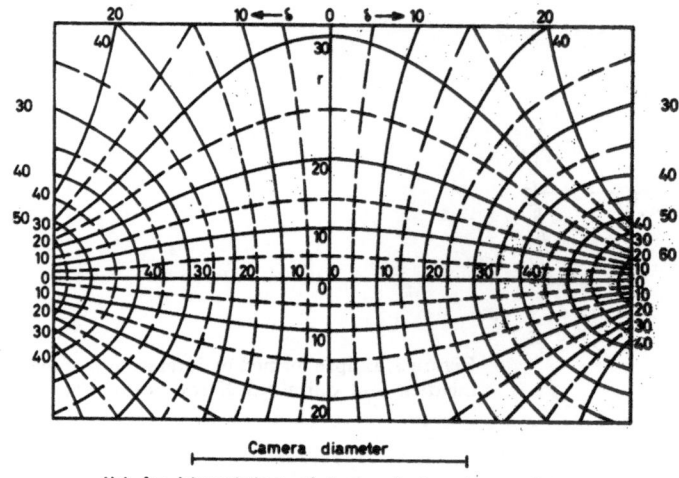

Net for interpretations of back-reflection Laue patterns

Figure 5. Net for cylindrical film equivalent to the Greninger net for flat film.

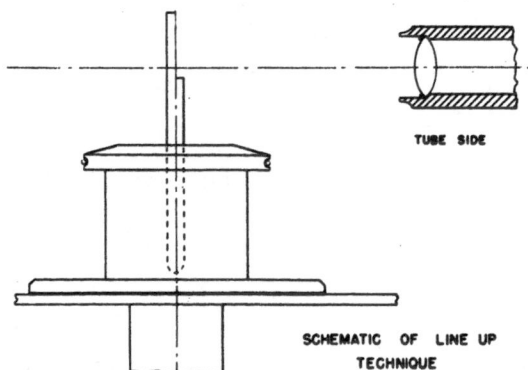

TUBE SIDE

SCHEMATIC OF LINE UP
TECHNIQUE

Figure 6. Instrumental arrangement for making cylindrical back-reflection Laue patterns.

EXPERIMENTAL

Before considering some examples, a few remarks should be made about procedure. Our work has been done on small crystals, mostly for cleavage or thermoelectric power studies; thus we have used a General Electric 10-cm-diameter single-crystal rotation camera for these studies. The procedure adopted here has been to carefully align a half-cylindrical metal pin and then focus on the flat surface the facing X-ray tube with a very shallow field microscope (Figure 6). The pin with its holder was then removed and, leaving the camera base and the microscope in this position, the sample was placed on the table of the goniometer and adjusted until it was in sharp focus. For larger samples, a larger table on the goniometer was substituted for the original one. For very large samples that will not fit into this camera a simple half cylindrical camera has been designed but thus far we have not found need for it.

The geometry resulting for the film curvatures in the one direction requires that the film be properly oriented relative to the net. Thus it is only possible to measure positions of plane normals, not of zone axes as is commonly done with a Greninger net.

Figure 7. Typical cylindrical back-reflection Laue pattern.

This disadvantage is offset by the greater amount of data, which allows the selection of enough important plane normals from the symmetry to rapidly determine the crystal orientation. It is even possible at times to determine the orientation visually and one needs only to measure a few angles to confirm it. This ease of selection of important planes should be especially useful in studies on lower symmetry crystals where tables of interplanar angles do not exist. With these systems, so many angles occur that even if they are calculated it is generally not possible to sort them out. Furthermore, any small change in one or more of the lattice parameters produces a different set of angles so that tabulation for all possible conditions is not feasible.

    A brief illustrative analysis of a cylindrical back-reflection Laue pattern by means of this cylindrical-film net should be quite useful. Figure 7 shows a typical cylindrical back-reflection pattern (Galena crystal) which was used for this illustration. A cursory

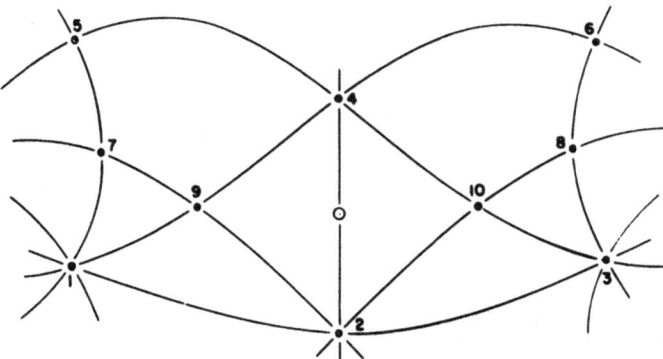

Figure 8. Arrangement of principle low-index reflections which are quite obvious in cylindrical back-reflection pattern.

examination of this pattern readily shows the positions and general arrangement of the principle low-index reflections. This arrangement as observed from the tube side (outside of film cylinder) of the film is shown in Figure 8. To measure the positions of the normals, the operator places the net on a viewing table and then places the back-reflection Laue pattern on the net so that the outside of the cylinder is viewed by the operator, keeping both pattern and net aligned horizontally. Table I shows the results of this measurement.

To measure the angles between the normals the operator now plots the positions shown in Table I on transparent paper by means of a Wulff net (N–S net axis horizontal). The result of this operation is shown in Figure 9. Now the operator turns the transparent overlay on the axis common to both overlay and the Wulff net so that the two positions representing the normals in question lie on a great circle of the Wulff net. In this position the angle between the normals can be read along the great circle in the usual manner. These results can be seen in Figure 9 together with tabulated data, Table II, for the cubic system. These tabulated data can be found in any of the standard books on X-ray metallography or X-ray diffraction for metallurgy.

Table I. Positions of Normals as Measured with a Cylindrical-Film Net

Reflection	$+\gamma°$	$-\gamma°$	$\leftarrow \delta°$	$\rightarrow \delta°$
1		16.5		46
2		17.5		0.5
3		19.0	44	
4	17+		0.5	
5	37			28.5
6	36+		30	
7	19.5			39
8	18−		39.5	
9	2.5			26
10	1.0		25+	

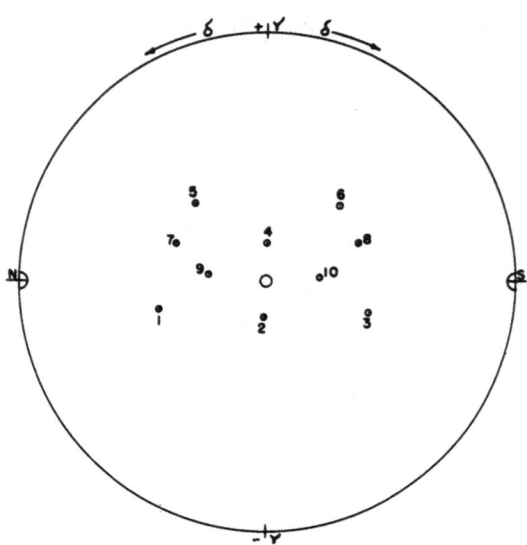

Figure 9. Arrangement of principle low-index reflections after measuring positions of normals and plotting by means of a Wulff net in the usual manner.

Table II. Angles Between Normals as Measured with a Wulff Net

Values of measured angles		Tabulated values		
Normals	Degrees	*hkl*	*hkl*	Degrees
1 2	45.5	100	110	45
2 3	44.5	110	010	45
2 4	34.5	110	111	35.3
4 5	35	111	101	35.3
4 6	35	111	011	35.3
1 7	27.5	100	210	26.6
3 8	27.5	010	021	26.6
2 9	31	110		
2 10	31	110		

DISCUSSION

N. C. Schieltz (Colorado School of Mines): You didn't mention the Wulff net.

R. J. Barton: The practice here with the cubics has been to make direct measurement. With the non-cubics, or sometimes with the cubics when things are extremely messy, we find that you take these readings off using a standard overlay on the Wulff net and proceed then to measure interplanar angles as you normally do. This brings up the one restriction then with this type of geometry, whereas with the flat film geometry you normally can rotate this, locate the poles of zones, with this geometry it is not possible to rotate the film on a net, it must be oriented in a fixed pattern. With the excess amount of data it doesn't prove to be any problem and once you go to the Wulff net it has axial symmetry just like any other Wulff net.

G. M. Gordon (University of California, Berkeley): The camera, you say, was a G.E. 10 cm powder pattern. Do you mean the rotating crystal camera?

R. J. Barton: Yes.

N. C. Schieltz: As a matter of fact, we use that for everything. It's a much better camera than the powder camera for powder work.

Chairman W. V. Cummings: I can buy a Greninger net or a Wulff net? Are yours available?

R. J. Barton: The nets are not. I can give you the points to be plotted. These I have. But the net I showed was a 5 degree net. We have calculations on 1 and 2 degree nets, also. The data exist if somebody wants to plot them, or if somebody has a plotter, they are on tape and can be simply plotted off a computer.

OBSERVATIONS OF G–P ZONE REVERSION IN Al–Zn–Mg ALLOYS BY SMALL-ANGLE X-RAY SCATTERING AND TRANSMISSION ELECTRON MICROSCOPY

R. W. Gould

University of Florida
Gainsville, Florida

and

E. A. Starke, Jr.

Georgia Institute of Technology
Atlanta, Georgia

ABSTRACT

A study of the reversion process in Al–Zn–Mg alloys has been made using small-angle X-ray scattering and transmission electron microscopy techniques. The rate and mode of Guinier–Preston zone dissolutions was investigated as a function of magnesium content, prior zone radius, and reversion temperature. Results indicate that in this system the reversion process is characterized by the preferential dissolution of the smallest G–P zones present after cold aging with a corresponding decrease in the volume fraction of zones. The amount of reversion at a specific temperature is dependent on magnesium content, however, the rate of reversion is independent of magnesium content.

INTRODUCTION

Precipitation-hardening alloys aged at low temperatures and subsequently heated for short periods at a temperature below the solvus often display a marked decrease in strength and hardness. This observation, made first by Gaylor,[1] has been called reversion or retrogression and is believed to be due to the solution of the hardening agents (G–P zones) at the elevated temperatures.[2,3] Earlier studies logically employed hardness measurements to monitor the changes occurring during the reversion process. More recent investigations[4] have utilized electrical resistivity to follow the solution of the zones. Both of these methods, however, involve indirect measurements and as such cannot provide a detailed picture of the changes occurring at the atomic level.

It seemed desirable, therefore, to undertake a study of the reversion process using methods by means of which the changes taking place in the alloy could be directly observed. Two such methods are currently available, i.e. small-angle X-ray scattering and transmission electron microscopy. Herman *et al.*[5] and Graf[6] have made some use of

---

[1] References are at the end of the paper.

the former method in reversion studies of Al–Zn and Al–Zn–Mg alloys, respectively, but a reversion study combining small-angle X-ray scattering and transmission electron microscopy has not been made.

Based upon early work of one of the present authors,[7] the system chosen for this investigation was Al–13 wt.%Zn containing small additions of magnesium. This system is similar to Al–Zn in that the G–P zones are spherical.

The present paper will describe the results of a study of the reversion process in two Al–Zn–Mg alloys using transmission microscopy and small-angle X-ray scattering. It is realized, as pointed out by Embury and Nicholson,[8] that the number of variables in such a study is quite large, but the present results should shed some light on the reversion process.

EXPERIMENTAL PROCEDURES

The alloys used in the present investigation were prepared with high-purity aluminum, zinc and magnesium by melting in a graphite crucible using an induction furnace. The 400 g ingots produced were homogenized for 10 hr at 250°C, 10 hr at 300°C, and 1 week at 450°C. The surface of the ingots was removed with a milling machine and foils 0.1 mm thick prepared by cold-rolling with intermediate anneals. The specimens were cut to the desired size, annealed 1 hr at 460°C, quenched in ice water, and aged at 50°C for 1 week, after which time the average zone radius remained constant. Two alloys were prepared in this manner: the first, designated R_3, was Al–12.8 wt.%Zn–0.066 wt.%Mg, and the second, designated R_2, was Al–12.9 wt.%Zn–0.24 wt.%Mg.* Alloy R_3 was treated for various times in a silicon oil bath at $160 \pm 1°C$ and $135 \pm 1°C$ and quenched into a mixture of dry ice and acetone. Alloy R_2 was given a similar treatment at $135 \pm 1°C$. By comparison with the Al–Zn system,[9,10] it was felt that these temperatures were suitable for a reversion study of these alloys. All samples were stored in dry ice until examined.

The small-angle X-ray examination was carried out at 1°C using a Kratky camera on a modified Norelco diffractometer as described elsewhere.[11]

Thin foils were prepared (at $-60°C$) from the same samples used in the X-ray investigation by the window technique of electropolishing.[12] These were subsequently examined by transmission in a Philips EM200 electron microscope.

EXPERIMENTAL RESULTS

Alloy R_2: Al–12.9 wt.%Zn–0.24 wt.%Mg

The sequence of reversion treatments at 135°C for this alloy and the corresponding X-ray data are given in Table I. The zone radii listed in this table were obtained by using Guinier's approximation utilizing the slope of an ln I versus ϵ^2 plot of the small-angle X-ray intensity data. Baur and Gerold[13] have shown that this method gives results comparable with the average radii obtained from electron microscopic observations. The ratio $Q_0(t)/Q_0(t=0)$ listed in Table I is the ratio of the integrated intensity after reversion time t at reversion temperature T_R to that of the integrated intensity before reversion. This ratio is proportional to the volume fraction of zones present.

The actual small-angle X-ray curves obtained after various treatments are shown for comparison in Figure 1. A plot of zone radius versus log time and integrated intensity versus log time obtained from the curves of Figure 1 is presented in Figure 2. Figures 3–8 are transmission electron micrographs obtained from the same foils used for the X-ray investigation.

* Chemical analysis kindly provided by Aluminum Company of America.

Table I. Small Angle X-Ray Data for the Reversion of Alloy R_2 (Al–12.9 wt.%Zn–0.24 wt.%Mg at 135°C)

Time t (min)	Zone radius before reversion (Å)	Zone radius after time t (Å)	$Q_0(t)/Q_0(t=0)$*
0	30.9	—	1.0
2	29.0	33.5	0.321
5	29.7	31.8	0.343
20	30.1	31.8	0.288
100	28.8	36.7	0.404
1000	27.7	41.6†	0.340

* $Q_0(t)$ is the integrated intensity after time t at reversion temperature T_R and $Q_0(t=0)$ is the integrated intensity before reversion.
† Scattering was probably influenced by precipitate particles.

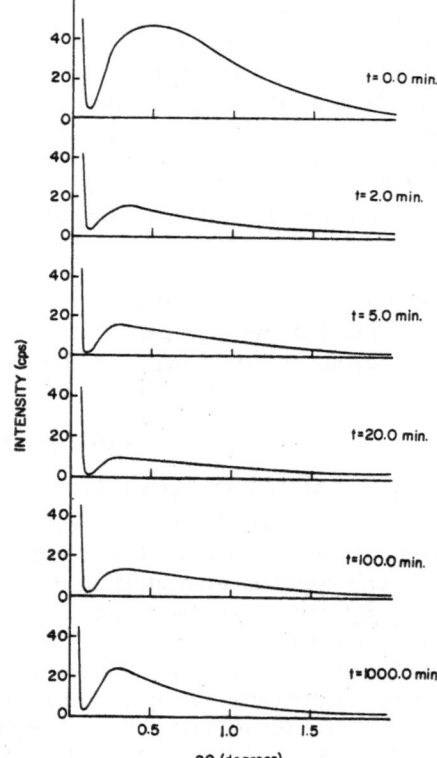

Figure 1. Small-angle scattering curves for Alloy R_2 (Al–12.9 wt.%Zn–0.24 wt.%Mg) reverted at 135°C for various times t.

Measurements of zone diameters were made with a graduated magnifying eye piece on magnified transparencies of the original microscope film. The center of the zone generally shows good contrast, however, the periphery gave poor contrast, possibly due to coherency strains and variance in zone composition between the center and edge.

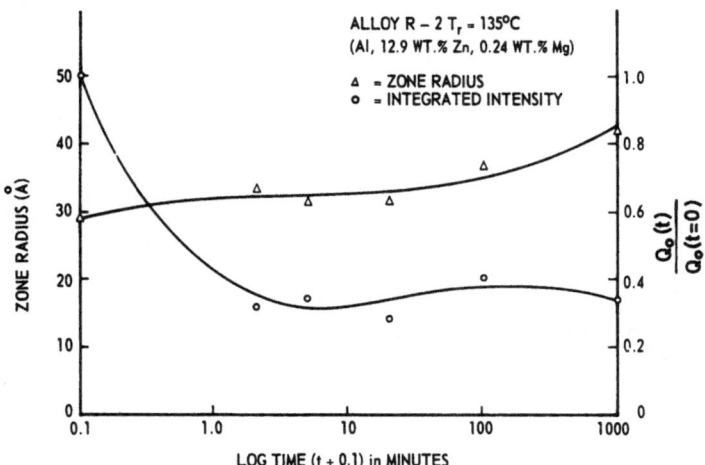

Figure 2. Zone radius and integrated intensity versus log revision time
for alloy R₂.

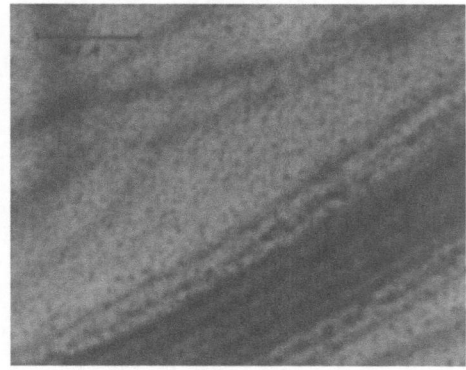

Figure 3. Transmission electron micrographs
of alloy R₂ reverted at 135°C for 0 min.

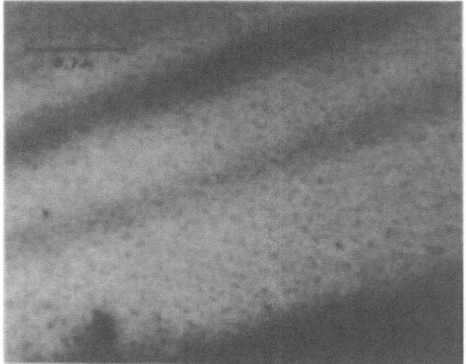

Figure 4. Transmission electron micrographs
of alloy R₂ reverted at 135°C for 2 min.

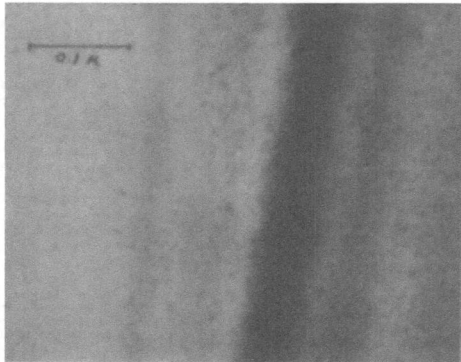

Figure 5. Transmission electron micrographs
of alloy R_2 reverted at 135°C for 5 min.

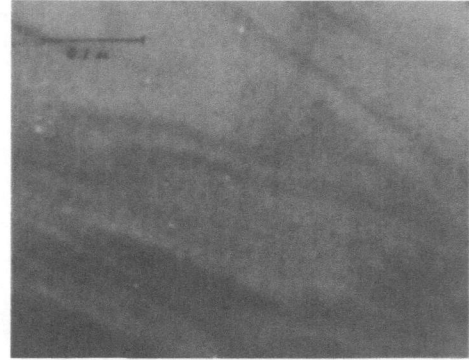

Figure 6. Transmission electron micrographs
of alloy R_2 reverted at 135°C for 20 min.

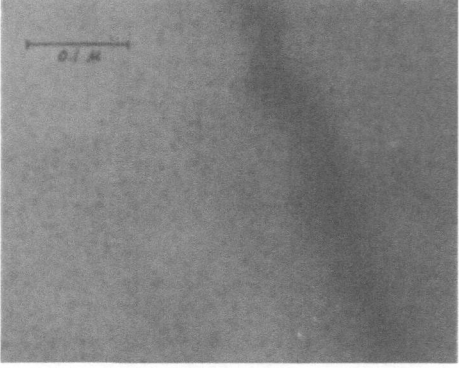

Figure 7. Transmission electron micrographs
of alloy R_2 reverted at 135°C for 20 min.

Consequently the measurements of zone diameter might be systematically underestimated by 20%. Figure 9 shows histograms of a large number of visual measurements of zone diameters made from Figures 3–8.

Alloy R_3: Al–12.8 wt.%Zn–0.066 wt.%Mg

The effect of temperature was introduced in the study of this alloy by observing the reversion both at 135 and 160°C. Table II contains information on the sequence of study at 135°C and Table III gives similar information on the 160°C treatment. The small-angle X-ray curves are given in Figure 10 for various times at 135°C and in Figure 11 for various times at 160°C. The main difference between the two treatments was the rapidity of the reversion process at 160°C. This can be seen quite clearly by comparing Figure 12, which is a plot of zone radius and integrated intensity versus log time at 135°C, with Figure 13, which is similar data obtained at 160°C.

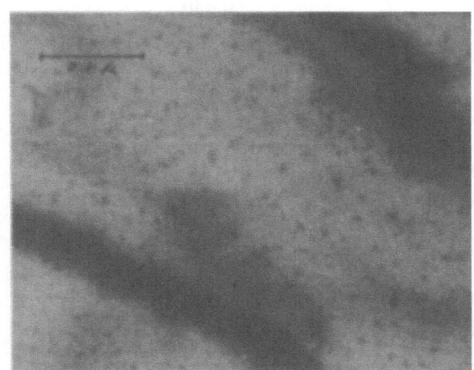

Figure 8. Transmission electron micrographs of alloy R_2 reverted at 135°C for 1000 min.

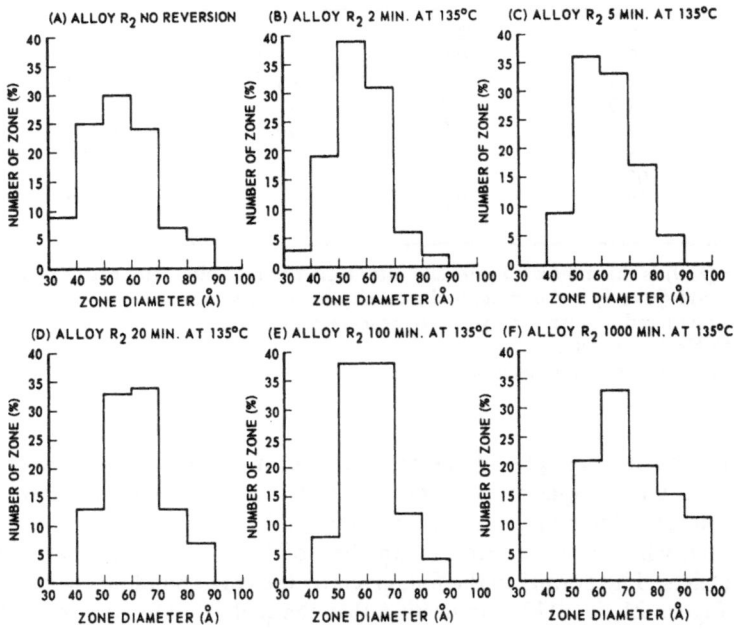

Figure 9. Histograms of zone diameters, measured on electron micrographs, of alloy R_2, reverted at 135°C for various times.

Table II. Small-Angle X-Ray Data for the Reversion of Alloy R_3
(Al–12.8 wt.%Zn–0.066 wt.%Mg at 135°C)

Time t (min)	Zone radius before reversion (Å)	Zone radius after time t (Å)	$Q_0(t)/Q_0(t=0)$*
0	39.0	—	1.0
2	38.0	38.1	0.628
5	40.0	36.7	0.344
10	38.0	39.6	0.285
40	38.0	37.8	0.220
100	39.0	†	0.336
400	39.0	†	0.360
1165	38.0	†	0.331
1550	38.0	†	0.284

* $Q_t(0)$ is the integrated intensity after time t at reversion temperature T_R and $Q_{t=0}(0)$ is the integrated intensity before reversion.
† No straight line portion on Guinier plot.

Table III. Small-Angle X-Ray Data for the Reversion of Alloy R_3
(Al–12.8 wt.%Zn–0.066 wt.%Mg at 160°C)

Time t (min)	Zone radius before reversion (Å)	Zone radius after time t (Å)	$Q_0(t)/Q_0(t=0)$*
0	42.3	—	1.0
0.25	40.0	43.0	0.482
0.50	42.3	43.0	0.298
1.0	42.3	42.4	0.240
2	42.3	40.0	0.105
5	42.3	†	0.056
100**			
1000	42.3	††	0.107

* $Q_t(0)$ is the integrated intensity after time t at reversion temperature T_R and $Q_{t=0}(0)$ is the integrated intensity before reversion.
† Scattered intensity too weak to measure zone radius.
** No X-ray data.
†† No straight line position on Guinier plot.

The electron micrographs of the foils treated at 160°C are presented in Figures 14–21. Electron micrographs taken from foils treated at 135°C are not shown since they are not significantly different from those obtained at 160°C.

Figure 20, which is an electron micrograph of R_3 after 1000 min at 160°C, shows oriented platelets. The normal to the foil, given by the corresponding electron diffraction pattern of Figure 21, is $\langle 110 \rangle$, indicating that the platelets are lying on {111} planes. Owing to the habit plane of these platelets and due to the low concentration of magnesium in this alloy, the platelets are tentatively identified as β-Zn.

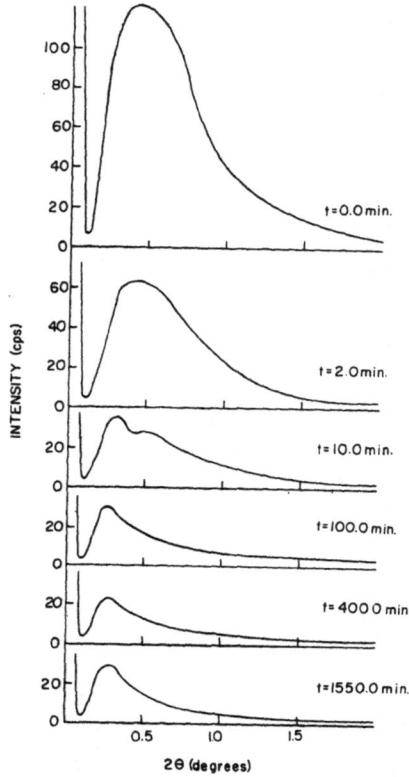

Figure 10. Small-angle scattering curves for alloy R_3 (Al–12.8 wt.%Zn –0.66 wt.%Mg) reverted at 135°C for various times t.

OBSERVATIONS

The results of the present study appear to substantiate the earlier conclusion of Belbeoch and Guinier[14] that the reversion process is characterized by preferential dissolution of the smallest G–P zones present after cold-aging. This is clearly illustrated by the results from alloy R_2. First, the X-ray results show that, while the volume fraction of zones decreases rather rapidly during the reversion treatment, the average zone size slightly increases. This in itself indicates that the small zones are being dissolved while the large ones are not. Second, the electron micrographs and the resulting histograms show that in the early stage of reversion the mean zone diameter stays approximately constant while the percent of small zones decreases rapidly. The change in size distribution might partially explain the shape change in the small-angle X-ray diffraction curves, shown in Figure 1.

The results obtained from alloy R_3 were also consistent with Belbeoch and Guinier's conclusions on dissolution, although the extent of reversion was different for the two alloys. This can be seen by comparing Figure 2 with Figures 12 and 13. For alloy R_2 (Al–12.9 wt.%Zn–0.24 wt.%Mg) the integrated intensity, which is a measure of volume fraction of zones present, leveled off after a few minutes at 135°C but never, even after 1000 min, appeared to approach zero. This indicates an irreversible portion in the reversion process of this alloy at this temperature.

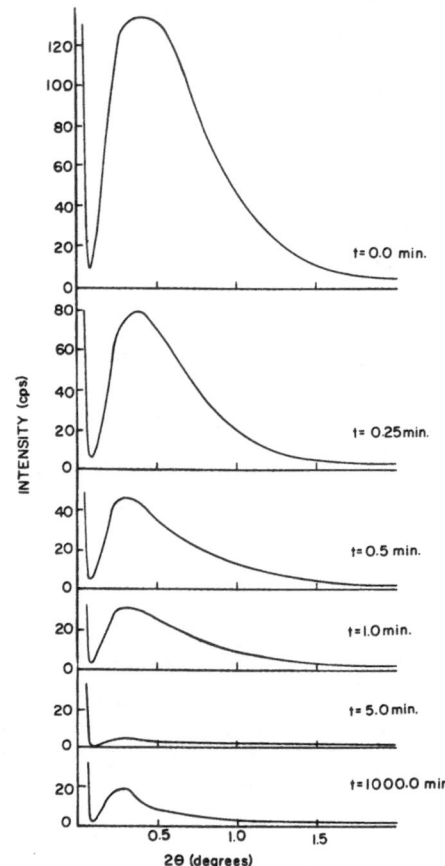

Figure 11. Small-angle scattering curves for alloy
R₃ reverted at 160°C for various times t.

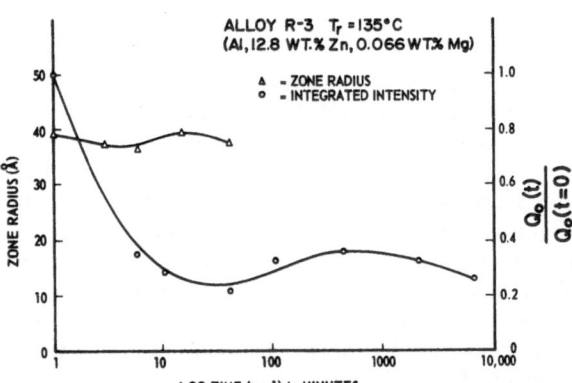

Figure 12. Zone radius and integrated intensity versus log reversion
time for alloy R₃ ($T_R = 135°C$).

X-ray data for alloy R₃, aged at 135°C, indicate that the volume fraction of zones
drops to a very low level even after 10 min at that temperature. The zone radius could not
be measured with any precision after 40 min of reversion. This was due to the increased
curvature of the Guinier plot, ln I versus ϵ^2. As Figure 13 shows, the reversion process
was essentially complete after 5 min at 160°C. The slight increase in integrated intensity
after 1000 min at 160°C might possibly be due to the precipitates shown in Figure 20.

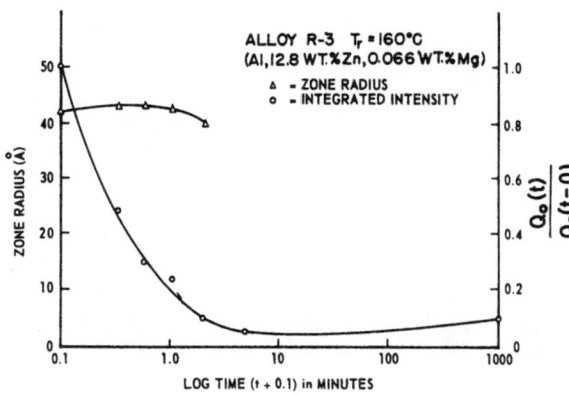

Figure 13. Zone radius and integrated intensity versus log reversion time for alloy R_3 ($T_R = 160°C$).

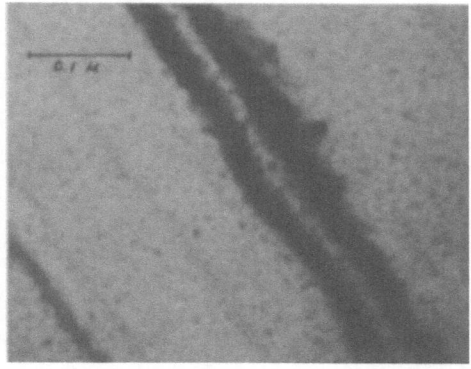

Figure 14. Transmission electron micrographs of alloy R_3 reverted at 160°C for 0 min.

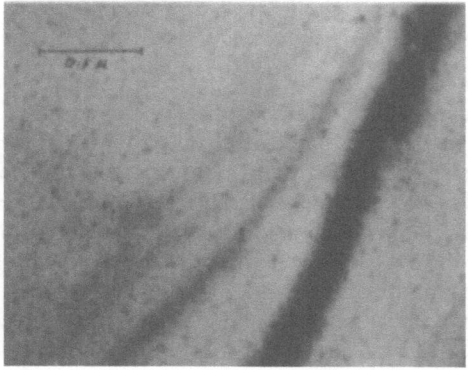

Figure 15. Transmission electron micrographs of alloy R_3 reverted at 160°C for 0.5 min.

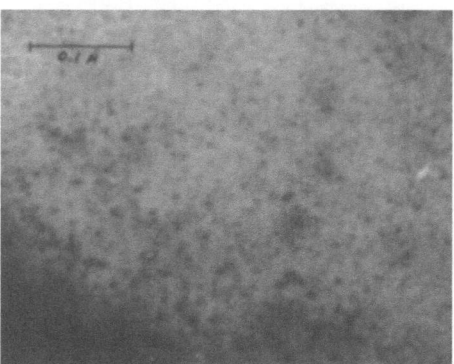

Figure 16. Transmission electron micrographs of alloy R_3 reverted at 160°C for 1.0 min.

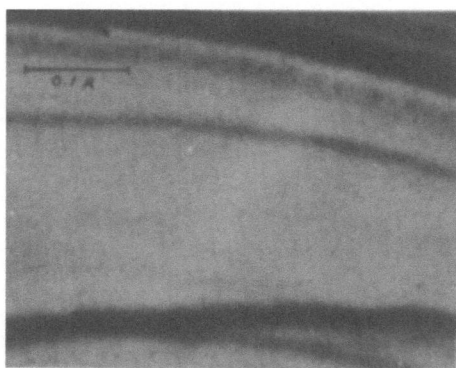

Figure 17. Transmission electron micrographs of alloy R_3 reverted at 160°C for 2 min.

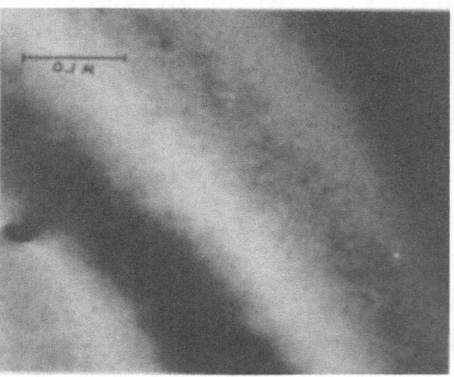

Figure 18. Transmission electron micrographs of alloy R_3 reverted at 160°C for 5 min.

The unreverted portion of the small-angle scattering intensity might be explained in several ways. It is possible that 135°C may be below the metastable miscibility gap for the ternary Al–Zn–Mg alloys studied. Polmear[15] has shown that the upper temperature limit of stability of G–P zones increases rapidly as one enters the ternary system from the Al–Zn binary. Graf's results[16] for an Al–9 wt.%Zn alloy containing 1 wt.% magnesium at a similar temperature would argue against this possibility. A second cause for the unreverted small-angle scattered intensity in R_2 might be the presence of two types of zones, one containing both magnesium and zinc and the other containing nearly pure zinc, analogous to the zones which form in Al–Zn. The presence of two types of zones has

Figure 19. Transmission electron micrographs of alloy R₃ reverted at 160°C for 100 min.

Figure 20. Transmission electron micrographs of alloy R₃ reverted at 160°C for 1000 min.

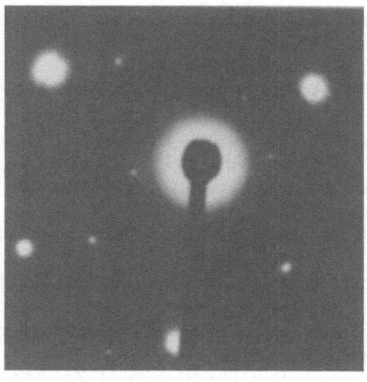

Figure 21. Transmission electron micrographs of alloy R₃ reverted at 160°C for selected area diffraction pattern for 1000 min.

previously been suggested by Guinier.[2] The complex Mg–Zn zones might be stable at this temperature while the zinc zones would dissolve. This is similar to the process observed by Graf[17] for an Al–7 wt.%Zn–3 wt.%Mg alloy. A third possibility for explaining the unreverted small-angle intensity is the presence of small precipitate particles. These particles, lying on {111} planes, were detected after long reversion times at 135°C, but their occurrence does not parallel the drop in integrated intensity. Thus, while they may explain the remaining small-angle intensity after long times, they cannot account for the low intensity found early in the reversion process. Therefore, one might conclude that the amount of reversion at a particular temperature is dependent on the magnesium content in this system. However, contrary to the conclusions drawn by Federighi and Thomas[18] the rate of reversion appears independent of magnesium content. This is illustrated quite clearly by comparing the integrated intensity versus long time plots of Figures 2 and 3.

CONCLUSIONS

1. The reversion process in the Al–Zn–Mg system is characterized by the preferential dissolution of the smallest G–P zones present after cold-aging with a corresponding decrease in the volume fraction of zones. Therefore, the strength loss of the alloy can be attributed to either the absence of small zones or decrease in total volume of zones or both.

2. The amount of reversion at a specific temperature is dependent on magnesium content for one or both of two reasons: (1) the change of the upper temperature limit of the metastable miscibility gap with magnesium content and (2) the presence of two types of zones, the temperature of stability being different for each type.

3. Contrary to the views of Federighi and Thomas, the rate of reversion seems to be independent of magnesium content.

ACKNOWLEDGMENTS

The authors wish to acknowledge the financial assistance given this research by the Army Research Office, Durham, North Carolina (Gould) and Georgia Institute of Technology Engineering Experiment Station and E. I. DuPont de Nemours Company (Starke). Appreciation is also expressed for the assistance of James Johnson (electron microscopy) and Guy Sheble (small-angle X-ray scattering measurements).

REFERENCES

1. M. L. V. Gaylor, "The Constitution and Age-Hardening of Alloys of Aluminum with Copper, Magnesium, and Silicon in the Solid State," *J. Inst. Metals* **28**: 213, 1922.
2. A. Guinier, "Heterogeneities in Solid Solutions," in: F. Seitz and D. Turnbull (eds.), *Solid State Physics: Advances in Research and Applications, Vol. 9*, Academic Press, New York, 1959, p. 293.
3. A. Kelly and R. B. Nicholson, "Precipitation Hardening," in: B. Chalmers (ed.), *Progress in Materials Science, Vol. 10*, The Macmillian Company, New York, 1963, p. 149.
4. C. Panseri and T. Federighi, "A Resistometric Study of Pre-Precipitation in Al–10% Zn," *Acta Met.* **8**: 217, 1960.
5. H. Herman, J. B. Cohen, and M. E. Fine, "Formation and Reversion of Guinier–Preston Zones in Al–5.3 wt.%Zn," *Acta Met.* **11**: 43, 1963.
6. R. Graf, "Some Observations of the Reversion Phenomenon in an Al–10%Zn Alloy," *Compt. Rend.* **249**: 1110, 1959.
7. R. W. Gould, "The Influence of Small Quantities of Magnesium Upon the Pre-Precipitation Behavior of Al–13 wt.%Zn Alloys," Dissertation, University of Florida, 1964.
8. J. D. Embury and R. B. Nicholson, "The Nucleation of Precipitates: The System Al–Zn–Mg," *Acta Met.* **13**: 403, 1965.
9. H. Herman and J. B. Cohen, "Resistivity Changes Due to the Formation of G–P Zones," *Nature* **191**: 63, 1961.
10. V. Gerold, "Zone Formation in Aluminum–Zinc Alloys," *Physics Status Solidi* **1**: 37, 1961.
11. R. W. Gould and V. K. Gerold, "Adaptation of the Norelco High Angle Diffractometer for Small-Angle Scattering Studies of Pre-Precipitation Phenomenon," *Norelco Reporter* **XII**: 7, 1965.
12. R. G. Nicholson, G. Thomas, and J. Nutting, "A Technique for Obtaining Thin Foils of Aluminum Alloys for Transmission Electron Metallography," *Brit. J. Appl. Phys.* **9**: 25, 1958.

13. R. Baur and V. Gerold, "Comparative X-ray and Electron Microscopic Measurements of the Size of Guinier–Preston Zones in an Aluminum–Silver Alloy," *Acta Met.* **12**: 1449 (1964).
14. B. Belbeoch and A. Guinier, "Relation Between Structures and Properties During Age Hardening in Al–Ag Alloys," *Acta Met.* **3**: 370, 1955.
15. I. J. Polmear, "The Upper Temperature Limit of Stability of Guinier–Preston Zones in Ternary Aluminium–Zinc–Magnesium Alloys," *J. Inst. Metals* **87**: 24, 1958–59.
16. R. Graf, "An X-Ray Study of the Precipitation Phenomenon in an Al–Zn–Mg Alloy with 9%Zn and 1%Mg (AZ 9 Gl)," *Compt. Rend.* **244**: 337, 1957.
17. R. Graf, "An X-Ray Study of the Precipitation Phenomenon in an Al–Zn–Mg Alloy Containing 7%Zn and 3%Mg," *Compt. Rend.* **242**: 1311, 1956.
18. T. Federighi and G. Thomas, "The Interaction Between Vacancies and Zones and the Kinetics of Pre-Precipitation in Al-Rich Alloys," *Phil. Mag.* **7**: 127, 1962.

DISCUSSION

L. S. Birks (Naval Research Laboratory): If I understood you correctly, you seem to be making a cause and effect relationship between the number of small zones and the hardness of the alloy. I claim ignorance in this field, so I can ask a stupid question. Are you sure that this is a cause and effect relation or could it be that whatever is causing these small zones to disappear is also changing the hardness, but the two are not necessarily related?

E. A. Starke: A very good point. We made the observation that the small zones disappear when the hardness decreases. Therefore, we attributed the hardness to the small zones being present. However, due to the dissolution of the small zones, the volume fraction decreased. Consequently the hardness might be related to the decrease in volume fraction.

A. Taylor (Westinghouse Research Laboratories): As an X-ray man, I want to ask a question which argues against the use of X-rays. What is the point of using small-angle scattering when you have the facilities to actually see the zones themselves and to make appraisals of the shapes and sizes?

E. A. Starke: Both techniques are useful. All previous workers in this field had been using small-angle scattering techniques. However, it is very difficult to interpret what is going on during the reversion process from the small-angle scattering curves. This is one point that should be emphasized. One could not tell from the small-angle scattering curves that the reversion process was accompanied by the dissolution of the small zones only. This, however, was obvious from the electron micrographs. From small-angle scattering one only obtains the fact that the volume fraction has decreased. It should be pointed out, however, that zones as small as 9 Å in diameter may be detected with the X-ray technique, whereas this would be difficult, though not impossible, with electron microscopy. Also, small-angle scattering is faster and gives the average zone size and volume percent. Therefore, one method compliments the other.

A. Taylor: Would electron microprobe analysis help in this work?

E. A. Starke: No, since the zones are from 10 to 100 Å in diameter, and the beam diameter of the electron microprobe is of the order of 10,000 Å.

B. S. Sanderson (National Lead Company): I wonder why you use the Guinier method of estimating the size. To interpret it as a radius of gyration one should get the slope of the plot that you indicate at zero angle.

E. A. Starke: The slope of the plot used for the radius calculation was determined from points taken from the curve at 80% and 40% maximum intensity. In the December 1964 issue of *Acta Metallurgica*, Baur and Gerold compare the different methods of measuring the zone radius from the small-angle scattering curves to that measured from transmission electron micrographs. They show three different methods which use the Guinier plot to determine the radius. One method, their R_3, gave the average zone size and this is the method we used in our calculations.

B. S. Sanderson: I would be willing to go along with your differences, but I would be hard pressed to look at it as an absolute size.

E. A. Starke: No, it is not an absolute size. That is why we did the histograms.

A. Taylor: Now that you have used the electron microscope on this, you can rule out, I suppose, the possibility of double Bragg scattering, because everything is so nicely separated.

E. A. Starke: Right, we can do that. Also, we are sure there is no double Bragg scattering from our kinetic studies. When we take a supersaturated solid that has no zones present and make a small-angle X-ray study we observe no intensity, i.e., no double Bragg scattering. As the sample ages and zones form we observe a peak which is initially low in intensity, but which increases in intensity with aging time. Therefore, we know that the intensity we are observing is not double Bragg scattering but due to the G–P zones.

E. F. Sturcken (E. I. DuPont Company): Were you able to make any observations about the distribution of the zones during this study?

E. A. Starke: This study is still under way. In fact, we have just begun it. We did not determine from our transmission micrographs the volume fraction present, only the zone distribution. We used X-rays to determine the volume fraction. But one can determine the volume fraction from the transmission micrographs. One has to know, however, the thickness which can be obtained in a number of different ways: from precipitates, measurement of dislocations that go all the way through the foil, etc.

E. T. Peters (Manlabs, Inc.): I am curious to learn a little bit more about your specimen preparation for transmission microscopy, and, also, what is your confidence that you did not change the specimen during specimen preparation?

E. A. Starke: The samples were electropolished with a nitric acid and methanol solution and cooled to $-60°C$ with acetone and dry ice. A large copper grip was used to hold the sample, which was efficient in conducting heat away from the sample during the polishing. The solution was stirred at all times and a thermometer placed near the sample indicated that the bath temperature remained constant. At one time a thermocouple was spot-welded to the sample and a temperature increase of only $7°C$ was detected during polishing. Since no changes could be expected at this low temperature $(-53°C)$ we are confident that our specimen was unaltered during preparation. A discussion on specimen temperature during electropolishing appeared in the January 1955 issue of *Transactions AIME*.

SUBSTRUCTURE MEASUREMENTS BY STATISTICAL FLUCTUATIONS IN X-RAY DIFFRACTION INTENSITY

E. F. Sturcken, W. E. Gettys,* and E. M. Bohn†

E. I. du Pont de Nemours & Co.
Aiken, South Carolina

ABSTRACT

The substructures of a beta-quenched and a recrystallized form of high-purity uranium were measured by a method based on statistical fluctuations in X-ray diffraction intensity. For these measurements, Warren's[1] statistical equation for determining grain size was modified to make the equation applicable to materials with high absorption coefficients or moderate-to-large grain size (> 20 microns) or both, since many metals fall into this category, and to allow for defocusing of the X-ray beam which occurs as a natural consequence of the experiment.

The beta-quenched uranium was found to have numerous subgrains with a range of misorientation angles that was smaller and larger than the limits of the X-ray measurements ($\Omega = 10^{-4}$ to 10^{-2} steradians). The presence of the large subgrains was corroborated by optical microscopy. The presence of very small subgrains was corroborated by transmission electron microscopy which showed 0.1- to 1-micron subgrains relatively free of dislocations bounded by dense dislocation networks, and by micro Laue diffraction patterns (30-micron beam diameter) which showed partial rings similar to a powder pattern.

The recrystallized uranium had no misorientation within the grains greater than 5.5×10^{-3} steradians. In contrast to the beta-quenched case, no subgrains were found either by transmission electron microscopy (TEM) or micro Laue diffraction patterns. The TEM micrographs showed a uniform distribution of dislocation networks. Since no other substructural elements were observed, the dislocations are believed to be the cause of the misorientation within the grains for solid angles of less than 5×10^{-3} steradians.

These preliminary experiments show that the statistical method may be used in conjunction with transmission electron microscopy and micro Laue diffraction for the study of substructure. The statistical method gives quantitative data on "bulk" specimens that can be given a meaningful interpretation with the aid of the other techniques.

INTRODUCTION

The use of photographic X-ray diffraction techniques for determining the perfection and size of grains dates back to at least 1935. For discussions and references to original papers on this subject, the reader is referred to the textbooks of C. S. Barrett[2] and A. Taylor[3] and to the papers of P. B. Hirsch.[4-7]

In the present study a powder diffractometer was employed. Diffractometer techniques are relatively new (1960) but have been discussed in papers by B. E. Warren[1] and

---

* Clemson College.
† Presently a graduate student at the University of Illinois.
[1] References are at the end of the paper.

C. S. Barrett.[8-9] By definition, a powder diffraction sample is one in which a large number of crystals (or grains) contribute to a given diffraction peak so that the integrated peak intensity remains constant for any position or any orientation of the sample (ignoring absorption effects). If only a small number of crystals contribute to a given peak (e.g., in a coarse-grained sample), then the sample is not an ideal powder, and changes in position or orientation of the sample cause large fluctuations in the integrated diffraction intensity.

Statistical fluctuations of the above type have in the past been a nuisance to the diffractionist; however, B. E. Warren[1] has shown that these intensity fluctuations may be used to determine the absolute grain size of the sample by the equation

$$\langle D^3 \rangle = \frac{3jA_0\Omega\gamma}{2\pi^2\mu} \frac{\langle (Y - \langle \bar{Y} \rangle)^2 \rangle}{(\langle \bar{Y} \rangle)^2} \tag{1}$$

where $\langle D^3 \rangle$ is the average cubed grain diameter, j is the multiplicity of the diffraction peak, A_0 is the X-ray beam area, Ω is the solid angle containing the plane normals that contribute to the diffraction peak, μ is the linear absorption coefficient, Y is the number of counts from any one sample position, and γ is the texture coefficient[10] if the sample has preferred orientation. The symbols, $\langle \ \rangle$ and $-$ signify average values of the variables.

If the grains of a material contain subgrains, that is, regions within the grain having small differences in orientation, then the solid angle Ω in equation (1) may be made small enough to permit measurement of the subgrain size. In this case, Ω is so small that two adjacent subgrains cannot contribute to the same measurement and hence are resolved into separate grains.

In the present report, the size of the solid angle was varied to measure the substructure of two differently heat-treated forms of high-purity uranium. The X-ray measurements give the average size of the misoriented regions within the grains. To deduce the nature of these regions, i.e., whether they are subgrains, dislocation networks, elastic strains, etc., requires correlation with transmission electron microscopy and other diffraction experiments, e.g., micro Laue diffraction patterns.

X-RAY MEASUREMENTS OF GRAIN SIZE

The Statistical Fluctuation Method

Theory. In equation (1) it is assumed that the X-rays penetrate many grain depths. Hence the theory is not applicable to materials having large absorption coefficients or large grain size or both. To cover these cases Warren's theory was developed in terms of surface diffraction (see Appendix I). The modified equation is as follows:

$$\langle D^2 \rangle = \frac{jA_0\Omega r\gamma}{\pi^2\sin\theta} \frac{\langle (Y - \langle \bar{Y} \rangle)^2 \rangle}{(\langle \bar{Y} \rangle)^2} \tag{2}$$

where $\langle D^2 \rangle$ is the average squared grain diameter, θ is the Bragg angle, $r = \rho'/\rho$, the relative density of the sample, ρ' is the density of the sample, ρ is the density of the crystal ($\rho = \rho'$ for a metal and $r = 1$). The other symbols have the same meanings as those given in equation (1).

In the present experiments, the solid angle was made small enough to measure the misorientation within the grains. In this case $\langle D^2 \rangle$ is the average diameter squared of the misoriented regions where it is assumed that the regions are circular. The assumption is made to permit one to calculate a diameter, $\langle D^2 \rangle^{\frac{1}{2}}$, for the regions. One could alternatively discuss only the average area, $\langle a_c \rangle$ (Appendix I), of the misoriented regions and thereby make no assumption about their shape.

At present, to deduce the nature of the misorientation within the grains it is necessary to employ other techniques in conjunction with the statistical measurements, such as optical and transmission electron microscopy and Laue diffraction patterns. However, if it turns out that each substructural element has a "characteristic misorientation distribution," then the nature of the element will be determined from the statistical measurement alone.

Apparatus and Procedure for Statistical Fluctuation Measurements. The object of the experiment is to measure the differences in diffraction intensity obtained, for a given reflection *hkl* by diffracting from various positions on the surface of a sample, in our case uranium. The area of the beam is made small enough compared to the grain size so that such differences occur and are large enough to be readily measured. These counts, called Y counts, along with the beam area and the solid angle through which the sample is rotated during a Y count, may be substituted in equation (2) to calculate the average squared grain diameter. Equation (2) is derived in Appendix I.

The diffraction geometry and the manner in which the experiment is performed are shown schematically in Figure 1. A Picker X-ray diffractometer with high intensity Cu K_α X-ray tube was used in conjunction with a scintillation counter and pulse height analyzer. The Picker diffractometer was particularly suited to these experiments since the

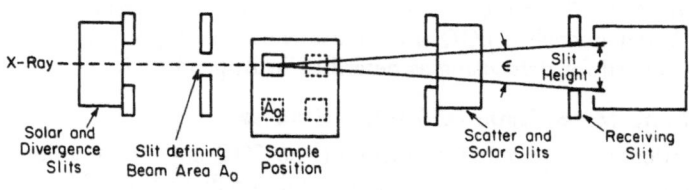

Front View

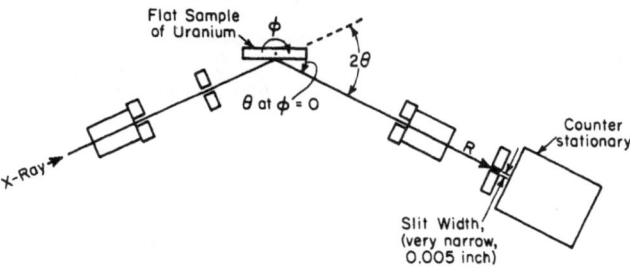

Top View

Figure 1. Diffraction geometry for uranium grain size. Measurements with Picker diffractometer. The ϕ rotation is about the same axis as the θ rotation. With the sample in some position (front view) and the counter stationary at the center of the (111) reflection, counts were taken while the sample was rotated at $1/16°$/min from $\phi = -4°$ to $\phi = +4°$. The counts were printed every 10 sec ($\phi = 0.0015°$). $\Omega = \epsilon\phi$ where $\epsilon = l/2R \sin \theta$, $l = 1.25$ cm, $R = 14.554$ cm, $\theta_{(111)} = 19.77°$, and $A_0 = 0.071$ cm^2.

ϕ motion (called omega motion on the instrument) is, like the θ motion, mechanically driven and marked in steps of 0.01°. In addition, a Picker digital printer was used to print the counts every 10 sec corresponding to a ϕ rotation of 0.0115°. Because of this automated feature, data for a nearly continuous range of solid angles, $\Omega(=\epsilon\phi)$, could be obtained. To minimize errors due to defocusing (see Appendix II) the rotations about ϕ were performed in the $+\phi$ and $-\phi$ directions, e.g., a rotation of 2° was performed by rotating from $-1°$ to $+1°$.

The solid angle Ω is computed from the relationship $\Omega = \epsilon\phi$, where ϕ is the angular distance in degrees through which the sample is rotated, $\epsilon = l/2R \sin \theta$ for which l is the height of the receiving slit, R is the distance from sample to receiving slit and θ is the Bragg angle. The solid angle can also be generated by leaving the sample at $\phi = 0$ and varying the width w of the receiving slit, in which case $\Omega = (wl)/(4R^2 \sin \theta)$. The latter method is limited in that the slit width must be so large for Ω values of the order of 10^{-2} that the counter may not be uniformly sensitive over the area of the receiving slit. In addition, it is easier to vary the solid angle by varying ϕ than changing the receiving slit width. The wide receiving slit geometry was used, as a check, at one small solid angle ($\Omega = 5 \times 10^{-5}$ steradians) and both methods gave $\langle D^2 \rangle^{\frac{1}{2}}$ values within 10% of one another.

Y counts [equation (2)] were measured for 50 locations[11] on each of the two recrystallized uranium QA samples and each of the two beta-quenched uranium βOQ samples. The preparation of the samples is described below. The area A_0 of the X-ray beam was 0.071 cm². Each location was measured for 1500 values of the solid angle Ω over a range of $\Omega = 10^{-4}$ to 10^{-2} steradians. A computer program was written to calculate $\langle D^2 \rangle^{\frac{1}{2}}$ from equation (2) for each Ω. A plot of solid angle Ω versus $\langle D^2 \rangle^{\frac{1}{2}}$ for the recrystallized QA and quenched βOQ uranium is shown in Figure 2. The X-ray beam area A_0 was corrected for defocusing as described in Appendix II.

Conventional and Micro Laue Diffraction Studies

Back-reflection photographs were made of the βOQ and QA uranium specimens. The Laue photographs were made with copper (white) radiation on a Norelco X-ray unit

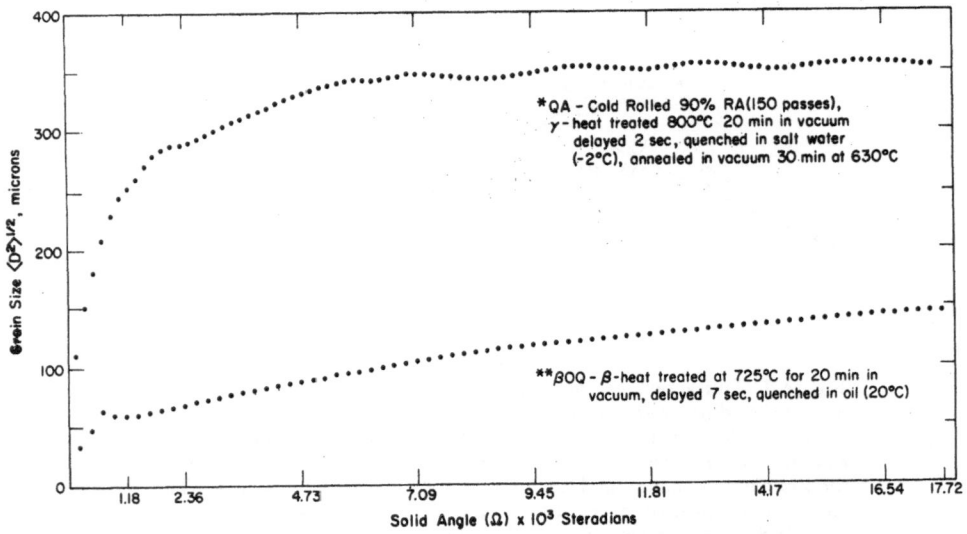

Figure 2. $\langle D^2 \rangle^{\frac{1}{2}}$ versus solid angle for QA and βOQ uranium.

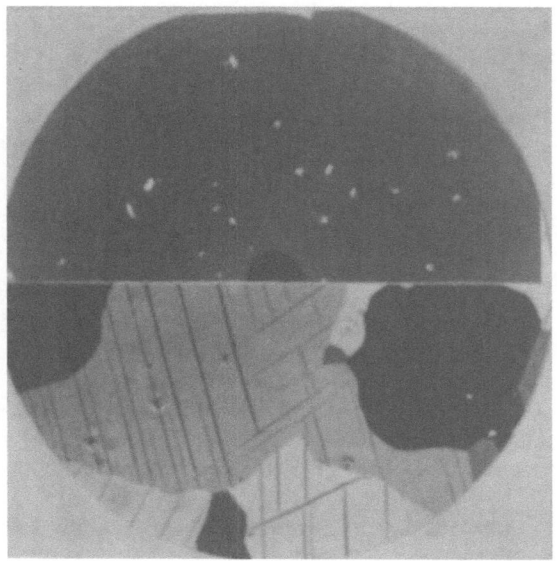

(a)

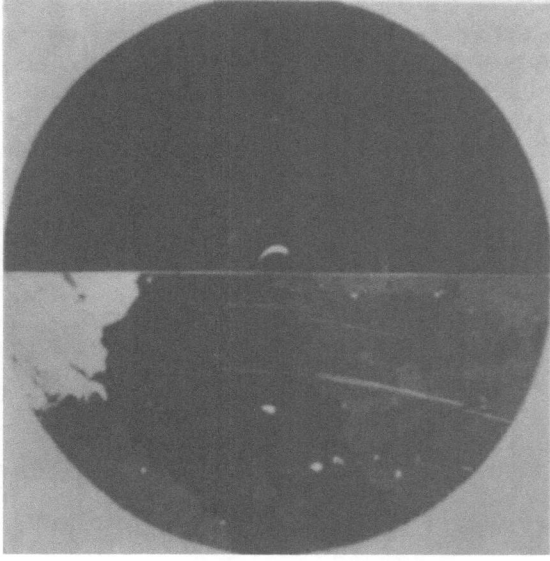

(b)

Figure 3. Optical micrographs and back-reflection Laue diffraction patterns of high-purity uranium. (a) QA uranium; optical macro-grain size 300–400 μ. (b) βOQ uranium; optical macrograin size 300–400 μ, subgrain size ~ 60 μ. (c) AE (α-extruded) uranium. (d) PM (powder metallurgy) uranium. (150×, reduced for reproduction 40%).

with 0.020-in. (~ 500 microns) collimators at a specimen-to-film distance of 3 cm. The Laue photographs for the βOQ and QA uranium are compared (Figure 3) with Laue photographs of alpha-extruded uranium and with very-fine-grained uranium prepared by a powder metallurgy process. Optical micrographs (polarized light) for these four types of uranium are also shown in Figure 3. The Laue patterns for the recrystallized uranium are similar to a single crystal photograph, whereas the Laue patterns for the quenched uranium have nearly continuous rings, suggesting much substructure, since the "macro" grains of the quenched uranium are the same size as those of the recrystallized uranium.

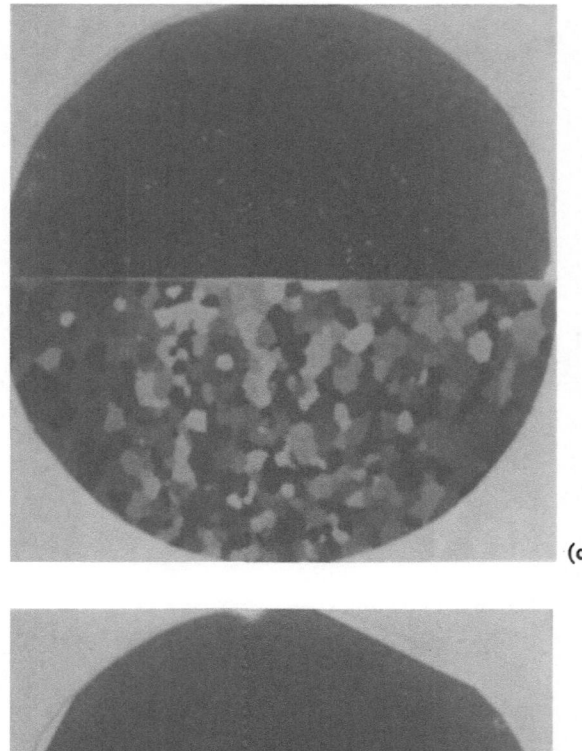

(c)

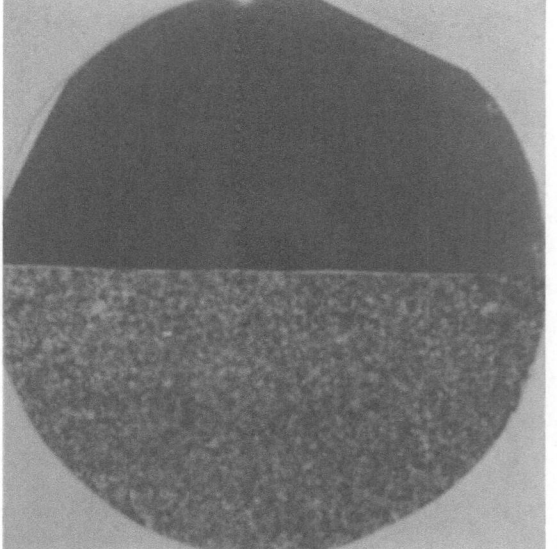

Figure 3 continued.

(d)

Micro Laue diffraction photographs were also made of the βOQ and QA uranium with Cu K_α radiation, using a Rigaku–Denki microfocus unit and associated micro Laue camera. The collimator size was 30 microns and the specimen-to-film distance was 0.6 cm. The micro Laue photographs are shown in Figure 4.

Berg-Barrett X-Ray Reflection Micrographs

Reflected X-rays may be used to form images of the grains on the polished surface of a metal. The images are formed by having a narrow, highly parallel beam incident nearly parallel to the surface of the sample and diffracted onto a film placed close to the sample (a few millimeters). The film should be fine-grained so that it can be enlarged. Several X-ray reflection techniques are described by Barrett.[2]

X-ray reflection micrographs (Figure 5) were taken of the βOQ and QA uranium. A fine focus X-ray beam [copper (white) radiation] from a Rigaku–Denki microfocus X-ray unit was used for the experiment. The beam was defined by two slits—a 0.05-cm slit near the beam and a 0.0038-cm slit 30 cm from the beam. The best results were obtained at 45 kV, 150 μA, for an 18-hr exposure. Nuclear track plates were used to record the grain images.

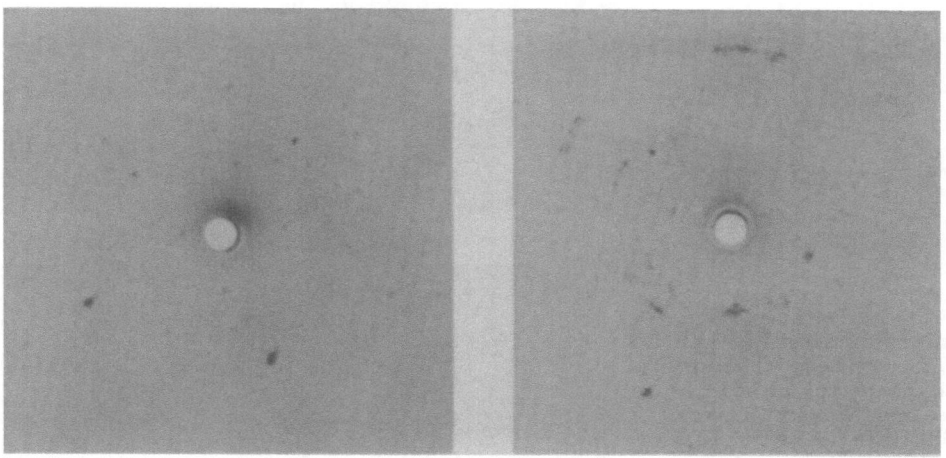

QA βOQ

Figure 4. Back-reflection micro Laue diffraction patterns of βOQ and QA uranium (30-micron beam diameter). QA uranium gives spots similar to those of a single crystal. Even within an X-ray beam diameter of 30 microns the βOQ uranium has a sufficient number of grains to give partial rings. (5 ×, reduced for reproduction 10%).

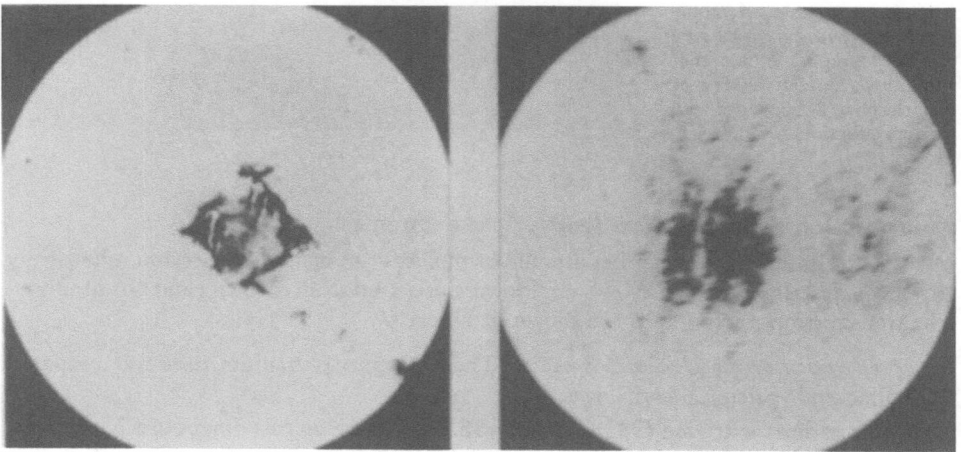

QA βOQ

Figure 5. Berg–Barrett X-ray reflection micrographs of QA and βOQ uranium. QA—Imperfection within grain prevents some portions from being reflected. βOQ—Note numerous subgrains and compare with optical microstructure of large grain of βOQ uranium in Figure 3. (50 ×, reduced for reproduction 30%).

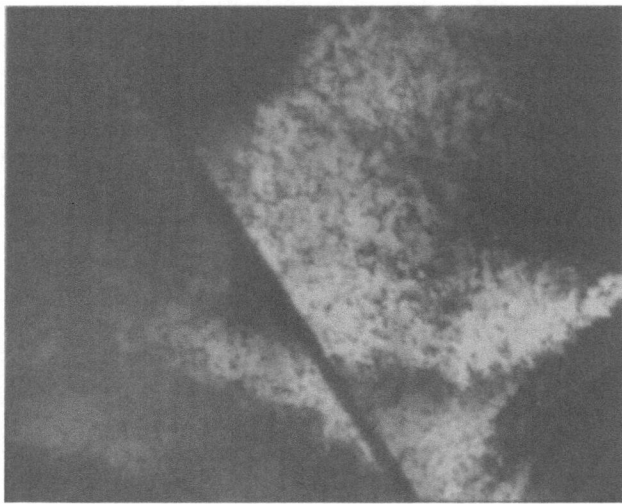

QA

Figure 6. Transmission electron micrographs of βOQ and
QA uranium. QA—Note
absence of subgrain boundaries and presence of dense
dislocation networks. The
dislocations give a continuous
range of misorientation as seen
from the statistical fluctuation
curve in Figure 2. (5200×,
reduced for reproduction
15%). βOQ—Note number
of small subgrains of 0.1 to
1 micron near center of photograph and twin on right side
of photograph. Note that
subgrains are relatively free of
dislocations. (8000×, reduced
for reproduction 15%).

βOQ

Transmission Electron Microscopy Measurements

The techniques for preparing uranium samples for transmission electron microscopy
have been described previously.[12] Samples were prepared from the βOQ and QA uranium.
The transmission micrographs are shown in Figure 6.

Preparation of the Uranium Samples. The high-purity uranium tube was prepared
at Mallinckrodt Chemical as follows:

(1) It was gamma-extruded (982°C) from a 18.9-in.-diameter cast dingot to a 7-in.-diam-
eter bar stock (ram speed 300 in./min), and (2) drilled and turned to 6.5-in. diameter,
then alpha-extruded (635°) to 3.12 in. OD × 1.84 in. ID (ram speed 15 in./min). The as-
extruded microstructure is shown in Figure 3. The impurity content of the uranium was
as follows (ppm): H, 6; B, 0.1; C, 35; N, 7; O, 15; Mg, 10; Al, 29; Si, 10; Cr, 1; Mn, 2;
Fe, 26; Ni, 18; and Cu, 8.

A section from the tube was rolled to 90% reduction in area at 325°C in 150 passes. Two of these samples. $1.5 \times 1.25 \times 0.125$ in., were heat-treated as follows: They were gamma-heat-treated in vacuum at 800°C for 20 min, delayed two sec, quenched in ice brine (-2°C), and annealed for 30 min in vacuum at 630°C. These two samples are referred to as QA uranium.

Two other samples were machined directly from the center wall of the alpha-extruded tube to dimensions of $2 \times 1.25 \times 0.25$ in. The samples were beta-heat-treated at 725°C for 20 min in vacuum, delayed 7 sec, then quenched in oil at 20°C. These two samples are referred to as βOQ uranium.

The samples were mechanically polished through No. 1 diamond dust and electropolished[13] to remove cold-worked metal. After the statistical fluctuation measurements were complete, the samples were used for Berg–Barrett reflection measurements, micro and conventional Laue back-reflection measurements, optical microscopy, and transmission electron microscopy.

RESULTS AND CONCLUSIONS

Comparison of the statistical measurements of grain size (Figure 2) for the QA uranium with the optical micrographs (Figure 3) shows that the macrograin sizes are in good agreement and hence the grain size equation derived in Appendix I appears quite valid for calculating grain size in cases where the absorption coefficient or the grain size or both is large. The QA uranium is a good sample for checking equation (2) at large solid angles because the misorientation within the grains is small. The twins present will, however, make the X-ray grain size a little smaller, since they register as separate grains.

The correction factor for defocusing which is derived in Appendix II is in general agreement with the experimentally measured correction factor (Figure 8). At the maximum ϕ rotation used (± 4°) it changes the effective area of the beam, A_0, by about 25%. However, at solid angles, corresponding to ± 10° rotation of ϕ, A_0 would change by 50%.

Consider the plot of Ω versus $\langle D^2 \rangle^{\frac{1}{2}}$ in Figure 2 for the QA uranium. There appears to be no misorientation within the grains greater than about 5.4×10^{-3} steradians. Above this solid angle the curve is nearly flat, oscillating with a small period because of the statistical nature of the measurements. Below the flat region there appear to be two ranges of misorientation, one between 2×10^{-3} and 5.5×10^{-3} steradians and one below 2×10^{-3} steradians. This conclusion is reached because of the two different rates of variation of $\langle D^2 \rangle^{\frac{1}{2}}$ with Ω. At first it was thought that the misorientation was due to subgrains. However, back-reflection micro Laue photographs with a 30-micron-diameter beam (Figure 4) show that there are no small subgrains. The transmission electron micrographs (TEM), (Figure 6) also show an absence of small subgrains but show a uniform distribution of dislocation networks. Since no other elements of substructure are observed, the dislocations are believed to be the cause of the misorientation within the grains.

Consider the plot Ω versus $\langle D^2 \rangle^{\frac{1}{2}}$ in Figure 2 for the βOQ uranium. The misorientation within the grains extends over a wide range, i.e., greater than 10^{-2} steradians and less than 10^{-4} steradians. It is apparent from the optical micrographs (Figure 3), that many subgrains are present. At the largest solid angle measured, $\Omega = 17.7 \times 10^{-3}$ steradians, and $\langle D^2 \rangle^{\frac{1}{2}}$ was 150 microns, was still increasing, and was considerably smaller than the "macro" grain size shown on the optical micrographs (Figure 3). As the solid angle is decreased below $\Omega = 2 \times 10^{-3}$ steradians, the constant $\langle D^2 \rangle^{\frac{1}{2}}$ between $\Omega = 2 \times 10^{-3}$ and $\Omega = 8 \times 10^{-4}$ suggests that all subgrains have been resolved. The sudden decrease at $\Omega = 8 \times 10^{-4}$ suggests that the "sub-sub-grains" are beginning to be resolved.

The TEM photographs for the βOQ uranium (Figure 6) show extremely small subgrains, relatively free of dislocations, bounded by dense dislocation networks. The subgrains are of the order of 0.1 to 1 micron. These data are corroborated by the back-reflection Laue photographs in Figures 3 and 4. The micro Laue photographs, Figure 4, cover a diameter of only 30 microns, and yet this area has sufficient grains to produce diffraction rings like a powder pattern. Further evidence of the extensive substructure in the βOQ uranium is shown by the conventional Laue photographs in Figure 3. They show that the βOQ uranium gives nearly continuous rings while the QA uranium gives a spotty pattern. According to the Laue photographs of Figure 3, the βOQ uranium has even finer subgrains than the alpha-extruded uranium.

Berg–Barrett photographs of typical "macro" grains of the recrystallized QA and quenched βOQ uranium are shown in Figure 6. These photographs support the other experimental evidence that the misorientation within the QA uranium is due to dislocation networks and that the misorientation within the quenched uranium is due to subgrains.

There are a number of twins (Figure 3) in both the βOQ and QA uranium. They are mainly of the {130} and {172} type and so have quite a different orientation than the grain itself. The effect of the twins is simply to reduce the average grain size $\langle D^2 \rangle^{\frac{1}{2}}$ since they simply add to the average as small grains. As shown in Figures 3 and 6, the twins vary from 30 microns to 0.1 micron in thickness.

APPENDIX I

Derivation of the Grain Size Equation for Materials with Large Absorption Coefficients or Grain Size or Both*

Assume a parallel beam of X-rays of area A_0 falling on a flat sample in a diffractometer (Figure 1). The sample and detector are aligned in the conventional θ, 2θ relationship. The receiving slit is wide compared to the 2θ width of the diffraction peak being studied. The length and width of the receiving slit subtend a solid angle Ω with the sample.

Alternatively, the solid angle Ω can be generated by employing a very narrow slit width and rotating the specimen through an angle θ about the axis of the diffractometer. In the former case, $\Omega = (wl)/4R^2 \sin \theta$ where w is the width of the receiving slit, l is the height or length of the receiving slit, R is the radius of the focal circle of the diffractometer, and θ is the Bragg angle. In the latter case, $\Omega = \epsilon\phi$ where $\epsilon = l/2R \sin \theta$.

To begin with, assume that all grains are of the same area a_c. The number of grains irradiated by the X-ray beam is given by

$$n = \frac{rA_0}{a_c \sin \theta} \tag{3}$$

where A_0 is the slit area defining the beam, $A_0/\sin \theta$ is the surface area irradiated, θ being half the Bragg angle, and r, the relative density of the sample $= \rho'/\rho$, where ρ' is the density of the sample and ρ is the density of the crystal ($r = 1$ for a solid sample).

The average number of grains in the irradiated area which contribute to the measured reflection is given by

$$\overline{m} = \frac{nj\Omega}{4\pi} = \frac{rA_0}{a_c \sin \theta} \frac{j\Omega}{4\pi} \tag{4}$$

* The following is a special case of B. E. Warren's derivation of grain size from measurements of statistical fluctuations in X-ray diffraction intensity. This treatment will follow closely the methods used in Warren's paper.[1]

Since $j\Omega/4\pi$ is small, the probability of a value m, $w(m)$, is given by the Poisson distribution law,[14]

$$w(m) = \frac{(\bar{m})^m e^{-\bar{m}}}{m!} \tag{5}$$

for which statistical variations follow the simple law

$$\langle (m - \bar{m})^2 \rangle = \bar{m} \tag{6}$$

Let Y be the integrated counts from the m crystals for one setting of the sample and $\bar{Y}$ be the average number of counts for all settings of the sample. The number of counts is equal to the number of grains diffracting into the solid angle Ω, allowing for attenuation of the beam by absorption, i.e.,

$$Y = \alpha_z m$$

and

$$\bar{Y} = \alpha_z \bar{m} \tag{7}$$

where $\alpha_z = K_1 \exp[-(2\mu rz)/\sin\theta]$, z is the depth of penetration, and μ is the linear absorption coefficient.

In Warren's case the X-ray penetrates through a number of grains until it finds one oriented in the correct position for diffraction. Hence, the attenuation of the X-ray beam due to absorption must be determined by integrating α_z over z as in Warren's[1] equations (6) and (7). However, when the depth of penetration is small compared to the grain dimensions, then the X-rays are either diffracted in the surface grains or not at all. The surface grains which diffract are all oriented the same way. So the attenuation due to absorption is constant for a given solid angle.* Hence α_z is constant and equation (7) may be written as

$$Y = Km$$

and

$$\bar{Y} = K\bar{m} \tag{8}$$

and from (6) and (8) we can write

$$\langle (Y - \bar{Y})^2 \rangle = K^2 \bar{m} \tag{9}$$

and

$$(\bar{Y})^2 = K^2 (\bar{m})^2 \tag{10}$$

or

$$\frac{\langle (Y - \bar{Y})^2 \rangle}{(\bar{Y})^2} = \frac{1}{\bar{m}} \tag{11}$$

Then from (4),

$$m = \frac{r A_0 j \Omega}{a_c \sin\theta \, 4\pi}$$

* The absorption does enter as an effect by reducing the beam intensity, and thereby A_0, as the sample is rotated about ϕ. The beam area, A_0, is corrected for both absorption and defocusing from the curve of Figure 10.

so that

$$\frac{\langle (Y - \bar{Y})^2 \rangle}{(\bar{Y})^2} = \frac{a_c \sin \theta \, 4\pi}{r A_0 j \Omega} \tag{12}$$

Approximating the grains as circles of diameter D, $a_c = \pi D^2/4$ and equation (12) becomes

$$D^2 = \frac{j A_0 \Omega r}{\pi^2 \sin \theta} \frac{\langle (Y - \bar{Y})^2 \rangle}{(\bar{Y})^2} \tag{13}$$

Equation (13) assumes that all grains are of the same size. If the grains are not the same size, the counts are a function of the area of the grains as well as the number of grains, so that equation (8) must be rewritten as

$$Y = K \sum_k a_k m_k$$

and

$$\bar{Y} = K \sum_k a_k \bar{m}_k \tag{14}$$

where a_k is the area of grains of size k, and m_k is the number of grains of size k which contribute to the reflection.

From equation (4),

$$\sum_k a_k \bar{m}_k = \sum_k a_k n_k \left(\frac{j\Omega}{4\pi} \right) \tag{15}$$

when n_k is the total number of crystals of size k being irradiated. The area of grains irradiated is $A_0 r/\sin \theta$, so

$$\sum_k a_k n_k = A_0 r/\sin \theta \tag{16}$$

and equation (15) becomes

$$\sum_k a_k \bar{m}_k = \frac{A_0 j \Omega r}{4\pi \sin \theta} \tag{17}$$

From equations (6) and (14),

$$\langle (Y - \bar{Y})^2 \rangle = K^2 \sum_k a_k^2 \bar{m}_k \tag{18}$$

and

$$(\bar{Y})^2 = K^2 (\sum_k a_k \bar{m}_k)^2 \tag{19}$$

After Warren we define a grain area weighted with respect to area as follows: If we have n_1 grains of area a_1, n_2 grains of area a_2, and n_3 grains of area a_3, then the area-weighted average $\langle a_c \rangle$ would be

$$[a_1(n_1 a_1) + a_2(n_2 a_2) + a_3(n_3 a_3)] \div [a_1 n_1 + a_2 n_2 + a_3 n_3]$$

Now, going back to Warren's notation,

$$\langle a_c \rangle = \frac{\sum\limits_k a_k^2 \overline{m}_k}{\sum\limits_k a_k \overline{m}_k} \tag{20}$$

and, from equation (18),

$$\frac{\langle (Y - \overline{Y})^2 \rangle}{(\overline{Y})^2} = \frac{\langle a_c \rangle}{\sum\limits_k a_k \overline{m}_k} \tag{21}$$

Substituting the value of

$$\sum\limits_k a_k \overline{m}_k$$

from equation (17),

$$\frac{\langle (Y - \overline{Y})^2 \rangle}{(\overline{Y})^2} = \frac{\langle a_c \rangle 4\pi \sin \theta}{A_0 j \Omega r} \tag{22}$$

which is identical to equation (12), except that a_c has been replaced by $\langle a_c \rangle$, defined in equation (20).

Approximating the grains as circles of diameter D, $a_c = \pi/4\, D^2$, which upon substitution into equation (22) yields

$$\langle D^2 \rangle = \frac{j A_0 \Omega r \gamma}{\pi^2 \sin \theta} \frac{\langle (Y - \overline{Y})^2 \rangle}{(\overline{Y})^2} \tag{23}$$

The value γ is introduced to correct for preferred orientation. It is the ratio of the intensity from the peak of the oriented sample to the same peak of a random sample.[10]

APPENDIX II

Correction Factor for Defocusing of the X-Ray Beam due to Rotation of the Sample about the Diffractometer Axis

The Picker diffractometer employed in the present studies has the parafocusing geometry[15,16] shown schematically in Figure 7. Normally the sample and counter rotate so as to maintain a θ, 2θ relationship. For the grain size measurements the counter is held stationary while the specimen is rotated about the diffractometer axis to generate a solid angle Ω. The rotation is referred to as $+ \phi$ and $- \phi$, however it has the same axis of rotation as θ except that the counter is stationary. R. E. Ogilvie has shown[17] that as the sample is rotated about ϕ, with the counter stationary, the focal point moves a distance X from the receiving slit given by the equation

$$X = R - R\left(\frac{\cos(\phi + \eta)}{\cos(\phi - \eta)} \right) \tag{24}$$

where R is the radius of the focal circle (14.554 cm) and $\eta = 90 - \theta$.

The focal point moves closer to the specimen for clockwise rotation and farther away for counterclockwise rotation. When the focal point moves, the beam becomes wider at the receiving slit as shown schematically in Figure 7. Hence a portion of the beam area

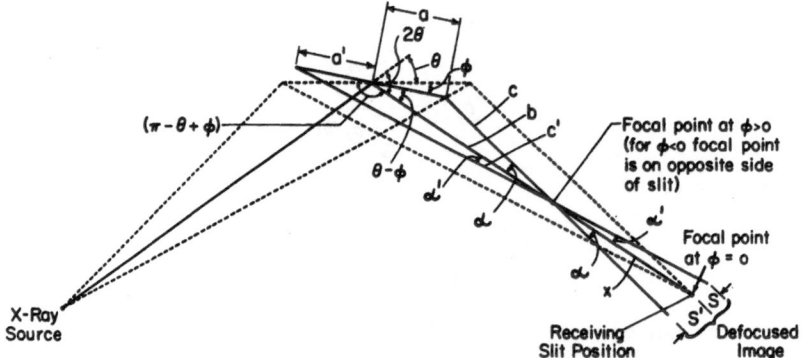

Figure 7. Schematic drawing showing defocusing caused by rotation about ϕ.

A_0 of equation (2) falls outside the slit, thereby decreasing the effective area of the beam from A_0 to a smaller value. An expression relating the width of the beam at the receiving slit to the angle ϕ may be derived from Figure 7 and equation (24) as follows: The beam width at the receiving slit equals $S + S'$ (Figure 7) where

$$S' = X \tan \alpha$$
$$S = X \tan \alpha' \tag{25}$$

The angles α and α' may be determined from the law of cosines by the equations

$$\cos \alpha = \frac{b^2 + c^2 - a^2}{2bc}$$

and

$$\cos \alpha' = \frac{b^2 + c'^2 - a^2}{2bc'} \tag{26}$$

where $b = b' = R - X$, $a = a' =$ beam width at sample$/2 \sin(\theta + \phi)$ and

and
$$c^2 = a^2 + b^2 - 2ab \cos \gamma$$
$$c'^2 = a^2 + b^2 + 2ab \cos \gamma \tag{27}$$

where $\gamma = \theta - \phi$. A plot of the fractional increase in beam width at the receiving slit versus ϕ is shown in Figure 8.

In order to check equations (25), (26), and (27), the fractional drop in (111) intensity of a section of powder metal uranium plate with very fine grains (6 microns) and random orientation (Figure 3) was measured during rotation from $\phi = -10°$ to $\phi = +10°$. The experimental conditions were the same as those employed in the grain size measurements. The results are also shown in Figure 8. The shape is the same as that of the theoretical curve; however, the diffraction intensity is not reduced as rapidly as suggested by the theoretical curve. Some reasons for the difference are as follows: (1) The intensity distribution across the beam width (Figure 9) is not constant. The central third of the beam is twice as intense as the outer two-thirds. Because the outer edges of the beam width are defocused first (Figure 7) the intensity drop in the experimental curve of Figure 8 is correspondingly small for small ϕ rotation. (2) Equations (25), (26), and (27) were derived for a "line" source (Figure 7) and a "zero width" receiving slit. In these experiments, the beam width at the source was 0.005 cm and the receiving slit width was 0.0127 cm.

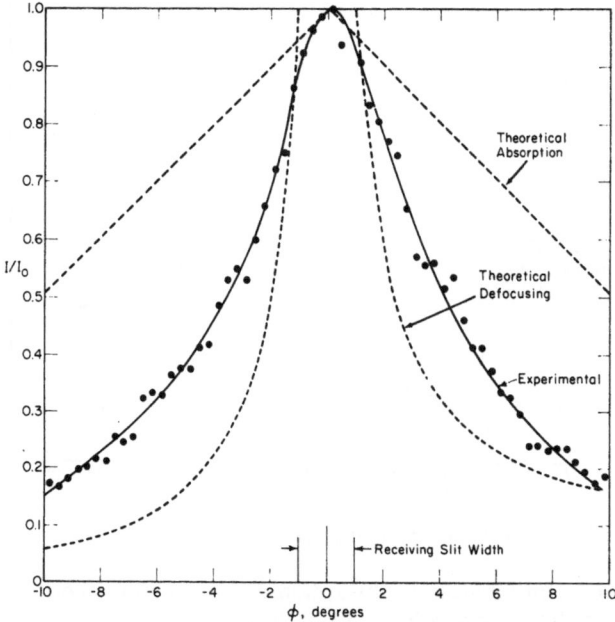

Figure 8. Decrease in diffraction intensity as a function of defocusing and absorption.

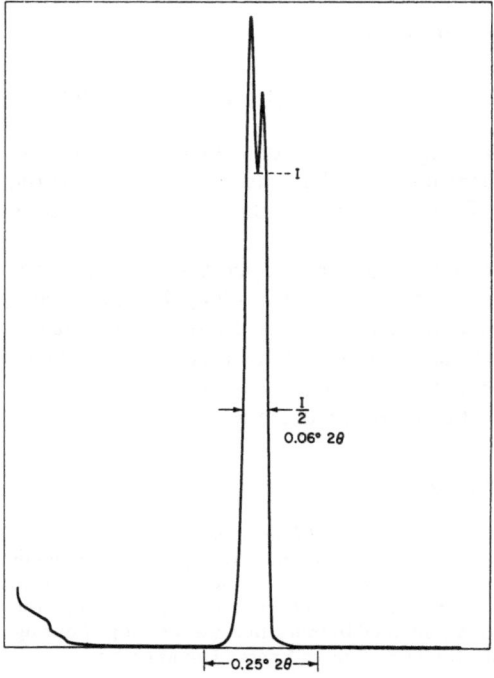

Figure 9. Intensity distribution across X-ray beam width.

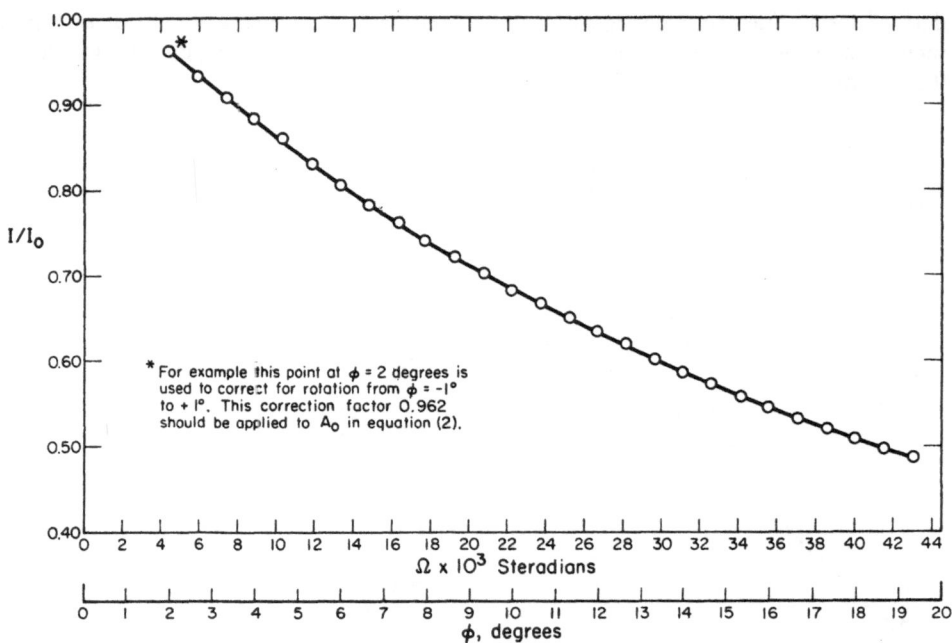

Figure 10. Decrease in diffraction intensity versus rotation from minus ϕ to plus ϕ.

Another factor that affects the beam intensity is absorption due to rotation about ϕ. The fractional decrease in intensity, I/I_0, is given by Schwartz[18] as

$$\left(\frac{I}{I_0}\right)_{\pm\phi} = 1 \mp \tan\phi \cot\theta \qquad (28)$$

where θ and ϕ have the same meaning as above. A plot of $(I/I_0)_{\pm\phi}$ vs. ϕ is given in Figure 8.

It should be noted that neither defocusing nor absorption affect the variance, $\langle(Y - Y)^2\rangle/(Y)^2$, since each of the 32 locations measured for a given solid angle has the same defocusing and the same absorption. These quantities reduce only the effective beam area A_0 of equation (2).

It is difficult to specify a correction factor of general application because each experimental setup will have differences in take-off angle, beam profile, absorption, and slit geometry. Hence it seems best to determine the A_0 correction factor experimentally as shown in Figure 8. To put the experimental data of Figure 8 in a usable form, we plot the average fractional decrease in intensity versus the solid angle Ω as shown in Figure 10. The fractional decrease in intensity is assumed equal to the decrease in A_0.

ACKNOWLEDGMENTS

The authors would like to thank Professor B. E. Warren of MIT for reviewing the derivation in Appendix I, C. L. Angerman of SRL for the transmission electron microscopy, R. B. Russell of Nuclear Metals, Inc., for the high-purity uranium, and Professor Sigmund Weissmann of Rutgers for helpful comments on the interpretation of the transmission electron micrographs and the Berg–Barrett reflection photographs. The information contained in this article was developed during the course of work under contract AT(07-2)-1 with the U.S. Atomic Energy Commission.

REFERENCES

1. B. E. Warren, "X-Ray Measurements of Grain Size," *J. Appl. Phys.* **31**: 2237–2239, 1960.
2. C. S. Barrett, *Structure of Metals*, McGraw-Hill Book Company, Inc., New York, 1952, Chapter V.
3. A. Taylor, *X-Ray Metallography*, John Wiley & Sons, Inc., New York, 1961, Chapter 14.
4. P. B. Hirsch and J. N. Kellar, " A Study of Cold-Worked Aluminum by an X-Ray Micro-Beam Technique. I. Measurement of Particle Volume and Misorientations," *Acta Cryst.* **5**: 162, 1952.
5. P. B. Hirsch, "A Study of Cold-Worked Aluminum by an X-Ray Micro-Beam Technique. II. Measurement of Shapes of Spots," *Acta Cryst.* **5**: 168, 1952.
6. P. B. Hirsch, "The Reflection and Transmission of X-Rays in Perfect Absorbing Crystals," *Acta Cryst.* **5**: 176, 1952.
7. P. B. Hirsch, "Mosaic Structures," in: B. Chalmers and R. King (eds.), *Progress in Metal Physics*, Vol. *6*, Pergamon Press Ltd., London and New York, 1956, pp. 236–339.
8. C. S. Barrett, "Determining Recrystallization by a Diffractometer Technique," in: J. B. Newkirk and J. H. Wernick (eds.), *Direct Observations of Imperfections in Crystals*, Interscience Publishers, Inc., New York, 1962, p. 395.
9. C. S. Barrett, "X-Ray Diffraction Studies at Low Temperatures," in: W. M. Mueller, G. R. Mallett, and M. J. Fay (eds.), *Advances in X-Ray Analysis*, Vol. *5*. Plenum Press, New York, 1961, p. 33.
10. E. F. Sturcken and J. W. Croach, "Predicting Physical Properties in Oriented Metals," *Trans. AIME* **227**: 934–940, 1963.
11. E. F. Sturcken and W. E. Gettys, *Determination of Grain Size in Uranium from Statistical Fluctuations in X-Ray Diffraction Intensity*, E. I. du Pont de Nemours and Co., DP-904, July 1964.
12. C. L. Angerman, "Transmission Electron Microscopy of Uranium," *J. Nucl. Mater.* **9**: 109–110, 1963.
13. N. Crank and R. N. Thudium, *Effects of Etching on Preferred Orientation Measurements*, Hanford Atomic Products Operation, HW-74429, August 1962.
14. H. Margenau and G. M. Murphy, *The Mathematics of Physics and Chemistry*, D. Van Nostrand Company, Inc., New York, 1943, pp. 422–425.
15. H. Seeman, *Ann. Physik* **59**: 455, 1919.
16. H. Bohlin, *Ann. Physik* **61**: 421, 1920.
17. R. E. Ogilvie, *Stress Measurement with the X-Ray Spectrometer*, M.S. Thesis, Department of Metallurgy, MIT, 1952.
18. M. Schwartz, *J. Appl. Phys.* **26**: 1507, 1955.

ANALYSIS OF THE BROADENING OF POWDER PATTERN PEAKS USING VARIANCE, INTEGRAL BREADTH, AND FOURIER COEFFICIENTS OF THE LINE PROFILE

N. C. Halder and C. N. J. Wagner

Yale University
New Haven, Connecticut

ABSTRACT

The broadening of powder pattern peaks has been studied by three methods—Fourier analysis, integral breadth measurements, and variance of the line profiles. The results obtained from the variances are compared with those obtained from the integral breadths and Fourier coefficients.

Tungsten filings were prepared at room temperature and their powder pattern peaks were recorded with a Norelco diffractometer using filtered Cu K_α radiation. The variances, integral breadths, and Fourier coefficients were calculated with the IBM 7094 computer. The results indicate that the variance is very sensitive to the range of integration $s_2 - s_1 = (2\theta_2 - 2\theta_1) \cos \theta_0 / \lambda$. An error of $\pm 10\%$ in this range due to the difficulty in choosing the correct background changes the values of the variance significantly and the integral breadth to a lesser extent. However, the same error does not affect the values of the Fourier coefficients.

Comparing the particle sizes and strains obtained by the three methods, it was found that the strains agreed remarkably well. The particle size calculated from the variance was smaller ($D_e{}^W = 150\text{Å}$) than that evaluated from the initial slope of the Fourier coefficients ($D_e = 210\text{Å}$) and from the integral breadths $2D_e \simeq D_I = 430\text{Å}$.

INTRODUCTION

The microstructural changes in metals can be studied by the analysis of powder pattern peaks which gives information about distortions, fragmentation, and faulting in the crystals. The measurement of the integral breadth based on Laue's definition[1] is very simple and useful if the broadening is due to particle size alone or an exact relationship between the integral breadths of particle size and strain is known when both are present. Since the powder pattern peak can be expressed as a Fourier series, it is possible to derive information about the particle sizes and strains directly from the Fourier coefficients. It was for some time a matter of discussion whether both of these methods would lead to similar results when applied to the study of the same material. Wagner and Aqua[2] analyzed the powder pattern peaks from cold-worked fcc and bcc metals by the Warren–Averbach[3] method of Fourier analysis and integral breadth measurements, and showed that there was reasonable agreement between the quantities: particle size, root-mean-square-strain, and stacking fault probability, measured by the two methods when the proper evaluation procedure is applied.

There is a third method which uses the variance or the second moment as a measure

---

[1] References are at the end of the paper.

of line broadening.[4] While the use of variance has been emphasized by Wilson,[5] its application has not been largely reported.[6-8] The methods of Fourier analysis and variance allow the correction of the instrumental broadening and the separation of particle size and strains without making any assumptions about the mathematical description of the line profile. Therefore, it seems of interest to investigate the applicability of the variance in describing the peak-broadening parameters in heavily deformed metals, and to compare the variance results with those obtained by Fourier analysis and integral breadth measurements.

X-RAY THEORY

The accuracy of any line-broadening analysis–integral breadth, Fourier coefficients, or variance–greatly rests on the exact measurement of the peak profile including its long tails, which are affected by the choice of the background level. The Fourier coefficients determined with respect to the experimentally determined background level are quite reliable and can be used with confidence for the evaluation of the microstructural parameters. Unlike Fourier coefficients, the variances are strongly sensitive to the tails of the peak and background level. The correct choice of an optimum range of integration and an appropriate background level has been stressed by Langford and Wilson.[6] An overestimated or underestimated range produces a significant error in the calculated value of the variances.

Since there exist good accounts of the Fourier analysis method in the literature,[9,10] no effort shall be made here to describe it in detail. However the following points are summarized: Jones[11] showed that the peak profile of a diffraction line is described by the convolution equation

$$h(s) = \int_{-\infty}^{\infty} f(s')g(s - s')ds' \tag{1}$$

where $f(s')$ is the peak profile due to the particle size and strain, $g(s - s')$ is that due to the geometry of the diffractometer, $s = 2\sin\theta/\lambda$, and $s' = 2\sin\theta'/\lambda$. It has been shown by Stokes[12] that

$$A(n) = \frac{H(n)}{G(n)} \tag{2}$$

where $A(n)$, $H(n)$, and $G(n)$ are the Fourier coefficients of $f(s')$, $h(s)$ and $g(s - s')$ profiles, respectively. The coefficients $A(n)$ can be relabeled as $A(L)$ for convenience, where $L = nd_{hkl}$ is a distance normal to the reflecting planes hkl of interplanar spacing d_{hkl}. The power distribution per unit arc length of a Debye–Scherrer line is given by

$$P'(s) = K \int_{-\infty}^{+\infty} A(L) \exp[- 2\pi i L(s - s_0)] \, dL \tag{3}$$

where $s_0 = 2\sin\theta_0/\lambda$, θ_0 being the peak maximum, and K is a constant depending only on the angle θ_0.[9] The coefficients $A(L)$ are given by[10]

$$A(L) = \frac{\displaystyle\int_{s_1}^{s_2} P'(s) \exp[2\pi i L(s - s_0)]ds}{\displaystyle\int_{s_1}^{s_2} P'(s)ds} \tag{4}$$

where $s_2 = 2 \sin \theta_2/\lambda$ and $s_1 = 2 \sin \theta_1/\lambda$ are the upper and lower integration limits, respectively. Equation (4) can be written as

$$A(L) = A^D(L) \, A^{PF}(L) \tag{5}$$

The distortion coefficients $A^D(L)$ and particle size coefficients $A^{PF}(L)$ are given for small values of L and ϵ_L as follows:[10]

$$A^D(L) = 1 - 2\pi^2 L^2(\langle \epsilon_L{}^2 \rangle - \langle \epsilon_L \rangle^2) \frac{h_0{}^2}{a^2} \tag{6}$$

and

$$A^{PF}(L) = 1 - L\left[\frac{1}{\bar{D}} + (1.5\alpha + \beta) \frac{V_{hkl}}{a}\right], \tag{7}$$

where $\langle \epsilon_L{}^2 \rangle$ is the mean-square strain, $\langle \epsilon_L \rangle$ is the mean strain, $h_0{}^2 = h^2 + k^2 + l^2$ for cubic crystals, a is the lattice parameter, $\bar{D}$ is the average particle size, α and β are the deformation and twin fault probability, respectively, and V_{hkl} is a constant which depends upon the crystal structure and the hkl planes.[10] One obtains directly from equations (4) and (6) the following:

$$\ln A(L) = \ln A^{PF}(L) - 2\pi^2 L^2(\langle \epsilon_L{}^2 \rangle - \langle \epsilon_L \rangle^2) \frac{h_0{}^2}{a^2} \tag{8}$$

Further,

$$nd_{hkl} = L = n'a_3' \tag{9}$$

and

$$a_3' = \frac{1}{s_2 - s_1} = \frac{1}{\Delta s} \tag{10}$$

where n' is the harmonic number and a_3' is a fictitious distance corresponding to the Fourier interval $s_2 - s_1$. The $\ln A(L)$ versus $h_0{}^2$ plots for small values of L will be straight lines, the slopes of which determine the difference between the mean-square strains and the squares of the mean strains, i.e., $\langle \epsilon_L{}^2 \rangle - \langle \epsilon_L \rangle^2$ (in filings, $\langle \epsilon_L \rangle = 0$, and we obtain the conventional definition of the rms strain $\langle \epsilon_L{}^2 \rangle^{\frac{1}{2}}$) and the intercepts give the coefficients $\ln A^{PF}(L)$. The $A^{PF}(L)$ versus L curves are linear for small values of L and the slopes are the measure of the negative reciprocal of the effective particle size given by

$$-\left[\frac{dA^{PF}(L)}{dL}\right]_{L=0} = \frac{1}{D_e} = \frac{1}{\bar{D}} + (1.5\alpha + \beta) \frac{V_{hkl}}{a} \tag{11}$$

A suitable selection of hkl planes allows the separation of the particle size $\bar{D}$ and the compound fault probability $(1.5\alpha + \beta)$.

The integral breadth in the reciprocal lattice space is defined by Laue[1] as follows:

$$b(s) = \frac{\displaystyle\int_{s_1}^{s_2} P'(s) \, ds}{P'(s_0)} \tag{12}$$

where the integral represents the area of the peak profile and $P'(s_0)$ is the intensity at the peak maximum $s_0 = 2 \sin \theta_0/\lambda$. Before being analyzed, the integral breadths are to be corrected for instrumental aberration. This can be achieved in either of two ways. The first is the use of an equation relating the true integral breadth $b(s)$, the measured integral breadth $B_c(s)$, and the instrumental breadth $b_a(s)$; the second is the use of Stokes[12] method for correction. Various analytic expressions have been used by many investigators; a good account of these is given elsewhere.[13] We will discuss here some of the very recent works on the use of analytic functions.

Wagner and Aqua[2] suggested a parabolic relation given by

$$b(s) = B_c\left[1 - \left(\frac{b_a}{B_c}\right)^2\right] \tag{13}$$

This relation was found to be a very good approximation in the case of several metals and alloys, namely, tungsten, SAP aluminum, niobium, silver–10% indium and silver–15% indium. Halder[14] has derived an exponential relationship between the integral breadths, i.e.,

$$b(s) = B_c\left\{\exp\left[-\left(\frac{b_a}{\sqrt{\pi}\,B_c}\right)^2\right]\right\} \tag{14}$$

This equation is applicable if the profile $g(s - s')$ due to the geometry of the diffractometer is represented by a Cauchy function and the profile $h(s)$ due to the cold-worked material is represented by a Gaussian function, and vice versa. Recently, Ruland,[15] by convoluting a Gaussian instrumental profile $g(s - s')$ with a Cauchy diffraction profile $f(s')$, has shown that

$$b(s) = \frac{b_a\exp[-(B_c/\sqrt{\pi}\,b_a)^2}{1 - \mathrm{erf}\,B_c/\sqrt{\pi}b_a} \tag{15}$$

where

$$\mathrm{erf}\frac{B_c}{\sqrt{\pi}b_a} = \int_0^{B_c/\sqrt{\pi}b_a}\exp(-t^2)\,dt$$

Equation (15) is too complicated and cannot be easily employed. However, to a good approximation it can be replaced by a parabolic equation as illustrated by equation (13), the deviation being at most +10%.

One can avoid the use of analytic functions and employ the Stokes method of correction for integral breadth measurements. The Stokes corrected integral breadths are shown to be

$$\frac{1}{b(s)} = \int_{-\infty}^{+\infty}A(L)\,dL \tag{16}$$

where $A(L)$ is given by equation (2). Thus we see that the integral breadths corrected using equation (15) or (16) are reliable and could be used for strain and particle size determination.

The separation of strain and particle size in the integral breadth measurements can be done if the shape of the particle size profile and the strain profile are known. Hall[16] assumed that both these profiles are described by Cauchy functions and obtained

$$b(s) = b^P + b^D \tag{17}$$

where $b^P = 1/D_I$ and $b^D = 2\epsilon_I s$. The value D_I is the integral breadth particle size and ϵ_I

is the integral breadth strain. Kurdyumov and Lysak[17] suggested that both these profiles are described by Gaussian functions. Therefore,

$$[b(s)]^2 = (b^P)^2 + (b^D)^2 \qquad (18)$$

But very recently, Schoening[18] and Halder and Wagner[19] showed that neither equation (17) nor equation (18) satisfy the actual particle size and strain distributions. They demonstrated that the particle size profile is approximately described by a Cauchy function and the strain profile is represented by a Gaussian function. Halder and Wagner[19] developed the approximate equation

$$\frac{b^P}{b(s)} = 1 - \left[\frac{b^D}{b(s)}\right]^2 \qquad (19)$$

which is very useful and gives particle size and strain values comparable to those obtained by Warren–Averbach method of Fourier analysis.

The use of variance, i.e., the second moment, as a measure of line broadening has been emphasized very recently.[5] The variance has attracted attention because of its additive property for the different causes of broadening. The variance $W(s)$ about the centroid s_g is given by

$$W(s) = \frac{\displaystyle\int_{s_1}^{s_2} (s - s_g)^2 P'(s)\, ds}{\displaystyle\int_{s_1}^{s_2} P'(s)\, ds} \qquad (20)$$

and

$$W(s) = W_0(s) + W^{PF}(s) + W^D(s)$$

where $W_0(s)$ is the variance due to the geometrical aberrations, $W^{PF}(s)$ is due to particle size and faulting, and $W^D(s)$ is due to strain. Wilson[5] showed that the variance of line profile is

$$W(s) = \frac{K\Delta s}{2\pi^2 D^W} - \frac{L}{4\pi^2 (D^W)^2} + (\langle \epsilon^2 \rangle - \langle \epsilon \rangle^2) s^2 \qquad (21)$$

where $\Delta s = s_2 - s_1 = \Delta 2\theta \cos \theta_0 / \lambda$, $\Delta 2\theta$ being the range of variance, D^W is the particle size, and K and L are constants defined elsewhere.[5] But one actually measures variance in the diffractometer coordinates given by

$$W(2\theta) = \frac{K\lambda\Delta 2\theta}{2\pi^2 D^W \cos \theta_0} - \frac{L\lambda^2}{4\pi^2 (D^W)^2 \cos^2 \theta_0} + 4(\langle \epsilon^2 \rangle - \langle \epsilon \rangle^2) \tan^2 \theta_0 \qquad (22)$$

The first and second terms are contributed by the smallness of the particle size and, as can be seen, one of them is range-dependent and the other is range-independent. As the second term is much less effective than the first and disappears for spherical particles, most of the particle-size broadening is shared by the first term. Thus an error in the measurement of Δs causes an error in the particle size, i.e., if Δs is overestimated, the particle size would be larger than the true value and if it is underestimated, the particle size would be smaller than the true value. The strain variance, however, remains unaffected for a small error in Δs. This is quite understandable from the fact that the effect of the strain is produced in the central part of the peak profile and the mistake is committed in the recovery of the tails which are characterized by the particle size only. The smaller the particle size, the longer are the tails and the greater is the uncertainty in the measurement of Δs.

Neglecting the range-independent part of the particle-size term and assuming the strain to be small, equation (21) may be written as

$$\frac{W(s)}{\Delta s} = \frac{1}{2\pi^2 D_e{}^W} + (\langle \epsilon^2 \rangle - \langle \epsilon \rangle^2) \frac{s^2}{\Delta s} \tag{23}$$

since $K/D^{W} = 1/D_e{}^W$. A plot of $W(s)/\Delta s$ against $s^2/\Delta s$ should therefore be linear and determine $D_e{}^W$ and $\langle \epsilon^2 \rangle - \langle \epsilon \rangle^2$ quite easily. In principle, $D_e{}^W$ should be equal to D_e, obtained from the initial slope of the Fourier coefficient $A^{PF}(L)$.

Pitts and Willets[20] have suggested that the standard deviation could also be used as a measure of the line broadening. By definition, the standard deviation is the square root of the variance and therefore the additive property of the variances will mean that the sum of the squares of the standard deviations are also additive:

$$[\sigma(\theta)]^2 = [\sigma^{PF}(\theta)]^2 + [\sigma^D(\theta)]^2 \tag{24}$$

Further, Willets[21] empirically suggested that the standard deviations due to both particle size and strain could be related by

$$\left[2\sigma(\theta) \frac{\cos \theta_0}{\lambda} \right]^2 = \left(\frac{K'}{D_e{}^P} \right)^2 + (\epsilon_W{}^P)^2 \left(\frac{\sin \theta_0}{\lambda} \right)^2 \tag{25}$$

where K' is a constant and $D_e{}^P$ and $\epsilon_W{}^P$ are the standard deviation particle size and strain, respectively. Although the theoretical basis for this equation was not discussed by Willets, it may be justified along the following lines.

In order to make the equations (23) and (25) identical and dimensionally the same, we substitute $1/a_3'$ for Δs [equation (10)] in equation (23) and obtain

$$W(s) = \frac{1}{2\pi^2 D_e{}^W a'_3} + (\langle \epsilon^2 \rangle - \langle \epsilon \rangle^2) s^2 \tag{26}$$

Therefore, the plot of $W(s)$ against s^2 should be linear. Writing equation (25) as

$$\left[2\sigma(\theta) \frac{\cos \theta_0}{\lambda} \right]^2 = \left(\frac{K'}{D_e{}^P} \right)^2 + (\epsilon_W{}^P)^2 \frac{s^2}{4} \tag{27}$$

and equation (23) as

$$W(2\theta) \left(\frac{\cos \theta_0}{\lambda} \right)^2 = \frac{1}{2\pi^2 D_e{}^W a'_3} + (\langle \epsilon^2 \rangle - \langle \epsilon \rangle^2) s^2 \tag{28}$$

it is easy to see that for $D_e{}^P = D_e{}^W$,

$$(K')^2 = \Delta s \frac{D_e{}^P}{2\pi^2} = \frac{D_e{}^P}{2\pi^2 a_3'} \tag{29}$$

and

$$\epsilon_W{}^P = 2(\langle \epsilon^2 \rangle - \langle \epsilon \rangle^2)^{\frac{1}{2}} \tag{30}$$

Willets used $K' = 1.44$ in the case of bromoiodide emulsions which he determined experimentally for 750Å crystals. But equation (29) shows that $K' = 1$ if $D_e{}^P \simeq 2\pi^2 a_3'$.

EXPERIMENTAL

Pure tungsten filings were prepared at room temperature (23°C) from a ductile specimen supplied in the platelet form and steel contamination was removed by magnetic separation. The filings were passed through 150 mesh and compacted into briquets using Duco cement as a binder and flattened with a glass plate. Annealed tungsten powder[2] was used for the correction of the instrumental aberration. The powder patterns were recorded for nickel-filtered copper radiation on a Norelco diffractometer provided with a Geiger counter. The chart recordings were conducted at 23°C.

The recorded profiles were examined first for constant background level before being analyzed. The background ratio between the cold-worked and annealed peaks was kept almost constant[22] for all hkl reflections except for the 222 and 400 peaks, which were not considered in the present analysis.

RESULTS

Fourier coefficients for 200 and 211 peaks for three different intervals of $\Delta s = s_2 - s_1$ were calculated. The change of Δs by $\pm 10\%$ does not produce any change in the Fourier coefficients. However, the variances, being more sensitive to the limits $s_2 - s_1$, are quite different for the change in range of $\pm 10\%$. The plots $A(L)$ versus L are shown in Figure 1. These Fourier coefficients yield a particle size of 210 Å and root-mean-square strain of 0.0043 (at $L = 0$). The true values of the variances and Δs were determined from the initial point of the horizontal part of the $W(s) - \Delta s$ curves as shown in Figure 2. The plot of $W(s)/\Delta s$ against $s^2/\Delta s$ is illustrated in Figure 3. Using a least square analysis, we obtain a particle size of 140 Å and root-mean-square strains of 0.0044.

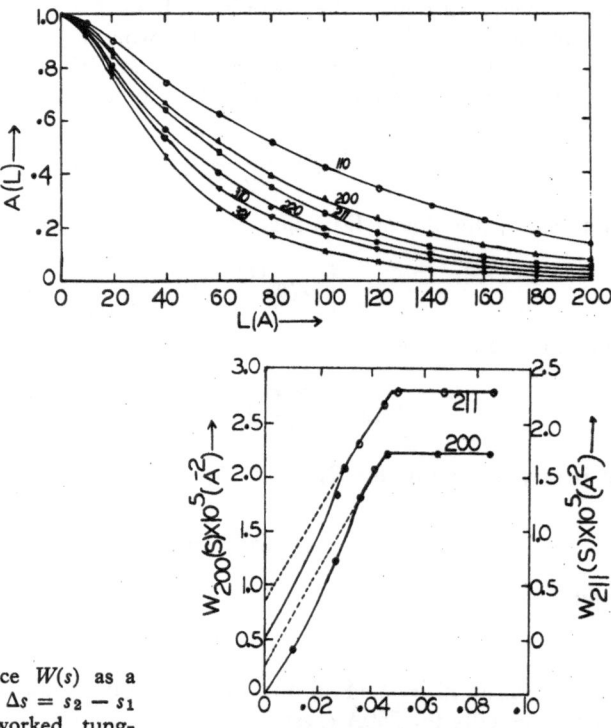

Figure 1. The plot of the corrected Fourier coefficients $A(L)$ of cold-worked tungsten filings as a function of the distance L normal to the reflecting planes hkl.

Figure 2. The plot of the variance $W(s)$ as a function of the integration range $\Delta s = s_2 - s_1$ for 200 and 211 peaks of coldworked tungsten filings.

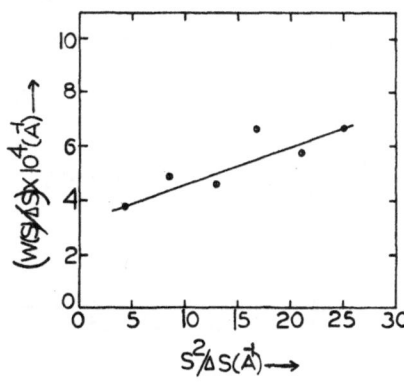

Figure 3. The plot of $W(s)$ versus $s^2/\Delta s$ for cold-worked tungsten filings.

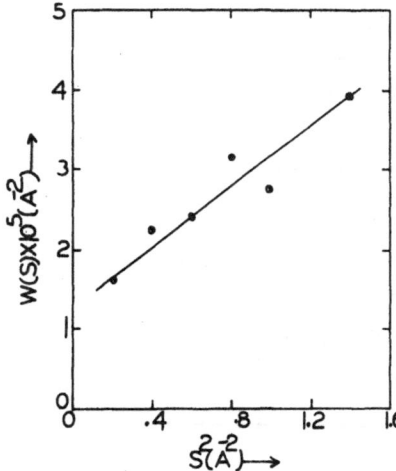

Figure 4. The plot of $W(s)$ versus s^2 for cold-worked tungsten filings.

According to equation (26), the values of the variances were also plotted against s^2 which is shown in Figure 4. The magnitude of a_3' was determined from a Δs versus s plot as shown in Figure 5, assuming it to be linear and the slow rise of the curve to be only a dispersion effect. The intercept of the straight line was the measure of $1/a_3'$. The effect of dispersion is to extend the line profile more towards the high-angle end than the low-angle end. The effect of dispersion on the variances of line profiles has been discussed by Wilson.[23] This set of calculations yields a particle size of 150 Å and root-mean-square strains of 0.0044, since in filings $\langle \epsilon_L \rangle \simeq 0$. Langford and Wilson[6] and Aqua[8] showed that the $W(s)$ versus Δs curve should be nonlinear at small values of Δs and linear after reaching the normal range. They suggested that an extrapolation of the straight line part would give strain and particle size from the intercept and slope respectively. Halder and Mitra[7] in their work on deformed tungsten wire determined strain and particle size from the variances of all the diffraction lines except (400). In their work $\Delta 2\theta$, which is related to Δs by $\Delta 2\theta = \Delta s \lambda / \cos \theta_0$, was obtained in a different manner. The value $\Delta 2\theta$ was experimentally estimated by making all the extrapolated lines of $W(s)$ versus Δs plots (Figure 2) parallel to one another because of the isotropic particle size in tungsten. The intercepts of the lines were the measure of $s^2 \langle \epsilon^2 \rangle$ and the slopes were equal to $1/(2\pi^2 D_e{}^W)$. Indeed, the results so obtained were in good agreement with those of Fourier analysis.

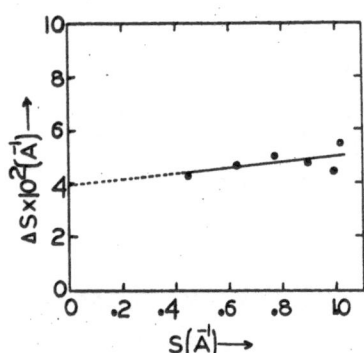

Figure 5. The plot of Δs versus s for cold-worked tungsten filings.

Table I. Particle Sizes and Strains for Cold-Worked Tungsten

Particle size		Strain	
Source	Size (Å)	Source	Strain
Warren–Averbach method, D_e	210	Warren–Averbach method, $\langle \epsilon_L{}^2 \rangle^{\frac{1}{2}}$	0.0040* and 0.0025†
Equation (19), D_I	430	Equation (19), ϵ_I	0.0037
Equation (26), $D_e{}^W$	140	Equation (26), $\langle \epsilon_L{}^2 \rangle^{\frac{1}{2}}$	0.0043
Equation (27), $D_e{}^P$	150	Equation (27), $\epsilon_W{}^P$	0.0044

* at $L = 0$.
† Averaged over $L = 0$ to $L = D_e$.

The particle size and strain were also calculated from integral breadths using equation (19). All the values of particle size and strain are shown in Table I.

It is clear from Table I that the particle size obtained by using equation (19) is twice as high as that obtained by Warren–Averbach analysis. As shown by Mitra and Halder[24] and Wagner and Aqua[2,25] in the case of metals containing faults, $D_I \simeq 2D_e$. It has been also shown[26] that for deformed metals, $D_I = \bar{D}^2/\bar{D} > \bar{D}$ where $D_e = \bar{D}$. The lattice strains of these two measurements are related by $\epsilon_I = 1.25 \langle \epsilon_L{}^2 \rangle^{\frac{1}{2}}$. If the root-mean-square strain of the Warren–Averbach analysis is considered over the average particle size, then $\epsilon_I \simeq 1.25 \times 0.0025 \simeq 0.0031$. This is in good agreement with the experimentally measured value $\epsilon_I = 0.0037$.

The variances give the lower value for particle size but fairly good value for strain. The reason for this is that the tail part of the line profile which determines the integration limits $s_2 - s_1$ and which is responsible for the particle size broadening has not been correctly evaluated. The strain broadening, which is dependent on the central part of the peak and mostly independent of its tail, has not been affected for a small error in the selection of the range of the integration limits.

Thus, we see that in the case of tungsten, the strain values are in very good agreement and are independent of the methods of calculations. A small error in the range is therefore not reflected in the strain values. Contrary to this, the particle size is strongly sensitive to this error. A smaller range Δs in the Fourier analysis produces a concavity upward in the $A^P(L) - L$ curves at small values of L. This is called the "hook effect" and is easily eliminated by considering the straight line portion of the curve for the determination of D_e. On the other hand, a larger Δs suppresses the hook, making the $A^P(L) - L$ curves more linear at small values of L. Therefore, it is possible to calculate the particle size by the Warren–Averbach method with equal confidence in either cases of positive or negative error in the measurement of true Δs. In the calculation of variances, however, a negative error or an underestimated Δs makes the quantity $W(s)/\Delta s$ greater, giving a smaller value for the particle size, and a positive error or an overestimated Δs will make the quantities $W(s)/\Delta s$ smaller, giving a larger value for the particle size. The measurement of Δs, is really disappointing and experimentally uncertain.

CONCLUSIONS

1. The Fourier method of line-broadening analysis by Warren–Averbach technique can be successfully employed for small values of strains, or when the strain distribution function can be approximated to a Gaussian function.

2. The integral breadth measurements lead to results comparable with those of Fourier analysis, when a parabolic relationship is used for instrumental correction as well as for particle size and strain separation.

3. The variances are extremely sensitive to the range of integration Δs. In general the quantity Δs is difficult to measure experimentally for powder pattern peaks having long tails. But a reasonable estimation of Δs is possible if one takes care of the choice of background level and the dispersion effect. The particle sizes are greatly affected by a small error in the measurement of Δs, whereas the strain values are almost independent of this error.

ACKNOWLEDGMENT

This investigation was supported by a contract from the Office of Naval Research.

REFERENCES

1. M. von Laue, "The Lorentz Factor and the Intensity Distribution in Debye–Scherrer Ring," *Z. Krist.* **64**: 115, 1926.
2. C. N. J. Wagner and E. N. Aqua, " Analysis of the Broadening of Powder Pattern Peaks from Cold-Worked Face-Centered and Body-Centered Cubic Metals," in: W. M. Mueller, G. R. Mallett, and M. J. Fay (eds.) *Advances in X-Ray Analysis*, Vol. 7, Plenum Press, New York, 1963, pp. 46–65.
3. B. E. Warren and B. L. Averbach, "The Effect of Cold-Work Distortion on X-Ray Patterns," *J. Appl. Phys.* **21**: 595, 1950.
4. A. J. C. Wilson, "Variance as a Measure of Line Broadening," *Nature* **193**: 568–569, 1962.
5. A. J. C. Wilson, "On Variance as a Measure of Line Broadening in Diffractometry, General Theory and Small Particle Size," *Proc. Phys. Soc. (London)* **80**: 286, 1962.
6. J. I. Langford and A. J. C. Wilson, *Crystallography and Crystal Perfection*, Academic Press, London, 1963, p. 207.
7. N. C. Halder and G. B. Mitra, "Particle Size and Strain in Deformed Tungsten," *Proc. Phys. Soc. (London)* **82**: 557–560, 1963.
8. E. N. Aqua, "The Separation of Particle Size and Strain by the Method of Variance," *Acta Cryst.* (in press).
9. B. E. Warren, "X-Ray Studies of Deformed Metals," *Progr. Metal Phys.* **8**: 147–202, 1959.
10. C. N. J. Wagner, "Analysis of the Broadening and Changes in Postiion of X-Ray Powder Pattern Peaks," in: J. B. Cohen and J. E. Hilliard (eds.), *Local Atomic Arrangements Studied by X-Ray Diffraction*, Gordon and Breach, New York, 1966, Chapter 6.

11. F. W. Jones, "Measurement of Particle Size by the X-Ray Method," *Proc. Roy. Soc. (London) Ser. A*: **166**: 16, 1938.
12. A. R. Stokes, " A Numerical Fourier-Analysis Method for the Correction of Widths and Shapes of Lines on X-Ray Powder Photographs," *Proc. Phys. Soc. (London) B* **61**: 382, 1948.
13. D. M. Vasil'ev and B. I. Smirnov, "Certain X-Ray Diffraction Methods of Investigating Cold-Worked Metals," *Soviet Phys.–Usp.* **4**: 226–259, 1961.
14. N. C. Halder, "A Study on the X-Ray Line Profile," *Physica* **30**: 1044–1050, 1964.
15. W. Ruland, "The Integral Width of the Convolution of a Gaussian and/or Cauchy Distribution," *Acta Cryst.* **18**: 581, 1965.
16. W. H. Hall, "X-Ray Line Broadening in Metals," *Proc. Phys. Soc. (London)* **62**: 741–743, 1949.
17. G. V. Kurdyumov and L. I. Lysak, *J. Tech. Phys. U.S.S.R.* **17**: 993, 1947.
18. F. R. L. Schoening, "Strain and Particle Size Values from X-Ray Line Breadths," *Acta Cryst.* **18**: 975–976, 1965.
19. N. C. Halder and C. N. J. Wagner, "Separation of Particle Size and Lattice Strain in Integral Breadth Measurements," *Acta. Cryst.* **20**: 312, 1966.
20. E. Pitts and F. W. Willets, "Determination of Crystallite Size from Diffraction Profiles by using Standard Deviation as a Measure of Breadth," *Acta Cryst.* **14**: 1302–1303, 1961.
21. F. W. Willets, "An Analysis of X-Ray Diffraction Line Profiles Using Standard Deviation as a Measure of Breadth," *Brit. J. Appl. Phys.* **16**: 323, 1965.
22. Shinichiv Sato, "Lattice Defects in Martensite by X-Ray Diffraction," *Japan J. Appl. Phys.* **1**: 210–217, 1962.
23. A. J. C. Wilson, *Mathematical Theory of X-Ray Powder Diffractometry*, Centrex Publishing Company, Eindhoven, 1963, p. 66.
24. G. B. Mitra and N. C. Halder, "Stacking Fault Probabilities in Hexagonal Cobalt," *Acta Cryst.* **17**: 817–822, 1964.
25. C. N. J. Wagner and E. N. Aqua, "X-Ray Diffraction Study of Imperfections in Rhenium," *J. Less-Common Metals*, **8**: 51–62, 1965.
26. E. F. Bertaut, "Debye–Scherrer Lines and Distribution of the Dimensions of the Bragg Domains in Polycrystalline Powders," *Acta Cryst.* **3**: 14–18, 1950.

DISCUSSION

A. Taylor (Westinghouse Research Labs): When you consider the errors involved in the determination of the background of a line, especially in the high orders, when you cannot even determine it properly because it runs out of the angular range of the photograph or diffractometer trace, is there any great advantage in using either Warren's method, Stokes' method, or the variance method over using the very simple early formula where they would just take the Laue integral breadth, use the Scherrer equation, and add on the term $\zeta \tan \xi$ for the broadening due to strain as well as the broadening due to particle size. Do the answers differ very greatly using the simple formula from using all these new complicated methods where you use the unfolding, etc.?

N. C. Halder: I will take up the first part of your question first. The background of the cold-worked profile is higher than the background of the annealed profile. We have followed the method used by Sato considering the ratio of the background levels between the cold-worked profile and the annealed profile and tried to keep this ratio nearly constant for all the reflections. In the case of the 400 reflection this ratio was reasonably high and that is why we didn't consider it. The 222 reflection, being very weak, we didn't consider either. Your other question was why we don't use the simple equation. The reason is that if we know exactly the strain profile and the particle size profile we could find out a definite relationship between the integral breadths. But these are not definitely known. As the line-broadening analysis (Warren–Averbach) shows, the particle size-distribution is approximately Cauchy type and the strain distribution is approximately Gaussian type for small strain, therefore, it would be unrealistic to use the simple equation that you mentioned. I showed on my slides that from the convolution method we get some other relationship which is not very easy to handle. That is why we used the parabolic equation, which is quite good. The accuracy is at most ±10%. We have published a paper on this subject, and if you are interested in that I can give you reference.

A. Taylor: You chose tungsten as your metal, but that I think has a very uniform distribution of Young's modulus, E_{hkl} is almost a sphere or a cube with very rounded-off corners. Now suppose you were checking something where E_{hkl} was very anisotropic with indices. What happens then to all your equations?

N. C. Halder: Tungsten is elastically isotropic and we do not see any anisotropy in the case of strain and particle size. We have also studied other metals which were anisotropic and which contained faults, for example, silver–indium. We examined two alloys, one with 10 at.% indium and the other with 15 at.% indium, and demonstrated that our equations could still be used without any difficulty. This we have added in the same paper to which I just referred.

LATTICE STRAIN MEASUREMENTS ON DEFORMED FCC METALS

Eckard Macherauch

*Max-Planck-Institut für Metallforschung
Stuttgart, Germany*

ABSTRACT

In this work experiments have been undertaken to determine the type of residual stresses of plastically deformed polycrystals of aluminum, copper, and nickel which cause a shift of the X-ray interference lines. In order to get accurate stress values, the lattice strain distribution has been measured in special planes of the deformed specimens, using the $\sin^2 \psi$ technique. The surface stresses were determined as a function of the macroscopic plastic strain of cylindrical specimens. With increasing strain, the residual surface stress component parallel to the direction of deformation increases. The stresses observed are compressive ones after tension and tensile ones after compression.

The changes in the residual lattice strain distributions which arise on progressive thinning of the plastically deformed specimens have been measured. From the stress values determined in each new surface layer, the stress distribution originally present in the undestroyed specimen was calculated. Tensile stresses in the interior of the specimens are in equilibrium with compressive stresses in the surface areas. The formation of residual stresses on specimens with a work-hardened surface layer is found to be quite different from that of well-annealed samples.

The results of this work seem to show that in uniaxially deformed cylindrical specimens, residual macrostresses arise which may be a consequence of a macroscopic inhomogeneity of work-hardening between the surface layers and the interior of the polycrystals.

INTRODUCTION

Polycrystals deformed plastically in a uniaxial tensile test show residual lattice strains, which broaden the X-ray line profiles and shift their positions. From the changes in the diffraction line positions the homogenous part of the lattice strains can be determined.[1-3]

In the past, a great number of residual lattice strain measurements on plastically deformed fcc metals and alloys have been undertaken. In addition to the determination of the magnitude of the surface residual strain on specimens (mainly parallel to the normal of the specimen surface) extended to a special amount, residual strains have also been measured using different X-ray wavelengths.[4,5] Furthermore, changes of the residual strains have been observed which arise on progressive thinning of deformed specimens.[6,7]

Several attempts have been made to explain the cause of these residual lattice strains existing in specimens deformed in a tensile test. From the uniaxiality of such a test it might be thought that no macrostresses would be produced. In spite of this fact two main lines of thought have been pursued in recent years. One of them leads to residual macrostresses, the other one to residual microstresses. Both types of explanation proceed from the common assumptions that under tensile loading in a specimen, "hard" and "weak"

---

[1] References are at the end of the paper.

regions occur, and that after unloading, the "hard" regions remain under tensile stresses, giving rise to a compression of the "weak" regions. On this basis the following suggestions have been proposed, to explain the residual lattice strains existing in homogeneous materials after tensile elongation:

1. The surface crystallites should have lower yield stress than the crystallites in the interior of the polycrystalline specimen (surface effect).[8,9]
2. Stresses in regions of coherent scattering are of opposite sign to those near and in highly distorted regions, subboundaries, and grain boundaries (coherent-area-effect).[10,11]
3. Due to the yield-point anisotropy intergranular stresses arise during plastic deformation. By means of X-rays lattice strains only of specially oriented crystallites are measured (orientation effect).[12]
4. After exceeding the yield point, differences in the work-hardened state between the surface layers and the interior of the specimen arise (hardening effect).[13]

Evidently proposals 1 and 4 lead to residual macrostresses, while proposals 2 and 3, on the other hand, give rise to microstresses. At present[6,7] there exists considerable confusion about the question which of the counted-up effects is mainly responsible for the measured residual lattice strains in plastically deformed polycrystals.

The purpose of this paper is to describe some critical experiments which may help to decide whether or not one of the suggestions mentioned above is able to explain the experimental results.

EXPERIMENTAL PROCEDURES

Cylindrical specimens of about 5 mm in diameter were used for all measurements. All specimens were carefully polished by standard electrolytic methods. The chemical composition of the materials used, the annealing treatment of the machined samples in a vacuum furnace, and their grain sizes are listed in Table I. The aluminum and copper specimens were deformed in a Polanyi-type tensile machine, and the nickel specimen in a Losenhausen-type tensile machine. The strain rate was about 10^{-4} sec.

The X-ray equipment used in these investigations consisted of a Seifert-Iso-Debyeflex apparatus equipped with Siemens-X-ray tubes with copper and cobalt targets. The arrangement for measuring residual strain distributions by means of the back-reflection film technique is shown in Figure 1. The specimen S can be translated and rotated relative to the fixed X-ray tube T and the collimating system C. On a shaft at the foot of the collimating tube the film chamber F is fixed. Thus the film is in a position

Table I. Chemical Composition, Heat Treatment and Grain Size of the Measured Materials

Material	Chemical composition	Heat treatment (furnace-cooled)	Grain size (μ)
Aluminum	99.99% Al	12 hr 250°C	130
Nickel-1	99.98% Ni	2 hr 550°C	15
Nickel-2	99.8% Ni	2 hr 700°C	30
Copper	99.98% Cu	2 hr 450°C	15
Aluminum–copper alloy	96% Al, 4% Cu	4 hr 350°C	25

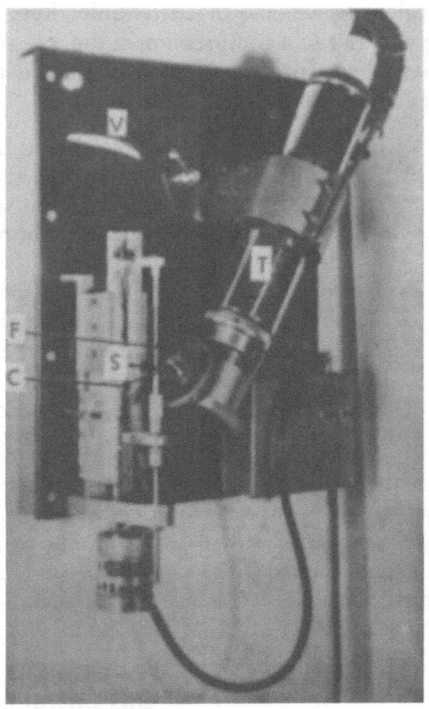

Figure 1. X-ray device for measurements of residual
stresses.

perpendicular to the axis of the incident X-ray beam. The inclination of the incident
X-ray beam to the surface normal of the specimen can be adjusted with the aid of the
screw V.

In order to determine the shifts of the interference lines due to the homogeneous
lattice strains the method of using a calibrating powder was applied. The $\sin^2 \psi$ method[14]
was used to get the best information about the interesting strain distributions. With this
method, five to six lattice-strain values in different directions ψ_i were determined. The
value ψ_i is the angle between the surface normal and the direction of measurement. In all
cases the lattice strains were investigated in a plane given by the direction of load and the
surface normal of the specimen. From the slope of the lattice strains e_ψ plotted versus
$\sin^2 \psi$,

$$m^* = \frac{\partial \epsilon_\psi}{\partial \sin^2 \psi} \tag{1}$$

stress values have been calculated using the equation

$$\sigma = \frac{m^*}{(\tfrac{1}{2}s_2)_{\text{X-ray}}} \tag{2}$$

which follows from the theory of elasticity and the basic principles of the X-ray stress
analysis. The value $(\tfrac{1}{2}s_2)_{\text{X ray}}$, one of the so-called X-ray elastic constants, is given by

$$(\tfrac{1}{2}s_2)_{\text{X-ray}} = \left(\frac{\nu + 1}{E}\right)_{\text{X-ray}} \tag{3}$$

where ν is Poisson's ratio and E is Young's modulus valid for X-ray measurements.

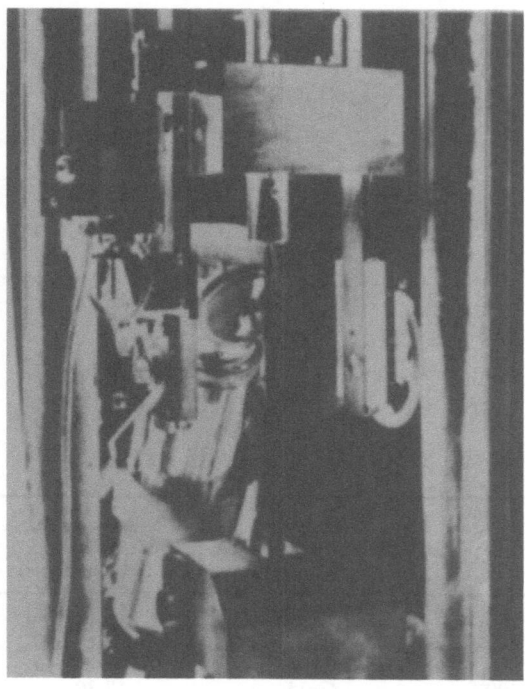

Figure 2. X-ray device for measurements of loading stresses.

In order to calculate stresses according to equation (2), the appropriate X-ray elastic constants were used. From equation (2) it follows that

$$\frac{dm^*}{d\sigma} = (\tfrac{1}{2}s_2)_{\text{X-ray}} \tag{4}$$

Therefore a tensile test in the elastic range has to be performed to measure the slopes m^* of the lattice strain distributions for different known values of tensile stresses. The measurements must be done in the plane determined by the surface normal and the direction of load. The experimental equipment for such measurements is shown in Figure 2. The specimen is held in a tension machine. During exposure, the tube with the collimating system is moved parallel to the fixed specimen. Experimental results determined in this way on prestrained specimens of pure copper and nickel are shown in Figure 3. The slopes of the straight lines lead to $(\tfrac{1}{2}s_2)_{\text{X-ray}}$ values of 13.3×10^{-5} kg/mm² for copper and 6.1×10^{-5} kg/mm² for nickel. The most important X-ray data of the materials used in the present investigations are listed in Table II.

To determine the distribution of residual stresses over the cross section of the deformed specimens a modification of the well-known turning-off method has been used.[15,16] The underlying principle is to measure the macroscopic strains of the specimens released by the removal of layers of the sample in which residual stresses were acting. In the case of a uniaxial stress state, the residual stress σ_{R_n} acting at the cross section area F_n of the undestroyed cylindrical specimen is given by the Heyn–Bauer differential equation,

$$\sigma_{R_n} = E\left[F_n \left(\frac{d\epsilon}{dF}\right)_n - \epsilon_n \right] \tag{5}$$

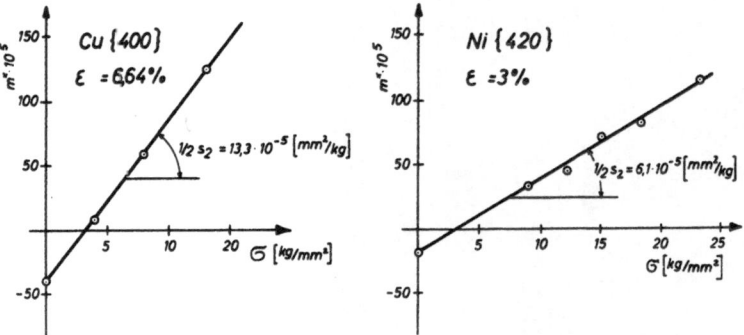

Figure 3. m^*-σ curves for determination of X-ray elastic constants.

Table II. X-Ray Date of the Measured Materials

Material	Radiation	Reflecting planes	Calibrating powder	X-ray elastic constant $\frac{1}{2}s_2$ ($\times 10^{-5}$ kg/mm^2)
Aluminum	Cu K_α	(511)/(333)	Silver	19.12
Nickel-1	Cu K_α	(420) and (313)	Germanium	6.10 and 5.90
Nickel-2	Cu K_α	(420) and (313)	Germanium	6.20 and 5.45
Copper	Co K_α	(400)	Gold	13.30
Aluminum–copper alloy	Cu K_α	(511)/(333)	Silver	19.74

where ϵ_n is the longitudinal macroscopic strain of the specimen after the cross section is reduced from the original F_0 to F_n and $(d\epsilon/dF)_n$ is the slope of the ϵ_n-F_n diagram. It can be shown that the term $EF_n(d\epsilon/dF)_n$ always represents the stress acting on the very surface of the specimen which has been partially turned off. This term can be determined by means of X-rays as a function of F_n. For that purpose the lattice strain distribution must be determined in each new surface layer. With the corresponding σ_x value, the integration of equation (5) leads to the following:

$$\sigma_{R_n} = \sigma_x - \int \frac{\sigma_x dF_n}{F_n} \tag{6}$$

From this equation the original residual stress distribution in the undestroyed specimen can be calculated. Instead of the usual turning-off method, an electrolytic etching technique was used to avoid residual stresses due to a mechanical removal of surface layers.

EXPERIMENTAL RESULTS

Typical lattice strain values are plotted in Figure 4 as a function of sin$^2 \psi$ for a nickel specimen extended to 3% resp. 26% in a tensile test. At small ψ values a lattice dilatation is observed, whereas at higher values a lattice compression is found. The lattice strains show approximately a linear dependence on sin$^2 \psi$. If least mean square lines are drawn through the measured points their slopes increase with increasing deformation. From

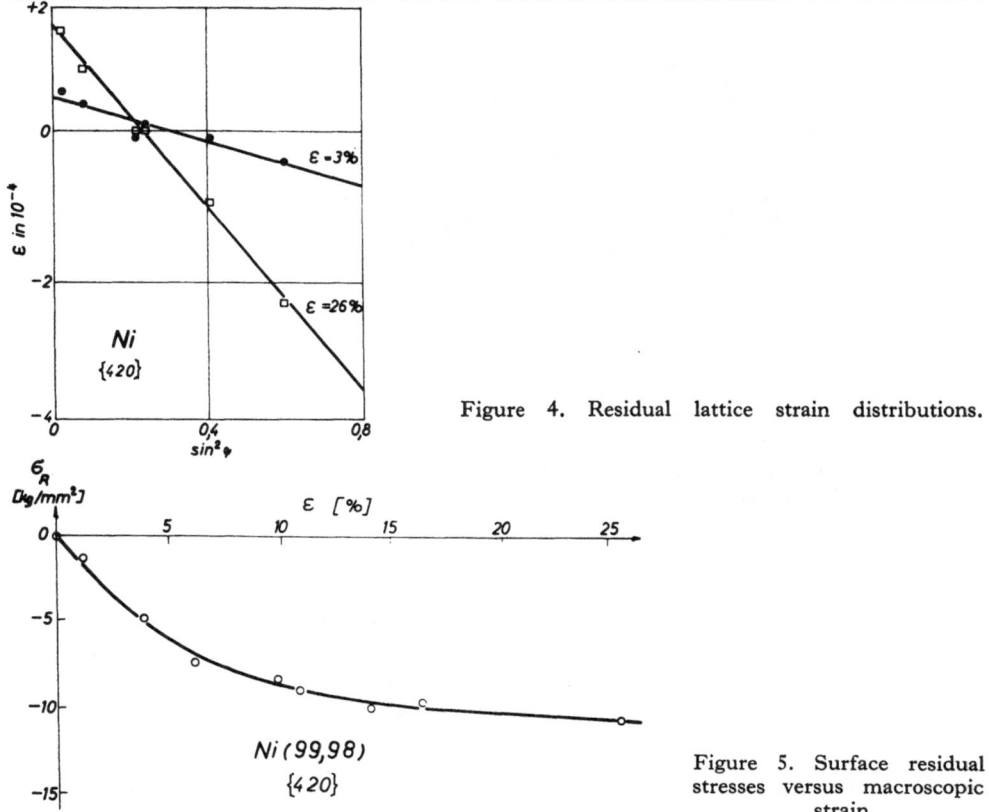

Figure 4. Residual lattice strain distributions.

Figure 5. Surface residual stresses versus macroscopic strain.

these slopes the actual surface stress can be determined using equation (2). As a result we get the dependence of the residual surface stresses on the degree of plastic deformation, which is shown in Figure 5. The surface stresses are of compressive nature. They increase with increasing plastic deformation of the specimen. After an extension of about 10% the compressive residual surface stress is about $- 8 \, kg/mm^2$.

By the investigation of other fcc metals similar results were obtained. Some of them are shown in Figure 6. As can be seen, the strain distribution curves from aluminum, copper, and nickel show similar characteristic features, at least after a sufficient amount of plastic deformation. In no case were distribution curves observed which were in agreement with theoretical calculations on the basis of the orientation effect discussed above. Those theoretical curves are shown as dotted lines in Figure 6. At this point it should be noted that measurements on different lattice planes lead to nearly the same result, e.g., (313) and (420) for nickel, (511)/(333), (420) and (222) for aluminum. As an example, Figure 7 shows the strain distributions which have been measured with different X-ray wavelengths (Co K_α, Cr K_α, and Cu K_α) on aluminum. The specimen has been deformed plastically by 4.6%. From the curves the residual stresses were calculated to be $- 1.7$, $- 1.6$, and $- 1.8 \, kg/mm^2$, respectively. Finally, it is shown in Figure 8 that an aluminum–copper alloy exhibits a behavior similar to the pure fcc metals as far as the dependence of the residual surface stresses on the degree of deformation is concerned. These stresses are plotted as a function of the plastic strain up to 7%. It can be seen that the negative stress values increase with increasing extension of the specimen. Measurements with other X-ray wavelengths confirm this dependence, too.

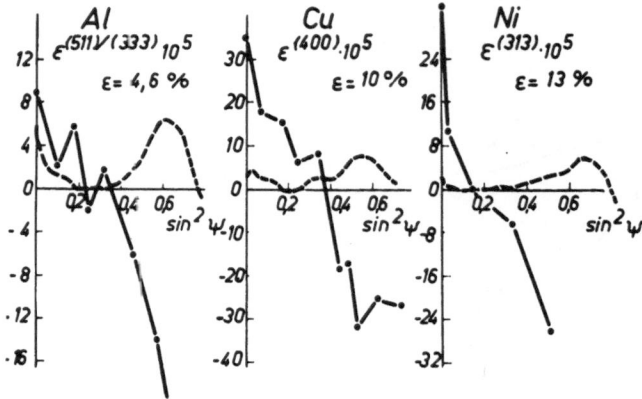

Figure 6. Residual lattice strain distributions. Measured values, full lines; calculated values according to the orientation effect, dashed lines.

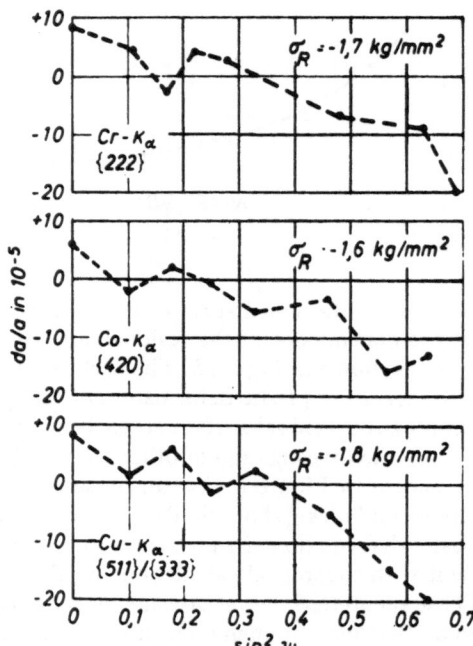

Figure 7. Dependence of residual strain distributions of a tensile-deformed aluminum specimen on X-ray wavelengths.

In order to determine the residual stress distribution over the cross section of the cylindrical samples, the following procedure was adopted: First the lattice strains as a function of $\sin^2 \psi$ were measured at the original surface of the specimen. Thereafter, a thin surface layer was removed electrolytically, which was followed by another determination of the lattice strains as a function of $\sin^2 \psi$, and so on. From each strain distribution the residual stress was calculated according to equation (2). The calculated stresses as a function of the remaining cross-sectional area are plotted in Figure 9, where typical results are given for aluminum, copper, and nickel strained plastically by 4.6, 10, and 13%, respectively.

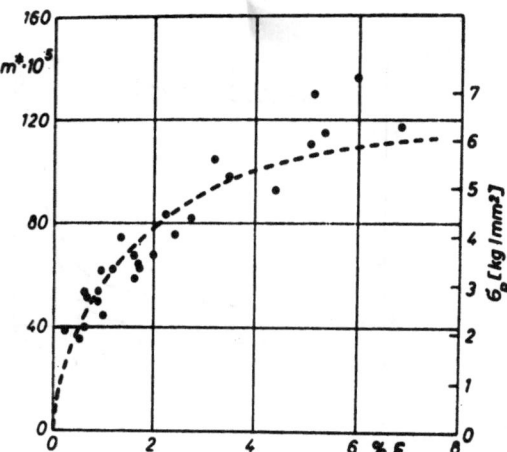

Figure 8. Surface residual stress of an Al-4% Cu alloy in dependence on the plastic strain.

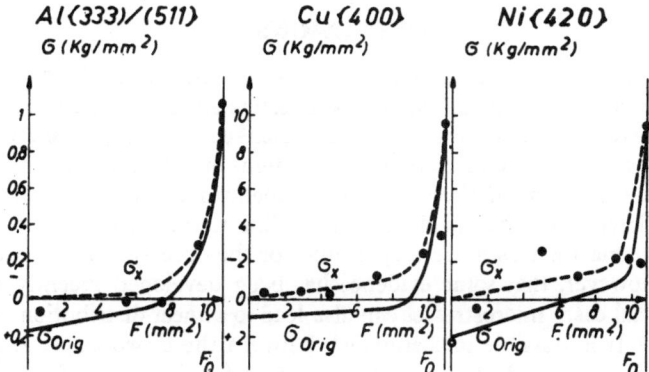

Figure 9. Residual stress distribution (full lines) over the cross section of uniaxially deformed fcc cubic metals. Dashed lines represent stress values determined after different amounts of electrolytic thinning.

The experimental values are marked by solid points. From the dashed curves the original stress distribution in the undestroyed specimen was calculated according to equation (6). The corresponding curves are shown as full lines in Figure 9. They drop from maximum negative values to zero within a small surface layer. Then the stresses change to positive values, which are present in the whole interior of the specimens. The compressive stresses in the surface area are in equilibrium with the tensile stresses of the remaining cross-sectional area. It should be emphasized that these tensile stresses never can be measured directly by the experimental method described above. Due to the thinning process the residual stresses relax in such a way that the actual surface layers always show residual stresses with negative sign. An aluminum–copper alloy deformed in a tensile test about 4% exhibits a similar residual stress distribution over the cross section. A typical result is shown in Figure 10.

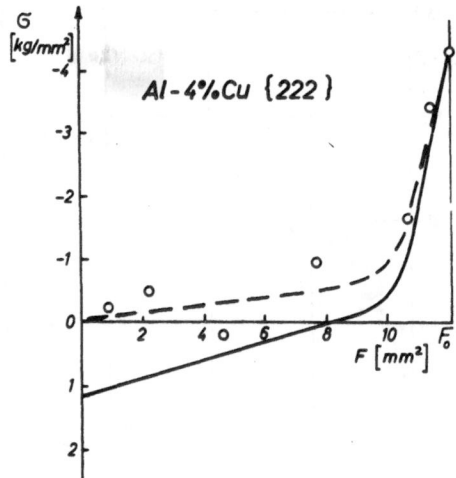

Figure 10. Residual stress distribution over the cross section of an Al–4%Cu alloy extended by 7%.

CONCLUSIONS

From the experimental results it can be deduced that residual lattice strains arise after passing the yield stress of the materials investigated. In a good approximation the lattice strains always show a linear dependence on $\sin^2 \psi$. During progressive thinning of the deformed specimens the lattice strain distributions are changed systematically.

What causes these results? From the linear dependence of the residual lattice strains on $\sin^2 \psi$ the following may be concluded: According to the basic rules of the theory of elasticity residual macrostresses are responsible for the observed lattice strains. It must be mentioned, however, that some evidence has been developed recently which shows that in some special cases microstresses can also lead to a linear distribution of the residual lattice strains.[5-7] It is possible to distinguish between these two cases. A pure state of microstress should not be affected by etching away surface layers of a deformed specimen in contrast to a state of macrostress, which may be influenced by such a procedure. The experimental facts mentioned above decide for the latter case. Therefore, the orientation effect and the coherent-area effect can be excluded from the factors which contribute to the residual lattice strains in uniaxially deformed fcc metals. Furthermore, the measurements on different lattice planes of the same metals lead to approximately the same results. This supports the conclusion that all crystallites in the measured areas have the same stress state.

All experimental facts agree with the idea that the residual lattice strains in uniaxially deformed specimens are due to macrostresses. The development of such stresses can be seen from Figure 11. After exceeding the yield point it is assumed that the flow stress σ_s of the surface layers never becomes equal to the average flow stress σ_M of the whole specimen. According to this assumption, an inhomogeneity can be expected for the stress distribution over the cross section of the polycrystalline sample during tensile deformation. With increasing deformation, the surface flow stress σ_s remains more and more behind the average flow stress σ_M and the flow stress σ_i in the interior of the specimen. It may be thought that this is a consequence of the exceptional position of the surface crystallites during plastic deformation. The surface crystallites as well as those which are close to the surface can deform much more easily than the crystallites in the interior of the

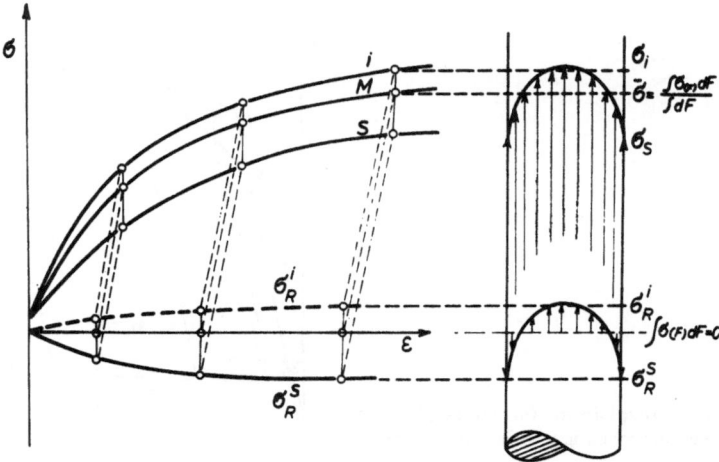

Figure 11. Schematic representation of work-hardening in different cross-sectional areas of a tensile·specimen.

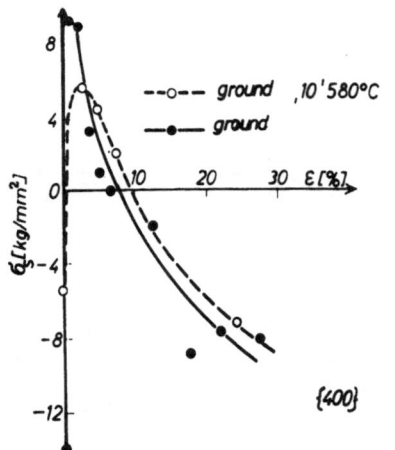

Figure 12. Surface residual stresses of ground specimen versus macroscopic strain.

specimen. Thus, work-hardening is different in both areas of the sample. This inhomogeneity of work-hardening causes the macroscopic inhomogeneous stress distribution postulated above. Consequently, after unloading, a residual macrostress state arises in the specimen as shown in the lower part of Figure 11. Compressive residual stresses in the surface layers balance tensile residual stresses in the interior. Indeed, flow stress measurements during a tensile work-hardening test of nickel and copper specimens have confirmed that the real stress acting in the surface layers is smaller than the average stress of the whole sample.

Further evidence can be given to support these conclusions. If the ideas proposed are valid, a difference in the dislocation density must be revealed between crystallites in the surface layers and in the interior of a specimen deformed in a tensile test. In fact, the etching method proposed by Guard[17] does show the right kind of difference in the dislocation density in the case of nickel. On the other hand we expect that well-annealed specimens with a work-hardened surface layer will reveal marked differences in the

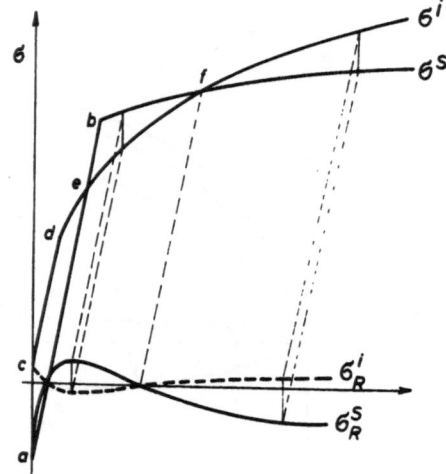

Figure 13. Model to explain the formation of surface residual stresses in specimen with work-hardened surface layers.

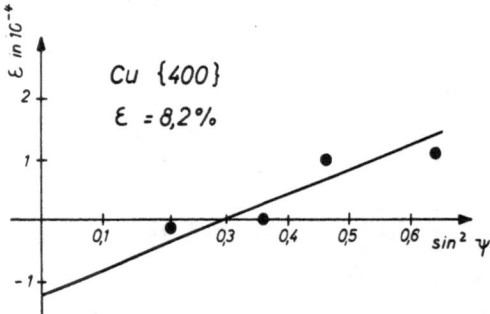

Figure 14. Residual lattice strain distribution after compression by 8.2%.

formation of surface residual stresses during a tensile test. As a typical example Figure 12 shows the behavior of ground specimens. One sample was only ground and the other one (dashed curve) was ground and then heat-treated. Due to the grinding process compressive residual stresses in the surface layers occur. It can be seen from Figure 12 that already after a small plastic deformation changes in the sign of the surface residual stresses appear. After tensile stresses reach a maximum, compressive stresses are again observed at higher deformations. This two-fold change of the sign of the surface residual stresses with increasing tensile deformation is independent of the measured *hkl* planes. Copper specimens with work-hardened surface layers revealed similar results.

A model which can explain these experimental results is illustrated in Figure 13. We suppose that the work-hardened surface layer has a more extended elastic range (a b) and a higher yield point than the well-annealed interior (c d) of the specimen. We further assume that the slope of the work-hardening curves of both areas of the deformed specimens differs in the suggested way. Then we can expect that the full line in the lower part of the picture describes the strain dependence of the surface residual stresses. The dashed line gives the residual stresses of the interior of the specimen. As can be seen, the two-fold change in the sign of the residual stresses is explained by this model. Unloading from the points e and f yields zero residual stresses.

A further proof to support our conclusions is the development of residual surface stresses after compressive deformation. As shown in Figure 14, after 8.2% compression, a residual lattice strain distribution occurs whose slope is opposite to that obtained from a specimen deformed in a tensile test. According to equation (2) it can be concluded that residual tensile stresses are acting in the surface layers. Furthermore, it has been proved that these tensile stresses increase with increasing compressive strains.

REFERENCES

1. C. S. Barrett, *Structure of Metals*, 2nd ed., McGraw Hill, New York, 1952.
2. B. D. Cullity, *Elements of X-Ray Diffraction*, Addison-Wesley Publishing Co., Reading, Mass., 1954.
3. R. Glocker, *Materialprüfung und Röntgenstrahlen*, 4th ed., Springer-Verlag, Berlin, 1958.
4. G. B. Greenough, "Quantitative X-Ray Diffraction Observations on Strained Metal Aggregates," *Progr. Metal Phys.* **3**: 176–219, 1952.
5. D. M. Vasil'ev and B. I. Smirnov, "Certain X-Ray Diffraction Methods of Investigating Cold Worked Metals," *Usp. fiz. Nauk* **73**: 503, 1961.
6. E. Macherauch, "Principles and Problems of the X-Ray Determination of Elastic Stresses," *Materialpruefung* **5**: 14–25, 1963.
7. V. Hauk, *Z. Metallk.* **55**: 626, 1964.
8. R. Glocker and H. Hasenmeier, "X-Ray Stress Measurements on the Beginning of the Flow Process in Carbon Steels," *Z. Ver. deut. Ing.* **84**: 825–828, 1940.
9. T. Nishihara and S. Taira, "Effect of Free Surface in the Yielding Resistance of Materials," *Mem. Fac. Eng. Kyoto Univ.* **12**: 90–118, 1950.
10. S. L. Smith and W. A. Wood, "Internal Stress Created by Plastic Flow in Mild Steel, and Stress-Strain Curves for the Atomic Lattice of Higher C Steels," *Proc. Roy. Soc. (London) Ser. A* **182**: 404–414, 1944.
11. D. M. Vasil'ev, "Character of the Initial Stage of Plastic Deformation in Polycrystalline Metals," *Fiz. Metal. i Metalloved.* **14**: 106–113, 1962.
12. G. B. Greenough, "Residual Lattice Strains in Plastically Deformed Polycrystalline Metal Aggregates," *Proc. Roy. Soc. (London) Ser. A* **197**: 556–567, 1949.
13. E. Macherauch and P. Müller, "Lattice Deformation and Residual Strains in Pure and Alloyed Aluminum Specimens after Stress Deformation," *Z. Metallk.* **49**: 324–331, 1958. "Influence of Wavelength in Lattice Strain X-Ray Measurements," *Z. Metallk.* **51**: 514–523, 1960.
14. E. Macherauch and P. Müller, *Z. Angew. Phys.* **13**: 305, 1961.
15. M. G. Moore and W. P. Evans, *SAE Trans.* **66**: 340, 1958.
16. K. Kolb and E. Macherauch, "X-Ray Studies of the Plastic Deformation of Sinter Nickel," **53**: 108–114, 1962.
17. R. W. Guard, "Etch Pit Method for Revealing Dislocation Sites in Nickel," *Trans. AIME* **218**: 573–574, 1960.

DISCUSSION

E. F. Sturcken (E. I. du Pont de Nemours & Co.): When you rotate the sample through the angle ψ you get an amount of defocusing. Do you have to move the counter in to measure your $\Delta d/d$?

E. Macherauch: Yes. The arrangement is such that the incident beam always hits the rotating specimen at the same point. The divergence is very small. Therefore we get only a small defocusing if we incline the specimen to the primary beam, and interference line as well as reference line are influenced in nearly the same way.

X-RAY STUDY OF WIRE-DRAWN NIOBIUM AND TANTALUM

R. P. I. Adler and H. M. Otte

Martin Company
Orlando, Florida

ABSTRACT

Deformation, introduced into niobium and tantalum specimens by wire drawing at room temperature, produced changes in the shape and position of X-ray diffraction peaks. The resultant peak profiles and locations of all available peaks were recorded using the Debye–Scherrer geometry on a modified diffractometer with crystal monochromated Cu K_{α_1} radiation. The amount of deformation in the surface layers of both metals was found to saturate essentially after only 20% reduction in area. The measured decrease in the lattice parameters of either material was attributed to a residual surface stress; the average value for the deformed saturated state for both tantalum and niobium wires corresponded to an equivalent longitudinal tensile stress of 35 ± 5 kg/mm². Integral breadth measurements revealed approximately equal X-ray particle sizes in the $\langle 100 \rangle$ and $\langle 110 \rangle$ directions; the minimum particle size for the microstructures of both metals was around 200 Å and occurred after the first few draws.

INTRODUCTION

The most significant mechanical effect produced by wire drawing metals is probably the attendant increase (two- to three-fold) in the tensile strength of the material (Table I). This increase is a consequence of the work-hardening associated with plastic deformation which, however, invariably reduces the ductility of the material. The drawing operation involves a number of processing variables such as the die angle, the type of lubrication, the reduction per draw, the draw direction with respect to the previous pass, and the rate of drawing. However, the drawing process basically consists of applying to the metal surface large shear forces which reduce the wire or rod diameter by causing an axial plastic flow of the surface; there is also with increasing severity of the draw (i.e., percent reduction in cross sectional area per pass), a progressive radial penetration of cold-work into the wire.

Extensive engineering studies of wire drawing have been made using techniques involving the measurement of dimensional changes on removing material[5] and also using analytic computations[6] of the flow arising from wire drawing. However, metallurgical investigations of lattice distortions and crystal imperfections have been limited. Since these play an important role in the plastic flow of the material and thus influence the effectiveness of the drawing operation, an X-ray investigation was undertaken to study the deformation of the surface layers of drawn wires. Specifically, the results for two bcc metals, tantalum and niobium, will be presented.

These two metals were selected because in many respects they are very similar. However, there is one interesting difference, which is that they have "opposite" elastic

5 References are at the end of the paper.

Table I. Comparison of the Room Temperature Strengths of Bulk and Wire Materials

Material	Crystal structure	Bulk material		Wire		
		Ultimate tensile strength (kpsi)	Reference	Ultimate tensile strength (kpsi)	Diameter (in.)	Reference
Type 302 stainless steel	fcc	85	1	316	0.006	3
Type 316 stainless steel	fcc	80	1	298	0.006	3
Tungsten	bcc	50–200†	2	476	0.005	3
Pearlitic steel (0.93%C)	two-phase	220	3	218	0.025* (0% RA)	4
				295	0.013 (73% RA)	4
				449	0.006 (94% RA)	4
				616	0.003 (98.5% RA)	4

* Original diameter of wire prior to cold-drawing.
† Swaged rod.

anisotropy as represented by the anisotropy factor $A = 2 (S_{11} - S_{12})/S_{44}$, where S_{ij} are the elastic constants (Tables II and III). For an elastically isotropic medium, $A = 1$, whereas for tantalum, $A = 1.6$ and for niobium, $A = 0.5$. Thus these two metals should show contrasting behavior toward the forces of wire drawing.

EXPERIMENTAL TECHNIQUES AND PROCEDURES

Preparation of Specimens

Rods 3.2 mm diameter of commercially pure niobium and tantalum (received from Fansteel Metallurgical Corporation) were used as the starting material. The niobium was annealed at 1200°C to yield a grain of size 0.01–0.03 mm and the tantalum at 1250°C to produce a grain size of about 0.05 mm.

All wire drawing was performed using "Vigor" (No. J or No. JA) drawplates at room temperature and a draw speed of 9 in./min. This rate was sufficiently low to avoid any noticeable heating. X-ray measurements were made between selected draws until a reduced diameter of approximately 0.5 mm had been reached. Smaller diameters than this yielded an insufficiently intense diffracted beam. The reduction in area (RA) was

Table II. Elastic Constants for Tantalum

Single-crystal elastic constants (Bolef[7]):

$C_{11} = 2.67 \times 10^{12}$ dynes/cm^2 $\qquad$ $S_{11} = 0.6855 \times 10^{-12}$ cm^2/dyne

$C_{12} = 1.61 \times 10^{12}$ dynes/cm^2 $\qquad$ $S_{12} = - 0.258 \times 10^{-12}$ cm^2/dyne

$C_{44} = 0.825 \times 10^{12}$ dynes/cm^2 $\qquad$ $S_{44} = 1.21 \times 10^{-12}$ cm^2/dyne

Mechanical (stress-strain) elastic constants (*Metals Handbook*[2]):

$E = 27 \times 10^6$ psi

$\nu = 0.35$

$\dfrac{\nu}{E} = 0.184 \times 10^{-4}$ mm^2/kg

Elastic moduli for X-ray stress measurements (Greenough[8]):

for isotropic stresses: $K_\sigma = S_{12} + (S_{11} - S_{12} - \tfrac{1}{2}S_{44})\Gamma$

where $\Gamma = \dfrac{h^2k^2 + k^2l^2 + l^2h^2}{(h^2 + k^2 + l^2)^2}$

for isotropic strains: $K_\epsilon = - \dfrac{(C_{11} + 4C_{12} - 2C_{44})}{2(C_{11} + 2C_{12})(C_{11} - C_{12} + 3C_{44})}$

for "average" case: $\bar{K} = \tfrac{1}{2}(K_\sigma + K_\epsilon) = - \nu/E$

	$-K_i (\times 10^{-4}$ mm^2/kg)			
hkl	Isotropic stress	Isotropic strain	Average	$\Delta a_{hkl}^{\text{stress}}(\text{Å})$*
110	0.173	0.179	0.176	$- 0.0029$
200	0.258	0.179	0.218	$- 0.0036$
211	0.173	0.179	0.176	$- 0.0029$
220	0.173	0.179	0.176	$- 0.0029$
310	0.228	0.179	0.203	$- 0.0033$
222	0.145	0.179	0.162	$- 0.0026$
321	0.173	0.179	0.176	$- 0.0029$
400	0.258	0.179	0.218	$- 0.0036$
411 330	0.207	0.179	0.193	$- 0.0032$
"Average"	0.199	0.179	0.189	$- 0.0031$

* $\Delta a_{hkl}^{\text{stress}} = a\bar{\bar{K}}\sigma$, where $a = 3.3$ Å and $\sigma = 50$ kg/mm^2.

defined in the standard way as $(A_0 - A_i)/A_0$ and the equivalent true (longitudinal) strain as

$$\epsilon = \ln\left(\frac{A_0}{A_i}\right) = \ln\left(\frac{1}{1 - RA}\right)$$

where A_0 and A_i are the original and reduced cross sectional areas. As basis of comparison, cold-worked filings of tantalum and niobium were also prepared and compacted with Duco cement into rods 2.2 mm in diameter.

X-ray Arrangement

Diffraction line profiles were recorded at $24 \pm 1°C$ using a Debye–Scherrer (D–S) arrangement on a General Electric XRD-5 diffractometer employing non-focused, crystal monochromated Cu K_{α_1} radiation ($\lambda = 1.5405$ Å) and a General Electric No. 2 proportional counter.[9] The wire specimens were spun on their symmetry axis at 150 rpm to provide better sampling over the entire cylindrical surface. Line profiles were taken from both negative and positive angles.

Table III. Elastic Constants for Niobium

Single crystal elastic constants (Bolef[7]):

$C_{11} = 2.46 \times 10^{12}$ dynes/cm^2 $S_{11} = 0.660 \times 10^{-12}$ cm^2/dyne

$C_{12} = 1.34 \times 10^{12}$ dynes/cm^2 $S_{12} = -0.233 \times 10^{-12}$ cm^2/dyne

$C_{44} = 0.287 \times 10^{12}$ dynes/cm^2 $S_{44} = 3.48 \times 10^{-12}$ cm^2/dyne

Mechanical (stress-strain) elastic constants (*Metals Handbook*[2]):

$E = 15 \times 10^6$ psi

$\nu = 0.38$

$\dfrac{\nu}{E} = 0.36 \times 10^{-4}$ mm^2/kg

Elastic moduli for X-ray stress measurements (Greenough[8]):

for isotropic stresses: $K_\sigma = S_{12} + (S_{11} - S_{12} - \tfrac{1}{2}S_{44})\Gamma$

where $\Gamma = \dfrac{h^2k^2 + k^2l^2 + l^2h^2}{(h^2 + k^2 + l^2)^2}$

for isotropic strains: $K_\epsilon = -\dfrac{(C_{11} + 4C_{12} - 2C_{44})}{2(C_{11} + 2C_{12})(C_{11} - C_{12} + 3C_{44})}$

for "average" case: $\bar{K} = \tfrac{1}{2}(K_\sigma + K_\epsilon) = \nu/E$

| | $-K_i$ ($\times 10^{-4}$ mm^2/kg) | | | |
hkl	Isotropic stress	Isotropic strain	Average	Δa_{hkl}^{stress} (Å)*
110	0.445	0.356	0.400	− 0.0066
200	0.233	0.356	0.295	− 0.0048
211	0.445	0.356	0.400	− 0.0066
220	0.445	0.356	0.400	− 0.0066
310	0.309	0.356	0.333	− 0.0055
222	0.515	0.356	0.435	− 0.0072
321	0.445	0.356	0.400	− 0.0066
400	0.233	0.356	0.295	− 0.0048
411⎫ 330⎭	0.360	0.356	0.358	− 0.0059
"Average"	0.381	0.356	0.368	− 0.0061

* $\Delta a_{hkl}^{stress} = a\bar{K}\sigma$, where $a = 3.3$ Å and $\sigma = 50$ kg/mm^2.

ANALYSIS OF THE DATA

The raw data consisted of the measurements of the peak position and of the line broadening. These were analyzed to obtain values for the lattice parameter, the root mean square (rms) strain, the domain size, and the probability of faults, if present.

Peak Position

The center of gravity of the truncated curve was taken as the position of the diffraction peak. The positive and negative values for a given *hkl* were averaged and corrected in most cases for a small measureable error due to the displacement of the spinning axis from the goniometer axis. The value of 2θ was then converted to an apparent lattice parameter a_{hkl} by use of Bragg's equation.

Table IV. Effects of Spacing Faults[12] on the Lattice Parameters of bcc Metals*

hkl	J'_{hkl}	$(\Delta a/a)_{hkl}^{fault}$†	$\Delta a_{hkl}^{fault}(Å)$ for $a = 3.3$ Å†‡
110	+ 0.208	+ 1.248 × 10⁻⁴	+ 0.0004
200	− 0.167	− 1.002	− 0.0003
211	+ 0.021	+ 0.126	~ 0
220	+ 0.208	+ 1.248	+ 0.0004
310	− 0.167	− 1.002	− 0.0003
222	− 0.167	− 1.002	− 0.0003
321	+ 0.048	+ 0.288	+ 0.0001
400	− 0.167	− 1.002	− 0.0003
330 411 }	+ 0.042	+ 0.252	+ 0.0001
420	− 0.167	− 1.002	− 0.0003

* $(\Delta a/a)_{hkl}^{fault} = J'_{hkl}\alpha\epsilon_f$.
† For $\alpha = 0.01$ and $\epsilon_f = 0.06$.
‡ Niobium and tantalum have about the same lattice parameter.

Systematic errors (e.g., specimen diameter, absorption, etc.) were removed by plotting a_{hkl} against the Nelson Riley[10] extrapolation function

$$\tfrac{1}{2}\left(\frac{\cos^2 \theta}{\sin \theta} + \frac{\cos^2 \theta}{\theta}\right)$$

for which a straight line can be drawn through all experimental points a_{hkl} obtained from the annealed metals. The extrapolated lattice parameter (i.e., a at $2\theta = 180°$) is the true value corrected for the most important systematic experimental effects. For deformed specimens, displacements of the experimental points a_{hkl} from a straight line can be attributed to the process of deformation, and may be associated with deformation (layer) faults or with residual stresses.

Deformation faulting. Faults on the {112} planes of body-centered-cubic (bcc) metals produce no peak shifts in the X-ray Debye–Scherrer patterns.[11–13] However, when there is a fractional change ϵ_f of the interplanar spacing at the fault, peak shifts are observed.[12] In terms of the present analysis, this shift may be expressed as

$$\left(\frac{\Delta a}{a}\right)_{hkl}^{fault} = J'_{hkl}\alpha\epsilon_f \tag{1}$$

where α is the fault probability and J'_{hkl} is a parameter. Table IV shows that these displacements are usually very small for representative values of α and ϵ_f. Figure 1 also graphically displays the theoretical relative peak shifts by means of a lattice-parameter Nelson–Riley plot.

Lattice Stresses. Stresses in elastically anisotropic materials can also cause peak shifts that vary with the indices of the hkl reflecting plane. The shifts (respectively strains) produced by a uniaxial stress are simple to analyze. However, in general, and certainly to some extent in wire drawing, a complex (triaxial) stress system is produced by deformation. Consequently, X-ray diffraction measurements made in only one direction, as

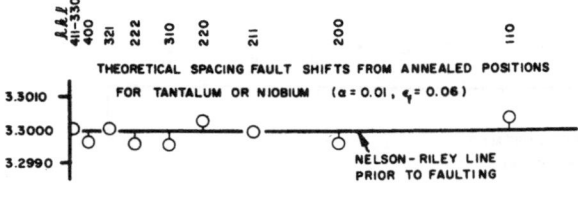

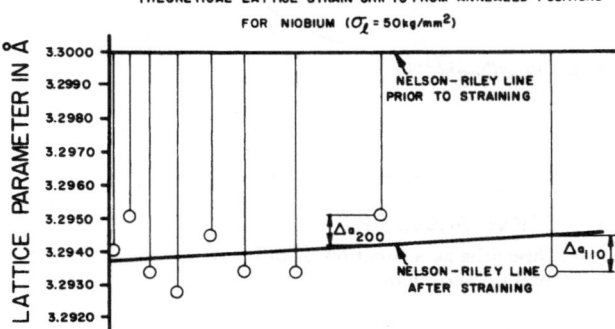

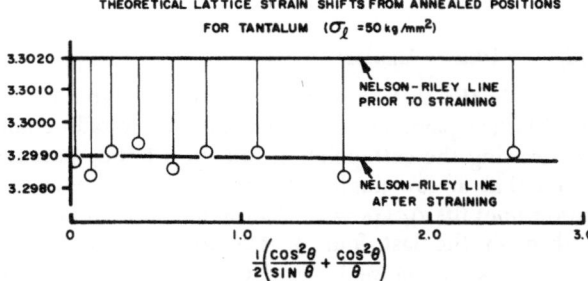

Figure 1. Theoretical changes in lattice parameter from various diffracting planes for tantalum and niobium due to spacing faults or lattice strains.

in the present case where the incident beam was perpendicular to the wire axis, are insufficient for a complete analysis unless another measurement at some other angle (to the wire axis) is also made. Since the drawn wires have radial symmetry, and in order to keep the analysis simple and tractable, we will interpret the change in a_0 measured perpendicular to the wire axis in terms of a radial Poisson strain ϵ_p due to an equivalent (uniaxial) tensile stress σ_l parallel to the wire axis. Thus we can retain the relationship

$$\sigma_l = \frac{1}{K}\,\epsilon_p \qquad (2)$$

where the strain $\epsilon_p = \Delta a/a$ and $\Delta a = a_{cw} - a_{ann}$ (the difference between the cold-worked *cw* and the annealed *ann* lattice parameters). For a given material, the magnitude of the modulus K is dependent on the crystallographic plane; the best experimental agreement[8] can be obtained by averaging the isotropic strain and isotropic stress moduli (Tables II and III). If instead the change in extrapolated lattice parameter a_0 is used to obtain strain, K can be approximated by a grand average $\langle \overline{K} \rangle = \frac{1}{2}(\overline{K}_\sigma + \overline{K}_\epsilon)$. Since $\langle \overline{K} \rangle$ is the weighted modulus for all the diffracting planes observed, there is excellent agreement (Tables II and III) between $\langle \overline{K} \rangle$ and the appropriate modulus $- \nu/E$ obtained from mechanical measurements.[2]

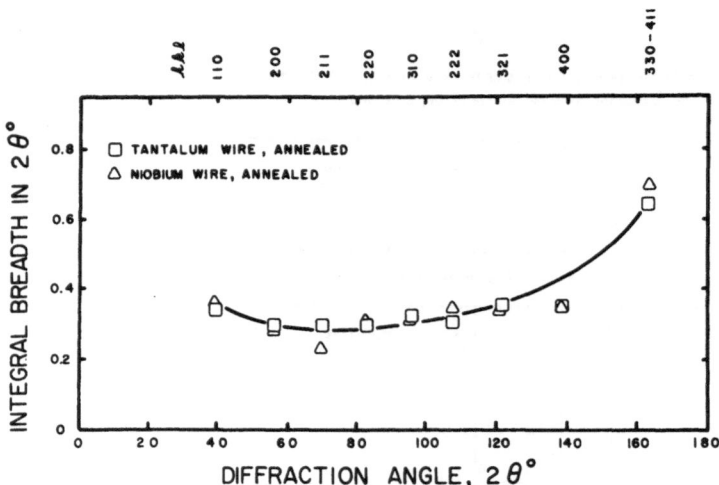

Figure 2. Variation of integral breadths as a function of diffraction angle for annealed tantalum and niobium wires.

Using the appropriate values of $\bar{K}$ and equation (2), one can calculate the individual lattice parameter changes,

$$(\Delta a)_{hkl}^{\text{stress}} = (a_{cw} - a_{ann})_{hkl}$$

corresponding to a longitudinal stress in the surface of a wire. For an arbitrary tensile stress one can find the extrapolated lattice parameter independent of crystalline aniso-tropy and experimental errors by drawing the extrapolation line on a lattice-parameter Nelson–Riley plot (Figure 1), such that at low angles the relative displacements (of a_{110} and a_{200} from the extrapolation line) fits $\Delta a_{110} = -\Delta a_{200}$, and at high angles fits a visually weighted placement through the last four or five a_{hkl} points. Selecting $\sigma_l = 50 \, \text{kg/mm}^2$ for illustration, we show in Figure 1 the decrease from the annealed values of all a_{hkl} for both niobium and tantalum; the magnitudes of the decreases depend, however, on the appropriate elastic moduli. To be noted is that the predicted relative displacements of the a_{hkl} around the extrapolation line for niobium with an anisotropy factor $A = 0.5$ is reversed from the pattern for tantalum, where $A = 1.6$, due to the "opposite" sense of the elastic anisotropy.

Peak Broadening

The true integral line breadths B, assuming Gaussian line shapes, were found by correcting for instrumental broadening[14] using the annealed wires as the geometrical standards. These annealed wires were found to be recrystallized and strain-free since all the a_{hkl} fell on the Nelson–Riley extrapolation line and the peaks were quite sharp. The corresponding integral breadths of both annealed metal wires were also found to be very similar (Figure 2), so that the broadening analyses were effectively taken from the same relative, if not ideal, geometrical standard.

The broadened peak profiles due to wire drawing, after correction for instrumental (geometric) effects, can be analytically described by[14,15]

$$(B_{hkl} \cos \theta)^2 = \left[\frac{\lambda}{D_e(hkl)} \right]^2 + (2\bar{\varepsilon}_{hkl})^2 \sin^2 \theta \qquad (3)$$

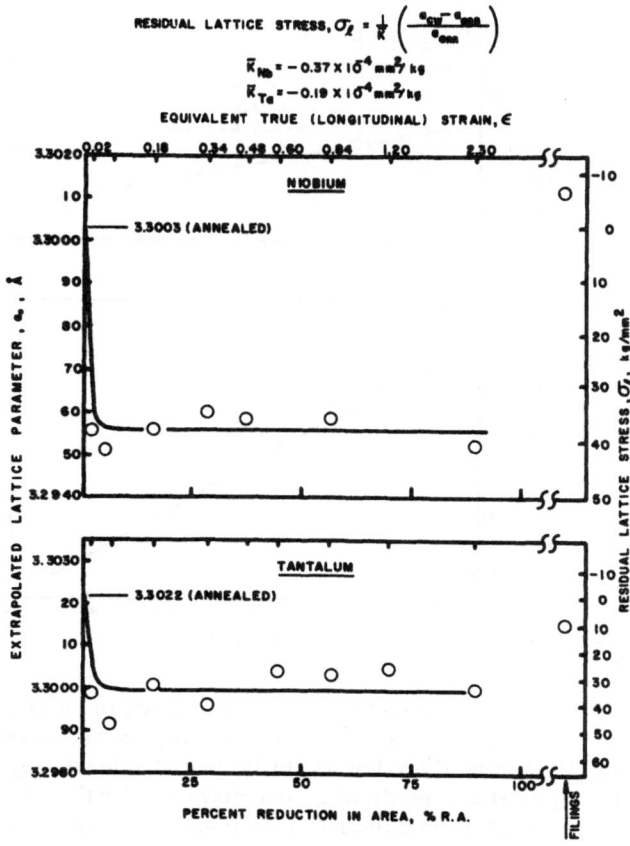

Figure 3. Variation of extrapolated lattice parameters and equivalent longitudinal residual stresses as a function of reduction in area by wire drawing for niobium and tantalum.

This equation assumes Gaussian line shapes which was a good approximation to the observed line shapes.

Thus, a plot of $(B_{hkl} \cos \theta)^2$ as a function of $\sin^2 \theta$ using sets of multiple-order peaks enables the rms strain $\bar{e}_{hkl}$ and the coherently diffracting domain size $D_e(hkl)$ to be determined. The value $D_e(hkl)$ also includes the effects from faulting if such faults are present.

RESULTS

Effects of Deformation on the Lattice Parameters of Niobium and Tantalum

After only 2% RA, a_0 for both the niobium and the tantalum wires dropped significantly and thereafter remained essentially at the reduced value (Figure 3); for niobium the decrease was approximately 0.0050 Å (0.15%) and for tantalum, 0.0020 Å (0.06%). The corresponding equivalent residual tensile stresses in the surface layers of the niobium and tantalum wires may be read from the right-hand scale of Figure 3. This scale was obtained using equation (2) and the appropriate $\langle \bar{K} \rangle$ from Tables II and III. Thus, for both metals fortuitously, the equivalent stresses σ_l had about the same constant magnitude of 35 ± 5 kg/mm². For the drawn niobium wire, the Nelson–Riley plots showed a definite and significant scatter pattern. This pattern and the reduced cold-worked lattice parameters were in accordance with the elastic anisotropy calculations

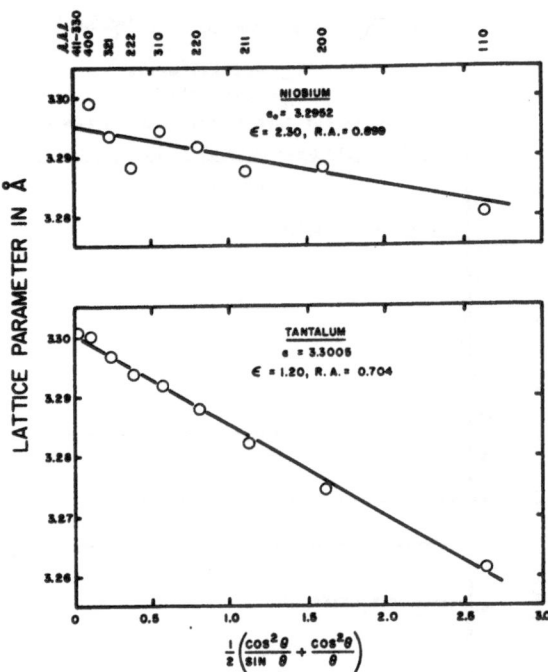

Figure 4. Typical lattice parameter analysis for niobium and tantalum wires.

(Table III), if the surface layers of the drawn niobium wire were considered to be in a state of residual, longitudinal tension (Figure 4). On the other hand, the a_{hkl} for tantalum displayed virtually no scatter and the Nelson Riley line could be placed immediately through most a_{hkl} (Figure 4). Since a_0 decreased on drawing, one may consider that the drawn tantalum wires are also under residual, longitudinal, tensile stress in the outer sections.

The a_{hkl} Nelson–Riley plots for cold-worked filings of both niobium and tantalum displayed virtually no scatter of the a_{hkl} points around the Nelson–Riley line.* Furthermore, only small differences not exceeding ± 0.0010 Å were observed between the extrapolated lattice parameters of the filings and the corresponding annealed wire values. These facts strongly suggest that there are no significant *net* residual macrostresses in the cold-worked filings, even though line broadening results indicate that these filings were deformed as least as much as the surface layers of the heavily drawn wires.

Effects of Deformation on the Line Broadening

The results of the line-broadening analyses are shown in Figures 5 and 6. The values of $D_e(hkl)$ systematically decreased to an apparent "saturation" point, which was reached after about 25% RA for niobium and only 5% RA for tantalum.

The strain-broadening behavior of the niobium wires indicated that the rms strains were anisotropic in good agreement with the isotropic stress model of polycrystalline deformation $(\bar{\varepsilon}_{110}/\bar{\varepsilon}_{100} = (K_\sigma)_{110}/(K_\sigma)_{100} \approx 1.9)$, and seemed to reach maximum values after 25% RA. The corresponding results for tantalum appear to be anomalous from this point of view.

---

* For niobium filings the small amount of scatter could be interpreted[12] by the presence of spacing faults ($\alpha = 0.02$, $\epsilon = 0.06$), but this explanation was not confirmed by line-broadening data evaluated in this study.

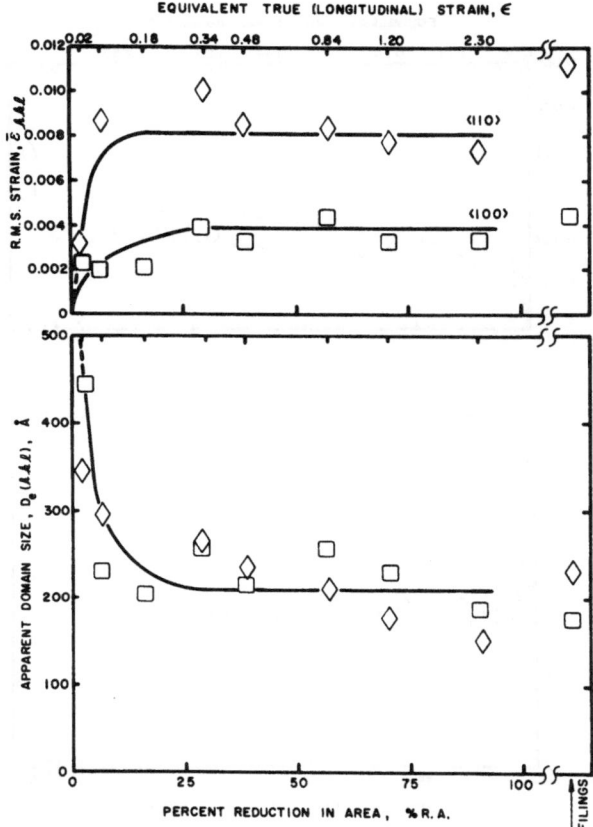

Figure 5. Variation of microscopic particle sizes and strains of niobium wire as a function of reduction in area by wire drawing, results from integral breadth analysis; $\diamond = \langle 110 \rangle$, $\square = \langle 100 \rangle$.

The broadening of individual peaks of cold-worked niobium and tantalum filings was greater than those of the heaviest drawn wires, respectively. The particle sizes were again isotropic and of the same magnitude as the minimum values found for the drawn wires; the larger integral breadth of the filings was due to a greater degree of strain broadening. However, the large anisotropy of rms strains was still present in the niobium filings, indicating that the detected elastic anisotropy was an inherent factor independent of deformation mode.

DISCUSSION

Microscopic Effects

The observation that (in the surface layers) the particle size $D_e(hkl)$ remains essentially constant after the first few percent RA is consistent with the formation of a stable cell size after the initial stages of deformation. Such cell structures have also been noted by electron microscopy during tensile deformation of tantalum[16] and niobium[17] while drawn tungsten wires[18] were found to have a minimum X-ray particle size of 300 Å. Furthermore, since the particle sizes of bcc refractory metal filings deformed at room temperature (see present work and Aqua and Wagner[19]) all lie around 200 ± 50 Å, this value appears to be a lower limit for these materials irrespective of the severity of the

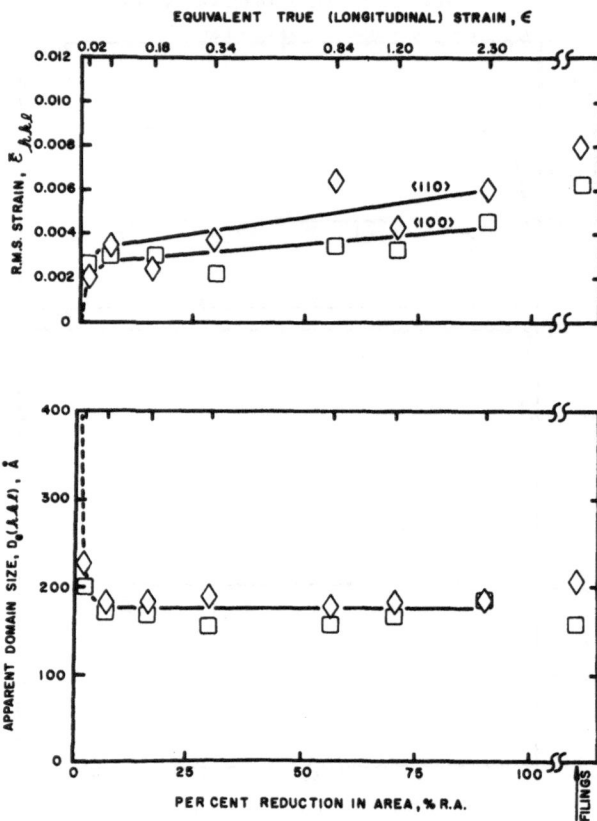

Figure 6. Variation of microscopic particle sizes and strains of tantalum wire as a function of reduction in area by wire drawing, results from integral breadth analysis; $\diamond = \langle 110 \rangle$, $\square = \langle 100 \rangle$.

deformation. Finally, the lack of any significant amount of stacking fault formation during wire drawing was indicated by the fact that $D_e(110) \approx D_e(100)$ (since $D_e(110/D_e(100) \to 2.8$ when stacking faults contribute to particle size broadening[11] and otherwise is generally close to unity). This result agrees essentially with the electron microscope observation[17] on tensile deformation of niobium.

The rms strains represent a measure of the internal strains in a deformed metal and can be taken as being proportional to the dislocation density. For tantalum the magnitudes of the rms strains increased with successive draws; for niobium increases only occurred up to 25% RA and then reached an apparent upper limit. Thus, since the $D_e(hkl)$ (≈ 200 Å for both niobium and tantalum wires) indicated a stable cell structure formed early in the deformation process, these differences between niobium and tantalum possibly are due to a continuing increase in the number of dislocations within cells in tantalum wires while the number within niobium cells are limited. However, for both niobium and tantalum, deformation by filing can produce a greater dislocation density (perhaps including partial or twin fault dislocations) within cells of this limiting size, since in the filings the rms strains are consistently greater than the largest rms strains for wires of either metal.

The observation that the ratio $\bar{\varepsilon}_{110}/\bar{\varepsilon}_{100} > 1$ for both niobium and tantalum is contrary to that made by Aqua and Wagner[19] on filings. They found that for tantalum the ratio was less than unity, as would be expected if the contributions to the rms strains were predominantly from long-range elastic strains in the (stressed) domains. Although

Table V. Mechanical Properties of Niobium and Tantalum

Material	History	Impurity content (%)	Yield strength (kg/mm²)	Ultimate tensile strength (kg/mm²)	Reference
Nb	Electron-beam melted and recrystallized	4×10^{-2}(a)	20(b)	28	21
Nb	Electron-beam melted and cold swaged, 95%RA	4×10^{-2}(a)	35(b)	42	21
Nb	Arc-melted and recrystallized	$\sim 11 \times 10^{-2}$(a)	28(b)	35	21
Nb	Arc-melted and cold swaged, 81%RA	$\sim 11 \times 10^{-2}$(a)	53(b)	64	21
Nb	Vacuum annealed	$2-7 \times 10^{-3}$	8–21(c)	—	17
Nb	Arc-melted	$\sim 5 \times 10^{-2}$	20(b)	29	22
Nb	Electron-beam melted	—	14(d)	—	23
Ta	Electron-beam melted and recrystallized	$\sim 1 \times 10^{-2}$	14(b)	21	24
Ta	Electron-beam melted and cold drawn, 98%RA	$\sim 1 \times 10^{-2}$	28(b)	35	24
Ta	Arc-melted and recrystallized	$\sim 10 \times 10^{-2}$	28(b)	42	24
Ta	Arc-melted and cold-worked 87%RA	$\sim 10 \times 10^{-2}$	77(b)	84	24
Ta	Arc-melted and recrystallized	$\sim 2 \times 10^{-2}$	28(d)	—	25

(a) Indicated interstitial impurity concentrations: as-received metallic impurities consist of Ta 0.1%, Fe, Si, Ti, Zr < 0.015% each.
(b) 0.2% Offset yield strength.
(c) Upper yield strength.
(d) Lower yield strength.

the discrepancy between this interpretation and the present results on the tantalum drawn wires is rather apparent, careful inspection shows that it also exists, but to a lesser extent, for the drawn niobium wires. In these, the ratio of the Young's moduli E_{100}/E_{110} (= 1.59 for niobium[19]) is lower than the observed $\bar{\varepsilon}_{110}/\bar{\varepsilon}_{100}$ ratio which agrees more closely, as already noted, with that expected from the isotropic stress model of polycrystalline deformation. The discrepancies may be attributed qualitatively to texture (or preferred orientation) for which evidence existed, as could be seen by examining the circumferential variation of intensity of the D–S rings on X-ray photograms. A quantitative assessment of the effect of texture on line broadening is not yet possible.

Results from tensile tests widely support[20] a relation of the form

$$\tau = \tau_0 + AGb\sqrt{\rho} \qquad (4)$$

where τ is the single-crystal resolved shear stress (on the slip plane in the slip direction), G is the shear modulus, b the Burgers vector and ρ the dislocation density; τ_0 and A are constants. Since the change Δa (Figure 3) in lattice parameter can be related to τ, and the rms strain $\bar{\varepsilon}_{hkl}$ to ρ, it follows that we would expect a correlation between Δa and $\bar{\varepsilon}_{hkl}$ as is observed for niobium. Thus Figures 3 and 5 show that both quantities level off after the first several percent RA. However, for tantalum this does not hold: $\bar{\varepsilon}_{hkl}$ continues to increase steadily, whereas Δa levels after about 5% RA. The situation appears to be one in which the dislocation density (in the surface layers) continues to increase with increasing RA, whereas the residual stress and the domain size remain essentially the same. One could, of course, readily enough propose models compatible with the observations for both niobium and tantalum, but at this stage further investigation is clearly desirable before embarking on a necessarily lengthy discussion of possible explanations.

Residual Stresses in Drawn Wires

As interpreted from radial strains, a residual surface stress exists in drawn wires; no significant residual strains were observed for niobium and tantalum filings. A primary component of this stress is probably a longitudinal one, produced by the preferential longitudinal extension of the surface layers with respect to the center of the wire. Macroscopic observations of an inverted conical lip on the trailing end of the wire also indicates such plastic flow. This stress state has been observed previously from analyses of dimensional changes.[5]

The tensile properties for both tantalum and niobium show a wide range of values (Table V), which can be attributed to variations in impurity elements and prior deformation. Note that differences in either impurity level or prior straining can each influence the mechanical properties by a factor of two. The observed value of 35 ± 5 kg/mm^2 from the X-ray analyses for either niobium or tantalum is generally greater than the yield stress and close to the ultimate tensile stress (with two exceptions for cold-worked niobium or tantalum.)

The residual stresses in the present work assume a constant value irrespective of the number of draws and are not detected in filings. A possible explanation is related to the occurrence of long-range inhomogeneous macroscopic flow in wires which is not present in filings and the observation[5] that any residual (surface) stresses developed during multiple-pass deformation processes must be substantially determined by the last pass. For wires the large localized amount of plastic flow on the surface produced during each successive pass effectively erases the "memory" of the previous stress system and imposes a new set of constraints (stress) on the surface. As the new set of constraints is unlikely to be significantly different than those imposed by the previous pass, the net stress system remaining from pass to pass is virtually constant. The effect of alternately reversed drawing directions on the microstructure and the stresses remain to be investigated.

Using destructive methods of analysis, the residual stresses in brass wires given single-pass drawings have been determined.[5] In particular, the magnitude of the longitudinal residual stress was found to reach an almost constant maximum value (where the magnitude itself depended on the die angle) in the approximate region of 5–35% RA and then fall off for larger RA. Such single-pass behavior of the stresses below 35% RA is similar and comparable to the present multipass results for all RA values tested. Any differences between multipass and single draws at $RA > 35\%$ can most probably be attributed to variations in plastic flow.

The Strength of Drawn Wires

The increase of strength of a drawn wire over that of the bulk-annealed material is proportional to the amount of deformation introduced or the associated reduction in actual diameter (Table I). The macroscopic tensile stresses in the wire surfaces are unlikely to contribute, per se, to the increased tensile strength of the wires. Thus the progressive strengthening produced by drawing must be primarily due to the increase in the effective volume containing a high dislocation density and associated with the severely cold-worked microstructure. This work-hardened zone initially forms at the wire surface due to the limited depth of plastic deformation arising from forces exerted by the die;[5] as the wire diameter is subsequently reduced, complete penetration of work-hardening occurs into the center of the wire and thereby gives rise to increased tensile strengths.

This concept of progressive nonhomogeneous work-hardening is also quite consistent with the observations of a gradual deterioration of ductility[2,4] which parallels the increase in tensile strength for wire drawing; the loss of ductility results from the proportionate increase in the volume of severely work-hardened material which is not capable of large plastic extensions. Furthermore, holding all other wire-drawing parameters constant and extending this concept to the limiting case where the entire cross section of the wire has the same stable cold-worked microstructure would lead to the conclusion that at such a deformation state the maximum inherent potential strengthening that can be produced by wire drawing has been attained.

CONCLUSIONS

1. Macroscopic residual radial strains corresponding to equivalent longitudinal tensile stresses of 35 ± 5 kg/mm² are produced in the outer layers of drawn wires of tantalum and niobium and remain of constant magnitude irrespective to the number of draws after the initial 2% RA.

2. For both wire-drawn niobium and wire-drawn tantalum the apparent X-ray particle sizes in the $\langle 100 \rangle$ and $\langle 110 \rangle$ directions reached a limiting value of approximately 200 Å after the first few draws; no evidence for the presence of stacking faults was found.

3. Filing tantalum and niobium metals produces higher rms strains (dislocation densities) than the maximum magnitudes in drawn wires but the resultant X-ray particle sizes (cell structure) were essentially the same for both metals deformed by either method. No residual macroscopic strains were observed in filings, indicating that residual stresses arise only when long-range inhomogeneous flow is produced by deformation processes.

ACKNOWLEDGMENTS

The authors would like to recognize H. Jordan and J. H. Colvin for assistance with experimental and analytic work and Dr. A. B. Michael, formerly of Fansteel Metallurgical Corporation, for supplying the material. This work was performed under a project sponsored by the Office of Naval Research.

REFERENCES

1. Latrobe Steel Company, Bulletin 100, Tech Topics: Stainless Steels, Latrobe, Pennsylvania 1961.
2. *Metals Handbook*, Vol. 1, 8th ed., American Society for Metals, Metals Park, Ohio, 1961, pp. 153, 1202, 1222, 1225.
3. W. O. Everling, "Super-High Strength Wire, A Component of Metallic Composites," in: *Proc. 6th Sagamore Ordnance Matls. Res. Conf., Composite Materials and Composite Structures*, Racquette Lake, N.Y., 1959.
4. J. D. Embury and R. M. Fisher, "The Structure and Properties of Drawn Pearlite," *Acta Met.* **14**: 147–159, 1966.

5. W. M. Baldwin, Jr., "Residual Stress in Metals," *Proceedings of the American Society for Testing Materials* **49**: 1–45, 1949.

6. T. A. Trozera, "On the Nonhomogeneous Work for Wire Drawing," *Trans. ASME* **57**: 309–323, 1964.

7. D. I. Bolef, "Elastic Constants of Single Crystals of the Body-Centered Cubic Transition Elements V, Nb, and Ta," *J. Appl. Phys.* **32**: 100–105, 1961.

8. G. B. Greenough, "Quantitative X-Ray Diffraction Observations in Strained Metal Aggregates," *Progr. Metal Phys.* **3**: 176–219, 1952.

9. H. M. Otte, "Lattice-Parameter Determinations with an X-Ray Spectrogoniometer by the Debye–Scherrer Method and the Effect of Specimen Condition," *J. Appl. Phys.* **32**: 1536–1346, 1961.

10. J. B. Nelson and D. P. Riley, "An Experimental Investigation of Extrapolation Methods in the Derivation of Accurate Unit Cell Dimensions of Crystals," *Proc. Phys. Soc. (London)* **57**: 160–177, 1945.

11. B. E. Warren, "X-Ray Studies of Deformed Metals," *Progr. Metal. Phys.* **8**: 147–202, 1958.

12. C. N. J. Wagner, A. S. Tetelman, and H. M. Otte, "Diffraction from Layer Faults in bcc and fcc Structure," *J. Appl. Phys.* **33**: 3080–3086, 1962.

13. C. N. J. Wagner, "Analysis of the Broadening and Changes in Position of X-ray Powder Pattern Peaks," in: J. B. Cohen and J. E. Hilliard (eds.), *Local Atomic Arrangements Studied by X-Ray Diffraction*, Gordon and Breach, New York, 1965, Chapt. 6.

14. A. Taylor, *X-Ray Metallography*, John Wiley & Sons, Inc., New York, 1961, pp. 605, 692, 788.

15. D. O. Welch and H. M. Otte, "The Effect of Cold-Work on the X-Ray Diffraction Pattern of a Copper–Silicon–Manganese Alloy," in: W. M. Mueller and M. J. Fay (eds.), *Advances in X-Ray Analysis, Vol. 6*, 1963, p. 96–120.

16. T. W. Barbee and R. A. Huggins, "Dislocation Structures in Deformed and Recovered Tantalum," *J. Less-Common Metals* **8**: 306–319, 1965.

17. L. I. van Torne and G. Thomas, "Yielding and Plastic Flow in Niobium," *Acta Met.* **11**: 881–893, 1963.

18. A. J. Opinsky, J. L. Orehotsky and C. W. W. Hoffman, "X-Ray Diffraction Analysis of Crystallite Size and Lattice Strain in Tungsten Wire," *J. Appl. Phys.* **33**: 708–712, 1962.

19. E. N. Aqua and C. N. J. Wagner, "X-Ray Diffraction Study of Deformation by Filing in bcc Refractory Metals," *Phil. Mag.* **9**: 565–589, 1964.

20. H. M. Otte and J. J. Hren, *Experimental Mechanics* **6**: 177–193, 1966.

21. A. L. Mincher and W. F. Sheely, "Effect of Structure and Purity on the Mechanical Properties of Niobium," *Trans AIME* **221**: 19–25, 1961.

22. E. S. Bartlett, D. N. Williams, H. R. Ogden, R. I. Jaffee, and E. F. Bradley, "High Temperature Solid-Solution-Strengthened Columbium Alloys," *Trans. Met. Soc. AIME* **227**: 459–467, 1963.

23. M. A. Adams, A C. Roberts, and R. E. Smallman, "Yield and Fracture in Polycrystalline Niobium," *Acta Met.* **8**: 328–337, 1960.

24. M. Schussler and J. S. Brunhouse, Jr., "Mechanical Properties of Tantalum Metal Consolidated by Melting," *Trans. AIME* **218**: 893–900, 1960.

25. C. S. Tedmon and D. P. Ferris, "The Dependence of Yield Stress on Grain Size for Tantalum and a 10% W–90% Ta Alloy," *Trans. AIME* **224**: 1079–1080, 1962.

DISCUSSION

S. R. Colberg (Naval Ordnance Test Station): Have you done any etching of the surfaces to see if that is the actual cause of this residual strain?

R. P. I. Adler: In this case we did not, but we are presently studying etched bronze wires. Here we do find that there is a decrease in the residual stresses (or strains) as you etch down.

N. C. Halder (Yale University): Do you not observe any preferred orientation after the wire has been drawn? If so, how did it affect the line-broadening analysis? In your talk you mentioned that the particle size ratio for the metals containing faults is of the order of 2.8. What do you mean by this? Is this the effective particle size ratio in two different directions?

R. P. I. Adler: I will answer the last question first. First of all, the particle sizes in two different directions, in this case the ⟨110⟩ and ⟨100⟩, were the same. If one has faulting and these contribute principally to the particle-size broadening, one would expect that the ratio would be anisotropic in the direction that I indicated.

N. C. Halder: What particle size do you mean? The effective?

R. P. I. Adler: Yes.

N. C. Halder: But recently Cohen of Northwestern University has shown that unless you correct for the particle anisotropy one will not get such; there is then some disagreement.

R. P. I. Adler: I don't think it has been solved at this particular point.

N. C. Halder: No, I know it definitely has not. I have just seen a paper which has been published in a book called Local Atomic Arrangements by Gordon and Breach Company of New York. I think it would have dealt with it.

R. P. I. Adler: Yes, I am aware of that work and I am not sure how to handle this particular problem when texture is present. In the second case, talking about preferred orientation, we made a quick qualitative check of the wires. We took Debye–Scherrer photograms of the wire; we found that there was indeed some ⟨110⟩ fiber texture. At present I don't think there is any analysis to handle this kind of effect. There is definitely a change in the broadening as you go around the Debye–Scherrer ring.

N. C. Halder: You said that you used the monochromator, right?

R. P. I. Adler: Yes.

N. C. Halder: Do you find any change in the lattice parameter against Nelson–Riley function plot for the low-angle reflections? Should you not, using a monochromator? You may get a straight line for all these points, but if you use a monochromator there will be a correction factor involved. Did you consider this correction factor?

R. P. I. Adler: This was taken care of in the Nelson–Riley function, because for the annealed material one can draw a straight line through it. I believe you are talking about flat samples or transmission samples where a correction factor may indeed have to be used. I think you are referring to the fact that you get a smooth curve, but this may not be a straight line. We didn't find this for the annealed materials, so no further correction was necessary.

THE STRUCTURE OF THE γ'-PHASE IN NICKEL-BASE SUPERALLOYS

S. Rosen and P. G. Sprang

Pratt and Whitney Aircraft
North Haven, Connecticut

ABSTRACT

Present day nickel-base superalloys are hardened in part by the precipitation of a phase which has variously been identified as Ni_3Al, $Ni_3(Al, Ti)$ and γ'. X-ray diffraction techniques which include precision lattice parameter measurements, intensity measurements, and phase identification are used to define the structural and chemical relationships upon which this phase is based.

These relationships are developed from the following considerations: crystal chemistry and atomic size factors which relate binary Cu_3Au-type T_3B phases (e.g., Ni_3Al) and ternary Perovskite-type T_3BC_x carbide phases (e.g., Y_3AlC), the determination of the number and kind of atoms in the unit cell of Ni_3Al and certain ternary phases, the crystallographic relationship between the structure of Y_3C and Y_3AlC, and phase relations in certain quaternary alloys.

From these considerations it is shown that the γ' phase may best be characterized as a Perovskite-type carbide phase having the chemical formula T_3BC_x. A model of the γ' structure is presented which indicates the position of the various atomic constituents based upon whether they are T or B elements. (An atomic component is considered of the T type if it is capable of substituting for nickel in Ni_3Al, of the B type if it can replace the aluminum. The essential features of this model are: T and B elements form an ordered T_3B lattice of the Cu_3Au type; carbon atoms are located only in octahedral holes in the centers of the Cu_3Au-type cells thereby establishing Perovskite-type T_3BC_x unit cells; the effective size of T and B atoms in the T_3BC_x unit cell is the same: hyperstoichiometric alloys, (ratio of B atoms to T atoms greater than one) will contain B atoms at face-centered positions in addition to a small amount of equilibrium vacant sites; in all alloys aluminum will preferentially occupy the cube corners of the unit cell; the amount of carbon which is soluble in T_3BC_x at any particular temperature is determined both by the distribution of the elements which are carbide-formers and the elements manganese, iron and cobalt. This model accounts for microstructural changes which occur in some nickel-base superalloys as a function of temperature and composition.

INTRODUCTION

The strength of present day nickel-base superalloys at high temperature is attributed in part to the presence of a dispersed phase (γ') in a solid solution matrix. The properties of these superalloys may be changed by modifying the gross composition so as to alter the relative amounts of the dispersed phase, the matrix, and the small amount of carbide phases which are present. Undoubtedly, these modifications also change the composition of these various phases and this change also greatly effects the properties of the alloys. Nevertheless, very little attention has been given to the composition and arrangement of the elements in the crystal lattice and to the influence of this on the properties of the alloy. It is known for example that crystal irregularities such as those produced by lattice vacancies, interstitial atoms, and substitutional atoms of different sizes greatly effect

131

mechanical properties. It therefore becomes desirable to investigate these structural features.

The dispersion hardening constituent as it appears in superalloys has variously been identified as Ni_3Al, $Ni_3(Al, Ti)$ and γ'. Where microprobe analysis or other means of analysis have been used to obtain the composition of the γ' phase, the following elements have been detected in concentrations greater than five atom percent: nickel, cobalt, titanium, aluminum, and chromium. It is known that the phase also contains a host of other elements in smaller amounts.

The structure of the stoichiometric binary Ni_3Al phase is of the Cu_3Au type, $L1_2$ Strukterbericht. A ternary carbide phase has lately been identified that bears a close relationship to the phases of the Cu_3Au type. In conjunction with observations to be made later, this similarity served as an indication that an understanding of the γ' phase might be obtained by characterizing the factors governing the nature and stability of the Ni_3Al and ternary carbide phases. In this paper, therefore, these phases are first characterized, then they are related to the γ' phase, and finally from this information a model of the γ' phase as it appears in superalloys is constructed.

EXPERIMENTAL

Alloys weighing about 10 g were prepared by arc-melting powder compacts made from various components on a water-cooled copper hearth using a tungsten electrode in a gettered argon atmosphere. With the exception of the rare earth metals, the purity of the starting materials was better than 99.9%. The purity of the rare earth elements used in this study was 99%. The experimental details have been described elsewhere.[1]

THE UNIT CELL OF Ni_3Al AND T_3BC_x

The cubic unit cell of Ni_3Al which is of the Cu_3Au-type structure is shown in Figure 1(a). The aluminum atoms occupy the cube corner positions and the nickel atoms the face-centered positions. The ternary carbide phases which are of interest in this investigation are characterized by the general formula T_3BC_x (e.g., $Co_3AlC_{0.11}$). The x value is calculated by dividing the atom percent of carbon by twenty. The arrangement of the metal atoms in the unit cell is the same as shown in Figure 1(a), the T and B atoms

[1] References are at the end of the paper.

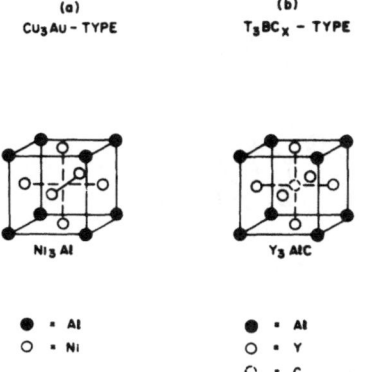

(a)
CU_3AU - TYPE

Ni_3Al

● = Al
○ = Ni

(b)
T_3BC_X - TYPE

Y_3AlC

● = Al
○ = Y
○ = C

Figure 1. Unit cell of Cu_3Au- and T_3BC_x-type structures.

forming an ordered lattice of the Cu₃Au type discussed above. Stadelmaier[1] reported on
the occurrence and stability of ternary carbide phases formed between B subgroup
elements with transition elements of group VII and group VIII. The present authors
reported carbide phases formed between aluminum and the group III elements. Here
it was shown that the carbon atom occupies the octahedral position in the center of the
unit cell. This places a maximum value of one on x. The unit cell of Y_3AlC is shown in
Figure 1(b).

SIZE FACTOR RELATIONSHIPS IN T_3BC_x PHASES

Dwight and Beck have discussed the occurrence of close-packed ordered binary
structures in alloys of transition elements. They observed that although the CN12 radius
of the components differed in hexagonal structures of the Cd_3Mg and Ni_3Ti type, the
experimentally determined axial ratios were in all cases quite close to the ideal value of
1.63 for spherical close packing of spheres of uniform size, indicating that some sort of
atomic size adjustment had occurred.

From the CN12 atomic radius of the component atoms of ordered binary phases,
a theoretical lattice parameter a' may be calculated. If ordered phases of the Cu_3Au type
are considered, the observed lattice parameter a is related to a' in a very interesting manner.
If the fractional lattice contraction or fractional lattice expansion $(a'-a)/a'$ or $(a-a')/a$ is
plotted against the fractional difference in Goldschmidt atomic radii, a linear relationship
occurs. If the a values observed for T_3BC_x phases are corrected for expansion due to
carbon solubility, the same relationship is obtained when the fractional lattice changes
of these hypothetical phases are plotted against the fractional difference in atomic radii.

The structures being considered may be looked upon as made up of the stacking of
ordered layers of atoms of the type shown in Figure 2. If the black atom in Figure 2(a)
is larger than the white atom it is clear that the atomic plane considered is no longer close
packed in the usual sense, since the smaller white atoms are no longer in contact with
each other. If the T atom in Figure 2(b) is larger than the B atom it is clear that the atomic
plane considered is again no longer close packed, since the B atoms would rattle around
if the requirement for universal contact of the T metals is maintained. For T_3B phases
where the B atom is smaller than the T atom, the fractional lattice expansion increases
with increasing relative size of the T atom, suggesting that in these instances, it is the
B atom that expands. Where the B atom is larger than the T atom, a fractional lattice
contraction is found which also increases with increasing relative size of the B atoms,
suggesting that in this case it is the B atoms that contracts. The radius calculated from

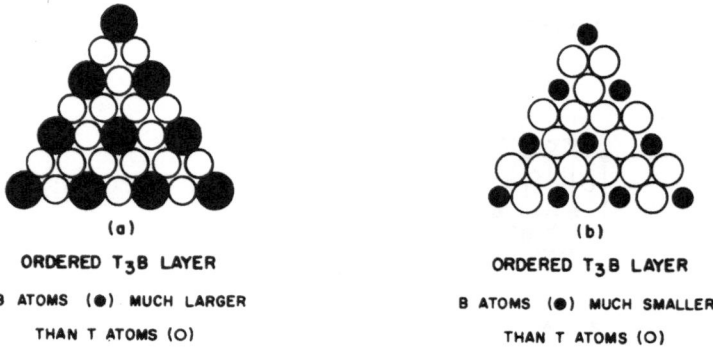

(a)
ORDERED T_3B LAYER
B ATOMS (●) MUCH LARGER
THAN T ATOMS (O)

(b)
ORDERED T_3B LAYER
B ATOMS (●) MUCH SMALLER
THAN T ATOMS (O)

Figure 2. Ordered atomic layers.

the observed cell size, assuming atoms of equal size, shows that when the B atom expands the T atom contracts somewhat and when the B atom contracts the T atom expands somewhat. It is noteworthy that the readjustments in size appear to be such as to produce effective radii which are more compatible with the close packing of uniform spheres.

Rosen and Sprang[2] have discussed these matters in detail. It is concluded that, from a geometrical point of view, ternary T_3BC_x phases resemble binary T_3B phases and are stabilized in part by efficient space filling. The component atoms undergo a mutual size adjustment which results in each atom obtaining a similar effective radius, thus minimizing the effect of strain energy when changes in composition are considered.

Although the chief criterion for the stability of the phases discussed is efficient space filling, the Hagg rule governing carbon solubility may be applied if, when using the Goldschmidt radii of the transition metal, the upper limit of the ratio r_C/r_T is extended to 0.62. In these instances, however, the slight enlargement of the T metal atom may reduce the radius ratio so as to be compatible with the original Hagg criteria.

It follows, therefore, that nitrogen, for example, which has a smaller radius than carbon, is expected to be soluble in Ni_3Al and form T_3BN_x nitrides. This, indeed, is found to be true.[2] Boron, on the other hand, has a radius considerably larger than carbon, and the simple radius criteria indicates that it should not be soluble in Ni_3Al or form ternary T_3AlB_x borides. Stadelmaier has found that boron in fact stabilizes ternary boride phases with the $Cr_{23}C_6$ structure in which the boron atom has a more suitable environment.

THE DEFECT Ni₃Al PHASE

In most phases changes of composition result in the substitution of one kind of atom for another, as in the case of substitutional primary solid solutions. In AlNi, however, it is found that certain changes in composition result in some of the nickel atoms dropping out of the structure, leaving lattice points unoccupied. This occurs in such a manner that the number of electrons per unit cell remains constant. Although Guard and Westbrook[3] made no explicit determination of the nature of the Ni_3Al phase, they obtained hardness data which indicated that defects may be present in Ni_3Al which contains excess aluminum.

In this investigation, the relationship between defects and composition in binary and ternary alloys was obtained from a combination of precision lattice parameter measurements and density determinations using the following equation:

$$T = \frac{(6.023 \times 10^{23})\,\rho(a)^3}{M_T + M_B/x + M_C/y}$$

where T is the number of T atoms in the unit cell, B is the number of B atoms in the unit cell, C is the number of C atoms in the unit cell, x is the atom ratio T/B, y is the atom ratio T/C, ρ is density in g/cm³, a is the lattice parameter in cm, and M is the molecular weight.

Lattice parameters were determined with a precision back-reflection camera and are considered accurate to 1 part in 7000. Densities, which were obtained by a liquid displacement method (using water as the displaced liquid), are the average of three measurements and are accurate to 1 part in 200. The maximum error in the calculated value of the number of atoms per unit cell should then be about 0.01. The results of these calculations for alloys which are annealed at 1000°C for five days are given in Table I. Figure 3 shows graphically the variation of lattice parameter and density with composition for the binary Ni_3Al phase.

Table I. The Number and Kind of Atoms per Unit Cell in Ni$_3$Al Alloys

Composition (at.%)	ρ(g/cm^3)	a(Å)	Atoms per unit cell					
			Total	Al	Ti	Ni	Cu	C
Ni–20.6 Al	7.67							
Ni–23.3 Al	7.52	3.5646	4.00	0.93		3.07		
Ni–23.8 Al	7.47	3.5656	4.00	0.95		3.05		
Ni–25.0 Al	7.41	3.5673	4.00	1.00		3.00		
Ni–25.8 Al	7.30	3.5684	3.96	1.03		2.93		
Ni–26.8 Al	7.22	3.5686	3.93	1.06		2.87		
Ni–30.0 Al	7.10							
Ni–22 Al–3 C	7.49	3.5830	4.12	0.91		3.09		0.12
Ni–24 Al–3 C	7.37	3.5848	4.10	0.98		2.99		0.13
Ni–27 Al–3 C	7.21	3.5915	4.13	1.11		2.89		0.13
Ni–13 Al–10 Ti	7.76	3.5811	4.01	0.52	0.40	3.09		
Ni–15 Al–10 Ti	7.66	3.5840	4.01	0.60	0.40	3.01		
Ni–17 Al–10 Ti	7.42	3.5868	3.95	0.67	0.40	2.88		
Ni–22 Al–13 Cu	7.72	3.5644	4.01	0.88		2.61	0.52	
Ni–24 Al–11 Cu	7.51	3.5757	4.00	0.96		2.60	0.44	
Ni–25 Al–10 Cu	7.48	3.5758	4.01	1.00		2.61	0.40	
Ni–26 Al–9 Cu	7.45	3.5715	4.01	1.04		2.61	0.36	
Ni–27 Al–8 Cu	7.28	3.5690	3.96	1.07		2.57	0.32	

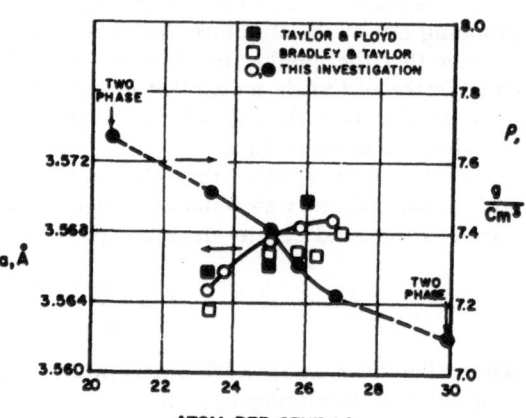

Figure 3. Lattice parameter and density versus composition for the Ni$_3$Al phase.

The results for the binary Ni$_3$Al phase show that, beginning with the nickel-rich composition at 23.3 at.% aluminum, and with increasing aluminum content, there is simple replacement of nickel atoms by aluminum atoms until the stoichiometric composition at 25 at.% aluminum is reached. At or slightly beyond this composition, nickel atoms begin to be rejected from certain lattice points and the defect structure begins. Unlike AlNi, however, aluminum atoms continued to replace nickel atoms at some of the face-centered positions. An alloy of composition Ni–26.8 at.% aluminum contains 6% of the aluminum atoms at face-centered positions beside those at the corners of the unit cell. About 2% of the face-centered positions are vacant. Hyperstoichiometric Ni_3Al may thus be described by the formula $(Ni, Al)_3Al$.

Bradley and Taylor[4] found that in AlNi the direct replacement of nickel by aluminum atoms does not proceed much beyond the stoichiometric composition. Only 1% of the nickel atoms are replaced. With a further increase of aluminum content, the number of aluminum atoms in the unit cell remains constant while nickel atoms are rejected from the unit cell. With increasing aluminum composition, the lattice parameter decreases as a result of the increasing concentration of vacant sites. In Ni_3Al, the aluminum content per unit cell continually increases along with the simultaneous creation of vacancies. This increase in aluminum content negates the contraction due to the creation of defects and results in a continually increasing lattice parameter. The ability of the large aluminum atom to replace the relatively small nickel atom is no doubt related to the mutual size adjustment which occurs in this phase.

Table I shows that alloying additions do not effect the defect structure of Ni_3Al in any predictable manner. Although it is expected that copper will cause an increase and titanium a decrease in the composition range for which defects exist, the results show no significant change. Table I also shows that in certain ternary alloys, aluminum atoms are present at face-centered positions in the unit cell in the same manner as in hyperstoichiometric binary Ni_3Al alloys. On the basis of lattice parameter changes, Arbuzov and Zelenkov[5] also have recently described the Ni_3Al phase which contains alloy elements (M) by the formula $(Ni, Al)_3 (Al, M)$.

At this point it is convenient to redefine what is meant by the terms B and T component in the T_3BC_x phases of interest. An atomic component will be considered of the B type if it can substitute for aluminum in Ni_3Al, of the T type if it can replace the nickel. For example, the Ni–Al–Ti ternary phase diagram at 1150°C[3] shows that the Ni_3Al phase field extends along an imaginary Ni_3Al–Ni_3Ti pseudobinary to a composition at which more than 60% of the aluminum has been replaced by titanium. Clearly titanium behaves as a B element. Similarly copper substitutes for more than 20% of the nickel in the Ni_3Al phase which extends in the direction of the imaginary Ni_3Al–Cu_3Al pseudobinary in the Ni–Cu–Al ternary system at 1150°C[3]. Copper obviously behaves as a T element. The Ni_3Al phase field in the Ni–Fe–Al ternary system at 1150°C extends in a direction bisecting the pseudobinary sections Ni_3Al–Ni_3Fe and Ni_3Al–Fe_3Al. Iron must behave as both a T and B element, substituting at the same time for nickel and aluminum (unless a disproportionally large number of aluminum vacancies occur).

Since in all hyperstoichiometric alloys (B greater than 25 at.%) B atoms must occupy face-centered positions in the unit cell, it becomes important to determine exactly which B atoms occupy the different sites.

THE POSITION OF THE B ATOM IN THE T_3BC_x UNIT CELL

The yttrium rich portion of the Al–Y–C phase diagram has been investigated by the present authors.[6] The features of interest are the occurrence of a ternary carbide phase with a small homogeneity range around the composition Y_3AlC and the negligible solubility of this phase in the binary NaCl-type Y_3C phase. This miscibility gap—which is rather surprising when one considers the similarity in structure and composition between the two phases—may be examined by comparing the arrangement of the different atoms in the unit cell of each phase.

The unit cell of Y_3C is of the NaCl type, however, whereas all the metal atom positions are occupied by yttrium atoms, vacant carbon sites exist which are distributed at random throughout the structure.[7] The unit cells of Y_3C and Y_3AlC, each drawn with

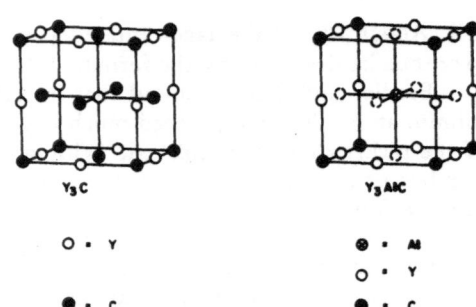

Figure 4. Unit cells of Y_3C and Y_3AlC.

the origin at a carbon atom position, are shown in Figure 4. The dashed circles represent lattice sites in the Y_3AlC structure which, although unoccupied, are the correct size to accept a carbon atom (they are, in fact, almost exactly the same size as the holes in the Y_3C structure which contain the carbon atoms). In this representation, the similarity between the unit cells of the two carbide phases is apparent. If the center yttrium atom in the unit cell of Y_3C is replaced by an aluminum atom and all the atoms which are nearest neighbors (these happen to be carbon atoms) are removed, the Y_3AlC unit cell is produced. The atom sites which are shown as dashed circles in the Y_3AlC unit cell represent the atoms which have been removed. The regular arrangement of carbon and aluminum atoms which results from the changes just discussed, produces a situation where an aluminum atom has no carbon atom as a nearest neighbor. In other words, the aluminum atoms are as far from a carbon atom as possible. However, the nearest neighbors of the remaining carbon atoms are unchanged. In simple terms, it is clear that the miscibility gap which exists between these phases is ultimately attributed to the fact that although there is some affinity between aluminum and yttrium atoms there is relatively little affinity between aluminum and carbon atoms. The carbon atom prefers to surround itself with yttrium atoms only, the aluminum atom moving as far from the carbon atom as possible. These considerations of mutual attraction and repulsion are fulfilled by the regular arrangement of atoms, which makes up the Perovskite structure.

If the matter of preferential location of the carbon and aluminum atoms in the Y_3AlC unit cell is correct, the tendency of the metalloid to avoid proximity to aluminum atoms must be reflected in the composition limits of the ternary phase. Contact of metalloid and aluminum atoms will be avoided if in composition each of these atoms are limited to a maximum of 20 at.%. For the isostructural phases which have been discovered to date, these limits are not transgressed.[1,2]

Let us return to the hyperstoichiometric T_3BC_x phases, where it has been shown that B atoms occupy the face-centered positions in the unit cell. In an alloy system containing aluminum and any of the B components from groups III, IV, and V, the aluminum atoms will occupy the cube corners of the unit cell preferentially, whereas the other B elements will fill the remaining cube corners and the required face-centered positions. This is clear when one considers the similarities between the monocarbide of yttrium and the monocarbides of the elements of groups III, IV, and V. All of these elements form stable carbides of the NaCl type, which within size limitations (15%) form complete solid solutions. Accordingly, the relative affinity of carbon for aluminum and for these elements will be much the same as the affinity of carbon for aluminum and for yttrium in Y_3AlC. Where a choice exists, the aluminum atom will move as far from the carbon atom as possible; any excess B element other than aluminum will enter the face-centered position surrounding the carbon atoms. It is probable that the chromium group elements, chromium, molybdenum, and tungsten, behave in a manner similar to the elements of the scandium, titanium, and vanadium group. Although these elements do not form stable monocarbides of the NaCl type at ordinary conditions of temperature and pressure, they are still under certain conditions stable in an octahedral environment of carbon atoms. For example, a large amount of Cr_3C_2 can be dissolved by TiC; thus, (Ti, Cr)C forms although no CrC exists.[8]

The fact that carbon will remain soluble in T_3BC_x phases when the carbide-forming elements are present is illustrated by the equilibrium between the Y_3C and the Y_3AlC phases and also the stabilization of the Ti_3AlC_x phase despite the existence of the very stable TiC phase.[9] Furthermore, in this investigation a single-phase alloy of composition 15 at.% Al–5 at.% Ti–10 at.% Fe–70 at.% Ni to which 3 at.% carbon was added shows different amounts of carbon in the as-cast and annealed (at 1300°C for 63 hr) condition. Incidentally, this latter observation is in agreement with the fact that a change in temperature will alter the unit cell dimensions, thereby changing the size of the octahedral hole at the center. The modified Hagg criteria, which has been shown to be valid for these phases, suggests that the solubility of carbon will show a temperature dependence. In this investigation, the solubility of carbon in Ni_3Al has in fact been found to increase from 5 at.% at 1000°C to approximately 7 at.% at 1300°C.

In a purely random arrangement of excess B atoms among the face-centered positions, the number of places in which two or more B atoms surround any given carbon atom is related to the composition by the simple theory of probability. In hyperstoichiometric T_3BC_x phases, however, a tendency for clustering probably exists in the sense that a carbon atom has a greater probability of being surrounded octahedrally by B metals than is indicated by a purely random distribution. Thus, in a hyperstoichiometric $Ni_3(Al, Ti)C_x$ alloy it is suggested that clusters of carbon surrounded by an octahedron of titanium may exist; for the environment that the carbon atom finds itself in, in this arrangement, is similar to that found in both the stable TiC phase and the ternary carbide phase T_3AlC_x. The size relationships which have already been discussed indicate that, to a limited extent, clustering may occur in hypostoichiometric alloys. The degree of clustering can be detected and measured by X-ray diffraction methods. However, a strictly monochromatic beam is required so as to be able to separate the scattering effect caused by clustering from other forms of diffuse scattering at low angles of θ.

THE SOLUBILITY OF CARBON IN Ni₃Al ALLOYS

Figure 5 shows the approximate phase fields at 1000°C of four T_3AlC_x carbides (solid lines) plotted on the same concentration triangle. The results for the manganese, iron, and cobalt phases are from the work of Stadelmaier. Stadelmaier found a number of other binary carbide phases which also take up carbon. For this reason and the others already discussed, these ternary carbide phases may be considered as carbon-stabilized T_3B compounds. Such an interpretation is especially justified because of the wide variations of stabilizer content which occur in the known T_3BC_x compounds. The similarity between the binary Ni_3Al phase and the ternary phases is quite apparent.

The dashed lines in Figure 5 show the effect of various additions on the solubility of carbon in Ni_3Al. Copper reduces the solubility of carbon to negligible proportions. There probably exists a continuous phase field between the Ni_3Al and the $Co_3AlC_{0.59}$ phase and also between the Ni_3Al and the $Fe_3AlC_{0.66}$ phase in the respective quarternary systems. Examination of many alloys would be required to define these phase fields accurately. In this investigation, a small number of alloys containing varying amounts of carbon but the same ratio of T metal to aluminum (3 : 1) were annealed at 1000°C for five days. Examination of the microstructures of these alloys reveals that the Ni₃Al phase field in alloys which contain cobalt extends to at least 8 at.% carbon and, in alloys which contain iron to at least 12 at.% carbon.

THE SOLUBILITY OF CARBON IN NICKEL-BASE SUPERALLOYS

A cursory examination of the partial Al–Ni–C ternary isothermal section at 1000°C has been made and the results are shown in Figure 6. The negligible solubility of carbon in the NiAl phase was determined from hardness measurements, the hardness of the NiAl phase in the alloys of varying composition agreeing almost exactly in all cases with the values reported by Westbrook for the binary phases.[10] The relatively small solubility of carbon in nickel at 1000°C has been assumed from the work of others,[11] and the one datum point shown.

Consider an alloy whose composition places it in the Ni₃AlC$_x$ + Ni two-phase field at 1000°C. Most of the carbon in this alloy will be contained in the Ni₃AlC$_x$ phase. It may be inferred that the solubility of carbon in Ni₃AlC$_x$ will be increased further relative to that in nickel by the addition of alloying elements such as manganese, iron,

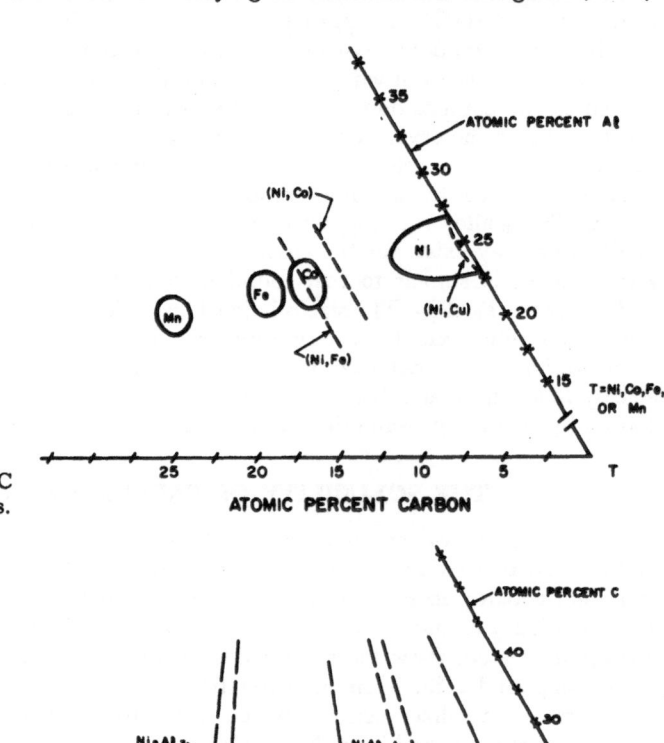

Figure 5. Phase fields at 1000°C of some ternary T_3AlC$_x$ carbides.

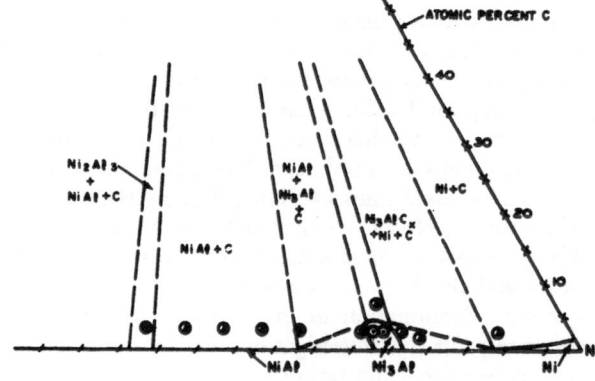

Figure 6. Partial Al–Ni–C isothermal section at 1000°C.

and cobalt. This may be seen by considering, for example, the element cobalt. As the solubility of carbon in cobalt at 1000°C is not much greater than the solubility of carbon in nickel at this temperature,[11] the addition of cobalt to nickel is not expected to result in any significant change in solubility. On the other hand, the solubility of carbon in Co_3AlC_x is 11.8 at.% at 1000°C, more than twice the value of 5 at.% for the solubility of carbon in Ni_3AlC_x at this temperature. Quite naturally, then, an addition of cobalt to Ni_3AlC_x should result in a significant increase in the solubility of carbon; this is precisely what has been observed. Consider now that nickel-base superalloys may be characterized as containing a Ni_3Al-phase in a nickel matrix. It follows that any carbon which is in solid solution in these alloys is most certainly contained in the Ni_3Al phase.

It has been shown in this laboratory that some carbon is tied up in this manner in the MAR-M200 superalloy. The addition of 2 wt.% aluminum to this superalloy results in an increase in the amount of carbides in the as-cast microstructure of the order of 50–100%.[12]

PROPOSED MODEL OF THE γ' STRUCTURE

The information which has been presented allows a model to be constructed of the $T_3BC_x(\gamma')$ structure. This model can be described as follows:

1. The effective size of all atoms in the T_3BC_x unit cell is the same.
2. In hyperstoichiometric alloys, B atoms replace T atoms at face-centered positions. Some face-centered sites are vacant due to processes which are probably electronic in nature.
3. Carbon atoms are located at the octahedral hole in the center of the unit cell.
4. In alloys which contain aluminum and any of the elements from group III, IV, V, and VI of the periodic table, aluminum will preferentially occupy the cube corners of the unit cell.
5. Hyperstoichiometric alloys may contain clusters of carbon surrounded octahedrally by elements of group III, IV, V, and VI. These clusters may also exist to a limited extent in hypostoichiometric alloys.
6. The amount of carbon which is soluble in T_3BC_x at a given temperature is determined in part by the presence of the elements manganese, iron, and cobalt.

SUMMARY

A number of results have been cited in this paper which suggest the presence of the γ' phase as T_3BC_x in superalloys. In concluding this paper the recent work of Wlodek[13] and Radavich and Couts[14] will be cited. Wlodek has tentatively identified a $Ni_3(Al, Ti)C$ phase in the massive γ' particles which are found in the nickel-base superalloy IN-100. The evidence for the presence of the carbide is based upon identification by X-ray and electron diffraction of a face-centered cubic phase with a slightly larger lattice parameter than $Ni_3(Al, Ti)$, whose chemical inertness in chloride solutions typifies the behavior of a carbide phase. Evidence is presented to show that aging of the as-cast alloy at 760°C (1400°F) for 300 hr produces a coherent precipitate of the carbide within the γ'. At 982°C (1800°F) the precipitate begins to dissolve along with the TiC phase.

In studies of a commercial nickel-base high-temperature alloy, Radavich and Couts also observed a precipitate in the γ' phase, although they were unable to identify it. These authors also noted a tendency for the γ' phase to coalesce [beginning at 927°C (1700°F)] and completely surround the TiC particles, whose rounded corners showed the beginning of the dissolution of the TiC phase. This latter observation indicates that the γ' phase contains carbon and is indeed a T_3BC_x carbide phase.

REFERENCES

1. H. H. Stadelmaier, "Ternary Compounds of Transition Metals, B-Metals and Metalloids," Z. Metallk. 52: 758–762, 1961.
2. S. Rosen and P. G. Sprang, "Ternary Carbide Phases Formed by Scandium-Group Elements with Aluminum and Carbon," in: W. M. Mueller, G. R. Mallett, and M. J. Fay (eds.), Advances in X-Ray Analysis, Vol. 8, Plenum Press, New York, 1965, pp. 91–101.
3. R. W. Guard and J. H. Westbrook, "Alloying Behavior of Ni_3Al (γ-Phase)," Trans. AIME 215: 807–814, 1959.
4. A. J. Bradley and A. Taylor, "X-Ray Analysis of the Nickel–Aluminum System," Proc. Roy. Soc. (London), Ser. A 159: 56–72, 1937.
5. M. P. Arbuzov and I. A. Zelenkov, "Structure of Ni_3Al Alloys with Additions of a Third Element," Fiz. Metal. i Metalloved. 15(5): 726–728, 1963.
6. S. Rosen and P. G. Sprang, Trans. AIME, To be published.
7. F. H. Spedding, K. A. Gschneidner, Jr., and A. H. Daane, "The Crystal Structures of some of the Rare Earth Carbides," Amer. Chem. Soc. 80: 4499–4503, 1958.
8. H. Novotny, in: Paul A. Beck (ed.), Electronic Structure and Alloy Chemistry of the Transition Elements, Interscience Publishers, Inc., New York, 1963, p. 189.
9. H. Von Philipsborn and F. Laves, "The Influence of Impurities on the Formation of the Cu_3Au-Type Structure from the Cr_3Si-Type Structure," Acta Cryst. 17: 213–214, 1964.
10. J. H. Westbrook, "Defect Structure and the Temperature Dependence of Hardness of an Intermetallic Compound," J. Electrochem. Soc. 104: 369–373, 1957.
11. M. Hansen, "Constitution of Binary Alloys," McGraw-Hill Book Company, Inc., New York, 1958, pp. 349–374.
12. B. J. Piercey, Private communication.
13. S. T. Wlodek, "The Structure of IN-100," Trans. Am. Soc. Metals 57: 110–119, 1964.
14. J. F. Radavich and W. H. Couts, Jr., "Effect of Temperature Exposure on the Microstructure of 4.5 Al–3.5 Ti Nickel-Base Alloy," Am. Soc. Metals, Trans. Quart. 54: 591–597, 1961.

DISCUSSION

A. Taylor (Westinghouse Research Laboratories): Can you please tell us something about the heat treatment and X-ray and density techniques you used to establish that there was in reality a defect structure in the Ni_3Al phase, and, also, whether you followed all your alloy making with very accurate chemical analyses of the material?

S. Rosen: It turns out that when you make the Ni–Al alloys by arc-melting, which was the process we used, you almost hit on the button the stoichiometry for the nominal composition. We have analyzed these and, if you notice, the errors for the differences between the analyzed compositions and the nominal ones were about 0.1 or 0.2 at.% aluminum. The alloys were arc-melted. They were annealed at about 1100°C for five days and when we looked at the microstructure they appeared homogeneous. These alloys have a tendency to core, due to the beta phase that appears, but we eliminated all of this with the annealing treatment. It turns out that when you calculate the number of atoms per unit cell this is very sensitive to the density measurements, unfortunately, and not to the lattice parameter measurements. The lattice parameter measurements were made with the back-reflection camera. These were more than adequate as far as the absolute value was concerned and the determination of atoms per unit cell. The density measurements were done by the liquid displacement method, which was just a matter of weighing these in water and out of water with appropriate corrections. We did these measurements three times and averaged the results. The estimated accuracy was one part in 200. It turns out that the error, as I mentioned before, was about 0.01 atoms per unit cell. That is about all I can say.

A. Taylor: Since the Ni_3Al phase is a very hard phase—if you forge it it is prone to crack—are you sure when you do a density on lumps on samples that it is absolutely free from tiny cracks and blowholes? Otherwise, you might misinterpret your results as being due to lattice defects.

S. Rosen: Probably the cracking of the Ni_3Al phase was due to the grain boundary hardness. What we did was section these innumerable times and we found that we had a sound alloy. We would take sections after we made the density measurements and take measurements on progressively smaller pieces and again we came out with consistent results. The alloys were not porous.

GENERATION OF A TWO-DIMENSIONAL SILICON CARBIDE LATTICE*

R. L. Prickett and R. L. Hough

Air Force Materials Laboratory
Wright-Patterson Air Force Base, Ohio

ABSTRACT

Silicon carbide was generated by pyrolysis of gas mixtures consisting of silicon tetrachloride, hydrogen, and organic vapors, such as acetone, on fine tungsten wires resistance-heated at 1500°C. Prominent two-dimensional structure was demonstrated for the 220 reflection. All other lines were of the normal three-dimensional lattice type.

Elevation of less than 100° in the pyrolysis temperature eliminated the two-dimensional reflection, and simultaneously changed the visible crystallite size.

Specialized techniques were used to generate the silicon carbide deposits and also to examine the structure of these deposits by X-ray diffraction to obtain lines from only the silicon carbide while ignoring the tungsten wire core. Diffraction techniques include offset collimation and vertical integration.

INTRODUCTION

The Air Force is concerned with refractory materials and their mechanical strength at elevated temperatures. Recent work includes a study of matrices containing refractory fiber substructures. One of the components of some of these fibers is silicon carbide.

Silicon carbide is being formed as a coating on a heated substrate such as pure tungsten wire or tungsten wire coated with boron by the interaction of gaseous compounds such as silicon tetrachloride, acetone, toluene, and a carrier gas, hydrogen. Because of the large number of compounds that may develop, X-ray identification of the final product is considered necessary. This paper describes some of the unusual results that are obtained and the variations from standard X-ray techniques that are used.

EXPERIMENTAL PROCEDURE

Figure 1 shows three typical silicon carbide deposits on tungsten wire substrates, number 1 formed at 2700°F, number 2 at 2900°F, and number 3 at 2800°F. Photomicrographs of these specimens are shown in Figures 2, 3, and 4 respectively. It can be seen that grain size varies directly with the temperature of formation, individual crystal facets being readily discernible at 2900°F. At somewhat lower temperatures, whisker-like growths are found, particularly if toluene vapor be substituted for acetone, as shown in Figure 5. Figure 6 shows a diagram of the apparatus used for forming β silicon carbide pyrolytically.[1]

---

[1] References are at the end of the paper.

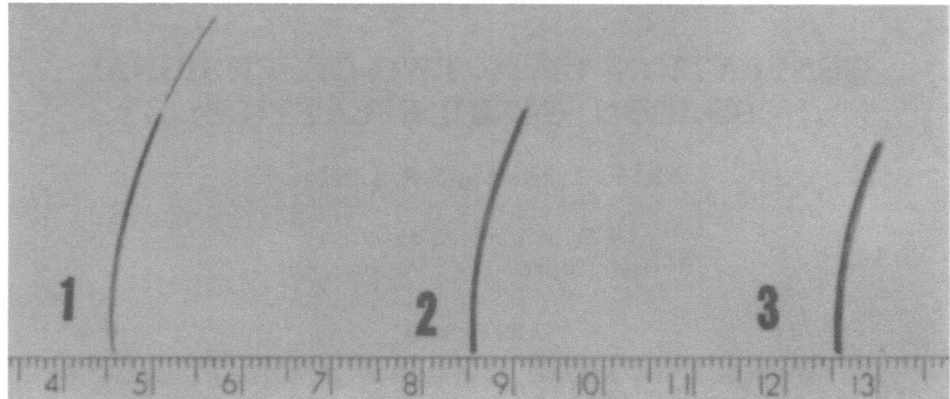

Figure 1. Silicon carbide specimens deposited on 10-mil tungsten wire.

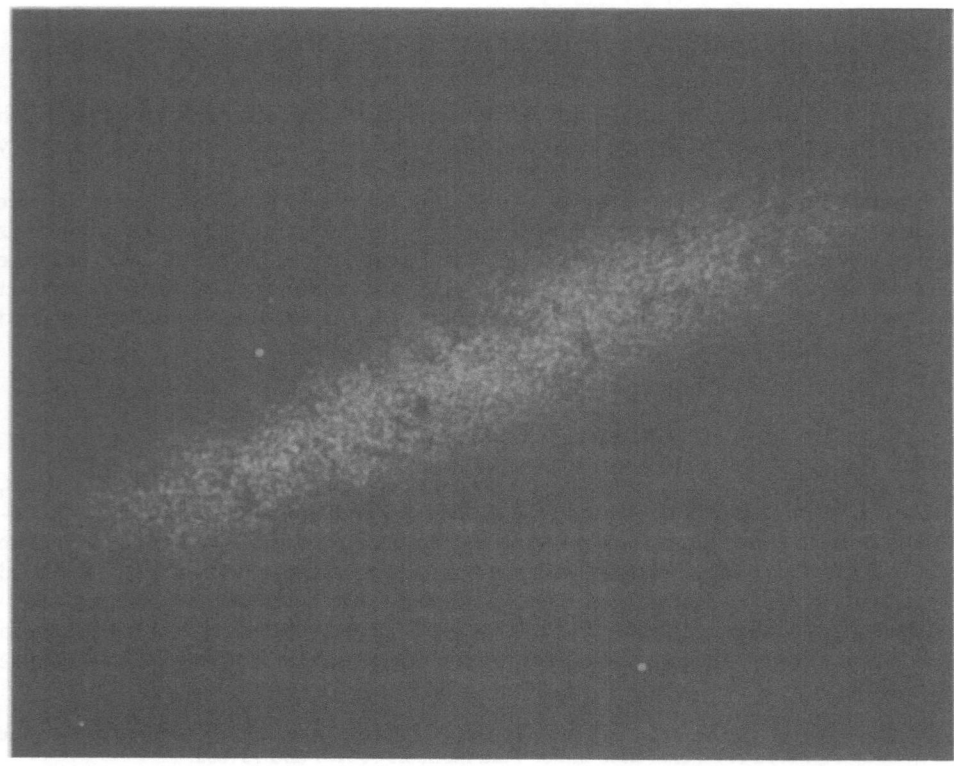

Figure 2. Photomicrograph of silicon carbide deposited at 2700°F.

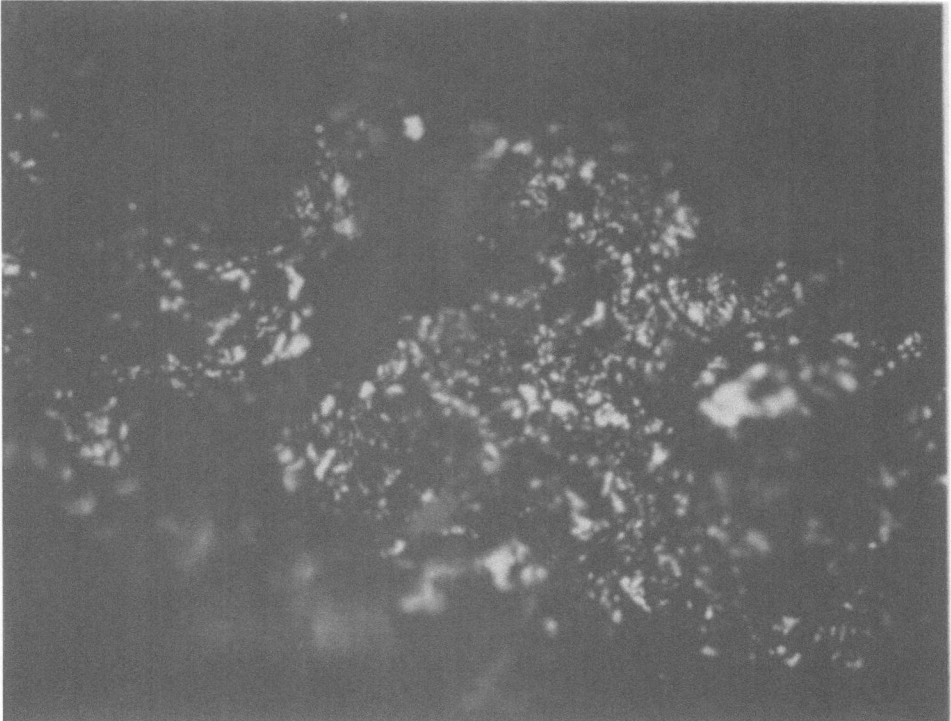

Figure 3. Photomicrograph of silicon carbide deposited at 2900°F.

There are three reasons why it is advisable not to follow strictly routine X-ray diffraction methods in analyzing these specimens.

1. The character of the core is not of critical importance in this research—whether or not tungsten is changed to a tungsten silicon carbon compound is not of interest. If standard diffraction techniques are applied, the complex compounds expected in the core area might produce a bewildering number of lines, making a study of the silicon carbide surface more difficult.
2. The specimens are somewhat thick for best resolution by regular techniques, Figure 1 showing their diameter to be in the vicinity of 1 mm; 0.2 to 0.5 mm is the recommended diameter.
3. The grainy and irregular surface character of the specimens might not give a representative diffraction pattern if studies are restricted to a small sector as is commonly done.

With the above ideas as guidelines, the samples are studied primarily by offset collimation techniques developed in this laboratory and, in a few cases, coupled with vertical integration.[2-7]

In Figures 7(A) and (B) are shown diagrams of the cross sections of X-ray powder cameras, with enlarged X-ray beams and powder samples drawn out of scale, to demonstrate how offsetting the X-ray beam to the side of a rather thick specimen (approximately 1 mm diameter) will increase resolution by decreasing the width of the diffraction lines generated. Figure 7(A) is the diagram of a standard camera, while Figure 7(B) is offset.

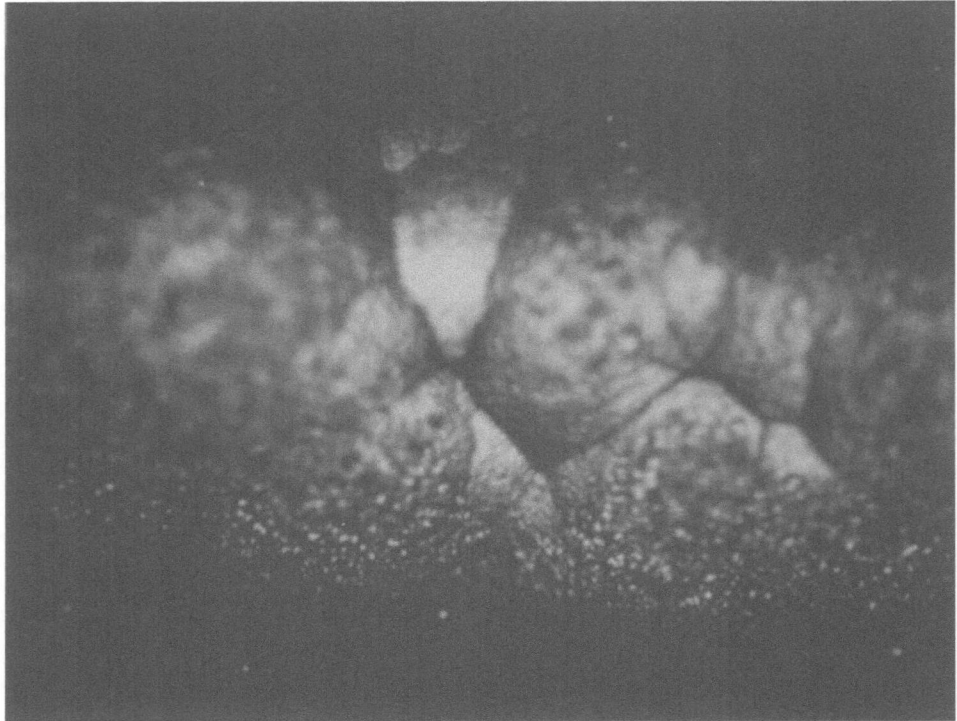

Figure 4. Photomicrograph of silicon carbide deposited at 2800°F.

Following the points marked within the cross section of the specimen, i.e., points P1 through P5 in Figure 7(A) and points P6 and P7 in Figure 7(B), allows one to visualize the origin of the individual diffraction spots D1, D3, and D5, the inherent substructure of which, represented by particulate points $D1_{P1}$, $D1_{P2}$, $D1_{P3}$, etc., helps explain how offset collimation increases resolution by decreasing diffraction line width.

There is no serious distortion of the X-ray diffraction pattern since the offset X-ray beam is still within the physical boundaries of the standard X-ray beam commonly used. A reduction in beam width alone without offsetting will give considerably less satisfactory results as may be demonstrated readily by the simple geometric construction used in Figures 7(A) and (B). The mechanical construction requirement for offsetting the X-ray beam is merely decentering the collimator bore by a few thousandths of an inch. Details of this technique have been given elsewhere.[2,3]

For elimination of spottiness on grossly crystalline fiber-like samples and to get a more representative analysis averaged over a large portion of the specimen, it is desirable to translate the specimen in a vertical direction, a technique termed vertical integration, during analysis. A special camera, designed by this laboratory and shown in Figure 8, is used for this purpose for some of the SiC specimens.

The one radian film cassette is shown suspended in the top of the photograph while the vertical integrating worm drive is controlled by the synchronous motor at the lower left of the photograph, and the sample rotating motor is at the lower right. Details of this camera have been given elsewhere.[4,5]

Figure 5. Whisker growth in silicon carbide deposited at 2600°F.

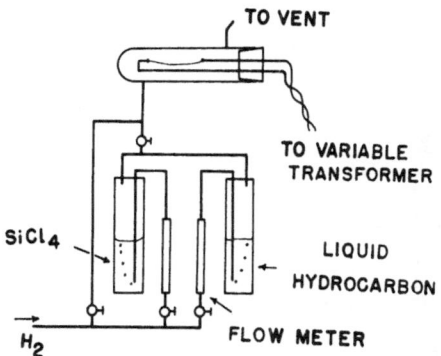

Figure 6. Silicon carbide preparation apparatus.

RESULTS

From Table I it will be seen that the samples are readily identifiable as β-silicon carbide. The expected (200) and (400) reflections are missing, indicating a strong orientation with the (100) planes parallel to the X-ray beam, that is, perpendicular to the fiber axis. Figures 9 and 10 show the diffraction patterns for samples 1 and 3, respectively; the hydrocarbon used for these fibers was acetone. The broadness of peaks (220) and (331) in silicon carbide formed at 2700°F, indicating a two-dimensional, random structure[7-24] primarily in the (220) direction, is almost completely removed on raising the formation temperature only 100°F, developing a sharp peak for the (220) reflection in sample 3, and sharpening the peak for the (331) reflection.

In Figure 11 is shown a typical silicon carbide fiber diffraction pattern taken without an offset collimator, the beam passing directly through the core. The lines from the

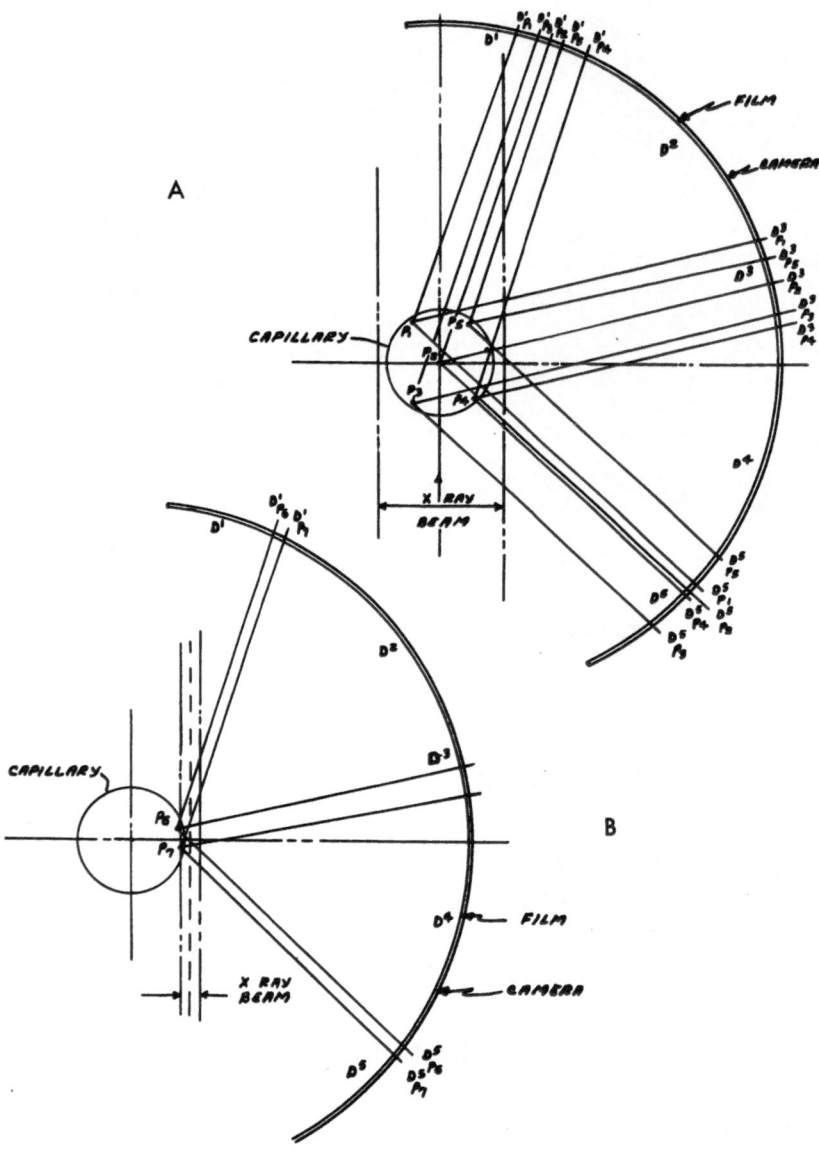

Figure 7. Cross sections of powder camera showing normal specimen and X-ray beam and diffraction lines drawn out of scale; (A) centered X-ray beam, (B) offset X-ray beam.

Figure 8. Structure-integrating
general-purpose camera.

Table I. Offset Diffraction Data for Silicon Carbide Samples 1 and 3

2θ (Copper radiation)	I/I_0	d	I.D.	hkl
Sample 1 (2700°F)				
32.0	20	2.53	β-Line	111
35.6	100	2.522	β-SiC	111
59.7*	20	1.549	β-SiC	220
71.5	40	1.320	β-SiC	311
75.3	70	1.262	β-SiC	222
100.4†	10	1.004	β-SiC	331
104.0	30	0.979	β-SiC	420
Sample 2 (2800°F)				
35.5	100	2.529	β-SiC	111
59.75	30	1.548	β-SiC	220
71.4	40	1.321	β-SiC	311
75.0	10	1.266	β-SiC	222
100.4	5	1.004	β-SiC	331
103.7	20	0.981	β-SiC	420

* Very broad.
† Broad and diffuse.

Figure 9. Offset diffraction pattern of SiC formed at 2700°F.

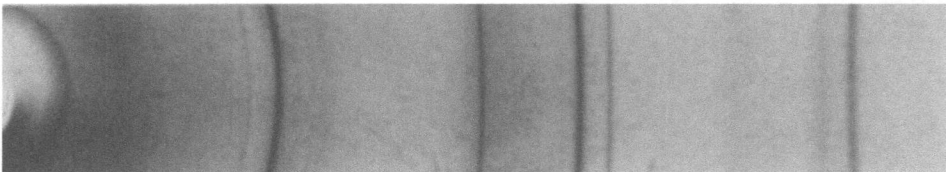

Figure 10. Offset diffraction pattern of SiC formed at 2800°F.

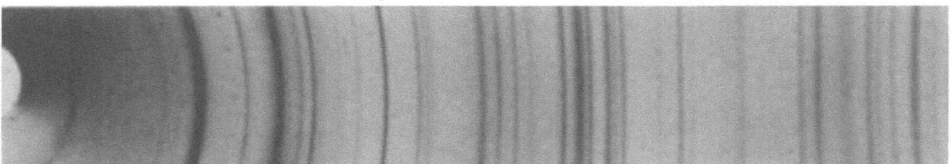

Figure 11. Silicon carbide diffraction pattern taken without offset collimation.

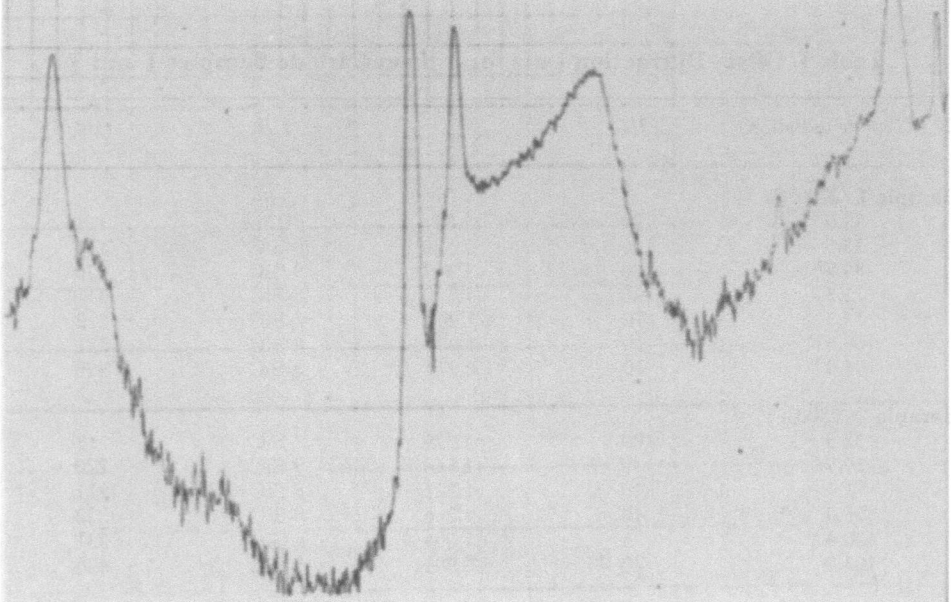

Figure 12. Densitometer scan of the powder pattern from β-silicon carbide formed at 2700°F.

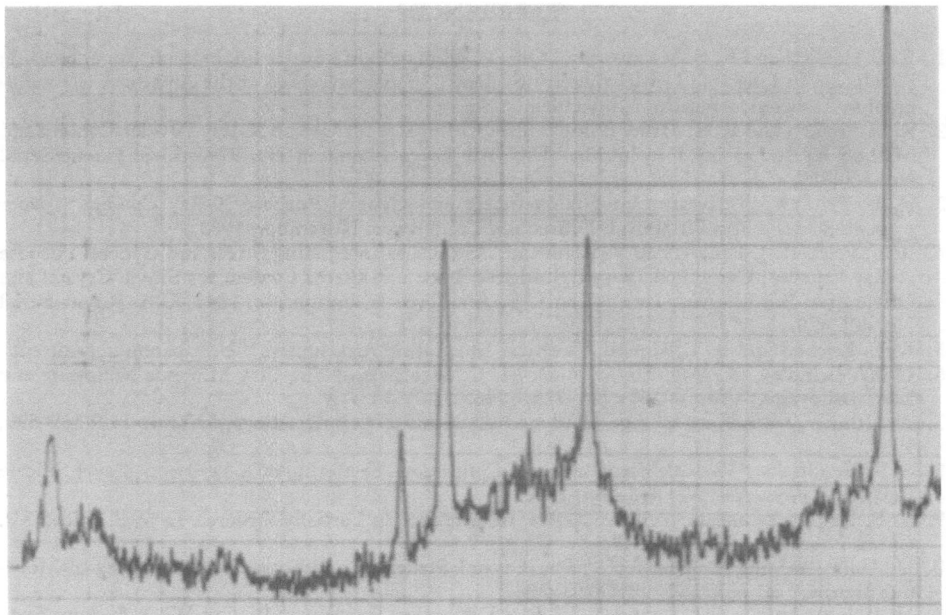

Figure 13. Densitometer scan of the powder pattern from β-silicon carbide formed at 2800°F.

additional compounds found in the core can be seen to interfere with the analysis. This particular silicon carbide sample showed no noticeable two-dimensional structure.

  Densitometer scans of these two diffraction patterns are shown in Figures 12 and 13. The (220) reflection in Figure 12 is almost too broad to be considered a peak in routine diffraction patterns, and shows considerable distortion, having a very gradual slope on the high angle side, a typical phenomenon in two-dimensional patterns. In Figure 13 the distortion is gone and a normal peak is present.

  The *hkl* indices in Table I indicate that there is no general restriction on *h*, *k*, or *l* for all planes, as was suggested by Brindley,[6] because of the two-dimensional character.

  Thibault's article[24] on SiC, an extract from his doctoral thesis, gives excellent coverage of the elementary crystallography. The refractive index of SiC exceeds that of diamond, and one might expect that a transparent grade might compete with the diamond, because of its relative hardness, as a gem stone in the future.

CONCLUSIONS

  Cubic silicon carbide can be formed by the pyrolysis of gaseous mixtures containing silicon and carbon compounds. Silicon carbide pyrolytically formed at 2700°F has an imperfect two-dimensional structure which is lost if higher temperatures are used.

ACKNOWLEDGMENTS

  The authors wish to thank members of the Technical Photographic Division at Wright-Patterson Air Force Base, particularly Mr. L. Riddle, for the excellent photographic assistance in a relatively difficult area.

REFERENCES

1. R. L. Prickett and K. S. Mazdiyasni, "Offset Collimation in Powder Photographs for Increasing Resolution in Unstable Low Symmetry Systems," paper presented at the American Crystallographic Association meeting, July, 1964.
2. R. L. Prickett and K. S. Mazdiyasni, "Offset Collimator for Use on X-Ray Powder Camera and Method for Increasing Resolution," U.S. Air Force Invention No. 9743, U.S. Patent Serial No. 380,959.
3. R. L. Prickett, "Structure Integrating Stress and General Purpose X-Ray Camera," paper presented at the 21st Pittsburgh Diffraction Conference, November, 1963.
4. R. L. Prickett, J. Fenter, and V. Robinson, "Structure Integrating Stress and General Purpose X-Ray Camera; Combined Long Cylindrical, Laue and Stress Camera for Single Crystal and Orientation Studies with Integration U.S. Air Force Invention No. 9842, U.S. Patent Serial No. 406,630.
5. R. L. Prickett and R. L. Hough, "Vertical Integrating Attachment," U.S. patent submitted.
6. G. W. Brindley, K. Robinson, and D. M. C. MacEwan, "The Clay Minerals Halloysite and Meta-halloysite," *Nature* **157**: 225–226, 1946.
7. A. Guinier, "Structure of Age-Hardened Aluminium-Copper Alloys," *Nature* **142**: 569–570, 1938.
8. S. B. Hendricks, "Variable Structures and Continuous Scattering of X-rays from Layer Silicate Lattices," *Phys. Rev.* **57**: 448–454, 1940.
9. S. B. Hendricks and E. Teller, "X-Ray Interference in Partially Ordered Layer Lattices," *J. Chem. Phys.* **10**: 147–167, 1942.
10. U. Hofmann and A. Hausdorf, "The Crystal Structure and Intracrystal Swelling of Montmorillonite," *Z. Krist.* **104**: 265–293, 1942.
11. H. P. Klug and L. E. Alexander, "X-Ray Diffraction Procedures," John Wiley & Sons, Inc., New York, 1954, p. 385.
12. L. Landau, "Scattering of X-Rays by Crystals with Variable Lamellar Structure," *Physika. Z. Sowjetunion* **12**: 579–585, 1937.
13. M. V. Laue, "Cross-Lattice Spectra," *Z. Krist.* **82**: 127–141, 1932.
14. I. M. Lifschits, "Scattering of X-Rays by Crystals of Variable Structure," *Physk. Z. Sowjetunion* **12**: 623–643, 1937.
15. E. Maegdefrau and U. Hofmann, "The Crystal Structure of Montmorillonite," *Z. Krist.* **98**: 299–323, 1937.
16. J. Méring, "Interference of X-Rays in Systems with Disordered Stacking," *Acta Cryst.* **2**: 371–377, 1949.
17. J. Méring and G. W. Brindley, "Banded X-Ray Reflections from Clay Minerals," *Nature* **161**: 774, 1948.
18. H. S. Peiser, H. P. Rooksby, and A. J. C. Wilson, "X-ray Diffraction by Polycrystalline Materials," The Institute of Physics (London), 1955, p. 418.
19. G. D. Preston, "Structure of Age-Hardened Aluminum-Copper Alloys," *Nature* **142**: 570, 1938.
20. B. E. Warren, "X-Ray Diffraction in Random Layer Lattices," *Phys. Rev.* **59**: 693–698, 1941.
21. A. J. C. Wilson, "X-Ray Optics," Methuen & Co., Ltd. (London), 1949.
22. A. J. C. Wilson, "X-Ray Diffraction by Random Layers: Ideal Line Profiles and Determination of Structure Amplitudes from Observed Line Profiles," *Acta Cryst.* **2**: 245, 1949.
23. R. L. Hough, "Method for the Pyrolytic Deposition of Silicon Carbide," U.S. Air Force Invention No. 10,114.
24. N. W. Thibault, "Morphological and Structural Crystallography and Optical Properties of Silicon Carbide (SiC)," *Am. Mineralogist* **29**: 249–278 and 327–362, 1944.

DISCUSSION

Chairman A. F. Berndt: I don't believe you mentioned the magnifications of the photomicrographs you showed.

R. L. Prickett: These were around 150×. They were not very high. They actually varied slightly from specimen to specimen to assist us in getting the portion out that we were interested in.

AN X-RAY DIFFRACTION STUDY OF THE PHASE TRANSFORMATION TEMPERATURE OF MnO

Charles P. Gazzara and R. M. Middleton

U.S. Army Materials Research Agency
Watertown, Massachusetts

ABSTRACT

X-ray diffraction measurements of MnO confirm the hypothesis that the structural temperature and the magnetic transition temperature, or Néel temperature, are the same. The usefulness of X-ray diffraction intensity data of MnO, with respect to an atomic structural refinement problem involving α-Mn powders, is discussed. Lattice constant values of MnO are listed between 100 and 310°K.

INTRODUCTION

Manganous oxide has been of interest to crystallographers because this compound undergoes a magnetic transformation at the Néel temperature, 122°K, going from a paramagnetic to an antiferromagnetic compound with decreasing temperature. Roth[1] found the low-temperature magnetic structure to be based on the formation of ferromagnetic sheets perpendicular to a rhombohedral axis. Another reason for concern over MnO is that in an X-ray diffraction study of α-manganese and β-manganese powders MnO was found to be the chief contaminant. This being the case, the effects of Mno on the physical properties of magnesium should be considered, especially when using powders.

The crystal structure of MnO is fcc rock-salt structure at room temperature, with a lattice constant of 4.4445Å.[2] The neutron diffraction work of Shull and Smart[3] indicated the necessity of a "magnetic" cell of twice the lattice spacing, namely 8.85Å at 80°K. Shull, Strauser and Wollan[4] determined from the broadening of the (111) neutron diffraction peak that the magnetic transition temperature is just below 124°K.

The first mention of a change in lattice spacing in MnO was reported by Ellefson and Taylor in 1934,[5] occurring at approximately 114°K. Ruhemann[6] noted a structural transformation at 115.9°K describing the low-temperature phase as an equilibrium of two near-cubic lattices. Toombs and Rooksby[7] reported a transition below 173°K from fcc to a symmetry approximating rhombohedral with a larger deformation occurring at 93°K. It was pointed out by Greenwald and Smart[8] that the temperatures at which the crystal structure changes in the oxides of the iron group, namely MnO, FeO, CoO, NiO, Cr_2O_3, are apparently the antiferromagnetic Curie temperatures.

The magnetic transition temperature of MnO is 122°K and Greenwald and Smart use Ruhemann's work as supporting evidence, indicating that the structural transition temperature found by Ruhemann to occur near 120°K.

The results of our work on MnO confirm the conclusion of Greenwald and Smart that the structural transition temperature and the Néel temperature are the same, and to

---

[1] References are at the end of the paper.

indicate our measurements of the rhombohedral unit cell of MnO just below the transition temperature.

RESULTS

The powder specimen of MnO was prepared by reducing Mn_2O_3 powder under hydrogen at 600°C for 1 hr. X-ray diffraction patterns, using Cr K_α radiation, were taken of the MnO sample in a low-temperature specimen holder,[9] and Figure 1 shows the transition of the fcc X-ray diffraction peaks to the rhombohedral peaks. The precise temperature of the structure transition is somewhat obscured by the fact that a temperature gradient exists across the specimen (approximately 5°K) as well as through the specimen, so that the temperatures shown are the specimen surface temperatures at the center of the specimen.

The temperature at which the structural transition from fcc to rhombohedral occurs is 123°K. This was concluded from the splitting of fcc (220) into the (1$\bar{1}$0) and (211) rhombohedral planes observed in Figure 2. In this case we were able to lower the temperature of the specimen an undetectable amount and yet initiate the structural change so that the peak splitting is observable.

The transition temperature (123°K) of MnO was confirmed by repeating this experiment using an MRC cryostat and measuring the surface temperature of the MnO powder.

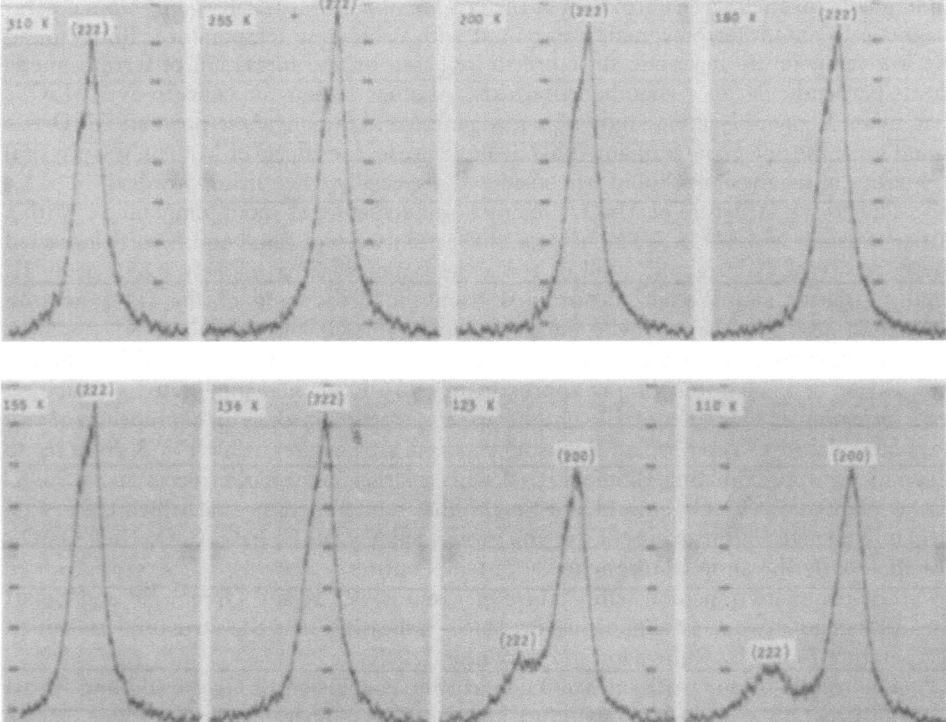

Figure 1. The effect of a change in temperature from 310 to 110°K on the (222) fcc diffraction peak of MnO.

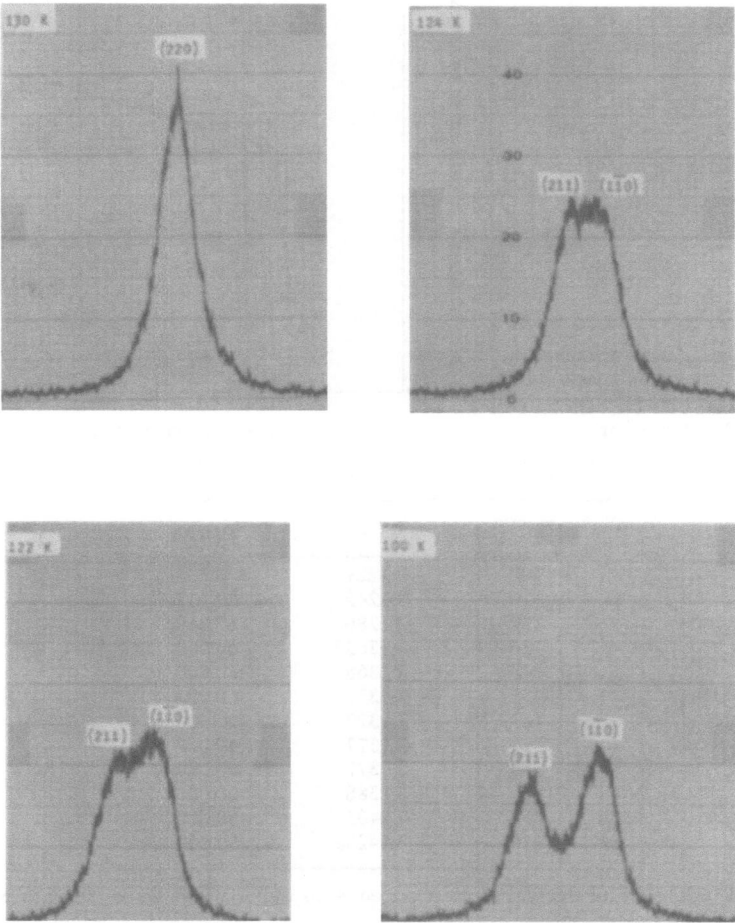

Figure 2. The splitting of the (220) fcc X-ray diffraction peak of MnO near
the transformation temperature.

The rhombohedral parameters α and a were computed from the angular separation of peaks (200)/(222) and (1$\bar{1}$0)/211), the former being weighted more heavily. The fcc lattice constants were computed by a least squares analysis and are in agreement with the room temperature value of Jay and Andrews.[2] Values of the diagonal lattice constants are given in Table I and plotted versus temperature in Figure 3.

The diffracted intensities are generally in agreement with those predicted for the rhombohedral cell just below the transition temperature of 123°K (see Table II). However, there appears to be a slight distortion of the rhombohedral cell below 100°K, but not enough to warrant structure refinement with the present intensity data. We were not able to detect any rhombohedral structural change in the lattice near 173°K as was reported by Toombs and Rooksby[7] or any other conclusive change in structure between 210 and 123°K.

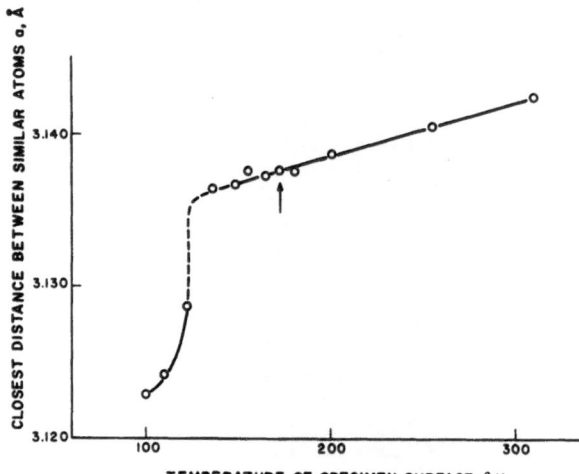

Figure 3. The closest distance (Å) between similar atoms a versus the temperature of the (K) specimen surface of MnO powder.

Table I. Values of Diagonal Lattice Constants

T (°K)	a (Å)	α (degrees)
100	3.1229	60.505_7
110	3.1242	60.437_9
122	3.1286	60.289_0
136	3.1363*	60.000
148	3.1366	60.000
155	3.1377	60.000
167	3.1372	60.000
173	3.1377	60.000
180	3.1377	60.000
200	3.1388	60.000
255	3.1405	60.000
310	3.1425_6	60.000

* The values of a listed for the fcc structure are equal to $a_0/\sqrt{2}$ where a_0 is the lattice constant of the fcc unit cell.

Table II. Diffraction Data on MnO

hkl	2θ	$|F|^{2*}$	$m|F|^2$(L.P.)†	I (measured)	$I/m|F|^2$(L.P.)
{ 100	50.03	138	6324	6750	1.07
{ 111	53.60	138	2064	2100	1.02
110	62.20	467	14950	16500	1.10
{ 1$\bar{1}$0	93.45	295	4894	6000	1.23
{ 211	94.35	292	4820	5700	1.18
{ 1$\bar{1}$1	117.37	79	1507	~1400‡	
{ 210	117.80	79	3037	~4000	1.10
{ 221	118.92	78	1527	~1400	
{ 200	126.45	208	4703	5640	1.20
{ 222	128.82	205	1623	1800	1.11

* $|F|^2$, structure factor magnitude $= FF^*$. $|F|^2 = (f_{Mn} - f_0)^2$, hkl odd, $|F|^2 = (f_{Mn} + f_0)^2$, hkl even, where f_{Mn} was taken as $f_{Mn^{++}}$ (corrected for dispersion) according to Roth,[1] and f_0 was taken as f_0^{--} (corrected for dispersion).

† m is multiplicity and L.P. is Lorentz Polarization factor.

‡ 3 peaks were difficult to resolve.

The usefulness of this X-ray diffraction data on MnO is illustrated in a problem involving a structural refinement of the atomic parameters of α-manganese. In measuring the X-ray diffracted integrated intensities from α-manganese powder, the diffraction peaks from the oxides of manganese could not be detected. This was unusual because vacuum fusion analyses revealed an oxygen contamination level of 0.40 wt.% O_2 while a neutron activation examination of the samples indicated 0.60 wt.% O_2 or greater. The presence of MnO was revealed by lowering the temperature of the specimen to 100°K, below the transformation temperature of MnO. The results of this procedure are shown in Figure 4. In one case, at 310°K, the (440) α-manganese diffraction peak is shown slightly broader than the (530)/(433) α-manganese peak, but at 100°K the previous hidden (220) MnO peak has split into the (211) and (1$\bar{1}$0), allowing the (211) MnO peak to become partly resolved. Also shown in Figure 4 is a diffraction scan of β-manganese at 100°K.

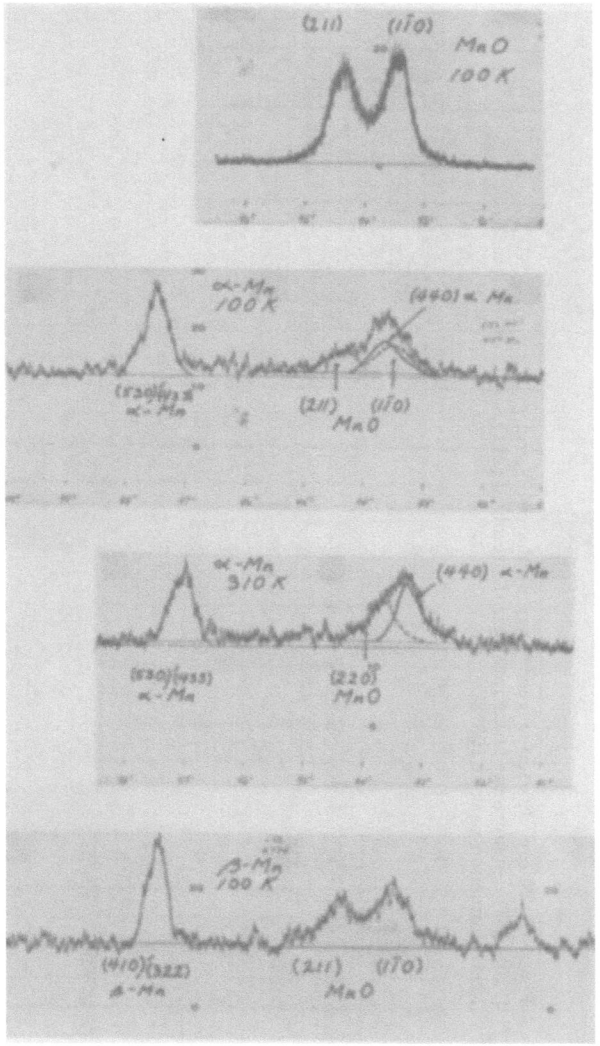

Figure 4. X-ray diffraction recordings of α-manganese and β-manganese peaks showing the effects of the low temperature (211) and (1$\bar{1}$0) MnO diffraction peaks.

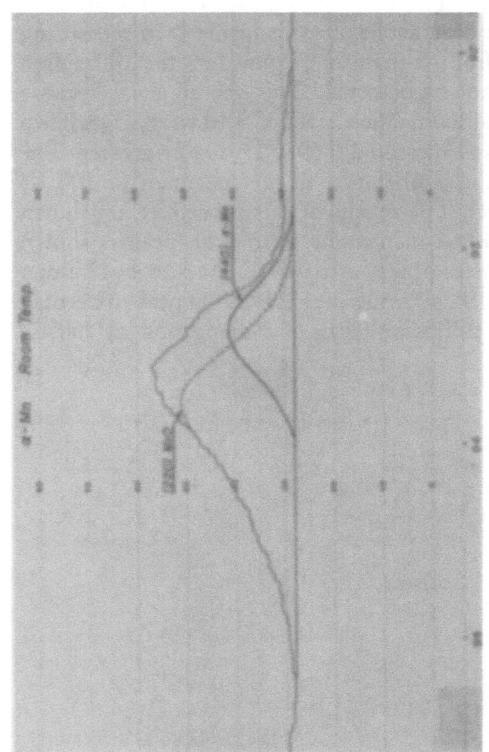

Figure 5. X-ray diffraction recordings of the (440) and (510)/(433) α-manganese peaks revealing the resolved (211) MnO diffraction peak at 100°K.

Here the (211)/(1$\bar{1}$0) MnO peaks are completely resolved, due to the fact that the (400) β-manganese peak, which occurs at the same 2θ angle as the (220) MnO peak, has a very low structure factor. The resolution of the (211) MnO peak and the (440) α-manganese peak is better shown in Figure 5 where a slower scanning speed, smaller receiving slit, smaller scale factor, and higher time constant were used. From the integrated intensity of the (211) MnO peak in Figure 5 we determined the amount of oxygen to be 1.1 wt.% O_2, which was in agreement with a neutron activation analysis of the same α-manganese powder. We therefore concluded that all of the oxygen in the α-manganese powder was present in the MnO form. Using the X-ray integrated intensities listed in Table II we could therefore correct the integrated intensities of the α-manganese diffraction peaks for MnO, thus yielding a more accurate structural refinement of the atomic parameters of α-manganese.

REFERENCES

1. W. L. Roth, "Magnetic Structures of MnO, FeO, CoO, and NiO," *Phys. Rev.* **110**: 1333–1341, 1958.
2. A. H. Jay and K. W. Andrews, "Note on Oxide Systems Pertaining to Steel Making Furnace Slags FeO–MnO, FeO–MgO, CaO–MnO, MgO–MnO," *J. Iron and Steel Inst. (London)* **152**: 15, 1945.
3. C. G. Shull and J. S. Smart, "Detection of Antiferromagnetism by Neutron Diffraction," *Phys. Rev.* **76**: 1256, 1949.
4. C. G. Shull, W. A. Strauser and E. O. Wollan, "Neutron Diffraction by Paramagnetic and Antiferromagnetic Substances," *Phys. Rev.* **83**: 333, 1951.
5. B. S. Ellefson and N. W. Taylor, "Crystal Structures and Expansion Anomalies of MnO, MnS, FeO, Fe$_3$O$_4$ between 100 and 200°K," *J. Chem. Phys.* **2**: 58, 1934.
6. F. Ruhemann, "Temperaturabhängigkeit der Gitterkonstanten von Manganoxyd," *Physik. Ber.* **16**: 2337, 1935.
7. N. C. Toombs and H. P. Rooksby, "Structure of Some Transition Elements at Low Temperatures," *Nature* **165**: 442, 1950.
8. S. Greenwald and J. S. Smart, "Deformations in the Crystal Structures of Anti-ferromagnetic Compounds," *Nature* **166**: 523–524, 1950.
9. C. P. Gazzara, "The Debye Temperature of Carbonyl Iron," in: W. M. Mueller, G. R. Mallett, and M. J. Fay (eds.), *Advances in X-Ray Analysis, Vol. 4*, Plenum Press (New York), 1960, pp. 93–107.

DISCUSSION

C. P. Gazzara: I might add that if you run into a similar problem, as we did with α-manganese (trying to detect the MnO diffraction peaks), e.g., involving FeO, CoO, NiO, and Cr$_2$O$_3$, you can perform essentially what we have done, that is, record the diffraction peaks of the oxides below and above the Néel temperature. This may help someone in the future to resolve hidden diffraction peaks.

CRYSTALLOGRAPHIC STUDIES OF NH_4Cl-NH_4Br SOLID SOLUTIONS

Sister Jane Edmund Callanan and Norman O. Smith

Fordham University
New York, New York

ABSTRACT

Both ammonium chloride and ammonium bromide undergo a transition, with rise in temperature, from an interpenetrating simple cubic (II) to a face-centered cubic (I) lattice at 183 and 137°C, respectively, and both the low- and high-temperature forms give a complete series of solid solutions. We have determined the lattice constants of the high-temperature solids at about 250° as a function of composition, and redetermined the lattice constants of the low-temperature solids at room temperature. The solutions were made by crystallization from water, followed by stirring in contact with mother liquor for at least three weeks at room temperature. Measurements were made with a Norelco-Philips diffractometer and recorder, with Cu K_α radiation. For the high-temperature work, a simple, inexpensive heating apparatus was developed. The only previous data reported for the high-temperature forms are the lattice constants of the pure components given by Bartlett and Langmuir.[13]

The low-temperature solutions showed negative deviations from Vegard's rule at both ends of the concentration range and a slight positive deviation elsewhere when high-angle data were used. The high-temperature solutions showed marked positive deviations from Vegard's rule over the whole compositions range. Values for the pure components agreed reasonably well with those of Bartlett and Langmuir.

The progress of the change II → I with time was followed for some of the solutions in the neighborhood of the transition temperature in an attempt to reveal the mechanism of the process.

INTRODUCTION

It has been demonstrated that both NH_4Cl and NH_4Br undergo a transition with rise in temperature from an interpenetrating simple cubic lattice (II) to a face-centered cubic lattice (I). This change takes place at about 183°C for the chloride and about 137°C for the bromide.[1,2] Both the low- and high-temperature forms of these salts give complete series of solid solutions as has been shown by the work of Flatt and Burkhardt[3] at 25°C and that of Rassow[4] near the melting point. The solutions also undergo a transition similar to that of the pure components. The relation between the composition of the solutions and the temperatures of the transition was determined in our laboratories by Costich, Maass and Smith[5] and is shown in Figure 1. It will be seen that the transition temperature varies smoothly with composition and exhibits a pronounced minimum.

X-ray diffraction measurements have been used in the study of solid solutions, particularly to demonstrate their existence and to give information with regard to the extent of miscibility. A gradual shift in the positions of the diffraction lines with changes in composition, which reflects the resulting change in lattice parameter, is accepted

---

[1] References are at the end of the paper.

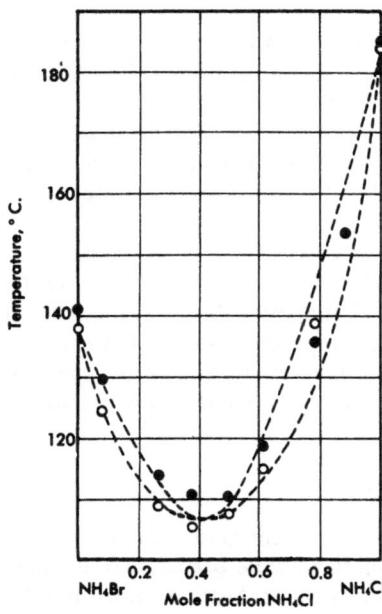

Figure 1. I $\rightleftharpoons$ II transition temperatures of NH_4Cl–NH_4Br solid solutions[5]; ● differential heating, ○ differential cooling, – – – – estimated equilibrium solidus lines. Reproduced with the permission of the American Chemical Society, from P. S. Costich, G. J. Maass, and N. O. Smith, *J. Chem. Eng. Data* **8**: 26, 1963.

generally as sufficient evidence for the existence of solid solutions. If the components are completely miscible there will be a shift from the positions in one pure component to those in the other. The interruption of this smooth change is indicative of immiscibility in the system.

In 1921 Vegard[6] proposed the following well-known additivity relationship between lattice constant and composition of the solution:

$$a^n = a_1{}^n + (a_2{}^n - a_1{}^n)x \tag{1}$$

with $n = 1$. Here a is the length of the side of the unit cube of the solution, a_1 the length of the side of the unit cube of pure component 1, a_2 the same for pure component 2, and x the mole fraction of component 2. This expression has been used to classify solid solutions as ideal or as exhibiting positive or negative deviations from ideality. The same forces considered responsible for deviations in other solutions, namely attraction or repulsion, are considered to be operative here. In 1923 Grimm and Herzfeld[7] developed a relation of similar form from ionic crystal lattice theory which would set the exponent in equation (1) equal to 8. Further considerations from calorimetric measurements led to the suggestion that $n = 3$; this would correspond to the thermodynamically sound additivity of molar volumes. Retgers,[8] in 1889, had proposed an additivity law based on specific volumes. Jette[9] suggested that density would be equally useful as a property for such a law. Through the measurements of lattice constants of several alkali halide solutions, Havighurst, Mack, and Blake[10,11] were able to rule out 8 as an exponent; their experimental error would not justify a choice between values of 1 and 3 for n. Zen,[12] in 1956, showed, by means of a short calculation, that the additivity of lattice constant and the additivity of molar volume are not equivalent unless the two volumes are very close in value.

Previous pertinent work with NH_4Cl and NH_4Br includes the determination at room temperature of lattice constants of the pure salts and of a single solution by Havighurst, Mack, and Blake,[10,11] the study of lattice constants and densities of the pure components at 20°C and 250°C by Bartlett and Langmuir,[13] and the investigation of the variation of lattice constant with composition for the low-temperature (II) solids by Anselmo and Smith.[14] Bridgman,[1] in his classic work, has determined the transition temperatures of the pure salts and the density changes accompanying the transition. Poyhönen[2] reports their transition temperatures and densities obtained through dilatometric studies. Wood, Secunda, and McBride[15] considered the mechanism of the transition for CsCl–CsBr solid solutions.

The work to be reported here confirms the existence of the high-temperature crystal form for the solid solutions of the system NH_4Cl–NH_4Br. For the first time the lattice constants for these, form I, have been determined at about 250°C and the relation between density and composition investigated. A time–temperature study of the transition provided insight into its mechanism. The study of the variation of lattice constant with composition for the low-temperature form (II) is more complete than any previously reported.

MATERIALS USED

The studies of the pure components were made with reagent grade salts which had been purified by recrystallization from water. The solutions were prepared from these materials by adding appropriate amounts of both salts to water, heating until the salts had dissolved, and allowing the solid solutions to precipitate by cooling to room temperature while stirring. The resulting suspensions were then stirred for at least three weeks to effect homogenization. The approximate proportions of salts and water to be used to give the desired solid-phase compositions were determined from the ternary phase diagrams of Flatt and Burkhardt.[3] The actual compositions of the resulting solids were determined by direct analysis.

EXPERIMENTAL

The analytic method found to be most satisfactory involved determinations of chloride and total halide content. Bromide ion was removed from some aliquots by oxidation to free bromine with potassium permanganate; these aliquots were allowed to stand for at least 36 hr to permit the escape of the bromine. The excess permanganate was destroyed with peroxide. The concentrations of chloride remaining after oxidation of the bromide and of the total halide in other aliquots were determined by potentiometric titration. Silver and saturated calomel electrodes were used with standard $AgNO_3$ solution as titrant. The mole fractions of NH_4Cl were found from the ratio of chloride ion to total halide.[5]

The validity of the analytical results was established through the analysis of synthetic mixtures. The results obtained are given in Table I. The accuracy is thus about 1% except for very low chloride content.

Table I. Analysis of Synthetic Mixtures

Mole fraction of NH_4Cl taken	Mole fraction of NH_4Cl found
0.103	0.099
0.490	0.496
0.777	0.768
0.930	0.920

X-ray powder patterns were taken with a Norelco-Philips diffractometer and recorder, using Cu K_α radiation and a nickel filter. The spectra for the pure salts and the solutions were obtained at room temperature for a range of 2θ values from 90 to 20°; at about 250°C the useful range was 50 to 20°. Because of the limitation of the useful range for the high-temperature work, the lattice constants for the low-temperature form of the solutions were determined using the same range. The work at the higher temperatures was performed using two types of sample heaters, both of which were simple and in-expensive. The first consisted of a ribbon heating element wound about flat sheets of mica in order to provide a heater of about 50 W. This was placed under the usual plate sample holder used with the diffractometer. The second consisted of a mica holder slipped over the ordinary aluminum plate; a convenient length of insulated strip heater was inserted into the holder.

As shown in Figure 2, a calibrated iron–constantan thermocouple was embedded in the sample. The leads from the heater and the thermocouple wire were brought through a carefully fitted opening in a scatter shield which had been modified for our purposes. Temperatures were read on a Leeds and Northrup thermocouple potentiometer. The current to the heater was controlled by a variable transformer and read on an ammeter. An infrared lamp was placed so that it shone directly on the sample from above to mini-mize the heat loss.

A second thermocouple placed in contact with the salt surface gave readings several degrees lower than the thermocouple which was placed in the sample. Repeated measure-ments indicated that this difference was essentially constant. Because of the experimental difficulties involved no attempt was made to measure the temperature of the surface while the spectra were being taken; however, the temperature given by the thermo-couple in the sample was sufficiently above the transition point that there could be no doubt that the surface layers were also above this point. It was necessary to take the spectra to be used for the determination of lattice constants at a single temperature to avoid undesirable irregularities arising from thermal expansion. A value of 253°C was chosen since at this temperature it had been possible to obtain a good spectrum for NH_4Cl (I), which has the highest transition point in the system.

For the work on lattice constants $\frac{1}{2}$ hr was allowed to ensure temperature equilibra-tion and completeness of the transition before beginning to take the spectra. That this was an unnecessary precaution was demonstrated by the time–temperature study which will be discussed subsequently.

Particular attention must be given to the condition of the surface in the high-temperature work. As the sample is heated it tends to expand and harden about air pockets. Once this has taken place, and the surface has been flattened and smoothed again, it remains flat and capable of giving good spectra for a considerable period of time.

Figure 2. Heating apparatus used
with diffractometer.

For the work at elevated temperatures reported here, the surface was smoothed just previous to the taking of the spectra.

In the time–temperature study, the behavior of the peaks in the neighborhood of 2θ values of 33° and 27° was followed closely: these are the areas in which are found the most intense peaks for the low- and high-temperature forms, respectively. The temperature was increased gradually from well below the transition point until there was evidence of reduction in intensity and sharpness of the peak at 33°. The temperature then was held steady and the changes in the low-temperature peak at 27° were noted. When there ceased to be further significant change the temperature was increased again.

The lines of the diffraction spectra were indexed according to the reciprocal lattice procedure given by Klug and Alexander.[16] Lattice constants and densities were then calculated.

RESULTS

The intensity and width of the peaks in the low-temperature form (II) of the solutions were entirely comparable to those of the pure salts. The lines were indexed independently and found to have the *hkl* values expected for the simple cubic crystalline form. By using a lower scale factor than had been used for form II, satisfactory spectra for form I could be obtained at elevated temperatures although these peaks were generally broader than would be expected at room temperature. Indexing of the lines of these spectra gave *hkl* values which indicated the existence of a face-centered crystal lattice.

The lattice constants (in Å) which were obtained in this study for NH_4Cl (II) and NH_4Br (II) are as follows (high-angle constant first, low-angle constant second): 3.878 ± 0.001, 3.879 ± 0.002; 4.064 ± 0.003, 4.056 ± 0.005. Selected values found by previous investigators are given in Table II. The agreement between the high-angle data and the literature values is good in view of the fact that this study was not designed to give precision lattice constants. High-angle here refers to 2θ values in the range 75–90°. The agreement is less good for those values obtained from low-angle lines; this is to be expected. These were calculated in order to have a basis of comparison with the high-temperature work.

The lattice constants (in Å) obtained for form I, 6.533 ± 0.023 for the chloride and 6.790 ± 0.026 for the bromide, may be compared with the values of Bartlett and Langmuir,[13] 6.533 and 6.90 for the chloride and bromide, respectively. In the light of the uncertainties in the actual elevated temperatures in both instances here, this agreement is satisfactory.

The variation of lattice constant with composition for the high-angle, low-temperature data is shown in Figure 3. There is negative deviation at both ends of the composition range and slight positive deviation elsewhere. Numerical values for these deviations are listed in Table III. Anselmo and Smith[14] found a similar trend at the NH_4Cl-rich end but had not investigated the NH_4Cl-poor region in sufficient detail to reveal the deviation. This is in contrast to the work of Havighurst, Mack and Blake,[11] who had found negative deviation for one solution of mole fraction of NH_4Cl equal to 0.345.

In an effort to determine whether possibly the molar volume is more nearly linear with composition than is lattice constant, this relationship was plotted and is illustrated also in Figure 3. The data are given in Table III. The trend shown by the lattice constants is evident here in the end regions, but the linear relationship is followed more closely elsewhere.

In Figure 4 the above relationships are illustrated for the low-temperature, low-angle data. Though the general pattern resembles that for the high-angle data, the positive

Table II. Selected Lattice Constants Found by Previous Investigators

Investigators	NH₄Cl (II) Å	NH₄Br (II) Å
Bartlett and Langmuir[13]	3.843	3.968
Havighurst, Mack, and Blake[10]	3.863	4.043
Anselmo and Smith[14]	3.875	4.059
Swanson and co-workers[17,18]	3.8756	4.0594

Table III. Deviations from Linearity

Mole fraction NH₄Cl	Lattice constant ($\times 10^3$ Å)			Unit cell volume ($\times 10^3$ Å³)			Density ($\times 10^3$ g/cm³)		
	A	B	C	A	B	C	A	B	C
0.083	− 6	2	37	− 305	54	5058	20	7	− 25
0.152	0	3	64	− 39	93	8671	17	12	− 43
0.272	4	10	42	92	416	5450	20	9	− 16
0.383	1	17	44	− 39	716	5695	31	5	− 12
0.518	4	12	24	102	459	2890	29	16	7
0.590	4	12	32	76	493	3883	25	14	2
0.710	6	15	25	197	606	3078	21	8	4
0.801	− 1	− 2	23	− 112	− 162	2774	25	25	2
0.885	− 4	− 4	11	− 242	− 236	1218	20	20	4
0.939	− 13	− 9	20	− 627	− 420	2548	25	19	− 6

A. Calculated from low-temperature, high-angle data.
B. Calculated from low-temperature, low-angle data.
C. Calculated from high-temperature data.

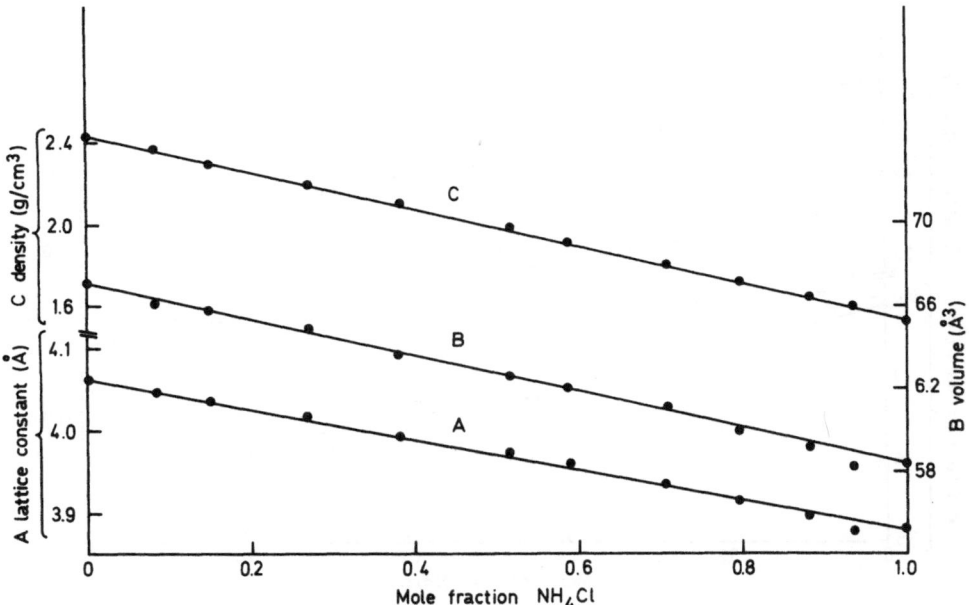

Figure 3. Variation of composition of solid solution with (A) lattice constant (Å), (B) unit cell volume (Å³), and (C) density (g/cm³); low-temperature, high-angle data only.

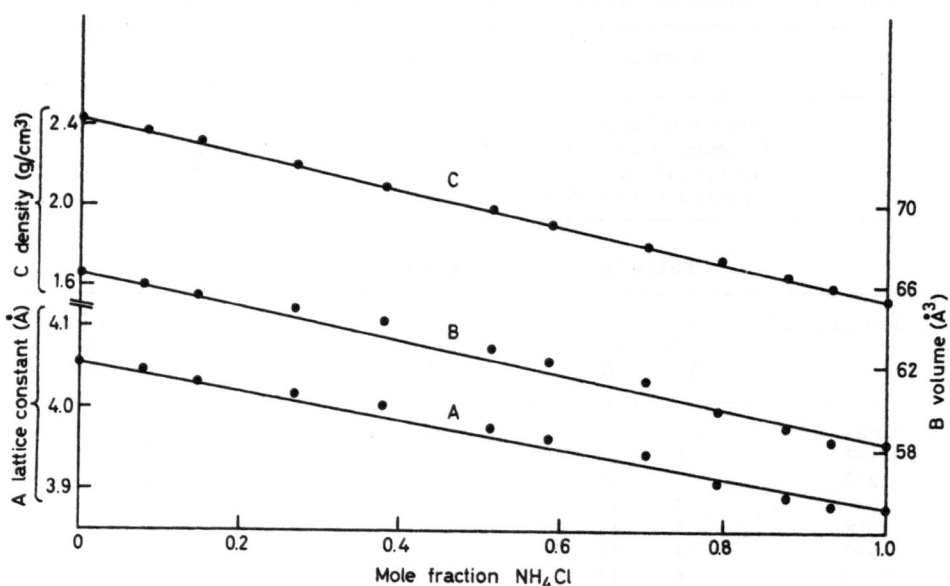

Figure 4. Variation of composition of solid solution with (A) lattice constant, (B) unit cell volume and (C) density; low-temperature, low-angle data only.

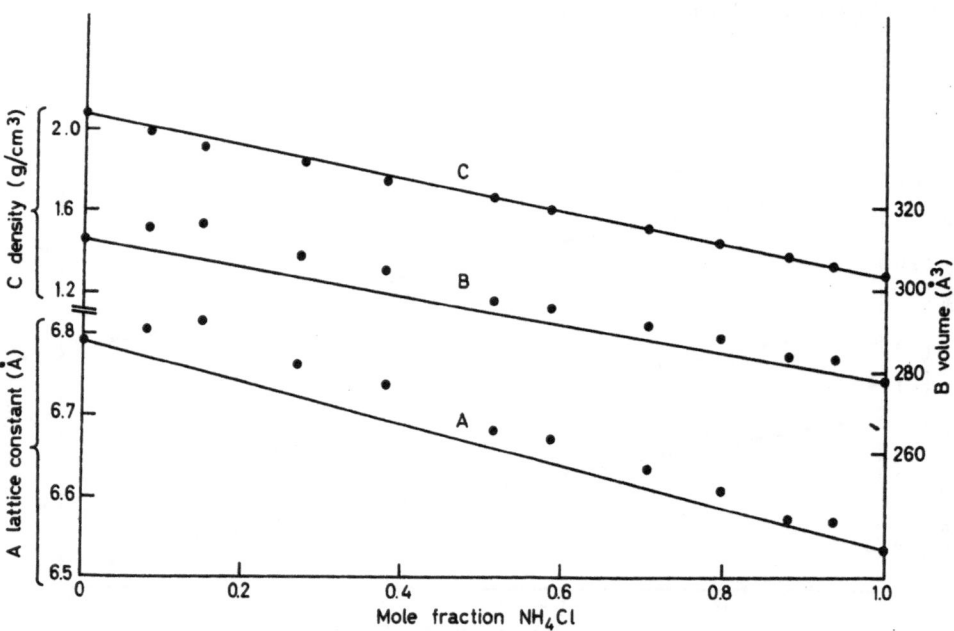

Figure 5. Variation of composition of solid solution with (A) lattice constant, (B) unit cell volume, and (C) density; high-temperature data only.

deviation is more pronounced in the curve for lattice constants. This causes the dip in the NH_4Cl-poor region to be at or above the straight line. The volume–composition relation here appears to be almost identical with the lattice-constant–composition curve.

Figure 5 gives the graphs for the high-temperature data. A pronounced positive deviation from linearity is everywhere observed; there is indication of even greater positive deviation at the ends of the diagram. The magnitudes of these deviations are listed in Table III.

The densities of forms I and II, as calculated from the lattice constants, are plotted as a function of composition as the upper curves of Figures 2, 3, and 4. In all three, the relation appears to be smoother and more nearly additive than either of the other relations. The actual deviations from additivity are given in Table III.

We may compare, first, the changes in density accompanying the transition for the pure components with available literature data and, secondly, the changes in density of the solutions with the changes in density of the pure components. These data are listed in Table IV. The changes in density reported by Bartlett and Langmuir[13] represent much the same kind of data, for they are considering the difference in the density of II at room temperature and I at about 250°C. Bridgman's[1] data were obtained from his monumental work on transitions at high pressure; the data of Poyhönen[2] refer to the change in density accompanying the transition, measured in the neighborhood of the transition point. Also, these last were obtained from dilatometric studies rather than from X-ray work.

One of the principal objects of this work was the elucidation of the mechanism of the transition process for the solid solutions. An examination of equilibrium diagrams for such transitions (cf. Figure 1.) shows that the heating of the low-temperature form of any solid solution (except one with a composition corresponding to the minimum in the curves) would result in the following sequence: As the lenticular region of the diagram is entered, the high-temperature solid solution (I) first appearing has a composition different from that of the material from which it is formed (II). As the temperature is slowly raised, I increases in quantity and II decreases, but the composition of both changes along the equilibrium solidus lines. At any one temperature all the material present in form I has a uniform composition and all of the material present in form II has a uniform composition, but these two compositions are different. When, with sufficient increase in temperature, the lenticular area has been crossed, all of form II has disappeared and nothing but form

Table IV. Density Changes Accompanying Transition

Study	Pure components		Solutions	
	% Change		Mole fraction	%
	NH_4Cl	NH_4Br	NH_4Cl	Change
This study	16.3	14.7	0.083	16.1
Bartlett and Langmuir[13]	15.7	20.8	0.152	17.2
Bridgman[1]	13	14	0.272	15.2
Poyhönen[2]	16.2	15.9	0.383	15.9
			0.518	15.7
			0.590	16.0
			0.710	15.9
			0.801	17.0
			0.885	16.8
			0.939	17.5

I, of composition corresponding to that of the original form II, remains. The above description applies only if equilibrium obtains at each temperature. The accompanying concentration changes and homogenization processes require rapid and extensive diffusion in the solid state, and should be revealed by a shifting of peak positions in the X-ray patterns caused by phase concentration changes, in addition to the shifts from mere lattice expansion. It was our desire to determine to what extent such equilibration does actually occur, and the following studies were made with this in mind.

An examination in the neighborhood of the high-temperature peak, 2θ equal to 27°, as soon as the temperature had reached 253°C showed that the high-temperature form was already present and that its position, intensity, and sharpness were the same at this time as they were after the temperature had been held at this level for a long period of time. In another instance, the heater was turned off and the spectrum taken immediately; the peak locations indicated that reversion to the low-temperature form had already occurred, although there were very slight shifts for some lines. These slight differences may be accounted for by the fact that the sample was cooling.

The time–temperature study described earlier was carried out for pure NH_4Br and for solutions of mole fraction of NH_4Cl equal to 0.5 and 0.8. As Figure 1 shows, at a mole fraction of 0.5 the two solidus lines are very close; at 0.8 they are more widely separated. In all three cases the low-temperature peaks tended to split as the temperature increased; they then became broad and ill-defined and very much reduced in intensity. Exactly the opposite changes occurred with the high-temperature peaks. They started out as broad, ill-defined ones and eventually, as the low-temperature peaks became more and more insignificant, the high-temperature ones developed in sharpness and intensity. Since the behavior of the solutions was identical to that of pure NH_4Br, where no separation into solutions of different composition is possible, it appears that no composition shift does occur as the system undergoes the transition. When changes had apparently ceased in the transition temperature region, the temperature was increased further. Only when the sample was somewhat above the transition region did the low-temperature form disappear completely. This behavior is in line with the results obtained by Wood, Secunda and McBride[15] for CsCl-CsBr solids solutions. They found a range of about 80° separating temperatures at which only one form could be detected. Within this range both forms were apparent. These authors also have evidence for a transition with no composition shift.

ACKNOWLEDGMENT

We are indebted to Dr. Luis Roldan and Mr. Forrest Rahl of Allied Chemical Corporation, Morristown, New Jersey, for suggesting the design of the sample heater.

REFERENCES

1. P. W. Bridgman, "Polymorphism at High Pressures," *Proc. Am. Acad. Arts. Sci.* **52**: 91, 1916.
2. J. Poyhönen, "Dilatometric Investigation of NH_4Cl at 183.1°C and NH_4Br at 137.2°C," *Ann. Acad. Sci. Fennicae, Ser. A VI*, No. 58, 1960.
3. R. Flatt and G. Burkhardt, "Investigation of Solid Solution Formation in Solutions," *Helv. Chim. Acta* **27**: 1605, 1944.
4. H. Rassow, "A Simple Method for the Determination of Melting Points and Critical Temperatures," *Z. anorg. u. allgem. Chem.* **114**: 117, 1920.
5. P. S. Costich, G. J. Maass, Jr., and N. O. Smith, "Transitions in Ammonium Chloride–Ammonium Bromide Solid Solutions," *J. Chem. Eng. Data* **8**: 26, 1963.
6. L. Vegard, "The Constitution of Mixed Crystals and the Space Occupied by Atoms," *Z. Physik* **5**: 17, 1921.

7. H. G. Grimm and K. F. Herzfeld, "The Lattice Energy and Dimensions of Mixed Crystals," *Z. Physik* **16**: 77, 1923.

8. J. W. Retgers, "The Specific Gravity of Isomorphic Mixtures," *Z. physik. Chem.* **3**: 497, 1889.

9. E. R. Jette, "Intermetallic Solid Solutions," *Trans. Am. Inst. Mining Met. Engrs.* **111**: 75, 1934.

10. R. J. Havighurst, E. Mack, Jr., and F. C. Blake, "Precision Crystal Measurements on Some Alkali and Ammonium Halides," *J. Am. Chem. Soc.* **46**: 2368, 1924.

11. R. J. Havighurst, E. Mack, Jr., and F. C. Blake, "Solid Solutions of the Alkali and Ammonium Halides," *J. Am. Chem. Soc.* **47**: 29, 1925.

12. E. Zen, "Validity of 'Vegard's Law'," *Am. Mineralogist* **41**: 523, 1956.

13. G. Bartlett and I. Langmuir, "The Crystal Structures of the Ammonium Halides Above and Below the Transition Temperatures," *J. Am. Chem. Soc.* **43**: 84, 1921.

14. V. C. Anselmo and N. O. Smith, "Lattice Constants of Ammonium Chloride–Ammonium Bromide Solid Solutions, *J. Phys. Chem.* **63**: 1344, 1959.

15. L. J. Wood, W. Secunda, and C. M. McBride, "Reactions Between Dry Inorganic Salts. IX. The Effect of Common Ions on the Transition Temperature of Cesium Chloride," *J. Am. Chem. Soc.* **80**: 307, 1958.

16. H. P. Klug and L. E. Alexander, *X-Ray Diffraction Procedures*, John Wiley & Sons, Inc., New York, 1954, p. 340.

17. H. E. Swanson and E. Tatge, "Standard X-Ray Diffraction Powder Patterns," NBS Circular No. 539, Vol. I: 59, 1953.

18. H. E. Swanson and R. Fuyat, "Standard X-Ray Diffraction Powder Patterns," NBS Circular No. 539, Vol. II: 49, 1953.

DISCUSSION

E. F. Sturcken (E. I. du Pont de Nemours & Co.): Sister, does this mean that the phase diagram showing these differences in composition is really not correct? That you really shouldn't draw it that way?

Sister J. E. Callanan: I wouldn't like to say it's not correct from other standpoints. I hadn't really thought about it from that standpoint. It's just that the X-ray did not show any evidence of such a transition. But if there is a composition shift it is not detectable, and we feel that we were able to heat the homogeneous solution through the transition without this composition shift occurring, which would seem to indicate what you say, that there is a discrepancy there as far as the phase diagram is concerned.

Chairman A. F. Berndt: Might this be explained merely by the fact that we don't have equilibrium when you do the seeding?

Sister J. E. Callanan: This could be and I think certainly that is very true. The phase diagram is an equilibrium diagram and in view of the behavior of the solid state it is very likely that equilibrium does not obtain as the transition occurs.

Chairman A. F. Berndt: I would think that you have shown this to be true by your measurements.

K. Aykan (E. I. du Pont de Nemours & Co.): If I understand you correctly, in the solid solution region there is a temperature region where you have low temperature and high temperature forms present. Am I correct in assuming that these concentration ratios do not change if you hold it longer at that temperature?

Sister J. E. Callanan: I'm sorry, I didn't quite get the question.

K. Aykan: For example, there was a temperature region at which I understand the low- and high-temperature forms were present. Does this say that this transition is a martensitic type? Like that discovered in zirconium dioxide, for example? So that it reaches a diffusion transformation, thus you can have low- and high-temperature forms present, a transformation independent of the time?

Sister J. E. Callanan: We do have a region in which both forms are present. If there is a composition shift occurring it is not detectable. Now the latter part of the question I didn't hear.

K. Aykan: If you hold at a longer time the ratios do not change at all?

Sister J. E. Callanan: No, they don't. We know that at this temperature at which the transition was occurring for a period of several hours. It just reached a point where no further change was occurring at all.

K. Aykan: What I was referring to was the zirconium dioxide transformation. There is a temperature interval where the monoclinic and tetragonal forms can be present together. This is independent of the time at which one holds at that temperature, so this is referred to as a martensitic type transformation, which is diffusionless. I was trying to see whether there was also correlation at this transformation.

Sister J. E. Callanan: That may very well be so. I am not familiar with the particular item you are mentioning, but certainly from the standpoint of the fact that there was no further change with time in this region, that very definitely is so.

ROTATIONAL POLYMORPHISM OF METHYL-SUBSTITUTED AMMONIUM PERCHLORATES

M. Stammler, R. Bruenner, W. Schmidt, and D. Orcutt

Aerojet-General Corporation
Sacramento, California

ABSTRACT

The thermal transformations which take place in solid methyl-substituted ammonium perchlorates have been studied using high-temperature X-ray diffraction and differential thermal analysis techniques. In the temperature range from 20°C to their decomposition temperature (above 300°C), ammonium perchlorate and tetramethyl ammonium perchlorate undergo only one enantiomorphic phase transition, namely at 240 and 340°C (with decomposition), respectively. This I–II transition is ascribed to the beginning of the free rotation of the ClO_4^- ions. The rotation of the cations, however, begins below room temperature. If the symmetry of the cation is lowered by having both methyl groups and hydrogens arranged around the nitrogen (as in monomethyl, dimethyl, and trimethyl ammonium perchlorates), there is an additional enantiomorphic phase transition. This II–III transformation is ascribed to the rotation of the cations which have, in the partially substituted ions, two sets of non-equivalent symmetry axes (different moments of inertia). The temperatures of transformation are discussed in terms of the space requirements for rotation. Symmetries and cell dimensions of some modifications were determined.

INTRODUCTION

High-temperature modifications of perchlorate salts of ammonium, tetramethyl ammonium, the alkaline metals, silver and thallium were first reported by Vorländer and Kaascht.[1] These salts have only one enantiotropic phase transition. For the alkaline perchlorates, the transition temperatures decrease with increasing atomic weight of the metal. An exception is lithium perchlorate which does not have a high-temperature modification. The transition temperature of ammonium perchlorate falls between rubidium and cesium perchlorate (Table I).

X-ray structure analysis of the room temperature modifications of some of the above mentioned perchlorates were made by Büssem and Herrmann,[2] who showed that the alkaline perchlorates, ammonium and thallous perchlorate, are isomorphic and crystallize in the orthorhombic system. The cubic high-temperature modifications of these perchlorates were analyzed by Braekken and Harang[3] and by Herrmann and Ilge[4] approximately at the same time. It was found that the high-temperature modifications are of cubic symmetry with a rock-salt type structure. Similar observations were made with other salts such as nitrates, fluoborates, and ammonium halides. The symmetry of the polymorphic phases in general increases with increasing temperature.

[1] References are at the end of the paper.

Table I. Transition Temperatures of Alkaline Perchlorates

Compound	Transition temperature (°C)	Atomic weight of metal
LiClO$_4$	—	—
NaClO$_4$	308	23
KClO$_4$	299–300	39
RbClO$_4$	279	85.5
CsClO$_4$	219	133
NH$_4$ClO$_4$	240	—

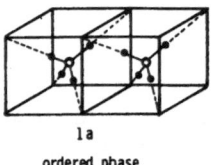
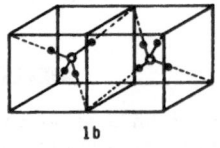

1 a
ordered phase

1 b
disordered phase

o = Nitrogen
• = Hydrogen

Figure 1. The orientation of the NH$_4^+$ ions in the ordered and disordered phase.

Two attempts were made to explain these observations: In the first, Pauling[5] attributed similar transitions to changes from the state in which most of the molecules are oscillating to that in which most of them are rotating. In the case of complex ions (AX_n) such as NH$_4^+$, ClO$_4^-$, BF$_4^-$, and MnO$_4^-$, it causes a substantial symmetry increase, since the initially nonequivalent X-atoms (with regard to lattice positions) become equivalent through rotation and the ion assumes spherical symmetry. This theory accounts for the symmetry increase of the high-temperature phases. However, as Eucken[6] pointed out, the above theory does not explain the abnormally high values of the heats of rotation $(C_R)^*$ above the II–III transition temperature of ammonium chloride. From the temperature function of the elasticity constants determined by Lawson,[7] the specific heat C_p of NH$_4$Cl just above the transition was found to be 19 cal/mole-°C from which a C_R value of 6.66 cal/mole is derived, thus eliminating free rotation, for which only 3 cal/mole are required. This result favours a theory suggested by Frenkel[8] which assumes a change in the orientation of the ammonium ions (order–disorder transition). In the ordered and in the disordered phase, the ammonium ions oscillate around an equilibrium orientation. In the ordered phase, the tetrahedral axes of oscillation of the ammonium ions, which coincide with the space diagonal of the elementary cell, are all parallel (Figure 1a). In the disordered phase, the two orientations (Figure 1b) appear at random in the crystal. Frenkel now assumes that below the transition temperature only the "ordered" phase exists, while random orientation exists above the transition temperature. Both theories have been repeatedly discussed in conjunction with phase transitions of various compounds, and it appears that both types of transitions exist.[9]

During this investigation, the phase transitions of a series of perchlorates were studied in which the symmetry of the cations was systematically changed by the substitution of methyl groups onto the ammonium ion.

* The heat of rotation C_R is defined as follows: $C_R = C_V - (C_M + C_S)$, where C_V is the specific heat at constant volume, C_M is the contribution of the lattice vibration to C_V and C_S is the contribution of the internal molecular vibration to C_V.

**Table II. Results of Tests for Handling Safety
on a Bureau of Mines Impact Machine**

Compound	Test* results (cm)
RDX	33
NH_4ClO_4	100
$MeNH_3ClO_4$	20
$(Me)_2NH_2ClO_4$	22
$(Me)_3NHClO_4$	25
$(Me)_4NClO_4$	35
$Et(AP)_2 \cdot \frac{1}{2}H_2O$	35

* 50% fire point with a 2 kg weight.

EXPERIMENTAL TECHNIQUES

Preparation of Compounds

The perchlorate salts of the substituted ammonium ions were prepared by neutralization of the free base with perchloric acid. The neutralization was performed in a round-bottom flask equipped with a dropping funnel, thermometer and teflon-coated magnetic stirring bar. The reaction flask was in an ice bath, and the rate of addition of perchloric acid was controlled so that the temperature did not rise above 25°C.

The methyl amine was a 40% water solution and was neutralized with concentrated perchloric acid. Dimethyl amine, trimethyl amine, and ethylene diamine were added to isopropyl alcohol to make approximately a 25% solution which was then neutralized with $6 N$ perchloric acid. The tetramethyl ammonium perchlorate was prepared by neutralizing a 10% water solution of tetramethyl ammonium hydroxide. After neutralization, the volume of solvent was reduced by evaporation in a rotary evaporator. The perchlorate salt was then allowed to crystallize out over a 24 hr period at 0°C.

All salts were recrystallized twice from isopropyl alcohol and dried over $CaSO_4$ in a vacuum desiccator. (With the ethylene diamine diperchlorate, this procedure yields the hemihydrate. To dehydrate this material at atmospheric pressure, it is necessary to heat it between 120 to 130°C.) Before drying the bulk of the material, small samples were dried and tested for handling safety on a Bureau of Mines impact machine. Table II lists the results of these tests and compares them to trimethylenetrinitramine (RDX).

High-Temperature Density Determination

The experimentally determined densities reported herein were obtained by conditioning the sample at the desired temperature and then weighing the solid immersed in a silicon oil of the same temperature. The temperature function of the density of the silicon oil was determined using an Invar metal bar which has a low thermal expansion coefficient. The sample holder was a basket made of 325 mesh stainless steel screen. Two to four determinations were made for each value reported. In some cases there was considerable scatter in the data which was mainly attributed to gas occlusions and voids within the crystals. Another source of error may have been a slight solubility of the compounds in the silicon oil particularly at elevated temperatures. Although all values determined by this method are too low, they indicate the approximate density of the materials.

High-Temperature X-Ray Diffraction Technique

All X-ray samples were run on a GE XRD-5 with nickel-filtered copper radiation or vanadium-filtered chromium radiation. A 3° dispersion slit, medium resolution collimator, and 0.2° receiving slit were standard for all runs. To obtain temperatures other than ambient, a 3-in.-long sample holder was utilized backed by coiled Nichrome wire which was in turn backed by aluminum foil and wire screen. The temperature was monitored by a millivolt potentiometer with an iron–constantan thermocouple inserted into the sample at one edge. Current was supplied to the Nichrome through a Variac transformer and held for several minutes before commencing a scan. The sample was ground to approximately 200 mesh and then carefully placed in the sample holder to hold any preferred orientation to a minimum.

Thermal Analysis

The differential thermal patterns were obtained using a DuPont 900 Differential Analyzer. Approximately 20 mg samples were used for each experiment. The heating rate employed was 20–30°C/min. The rate of the temperature drop, while taking the cooling curves, was uncontrolled and rather rapid. In some cases, the thermal stability of the compounds was tested using thermogravimetric techniques. The perchlorates were kept at temperatures close to their melting points for prolonged periods of time. Decomposition was measured by determination of the weight loss as a function of time.

RESULTS

During this investigation the polymorphic phase transition of ammonium perchlorate (AP), monomethyl ammonium perchlorate, dimethyl ammonium perchlorate, trimethyl ammonium perchlorate, and tetramethyl ammonium perchlorate were studied using the above outlined techniques. The only difunctional salt included in this study was ethylene diammonium diperchlorate. In each case, the modification stable at the highest temperature is identified as Phase I, the modification stable at the next lowest temperature is Phase II, etc.

Ammonium Perchlorate (AP)

Commercially available AP which was twice recrystallized from methyl alcohol was used for the experiments performed with this compound. Since the properties of this material are well known, it was used as a standard in some experiments [such as for the differential thermal analysis (DTA)]. The pertinent properties of AP are briefly reviewed. AP exists in three modifications, a low-temperature form which is stable below −190°C (Phase III),[10] the orthorhombic modification which is stable from −190°C to +240°C (Phase II), and the cubic high-temperature modification (Phase I) which is stable above 240°C.

$$\text{(III)} \underset{\text{Unknown}}{\text{NH}_4\text{ClO}_4} \overset{-190°C}{\rightleftharpoons} \text{(II)} \underset{\text{Orthorhombic}}{\text{NH}_4\text{ClO}_4} \overset{240°C}{\rightleftharpoons} \text{(I)} \underset{\text{Cubic}}{\text{NH}_4\text{ClO}_4}$$

Phase II has a $BaSO_4$ type structure in which the NH_4^+ ions are rotating. The modification which is stable above 240°C has rock-salt type structure which is believed to result from the free rotation of the ClO_4^- ion. The density of the orthorhombic form (Phase II) at 25°C is 1.95 g/cm³ while the density of the cubic modification (Phase I) is 1.71 g/cm³.

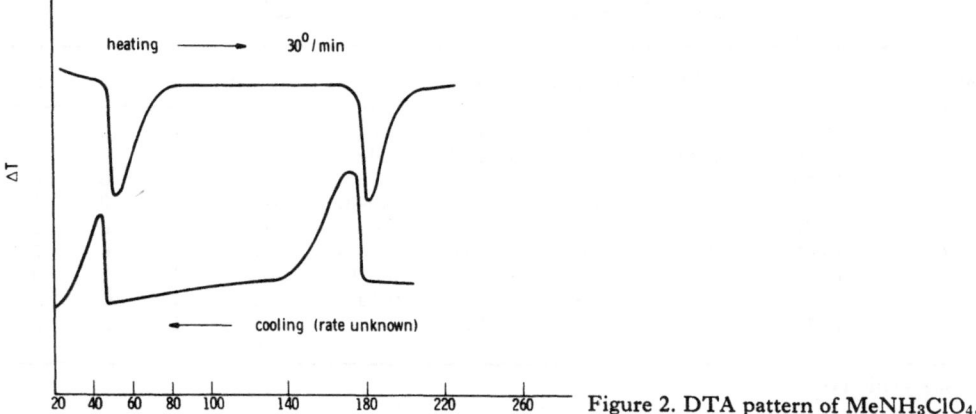

Figure 2. DTA pattern of MeNH₃ClO₄.

Table III. X-Ray Diffraction Pattern of MeNH₃ · ClO₄ at 200°C, Cu K_α Radiation; Phase I (Cubic)

No.	$\theta_{Cu\,K_\alpha}$	d (Å)	Relative intensity	$\sin^2\theta_{exp}$*	$\sin^2\theta_{calc}$†	$\Delta \times 10^{-4}$‡	h	k	l
1	8.61	5.150	71.5	0.0224	0.0222	−2	1	0	0
2	12.20	3.640	100	0.0447	0.0444	−3	1	1	0
3	14.90	2.996	5.7	0.0661	0.0666	+5	1	1	1
4	17.35	2.580	4.5	0.0889	0.0888	−1	2	0	0
5	19.40	2.319	1.2	0.1105	0.1110	+5	2	1	0
							2	0	1
							1	2	0
							1	0	2
6	21.35	2.110	1.1	0.1327	0.1332	+5	2	1	1
							1	2	1
							1	1	2

* Experimental values.
† Calculated values.
‡ $\Delta \times 10^{-4}$ represents the differences between $\sin^2\theta_{exp}$ and $\sin^2\theta_{calc}$. The negative sign results from smaller calculated values, the positive sign from smaller experimental values.

When AP is thermally decomposed by applying heat at a rate of 20°C/min (or higher) deflagration occurs, without melting, at approximately 460°C. Petricciani et al.[11] have shown that Phase I decomposes if small amounts of $KClO_3$ are in the AP lattice, becoming completely unstable if the concentration of $KClO_3$ exceeds 0.08%. $KClO_3$ has little, if any, effect on the decomposition of Phase II.

Monomethyl Ammonium Perchlorate (MeNH₃ · ClO₄)

The thermogram of MeNH₃ · ClO₄ obtained between room temperature and the decomposition temperature (330°C) is shown in Figure 2. The heating and cooling curves are marked by arrows pointing toward right or left, respectively. The two endothermic reactions (heating curve) beginning at 48°C and 178°C are reversible, as shown by the cooling curve (Figure 2). The endotherm (heating curve) beginning at 255°C (Figure 4)

Table IV. X-Ray Diffraction Pattern of $MeNH_3 \cdot NH_3 \cdot ClO_4$ at 77°C, Cu K_α Radiation; Phase II (Tetragonal)

No.	$\theta_{Cu\,K_a}$	d (Å)	Intensity	$\sin^2 \theta_{exp}$	$\sin^2 \theta_{calc}$	$\Delta \times 10^{-4*}$	h	k	l
1	4.81	9.205	54	0.0070	0.0069	−1	0	0	1
2	9.50	4.670	100	0.0272	0.0276	+4	0	0	0
3	11.56	3.850	45	0.0401	0.0398	−3	1	0	0
4	12.50	3.550	15.5	0.0469	0.0467	−2	1	0	1
5	14.38	3.100	7.8	0.0617	0.0620	+3	0	0	3
6	15.05	2.970	14.0	0.0674	0.0674	0	1	0	2
7	16.45	2.720	9.0	0.0802	0.0796	−6	1	1	0
8	18.59	2.410	2.5	0.1017	0.1018	+1	1	0	3
9	19.15	2.350	4.1	0.1077	0.1072	−5	1	1	2

* See Table III.

is due to the melting of the sample. At 178°C the material undergoes an enantiotropic phase transition from Phase I to Phase II. This transition is related to the I–II transition of AP in that small amounts of ClO_3^- ion in the lattice of $MeNH_3 \cdot ClO_4$ cause Phase I to decompose rapidly while Phase II is unaffected.

The X-ray diffraction pattern of $MeNH_3 \cdot ClO_4$ (Phase I) was taken at 200°C. The reflections are listed in Table III and are indexed using a cubic primitive unit cell:

Cubic primitive . . . . . $a_0 = 5.18$ Å
Volume of elementary cell . . . $V = a^3 = 138$ Å^3
Number of moles per unit cell . . $Z = 1$
X-ray density . . . . . $d_x = 1.58$ g/cm^3 (at 200°C)
Determined density . . . . $d_p = 1.52$ g/cm^3 (at 187°C)
Molecular weight 131.5

Below 178°C, the X-ray pattern of $MeNH_3 \cdot ClO_4$ changes drastically. The reflections obtained at 77°C are listed in Table IV and are indexed using a tetragonal unit cell (Phase II):

Tetragonal . . . . . . $c/a = 2.40$; $a_0 = 3.87$ Å; $c_0 = 9.28$ Å
Volume of elementary cell . . . $V = a^2 \cdot c = 138$ Å^3
Number of moles per cell . . . $Z = 1$
X-ray density . . . . . $d_x = 1.58$ g/cm^3 (at 77°C)
Determined density . . . . $d_p = 1.56$ g/cm^3 (at 72°C)
Molecular weight 131.5

It is worth mentioning that during this I → II transition, the cell volume remains constant within the experimental error. Since the number of formula units per elementary cell remains constant also during this transition, Phase II may be considered as a tetragonally deformed Phase I.

The Phase III stable below 48°C is of monoclinic or triclinic symmetry. Based on the assumption that the size and bond angles of CH_3 and NH_2 groups are very similar, and that the perchlorates are isomorphous with fluoborates,

$[CH_3–NH_3]^+$ $[NH_2–NH_3]^+$
Monomethyl ammonium ion   Hydrazinium ion

Table V. Reflections and Intensities of Phase III of $MeNH_3 \cdot ClO_4$ and of $NH_2-NH_2 \cdot HBF_4$[12]

| No. | Phase III | | $NH_2-NH_3 \cdot BF_4$[11] | | | | | |
	d (Å)	Intensity	d_{obs}	d_{calc}	Intensity	h	k	l
1	9.817	16.0	—	—				
2	9.16	38.0	—	—				
3	8.838	75.0	—	—				
4	6.146	3.8	—	—				
5	5.824	1.1	5.78	5.74	W	0	0	2
6	5.50	1.1	4.89	4.91	S	2	0	0
7	5.368	3.2	4.04	4.06	W	1	1	2
8	5.241	5.4	3.61	3.62	MW	3	0	3
9	5.092	2.7	3.48	3.46	VS	4	0	1
10	4.92	24.0	3.35	3.35	VS	3	1	2
11	4.85	19.0	3.24	3.24	M	3	1	1
12	4.56	13.5	3.09	3.09	M	2	2	2
13	4.436	100.0	2.98	2.99	M	3	1	3
14	4.33	15.0	2.86	2.86	W	0	0	4
15	4.227	2.7	2.72	2.70	MS	4	1	1
16	4.13	4.3	2.65	2.66	MW	$\bar{4}$	1	1
				2.65				
17	3.81	12.5	2.46	2.44	S	4	0	2
18	3.72	19.0						

an attempt was made to index Phase III using the monoclinic unit cell of hydrazinium fluoborate.[12,13] This attempt failed; the unit cell of the methyl derivative is much larger than the unit cell of the hydrazinium salt. The d spacings of Phase III are listed in Table V. This table also contains the reflections and indices of hydrazinium fluoborate as reported earlier[12] in order to show the difference of both patterns.

That these compounds are not isomorphous may be due to the hydrazinium ion having two equivalent nitrogens and the hydrogens around these nitrogens being able to form hydrogen bridges. The proton from the acid cannot be assigned to one nitrogen of the hydrazinium ion.[13] This is different for the methyl ammonium ion, where the acid proton is localized and the tighter carbon-hydrogen bond does not permit hydrogen bridging to any extent. The phases observed between room temperature and the decomposition temperature of $MeNH_3 \cdot ClO_4$ are summarized in the following schematic:

$$\text{Phase III} \underset{\approx \text{ Monoclinic}}{\overset{48°C}{\rightleftharpoons}} \text{Phase II} \underset{\text{Tetragonal}}{\overset{178°C}{\rightleftharpoons}} \text{Phase I} \underset{\text{Cubic}}{\overset{255°C}{\rightleftharpoons}} \text{Melt}$$

Above 255°C slow decomposition begins. No weight loss was observed by heating the material at 240°C over a 90-minute period.

Dimethylammonium Perchlorate ($Me_2NH_2 \cdot ClO_4$)

Two reversible endothermic phase transitions were observed in the differential thermogram of $Me_2NH_2 \cdot ClO_4$ (Figure 3). The endotherm occurring at 180°C is due to the melting of the sample. Phase I is stable from 180°C down to 38°C, while below 38°C Phase II is the stable modification. Above approximately 300°C, decomposition becomes

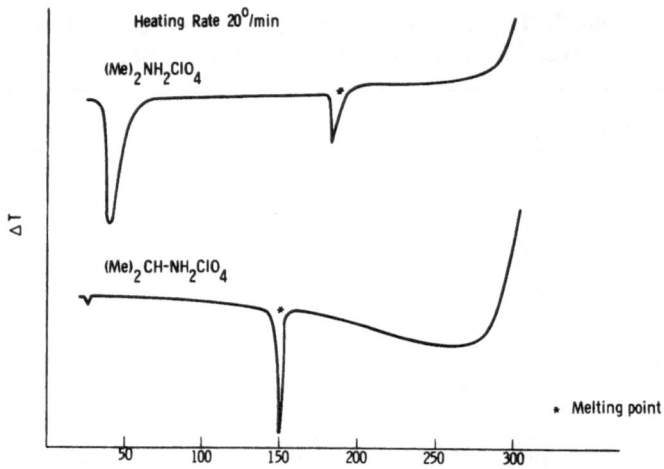

Figure 3. DTA patterns of (Me)₂NH₂ClO₄ and (Me)₂CH-NH₂ClO₄.

Table VI. X-Ray Diffraction Pattern of Me₂NH₂ · ClO₄ at 45°C, Cu K_α Radiation; Phase I (Tetragonal)

No.	$\theta_{Cu\,K_a}$	d (Å)	Intensity	$\sin^2 \theta_{exp}$	$\sin^2 \theta_{calc}$	$\Delta \times 10^{-4*}$	h	k	l
1	8.13	5.44	90	0.0200	0.0201	+1	1	0	0
2	11.60	3.84	100	0.0404	0.0402	−2	1	1	0
3	14.15	3.155	8	0.0597	0.0591	−6	1	1	2
4	16.45	2.722	12	0.0802	0.0804	+2	2	0	0
5	18.43	2.436	4	0.0999	0.0993	−6	2	0	2
6	20.25	2.227	3	0.1198	0.1994	−4	2	1	2
7	21.88	2.067	1	0.1389	0.1381	−8	1	0	5

* See Table III.

rapid. The X-ray diffraction pattern of Phase I was taken at approximately 45°C. Reflections and indices are listed in Table VI for a tetragonal unit cell:

Tetragonal . . . . $c/a = 2.08$; $a = 5.44$ Å; $c = 11.22$ Å

Volume of elementary cell . . . . $V = a^2 \cdot c = 331.5$ Å³
Number of moles per unit cell . . . $Z = 2$
X-ray density . . . . . $d_x = 1.46$ g/cm³ (at 45°C)
Determined density . . . . $d_p = 1.41$ g/cm³ (at 71°C)
Molecular weight 145.6

The modification of Me₂NH₂ · ClO₄ below 38°C has a fairly complex lattice of monoclinic or lower symmetry. The reflections of this phase (designed as Phase II) are listed in Table VII.

The transitions of Me₂NH₂ · ClO₄ between 25°C and the decomposition temperature are summarized below:

$$\text{Phase II} \underset{\sim \text{ Monoclinic}}{\overset{38°C}{\rightleftharpoons}} \text{Phase I} \underset{\text{Tetragonal}}{\overset{180°C}{\rightleftharpoons}} \text{Melt}$$

Table VII. Reflections and Intensities of Phase II of $Me_2NH_2 \cdot ClO_4$ at 25°C, Cu K_α Radiation

No.	$2\theta_{Cu\,K_\alpha}$	d (Å)	Intensity
1	14.20	6.24	16
2	15.00	5.91	37
3	16.25	5.45	23
4	17.00	5.21	42
5	18.30	4.85	14
6	19.20	4.62	25
7	20.10	4.41	9
8	20.65	4.30	20
9	21.60	4.11	6
10	22.30	3.97	12
11	23.10	3.85	11
12	23.70	3.76	32
13	24.40	3.65	100
14	25.0	3.56	30
15	25.8	3.45	18
16	26.3	3.39	65
17	28.8	3.10	6
18	30.2	2.96	5

It is interesting to note that the crystals of this compound reveal high attractive forces for each other when they are in the tetragonal (Phase I) modification. They can be pressed together to form a waxlike material with a tacky surface. This process is reversible and upon cooling the material below the transition temperature, the brittle state is regained. While the transition from the brittle state to the tacky state is rather rapid, the reverse reaction appears to be slower. The isomorphous dimethyl ammonium fluoborate behaves in a similar manner. Another compound, 1,1 dimethyl hydrazine fluoborate, has also been reported[12] to exhibit tacky properties, which are comparable with those of the dimethyl ammonium fluoborate and perchlorate. This property was attributed[12] to a small CH_3,CH_3 separation which—in the case of the hydrazinium salt—was estimated to about 2.4 Å. Similar reasoning may be applied for the dimethyl ammonium perchlorate and fluoborate; the isopropylammonium perchlorate was also prepared in order to investigate whether the tacky properties are associated with a particular structure of the cation, such as the following:

$$\left[\begin{array}{c} CH_3 \\ > N—NH_3 \\ CH_3 \end{array} \right]^+$$
1, 1 dimethylhydrazinium

$$\left[\begin{array}{c} CH_3 \\ > NH \cdot CH_3 \\ CH_3 \end{array} \right]^+$$
trimethylammonium

$$\left[\begin{array}{c} CH_3 \\ > CH \cdot NH_3 \\ CH_3 \end{array} \right]^+$$
i-propyl ammonium

Table VIII. Diffraction Data of i-Propyl $NH_3 \cdot ClO_4$ at $\sim 30°C$, Cu K_α (Ni) Radiation; Phase I (Tetragonal)

No.	d (Å)	Intensity	$\sin^2 \theta_{exp}$	$\sin^2 \theta_{calc}$	$\Delta \times 10^{-4*}$	h	k	l
1	7.00	100	0.0121	0.0121	0	0	0	1
2	5.20	3	0.0219	0.0218	−1	1	1	0
3	5.05	6	0.0234	0.0230	−4	1	0	1
4	4.20	20	0.0335	0.0339	+4	1	1	1
5	3.68	9	0.0435	0.0436	+1	2	0	0
6	3.23	35	0.0566	0.0557	−9	2	0	1
7	2.96	5	0.0675	0.0666	−9	2	1	1
8	2.88	2	0.0710	0.0702	−8	1	1	2
9	2.60	3	0.0872	0.0872	0	2	2	0
10	2.46	1	0.0979	0.0981	+2	3	0	0
				0.0993				
11	2.39	2	0.1039	0.1037	−2	2	1	2

* See Table III.

This compound undergoes a very low-energy phase transition at 25°C (Figure 3). Above 25°C up to 150°C where melting occurs Phase I is stable. Phase II, stable below 25°C, gives the same X-ray diffraction pattern as Phase I but with different intensities. The X-ray diffraction pattern (Table VIII) taken at 30°C could be indexed using a tetragonal unit cell:

I-propyl ammonium perchlorate at 30°C

Tetragonal	$c/a = 0.951, a = 7.34$ Å; $c = 6.98$ Å
Volume of elementary cell	$V = a^2 \cdot c = 375$ Å^3
Number of moles per unit cell . . .	$Z = 2$
Molecular weight	159.57
X-ray density	$d_x = 1.42$ g/cm^3 (at 30°C)
Determined density	$d_p = 1.41$ g/cm^3 (at 25°C)

Neither Phase I nor Phase II of this compound reveals the tacky properties of the $Me_2NH_2 \cdot ClO_4$ or 1,1 dimethylhydrazinium perchlorate.

Trimethyl Ammonium Perchlorate ($Me_3NH \cdot ClO_4$)

The transition temperatures of $Me_3NH \cdot ClO_4$ (including the melting point) are considerably higher than for the two methyl ammonium perchlorates already discussed (Figure 4). This material melts, while slowly decomposing, at 275°C. Phase I is stable below 275°C and above 207°C. The temperature range in which Phase II is the stable modification is between 116 and 207°C. Below 116°C, Phase III appears which has the lowest symmetry among the polymorphous phases of $Me_3NH \cdot ClO_4$. These phase transitions appear in the thermogram (Figure 4) as endotherms. All three phases have a rather large thermal expansion coefficient which made it difficult to obtain clear-cut high-temperature X-ray diffraction patterns because of temperature fluctuations during

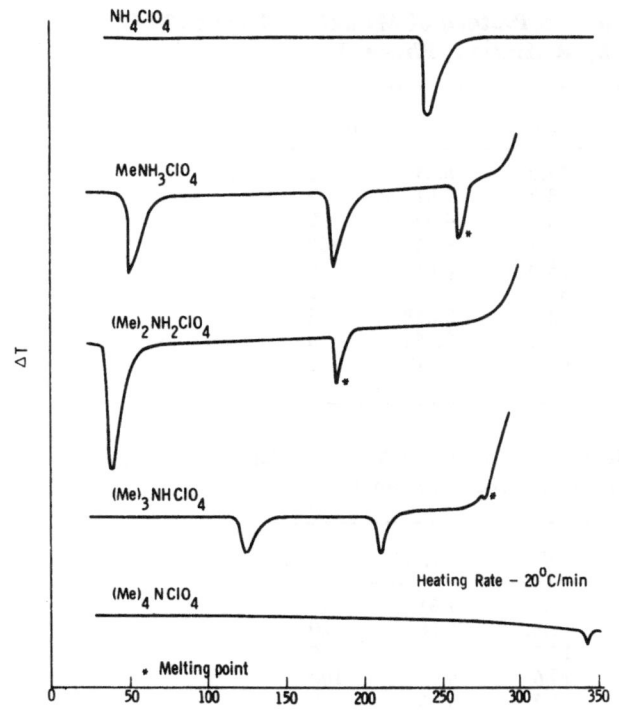

Figure 4. DTA patterns of ammonium and methyl substituted ammonium perchlorate salts.

Table IX. X-Ray Diffraction Pattern of Me₃NH · ClO₄ at ~ 240°C, Cu K_α Radiation; Phase I (Tetragonal)

No.	$\theta_{Cr\,K_a}$	d (Å)	Intensity	$\sin^2 \theta_{exp}$	$\sin^2 \theta_{calc}$	$\Delta \times 10^{-4}$	h	k	l
1	11.4	5.81	16	0.0391	0.0391	0	1	1	0
2	15.1	4.40	100	0.0679	0.0697	+18	1	1	1
3	16.3	4.08	37	0.0788	0.0783	−5	2	0	0
4	17.9	3.73	13	0.0945	0.0976	+31	2	1	0
5	19.3	3.470	10	0.1092	0.1089	−3	2	0	1
6	23.6	2.861	6	0.1603	0.1616	+13	1	1	2
7	24.7	2.741	< 6	0.1746	0.1754	+8	3	0	0

* See Table III.

these runs. In Table IX the reflections observed above 212°C are listed and indexed tetragonally:

Tetragonal	.	.	.	.	. $c/a = 0.8$; $a = 8.20$ Å; $c = 6.56$ Å
Volume of elementary cell	.	.	.	. $V = a^2 \cdot c = 439$ Å³
Number of moles per cell	.	.	.	. $Z = 2$
X-ray density	.	.	.	.	. $d_x = 1.21$ g/cm³ (≈ 245°C)
Determined density	.	.	.	. $d_p = 1.27$ g/cm³ (223°C)
Molecular weight	.	.	.	.	. 159.5

Table X. X-Ray Diffraction Pattern of $Me_3NH \cdot ClO_4$ at 180°C, Cr K_α Radiation; Phase II

No.	2θ	d (Å)	Intensity
1	14.9	8.84	85
2	18.8	7.01	2
3	26.5	4.99	2
4	29.1	4.56	100
5	35.0	3.81	20
6	38.2	3.50	9
7	43.8	3.078	12
8	45.6	2.958	6
9	49.8	2.717	1

Table XI. X-Ray Diffraction Pattern of $Me_3NH \cdot ClO_4$ at 25°C, Cu K_α Radiation; Phase III

No.	2θ	d (Å)	Intensity
1	12.1	7.31	34
2	15.85	5.58	20
3	17.7	5.01	5
4	20.6	4.31	100
5	24.3	3.66	95
6	26.1	3.42	14
7	27.2	3.28	5
8	27.9	3.195	1
9	31.9	2.805	7
10	33.5	2.675	8
11	35.0	2.564	6
12	35.8	2.508	2
13	36.5	2.462	1
14	38.1	2.362	4
15	39.7	2.270	2

The reflections observed for Phase II at 180°C are listed in Table X, and the reflections obtained for Phase III at 25°C are shown in Table XI. The stability ranges of the three phases existing between room temperature and the decomposition temperature are schematically shown as follows:

$$\text{Phase III} \underset{}{\overset{116°C}{\rightleftharpoons}} \text{Phase II} \underset{}{\overset{207°C}{\rightleftharpoons}} \underset{\text{Tetragonal}}{\overset{275°C}{\text{Phase I}}} \rightarrow \text{Melting with decomposition}$$

Phase I begins to decompose slowly at about 270°C. At 295°C, trimethyl ammonium perchlorate decomposed completely within 20 hr.

Tetramethyl Ammonium Perchlorate ($Me_4N \cdot ClO_4$) and Fluoborate ($Me_4N \cdot BF_4$)

Tetramethyl ammonium perchlorate was prepared by Vorländer and Kaascht[1] and was reported to explode at about 350°C. The temperature at which the phase transition occurred was close to the explosion temperature and could not be determined accurately.

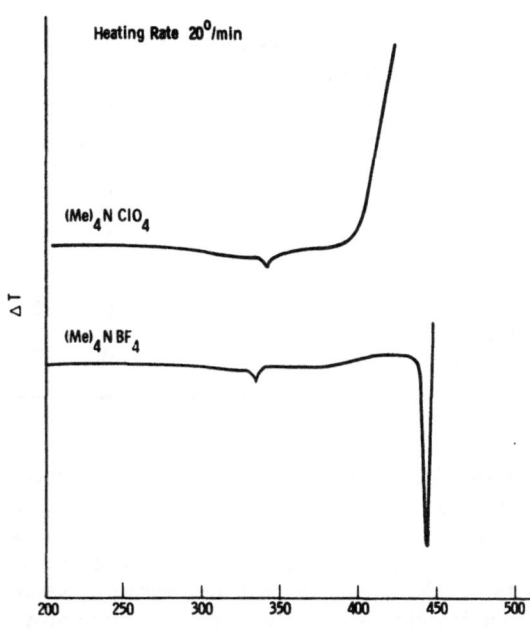

Figure 5. DTA patterns of tetramethyl ammonium perchlorate and fluoborate.

Table XII. X-Ray Diffraction Data of Me₄N · ClO₄ at 25°C, Cu K_α Radiation; Phase II (Tetragonal)

No.	θ	d (Å)	Intensity	$\sin^2 \theta_{exp}$	$\sin^2 \theta_{calc}$	$\Delta \times 10^{-4}$*	h	k	l
1	7.45	5.950	5	0.0168	0.0168	0	1	0	0
2	9.10	4.875	4	0.0250	0.0253	+3	0	0	2
3	10.55	4.215	100	0.0336	0.0336	0	1	1	0
4	14.00	3.185	3	0.0585	0.0570	−15	0	0	3
					0.0589	+4	1	1	2
5	15.10	2.960	2	0.0679	0.0672	−7	2	0	0
6	15.85	2.822	1.3	0.0746	0.0735	−11	2	0	1
					0.0738	−8	1	0	3
7	16.90	2.652	0.8	0.0845	0.0840	−5	2	1	0
8	17.70	2.535	0.4	0.0924	0.0925	+1	2	0	2
					0.0903	−21	2	1	1
					0.0906	−18	1	1	3
9	18.55	2.423	0.6	0.1012	0.1012	0	0	0	4
10	21.60	2.094	0.2	0.1355	0.1345	−10	2	2	0
					0.1348	−7	1	1	4
11	22.15	2.045	0.1	<0.1421	0.1410	−11	2	1	3
					0.1408	−13	2	2	1
12	23.0	1.973	0.3	0.1527	0.1510	−17	3	0	0
					0.1573	+46	2	0	1
13	24.2	1.881	0.1	<0.1680	0.1680	0	3	1	0
					0.1684	+4	2	0	4
14	24.85	1.835	0.1	<0.1766	0.1763	−3	3	0	2
					0.1743	−23	3	1	1
15	25.55	1.788	0.1	0.1860	0.1852	−8	2	1	4
16	25.95	1.760	0.1	<0.1915	0.1920	+5	1	1	5
					0.1915	0	2	2	3
					0.1933	+18	3	1	2
17	29.1	1.585	0.1	<0.2365	0.2357	−8	2	2	4

* See Table III.

Table XIII. X-Ray Diffraction Data of $Me_4N \cdot BF_4$ at 25°C, Cu K_α Radiation; Phase II (Tetragonal)

No.	θ	d (Å)	Intensity	$\sin^2 \theta_{exp}$	$\sin^2 \theta_{calc}$	$\Delta \times 10^{-4*}$	h	k	l
1	11.35	5.82	3	0.0387	0.0387	0	1	0	0
2	13.8	4.80	10	0.0569	0.0569	0	0	0	2
3	16.15	4.12	100	0.0773	0.0774	+1	1	1	0
4	21.5	3.125	6	0.1343	0.1343	0	1	1	2
5	24.4	2.773	4	0.1707	0.1697	−10	2	0	1
6	27.4	2.490	8	0.2118	0.2124	+6	2	0	2
7	29.85	2.300	2	0.2478	0.2505	+27	2	1	2
8	32.30	2.142	1	0.2855	0.2837	−18	2	0	3

* See Table III.

Therefore, both the perchlorate salt and the isomorphous tetrafluoborate salt were investigated. The thermograms of both compounds are shown in Figure 5. According to these curves, the perchlorate undergoes a phase transition at 340°C and detonates at approximately 430°C. The fluoborate salt shows a similar phase transition to occur at 333°C, and melts with decomposition at 435°C. No clear-cut high-temperature X-ray pattern could be obtained from either compound, which is mainly due to poor temperature control of the specimen and the narrow temperature range in which Phase I is stable. An improvement in the high-temperature X-ray technique (used for this study) is needed in order to obtain a diffractogram without appreciable decomposition. It is noteworthy that in the case of both tetramethyl compounds, the sensitivity of the thermal analyzer had to be increased tenfold in order to detect the II → I transition clearly. This indicates that very little energy is required for this transition.

The room-temperature diffractograms (taken at 25°C) for both the perchlorate and the fluoborate are listed and indexed tetragonally in Tables XII and XIII:

Cell dimensions of $Me_4N \cdot ClO_4$ at 25°C (Phase II)

Tegragonal	$c/a = 1.63$; $a = 5.95$ Å; $a^2 = 35.4$ Å^2; $c = 9.70$ Å
Volume of elementary cell	$V = a^2 \cdot c = 343.5$ Å^3
Number of moles per cell . . .	$Z = 2$
X-ray density	$d_x = 1.67$ g/cm^3
Determined density	$d_p = 1.35$ g/cm^3
Molecular weight	173.5

Cell dimensions of $Me_4N \cdot BF_4$ at 25°C (Phase II)

Tetragonal	$c/a = 1.65$; $a = 5.82$ Å; $c = 9.60$ Å
Volume of elementary cell	$a^2 \cdot c = 324$ Å^3
Number of moles per cell . . .	$Z = 2$
X-ray density	$d_x = 1.175$ g/cm^3
Molecular weight	115

Herrmann and Ilge[4] have reported that the tetramethyl ammonium permanganate, $(CH_3)_4N \cdot MnO_4$, also has tetragonal symmetry. This would be expected, since the MnO_4^--ion (like ClO_4^- and BF_4^-) is tetrahedral, with the manganese in the center.

However, no high-temperature modification was observed for the permanganate, due to the low decomposition temperature. The d spacings which have been reported by Moss and Sharp[14] for the $Me_4N \cdot BF_4$ agree with those obtained from a sample which was prepared by neutralization of $(CH_3)_4 N \cdot OH$ with aqueous HBF_4 except for the first three lines. It is possible that the sample of Moss and Sharp contained some hydrolysis products which could cause an expansion of the unit cell. Using Moss and Sharp's data,[14] the unit cell for the fluoborate is larger than for the perchlorate. This would be in contrast to the findings in other perchlorate and fluoborate salts (NH_4BF_4-NH_4ClO_4 or $N_2H_5BF_4$-$N_2H_5ClO_4$).[12]

In order to obtain as many reflections for the $Me_4N \cdot ClO_4$ (Phase II) as are listed in Table XII it was necessary to increase the sensitivity of the proportional counter eightfold after reflection No. 3. The strong intensity of the (110) reflection suggests that this plane contains the highest concentration of ClO_4^- ions. A similar observation is made with the fluoborate salt, in that there is a steep drop in the intensity at both sides of the (110) reflex.

The phase transitions of the tetramethyl ammonium perchlorate and fluoborate are summarized below:

$$Me_4N \cdot ClO_4: \text{Phase II} \overset{340°C}{\rightleftharpoons} \text{Phase I} \overset{430°C}{\rightarrow} \text{Detonation at } 430°C$$
$$\text{Tetragonal}$$

$$Me_4N \cdot BF_4: \text{Phase II} \overset{333°C}{\rightleftharpoons} \text{Phase I} \overset{430°C}{\rightarrow} \text{Melting with decomposition}$$
$$\text{Tetragonal}$$

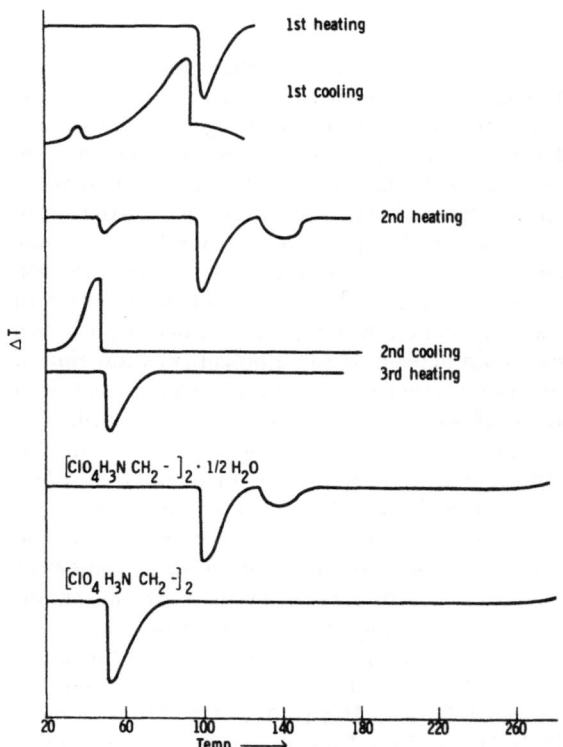

Figure 6. Ethylenediaminediperchlorate.

Ethylene Diammonium Diperchlorate [Et(NH₃ · ClO₄)₂]

This compound was the only difunctional perchlorate which was investigated during this study. It exists as a hemi-hydrate and in the anhydrous form. The heating and cooling curves of both forms are shown in Figure 6. The hemi-hydrate undergoes a reversible phase transition at 95°C and loses $\frac{1}{2}$ mole of water between 120 and 130°C (at a heating rate of 20°C/min). The anhydrous material has a reversible phase transition at 52°C and decomposes near 290°C.

$$\text{Hemi-hydrate: Phase II} \underset{}{\overset{95°C}{\rightleftharpoons}} \text{Phase I} \underset{-\frac{1}{2}H_2O}{\overset{120°C}{\rightarrow}} \text{Phase I (anhydrous)}$$

$$\text{Anhydrous: Phase II} \overset{52°C}{\rightleftharpoons} \text{Phase I} \overset{290°C}{\rightarrow} \text{Phase I (decomposition)}$$

X-ray diffraction data of Phase II of the hemi-hydrate are listed in Table XIV.

No diffraction pattern of Phase I of the hemi-hydrate was obtained. The reflections of Phase I and II of the anhydrous material obtained with nickel-filtered copper radiation are listed in Tables XV and XVI. The dimensions of the tetragonal unit cell of the anhydrous Phase I are:

Tetragonal	$c/a = 0.46; a = 11.35$ Å; $c = 5.23$ Å
Volume of elementary cell	$V = a^2 \cdot c = 670$ Å³
Number of moles per cell	$Z = 3$
X-ray density	$d_x = 1.94$ g/cm³

Summary of Experimental Results

All the experimentally determined data obtained for the substituted ammonium perchlorates during the course of this study are summarized in Table XVII. Phase transitions like those described above, where the crystal structure changes with increasing temperatures to higher symmetries, have been interpreted in terms of the onset of rotation of molecules (Pauling[5]) or ions, or in terms of the onset of disorder (Frenkel,[8] Lawson,[7] Nagamiya[15]), i.e., distribution of rotational oscillation axes over a fixed set of possible positions. Levy and Peterson[9] have used both concepts to explain different phase transitions and the corresponding crystal structures of ND₄Br (as determined by neutron diffraction): Phase III → Phase II, onset of disorder, Phase II → Phase I, onset of rotation with a random distribution of axes of rotation over a given number of positions. The energy requirements for above cases would depend on some potential as function of the polar angle $V(\phi)$ as to the axis of rotation (or rotational oscillation) and one equilibrium position (potential minimum); the moment of inertia I would be involved as well. The application of these concepts to the phase transitions of the differently substituted ammonium perchlorates can be only qualitative in view of the uncertainty of $V(\phi)$. The more readily accessible moments of inertia do not show a clear correlation with transition temperatures. The energy take-up for a simple harmonic oscillator would be inversely proportional to the square root of I. Table VII, however, shows a somewhat irregular increase of transition temperatures T_2 with increasing mass of the cation (the corresponding moments of inertia I increase in the same order). From this it is quite clear that $V(\phi)$ is most important for the transition temperatures. This influence is readily seen with the ammonium halides: The Phase II → Phase I transition temperatures show a substantial decrease from NH₄Cl (184°C) to NH₄I (−17.6°C). The low-temperature transition of NH₄ClO₄ at −190°C[10] may be related to this series. The space

Table XIV. X-Ray Diffraction Data of Et(NH$_3$ · ClO$_4$)$_2$ · $\frac{1}{2}$ H$_2$O at 25°C, Cr K_α Radiation; Phase II (Hemi-Hydrate)

No.	$2\theta_{Cr}$	d (Å)	Intensity
1	21.0	6.28	31
2	21.4	6.17	50
3	23.2	5.70	12
4	23.5	5.63	24
5	27.1	4.89	12
6	27.7	4.78	5
7	28.6	4.64	7
8	29.9	4.44	7
9	32.3	4.12	100
10	34.0	3.92	80
11	34.65	3.85	40
12	35.1	3.80	20
13	35.5	3.76	24
14	37.2	3.590	52
15	38.1	3.510	5
16	38.4	3.482	7
17	39.9	3.358	33
18	40.5	3.345	19
19	41.4	3.241	22
20	41.6	3.226	22
21	43.5	3.091	24
22	45.6	2.956	5
23	47.0	2.873	7
24	48.1	2.811	9
25	48.9	2.768	9

Table XV. Diffraction Data of Et(NH$_3$ · ClO$_4$)$_2$ at 80°C, Cu K_α Radiation; Phase I (Anhydrous, Tetragonal)

No.	$\theta_{Cu\,K\alpha}$	d (Å)	Intensity	$\sin^2\theta_{exp}$	$\sin^2\theta_{calc}$	$\Delta \times 10^{-4}$*	h	k	l
1	4.0	11.05	1	0.0048	0.0046	−2	1	0	0
2	5.51	8.036	2	0.0093	0.0093	0	1	1	0
3	8.70	5.09	37	0.0229	0.0217	−12	0	0	1
					0.0230	+1	2	1	0
4	10.99	4.04	32	0.0362	0.0365	+3	2	2	0
5	11.55	3.81	100	0.0411	0.0415	+4	3	0	0
					0.0401	−10	2	0	1
6	16.10	2.78	14	0.0769	0.0739	−30	4	0	0
					0.0784	+15	4	1	0
7	16.60	2.70	32	0.0826	0.0830	+4	3	3	0
					0.0817	−9	3	2	1
8	18.85	2.385	2	0.1044	0.1047	+3	3	3	1
9	20.05	2.25	7	0.1175	0.1150	−25	5	0	0
					0.1150	−25	4	3	0
					0.1140	−35	4	2	1
10	20.45	2.20	4	0.1220	0.1237	+17	2	2	2
					0.1200	−20	5	1	0
11	26.80	1.71	5	0.2033	0.1998	−35	1	0	4
					0.2019	−14	5	0	2
					0.2019	−14	4	3	2
					0.2045	+12	1	1	3
12	27.30	1.07	2	0.2132	0.2136	+4	2	0	3

*See Table III.

Table XVI. X-Ray Diffraction Data of Et(NH₃ · ClO₄)₂ at 25°C (Anhydrous), Cu K_α Radiation; Phase II

No.	$2\theta_{Cu}$	d (Å)	Intensity
1	7.34	12.05	12
2	11.60	7.63	2
3	14.50	6.11	2
4	15.90	5.57	10
5	17.10	5.18	3
6	18.24	4.87	6
7	19.30	4.60	6
8	20.20	4.39	3
9	21.74	4.08	100
10	22.94	3.87	52
11	23.92	3.72	28
12	24.98	3.565	37
13	25.72	3.462	4
14	26.80	3.325	22
15	27.82	3.205	35
16	29.04	3.073	26
17	30.40	2.940	8
18	31.30	2.858	5
19	31.90	2.805	6
20	32.40	2.764	5
21	34.20	2.622	5
22	35.40	2.535	15
23	37.50	2.398	23
24	39.10	2.304	5
25	40.60	2.222	3
26	41.80	2.161	3
27	44.20	2.049	3
28	46.20	1.965	4

requirements for the rotation of the NH_4^+ ion are such that no substantial lattice expansion is necessary; the potential barrier to be overcome for the onset of rotation is mainly governed by hydrogen bonding to the oxygens of the ClO_4^- ions. With increasing substitution of hydrogens by methyl groups, the space requirements for cationic rotation (II → III transition) increase simultaneously. Therefore, one would expect an increase of the II → III transition temperatures, since a larger thermal expansion of the lattice is necessary to allow the onset of cationic rotation. This indeed was observed for the II → III transition temperatures of $NH_4^+ClO_4$ and $(CH_3 \cdot NH_3)^+ \cdot ClO_4^-$, which increase from − 190°C to + 48°C, respectively. Comparing the II → III transition temperatures of $(CH_3 \cdot NH_3)^+ \cdot ClO_4^-$ and $[(CH_3)_2 \cdot NH_2]^+ \cdot ClO_4^-$, however, one observes a slight decrease from + 48°C to + 38°C respectively. This inversion may be explained by relatively smaller space requirements for the rotation of the $(CH_3)_2NH_2^+$, cation if one compares an axis of rotation for the $(CH_3 \cdot NH_3)^+$ ion, which is perpendicular to the C-N-axis, and for the $[(CH_3)_2 \cdot NH_2]^+$ ion, an axis of rotation which is perpendicular to the plane determined by the nitrogen and the two carbon atoms. The steep increase of the II → III transition temperature from the dimethyl ammonium perchlorate to the trimethyl ammonium perchlorate, however, may be explained by the existence of four nearly equivalent axes of rotation. The onset of rotation with a random distribution over these four axes increases the relative space requirement over those of the mono- and dimethyl ammonium ions. The tetramethyl ammonium ion rotation would then be

Table XVII. Summary of Experimental Data obtained for Substituted Ammonium Perchlorates

No.	Compound	Melting point (°C)	T_1 I → II transition temperature (°C)	Phase I Unit cell parameter (Å)	Phase I density Temperature (°C)	d_x (g/cm^3)	d_p (g/cm^3)	T_2 II → III transition temperature (°C)	Phase II unit cell parameter (Å)	Phase II density Temperature (°C)	d_x (g/cm^3)	d_p (g/cm^3)
1	NH_4ClO_4	450 decomposes	240	$a = 7.67$	250	1.73	1.7	≈ −190	$a = 9.231$ $b = 5.813$ $c = 7.453$	25	1.95	1.9
2	$CH_3NH_3 \cdot ClO_4$	255 melts and decomposes	178	$a = 5.18$	200 187	1.58 —	— 1.52	48	$a = 3.87$ $c = 9.28$	77 72 25	1.58 —	— 1.56 1.65
3	$(CH_3)_2NH_2 \cdot ClO_4$	180	180	$a = 5.44$ $c = 11.22$	45 71	1.46 —	— 1.4	38	$c = 9.28$	25	—	1.40
4	$(CH_3)_3NH \cdot ClO_4$	275 melts and decomposes	207	$a = 8.20$ $c = 6.56$	245 223	1.21 —	— 1.27	116	—	133 25	— —	1.35 1.41
5	$(CH_3)_4N \cdot ClO_4$	detonates at 380	340	—	—	—	—	—	$a = 5.95$ $c = 9.70$	25	1.67	1.35
6	$(CH_3)_4N \cdot BF_4$	430 melts and decomposes	335	—	—	—	—	—	$a = 5.82$ $c = 9.60$	25	1.18	—
7	$(CH_2NH_3 \cdot ClO_4)_2$	290 decomposes	54	$a = 11.35$ $c = 5.23$	80 70	1.94 —	— 1.8	—	—	25	—	—
8	$(CH_2NH_3 \cdot ClO_4)_2 \cdot \tfrac{1}{2}H_2O$	— $\tfrac{1}{2}H_2O$ at 120	95	—	70	—	1.7	—	—	25	—	1.80
9	i propyl$NH_3 \cdot ClO_4$	150 melting	25	$a = 7.34$ $c = 6.98$	30 25	1.42 —	— 1.41	—	—	—	—	—

expected at even higher temperatures, which may be close to transition temperatures ascribed to the beginning of anion rotation (ClO_4^- and BF_4^-). Only one phase transition has been observed for tetramethyl ammonium perchlorate and fluoborate. This transition is associated with only a very small energy of transformation. It is most likely that here the rotation of cations and anions occurs simultaneously, since both ions are of comparable size. The onset of cationic rotation, for example, would cause (in addition to the thermal expansion) the lattice parameter to increase sufficiently, to permit anion rotation.

ACKNOWLEDGMENT

The authors are indebted to Dr. A. O. Dekker of Aerojet-General Corporation, whose continued interest and encouragement made this work possible, and to Aerojet-General Corporation for the permission to publish. The density determinations at elevated temperatures were performed by A. Lange.

REFERENCES

1. D. Vorländer and E. Kaascht, "Neue Formen der überchlorsauren Salze," *Ber. deut. chem. Ges.* **56B**: 1157, 1923.
2. W. Büssem and K. Herrmann, "Röntgenographische Untersuchung der einwertigen Perchlorate, *Z. Krist.* **67**: 405, 1928.
3. H. Braekken and L. Harang, "Die kubische Hochtemperatur struktur einiger Perchlorate." *Z. Krist.* **75**: 538, 1930.
4. K. Herrmann and W. Ilge, "Röntgenographische Strukturerforschung der kubischen Modifikation der Perchlorate," *Z. Krist.* **75**: 41, 1930.
5. L. Pauling, "The Rotational Motion of Molecules in Crystals," *Phys. Rev.* **36**: 430, 1930.
6. A. Eucken, "Rotation von Molekeln und Ionengruppen in Kristallen," *Z. Electrochem.* **45**: 126, 1939.
7. A. W. Lawson, "The Variation of the Adiabatic and Isothermal Elastic Moduli and Coefficient of Thermal Expansion with Temperature through the λ-Point Transition of NH_4Cl," *Phys. Rev.* **57**: 417, 1940.
8. J. Frenkel, "Über die Drehung von Dipolmolekülen in festen Körpern," *Acta Physicochem. U.R.S.S.* **3**: 23, 1935.
9. H. A. Levy and S. W. Peterson, presentation before the International Congress for Pure and Applied Chemistry, New York, 1951.
10. M. Stammler, D. Orcutt, and P. C. Colodny in: W. M. Mueller, G. R. Mallet, and M. J. Fay (eds.), *Advances in X-Ray Analysis*, Vol. 6, Plenum Press, New York, 1962, p. 202.
11. J. C. Petricciani, S. E. Weberly, W. H. Bauer and T. W. Clapper, *J. Phys. Chem.* **64**: 1309, 1960.
12. M. Stammler and D. Orcutt, in: W. M. Mueller, G. R. Mallett, and M. J. Fay (eds.), *Advances in X-Ray Analysis*, Vol. 5, Plenum Press, New York, 1961, p. 133.
13. J. W. Conant, L. L. Corrigan, and A. A. Sparks, "The Structure of Hydrazinium Fluoborate", *Acta Cryst.* **17**: 1085, 1964.
14. K. C. Moss and D. W. A. Sharp, letter to the editor, *J. Inorg. Nucl. Chem.* **13**: 328, 1960.
15. Takeo Nagamiya, "Zur Theorie der Umwandlung der festen Ammoniumhalogenide bei tiefen Temperaturen," *Proc. Phys. Math. Soc. Japan*, **25**: 540, 1943.

ORIENTED CRYSTALLIZATION OF UREA IN MEMBRANES

Edwin H. Shaw, Jr.

University of South Dakota
Vermillion, South Dakota

ABSTRACT

Inner egg shell membrane was soaked in 2 M urea solution and dried. Cu K_α X-rays directed perpendular to the membrane yielded an unsymmetrical diffraction pattern with urea 110, 111, 102, 210 and 211 planes diffracting to the right and urea 101, 200 and 201 planes diffracting to the left. The urea appears to be hydrogen-bonded diagonally between the lamellae of β-keratin at an angle of 33.5° to the 6.68 Å b-axis (fiber axis) of the parallel pleated sheet structure, shortening the half-axis lengths in the a direction from 4.73 Å to 4.41 Å. Similar results were obtained with formalinized dog mesentery. Keratinous outer egg shell membrane, formalinized collagenous dog renal capsule, formalinized collagenous pheasant pericardium, and formalinized callogenous human arachnoid yielded symmetrically oriented urea layer lines. Formalinized dog renal fascia and pleura yielded random urea lines.

INTRODUCTION

In previous work,[1,2] the author has demonstrated that the hydrogen bonding of urea derivatives and of resorcinol on collagen can form oriented crystals in this fibrous material, yielding information as to the spacing of hydrogen bond acceptors along the collagen fiber and as to the diameter of the core of the collagen spiral. Because of its high dipole moment, 4.20 D from the water,[3] and good crystallinity, urea is an excellent tool for the exploration of structures by the method of oriented crystallization.

EXPERIMENTAL PROCEDURES AND RESULTS

Biological membranous materials were soaked in 2 M urea solutions, pulled smooth over an open glass cylinder which fits snugly over the collimator, dried, and Cu K_α X-ray diffractograms made with the X-ray beam perpendicular to the dried membrane surface. In the case of the inner egg shell membrane, which consists of keratin[4] with some mucin as a binder,[5] the procedure yielded an unsymmetrical layer line pattern as indicated in Figure 1.

The layer line spots are spread to an angle of approximately 45°, indicating a tilting spread of approximately 22.5° in the plane of the membrane. There is probably also enough variation in the direction normal to the membrane to produce the equivalent of an oscillation diffractogram from the stationary membrane. In this pattern, we can explain the alteration of spots to the right (110, 111, 102) and to the left (101, 200, 201) of the center in terms of the diagram in Figure 2, where unsymmetrical diffraction by the urea 110 and 200 planes is illustrated in terms of the angle of rotation of the urea crystal

[1] References are at the end of the paper.

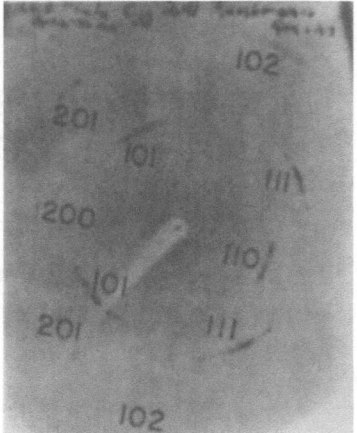

Figure 1. Unsymmetrical crystallization of urea in inner egg shell membrane.

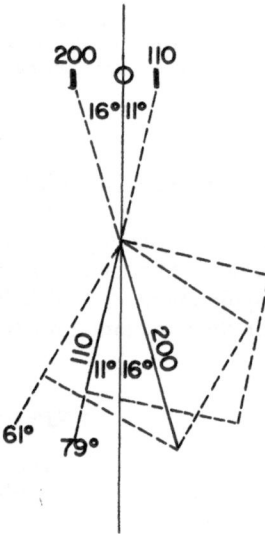

Figure 2. Angular spread for unsymmetrical diffraction by urea planes.

from the position in which the 110 planes, the sides of the crystal, were at the 90° position, parallel to the X-ray beam. In order to verify the unsymmetrical diffraction of the spots involved, a crystal of urea was oscillated over 5° ranges of rotation and the spots produced identified, with the results noted in Table I. The difference between 61° and 79° (18°) represents the minimum oscillation angle at which 110 planes diffract to the right and 200 planes diffract to the left as indicated in Figure 2.

Figure 3 illustrates the insertion of the urea molecule diagonally across the pleated sheets of the parallel arrangement of β-keratin in the orthogonal form proposed by Frazer and MacRae.[6] The arrangement shortens the half-length of the 9.46 Å a-axis from 4.73 Å to 4.41 Å with an angle of 33.5° between the urea molecules and the membrane surface.

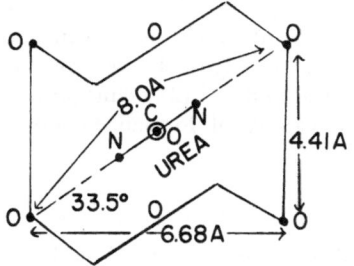

Figure 3. Projection of urea in the interlamellar space of keratin, diagonally across the layers of the 6.68 Å fiber *b* axis on the *a–b* projection. The central oxygens are above the plane and the corner oxygens are below it.

Table I. Angular Positions of Crystal where Urea Planes Diffract

Index	Angle (deg)	Direction
101	65–70	Left
110	75–85	Right*
111	70–75	Right
200	60–65	Left†
201	65–70	Left
102	65–75	Right
210	50–60	Right
211	45–55	Right

* 79° calculated.
† 61° calculated.

Table II. Action of Urea on Membranes

Membrane	Composition	Type of urea X-ray pattern
Inner egg shell	Keratin and mucin[5]	Unsymmetrical oriented
Mesentery (dog)	Unknown	Unsymmetrical oriented
Outer egg shell	Keratin[4]	Symmetrical oriented
Renal capsule (dog)	Collagen	Symmetrical oriented
Pericardium (pheasant)	Collagen	Symmetrical oriented
Arachnoid (human)	Collagen	Symmetrical oriented
Renal fascia (dog)	Unknown	Random
Pleura (dog)	Collagen	Random
Pericardium (dog)	Collagen	None*
Inner corneal (dog)	Collagen	None

* Interspiral spacing of collagen increased from 13.6 Å to 16.5 Å.

With the central oxygens above the plane of the nitrogens, they may also be hydrogen-bonded to the nitrogens, contributing to the shortening of the *a*-axis. Since the urea molecules parallel the 110 planes in the crystal, which are the bounding planes of the tetragonal crystals, this means that the 110 crystal faces of urea form at approximately 56.5° (or 33.5°) from surface and diffract at 11° from the beam or 79° from the surface of the membrane. The spread of the membrane plane of approximately 22.5° from this 56.5° center yield an "oscillation range" of from 34° to 79° from the membrane surface, which covers the diffraction angles noted in Table I. In this range, the 210 and 211 planes would be expected to scatter to the right, as has been observed in additional runs. An unsymmetrical oriented crystallization pattern similar to that in Figure 1 was produced when a crystal of urea was oscillated between the angles 44° and 79° (0° being perpendicular to the beam). The procedure was applied to a variety of formalized animal membranes with the results indicated in Table II.

SUMMARY

When biological membranes are soaked in strong solutions of urea, dried, and subjected to X-ray diffraction with the beam perpendicular to the membrane, there are four possible results.

1. Crystallization may be oriented in two directions by hydrogen bonding to surface components (the carbonyl group in proteins), leading to unsymmetrical diffraction patterns similar to oscillation patterns, from which information as to the structure of the membrane may be obtained.
2. When crystallization is oriented in only one direction, symmetrical diffraction patterns are obtained similar to rotating crystal runs.
3. With random orientation of protein in the membrane, random crystallization of urea occurs.
4. In some cases urea crystallization is suppressed.

Attempts to use this procedure with commercial films failed, presumably because of lack of fit of the urea hydrogen bonds to acceptors in the films.

ACKNOWLEDGMENTS

The author acknowledges the technical assistance of Inara Zarins and Marcia Mills. This investigation was supported in part by grant (USPH 1-SO1-FR-05421-01) from the National Institute of Health, U.S. Public Health Service.

REFERENCES

1. E. H. Shaw, Jr., "Oriented Crystallization of Amides on Collagen with Modification of the Collagen Lattice," in: W. M. Mueller, G. R. Mallett, and M. J. Fay (eds.), *Advances in X-ray Analysis, Vol. 7*, Plenum Press, New York, 1964, p. 252.
2. E. H. Shaw, J. and Alton R. Christensen, "Modification of Collagen and Nylon Lattices by Resorcinol," in: W. M. Mueller, G. R. Mallett, and M. J. Fay (eds.), *Advances in X-Ray Analysis, Vol. 8*, Plenum Press, New York, 1965, p. 175.
3. W. R. Gilkerson and K. K. Srivastava, "The Dipole Moment of Urea," *J. Phys. Chem.* **64**: 1485, 1960.
4. H. O. Calvery, "Some Analyses of Egg Shell Keratin," *J. Biol. Chem.* **100**: 183, 1933.
5. T. Moran and H. P. Hale, "Physics of the Hen's Egg. I. Membranes in the Egg," *J. Exptl. Biol.* **13**: 35, 1936.
6. R. D. B. Fraser and T. P. MacRae, "An Investigation of the Structure of β-Keratin," *J. Mol. Biol.* **5**: 457, 1962.

HIGH-INTENSITY ROTATING ANODE X-RAY TUBES

A. Taylor

Westinghouse Electrical Corporation
Pittsburgh, Pennsylvania

ABSTRACT

Demountable rotating anode X-ray tubes with a $7\frac{1}{2}$ kW power dissipation have been built for conventional diffraction work with powder cameras and equi-inclination Weissenberg goniometers, and for use with a tetrahedral press for studying crystalline matter at ultra-high pressures. The tubes employ a highly compact cooling and sealing arrangement on the rotating anode which enables four windows to be used with the focal spot close to the specimen. A rotational speed of 1750 rpm with a focal spot size of 10 × 1 mm enables the tubes to be operated at 250–275 mA at 30 kV DC or at 150 mA, 50 kV DC.

GENERAL DESCRIPTION OF THE X-RAY TUBES

Two X-ray tubes have been built on somewhat similar principles. One of these is conventional in that it is fixed rigidly in position and is mounted vertically in a cabinet with four beryllium windows standing approximately 16 in. from the cabinet top (Figure 1). This tube is arranged to be used with standard types of crystallographic equipment such as powder cameras, Weissenberg goniometers, and goniostats. The second tube is separated from the cabinet and becomes, in effect, an accessory to a much larger installation, in this case a tetrahedral anvil press. In this second arrangement the beam must be directed with a high degree of accuracy and positional stability through the centroid of the tetrahedral gasket surrounding the specimen to be compressed, and through the narrow gap between the anvils. To achieve this, the X-ray tube and vacuum manifold form an articulated system of sliding and rotating vacuum joints which permits translation and rotation of the tube with respect to vertical and horizontal axes.

Because the construction of the anode and the cathode assemblies of both tubes is essentially the same, the more conventional vertically arranged tube will be described in detail, and any changes made to produce the articulated version will only be given a brief mention. The vertical tube, which is shown in Figure 1, is mounted on a cabinet which houses the control instruments, powerstat, and vacuum pumps, the high-voltage transformer being in a separate unit. The water-cooled body of the tube is made of a cupro-nickel alloy, the cross section of the tube being $6 \times 2\frac{3}{4}$ in. An important feature of the design is that the corners of one of the larger faces are bevelled at approximately 45° to permit a horizontally mounted Nonius Weissenberg goniometer to swing into an equi-inclination position of roughly 43°. (The same shape also enables the body of the tube to fit snugly between the rams of the tetrahedral anvil press, thus allowing the focus to come within a reasonable distance of the sample.) The tube is fitted with four beryllium windows which are easily removed for cleaning. These windows, which are approximately 16 in. from the cabinet top, enable the 10 × 1 mm focal spot to be seen from the "long" or "foreshortened" positions, with the beams emerging at a downward angle of approximately 5–6° with respect to the horizontal, as in conventional tubes. The flexible

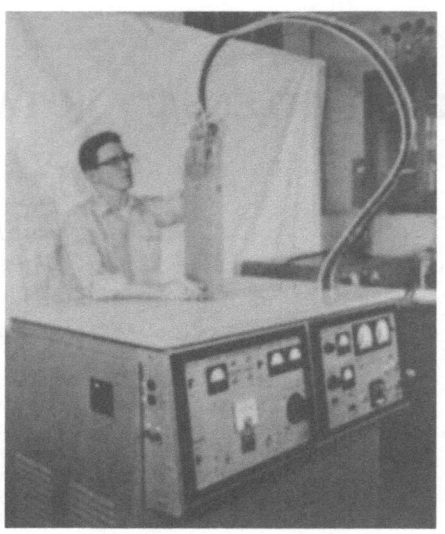

Figure 1. General view of X-ray tube and
control cabinet.

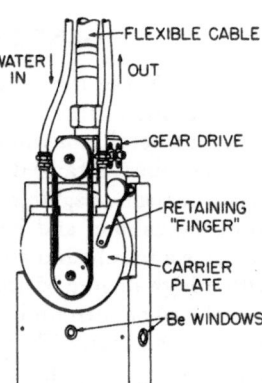

Figure 2. Close-up view of anode drive, water cooling pipes, and
carrier plate.

cable leaving the cabinet top connects with a right-angled gear drive at the top of the
X-ray tube (Figure 2), the drive being fitted with a pulley wheel which rotates the anode
via an orlon belt. Flexible Teflon inlet and outlet pipes for the anode-cooling water are
clipped to the cable for convenience.

The X-ray tube body is actually mounted with an O-ring vacuum seal on top of a
large vacuum manifold which is located horizontally beneath the cabinet top, with the
cathode insulator and cathode assembly mounted on the manifold from below. This
arrangement makes it possible to rotate the X-ray tube about its vertical axis so that the
beam from the foreshortened focal spot can be viewed either from the front of the cabinet
or from the side.

CARRIER-PLATE ASSEMBLY AND ANODE

The general arrangement of the carrier plate with its associated bearings, hollow
drive shaft, pulley wheel, and seals is shown in Figure 3, along with the anode. A special
feature of the design is the re-entrant shape of the drive shaft, which enables the vacuum

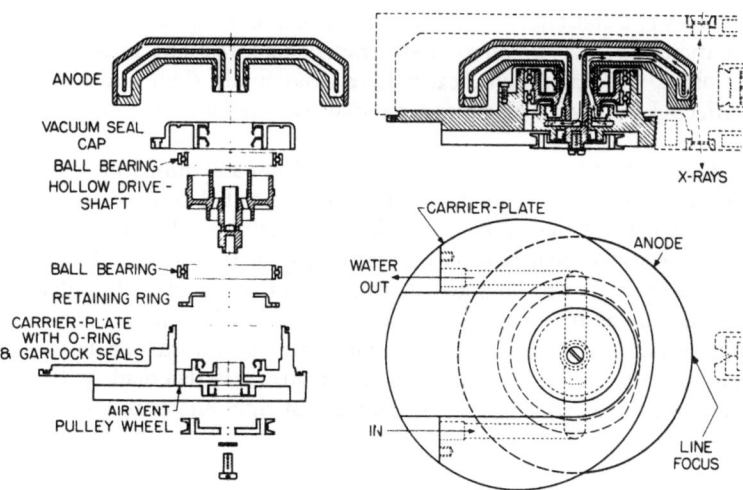

Figure 3. Anode and carrier-plate assemblies.

seals and one of the water seals to be housed within the space occupied by the ball bearings. This makes for an extremely compact arrangement which allows unimpeded access to all windows of the tube. In previously described rotating anode tubes,[1,2] the assembly was such that the cooling water entered and left the anode via a stationary union which was pushed through a rotating water seal and into the rotating hollow shaft of the anode. Because of its necessarily small size, the union was difficult to fabricate and there was always the possibility of seizure if the lubrication on its surface failed. Moreover, its small bore limited the flow of water to the anode. The present arrangement completely eliminates the water union by enabling the cooling water to enter and leave the hollow drive shaft from the side, rather than from the end, via the stationary carrier plate. Not only can we increase the sizes of the openings, thus permitting a larger flow of water and consequently greater tube power, but the lips of the vacuum seals rubbing on the rotating shaft can in effect be cooled by the incoming cold water, instead of being heated by the outflowing water as in previous arrangements. In the present design, the vacuum seals are housed in a separate and easily detachable cap. This enables them to be replaced as a unit, together with the cap, in a few minutes, so that the down-time is negligible in the event of a seal wearing out. The worn seals can then be replaced at leisure.

As may be seen from Figure 3, the carrier plate is fitted with two chevron-type seals,* which are shown in place. These bear on the hollow drive shaft to prevent leakage of water which enters and leaves the side of the shaft via the passages in the carrier plate. The inner seal, which takes the full pressure of the incoming water, is held in place by a retaining ring; otherwise it would creep out of position and rub against the lower face of the hollow drive shaft, thus impeding its rotation. The drive shaft is fitted with two sets of stainless steel precision high-speed ball bearings and pushed into place in the boss of the carrier plate, the bearings being held in place by a special adhesive† to avoid any undue distortion. The pulley wheel is then placed on the square-shaped end of the drive shaft and held in position with the retaining screw and lock washer. The vacuum-seal cap with its two chevron seals is now slid into position over the anode-end of the drive shaft and held

[1] References are at the end of the paper.
* Manufactured by the Garlock Company, Palmyra, New York.
† Loctite Type A; Loctite Corporation, Newington, Connecticut.

in place with screws, a stationary vacuum seal between the cap and the boss being made with a conventional O-ring which is under radial compression. Finally, the anode is screwed into place, the water-tight seal between the anode shaft and the hollow drive shaft being effected by two O-rings under radial compression in grooves on the anode shaft.

The anode, which is 5 in. in diameter, is made from a spinning of O.F.H.C. copper and brazed to a copper base. A copper septum plate and radial grooves directs the cooling radially across the base of the anode and behind the cylindrical surface where the focal spot is located, the water then flowing across the top of the plate and out through the center hole in the shaft. It will be noted that the top of the anode is bevelled to conform to the shape of the X-ray tube which must be compatible with the requirements of equi-inclination Weissenberg goniometers, precession cameras, and the peculiar geometry of a tetrahedral angle press. In the latter case, the focal surface is also made with a 6° slope so that the beam from the foreshortened spot emerges at right angles to a face of the X-ray tube.

THE CATHODE ASSEMBLY

The biased, adjustable cathode assembly described in previous publications[1,2] has been completely redesigned, sliding O-ring seals now being used instead of a sylphon bellows arrangement. The new design makes it much easier to replace the filament and maintain the alignment of the cathode assembly with respect to the anode.

The cathode assembly shown in Figure 4 comprises a flanged porcelain insulator which carries an accurately ground bronze sleeve. This sleeve makes a vacuum-tight sliding O-ring seal with a stainless steel tube, the far end of which is terminated by the focusing cup (Wehnelt cylinder). Sliding inside the stainless steel tube and separated from it by mica spacers and by a Teflon insulator is a second stainless steel tube which is terminated by the support to which the 12-turn helical filament is attached. The distance between the focusing cup and the anode may be adjusted without breaking the vacuum simply by rotating the external focusing cup positioning nut, the pressure of the atmosphere holding the nut firmly pressed against the outer face of the bronze sleeve. The filament may likewise be moved independently with respect to the slot in the focusing cup by means of its own externally mounted filament-adjusting nut.

The distance between the focusing cup and the anode is not very critical, a distance of 12–14 mm being suitable. On the other hand, the position of the filament with respect to the slot in the focusing cup is extremely critical and must be set every time the filament is renewed. To obtain the sharpest focus with minimum X-ray emission from its immediate environs on the anode face, a bias of up to 500 V may be applied between the

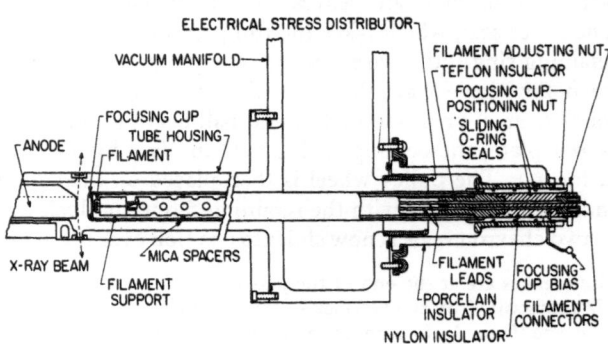

Figure 4. Cathode assembly.

filament and the focusing cup, the potential being provided by a potentiometer housed in the high-voltage transformer tank.

The focal spot size is 10×1.0 mm, obtained with a tungsten helix consisting of 12 turns of 11.5 mil wire wound on a mandrel of a diameter of 48 mils. With a rotational speed of 1750 rpm, the tube could be operated at 250–275 mA at 30 kV DC using a transistorized-current stabilizer[3] and a three-phase six solid state diode rectified HT supply, the rise in temperature of the cooling water being only 19°C at a flow rate of about 1.8 gal/min. By modifying the shape of the focusing cup and the filament size, it would of course be possible to obtain either a smaller or a larger focus to suit the particular crystallographic problem, but the specific loading of the anode could not be greatly increased with an anode speed of 1750 rpm because of the possibility of surface erosion caused by the electron bombardment.

X-RAY TUBE FOR TETRAHEDRAL ANVIL PRESS

The main design features of the anode–carrier plate assembly and of the anode are similar to those of the vertical tube and need not be described again. In this case, the tube is mounted horizontally as shown in Figures 5 and 6, the vacuum manifold being held in a clamp which is attached to the frame of the press. The central portion of the tube is, in effect, a large rotatable vacuum joint made with radially compressed O-rings, which allows the tube to be rotated about a horizontal axis, and also to be translated horizontally and locked in position. The clamp allows both rotation and sliding with respect to the vertical axis, with an additional sliding motion at right angles, thus making it possible to direct the beam from the foreshortened focal spot accurately through the center of the press. However, the X-ray tube is fitted with two beryllium windows and it may be used as an orthodox horizontal X-ray tube when not used in conjunction with the high-pressure apparatus.

When used in conjunction with the press, the beam from the X-ray tube must be capable of traversing a total distance of about 2 cm in a beryllium or LiH gasket without too great an intensity loss by absorption. For this reason, a rhodium-plated anode is used, the tube being then operated at 150 mA and 50 kV DC using a three-phase system with six solid state rectifier diodes. As is the case with the more conventional vertical unit, the anode can be changed within a few minutes for other types of work. It will be noticed, in the present case, that the anode is driven by a belt at approximately 45° to the body of the tube. This is to prevent the flexible drive shaft and gear drive from interfering with the diffractometer counter and its associated tracks in the back-reflection region.

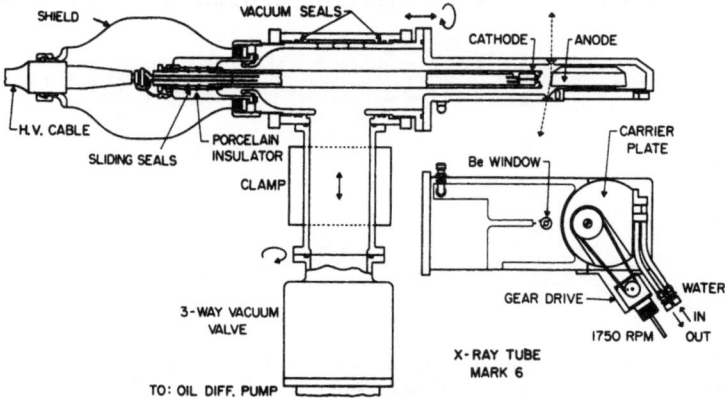

Figure 5. Diagrammatic view of articulated X-ray tube.

Figure 6. X-ray tube, three-way valve, and diffusion pump mounted on tetrahedral anvil press.

FIGURE OF MERIT

To obtain a "figure of merit" for the vertical X-ray tube, a series of Weissenberg patterns were taken of a minute zinc glycenate crystal. Operating the tube at a stabilized current of 250 mA and 30 kV, the approximate exposure time for a zero layer line was $1\frac{1}{2}$ min, while the time to take a complete traverse was 1 hr. These exposure times are approximately $\frac{1}{20}$ to $\frac{1}{25}$ of the time required with a standard sealed-off tube operated with bi-phase rectification at 20 mA and 30 kV peak. Moreover, the background was clean and the spots perfectly sharp. Under the same conditions, using the identical 11.46-cm-diameter Debye–Scherrer camera and a tungsten sample, a powder photograph which took 4 hr with the conventional sealed-off tube took only 10 min with the rotating anode tube. These patterns are shown in Figure 7. Thus, if we take the figure of merit of a conventional sealed-off tube as unity, that of the rotating anode tube would lie between 20 and 25. This latter figure could even be improved upon by sharpening the focus still further and by reducing the thickness of the beryllium windows from their present value

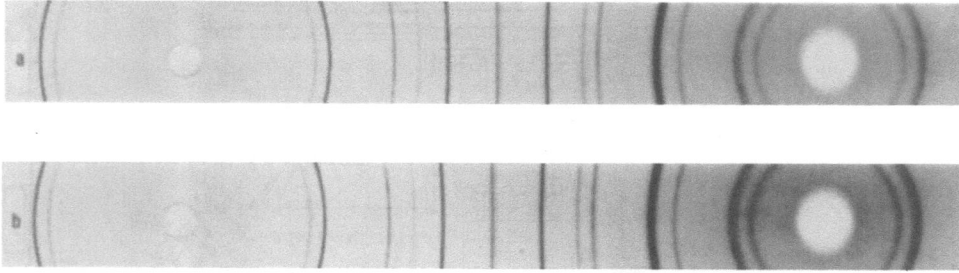

Figure 7. Debye–Scherrer patterns of tungsten powder taken with $\mathrm{Cu}\,K_{\alpha+\beta}$ radiation in a 114.6 mm diameter Philips camera. (a) With sealed-off tube at 20 mA, 35 kV, 4 hr; (b) With rotating anode tube, 250 mA, 30 kV, 10 min.

of 30 mils to 10 mils. Furthermore, the figure of merit of the rotating anode can be maintained indefinitely by cleaning the anode and windows as required, whereas this is impossible with sealed-off units which show spectral impurities and a marked falling off in performance after only 1000 hr of service.

ACKNOWLEDGMENT

Among the many involved in constructing the two X-ray tubes described above, the author wishes to thank F. W. Senchur and H. B. Ryden for their assistance in drafting and design and Norman Doyle for his help in assembling and operating the equipment.

REFERENCES

1. A. Taylor, "A 5 kW Crystallographic X-Ray Tube with a Rotating Anode," *J. Sci. Instr.* **26**: 225, 1949.
2. A. Taylor, "Improved Demountable X-Ray Tube," *Rev. Sci. Instr.* **27**: 757, 1956.
3. A. Taylor and K. H. Sueker, "Transistor AC Regulator for X-Ray Tube Current," *Electronic Industries*, May, 1963, p. 121.

DISCUSSION

G. A. Brown (Simens America Inc.): I would like to know when you use this tube in a stationary application, do you find that the plating of the target is homogenous enough to get the same intensity if you switch from one target to another when you have several plated on the same disc?

A. Taylor: What you have to do is adjust the milliamperage for each focal spot on each different material according to the load that the particular material will take. Depending on spot size, a stationary target of copper would take up to 20 mA, but zinc would only take up to 8–10 mA, and cobalt would be about 15 mA. On account of phase transitions, manganese flakes off when rotating the anode, but with the anode stationary a 10 × 1 mm focal spot can be operated at 10–12 mA. However, if a particular focal area should become eroded, the target is rotated a few degrees to expose a new area so you have, as it were, an infinite number of focal spots around the periphery. Alternatively, if you have patches of different metals, as would be required for microradiography, then you could have as many as 50 focal spot locations. As a result, the plated target would last indefinitely.

G. A. Brown: And are you able to balance the intensity relatively by adjusting the milliamps?

A. Taylor: Yes.

T. Barber (White Sands Missile Range): What sort of power were you able to get out of the chrome when you were using your rotating anode?

A. Taylor: We would normally operate the chromium target at roughly four to five times the excitation potential, namely at 25 to 30 kV. This keeps the background down. With such a voltage the chromium target could be operated at the full output of the tube, namely 275 to 300 mA, on a 10 × 1.0 mm focal spot.

R. P. I. Adler (Martin Company, Orlando): Have you considered using a filter material instead of a beryllium window to increase your intensity?

A. Taylor: There is no real point in doing that, because a beryllium window of thickness 20 mils makes very little difference to the intensity of most radiations. Chromium is beginning to show some change. If you want aluminum radiation, then you would perhaps even put a mylar window on. The danger then is that the splashback of electrons would burn a hole in the mylar pretty quickly and so you would have to cut your milliamps down. Incidentally, if the focus were enlarged from its conventional size of 10 × 1 mm to a circle of diameter 1.0 cm, the output of the tube could be greatly increased and fluorescence analysis could be carried out with an output of 1 A or more. I am planning an X-ray tube—it is still on the drawing board—where 25 kW could be continuously applied to the focal spot for fluorescence analysis. There is no reason why you shouldn't be able to do these things provided your target is big enough and rotating fast enough to dissipate the heat.

The cooling is very much more efficient than in a stationary target tube, simply because you are moving the target. The difficulty with a stationary anode tube is getting the flow of water fast enough. That is what is holding down the output. But here, by having the rotation, as well as a higher water flow, you can increase the power enormously—far more than you would think at first sight. So that the 25 kW fluorescence analysis tube is not out of the question. Thus, if you were using aluminum radiation for the softer elements, you could operate at 25 kW and, by gating the window, you could have your analyzing crystal in an adjacent vacuum chamber in direct communication with the primary X-ray source. Once you have this type of experimental equipment, almost anything is possible. You cannot do this type of work with standard, sealed-off apparatus.

A. Paddock (Denver Research Institute): When you have a number of targets on the rotating wheel, how do you index this from the inside when you want to switch from one sample to the next?

A. Taylor: That's very easy because the little pulley wheel on the outside has a one-to-one correspondence with the target face. You just mark it off into sections and label them cobalt, nickel, iron, and copper, corresponding to the plated regions on the target face. There is no problem there.

E. T. Peters (Manlabs, Inc.): How do you determine the focal spot size and shape?

A. Taylor: That is done by a pinhole camera arrangement. A lead disc with a tiny hole drilled in it is placed over the window of the X-ray tube and a pinhole camera image of the focal spot is observed on a fluorescent screen.

A. Vaciago: (Catholic University, Rome): Has any analysis been made of the degree of homogeneity of the cross section of your X-ray beam? You are certainly aware of the discussion in London last winter when it came out that when you go in for precision measurements on single-crystal X-ray diffraction you run into trouble with the fact that the cross section of the standard X-ray beam even after collimation is not quite homogenous. Has any measurement or any analysis of this sort been made on these beams?

A. Taylor: I haven't made an analysis, but commercial sealed-off X-ray tubes frequently have a "double-hump" for better dissipation of the heat. Some commercial tubes may exhibit two or even three focal lines. With the demountable rotating anode X-ray tube the focus can be adjusted both by mechanical movement of the filament and by altering the bias. I am quite sure that the focus is Gaussian rather than double-humped. But there is the danger that if you don't make a uniformly wound filament, one or two hot spots may form along the focal line. I have found that if you use six diode rectification instead of half-wave rectified AC one seems to get a very much more uniform spot. I can't explain it. I would have to do a complete analysis with an electrolytic trough to see how the electrons focus. But there isn't time for that. There are enough practical problems in developing the tube quite apart from these things. As far as we can tell from Weissenberg camera photographs of lead glycenate and from the work being done at Brookhaven, the tube is functioning just as steadily as a sealed-off tube, but with an X-ray intensity stepped up by a factor of 20 to 25.

A DUAL COUNTER X-RAY ANALYZER FOR THE RAPID QUANTITATIVE ANALYSIS OF TWO-PHASE SYSTEMS

B. S. Sanderson and L. E. MacCardle

National Lead Company
South Amboy, New Jersey

ABSTRACT

The normal procedure for a quantitative measurement of two phases with an X-ray diffractometer is to scan or count the intensities of two diffraction peaks and then to calculate the ratio of the two phases using an appropriate equation. This paper will describe an improved method of quantitatively measuring two phases in a sample. The diffractometer, which was constructed by Philips Electronics Instruments, consists of two fixed scintillation counters mounted at the Seeman–Bohlin focusing positions for the two desired diffraction lines. Each counter is preceded by a lithium fluoride curved crystal monochromator. The output from each counter feeds to a separate scaler which can be arranged to gate both scalers at a preselected number of counts. At this point a tape printer prints out the counts for both scalers. These counts are then related to the proportions of each phase present by an appropriate equation. Most of the electronics are solid-state designs. This paper will describe the instrumentation and show how it can be used to measure rapidly and precisely the ratio of anatase to rutile in a titanium dioxide system. The savings in time over the conventional method can be as much as 100 to 1.

INTRODUCTION

In order to quantitatively measure the relative amounts of the phases in a two-phase system, it is necessary to measure at least two diffraction peaks which are characteristic of the two phases. This is usually done on a diffractometer by scanning each peak and measuring the peak intensity after subtracting the background. The precision of a rate-meter recording is dependent on the speed of scanning; therefore, in order to get good precisions it is necessary to scan slowly. When one of the phases is present at a very low level it is necessary to increase the time constant of the ratemeter and scan very slowly in order to distinguish this peak from the background. A fixed count procedure on the two peaks would eliminate the slow scanning but it is still necessary to position the diffractometer at the proper 2θ positions. A new X-ray analyzer will be described which will take advantage of the accuracy of the fixed count method and will avoid the necessity of setting the 2θ angles for each peak during an analysis. This is accomplished by using two counters mounted at the appropriate Seeman–Bohlin focusing positions for the two diffraction peaks. A scaler is used to count the output of each counter simultaneously. One of the scalers is used to gate both counters by setting in a preselected count on the proper scaler. The two counts are then printed out on a tape printer.

X-RAY OPTICS

Figure 1 gives a diagram of the Seeman–Bohlin focusing conditions for two diffraction peaks. Note that for a focusing circle of radius R all the focusing positions for any

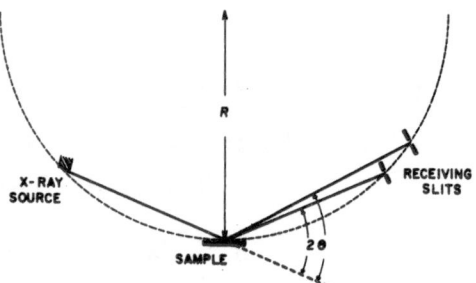

Figure 1. Seeman–Bohlin focusing for two diffraction peaks.

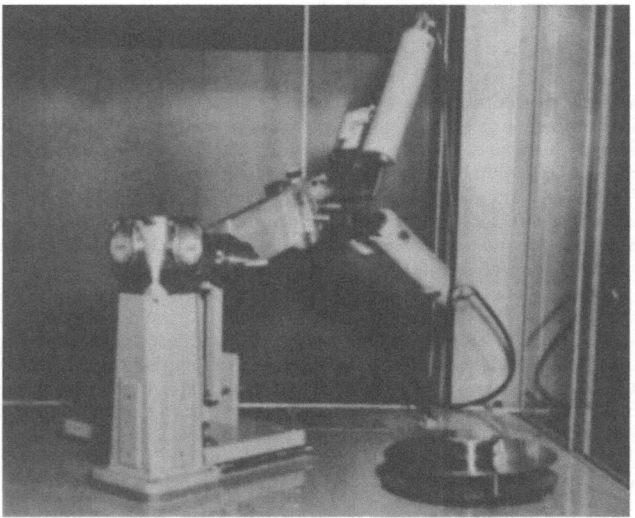

Figure 2. General view of the goniometer.

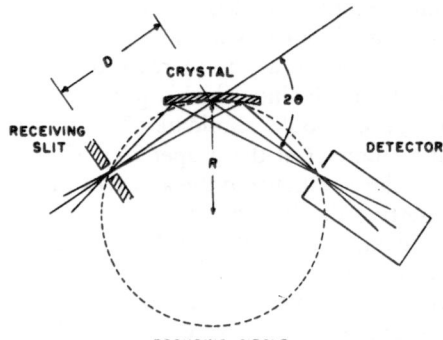

Figure 3. Arrangement of the monochromator.

value of 2θ lie on this circle. If the sample is a powder with a surface having a radius of curvature equal to R, the beam will be exactly in focus. If the sample has a plane surface and if R is large enough the beam will be approximately focused. Once the apparatus has been lined up properly, there is no further movement and so alignment is maintained.

A general view of the goniometer is shown in Figure 2. The slit for the higher 2θ value is first focused by moving the arm which holds both slits. Then the fine adjustment is used to focus the slit for the other peak. The angle of the sample can also be adjusted independently of the 2θ adjustment.

After the beams have passed through the receiving slits, they are monochromatized by adaptations of the curved crystal monochromator, made by the Advanced Metals Research Corporation.[1] This monochromator uses a LiF crystal, elastically bent to a radius of 132.6 mm using Johann-type focusing.[2] Figure 3 shows the arrangement of the monochromator. This arrangement, with $D = 47.0$ mm and $R = 66.1$ mm, gives excellent intensities with high peak-to-background ratios. There is no need for a beta filter with this arrangement.

DETECTORS

The detectors used are Philips transistorized scintillation counters, Cat. No. 52572. These counters have lower noise levels than most scintillation counters.

ELECTRONIC CIRCUITS

The signals from the counter preamplifiers are fed to separate Philips type PW 4022 pulse amplifiers. These units are made up of a stabilized high-voltage supply and an amplifier. The output pulses can be attenuated by two switches in steps of 1, 2, 4, 6, and 8 and 1, 10, 100, 1000 so that a threshold discrimination results. This prevents noise pulses from going to the amplifier. The amplified pulses are then fed to individual scalers. These scalers are Philips type PW 4230 and are solid state devices. A preset count switch makes it possible to use a signal from the preselected decade and to stop the counting of both scalers. The highest preset count which can be set is 4×10^7. If the preset count selector switch is set at infinity the scalers will continue counting until stopped by an external signal. It is also possible to use an internal calibrating signal of 120 cps to give a fixed time count for one of the scalers. For this mode of operation, one scaler acts as a timer for the other. A Philips type PW 4242 ratemeter can be connected in parallel to either one of the two scalers by a jacked cable. The counting rate can be monitored on a meter or by plugging in a strip chart recorder to the ratemeter. The ratemeter is also of all solid-state design. It is particularly useful in the alignment of the instrument. The stop pulse from the scaler goes to a Philips type PW 4200 printer control to initiate the print-out process. This unit generates interrogation pulses which are fed to each decade of the scaler and also, in parallel, to a memory unit in the printer control. When each decade is filled, the output pulse generates a signal in the printer control which initiates the selection of the proper printer key in a Victor printer. It is possible to select a printing mode by a switch, which will print out the counts in either scaler alone, or print the counts in both scalers on successive lines of the printer.

APPLICATION TO ANATASE–RUTILE SYSTEM

This instrument has been designed for the analysis of anatase and rutile in titanium dioxide pigments. The 101 line of anatase is found at $2\theta = 25.3°$ for Cu K_α radiation and the 110 line for rutile is at 27.4°. The close angular proximity of these two lines make the placement of the counters very difficult. However, by inverting one of the curved crystal monochromators, the two counters can be easily mounted.

[1] References are at the end of the paper.

Table I. Results of Calculations on Rutile

Standard	R	$\bar{A}$	$\Sigma(A - \bar{A})^2$	Degrees of Freedom	% Rutile Standard	% Rutile Calculated	% Rutile Difference
1	10,000	930.71	20,642.96	23	99.350	99.465	−0.115
3	10,000	1462.67	37,839.33	23	95.465	93.996	1.469
4	10,000	1015.79	26,009.96	23	97.735	98.548	−0.813
5	10,000	903.21	25,913.96	23	99.435	99.765	−0.330
6	10,000	2313.54	64.597.96	23	86.140	86.397	−0.257
	Standard deviation	0.429		115			

$$\% \text{ rutile} = 100/(0.903 + 0.000110 \, A)$$

$$\text{Standard error (calculated \% rutile)} = 0.476 \sqrt{\frac{1}{M} + 0.20 + \frac{(A - 1325.2)^2}{1395078}}$$

The greatest interest in titanium dioxide pigment analysis is in the region of high rutile. A series of standards was prepared in the range of 85–100% rutile. In this range, the relationship between the percent rutile and the intensities of anatase and of rutile can be fitted well to the following equation:

$$\% \text{ Rutile} = \frac{100 I_R}{I_R + k I_A} \tag{1}$$

where I_R and I_A are the intensities above background of the rutile peak and the anatase peak, respectively, and k is a constant. To take advantage of the simultaneous counting ability of the analyzer, a method is used in which a count for rutile is preselected (i.e., 10,000 or 20,000). When the analysis is initiated, both scalers will accumulate counts from their respective counters until the rutile count reaches the preselected value. At this time, both scalers are stopped and the results printed out on a Victor tape printer. If there were no background the anatase count would be a direct ratio of the rutile. In the presence of background the following equation represents the relationship between the observed counts and the percent rutile:

$$\% \text{ Rutile} = \frac{100(R - B_R)}{R - B_R + k(A - B_A)} \tag{2}$$

where R and A are, respectively, the observed rutile and anatase counts and B_R and B_A are the backgrounds for these two peaks. Rearranging equation (2) we have:

$$\frac{1}{\% \text{ Rutile}} = \frac{R - B_R - 100 B_A k}{100(R - B_R)} + \frac{(k)}{100(R - B_R)} A \tag{3}$$

where k, B_R, B_A, and R are all constants in the range of rutile from approximately 85–100%, so that this equation is linear when $1/\%$ Rutile is plotted versus A. The two constants for this equation were fitted by least squares and the results are shown in Table I. The degree to which equation (3) approaches linearity is dependent on keeping the background as low as possible. For a sample of approximately 100% rutile and 20% by weight TiO_2 in polystryrene, the B_A is about 7% of the rutile peak. The counting rate for the rutile peak is about 1670 counts/sec. This means that a 10,000 count analysis takes about 6 sec.

The uncertainty of the results comes from two different sources. First there is the variation due to the lack of reproducibility of the samples when replicated. This is designated as the precision and for these data (i.e., 10,000 rutile counts) is measured by a standard deviation of 0.4% rutile with 115 degrees of freedom. A second source of error is due to the lack of agreement of the standards to equation (3). This could arise either because the mathematical model is not exact or because the value of the standard is not known exactly. This error is called accuracy. The overall error of a value predicted from equation (3) is given by the following equation:

$$SE(\% \text{ Rutile}) = \frac{s}{B}\sqrt{\frac{1}{M} + \frac{1}{N} + \frac{(A - \bar{A})^2}{Z}} \tag{4}$$

where M is the number of replications and N is the number of standards used to fit equation (3), s is the standard deviation of A which is the anatase counts including background, $\bar{A}$ is the average value of the counts used to estimate equation (3), Z is a constant used in the estimation of equation (3), and B is the constant which transforms s to units of percent rutile.

It should be noted that when an estimate is made at the mean value of the standards, the error is a minimum, since the last term in equation (4) becomes zero. As an estimate is made farther from the mean value of the standards, the reliability becomes poorer. Note also that the term $1/N$ is fixed by the number of standards used and the value of the expression under the square root sign can not be reduced below the value of $1/N$.

The expected value of the standard deviation of A is $\sqrt{R + A}$, which is about 1000 for an R of 10,000. The average observed value of the standards was found to be about 1200, indicating that only a small improvement can be made without increasing the number of rutile counts. A time of 6 sec to accumulate 10,000 rutile counts for a 20% TiO_2 sample in lucite gives an overall error of about 1%. This is comparable to the error obtained with the standard ratemeter scan, which takes about 5 min. The saving in time and convenience is quite obvious.

ACKNOWLEDGMENTS

The authors would like to acknowledge the contributions of Dr. R. Westberg, C. Zalkin, and R. Deichert of Philips Electronic Instruments, and of Dr. J. A. Dunne, formerly with Philips Electronic Instruments, in the development and design of this analyzer.

REFERENCES

1. D. M. Koffman and S. H. Moll, "A Curved Crystal Monochromator for the X-Ray Diffracto-meter," *Norelco Reporter* 11: 95, 1964.
2. H. H. Johann, "More Intense X-Ray Spectra Obtained with Concave Crystals," *Z. Physik* 69: 185, 1931.

DISCUSSION

A. Taylor (Westinghouse Research Labs.): You've got percent rutile listed as 99.465. How many decimal places can you carry? Are they genuine places or are they just machine calculations?

B. S. Sanderson: They are just machine calculations. This is a computer output.

A. Taylor: Well then, what do the figures mean?

B. S. Sanderson: The precision of the numbers is represented by the standard deviation which, of course, is not anywhere comparable to three figures. However, by merely increasing your count or by using other standards, one can bring the standard deviation down. I have reduced this number down to 0.1 now, and by increasing the count one can bring this down to 0.01. I like to report one more than a significant figure in order to determine where the variation lies. The computer program is written very simply to print out all of these decimal places because potentially we can get numbers of this sort. We don't show them in here, but we have had some standardization procedures in which we do. I might point out that by adjusting the slit sizes and making some more careful adjustments we can reduce the background. The background is essentially the thing that hurts you. I might also point out that the standard deviation expressed as percent rutile, obtained by going through a standard error analysis and differentiating the equation for the percent rutile, comes out to be about $(100k/R)[E(A)]$. Now in our case R is 10,000 and k comes out approximately 1. You have a 100 times reduction in the error in percent rutile over the error in A. So even though you seem to be getting large fluctuations in the value of A that is measured here, it does not show up as a very large value in the percent rutile.

E. F. Sturcken (E. I. du Pont de Nemours & Co.): With regard to other factors which affect the intensity, such as preferred orientation or large grain size, it would seem that you would need to make sure that you had enough beam area on the sample. My question is, if you wanted to extend your method to peaks which were farther apart on your Seeman–Bohlin circle, would you then have to go to a curved specimen so that you would get enough beam area in order to avoid these intensity errors?

B. S. Sanderson: This could be very easily done. In fact, we have tentatively looked at some curved specimens. Since you have a fixed circle, there is no problem in making up some kind of a jig for powder samples, or if you are going to mount them in plastic, making your die curved to that radius. So you can make a curved sample almost as easily as you can a flat one by merely changing your sample arrangement. As far as preferred orientation in this case, we are talking about powders that have particle sizes of the range of 0.2 μ so this is no problem.

CORRECTION FOR NON-LINEARITY OF PROPORTIONAL COUNTER SYSTEMS IN ELECTRON PROBE X-RAY MICROANALYSIS

Kurt F. J. Heinrich, Donald Vieth, and Harvey Yakowitz

National Bureau of Standards
Washington, D.C.

ABSTRACT

While the theoretical basis for the correction of non-linearity of detector systems is well known, methods for the determination of dead-time effects must be adapted to electron probe microanalyzer systems. Two such methods, one employing both X-ray and current measurements and the other employing simultaneous X-ray measurements on two spectrometers, are described. The effect of pulse-height shrinkage at high counting rates on the linearity of the detector system is discussed. When the proposed corrections for the dead-time of X-ray detector systems employing proportional counters are applied to the X-ray intensity measurements obtained with the electron probe microanalyzer, count rates as high as 50,000 counts/sec can be used.

INTRODUCTION

The response of X-ray detector systems is non-linear due to the limited capability of detectors as well as of the associated circuitry to register photons received within very short time intervals.[1,2] The time interval after the arrival of a photon in the detector during which the system is incapable of registering the arrival of another photon is called dead time (τ). Schiff [3] proposed in 1936 to correct for the resulting counting loss by the equation

$$N = N' \exp(N\tau) \tag{1}$$

in which N' is the observed counting rate and N is the true counting rate. This equation is valid for proportional and scintillation counters, provided that the intrinsic resolving time of the detector is large compared with the intrinsic resolving time of the associated electronic recording circuits. Ruark and Brammer[4] noted that Schiff's equation is not applicable when the counting loss is caused by a large non-extendible dead time in the electronic recording apparatus, or in the detector (in the case of Geiger counters), if this dead time is long compared with the charge collection time of the detector. Ruark and Brammer derived for this case the following correction equation:

$$N = \frac{N'}{1 - N'\tau} \tag{2}$$

The theory and practice of dead-time correction is exhaustively treated in the literature,[5-8] and dead-time corrections are made as a matter of routine in the branches of physics and chemistry associated with the measurements of radiation by the detector devices mentioned above.

---

[1] References are at the end of the paper.

This correction, however, has not received equal attention in spectrochemical analysis by X-ray fluorescence and in electron probe microanalysis, although the use of equation (2) was advocated by Wittry.[9] X-ray fluorescence analysts usually assume[10] that their detection systems are linear up to 10^4 counts/sec; since the analysis is based solely upon empirical calibration curves, this assumption is not tested in practice.

In electron probe microanalysis it is frequently impossible to obtain standards of composition close to that of the specimens to be analyzed. The analysis is therefore usually performed by relating the X-ray intensity observed on the specimen to that obtained from a pure element. Corrections are applied in order to account for X-ray absorption within the specimen, for fluorescent X-ray emission, and for the variation of X-ray emission as a function of atomic number (atomic number effect). Seldom, however, is the dead time of the detector system determined, and the corresponding correction applied to the observed counting rates. Rates up to 5×10^4 counts/sec can be obtained on modern microprobes under typical conditions. Such counting rates are desirable when a great number of analyses must be performed on a single specimen. In such cases, proper dead time corrections are mandatory.

Most microprobes are provided with two or more curved crystal X-ray spectrometers. Usually, one or more spectrometers are located in an evacuated enclosure. This arrangement renders the common techniques of dead time determination by the use of X-ray filters or radioactive emitters impractical. Instead, it can be assumed that, for a given set of conditions, the flux of X-ray photons reaching the detector is proportional to the current of the electron beam incident upon the specimen:

$$N = ki \tag{3}$$

where i is the beam current. If equation (2) is applicable, it follows that

$$\frac{N'}{i} = k(1 - N'\tau) \tag{4}$$

Therefore, a plot of N'/i as a function of N' will produce a straight line. The value of k can be obtained by extrapolating N'/i to $N' = 0$, and τ can be determined by transforming equation (4) into

$$\tau = \frac{1}{N'} - \frac{1}{ki} = \frac{1 - (N'/i)/k}{N'} \tag{5}$$

Once it is proven for a given set of conditions that equation (2) is applicable, the dead time can be determined by measuring X-ray counting rates at two beam-current values with the ratio

$$R = \frac{i_2}{i_1}$$

Then the dead time can be determined by the equation

$$\tau = \frac{1}{R - 1}\left(\frac{R}{N_2'} - \frac{1}{N_1'}\right) \tag{6}$$

In a system governed by equation (1), a slightly bending curve would be obtained in the above mentioned plot. For counting rates up to 4×10^4 and dead times up to 3 μsec, the differences resulting from the use of equation (1) rather than equation (2) will be within experimental error as shown in Figure 2.

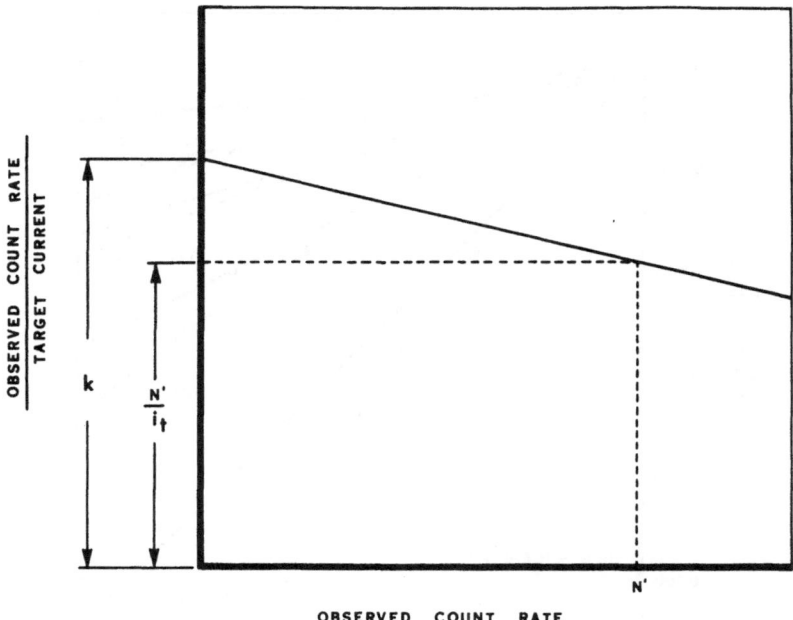

Figure 1. Schematic plot for the determination of dead time by current measurement.

Serious bending of the curve can be caused, however, as a consequence of the unexpectedly large shrinkage of pulse height at high counting rates[11] observed with several electron probes equipped with proportional counters. If such an effect is present, a pulse-height selector setting which is adequate for lower counting rates can result in serious nonlinearity at high count-rate levels.

Most microprobes are not equipped with Faraday cages for measuring beam current. Even when such a device is available, the beam current cannot be measured at the same time as the X-ray intensity. It is therefore preferable to replace the beam current in equation (3) by the target current (Figure 1), or by a monitor current, which is proportional to the beam current, provided by some instruments. In this case, the proportionality between beam current and target or monitor current over the range of measurement must be confirmed experimentally (Figure 3).

An accuracy of current measurements of 1–2% must be maintained over the measured range, which typically covers from 10^{-7} to 10^{-9} A. This is not obtainable in many current-measuring systems installed in microprobes. If the target current (specimen current) is employed, the impedance of the meter input must be low enough to avoid the effects of specimen biasing at higher current levels.[12]

If the measurement of currents is difficult or impossible to perform with the available instrumentation, the dead time can be determined by making simultaneous X-ray measurements on two independent detector circuits.[13] This method is well suited to electron probe instruments having more than one spectrometer. A series of measurements is made in which the counting rate in one detector is higher than in the other, so that dead-time effects appear more prominently in the data produced with the first counter. In a second series the ratio of counting rates is reversed. The evaluation of the two series yields the dead time of both detector systems.

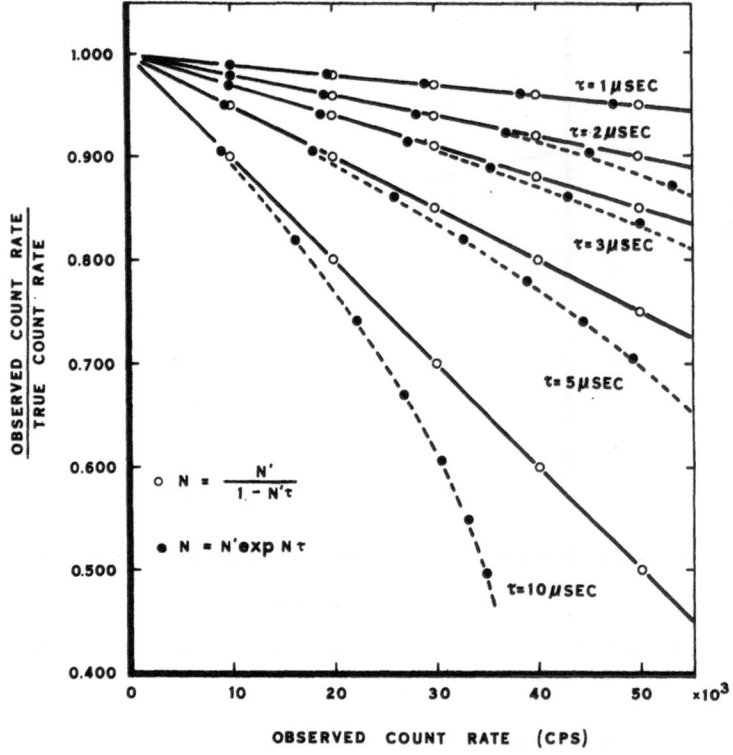

Figure 2. Comparison of two proposed expressions for calculating dead time.

Using subscripts 1 and 2 to indicate the two detector circuits, we obtain from equation (2) the following:

$$N_1 = \frac{N_1'}{1 - N_1'\tau_1}$$

$$N_2 = \frac{N_2'}{1 - N_2'\tau_2}$$

Designating $N_1/N_2 = C$, we obtain

$$\frac{1}{N_2'} - \tau_2 = C\left(\frac{1}{N_1'} - \tau_1\right)$$

$$\frac{N_1'}{N_2'} = C + N_1'(\tau_2 - C\tau_1) \tag{7}$$

By plotting N_1'/N_2' as a function of N_1', a straight line is obtained. Its extrapolation to $N_1' = 0$ yields the value of C.

To determine τ_1 and τ_2, two sets of measurements, with substantially different values for C, are necessary. Where C_1 is the first set, obtained from the plot of N_{11}' versus N_{11}'/N_{21}', C_2 is the second set, obtained from plot of N_{12}' versus N_{12}'/N_{22}'

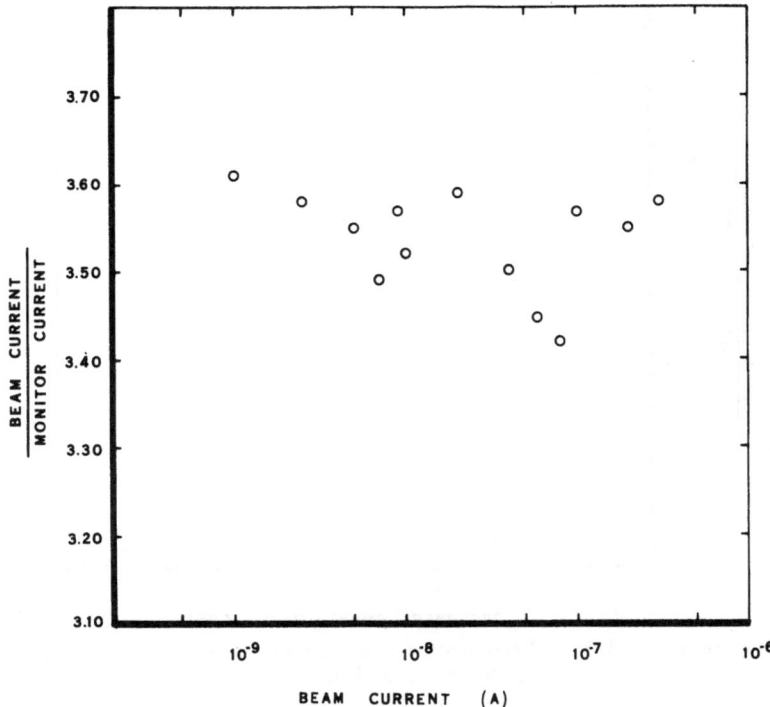

Figure 3. Constancy of the ratio of beam current to monitor current with respect to beam current.

N'_{11} is the observed counting rate, spectrometer 1, first set, N'_{12} is the observed counting rate, spectrometer 1, second set, N'_{21} is the observed counting rate, spectrometer 2, first set, and N'_{22} is the observed counting rate, spectrometer 2, second set, we obtain:

$$\tau_1 = \frac{1}{C_2 - C_1}\left(\frac{1}{N'_{21}} - \frac{1}{N'_{22}} - \frac{C_1}{N'_{11}} + \frac{C_2}{N'_{12}}\right)$$

$$= \frac{1}{C_2 - C_1}\left[\frac{1}{N'_{11}}\left(\frac{N'_{11}}{N'_{21}} - C_1\right) - \frac{1}{N'_{12}}\left(\frac{N'_{12}}{N'_{22}} - C_2\right)\right]$$

$$\tau_2 = \frac{1}{C_2 - C_1}\left(\frac{C_2}{N'_{21}} - \frac{C_1}{N'_{22}} - \frac{C_1 C_2}{N'_{11}} + \frac{C_1 C_2}{N'_{12}}\right) \qquad (8)$$

$$= \frac{1}{C_2 - C_1}\left[\frac{C_2}{N'_{11}}\left(\frac{N'_{11}}{N'_{21}} - C_1\right) - \frac{C_1}{N'_{12}}\left(\frac{N'_{12}}{N'_{22}} - C_2\right)\right]$$

The values for C_1, C_2, N'_{11}, N'_{12}, N'_{11}/N'_{12} and N'_{12}/N'_{22} can be read directly from the graphs (Figures 7 and 8), so that the calculation is quite simple.

EXPERIMENTAL

Figure 4 shows a plot of the observed counting rate for unit target current N'/i_t as a function of the observed count rate N'. The experimental conditions were as follows:

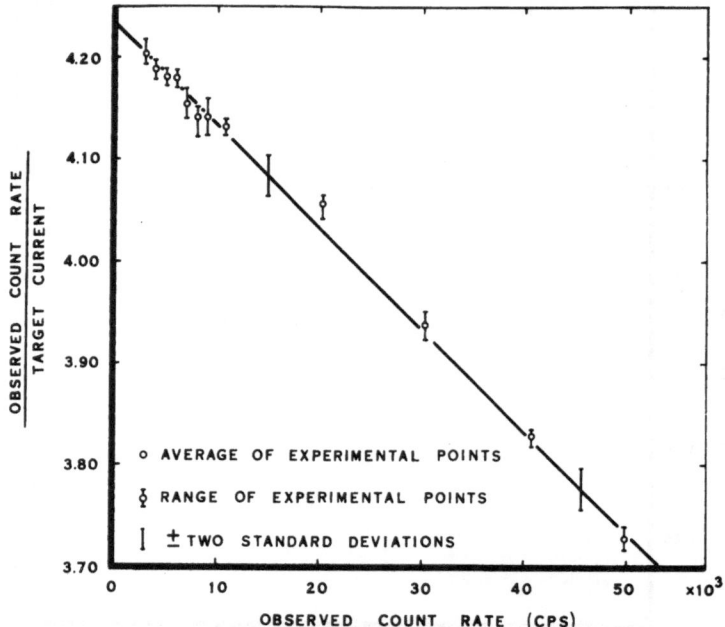

Figure 4. Dead-time determination using target current measurement.
Spectrometer no. 1, Cu K_α.

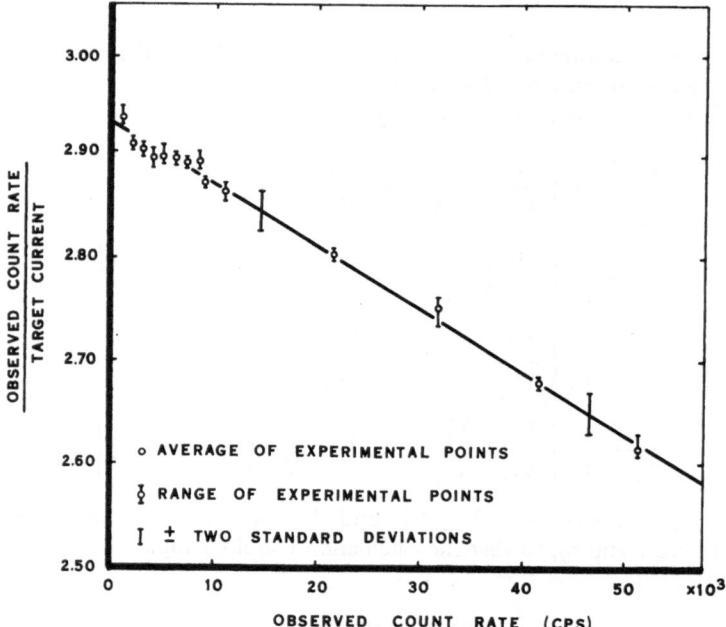

Figure 5. Dead-time determination using target current measurement.
Spectrometer no. 2, Cu K_α.

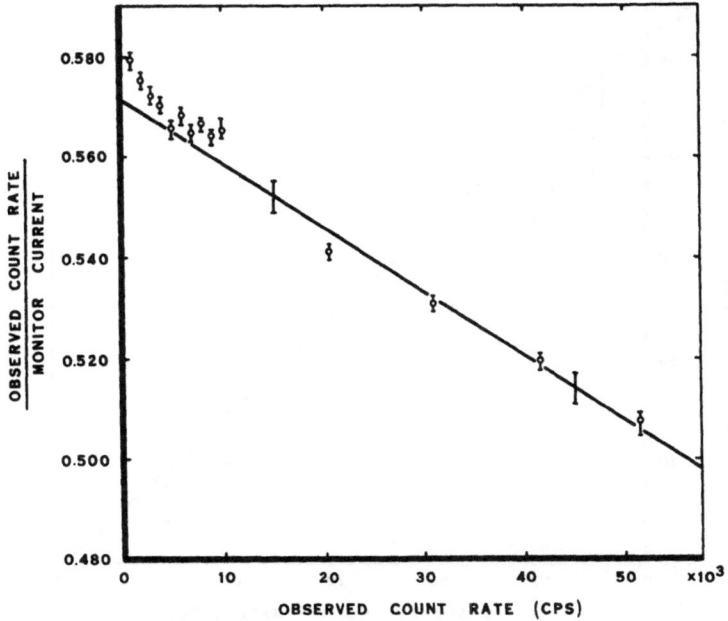

Figure 6. Dead-time determination using monitor current measurement. Spectrometer no. 2, Cu K_α.

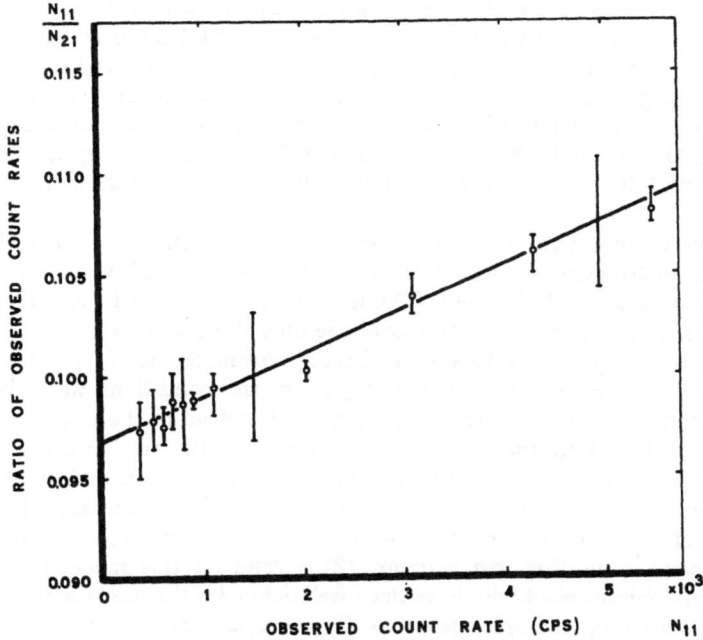

Figure 7. Dead-time determination using the count-rate method. Measurement series 1.

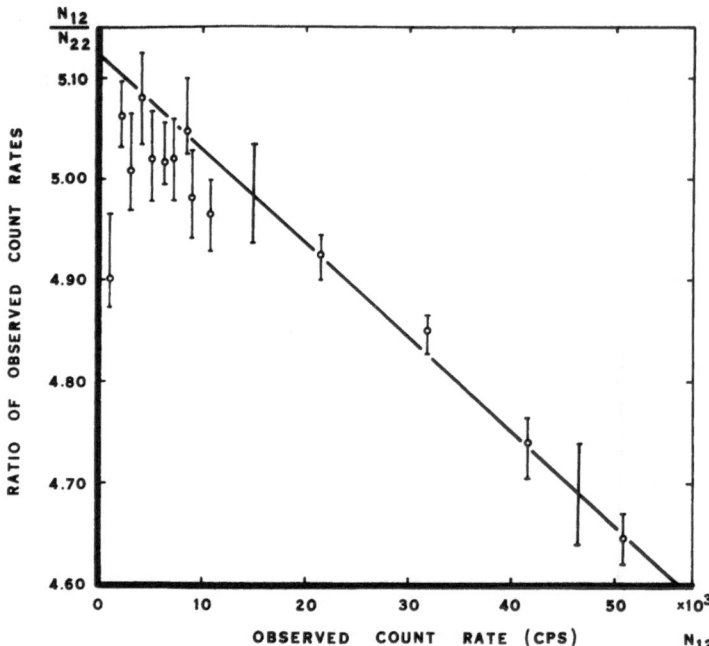

Figure 8. Dead-time determination using the count rate method. Measurement series 2.

The instrument used was the Applied Research Laboratories electron probe micro-analyzer EMX. Spectrometer no. 1, with a sealed argon–methane proportional detector, was employed in the experiment, registering the Cu K_α line produced on pure copper, with an accelerating voltage of 30 keV. The detector voltage was 2100 V, and the pulses were amplified to a mean pulse height of 40 V (monitored on an oscilloscope), at low counting rates. Hamner pulse height analyzer N-302 was used in integral position, with the base line set at 10 V. Between 100,000 and 200,000 pulses were collected for each point.

The target current was fed into a Keithley 610B electrometer, and the voltage output of this instrument connected to a Dymec model 2210 voltage-to-frequency converter, operating at the 10 V range. The frequency so generated was scaled with the ARL scaler at the same time as the output of the pulse height analyzer.

Figure 5 shows similar results obtained on spectrometer no. 2, using an identical detector. In the experiment illustrated in Figure 6, the conditions were the same as above, except that the monitor current was measured instead of the target current.

Figures 7 and 8 show the curves corresponding to the two measurement series needed for the X-ray signal ratio method explained previously. In each of these series, Cu K_α radiation was registered in one spectrometer, and Cu K_β radiation in the other. In all experiments, the curves obtained were linear up to observed counting rates of 5×10^4 counts/sec, indicating that equation (2) is valid for this range. In most cases, the slopes of these curves would not have been well defined if the measurements had been restricted to the usual operating range up to 10^4 counts/sec. Since in the lower counting range, time intervals up to 100 sec were employed, it is doubtful that accumulation of more counts would have increased the precision, in view of slow drifts of signal intensity.

Table I. Results of Dead Time Determinations

Method	Spectrometer No. 1 $\tau(\mu \text{ sec})$	Spectrometer No. 2 $\tau(\mu \text{ sec})$
counts/target current	2.33	2.07
counts/monitor current		2.22
counts/counts	2.40	2.29

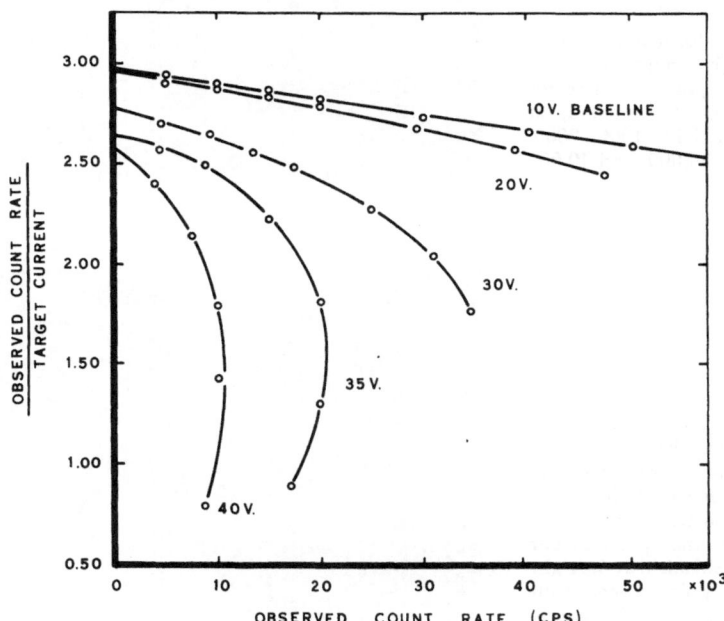

Figure 9. Effect of pulse-height selector baseline position on the determination of dead time.

Table I summarizes the numerical values for dead time obtained in the above experiments. For evaluation of the usefulness of the data contained in Table I, it should be considered that the effect of an error in the dead time σ_τ will be a relative error in the counting rate such that

$$\frac{\delta N}{N} = N\delta\tau$$

Thus an error of 0.22 μsec in the value of τ (equal to the range of value under spectrometer no. 2, Table I) would produce, at a counting rate of 4×10^4 counts/sec, a relative error of 0.9% in the determination of the true counting rate N.

The effect of pulse-height shrinkage is illustrated in a series of measurements similar to that shown in Figure 2, except that the baseline setting of the pulse-height analyzer was varied (Figure 9). When the baseline is raised sufficiently, the apparent dead time of the system increases and becomes a function of the counting rate. In extreme cases, an increase of true counting rate beyond a limiting value will produce a decreasing

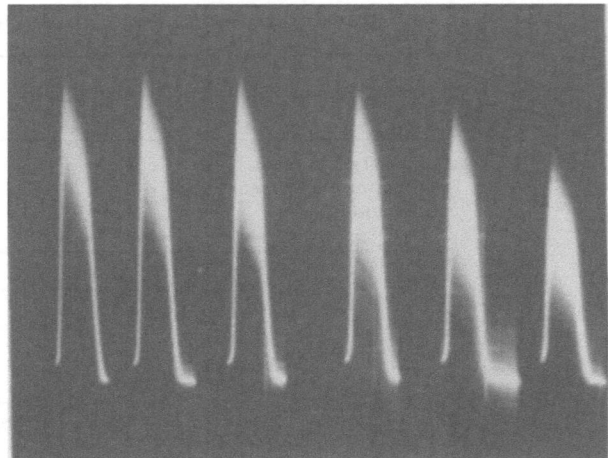

Figure 10. Pulse-height shrink-age at high counting rates. Amplified proportional detector pulses of Au $M\alpha$ radiated, at (from left to right) 1000, 2000, 4000, 10,000, 20,000, and 40,000 counts/sec.

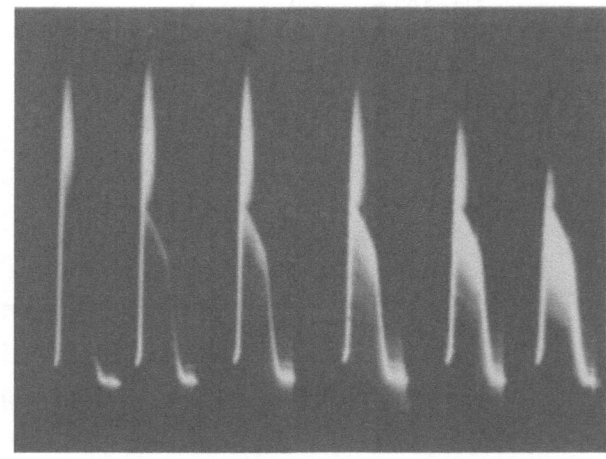

Figure 11. Same pulses as on Figure 10. The baseline of the pulse-height selector has been set close to the lower limit of the pulse-height distribution. Pulses accepted by the pulse-height selector are unblanked at the right side of each pulse trace. With rising counting rate, an increasing fraction of the pulses is rejected by the pulse-height selector. The rejected pulses appear bright at the right half of each pulse trace.

apparent counting rate. As the average pulse height decreases, the width of the pulse-height distribution increases. Hence, at high counting rates some of the lower pulses fall below the baseline. As these pulses are not accepted by the pulse-height analyzer, the coincidence loss increases. This is illustrated by a series of oscilloscope photographs similar to those of Figure 10, except that the output of the pulse height analyzer is allowed to unblank the oscilloscope brightness control, by means of a device to be described in more detail elsewhere.[14] Thus, the pulses accepted by the pulse-height analyzer are marked by interrupted traces, while the pulses which are rejected are uninterrupted. It can be clearly observed in Figure 11 that the number of rejected pulses below the baseline increases rapidly with increasing counting rate.

A curious effect produced by the combination of high counting rates and close baseline setting can be seen in Figure 12. In the passage through an intense peak during a wavelength scan, excessive intensity produced a decrease of the apparent counting rate in the zone of highest intensity, so that instead of a single peak, an apparent line doublet was obtained. This effect did not appear when the baseline was lowered sufficiently.

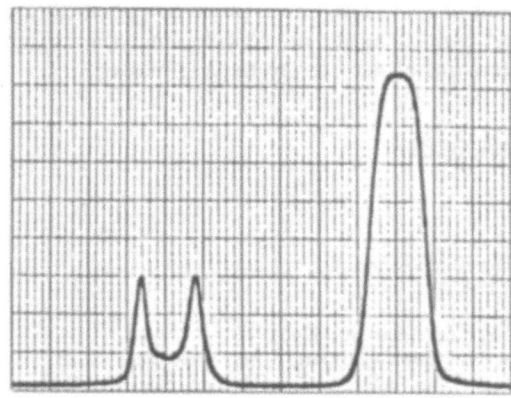

Figure 12. X-ray wavelength scan through Cu K_α, at a count rate of 50,000 counts/sec. High baseline setting (left side) results in suppression of the center of the peak. Lowering of the baseline (right side) restores the normal tracing of the peak.

SUMMARY

The experimental results here presented show that, under appropriate conditions, equation (2) describes effectively the coincidence losses occurring with gas proportional counting systems in the microprobe, at counting rates up to 50,000 counts/sec. The dead time can, and should, be determined using these high counting rates; these rates can be used for analytical purposes where rapid accumulation of statistically meaningful data is desired.

The dead time can be determined in electron-probe microanalyzers by combining current measurements with X-ray measurements, or, in multispectrometer instruments, by combination of X-ray measurement on two independent circuits. Where access to the X-ray path is available, it is suggested that measurements with a single filter in and out be performed, as suggested by Short.[10] In this case, in equation (7), $\tau_1 = \tau_2$, so that

$$\frac{N_1'}{N_2'} = C + N_1'\tau(1 - C)$$

After determining C at the intercept $N_1' = 0$ of the plot of N_1'/N_2' versus N_1', the value of τ can easily be calculated. For a quick estimation of the magnitude of the coincidence loss, it is useful to remember that equation (2) can be rewritten as

$$\frac{N - N'}{N} = N'\tau$$

so that the observed counting rate (in units of 10^4 counts/sec) multiplied with the dead time (in μsec) gives the percentage of coincidence loss.

The methods described in this paper are preferable to the empirical method of Short, in which a non-linear correction curve must be constructed, and great weight is given to the data obtained at low counting rates, usually with poor statistical precision. The underlying concept of a linear range of negligible coincidence loss is not deemed useful, and can in fact be misleading. Measurable coincidence losses do indeed occur even at counting rates well below 10^4 counts/sec.

Fit of data to equation (2) cannot, however, be taken for granted unless the experimental conditions are properly chosen. The dead time of a system is determined by the whole system rather than by the detector only. Improper settings of the pulse-height analyzer with respect to mean pulse height may affect the fit, due to pulse-height shrinkage at high counting rates. To establish the best conditions for microprobe analysis at high counting rates, further study of the characteristics of detector systems is needed.

REFERENCES

1. H. Friedman, *Electronics* **18**: 132, 1945.
2. K. Lonsdale, *Acta Cryst.* **1**: 12, 1948.
3. L. J. Schiff, *Phys. Rev.* **50**: 88, 1936.
4. A. Ruark and F. E. Brammer, *Phys. Rev.*, **52**: 322, 1937.
5. N. Hole, *Arkiv. Mat., Astro. Fys.* **33A** (11), 1947.
6. R. Jost, *Helv. Phys. Acta* **20**: 173, 1947.
7. M. Blackman and J. L. Michiels, *Proc. Phys. Soc. (London)* **60**: 594, 1948.
8. R. D. Evans, *The Atomic Nucleus*, MacGraw-Hill Book Company, Inc., New York, 1955.
9. D. B. Wittry, in: *X-Ray and Electron Probe Analysis*, ASTM Special Technical Publication No. 349, ASTM, 1963.
10. M. A. Short, *Rev. Sci. Inst.* **31**: 618, 1960.
11. S. L. Bender and E. J. Rapperport, "The Electron Microprobe," in: *Proceedings of the Symposium on Electron Probe Microanalysis*, Electrochemical Society, John Wiley & Sons, Inc., New York, 1966.
12. K. F. J. Heinrich, in: W. M. Mueller, G. R. Mallett, and M. J. Fay (eds.), *Advances in X-Ray Analysis, Vol. 7*, Plenum Press, New York, 1963, p. 325.
13. W. Petzold, *Z. Angew. Phys.* **15**: 158, 1963.
14. K. F. J. Heinrich, to be published.

DISCUSSION

J. Merritt (Shell Development Company): You weren't explicit about this, but I assume you have to make the dead time correction with a new calibration every time you have a new type of target. If I go from one target sample to a different sample type, do I have to make a new calibration to determine the dead time, or can I transfer it from one sample to another?

K. F. J. Heinrich: You mean if you go from one wavelength to the other, for instance?

J. Merritt: Well, all right, take that as an example.

K. F. J. Heinrich: This aspect we have not yet determined. One would predict, of course, that if, as probably should be done, they have built a well-determined dead time into the electronic system of your instrument, this should not be necessary. On the other side, you see the fact that we have a tremendous pulse shrinkage over the range from 0 to 50,000 counts, and we still get linearity on the plots, which seems to indicate that the value of the dead time is not seriously affected by the pulse height. Therefore I would think that under these conditions you would not have to determine your dead time for every situation you run into. This, of course, is only true if the pulse-height analyzer is properly set, because if you have a high-energy pulse in one side where your baseline is far below your distribution and then you go to a soft X-ray, say aluminum, the situation might be completely different. I would predict that under the conditions we have used on copper we would not have had much luck with aluminium at 50,000 cps. But this must be investigated further.

L. A. Fergason (Mallinckrodt Chemical Works): I am glad to hear you talk about this shift. We have been using a xenon counter, and we observed three phenomena on our multichannel analyzer: (1) a large shift with counting rate, (2) the broadening of the peak as it goes to a higher counting rate and (3) a change in shape of the peak. It goes from a nice Gaussian distribution to a very skewed peak at a high counting rate.

K. F. J. Heinrich: Yes, we have observed the same thing. I was very glad to hear you mention it today. We have observed this, in fact, also on fluorescent instruments. One argument is that most of our detectors in the probe have a rather narrow slit, and if this slit is then normal to the wire of the detector, you get all your charge in through this area, but I would certainly not try to give this as a definite explanation. There must be other effects, because apparently similar detectors will give similar amounts of this effect. This is certainly a factor that should be studied very much. I might point out that if this shift is not corrected for, pulse-height analyzers have very little value on this situation. Because, if you set the window at 50,000 counts, you will run out of your window in the other direction; this has to be very carefully considered. There is one point on which I would caution those who use multichannel analyzers—I don't know if you have done it—and that is that you also must check to see if the channel distribution does not vary with the intensity of the count. This admittedly exists in some of the multichannel analyzers; when testing here you pair the intensity which you inject with a standard pulse of fixed height, and this particular peak on your pulse distribution then should not shift if you increase the counting rates of the others. But all of the phenomena that you have mentioned—the change of shift, the change of width, and the change of height—have been noted by several observers. These things are certainly to be taken into account in X-ray analysis.

A COMBINED FOCUSING X-RAY DIFFRACTOM-ETER AND NONDISPERSIVE X-RAY SPECTROMETER FOR LUNAR AND PLANETARY ANALYSIS

K. Das Gupta

California Institute of Technology, Pasadena, California

Herbert W. Schnopper

Cornell University, Ithaca, New York

Albert E. Metzger and Rex A. Shields

California Institute of Technology, Pasadena, California

ABSTRACT

An instrument is described which is intended to perform a dual purpose (elemental–structural) analysis consistent with the environmental conditions implied by lunar or planetary operation. The diffractometer section is based on a modified Seeman–Bohlin focusing principle in which a sharp-line focus target, a powdered sample, and a movable detector slit all lie on the focusing circle. The convolution of the projections on the focal circle, of a narrow receiving slit on the detector, the line focus target, combined with a high dispersion produce higher resolution and intensity than is common with Bragg focusing diffractometers with similar instrumental parameters. The range of *d*-spacings covered is from 1 to 7 Å (chromium target). The chemical analysis section of the instrument utilizes the fluorescent X-rays produced in the specimen by the primary beam. A proportional counter and pulse-height analyzer accomplish detection and energy discrimination. Resolution is low, but the analysis can distinguish between elements in the range of atomic numbers 11 to 29. Data from a breadboard model is presented. The entire unit, although primarily intended to meet the requirements of space, performs equally well as a routine laboratory analyzer. The horizontal, stationary nature of the specimen holder suggests several specific applications.

INTRODUCTION

A determination of the elemental and mineralogical composition of the moon and planets is one of the most important scientific objectives in the exploration of the solar system. X-ray diffraction and X-ray spectroscopy have a long and successful history of development and application in the laboratory. It has therefore been natural to give careful consideration to the use of these techniques for the *in situ* analysis of samples of the lunar surface.

The instrument to be described in this report is intended to perform a simultaneous elemental and mineralogical analysis. The design of a possible experiment for remote operation on an unmanned or manned spacecraft has been derived, subject to the following stringent boundary conditions: (1) reasonably high sensitivity, (2) good lattice spacing resolution, (3) atomic number resolution, (4) rapid data accumulation, (5) low power

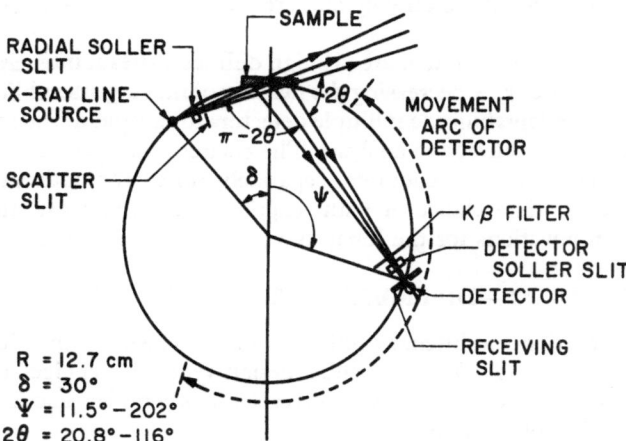

R = 12.7 cm
δ = 30°
Ψ = 11.5° − 202°
2θ = 20.8° − 116°

Figure 1. Schematic of the focusing diffractometer.

requirements, (6) moderate weight, and (7) simplicity of mechanical construction and operation. Although the primary objective has been the analysis of lunar material, this instrument may also have application in the laboratory.

DESCRIPTION

Based on the method of focusing, diffractometers are classified into two categories:

1. Bragg focusing: The distance from the sample to the source slit is equal to its distance from the detector slit.
2. Seeman–Bohlin focusing: The source slit, the powdered sample, and the detector slit all lie on the circumference of the focusing circle.

In Bragg focusing, the detector moves in a circle with the sample at the center, while in the Seeman–Bohlin method the detector moves in a circle with the sample on the circumference of the circle.

A lunar diffractometer utilizing the Bragg focusing approach was designed by W. Parrish of Philips Laboratories, developed to the pre-prototype stage by Philips Electronic Instruments, and has undergone extensive modification and evaluation at the Jet Propulsion Laboratory at the California Institute of Technology.[1,2] Although this instrument is notably successful in retaining a high level of intensity and resolution with only a fraction of the power and weight required by a laboratory diffractometer, the Seeman–Bohlin focusing principle lends itself to more efficient power use, a simpler mechanical configuration, and a more rapid analysis. In order to take advantage of these potentialities, a new instrument for lunar and planetary application, based on the Seeman–Bohlin focusing principle, was designed and breadboarded at the Jet Propulsion Laboratory.

The JPL-designed diffractometer (Figure 1) is based on a fixed source–sample relationship which allows a simple focusing design. The conventional Seeman–Bohlin optics have been modified, however, by placing a line source directly on the focusing circle.[3,4] This arrangement provides the most efficient utilization of the source. The line source is imaged at points appropriate to the various lattice spacings by diffraction from a curved sample. By means of a simple linkage, a detector is made to rotate about the center of the focal circle and point a narrow receiving slit at the center of the sample.

[1] References are at the end of the paper.

One or more fixed counters observe the source directly and are used for a nondispersive chemical analysis.

The advantages offered by the JPL instrument are (1) high diffracted-beam intensity per power input, (2) mechanical simplicity by decreasing the number of moving assemblies from two to one, (3) light weight, (4) adaptability to multiple detectors for a rapid diffracto-meter scan, (5) combined chemical and diffraction analysis. The chemical analysis can be extended by the use of several nondispersive counters, separately tuned for optimum spectral response through an appropriate choice of path length, window, and gas fill. Items (4) and (5) are important innovations for the lunar and planetary application.

DESIGN CONSIDERATIONS—DIFFRACTOMETER PORTION

The requirements of line resolution and intensity follow a quasi-exclusion principle, i.e., their product is roughly constant. Given a particular design type, the choices of instrumental parameters which tend to increase one will decrease the others. In practice, a compromise is reached; this philosophy is applied here.

The angular extension of the image of the source and the angular projection of the detector entrance window determine the resolution (defined here as $D/\Delta D$, where D is the lattice spacing). These contributions are augmented somewhat by vertical divergence (divergence normal to the plane of dispersion), various alignment errors, sample pene-tration, etc. To begin with, these latter effects will be considered negligible on the assumption that the overall resolution required will not be extremely high. They will be treated empirically from an experimental viewpoint when necessary.

The parameters to be treated are R, the radius of the focal circle; ds the linear extension of the source along the focal circle; dl the width of the receiving slit; δ, the angular position of the source with respect to the sample; and λ, the wavelength of the incident X-ray beam. These parameters are discussed in terms of the dispersion and the resolution of the instrument. The instrumental variable is ψ, the angle from the center of the sample to a diffraction peak. It can be shown from Figure 1 that

$$\psi = 2\left(\theta + \theta - \frac{\delta}{2}\right) = 4\theta - \delta \tag{1}$$

where θ is the Bragg angle of interest corresponding to a diffraction peak. Thus, θ is given by

$$\theta = \frac{\psi + \delta}{4} \tag{2}$$

From Bragg's law, the lattice spacing D is given by

$$D = \frac{n\lambda}{2 \sin \theta} = \frac{n\lambda}{2 \sin \left[(1/4)(\psi + \delta)\right]} \tag{3}$$

The dispersion of the instrument is then

$$\frac{\partial D}{\partial \psi} = -\frac{n\lambda \cos \left[(1/4)(\psi + \delta)\right]}{8 \sin^2 \left[(1/4)(\psi + \delta)\right]} = -\frac{n\lambda}{8} \frac{\cos \theta}{\sin^2 \theta} \tag{4}$$

Since $\psi = a/R$, where a is the circumferential distance from the center of the sample to the diffraction peak, then

$$\partial \psi = \frac{1}{R} \partial a \tag{5}$$

and

$$\frac{\partial D}{\partial a} = \frac{1}{R}\frac{\partial D}{\partial \psi} = -\frac{n\lambda}{8R}\frac{\cos\,[(1/4)(\psi + \delta)]}{\sin^2\,[(1/4)(\psi + \delta)]} = -\frac{D}{4R}\cot\theta \qquad (6)$$

The following value is then given:

$$\left[\frac{\partial D}{\partial a}\right]_{\text{Bragg}} = -\frac{D}{2R}\cot\theta = 2\left[\frac{\partial D}{\partial a}\right]_{\text{Focusing}}$$

This factor of two is significant as it means that, with the same radius for both types of focusing, the diffraction peaks are more widely spaced out along the circle in the Seeman–Bohlin (focusing) design. Alternatively, in order to give the same circumferential spacing, the Bragg instrument must have twice the radius of the JPL instrument with a subsequent loss of intensity because of the increased source-to-detector distance. This argument is important since the resolution of both instruments is determined by extensions along the focal circle which cannot be made arbitrarily small. The advantage here is that, for the same size of receiving slit on the detector, the resolution with the JPL instrument is greater. Note, also, that for a given λ, as D increases, θ increases and $\cot\theta$ decreases. The finite extension of the sample along the focal circle imposes a lower limit on the θ, and hence an upper limit on the D spacing which can be observed. These limits are eased by going to a longer wavelength of irradiation, and by bringing the source as close as possible to the sample. However, these measures lead to increased scattering at low diffraction angles, and the useful scanning range is also determined by the degree to which this scattering can be limited.

The resolution of the instrument imposes a limit on the precision of the D-value measurement. For present purposes the resolution is defined as D/dD. In the simplest approximation, D is a function of only ψ and δ given by:

$$D = D(\psi, \delta) \qquad (8)$$

Thus,

$$dD = \frac{\partial D}{\partial \psi}d\psi + \frac{\partial D}{\partial \delta}d\delta \qquad (9)$$

Substituting from equation (2) and equation (6),

$$dD = -\frac{D}{4}\cot\theta(d\psi + d\delta) \qquad (10)$$

More detailed discussions of the general instrument window problem, the effect of the line profile, etc., are given elsewhere.[5,6]

The angular extension of the receiving slit along the focal circle is given by

$$d\psi = \frac{dl}{R\sin\psi/2} \qquad (11)$$

and the angular extension of the source along the focal circle is given by

$$d\delta = \frac{ds}{R} \qquad (12)$$

Table I. Expected Instrument Performance

ψ (deg)	θ (deg)	Cu K_α		Cr K_α		Ti K_α		All λ
		D (Å)	dD (Å)	D (Å)	dD (Å)	D (Å)	dD (Å)	D/dD
12.0	10.5	4.35	0.489	6.29	0.707	7.53	0.846	8.9
22.9	13.2	3.37	0.174	5.03	0.259	5.98	0.307	19.4
45.8	19.0	2.37	0.0489	3.53	0.0729	4.22	0.0872	48.5
68.8	24.7	1.85	0.0226	2.75	0.0335	3.28	0.0397	82.0
91.7	30.4	1.53	0.0137	2.27	0.0203	2.71	0.0221	112
114.6	36.2	1.31	0.0080	1.95	0.0118	2.32	0.0141	165
137.5	41.9	1.16	0.0060	1.73	0.0082	2.05	0.0095	210
160.4	47.6			1.56	0.0057	1.86	0.0068	272
183.4	53.3			1.44	0.0043	1.71	0.0051	333
206.3	59.1			1.34	0.0033	1.60	0.0039	413
229.2	64.8			1.28	0.0025	1.51	0.0030	505
252.1	70.5			1.22	0.0019	1.45	0.0023	641
275.0	76.3			1.18	0.0014	1.41	0.0017	826

By differentiating equation (2) and using the last three equations, the resolution can be expressed as

$$\frac{D}{dD} = -\left(\frac{4}{d\psi + d\delta}\right)\tan\theta = -\frac{\tan\theta}{d\theta} = -\frac{4R}{dl/\sin(\psi/2) + ds}\tan\theta \qquad (13)$$

From equation (13), it is evident that high resolution is achieved with large R, large θ (or λ), and small dl and ds, as expected.

The approach to the design of the instrument has been somewhat empirical. Definite limitations are imposed by the necessities of lunar operation; the objective has been to observe these limitations and still retain performance approaching that of an ordinary laboratory instrument. As a first approximation, the following parameters were chosen for the instrument: $dl = ds = 1$ mm $R = 12.7$ cm (5 in.) and $\delta = 30°$. Three possible wavelengths have been considered: Cu K_α, $\lambda = 1.54$ Å; Cr K_α, $\lambda = 2.29$ Å; and Ti K_α, $\lambda = 2.75$ Å. Table I, obtained with the help of equations (2), (3), (10), and (13), illustrates the expected performance of the instrument for these parameters. Note that neither the resolution D/dD, nor the line width dD, of the instrument are constant, and that both become poorer at low 2θ values.

The three wavelengths show some overlap in D value ranges. The choice of either Cr K_α or Ti K_α seems to be a reasonable compromise between the range of D value to be covered and the cutoff at low values of θ.

CONSTRUCTION AND ALIGNMENT

The breadboard instrument shown in Figure 2 was constructed using the results outlined in the previous section as a guide. The aluminum plate on which it is mounted also provides shielding protection. It was decided to make all preliminary alignments and measurements using 400-mesh quartz powder as a standard. This material has a wealth of diffraction structure. Although the instrument was planned for radiation of longer wavelength than copper, the time required for procurement made it necessary to use a copper target tube in the preliminary work. A copper target tube built by N. V. Philips Gloelampfabriken for the Bragg lunar instrument was used, although this meant a loss of the low-angle region. However, this made it possible to make more direct comparisons with the performance of the Bragg lunar instrument as well as with a standard laboratory instrument.

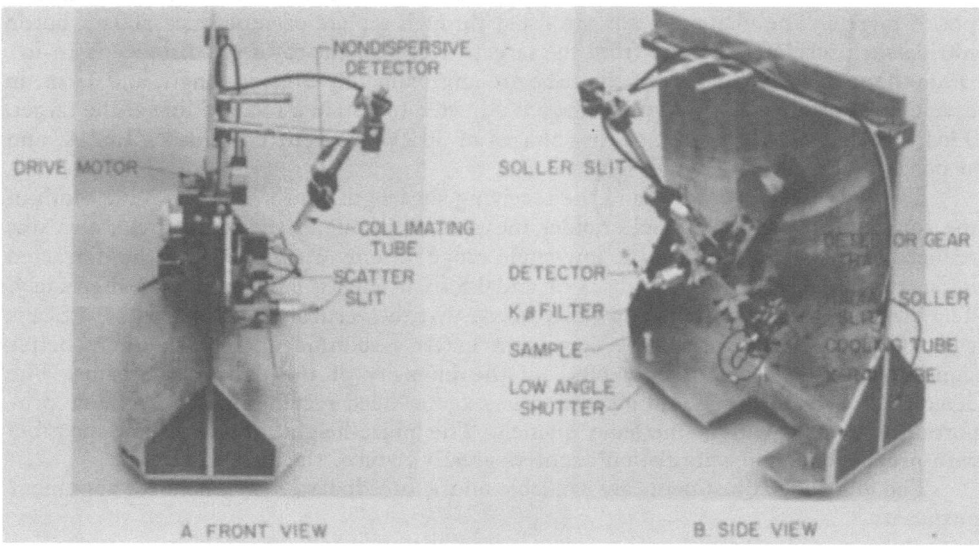

Figure 2. Combined diffractometer–spectrometer breadboard.

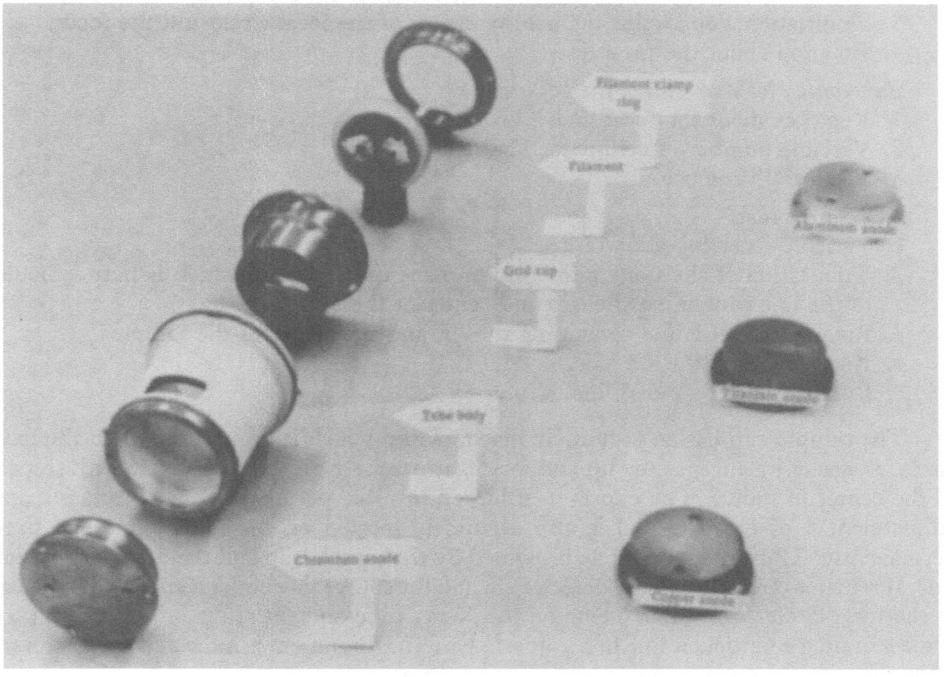

Figure 3. Disassembled view of the demountable X-ray tube.

Operation of the focusing diffractometer with radiation other than Cu K_α was made possible by construction of an unsealed X-ray tube, based on the design of a point-source electron gun.[7] The tube, shown in Figure 3, consists of a line-source filament made of 0.010-in. tungsten wire $\frac{3}{4}$-in. in length, an eliptical focus grid of copper which is placed just below the filament, and an interchangeable assortment of machined and electroplated targets. The filament leads are fitted through a plate of supermica 500. A boron nitride shell insulates the gun from the target. The filament-to-target distance is $1\frac{1}{8}$-in., while the overall dimensions of the tube are approximately 3 in. in length and $1\frac{3}{4}$ in. in diameter. A positive control grid voltage is adjusted to obtain a focused line at the target. This tube has been operated up to voltages of 30 kV, filament currents of 1.67 A, and at power levels of up to 25 W.

The sample holder width and the receiving slit length conform to the $\frac{3}{4}$-in. width of the source. Initially, the sample holder measured 1 in. along the focal circle; this was reduced to 0.5 in. to increase the low-angle range, at only a small cost to the diffracted intensity. Two receiving Soller slits, 1 in. and 0.5 in. in length, have been used alternately. The blade spacing of both slits is 0.028 in., so that the vertical divergence is 1.8°. Scans with the longer collimator have somewhat better resolution and considerably better signal-to-noise values, but only 60% of the intensity of the 0.5-in. collimator. The detector type used throughout all testing is a xenon-filled rectangular side window proportional counter with a methane quench. The pulse-height discrimination possible with proportional and scintillation counters greatly reduces the background.

The following adjustments are available on the breadboard instrument for alignment purposes:

X-ray tube motions
1. Translation along a line normal to the ideal focal plane.
2. Translation along a line defined by the center of the sample and the focus.
3. Translation along a line defined by center of the focal circle and the focus.
4. Rotation about the focal line.

Exit collimator and scatter-slit adjustments
1. Rotation about the focal line.
2. Variable number of collimator blades.
3. Vertical divergence.

Detector motions
1. Variable receiving-slit width.
2. Translation of slit along a line in the plane of dispersion which is perpendicular to the line containing the slip and center of the sample.
3. Rotation about a line containing the slit and the center of the sample.
4. Rotation about the slit axis.
5. Rotation about a radial line.

The sample cup has an accurately machined lip which is ground to match the focal circle. A spacer jig engages the lip and locates it precisely on the focal circle with respect to the center of radius. The cup is then locked in place permanently. The rotation axis (adjustment 4 of the detector) is also accurately located on the focal circle and fixed permanently. Thus, the focal circle is defined by the sample cup and the detector rotation axis. It remains only to bring the detector slit and the source line both on and perpendicular to the focal circle. The adjustments on the X-ray tube and slit are manipulated in turn while scanning a particular line in the quartz pattern. Alignment is indicated by maximum resolution, peak intensity, and signal-to-noise ratio.

Table II. Operating Conditions for Scans

	Tube target	Voltage (kV)	Current (mA)	Filter (mil)	Receiving slit (in.)	Scan speed (deg/min)
Laboratory diffractometer	Cu	40	10	Ni–0.5	0.005	2
PEI diffractometer	Cu	25	1	Ni–0.6	0.006	0.5
Focusing diffractometer	Cu (Eindhoven tube)	25	1	Ni–0.5	0.004 0.008 0.014	2
Focusing diffractometer	Cr (JPL tube)	25	1	V–0.5	0.008	4

A systematic approach to alignment was developed as testing proceeded. Experience has shown that only a few of the adjustments built into the breadboard model have been required for routine alignment. After initial installation of the detector and tube assemblies, the only adjustments of significance have been (1) a detector-slit scan at right angles to the diffraction line, (2) scanning of the X-ray tube along the 30° line perpendicular to the focal circle for maximum resolution, (3) rotation of the X-ray collimator for maximum intensity and resolution, and (4) narrowing the width of the scatter slit at the exit of the X-ray Soller slit, to increase the signal-to-noise ratio to a point where improvement is outweighed by the loss in intensity.

RESULTS–DIFFRACTOMETER PORTION

The initial results indicated a rather asymmetric line shape, poor resolution, unsatisfactory intensity, and a low signal-to-noise ratio. These undesirable features were eliminated by (1) controlling with Soller slits the beam divergence in the plane perpendicular to the plane of dispersion, (2) installing a radial Soller slit at the X-ray tube to give a better definition to the line focus, (3) placing scatter slits to eliminate extraneous signals, and (4) using a variable-width receiving slit to optimize resolution, intensity, and signal-to-noise ratio.

With the focusing diffractometer modified and aligned, studies were made to determine the performance characteristics of the instrument. Unless otherwise stated, operating conditions for the scans to be reported are as listed in Table II. In each case, the detector used was a xenon-filled side-window proportional counter.

Early in the test program, a set of scans was obtained with three values of receiving-slit width. The incident radiation was Cu K_α, and the sample was 400-mesh quartz. The values of intensity, signal-to-noise ratio, and resolution are given in Table III. The results indicate that the increased intensity with a receiving slit width of 0.014 in. sacrifices too much resolution to be acceptable. A slit width of 0.008 in. appears to be a good compromise and has been taken as standard for subsequent scans.

The signal-to-noise ratio of the quartz 101 peak is poor because at low angles of scan, the receiving slit begins to see radiation coming directly from the mouth of the X-ray tube. A special shield was installed as a low-angle shutter between the sample and the receiving slit to eliminate this line-of-sight effect, at no cost to the intensity of line radiation diffracted from the sample. An arm and track arrangement linking the shield to the detector allows the shield to drop out at the value of ψ at which the detector no longer can see the X-ray tube. Figure 4 illustrates the effectiveness of this shutter in reducing the background at low scanning angles.

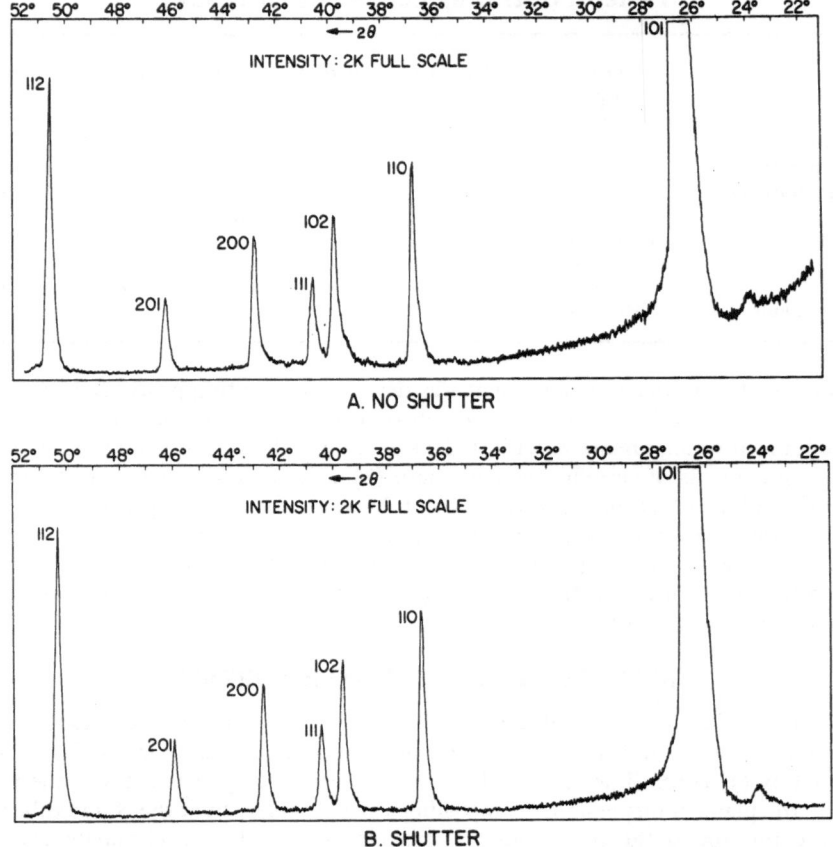

Figure 4. Effect of the shutter in reducing background at low scanning angles.

Scans were made with the demountable tube using anodes of titanium, chromium, and copper. Initial results with Ti K_α radiation were promising in terms of resolution and intensity, although the dispersion around the focal circle was somewhat greater than desired. A more serious obstacle to the use of Ti K_α is lack of a suitable $K\beta$ absorption filter. Cr K_α was therefore chosen as an optimum compromise to obtain good performance with adequate dispersion. Tests of the demountable tube with a copper target provided a point of reference with the sealed Philips tube. A comparison of the 112 peak of quartz at the same operating voltage and current indicated that the unsealed tube provided equivalent resolution and signal-to-noise ratio, but is 25% less efficient (after correcting for window absorption and a small difference in source-to-sample distance).

Four comparative scans are shown in Figure 5. All were run with a < 400-mesh sample of quartz. These include (A) a good quality scan with a laboratory diffractometer, (B) a scan obtained with an engineering model of the Bragg lunar diffractometer, (C), and (D) two scans by the focusing diffractometer using copper and chromium targets, respectively. Table II gives the operating conditions of these scans. The results are summarized for certain of the more prominent peaks over a wide range of D values, in Tables IV, V, and VI.

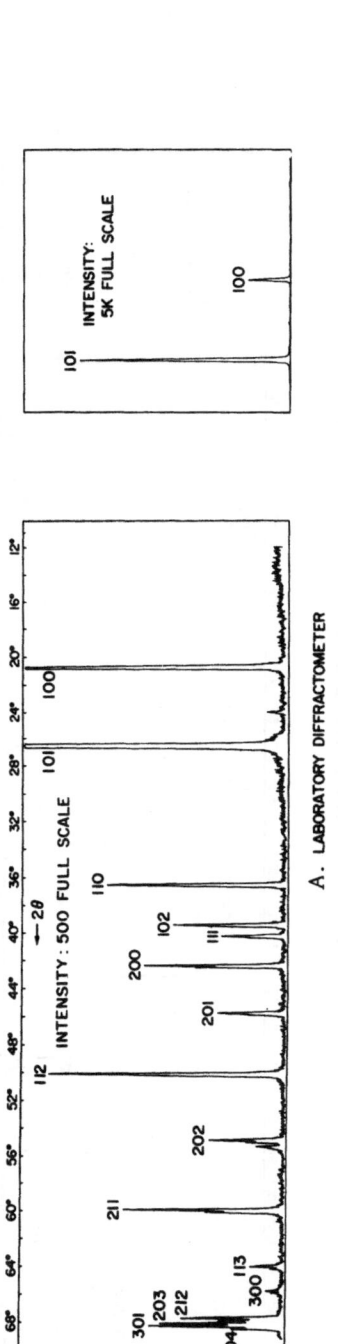

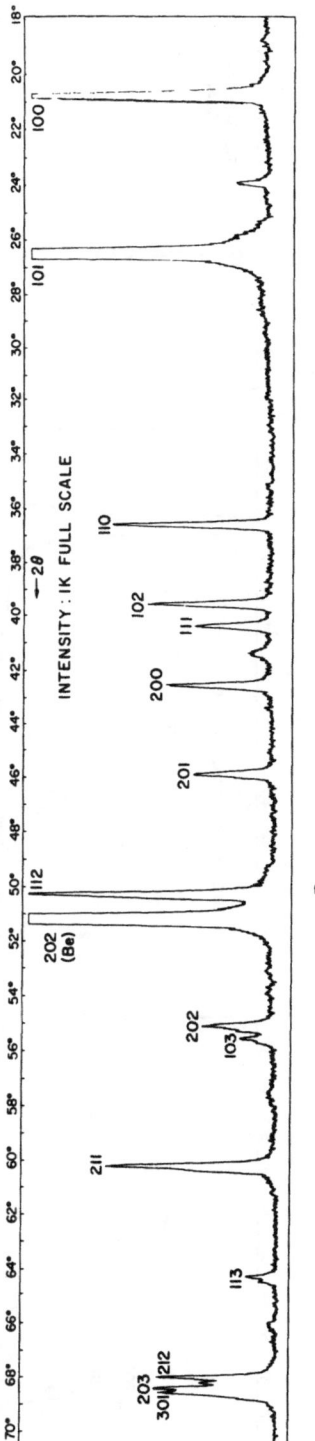

Figure 5. Comparative quartz diffraction patterns.

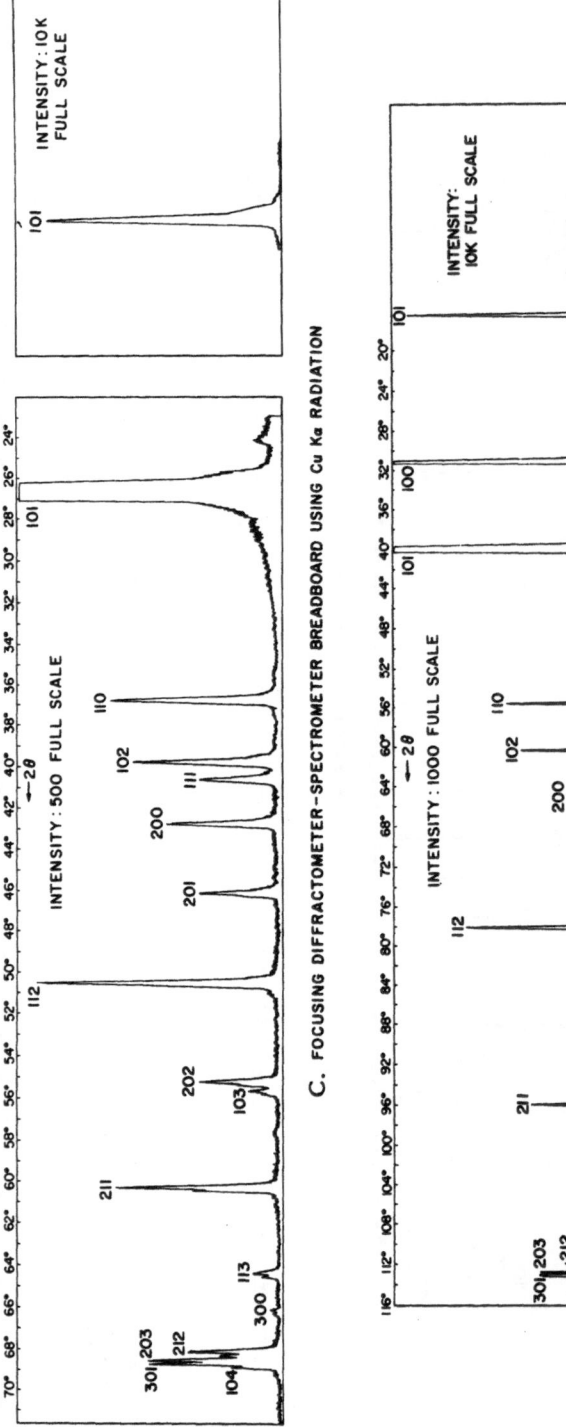

C. FOCUSING DIFFRACTOMETER–SPECTROMETER BREADBOARD USING Cu Kα RADIATION

D. FOCUSING DIFFRACTOMETER–SPECTROMETER BREADBOARD USING Cr Kα RADIATION

Figure 5 continued.

Table III. Selected Characteristics of Diffraction Patterns
for Three Values of Receiving Slit Width *dl*
with Cu K_α Radiation

Quartz peak	Intensity (counts/sec)			Signal-to-noise			Resolution		
hkl (2θ)	0.004 in.	0.008 in.	0.014 in.	0.004 in.	0.008 in.	0.014 in.	0.004 in.	0.008 in.	0.014 in.
101 (26.6°)	8,200	10,400	12,200	41	26	16	0.29	0.42	0.61
110 (36.4°)	830	1,390	2,020	10.4	9.9	8.2	0.21	0.24	0.32
112 (50.2°)	1,000	1,640	2,600	30.8	27.4	26.0	0.19	0.22	0.30
211 (60.0°)	—	—	—	—	—	—	0.18	0.20	0.28
203 (68.1°)	430	720	1,280	23.0	24.0	21.2	—	—	—

Table IV. Intensity (counts/sec)

Quartz peak *hkl* D(Å)	Laboratory diffractometer	Bragg lunar diffractometer	Focusing diffractometer	
			Cu target*	Cr target
100 4.26	790	810	—	2,200
101 3.34	3,880	4,420	9,060	9,630
110 2.46	330	305	665	575
112 1.82	440	520	950	730
211 1.54	300	330	645	490
212 1.38	190	230	370	350

*$dl = 0.008$ in.

Table V. Signal-to-Noise Ratio

Quartz peak *hkl* D(Å)	Laboratory diffractometer	Bragg lunar diffractometer	Focusing diffractometer	
			Cu target*	Cr target
100 4.26	65	16	—	34
101 3.34	320	60	113	148
110 2.46	36	8.5	22	39
112 1.82	53	14.5	48	53
211 1.54	41	12.9	36	49
212 1.38	23	10.2	25	35

*$dl = 0.008$ in.

Table VI. Resolution (Half-Height Width)

Quartz peak *hkl* D(Å)	Laboratory diffractometer	Bragg lunar diffractometer	Focusing diffractometer	
			Cu target*	Cr target
100 4.26	0.15	0.20	—	0.20
101 3.34	0.16	0.19	0.31	0.16
110 2.46	0.17	0.19	0.28	0.14
112 1.82	0.16	0.22	0.23	0.15
211 1.54	0.13	0.26	0.17	0.18
212 1.38	0.14	0.19	0.17	0.12

* $dl = 0.008$ in.

The use of chromium radiation has improved the performance of the focusing diffractometer by extending the scanning range, enhancing peak resolution, and, especially at low values of 2θ, by reducing peak asymmetry. Resolutions have become comparable to those obtained from the laboratory diffractometer. Resolution of the $\alpha_1\alpha_2$ doublet does not extend to 2θ values as low as the laboratory diffractometer scan because the percentage $K_{\alpha_1}-K_{\alpha_2}$ separation is only two-thirds as great for chromium as for copper. The signal-to-noise values of the focusing diffractometer show up increasingly well at higher values of ψ. Peak intensities with the focusing diffractometer average twice those of the laboratory diffractometer even though the operating power is only $\frac{1}{16}$ as great. Occasional instability of the demountable X-ray tube and the fact that resolutions up to 20% better than the diffractometer pattern of Figure 5(D) were obtained in the alignment procedure suggest that still better results can be achieved with the same operating conditions.

The beryllium line in the Bragg lunar diffractometer pattern is generated by a 0.002-in. foil positioned above the specimen as part of the mounting system for this instrument.

Both focusing diffractometer scans were run under vacuum, a characteristic of the lunar environment. A comparison of vacuum and atmospheric operation with copper radiation showed the increased intensity in vacuum to be about 10% at $\psi = 110°$, and proportionately less at smaller sample-to-detector distances.

The focusing diffractometer patterns of Figure 5 and Tables IV–VI were obtained using the 1-in. receiving Soller slit for copper radiation, and the 0.5-in. receiving Soller slit for chromium radiation. With the 0.5-in. collimator and the copper target, an intensity in the 101 peak was over 14,000 counts/sec, but with some loss in resolution. The reasons for the lower counting rate with chromium radiation have not been fully explored; they can be ascribed, in part, to absorption losses in the sample, the K_β filter, and the detector window. A K_β filter of optimum thickness is expected to recover some of this intensity.

Use of the short receiving Soller slit and the 0.5-in. sample cup decreases the low-angle scan limit to 20.8° 2θ; with chromium radiation this corresponds to a 6.36 Å spacing. A further increase to 7.2 Å will be possible by taking advantage of the smaller sample cup to decrease δ—the angle of the source with respect to the sample—to 25°. Reference to the Hanawalt index shows that very few inorganic materials are known for which the three strongest lines of the diffraction pattern all occur above 7.2 Å.[8] The upper limit of the scanning range is 140° 2θ.

The effect of the shutter improving the signal-to-noise ratio of peaks at low 2θ angles is shown quantitatively by comparing the values for the 101 peak in Tables II and V. The ratio has increased by better than a factor of four.

Figures 6 and 7 show the results of scans made with the focusing diffractometer for several minerals other than quartz. The sealed copper target tube was used for the patterns of Figure 6, which include calcite ($CaCO_3$); forsterite, a magnesium-rich olivine common to chondritic meteorites; and microcline, one of the potash feldspar group ($KAlSi_3O_8$). Figure 7 is a diffraction pattern of malachite with chromium radiation. Malachite, a copper mineral, was chosen because it is rich in structure at high D-spacings (the 020 peak has a spacing of 5.99 Å). Good resolution and signal-to-noise ratios have been retained out of the limit of scanning. Peak resolutions at low 2θ values are much better than those tabulated in Table I; the calculated resolutions were based on a receiving-slit width of 1 mm (0.025 in.), whereas a much smaller slit width (0.008 in.) has been used in practice. All samples were ground and sifted through a 400-mesh screen.

The particle-size distribution of the specimen is critical to the quality of diffraction response. The intensity and reproducibility of a diffraction line depend on the number of particles irradiated by the incident beam as well as on the geometrical factors which

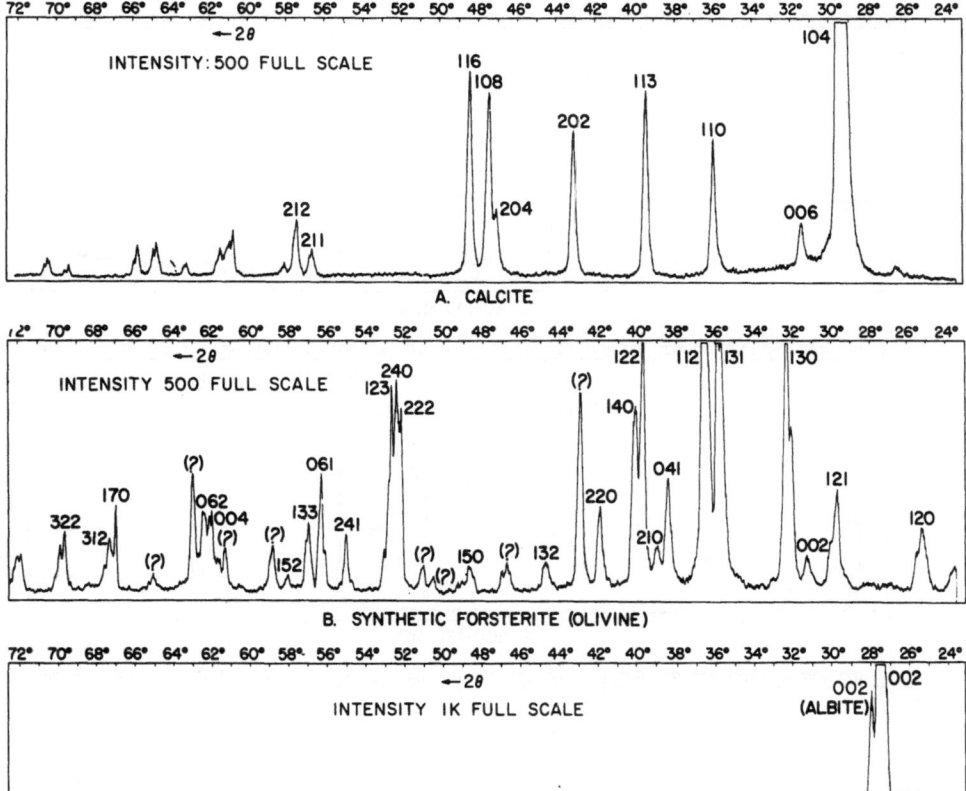

Figure 6. Diffraction patterns from the combined diffractometer–spectrometer breadboard.

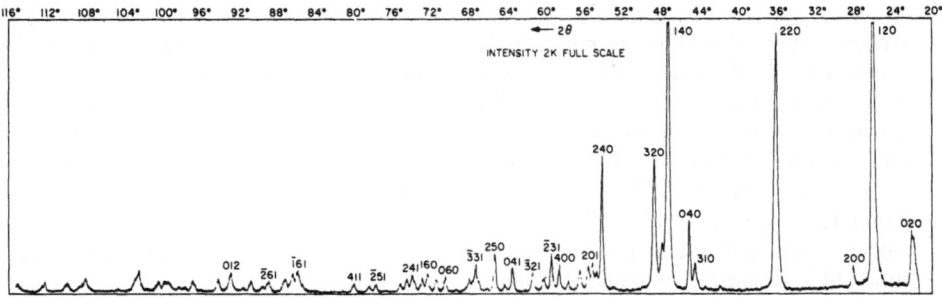

Figure 7. Diffraction pattern of malachite with Cr K_α radiation.

determine the fraction of favourably oriented crystallines.[9] The number of particles is inversely proportional to the cube of the mean particle-edge size. A fairly smooth piece of consolidated quartz was examined briefly on the breadboard instrument and, as

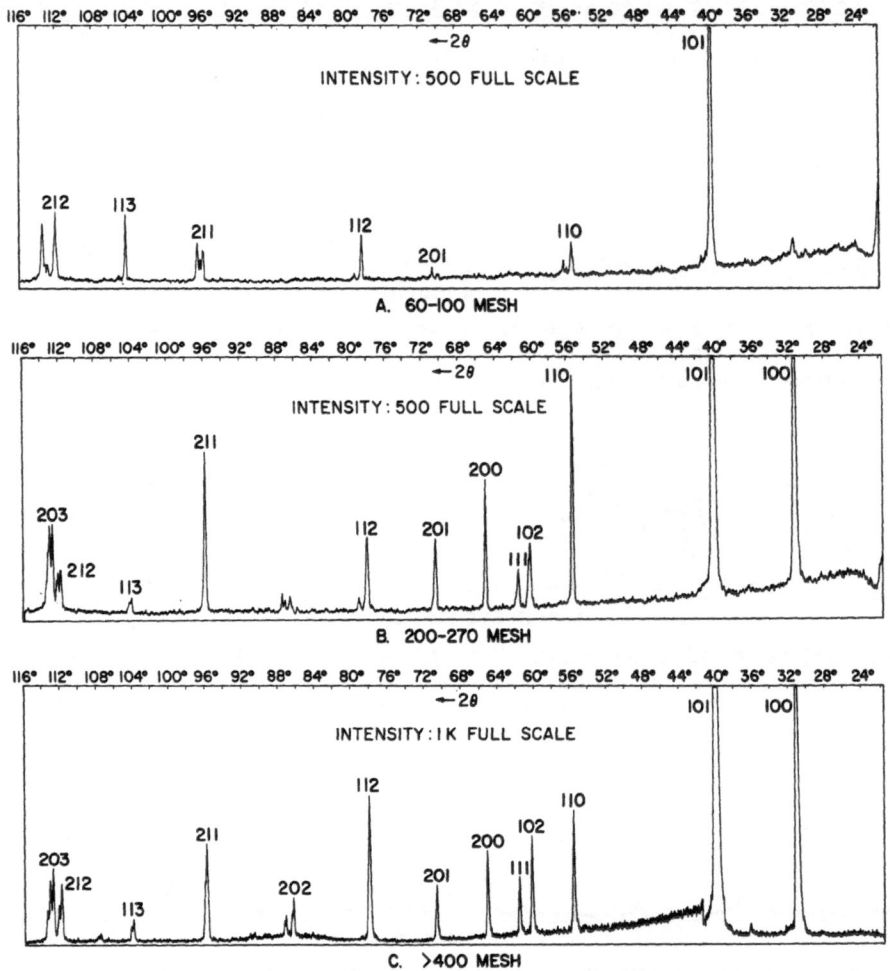

Figure 8. Effect of particle size distribution pattern of quartz with Cr K_α radiation.

expected, gave almost no recognizable response. A set of scans with quartz powders having widely different particle-size distributions was obtained with the chromiun target under identical operating conditions. The results are shown in Figure 8. Compared to that of the < 400-mesh sample, the 200–270-mesh sample shows considerable degradation, an effect which becomes extreme for the 60–100-mesh sample. When the same samples were run on the laboratory diffractometer (copper target), the degradation was found to be qualitatively similar, though not identical, demonstrating that the use of longer wavelength radiation has not made the quality of response appreciably more susceptible to particle-size distribution.

It has been pointed out that the presence of an appreciable amount of iron in the sample will produce a significant background fluorescence under copper radiation. This effect should not occur with chromium radiation. Accordingly, a sample mixture containing 10% iron, as Fe_2O_3, and quartz was scanned successively with the copper and chromium targets. The results are shown in Figure 9 and can also be compared to the 100% quartz

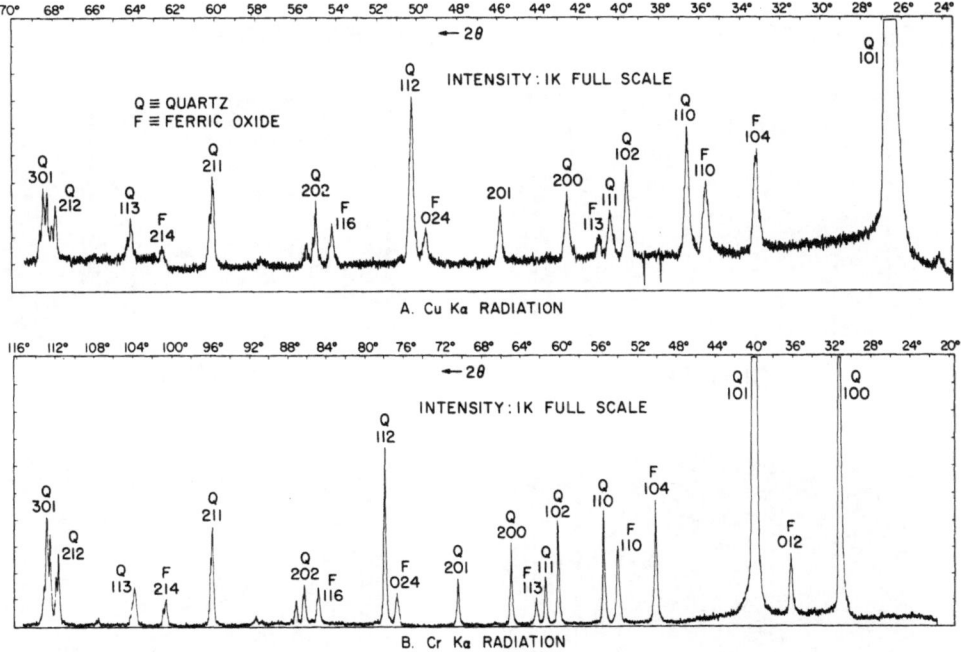

Figure 9. Diffraction patterns of 14.3% Fe_2O_3 (10%)–quartz mixture.

scans for Figures 5(C) and 5(D). It is seen that with copper radiation, fluorescent emission from the iron has produced a very significant increase in background with a resultant degradation in the signal-to-noise ratio, and that this continuum is absent when chromium radiation is used. The interference could be substantially reduced with a copper target by narrower pulse-height discrimination at the detector, but the single-channel threshold and window-width settings used in these scans were found to optimize intensity and signal-to-noise ratios. Furthermore, narrow pulse-height discrimination in space is undesirable because of the susceptibility of proportional counter gain to changes in applied voltage and counter temperature.

NONDISPERSIVE SPECTROMETER

The feasibility of nondispersive elemental analysis has been discussed elsewhere.[10,11] In this combined design, the nondispersive X-ray spectrometer shares the X-ray tube, high-voltage supply, and sample in common with the diffractometer. Gas-filled proportional counters are currently the most suitable pulse-amplitude energy-dependent detectors of low-energy X-rays. One or more such detectors placed outside the plane of the focal circle (to avoid interference with the diffractometer) will respond to fluorescent radiation emitted by the sample. The variation of fluorescence yield with atomic number and the inverse proportionality of range with energy favor the detection of heavier elements. Use of as thin a window as possible is desirable; e.g., the transmission factors of a 1.5 mg/cm² mica window and a 0.88 mg/cm² (0.25 mil) Mylar window for Si K_α radiation are 13% and 60%, respectively. A short, effective path length in the counter and the use of a light gas having a low mass-absorption coefficient will also tend to equalize the response as a function of atomic number.

A. Cu Kα RADIATION   B. Cr Kα RADIATION

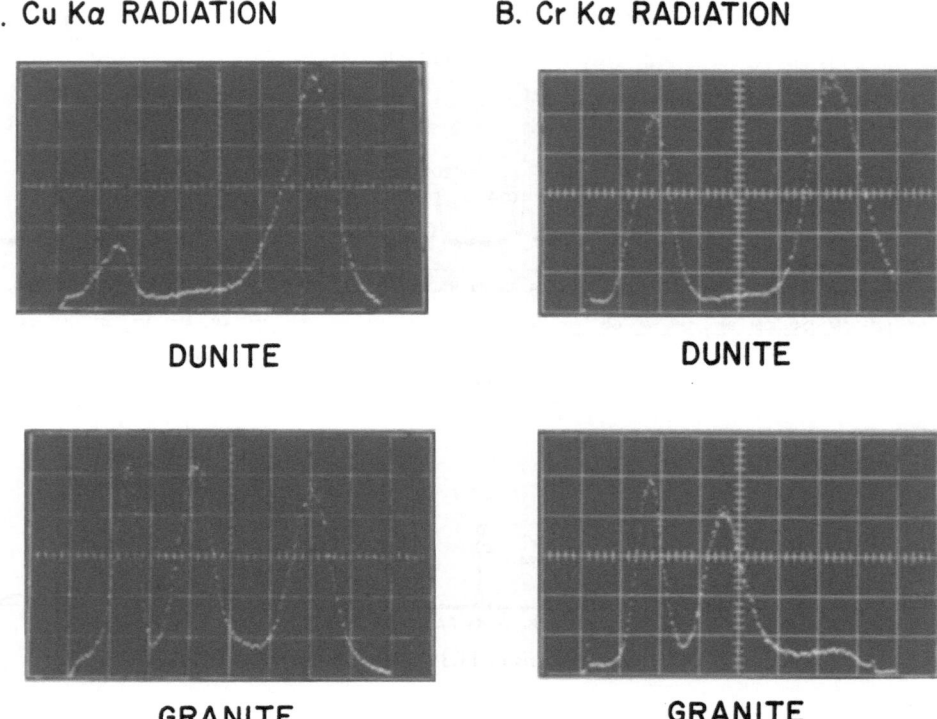

DUNITE   DUNITE

GRANITE   GRANITE

Figure 10. Nondispersive spectra from the combined diffractometer–spectrometer breadboard.

Following the proportional counter and suitable amplification, a system of pulse-height analysis is required for energy discrimination. An analyzer of 128 channels is desirable if the spectrometer is to detect all elements present as major constituents between atomic number 11 (Na) and 27 (Ni). Figure 10 shows four nondispersive spectra from two contrasting rock types obtained by the instrument. The detector was a neon-filled, sealed, thin-window, aluminized Mylar proportional counter. Peaks of the display represent Mg + Al + Si, K + Ca (almost absent in the dunite), and Fe. The ratio of peak heights shows that iron excitation is less predominant when excited by the copper continuum.

The most promising approach to reducing these multichannel spectra will employ computer techniques of sequential spectrum stripping or least squares analysis. The latter approach requires a library of standard spectra obtained from pure elements, or simple compounds measured under the identical experimental conditions as the "unknown" spectra. The solution to be obtained is the relative intensity of each component making up the composite spectrum with respect to its library function. Such techniques have been used to resolve complex gamma-ray pulse-height displays, and should be applicable to nondispersive X-ray spectra. Resolution of adjacent elements will be difficult but can be aided in certain cases by the use of selective absorption filters at the cost of some increase in mechanical complexity. A relatively crude but simple hand reduction of the data can be performed by simply integrating between peak limits. An example is shown in Figure 11, in which the concentration by weight of the combined response due to K + Ca is plotted as a function of fluorescent intensity.

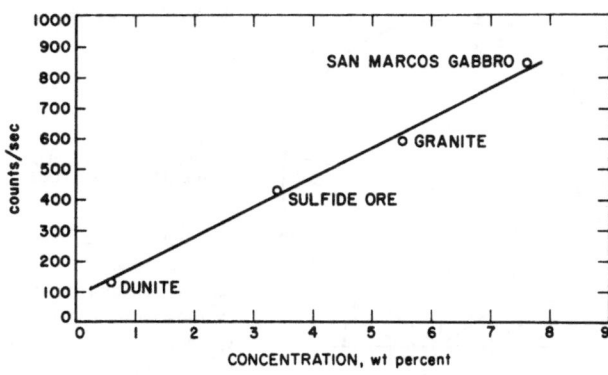

Figure 11. Nondispersive response of emitted X-ray intensity versus concentration from K + Ca in rock material.

SUMMARY AND APPLICATIONS

The combined focusing diffractometer–nondispersive-spectrometer has performed well and confirmed all major design objectives. Only the use of an aluminum target for extremely large D-spacings has failed to measure up to expectations. Chromium and copper provide resolution and signal-to-noise values which approach that available from laboratory diffractometers, and do so within the constraints of weight, volume, power, and shapes suitable for lunar or planetary use. The instantaneous power requirement is approximately $\frac{1}{50}$ of that for a laboratory instrument when copper is used as the target, and somewhat less than this for a chromium target with its wider D-spacing range. The feasibility of combining elemental and mineralogical analysis has been demonstrated.

The width of the lines recorded with the JPL-designed instrument is not constant because the projection of the slit width on the focal circle is a function of ψ (see Figure 11). The effect becomes significant at low angles of the scan. It can be remedied by leaving the slit fixed upon the detector arm and pointing only the detector during the scan. Then the low-angle line widths would decrease to a value comparable with the high-angle values. Peak intensity and background would also decrease, but the net result will be improved performance at these low angles.

Time and energy are at a premium in spacecraft missions. The ability to use multiple detectors is therefore a significant feature of the JPL-designed instrument. The focusing diffractometer is scanned by moving only the detector. The breadboard instrument uses only one detector, but it is possible by a straightforward modification of the present system to permit mounting of a large number of detectors on the focal circle, each one covering a portion of the scanning range with some overlap with its neighbor. The scan time of the instrument out to a lattice spacing of 1.2 Å is about 45 min, for the chromium target ($\psi = 200°$), and about 50 min for copper ($\psi = 110°$), at scanning speeds of 4°/min and 2°/min, respectively, speeds which have been found to only slightly degrade the quality of scan. If eight detectors are used, the focusing diffractometer will be able to obtain a pattern in 6 to 8 min, approximately $\frac{1}{20}$ of the time which has been specified for the Bragg lunar diffractometer. Additional power will be required by the added detectors, but the corresponding total energy (w-hr) saving in spacecraft power (not including the spectrometer portion) should be of at least a factor of 10.

It is estimated that a space-hardened model of the combined instrument could be built weighing no more than 20 lb, while the focusing diffractometer alone would require about 70% of the combined total.

As a more advanced development, a dispersive X-ray spectrometer with much greater sensitivity could be designed as a companion unit for the focusing diffractometer. This

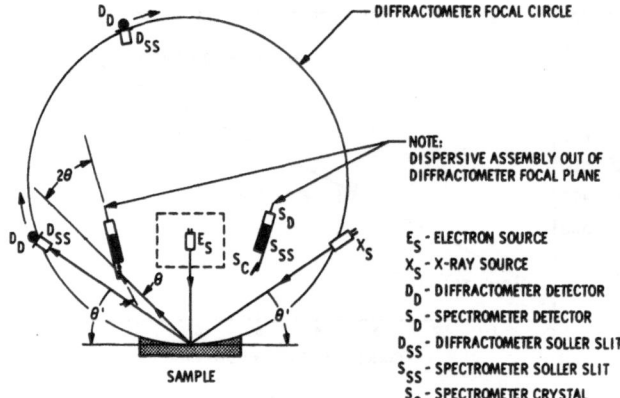

Figure 12. Proposed combined focusing diffractometer and dispersive spectrometer for remote analysis.

E_S - ELECTRON SOURCE
X_S - X-RAY SOURCE
D_D - DIFFRACTOMETER DETECTOR
S_D - SPECTROMETER DETECTOR
D_{SS} - DIFFRACTOMETER SOLLER SLIT
S_{SS} - SPECTROMETER SOLLER SLIT
S_C - SPECTROMETER CRYSTAL

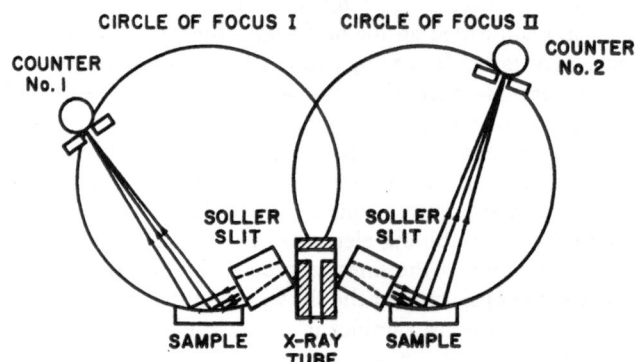

Figure 13. Siamese-twinned focusing diffractometers.

can be done by the simple application of conventional crystal optics as shown in Figure 12. The sample would become the source of fluorescent radiation, and if a fixed, nonscanning system is employed, a crystal-collimator-detector unit for each element would be positioned at the proper angle θ to satisfy the Bragg equation for its characteristic wavelength. It would be desirable for the X-ray source to be common to both modes so that the elemental and mineralogical analyses would be simultaneous. If this arrangement produces an insufficient response from the light elements, an electron source of excitation could be employed.

This newly designed diffractometer has potential applications other than *in situ* extraterrestrial analysis. The same characteristic rapidity through multidetector scanning and low power consumption suggest a place in the laboratory for routine analysis and possible adaptation as a battery-operated semi-portable field unit. High resolution can be achieved with this instrument by using a fine convergent Soller slit system with a minimum horizontal and vertical divergence. The factor of more than an order of magnitude advantage in intensity over a conventional laboratory unit, gained by aligning the target directly on the circle of focus, allows the use of fine collimation while retaining a reasonable intensity of the diffraction peaks.

Since the configuration of the instrument allows the sample holder to remain in a fixed horizontal position, a variety of problems involving loose powders and liquids can be investigated without the usual container problem. Furthermore, the sample holder can easily be incorporated into a variety of adiabatic enclosures without the need of a special moving vacuum seal and complicated heating or cooling lines. Any special purpose environmental rig can be designed with the sample holder as an integral part, as long as the spatial location of the sample with respect to the center of the focal circle is maintained. This unique feature of the fixed sample instrument allows diffraction studies to be made *in situ* over the whole range of high and low temperatures involving such phenomena, for example, as phase change, order and disorder transformation, strain, and radiation damage.

A rather unique capability of this focusing geometry is illustrated in Figure 13. Two focusing circles with two different samples and detectors could operate simultaneously with one fine focus target adjusted to have a position at the intersection point of the two circles.

ACKNOWLEDGMENTS

The authors wish to thank Neil L. Nickle and Dr. James A. Dunne for providing the laboratory diffractometer data and making many of the specimens available. This work was supported at the Jet Propulsion Laboratory, California Institute of Technology, under Contract No. NAS 7-100, sponsored by the National Aeronautics and Space Administration.

REFERENCES

1. W. Parrish, *Lunar Diffractometer Geometry*, in press (1965).
2. R. C. Speed, D. B. Nash, and N. L. Nickle, " A Lunar X-ray Diffraction Experiment," in: W. M. Mueller, G. R. Mallett, and M. J. Fay (eds.), *Advances in X-Ray Analysis, Vol. 8*, Plenum Press, New York, 1964, pp. 400–419.
3. G. W. Brindley, in: H. S. Peiser, H. P. Rooksby, and A. J. C. Wilson (eds.), *Monochromators and Focusing Cameras from X-Ray Diffraction by Polycrystalline Materials*, Institute of Physics, London, 1955.
4. K. Das Gupta and N. Pan, "A New Experimental Technique for X-Ray Diffraction Study," *Sci. Ind. Res. (India)* **17B**: 131–133, 1958.
5. A. J. C. Wilson, *The Mathematical Theory of X-Ray Powdered Diffractometry*, Philips Technical Library, Eindhoven, 1963.
6. G. Kunze, "Intensitäs-, Absorptions- und Verschiebungsfaktoren von Interferenzlinien bei Bragg–Brentano und Seeman–Bohlin Diffraktometern I," *Z. Angew. Phys.*, Dec. 24. 1964, p. 522.
7. *X-Ray Spectrometer Final Report* prepared by Philips Electronic Instruments, JPL Contract No. 950159.
8. *Powder Diffraction File*, J. V. Smith (ed.), American Society for Testing and Materials, 1964, Index (inorganic).
9. L. Alexander, H. P. Klug, and E. Kummer, "Statistical Factors Affecting the Intensity of X-Rays Diffracted by Crystalline Powders," *J. Appl. Phys.* **19**: 742, 1948.
10. L. S. Birks, *Electron Probe Micro-Analysis*, Interscience Publishers, Inc., New York, 1963, p. 95.
11. A. E. Metzger, *A Nondispersive X-Ray Spectrometer for Extraterrestrial Geochemical Analysis*, JPL Space Programs Summary, No. 37–31, Vol. IV, 1965, pp. 273–277.

DISCUSSION

G. W. Martin (Stanford University): How do you obtain a curved sample and how critical is this curvature? How would you do it for a liquid?

A. E. Metzger: We obtained the curved sample by curving our sample holder and filling it to the level of curvature. Although we haven't done any tests of this type as yet, a curved sample is quite important for good resolution. The larger the sample the more important it is to place the entire sample on the focal circle.

For liquids, a small amount of specimen would be best in order to minimize the defocusing at the center of the holder. The better the wetting action of the specimen material the less defocusing there will be and the greater will be the area exposed to the beam. The stationary upright position of the specimen holder is the most significant advantage for study of liquids.

G. W. Martin: For a lunar application, how would you do this? How would you arrange for the sample to be curved?

A. E. Metzger: This is a design feature in common with the earlier lunar diffractometer; the systems approach to this—and I am not saying that this is the best or the only approach—was to insert the sample into a holder covered by a thin, curved beryllium window and to press the sample up against that beryllium window. We would probably prefer to do without the window, if a satisfactory open surface could be provided and the calibration peak produced by the beryllium was not necessary.

H. K. Herglotz (E. I. du Pont de Nemours & Co.): A symmetrical version of the Seemann–Bohlin method was published in the 1940's.* The primary beam is perpendicular to the sample surface in this method. It allows a flat specimen as an acceptable approximation. Sharp lines are obtained from this flat specimen.

A. E. Metzger: Would that be a sample with as much extension relative to the diameter of the circle as we are using here? Half an inch? A flat specimen can be optimally focused for any one point on the focusing circle. The region around that point will show better line resolution than elsewhere on the circle.

H. K. Herglotz: One must make a compromise between conflicting requirements of large specimen area and achievable resolution. At angles θ close to 90°, the geometrical deviation of the sample from the focusing circle is less important.

A. E. Metzger: In comparing the resolutions that are attained with the focusing instrument versus the laboratory instrument, the focusing instrument shows up increasingly well at the higher angles, more poorly at the lower angles.

H. K. Herglotz: This is expected from the geometry of the Seemann–Bohlin method. The symmetrical version brings some advantages and some disadvantages.

L. S. Birks (Naval Research Laboratory): I would like to comment on the nondispersive analysis. We do have a computer program running to unfold the energy spectra. It gives the relative X-ray intensity and also calculates the standard deviation, which is no longer so easy to obtain because of the interrelationships between the energies and we would be happy to send this program to anyone who wants to write to us. I think the instructions for this program have not been prepared yet, but should be within the next few months. If you want to write us a letter we will be glad to put your name on the list to get a copy of the computer program when it is available.

A. E. Metzger: I hope you will put us on your mailing list. We are working on such a program now and made initial attempts to use a rather sophisticated gamma-ray least-squares analysis technique, but without much success. It looks as though this approach is oversophisticated.

W. Reuter (IBM Research Center): It appears that there will probably be a considerable degree of ambiguity in the interpretation of the nondispersive fluorescence results, particularly if you have situations like magnesium–aluminum–silicon, which has probably a high degree of probability of occurring on the Moon as well as on Mars. It would appear that there really is no considerable advantage over the method of a α-particle back-scattering which is particularly strong for oxygen and nitrogen, which you can't get with X-ray fluorescence. I wonder if there has ever been an attempt to take the original suggestion of Phillips, using the fixed channel geometry. What kind of sensitivity could one expect if a power supply of 25 kV and 1 mA were used?

A. E. Metzger: A nondispersive analysis can be performed much more rapidly than an α back-scattering analysis, and in any application in which operating time is limited or many samples are to be examined, this gives the fluorescence technique a definite advantage.

We did a considerable amount of work with Philips on the fixed channel geometry system, using a breadboard which they constructed. The concept looks entirely sound. It is a considerably more complex instrument—a nondispersive spectrometer with a fixed-crystal collimated detector system for each element. The dispersive spectrometer combined with the focusing diffractometer illustrated in Figure 12 is essentially the same approach. The breadboard experiments that we performed indicated sensitivities down to 0.1%, and for the heavier elements, considerably better than that.

W. Reuter: Was this work with the 25 kV, 1 mA source?

A. E. Metzger: No, an electron gun was used. The beam current required for a good response when you use electron excitation is considerably less than that for X-ray excitation, and our nominal currents were 10–50 μA. It will be difficult to get an adequate dispersive response from the lighter elements with an X-ray tube operating at 25 kV and 1 mA.

* F. Regler, *Z. Tech. Physik* **24**: 291, 1943.

TWO-CRYSTAL X-RAY SPECTROMETER ATTACHMENT

Leonid V. Azároff*

Illinois Institute of Technology
Chicago, Illinois

ABSTRACT

A two-crystal spectrometer that makes full use of conventional diffractometer or spectrometer motions has been constructed so that it can be easily attached to a commercial instrument. The requisite degree of parallelism of the two crystal-rotation axes is maintained by keeping both parallel to a single ground surface. Special alignment aids have been constructed to facilitate the rapid alignment of the instrument, which can be operated manually in 1″ intervals or used for automatic scanning in steps of 0.01°. Some guide lines for comparing the relative performance of spectrometers in X-ray absorption spectroscopy are suggested.

INTRODUCTION

The first two-crystal spectrometer arrangement was employed by A. H. Compton[1] about fifty years ago. Since then, numerous instruments have been designed and constructed[2] without altering the original principles of its operation[3] in any significant way. A reliable accurate two-crystal instrument is relatively difficult and, therefore, costly to construct so that the recent resurgence of interest in the X-ray absorption spectroscopy of chemical compounds and of alloys has been fostered by the utilization of one-crystal spectrometers. Although such spectrometers are more easily available, since commerical X-ray instruments can be used, they have two serious limitations: Their resolving power is inherently lower than that of a two-crystal instrument and, for quantitative work, there do not exist at present any means for correcting the instrumental distortions introduced in the curves recorded.

In order to obtain the superior resolution of a two-crystal spectrometer that is needed for the study of the fine structure near the main absorption edge of an atom without constructing an elaborate instrument, it was decided to devise an attachment to a commercial diffractometer available in our laboratory. First, such an attachment was procured from Otto von der Heyde[4,5] and, although it was found to be adequate, it was decided to construct a new spectrometer that yielded larger intensities, was easier to align, and could be operated in an automatic scanning mode for extended fine-structure studies. The constructional features and alignment procedures for this instrument are described below.

TWO-CRYSTAL ATTACHMENT

The two-crystal attachment is shown separately in Figure 1. It consists of a stainless steel backing plate on which two elevated flats have been ground in a single operation.

---

*Present address: Institute of Materials Science, University of Connecticut, Storrs, Connecticut.
[1]References are at the end of the paper.

At right angles to this ground flat, two holes have been bored through the plate, thus establishing the two rotation axes for the two crystals. The crystal holders are attached at right angles to two mounting plates, which can be rotated about these two axes; however, their parallelism is assured not only by the parallelism of the two bored holes but also by maintaining the two mounting plates parallel to the ground surfaces on the steel backing plate. The crystal holders can accommodate circular crystals of up to 1.25 in. in diameter, which are held in position against a circular retaining ring by a spring-loaded

Figure 1. Two-crystal spectrometer attachment.

Figure 2. Attachment mounted on Norelco diffractometer with alignment rod in position.

plate. The crystal can be rotated about an axis normal to its surface by rotating the retainer ring, about the rotation axis of the crystal holder (crystal-rotation axis), and about a third, mutually orthogonal direction by means of a small screw at the front of each holder that tilts the plate housing the retainer ring and crystal. Thus is is possible to align the reflecting planes of the crystal parallel to the crystal-rotation axis quite easily. When a small amount of nonparallelism between the ground surface of the crystal and its reflecting planes can be tolerated, the ability to rotate the crystal within its own plane allows one to optimize the crystal setting so as to minimize any spectral distortions such nonparallelism may introduce.

The first crystal holder bears angular graduations and is fastened to a long shaft (perpendicular to the ground surface) which passes through the center of the commerical diffractometer. As can be seen in subsequent figures, this attachment was devised to fit on a Norelco diffractometer, although similarly constructed instruments can be built to fit other commerical diffractometers as well. The second crystal holder is attached to a graduated circular plate bearing an arm which is joined through a pivot block to a parallel arm extending from the back of the steel backing plate (Figure 2). This second arm rotates about a shaft passing through the second hole bored in the backing plate. Since the second crystal is thus attached to the second arm, its rotation is accomplished by rotating this entire assembly through large angles or by rotating the front arm relative to the rear arm through seconds of arc by means of the micrometer tangent screw shown attached to the two arms below the pivot block. An indicator can be affixed to the front arm to measure very precisely its actual displacement. This serves as a check on the micrometer readings.

The employment of this particular arrangement has several noteworthy advantages. The crystal holder can be set at arbitrary angles relative to the graduated circular plate. As shown later, this is most convenient during the initial alignment of the spectrometer. The parallelism of the two crystals during its operation is assured by maintaining the parallelism of the two graduated circular plates with respect to the ground surface on the steel backing plate. Thus the second crystal can be rotated through large angles by loosening a nut locking the rear arm to the second rotation shaft without relying solely on the shaft to maintain parallelism. Similarly, the small angular adjustments accomplished by the tangent screw do not disturb the parallelism in the alignment.

An additional arm is also affixed to the rear of the second crystal so as to rotate about its rotation shaft. This arm protrudes from a graduated ring and includes the detector mounting bracket, which can be rocked through small angles by means of an independent tangent screw. This permits precise positioning of the detector relative to the second crystal after the approximate setting has been made by reading the graduated circle.

The center-to-center distance of the two bored holes through which the two crystal rotation shafts pass was chosen to fit the specimen-to-detector shaft distance in the diffractometer employed to support the attachment. Thus the second rotation ·shaft is affixed to the pin on which the detector arm of the diffractometer normally is mounted. This is deliberately made a loose junction to prevent any binding when the spectrometer is operated automatically. It has the important virtue of supporting the attachment at two points so that its weight distribution is better balanced. The loose fit does require, however, that the scanning direction must always be in the same sense to assure reproducibility. On the plus side, it permits one to utilize the gearing and stepping devices built into the commerical diffractometer.

INSTRUMENTAL ALIGNMENT

Since the two-crystal attachment is scaled so that the θ-to-2θ relation of the diffractometer motion is not disturbed, the alignment procedure is relatively straightforward.

A metal disk 1.25 in. in diameter is inserted in the position of the first crystal. Through a small hole in its center, a $\frac{1}{8}$-in. drill rod with tapered ends is inserted, as shown in Figure 2. For proper alignment, this drill rod should extend from the center of the first diffractometer slit to the center of the detector slits when the diffractometer is set at $\theta = 2\theta = 0°$. Thus, after the diffractometer has been set to the appropriate take-off angle,* the arrangement shown in Figure 2 permits the location of the approximate zero settings of the first crystal and the detector. After the alignment rod with its plate is removed, the zero alignment can be refined by inserting a very narrow slit in the first crystal holder, shown in Figure 3. By rotating the slit by 180° about the crystal-rotation axis, the zero setting of the detector ($2\theta = 0°$) can be adjusted by means of the tangent screw in the detector mounting bracket. This step completes the mechanical alignment of the instrument.

After the first crystal has been inserted in its holder, the diffractometer is cranked to the appropriate Bragg angle for the reflection chosen. The first crystal now should be at the correct θ angle to diffract X rays to the detector, which concurrently should be at the correct 2θ angle, as in Figure 4. (The second crystal holder is empty and positioned so as not to obstruct the diffracted beam.) With the X-ray beam on, the first crystal is now adjusted to yield maximum intensity and minimum diffraction halfwidth. This is carried out primarily with the aid of the small rocking screw which aligns the reflecting planes parallel to the rotation axis. A simple check of the parallelism between the crystal surface and the diffracting planes is afforded by a rotation of the retaining ring about its own axis. In this connection, it should be noted that rotations by arbitrary amounts are not reliable, since multiple reflections (Renninger effect) and variations in crystal perfection may cause considerable intensity variation. To check the parallelism, the crystal must be rotated through a full 180° within its own plane. The ability to make smaller rotations, however, affords an opportunity to optimize the crystal setting during its alignment.

Figure 3. Zero alignment using special alignment slit in first-crystal position.

*As is well known, another important advantage of the two-crystal instrument is that slits are not required to define the beam dimensions (resolution), so that larger take-off angles can be utilized than those in one-crystal instruments.

Figure 4. One-crystal operating
mode used for aligning first crystal.

In view of the design of the spectrometer, the second rotation shaft is located at the correct 2θ-angle after the first crystal has been set at θ. This means that the detector arm now can be rotated by θ degrees in either direction to accommodate the parallel or anti-parallel setting of the second crystal. With the second crystal inserted in its holder, it is then aligned in a manner quite analogous to that used in aligning the first crystal. After the second crystal has been adjusted, the spectrometer is ready for operation.

SPECTROMETER OPERATION

The two modes of operating the two-crystal instrument are shown in Figures 5 and 6. The parallel arrangement, Figure 5, can be used to measure the rocking curve of the second crystal or for crystal perfection studies, quantitative intensity measurements, and so forth. For this reason, this mode of operation should be called two-crystal diffractometry, following a suggestion first put forth by S. Weissmann.[6] The increased dispersion of the antiparallel arrangement, Figure 6, makes this mode of operation preferable in spectroscopy. The various ways that these two arrangements can be utilized in practice have been described in detail elsewhere[2,3] so that they will not be considered further here. In either mode of operation, the first crystal is set at the "center" of the spectral region being examined by the second crystal, which is then "rocked" by means of the tangent screw in steps that can be measured in seconds of arc.

Because the above-described attachment fits exactly the two moving "axes" of the diffractometer, it is also possible to employ the mechanical drive of the diffractometer to scan through the spectral "window" defined by the stationary crystal. In this operating mode, the second crystal is held fixed relative to the detector while the first crystal is rotated by the mechanical drive. (The second crystal's position is automatically rotated through twice this angle.) This arrangement is particularly useful when the extended fine structure of an absorption edge is desired. It may be necessary to reset the stationary crystal several times during such a scan, since its spectral width is limited by its perfection,

Figure 5. Parallel arrangement for
two-crystal diffractometry.

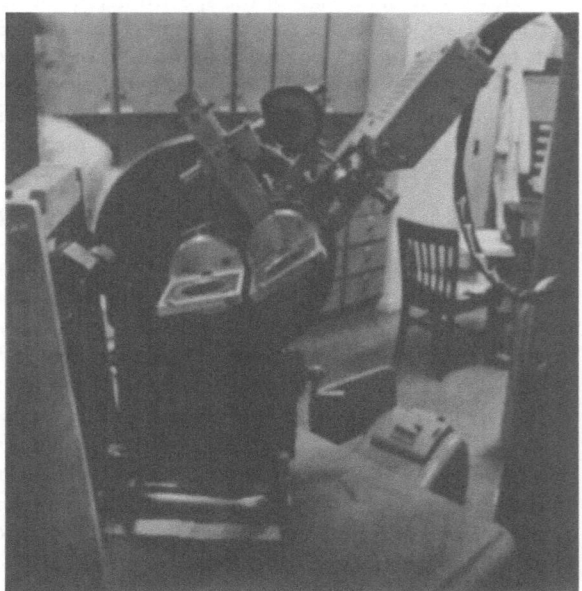

Figure 6. Antiparallel arrangement
for two-crystal spectroscopy.

depending on the desired scanning range. Thus it is possible to obtain the extended fine structure with the same reliability that has been limited hitherto to the near-in fine structure. It is hoped that a more regular application of two-crystal spectrometry to such studies will serve to increase the quality of the fine structures discussed in the literature.

This naturally leads to the discussion of the accuracy and resolving power of this instrument. To begin, it is not possible to discuss accuracy without some absolute basis, not presently available in X-ray spectroscopy, so that only the relative precision of two instruments can be compared. Even here, a certain amount of arbitrariness exists which is occasionally inflated by sheer wishful thinking. To avoid misunderstandings, only the reproducibility of a single measurement will be considered here. Even this depends on several unrelated factors, including mechanical stability of the spectrometer, the stability of the X-ray source intensity, detector response, and its attendant electronic circuitry. By using counting statistics assuring a statistical reproducibility of approximately 0.1%, the above attachment was tested in an air-conditioned laboratory, using a commercial constant-potential generator, diffraction tube, scintillation counter, and attendant circuitry, including pulse-height analysis. It was found that the intensity of the fine structure in the absorption edge of the same metal foil could be reproduced within 1% while reproducing the same angular (spectral) position to within 3 seconds of arc.

Actually, the resolving power of a spectrometer may be even more important than its reproducibility, particularly in X-ray absorption spectroscopy. This is so because it is not presently possible to interpret rigorously the observed fine structures, so that the necessarily more qualitative interpretations depend markedly on how much detail can be resolved, as well as on how reliably such details can be reproduced. When comparing the relative performance of two instruments, therefore, it is again necessary to define carefully the criteria to be employed. In a previous comparison of one-crystal spectrometers used in X-ray absorption spectroscopy,[5] it was suggested that the fine structure appearing in the copper K edge could serve as a useful "standard." This is so because copper foils can be easily and uniformly manufactured into suitable samples. In the present context, the first maximum in this copper absorption curve, lying approximately 4 eV beyond the Fermi level, affords a convenient structural detail for comparing spectrometer resolving powers. Immediately following this maximum, labeled A in Figure 7, there is a small dip following which the curve again rises. Measuring the heights of the curve at these two points above a common background, shown by the dashed line in Figure 7, it is possible to define a ratio between the difference in the linear absorption coefficient μ at A and α and the total height of the maximum at A by

$$\frac{\mu_A - \mu_\alpha}{\mu_A}$$

The advantage that this definition has over some others is that it is independent of the units employed and has a practical significance in absorption spectroscopy, since it measures the kind of fine structure that is usually sought. It is, of course, possible to vary even this simple ratio in a single instrument by altering the thickness of the absorber or by using different kinds of crystals. If one standardizes on the optimum thickness for a copper foil, it is found that the above ratio does not vary significantly for thicknesses of $7 \pm 1.5\,\mu$, so that the relative resolving powers of two different spectrometers, or of different crystals in the same instrument, can be meaningfully compared. Using this criterion, it was found that the above ratio ranged from 1.5% to 2.0%, depending on which pair of crystals, silicon or germanium set to reflect from (111), were used. These ratios were measured after a suitable correction had been made for instrumental broadening effects.[7] Since such corrections cannot be made for one-crystal spectrometers, the absence of a dip at α in corresponding one-crystal curves clearly demonstrates that the resolving power of two-crystal spectrometers is sufficiently greater to justify their exclusive use in X-ray absorption spectroscopy whenever really meaningful data are desired.

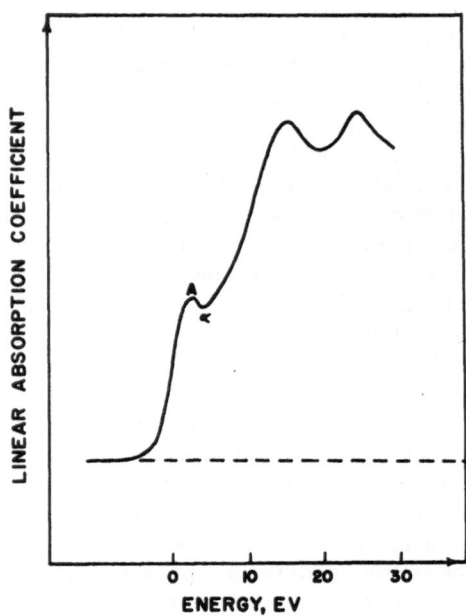

Figure 7. Absorption-edge fine structure near the K edge of copper. (Linear absorption coefficient expressed in relative units.)

ACKNOWLEDGMENTS

The construction of the above instrument was greatly aided by the diligence of Messrs. Deptolla and Leu, while its testing was carried out independently by R. J. Donahue, H. V. L. Narasimhamurty, and H. C. Yeh. This research was supported by a grant from the National Science Foundation.

REFERENCES

1. A. H. Compton, "The Intensity of X-ray Reflection and the Distribution of Electrons in Atoms," *Phys. Rev.* **9**: 29, 1917.
2. A. E. Sandström, "Experimental Methods of X-ray Spectroscopy: Ordinary Wavelengths," in: S. Flügge (ed.) *Handbuck der Physik, Vol. 30*, Springer-Verlag, Berlin, 1957, pp. 79–245.
3. A. H. Compton and S. K. Allison, *X-rays in Theory and Experiment*, 2nd ed., D. Van Nostrand Company, Inc., Princeton, 1935.
4. Otto von der Heyde, Newton Highlands 61, Massachusetts.
5. G. J. Klems, B. N. Das, and L. V. Azároff, "Single-Crystal Spectrometers for X-Ray Absorption Spectroscopy," in: *Developments in Applied Spectroscopy, Vol. 2*, Plenum Press, New York, 1963, p. 275–284.
6. S. Weissmann, *Method for the Study of Lattice Inhomogeneities Combining X-Ray Microscopy and Diffraction Analysis*, Rutgers University Rearch Bulletin, No. 39, 1956.
7. L. G. Parratt and C. F. Hempstead, *Correction of Complex Spectra for Instrumental Resolving Power, Part I: Model Windows*, Air Force Office of Scientific Research Technical Report AFOSR TN-56-388, 1965; ASTIA No. AD 96046.

DISCUSSION

H. W. Pickett (General Electric Company): What is the range in maximum obtainable θ, at which the counter or the arm might begin to conflict?

L. V. Azároff: I have no idea because we are working strictly with the K edges of only a few transition elements; they are all around 20°. The whole measurement is only about a half a degree in angle, so we never even worried about that problem. But it's not going to be very large.

H. W. Pickett: It might go to 40°?

L. V. Azároff: Something like that.

N. Spielberg (Philips Laboratories): This is a very nice little instrument. I was a little concerned about your statement that you can only get a 3° take-off angle on the vertical tube mounting. Ordinarily these are used quite often at 6°; you can get this either by shimming the tube tower or by dropping the goniometer. I think the limitation is primarily the size of the window, particularly in this case, since you have no divergence of the beam. You are taking essentially parallel radiation. I wondered also about something I believe you mentioned in the abstract. This was a precision of a few seconds on the positioning. With some of the high-quality semiconductor crystals, if the rocking curve with a two-crystal width is of the order of 10 or 12″ and a precision of 3″ it might slip you off the peak of the diffraction pattern as indicated by the rocking curves. You are still having a much better resolution than the single-crystal instrument, but it does limit the resolution of this one.

L. V. Azároff: Yes and no, if I may put it that way. You are right, that this makes the alignment of the instrument a really hair-raising operation. Actually I am talking about running two consecutive runs, such as two absorption curves, as I showed, and then comparing point by point the intensities versus the angular readings which we have read off our instrument. We find that those vary by no more than 3 seconds of arc. Of course, once they go by 3″ they are systematically all displaced by 3″. This is just a matter of the readings on the instrument versus what actually shows up on the final plot.

N. Spielberg: This is a general shift of the whole plot?

L. V. Azároff: Essentially it is a general shift. This merely means that we can't go back and set the instrument as far as our dials are concerned to better than ±3″ of where we thought we were. The disadvantage that this has is that if we want to go back to check a particular point, we can't do it. We have to go through a range of points and cover the whole territory.

N. Spielberg: One other point. I think that was a proportional counter shown in one of your slides, rather than a scintillation counter.

L. V. Azároff: We have two instruments, actually. We are using scintillation on one and proportional on the other. So this may have been a proportional counter.

PLANE-POLARIZED, TWO-CRYSTAL X-RAY SPECTROMETER

Raymond J. Donahue and Leonid V. Azároff

Illinois Institute of Technology
Chicago, Illinois

ABSTRACT

A two-crystal spectrometer employing a Cole polarizer made from a germanium crystal as one of the crystals is described. Experimental tests show that placing the polarizer crystal in the second position markedly improves the degree of plane polarization, in full agreement with the predictions of the dynamical theory of X-ray diffraction. Tests indicate that the twice-diffracted beam is better than 95% plane polarized. The angular resolving power of the instrument is comparable to that of conventional two-crystal spectrometers. The almost wholly plane-polarized monochromatic X-radiation, moreover, permits certain tests to be carried out far more effectively than is possible with partly polarized X-ray beams.

INTRODUCTION

A fine structure, extending for hundreds of electron volts from the main edge, is observed when the X-ray absorption spectra of crystalline materials are recorded with a high-resolution spectrometer. According to the Kronig theory,[1] a more pronounced extended fine structure should be observed when a single-crystal absorber, instead of a polycrystalline absorber, is used, and when plane-polarized X-rays are substituted for unpolarized X-rays. Experimental verification of this has not been very successful to date. Stephenson[2] was the first to use partially polarized X-rays, in 1933, to investigate the X-ray K absorption edge of bromine in a single crystal of KBr, without really conclusive results. Since then several experimental studies [3-6] using partially polarized X-rays have reported some structure variations, but none have been of sufficient magnitude to allow definite conclusions. More recently, Boster and Edwards,[7] using 82.4%-polarized X-rays and a single-crystal copper foil, compared the extended fine structures for several different angular positions of the foil relative to the plane of polarization. Although differences as large as 5% in the positions of some peaks were reported, Azároff[8] pointed out that it is not possible to assign a real meaning to these observations because the experimental errors in locating these positions are at least as large as $\pm 2\%$. Another recent study[9] investigating the K-absorption spectrum of a germanium single crystal with 90%-polarized radiation reported no changes in the maxima or minima in the extended fine structure when the direction of the polarization vector was changed relative to the absorber. An accompanying theoretical analysis of the relation between X-ray absorption spectra, specifically the extended fine structure, and the state of polarization of the incident X-ray beam carried out by Alexander, Fraenkel, Perel, and Rabinovich,[9] indicated that no changes in the fine structure should be expected from cubic single crystals.

---

[1]References are at the end of the paper.

In order to experimentally test the existence or absence of a dependence on polarization, it was decided to carry out a series of measurements on single-crystal foils of copper, grown parallel to different crystal planes. This would allow a test of Kronig's theory, as well as that of the later analyses, because both the plane normal to the X-ray beam and the direction of the polarization vector within each plane could be varied. To make the test a really decisive one, it was concluded that a truly 100% polarized beam would be desirable. It was decided, therefore, to make use of the anomalous transmission effect of X-ray diffraction which, as had been previously demonstrated by Cole, Chambers, and Wood,[10] could be utilized to construct an X-ray polarizer. The present paper describes a two-crystal spectrometer, built in our laboratory, in which one of the two crystals is a Cole polarizer. Partly because of the low intensity of the X-ray source on which its operation was tested, the results of the verification of the Kronig theory will be discussed in a later publication.

GENERAL INSTRUMENT

The apparatus assembled to carry out the proposed investigation, in addition to the specially built X-ray polarizer, included a standard Norelco diffractometer equipped with geneva gears for continuous or point-by-point scanning (in 0.01° steps), a scintillation counter followed by a pulse-height analyzer, a single-crystal holder built to provide three mutually perpendicular rotation motions, a silicon (111) crystal with a (1, −1) halfwidth of 9″, a topaz (101) crystal with a (2, −1) halfwidth of 17″, a dislocation-free germanium slab 1 in. × ¾ in. × 0.5 mm [cut with the (220) planes perpendicular to the faces 1 in. × ¾ in. and ¾ in. × 0.5 mm and parallel to the face 1 in. × 0.5 mm] for the X-ray polarizer, and three copper crystal foils respectively parallel to (100), (110) and (111). The germanium wafer serves as the polarizer when set to reflect the incident X-ray beam from its (220) planes.

It is possible to locate the polarizer in essentially two ways. At first it was decided to mount the polarizer directly on the X-ray tube so that the plane-polarized X-ray beam would fall on the second crystal, which is mounted at the center of rotation of the diffractometer.

The X-ray polarizer crystal and the crystal in the diffractomer center can be aligned so as to constitute a parallel double-crystal arrangement or an antiparallel double-crystal arrangement. Figures 1 and 2 illustrate schematically these two arrangements. An alternative arrangement is also possible in which the X-ray beam from the tube falls directly on the crystal at the diffractometer center so that the polarizer is placed in the

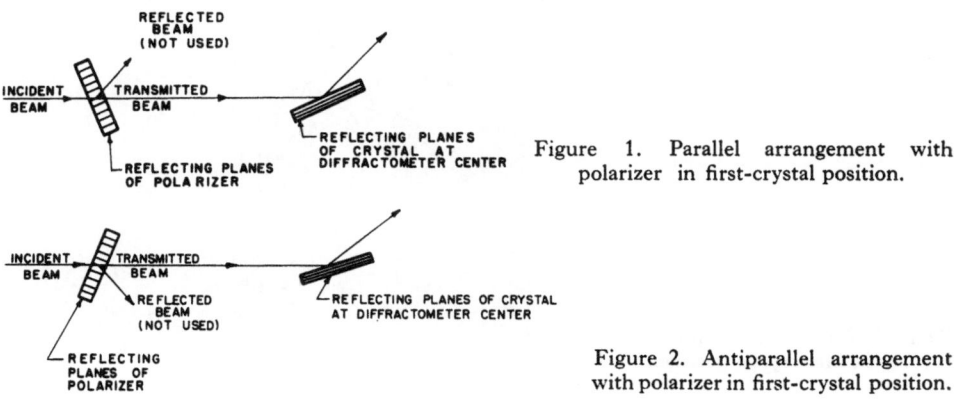

Figure 1. Parallel arrangement with polarizer in first-crystal position.

Figure 2. Antiparallel arrangement with polarizer in first-crystal position.

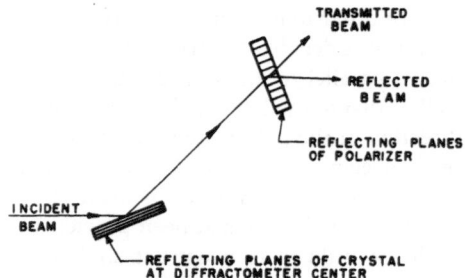

Figure 3. Parallel arrangement with polarizer in second-crystal position.

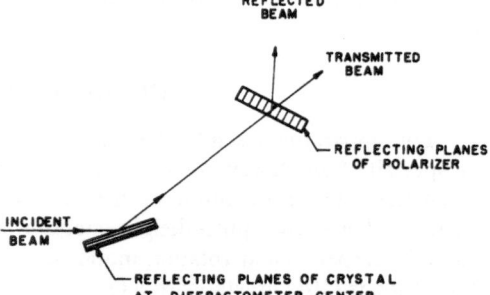

Figure 4. Antiparallel arrangement with polarizer in second-crystal position.

Figure 5. Two-crystal with polarizer in first-crystal position.

second-crystal position, in front of the detector. The two spectrometer configurations for this disposition are shown in Figures 3 and 4. The actual instrument is shown in Figure 5, where the polarizer is mounted on the X-ray tube tower, the second crystal or analyzer is at the diffractometer center, and a holder in which an absorbing foil has been inserted is placed in front of the detector.

X-RAY POLARIZER

An expanded view of the polarizer assembly is shown in Figure 6. The polarizer crystal is mounted on the cradle marked part A at the top of the figure. It is so mounted in cradle B that it can rotate freely about line O'–O', which is designated the θ axis of the polarizer crystal. The cradle B in turn fits into a dovetailed slot in part C which is mounted so as to rotate about the polarizer axis A'–A' (vertical line in Figure 6). This rotation allows the alignment of the plane polarization. Finally, the assembly fits onto the base support, part D, which also has a graduated circle about which the θ rotation of the polarizer cradle (part A) can be noted. All rotations are accomplished by means of micrometer screws.

The X-ray beam passes along the polarizer axis A'–A'. After the dislocation-free germanium crystal is positioned in the cradle, the cradle is rotated to the approximately correct θ angle, which is read on the dial of part D. The reflection can then be tuned by adjusting this angle slightly with the X-ray beam turned on. When the anomalously transmitted beam is desired, it can be detected by noting an increase in the transmitted beam as the crystal is rotated into reflection position as described in the next section.

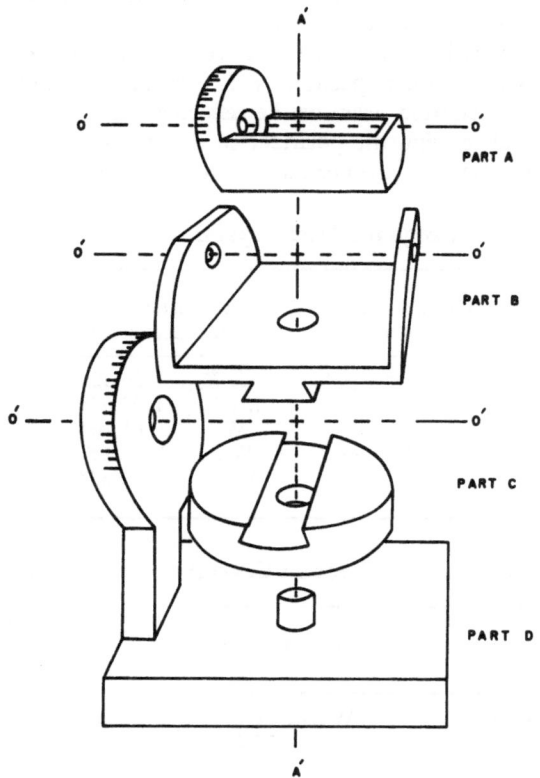

Figure 6. Expanded view of polarizer assembly.

Rotating the polarizer about its axis, A′–A′, has the effect of rotating the plane of polarization about the direct-beam direction. The diffracted beam then moves along a cone about the direct-beam path. When the polarizer is in the first-crystal position (Figures 1 and 2), the polarization direction can be set arbitrarily. When it is in the second-crystal position, then it is necessary, for maximum transmitted intensity, to rotate so that the reflecting planes are as nearly parallel (or perpendicular) to those of the first crystal as possible. This assures that one of the polarization directions of the polarizer is parallel to the electric vector having maximum amplitude. The choice of one arrangement over the other and some of the consequences are discussed further below.

SPECTROMETER ALIGNMENT

The alignment of the two-crystal spectrometer equipped with an X-ray polarizer is essentially quite similar to the alignment of any double-crystal spectrometer. First the diffractometer (Figure 5) is aligned with the single crystal at its center to yield the maximum resolution for an emission peak of the X-ray tube. Table I gives the pertinent one-crystal information for the arrangements tested in our laboratory. For the two crystals used, the reflecting planes and their d values, as well as the angular settings for the various characteristic lines and the order of reflection employed, are listed.

After the first crystal has been correctly aligned, the polarizer crystal is mounted in position and tuned to its reflection. With the polarizer mounted on the X-ray tube tower, this is done by rotating the crystal until an anomalous increase in the X-ray beam reflected by the analyzer crystal is observed. When the polarizer is mounted directly ahead of the detector, it is convenient to remove the diffractometer arm and to replace it with a special arm which not only allows the rotation of the polarizer about the arm's pivot, but also permits the detector to be rotated about the same axis (O′–O′ in Figure 6). Thus, when the polarizer is set in the second-crystal position, it is a relatively simple matter to utilize either the anomalously transmitted beam or the reflected beam. When the latter is used, it is, of course, essential to align the polarizer so that the diffracted beam lies in the plane of the diffractometer. Some of the pertinent settings for the second

Table I. Diffractometer Settings for First Crystal

Crystal, hkl, $2d_{hkl}$	X-ray target	Diffractometer slits	Characteristic line, order of reflection, 2θ
Topaz, (101), 2.7120 Å	Mo	$\frac{1}{30}°$, $\frac{1}{30}°$, $\frac{1}{12}°$	Mo K_{α_1}, I, 30.32° Mo K_{α_1}, II, 63.07° W L_{α_1}, I, 65.96°
Silicon, (111), 6.2706 Å	W	$\frac{1}{30}°$, $\frac{1}{30}°$, $\frac{1}{12}°$	W L_{α_1}, I, 27.24° W L_{α_1}, III, 89.87°

Table II. Diffractometer Settings for Second Crystal

Crystal, d_{220}	Diffractometer slits	Characteristic line, θ
Ge, 2.000 Å	1°, 1°, 4°	W L_{α_1}, 21.7° Mo K_{α_1}, 10.2°

crystal are listed in Table II. Note that the use of a double-crystal spectrometer eliminates the need for slits to define the resolution, so that the larger slits indicated serve primarily to minimize random scattering.

OPERATION

In principle, it appears that the orders of placement illustrated in Figures 1–4 should not affect the operation of this instrument. As pointed out by Professor N. Kato during a visit in our laboratory, when the polarizer follows the first crystal, its polarizing action is more "efficient", because the X-ray beam leaving the first crystal is mono-chromatic and more nearly parallel. The distinction between these two arrangements, however, is not very noticeable when characteristic lines are examined. Using the arrangements indicated in Figures 1 and 2, with the polarizer positioned as indicated in Figure 5, the Mo $K_{\alpha_{1,2}}$ peaks are shown for the parallel and antiparallel arrangements in Figure 7. The two sets of curves are the actual spectrometer tracings recorded on the pen recorder, and clearly demonstrate the increased dispersion and resolving power of the antiparallel arrangement. It should be noted that virtually identical curves are obtained when the polarizer is the second crystal, both as to intensity and angular widths.

The importance of the polarizer's position becomes significant when the spectrometer is set at the wavelengths corresponding to an absorption edge, for example, the K edge of copper. At these wavelengths, the total intensity in the appropriate portion of the continuous spectrum is considerably reduced, so that small intensity changes become most significant. In the parallel position (Figure 1), the anomalously transmitted beam

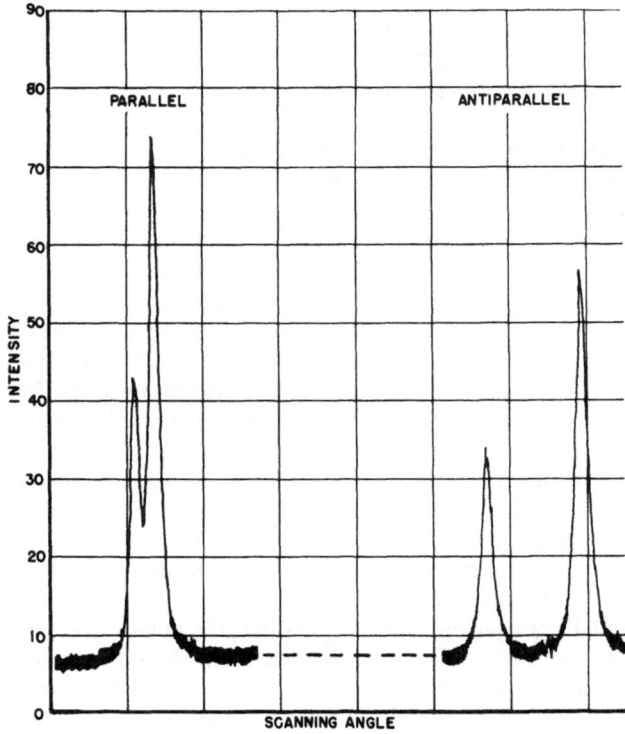

Figure 7. First-order Mo $K_{\alpha_{1,2}}$ reflection for parallel and antiparallel arrangements (the dashed line represents background intensity prior to alignment of polarizer crystal).

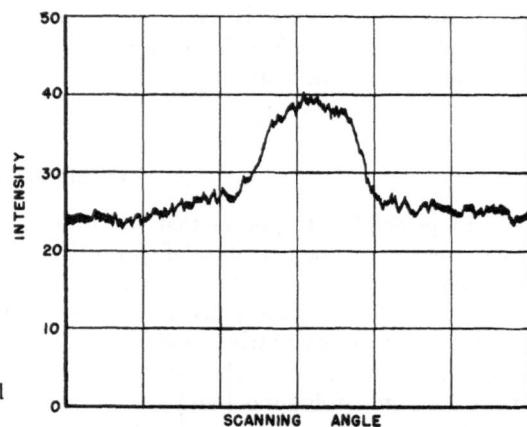

Figure 8. Copper edge region in parallel arrangement.

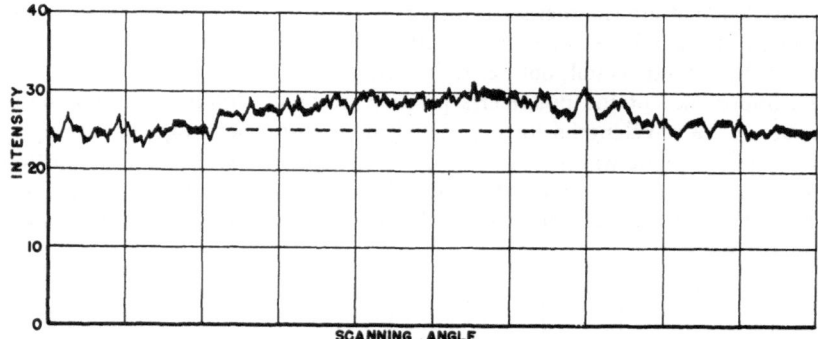

Figure 9. Copper edge region in antiparallel arrangement.

rises above background as shown in an actual tracing reproduced in Figure 8. It is clearly seen that this rise above the background is quite small, and because of the dispersion in the antiparallel arrangement (Figure 2) it is even further reduced (Figure 9), and becomes barely distinguishable from the general background. The background intensity on the chart in Figure 9 is at 25, while the anomalously transmitted beam has a relative intensity of 29. Some of the general "background" detected in both arrangements is caused by parasitic scattering, while the rest stems from the "unaffected" components transmitted by the polarizer in the forward-beam direction. Although this component can be minimized by increasing the crystal's thickness, there is an optimum thickness for a specific type of crystal, as pointed out by Cole et al.[10] For the germanium crystal actually employed, it was estimated that the anomalously transmitted beam was 70% polarized when topaz was placed in the first-crystal position and set to reflect at the K edge of copper.

In order to obtain the 100%-polarized beam originally desired, it was decided, therefore, to utilize the diffracted beam rather than the transmitted beam from the polarizer (Figures 3 and 4). Except for swinging the detector about the polarizer θ axis, by an amount 2θ, the alignment of the spectrometer remains as before. The diffracted beam cannot contain an "unaffected" component so that, provided the polarizer crystal is sufficiently perfect to diffract anomalously only, the diffracted beam should be 100% polarized. Unfortunately, the total diffracted-beam intensity in the absorption-edge region is still extremely low. Even though undesired background radiation is virtually eliminated in this setting, it was found that the finally transmitted intensity was too low to permit the accumulation of statistically significant data when an absorbing metal foil was placed in its path. Thus, with the conventional X-ray generator employed in this analysis, it did not prove possible to carry out the initially assigned task.

CONCLUSION

The two-crystal spectrometer, employing a Cole polarizer as one of its crystals, was found to have the necessary dispersion and resolution for the study of X-ray absorption spectra with plane-polarized X-radiation. Unfortunately, the X-ray generator employed did not produce a sufficiently high intensity to permit the application of this instrument to the study of possible changes in X-ray absorption-edge fine structures, as predicted by Kronig. It is hoped to obtain a more intense X-ray source in the very near future and to carry out the contemplated studies at that time. As clearly demonstrated above, the intensity is quite adequate, even with a conventional generator, provided characteristic radiation emanating from the X-ray tube can be utilized.

ACKNOWLEDGMENTS

The author would like to thank H. Deptolla for his aid in the construction of the polarizer attachment. This research was supported by a grant from the National Science Foundation.

REFERENCES

1. R. de L. Kronig, *Z. Physik* **70**: 317, 1931; **75**: 191, 1932.
2. S. T. Stephenson, *Phys. Ref.* **44**: 349, 1933.
3. R. Krogstad, W. Nelson, and S. T. Stephenson, *Phys. Rev.* **92**: 1394, 1954.
4. W. F. Nelson, Ph.D. Thesis, Washington State College, 1956 (unpublished).
5. J. M. El-Hassaini and S. T. Stephenson, *Phys. Rev.* **109**: 51, 1958.
6. J. N. Singh, *Phys. Rev.* **123**: 1724, 1961.
7. T. A. Boster and J. E. Edwards, *J. Chem. Phys.* **36**: 3031, 1962.
8. L. V. Azároff, *Rev. Mod. Phys.* **35**: 1012, 1963.
9. E. Alexander, B. S. Fraenkel, J. Perel, and K. Rabinovich, *Phys. Rev.* **132**: 1554, 1963.
10. H. Cole, F. W. Chambers, and C. Wood, *J. App. Phys.* **32**: 1942, 1961.

DISCUSSION

H. K. Herglotz (E. I. du Pont de Nemours & Co.): The short wavelength part of the continuous spectrum is partially polarized, anyway. So you should have dependence on θ.

R. J. Donohue: Nailing the copper edge down to a particular place where we are going to examine it, if we tune the polarizer to this position, we will get anomalous increase above this general background.

EVALUATION OF QUANTITATIVE ELECTRON MICROPROBE ANALYSES OF MULTIPHASE MICROCRYSTALLINE REFRACTORY MATERIALS*

David H. Speidel

The Pennsylvania State University
University Park, Pennsylvania

ABSTRACT

An appropriate test for homogeneity of composition of standards and of unknown samples is an analysis of variance of the intensity ratios of the components within the grains opposed to the variance between the grains under conditions of constant beam current capacitor change. Replication of analyses is necessary to distinguish inherent instrumental variation from actual chemical inhomogeneities present in the samples. This replication permits an estimate of the imprecision caused by inhomogeneities and the statistical variation of X-ray quanta. The total uncertainty of an analysis is the sum of systematic error—primarily reference standard uncertainty—and the imprecision. For multiphase mixtures of microcrystalline refractory oxide material, an uncertainty of $\pm 2\%$ of the amount of oxide component present in a given phase should be considered a good analysis. This value is based on experimental work in the system $MgO-FeO-Fe_2O_3-SiO_2$.

INTRODUCTION

The determination of chemical composition of individual phases present in microcrystalline refractory oxide bodies often involves problems which prohibit the use of conventional analytical techniques. For instance, the cell dimensions may not change significantly with changing composition. A good example of this is the solid-solution series magnesioferrite–$(MgFe_2O_4)$–magnetite(Fe_3O_4). In some cases, the optical properties of a phase may not be distinguishable if the phase is masked by other phases that may be opaque or just much more abundant. In synthetic refractory oxide systems, a common difficulty is the extremely small size of the grains, making optical measurements difficult. The usual intimate intergrowths of the phases precludes mechanical separation for a "wet" chemical analysis. For cases in which the phase composition is not otherwise determinable, the electron microprobe provides rapid analyses of selected grains in an inseparable matrix. The present paper presents some of the inherent problems and procedural techniques involved in determining the quantitative value of electron microprobe analyses of multiphase microcrystalline mixtures of refractory oxide materials.[1,2]

---

*Contribution No. 64–44 from the College of Mineral Industries, The Pennsylvania State University, University Park, Pennsylvania.
[1]References are at the end of the paper.

EXPERIMENTAL PROCEDURES AND THEORETICAL CONSIDERATIONS FOR THE PRESENT INVESTIGATION

The following phase assemblages in the system $MgO-FeO-Fe_2O_3-SiO_2$ were studied over a range of compositions and oxygen fugacities (0.21 to $10^{-10.0}$ atm) at subsolidus temperatures: olivine + pyroxene + magnesioferrite, olivine + tridymite + magnesioferrite, and pyroxene + tridymite + magnesioferrite. The primary aim of the work was to determine the changes in composition of the coexisting phases with changes in temperature and oxygen fugacity.[3,4]

The conventional quenching technique was utilized under conditions of controlled atmosphere.[5] Mechanically mixed and sintered oxides of a composition known to give the desired assemblage were held at constant temperature under chosen atmospheric conditions for 10 to 21 days and then quenched to room temperature. Samples were ground and replaced two or three times during the run's duration to facilitate the attainment of equilibrium. The quench products had grains of 5 to 30 μ in size, each grain usually of several different crystals. Small portions of the sample were set in DOW Epoxy 334 and hardened by the addition of diethylene-tetraethylamine. Polishing was done by the sequential use of 6-,3-,1-, and $\frac{1}{4}$-μ diamond paste on a mechanical lap wheel. The samples were carbon-coated to a thickness of about 200 to 400 Å and placed in a brass holder in the microprobe chamber for analysis. The particular grain to be analyzed was approximately located by reflected light optics and exactly located by the use of the electron back-scatter image. The beam current was adjusted to read a constant value on the specimen to be analyzed, in this case, 0.05 μA with a voltage of 20 kV. The beam-current capacitors were charged to capacity, and the relative charge of the elements on the X-ray channel capacitors was recorded on a strip chart. Because the amount of magnesium in the spinel phase was quite low in some samples, a 20-sec scaler count of magnesium was taken.

The resulting relative intensities were transformed into relative amounts of the oxides present by comparison with known standards of similar chemical composition and structure.

A series of five magnesioferrite-magnetite solid solutions were prepared in air at 1300–1400°C using the data of Phillips, Somiya, and Muan.[6] These solid solutions were used as standards for the determination of the magnesium/iron ratios of the unknown spinel phases. The standards were calibrated in number of magnesium counts per 20 sec over chart intensity of iron. All iron was calculated as FeO. Two natural olivines, one synthetic olivine, and one natural amphibole were used as standards for the iron magnesium silicates. They were calibrated in terms of chart intensity of magnesium over chart intensity of iron.

ASSIGNMENT OF UNCERTAINTY TO QUANTITATIVE ELECTRON MICROPROBE ANALYSES

To evaluate a report on quantitative electron microprobe analyses, it is necessary to know the sources of error, with respect to systematic error and precision, and the numerical importance attached to them. Various sources of systematic error, such as instrumental counter drift, long-term beam-current variations, carbon coating, surface irregularities and contamination, and material loss can be minimized by comparison of known standards and the unknown sample prepared in the same manner and run in the same holder within a relatively short period of time of approximately 1 hr. The degree of dependence of unknown samples on known standards means that reference uncertainty is the major

Table I. Sources of Variation Contributing to the Variance

Source	Factors contributing to variance
Grains	Chemical inhomogeneities between grains
	Chemical inhomogeneities within grains
	Machine variation
	Material loss
Points within grains	Chemical inhomogeneities within grains
	Machine variation
	Material loss

source of systematic error. A definite requirement of acceptable standards is that the reference to any portion of the standard will give identical results. The homogeneity is difficult to measure and even more difficult to achieve on the micron level of sensitivity of the electron microprobe. Common procedural techniques to achieve equilibrium in synthetic systems have been to approach a given temperature from both higher and lower values and to use different starting materials. Equilibrium has been assumed if there is a lack of change of physical properties, a positive test only for nonequilibrium. Moreover, changes in the physical properties are usually not sensitive enough to indicate that all portions of the standard are of the same composition, and that there are no local enrichments, several microns or less in size, of a particular component.

Many electron microprobe analysts have assumed that an extremely low standard deviation for points over several grains indicates no significant chemical variation. As can be seen from Table I, a low standard deviation would indicate only that there is no chemical variation within the combined factor of inherent machine variation and material loss. It is conceivable that a real chemical variation could be disguised. If the instrumental variation were giving high readings on points of low concentration and fluctuated to give low readings on points of high concentration, the above tests would show an extremely small standard deviation, even though there were two distinct composition populations. This type of population variation could easily occur in synthetic systems, where subsolidus reaction rates are extremely sluggish. It is possible to test, within the limits of material loss, the counter effects of machine variation and chemical inhomogeneity by a replication of analyses.* The factors contributing to the variance for the different sources of variation with replication of determinations are given in Table II.

Internal equilibrium is assumed if the variance of the intensity values of the points within the grains is not significantly different from the intensity values between the grains. This means that the compositional variations within and between grains are being tested against the inherent instrumental variation for significant differences. Table III tabulates the data points for a particular spinel standard. There are two determinations within each point, four points within each grain, and eight grains for the sample. Table IV illustrates the analysis of variance and shows the acceptability of the sample as a standard in terms of its homogeneity. Similar procedure should be used, at least to some degree, to establish the homogeneity of unknown samples.

If the described tests indicate homogeneous composition for a particular phase, the replication of analyses enables an estimate of the precision to be made. This statistical

---

*By limiting the composition to refractory oxides, material loss is negligible and this limit can be removed.

Table II. Sources of Variation Contributing to the Variance (Replicated Analyses)

Source	Factors contributing to the variance
Grains	Chemical inhomogeneities between grains Chemical inhomogeneities within grains Machine variation Material loss
Points within grains	Chemical inhomogeneities within grains Machine variation Material loss
Replication of points within grains	Machine variation Material loss

Table III. Data for Spinel Standard [(magnesium count) (chart intensity of iron)], $X \pm 1s = 11.04 \pm 0.73*$

Point	Grain 1	Grain 2	Grain 3	Grain 4	Grain 5	Grain 6	Grain 7	Grain 8
1	10.75	11.13	10.53	11.37	10.52	10.46	11.34	11.81
	11.46	11.85	11.41	10.34	10.64	11.16	10.82	11.43
2	11.53	11.05	11.03	10.99	10.60	10.99	10.75	11.76
	11.51	10.60	11.26	11.49	11.62	10.86	10.26	11.07
3	11.11	11.64	10.70	11.11	10.16	11.31	10.97	10.44
	10.62	11.31	11.45	10.96	9.75	11.57	10.97	10.83
4	10.69	11.60	11.06	11.68	10.86	11.52	11.43	10.77
	10.85	11.46	10.95	11.34	10.05	11.84	11.14	10.37

*Two replications per point, four points per grain.

Table IV. Analysis of Variance for Spinel Standard of Table III

Item	Source of variance	Degrees of freedom	Sum of square	Mean square variance	$F*$
$8 = g$	grains	$g - 1 = 7$	32.142	4.5917	
					1.66
$4 = p$	points	$g(p - 1) = 24$	66.350	2.7646	
					2.07
$2 = d$	determinations	$gp(d - 1) = 32$	42.702	1.3344	
Total $64 = t$		$gpd - 1 = 63$	141.194		

$P_{05,7,24} = 2.43$ $P_{05,24,32} = 1.86$ $P_{01,24,32} = 2.42$

*If the F value, the Fisher function, exceeds the P_{05} and P_{01} values given (Arkin and Colton[7]) there is less than a 5% chance or 1% chance, respectively, that the difference is accidental, i.e., the probability of being wrong in claiming inhomogeneity of composition is less than 5% (or 1%).

estimate is impossible without replication of analyses. The statistical error of X-ray quanta was found to be as high as 10% of the amount for small scaler counts and low chart readings. It was discovered that a wide range in the intensity ratios is usually necessary to give significant difference in the calculated weight percent of the oxide components.

Some estimate of error based on systematic error, mainly reference-standard uncertainty, and on precision, caused by inhomogeneity of the samples and statistical X-ray quanta variations, should be included in any quantitative analysis by the electron microprobe. For the present example,[3,4] the uncertainties in FeO values is 0.5 wt.% for 97 wt.% FeO and 2.5 wt.% for 80 wt.% FeO for the spinel phase. The uncertainties were based on an imprecision of 10% for the experimentally determined ratio (magnesium counts)/(intensity of iron).

The uncertainty in FeO values for the silicate phases does not exceed 5 to 7% of the amount of FeO present. This is based on a general imprecision of 8% for the ratio (intensity of magnesium)/(intensity of iron) and a 3% variation caused by reference uncertainty.

CONCLUSION

Replication of determinations is necessary to distinguish inherent instrumental variations from inhomogeneities of sample composition in work with an electron microprobe analyzer. It is meaningless to discuss the particular composition of a phase as determined by the microprobe if statistical evaluation shows a continuous variation of compositions or shows two or more distinct populations of compositions. If there is little or no systematic error, primarily reference-standard uncertainty, and if the sample has been shown to be one population by an analysis of variance, the main source of error is the imprecision caused by the statistical variation of X-ray quanta. The example given shows that claims of $\pm 2\%$ of a refractory oxide component in a quantitative analysis by the electron microprobe of a multiphase microcrystalline (30 μ or less) mixture should be considered as a good analysis. Claims of less uncertainty on grains of this size for oxide components should be regarded with suspicion unless justification is spelled out.

ACKNOWLEDGMENTS

This research was sponsored by the American Iron and Steel Institute. The electron microprobe analytical technique was set up with the aid of E. W. White of the Mineral Constitution Laboratory, Pennsylvania State University. J. C. Griffiths suggested some of the statistical evaluation techniques. Arnulf Muan, E. F. Osborn, and G. Gibbs critically read the manuscript.

REFERENCES

1. "3rd International Symposium on X-Ray Optics and X-Ray Microanalysis", 1962, Stanford, California.
2. "Abstracts of the 1964 Fall Meeting, The Electrochemical Society, Inc., 2, no. 2, Electrothermics and Metallurgy Division, Electron Probe Techniques", 1964, Washington, D.C.
3. D. H. Speidel, Thesis, "Element Distribution Among Coexisting Phases in the System MgO–FeO–Fe$_2$O$_3$–SiO$_2$–TiO$_2$ as a Function of Temperature, Oxygen Fugacity, and Bulk Composition," 1964.
4. D. H. Speidel and E. F. Osborn, "Element Distribution Among Coexisting Phases in the System MgO–FeO–Fe$_2$O$_3$–SiO$_2$–TiO$_2$ as a Function of Temperature, Oxygen Fugacity, and Bulk Composition," Geol. Soc. Am., Miami Beach, Florida, November, 1964 (abstract).

5. A. Muan and E. F. Osborn, "Phase Equilibria at Liquidus Temperatures in the System MgO–FeO–Fe$_2$O$_3$–SiO$_2$," *J. Am. Ceram. Soc.* **39**: 121–140, 1956.
6. B. Phillips, S. Somiya, and A. Muan, "Melting Relations of Magnesium Oxide–Iron Oxide Mixtures in Air," *J. Am. Ceram. Soc.* **44**: 167–169, 1961.
7. H. Arkin and R. R. Colton, *Tables for Statisticians*, College Outline Series, Barnes and Noble, Inc., New York, 1963, pp. 122–125.

DISCUSSION

E. T. Peters (Manlabs, Inc.): Could you describe the method you use for carbon coating? How do you determine thickness and do you feel that any other material would be a better electrical conductor than carbon?

D. H. Speidel: We put the sample in a spark chamber and estimate the thickness. It was done by Dr. White through a series of experiments where he varied the time on an oil drop to see just what the thickness was for a particular length of time for a particular voltage through the carbon arc. We tried nothing else, for no other reason than that this was convenient.

A METHOD FOR TRACE ANALYSIS WITH AN ELECTRON MICROPROBE

L. A. Fergason

Mallinckrodt Chemical Works
Saint Charles, Missouri

ABSTRACT

A "chopped beam" system of analysis has been devised for a scanning electron microprobe equipped with a 400-channel analyzer. The system has been programed with circuitry which places a signal onto the X or Y scanning coils of the probe, so that the beam jumps back and forth between two analytical areas. The same signal is used to activate alternate halves or quadrants on the 400-channel analyzer in a synchronous manner. The analyzer accumulates X-ray intensity data in the appropriate halves and quadrants as the electron beam oscillates between sample and standard or, in the case of trace analysis, between the unknown and the pure major constituent for background correction. The probe may be left unattended while it is gathering information in this manner. The dwell time of the probe on a given analytical area is 6 sec "live" time.

Errors due to instrumental drift and sample contamination are nullified or minimized by this technique. Consequently, theoretical precision is closely approached for extended counting times. 100 ppm levels of aluminum, silicon, nickel, and iron in uranium have been determined to precisions as good as $\pm$ 10 ppm at the 95% confidence level.

INTRODUCTION

The demands on any analytical instrument usually exceed its capabilities and the microprobe, partly because of its versatility, is certainly no exception. Although referred to as a microanalyzer, the electron microprobe is not a particularly sensitive instrument,[1] sensitive, that is, in its ability to detect a very low concentration of one element in another. Ziebold,[2] by numerical evaluation, showed that the probe can be used to obtain trace analysis below 100 ppm, and further stated that a more elaborate setup should allow the analyst to determine 10 ppm or less. Birks[3] calculated the practical theoretical limits of sensitivity to be 30 ppm under specified conditions and stated that such a limit had not been achieved in practice, but for a Cu K_α peak-to-background ratio of 1300 : 1 using a quartz crystal and 1100 : 1 using a LiF crystal, a limit of about 100 ppm might be expected and such values have actually been observed by several analysts.

The determination of the phase diagram of an alloy with elements showing limited solid solubility requires maximum sensitivity and precision. For example, the solubilities of iron, nickel, silicon, and aluminum in α-uranium require quantitative determinations well below 100 ppm. This requirement demands precisions an order of magnitude better than has been achieved by the probe in any matrix, and unfortunately, the analysis of elements in uranium is comparatively inefficient. For example, with the LiF crystal of the Cambridge microprobe and using Soller slits, nickel in an iron matrix gives a

---

[1] References are at the end of the paper.

peak-to-background ratio of 700 : 1. However, nickel in a uranium matrix gives a ratio of only 230 : 1. In addition, the mass absorption coefficients for X-rays passing through uranium are relatively high and the electron back-scatter coefficient is about 0.5. All of these things combine to lower the efficiency of analysis of these elements in uranium.

METHODS OF TRACE ANALYSIS

Theoretically, precision is a statistical matter, that is, a function of peak-to-background ratio and total counts that have been accumulated.[4] Thus, it would seem to follow that increased precision without limit could be obtained by extending the time of analysis. In actual practice, however, theoretical precision is never attained, but is limited by the stability of the instrument and other factors. Because of the instabilities of the instrument, the total time of analysis is limited to about 20 or 30 min, beyond which time instrumental errors can become greater than those arising from random counting statistics.

To date, there have been two general methods of attack in the field of trace analysis. One is the fabrication of more sophisticated crystal spectrometers to improve the peak-to-background ratios. The other approach is to improve the electronic stability with more elaborate regulator circuitry. In an attempt to achieve the maximum performance with the existing equipment at Mallinckrodt, a method of "chopped-beam" analysis has been devised. Fundamentally, this method breaks up the comparison of two analytical areas into small time increments. When analyses are conducted in this manner, effects of instrumental drift and carbon contamination are the same for the sample and for the standard or the pure elements used for background correction. Consequently, extended counting times can be used to improve the statistical population, and to thereby enhance the ability to measure a peak in the presence of a high background.

INSTRUMENTATION

Figure 1 is a skeletal drawing of the integrated system of a multichannel analyzer (MCA) and a Cambridge microprobe. This system has been described in another report.[5] The units of primary importance to the chopped-beam mode of analysis are outlined with heavy lines. The crystal spectrometer and proportional detector, as supplied by Cambridge, are used in this method. The pulses from the detector, which are proportional in height to the quantum energy of the diffracted X-rays, are amplified and routed to the analog-to-digital convertor (ADC). The ADC measures the pulse heights and stores the pulses in the appropriate channels of the ferrite core memory, which is incorporated in the MCA. As the data accumulates in the memory, it may be monitored with an oscilloscope (Figure 2). The channels or points on the abscissa represent increments of energy or pulse height. The ordinate represents the number of pulses counted and stored in each of the energy increments.

The 400-channel memory is divisible into halves or quadrants. These half and quadrant groups may be selected manually at will, or they may be automatically selected by means of a signal pulse. Switching from one half to the other half and back again, or back and forth between two quadrants, may be achieved by means of a pulse coming from the clock in the timing circuitry. 100 channel quadrants are usually used because the time for data read-out with the digital printer is one half that of the 200 channel halves. The timing circuitry has several functions and may register clock or live time. Live time represents only that time in which the equipment is not processing data and is ready to accept a pulse. Thus, when live time is used, the usual calculations to correct for the bias due to pulse losses at higher counting rates are not necessary. The timer may

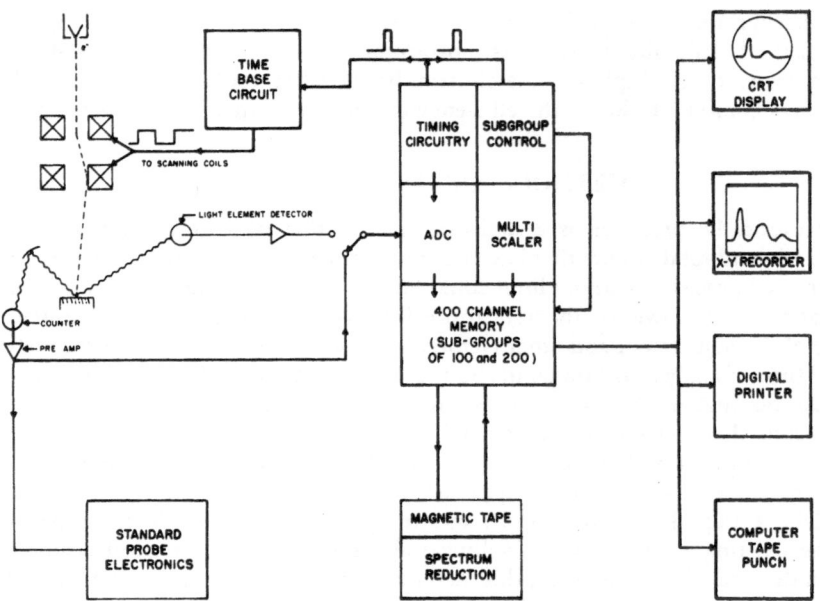

Figure 1. Diagram of system combining multichannel analyzer and electron micro-probe.

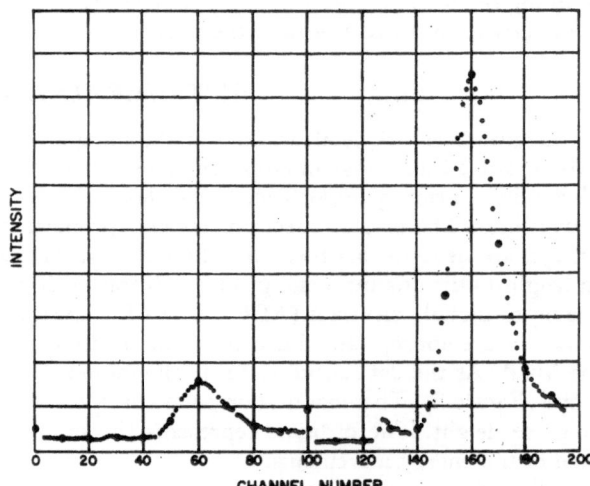

Figure 2. Oscillograph of two spectra at Fe K_α wavelength representing background from pure uranium and U_6Fe, respectively.

be preset so that after a given interval of time, accumulation in the memory is stopped and the results are typed out. The memory is then cleared in preparation for the next analysis and the analytical cycle is started again.

Meanwhile, a pulse is produced by the clock every 6 sec of either clock or live time. This pulse is used to trigger a flip-flop circuit, which in turn selects and activates the memory subgroups. In this manner, subgroups are interchanged every 6 sec. The same pulse is routed to the time-base circuit on the Cambridge microprobe. There it is applied to a Schmidt trigger, which produces a square wave which is applied to the X or Y scanning coils. Thus, in synchronization with the subgroup exchange, the analytical areas exchange every 6 sec. The amplitude of the square wave on the X and Y scanning coils is adjustable, so that the probe may be manually controlled to oscillate between any two points within the area of observation (i.e., 400 μ^2). The basic circuitry of the Cambridge microprobe is used in positioning the probe; the stability of the beam, relative to normal operation, is not affected in any way.

The square wave from the Schmidt trigger has a fast rise time. Consequently, the transient time between the analytical areas is as rapid as the scanning coils will permit. No measurable interference has been encountered because of the scanning of the matrix between the analytical areas. Should interference of this nature occur, however, the effect would be equal for the two areas.

ANALYTICAL PROCEDURE

The samples and standards for this method of analysis were usually machined into thin strips and clamped in a laminar fashion, as shown in Figure 3. Conductive bakelite was used as a mounting material to prepare a 1-in.-diameter specimen. Some care was exercised to minimize the spacing between the strips, so that the distance required to oscillate the probe between the standards and samples would not be greater than 100–200 μ. When Ziebold's empirical correction was to be applied, intermetallic compounds of known stoichiometry were mounted in the specimen so their X-ray emissions could be compared to that of the pure elements.[6] These compounds were U_6Fe and UFe_2 for iron, U_3Si and U_3Si_2 for silicon, and UAl_2 and UAl_4 for aluminum.

When trace constituents were to be analyzed, it was assumed that all of the contributions to background X-ray intensity came from the matrix—for example, pure uranium.

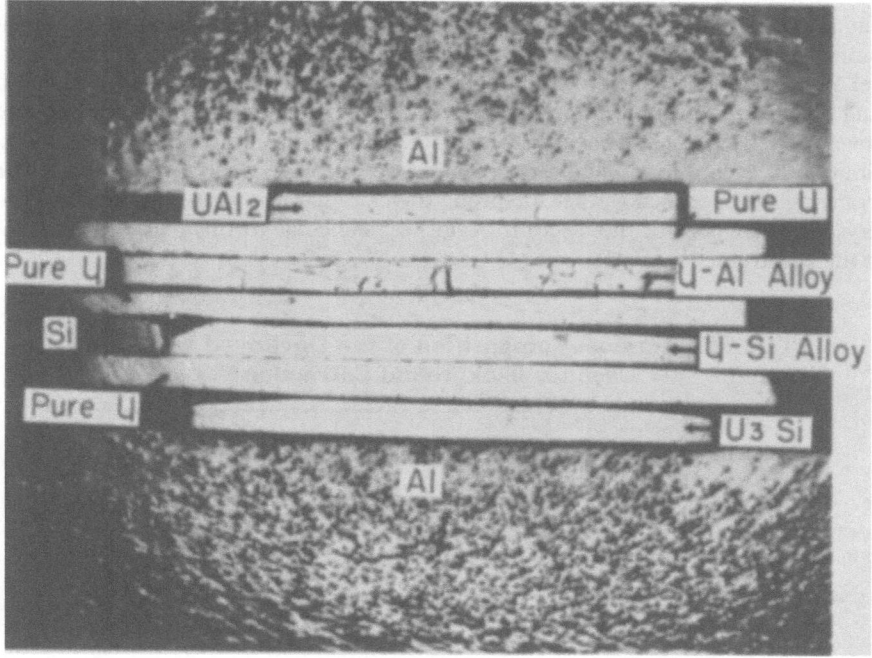

Figure 3. Typical specimen mount for trace element analysis (etched for optical observation, 50×).

Thus the beam was allowed to oscillate between the pure uranium and the unknown. Consequently, the difference in integrated counts between these areas was related to mass composition. The impurity levels for the uranium used in background corrections are listed in Table I. These values were added to the results of the analysis.

So that subsequent statistical treatment could be done and trends in the analysis could be observed, the data was automatically read out at regular preset intervals during the analyses. The time selected was more or less arbitrary, but for the elements aluminum and silicon, comparatively short intervals were selected so as to better evaluate the effect of build-up of carbon contamination upon the lower energy X-ray intensities. The equipment was then allowed to gather data for an extended time, alternately from the two analytical areas. Statistically, the interval readings were treated as replicas, and the precision was calculated by normal statistics at the 95% confidence level. The result of this treatment was then compared to the precision expected of a perfect instrument, as determined by counting statistics.

A possible source of error in quantitative analysis with the microprobe is caused by a shift in gain of the proportional counter with counting rate. A very large error can be incurred by this phenomenon when a xenon counter with a pulse-height analyzer is used to compare the X-ray intensity of a pure elemental standard to that of the element at low concentrations. The nature of MCA data presentation allows the operator to determine true peak position with considerable precision. This cannot be achieved with the pulse-height analyzer. Another advantage of the digital-type data collection of the multichannel analyzer is that the optimum window width or number of channels selected for calculations may be accurately determined and reproduced. A computer program has been constructed to calculate true peak position and optimum window width for these analyses.

The column of a scanning microprobe must be carefully aligned for precision analysis, so that all points in the area of observation are equal in sample current on a homogeneous specimen. The sample current is read to four significant figures by means of a Beckman Model 910-2 digital voltmeter. The alignment of the spectrometer is also checked by causing the electron beam to jump between two points on a pure element. X-ray intensities measured from these two areas should be within the limits of normal statistical distribution of each other. The best agreement between two analytical areas is achieved by arranging the specimen so that the probe traverses parallel to the axis of curvature of the crystal. Movement of the electron beam perpendicular to this axis will tend to defocus the spectrometer.

Table I. Chemical Composition of the Unalloyed Uranium Used for Background Correction

Element	Impurity level (ppm)
Al	8
Fe	18
Ni	5
Si	16
C	24
Total impurities	< 100

RESULTS AND CONCLUSIONS

The results of analysis of aluminum, silicon, iron, and nickel in uranium are presented in Table II, along with the statistical evaluation. Although the peak-to-background ratio of aluminum and silicon was more favorable than that of the iron and nickel, equivalent precision was not obtained, because the duration of the counting had to be limited due to the effects of carbon contamination. The use of a liquid-nitrogen cold finger in the immediate vicinity of the specimen seemed to mitigate this problem.

Figure 4 shows the effect of time versus X-ray intensity at the Ni K_α wavelength. Nickel and iron show a similar trend in the total reading of background and unknown intensity; however, the difference between background and unknown intensity remains constant within normal statistical variation at the 95% confidence level. The exact nature of the cause of this overall decrease in X-ray intensity has not been determined. No trend or bias of the results with time could be detected at the 95% confidence level within the 15-hr counting period. These data were collected overnight with an unattended probe. The total counts accumulated for peak and background were 1,902,178 and 1,866,825, respectively. The overall precision was ± 10.0 ppm.

Table II. Determination of Trace Elements in Uranium

Element analyzed	Known concentration (ppm)	Results of analysis (ppm)	Theoretical precision (ppm)	Actual precision (ppm)	Number of data points	Total counting time	Detector used	Probe voltage (keV)
Fe	154.3 ± 1.1	170[a]	± 16	± 20	7	14 hr	P-10	25
Fe	154.3 ± 1.1	180[a]	± 19	± 19	6	6 hr	Xenon	25
Fe	165.6 ± 0.6	185[a]	± 13	± 13	5	10 hr	Xenon	25
Ni	43	42[a]	± 15	± 17	9	9 hr	Xenon	25
Ni	108 ± 4	115[a]	± 8	± 10	15	15 hr	Xenon	25
Al	743	702[b]	± 25	± 43	10	200 min	P-10	12
Si	111.2 ± 1.2	86[b]	± 29	± 49	10	200 min	P-10	13
Si	111.2 ± 1.2	100[b]	± 42	± 40	6	120 min	P-10	13

[a] Compared directly to standards (U_6Fe and U_6Ni).
[b] Ziebold's empirical correction applied.

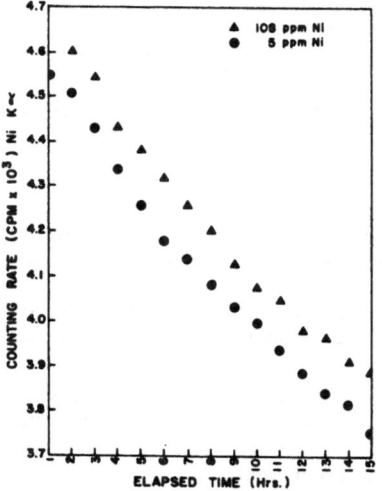

Figure 4. Trace analysis of nickel in uranium, showing the effect of time on replica determinations.

It is expected that the aluminum analysis can be improved to yield greater sensitivity, because of a favorable signal-to-noise ratio. Silicon, however, is a particularly troublesome element. Ca K_α and S K_α lines resulting from the fluorescence of the gypsum crystal interfere with the analysis. Here, again, may be seen the advantage of a multichannel analyzer read-out system. By proper selection of window width, this type of interference can be minimized.

Perhaps the most significant information that can be derived from Table II is that actual precision is fairly close to theoretical. Under these conditions, the attainment of trace analysis is limited only by the resolution of the spectrometer and by the condition of the sample surface.

From considerations of peak-to-background ratio, as mentioned earlier, it has been calculated that nickel in a matrix of iron could be determined to a precision of at least ± 4 ppm.

SUMMARY

A method for trace analysis with the microprobe has been described. A multichannel analyzer is used to help improve the precision of measurement of relative X-ray intensities and to allow a more precise use of the data collected. Trace analysis at 100 ppm levels extending down to ± 10 ppm precision for iron, nickel, silicon, and aluminum in uranium has been achieved by this method.

ACKNOWLEDGMENT

I wish to thank Dr. Norbert Neumann for valuable advice and encouragement and Glen Gower for his assistance with the electronic design. This work was performed under the auspices of the U.S. Atomic Energy Commission.

REFERENCES

1. R. Theisen, "Detection Limits of Electron Microprobe Analysis," presented at E.C.S.–A.I.M.E. Meeting, Toronto, May 1964.
2. T. O. Ziebold, "Trace Determinations in Microanalysis," presented as a lecture at M.I.T. Special Summer Program on the Electron Microanalyzer and its Application, August 3–4, 1964.
3. L. S. Birks, *Electron Probe Microanalysis*, Interscience Publishers, Inc., New York, 1963, p. 136.
4. A. Jarret, *Statistical Methods Used in the Measurement of Radioactivity*, Technical Information Service Extension, U.S. Atomic Energy Commission, *AECU-262*, June 1946.
5. L. A. Fergason and N. F. Neumann, "An Energy Dispersion System for Light Element Analysis with the Microprobe," to be published.
6. T. O. Ziebold and R. E. Ogilvie, "Quantitative Analysis with the Electron Microanalyzer," *Anal. Chem.* **35**: 621, 1963.

DISCUSSION

K. F. J. Heinrich (National Bureau of Standards): I think this is a very elegant method, and wish you could enlighten me a bit more as to why you use the multichannel analyzers rather than, for instance, switching from one scaler to another.

L. A. Fergason: For one thing, we had the equipment for another purpose, and it's a natural to program on something like this. The timing circuitry is there; the read-out is convenient. I suppose it could be done on any type of circuitry with switching.

J. Merritt (Shell Development): There were a couple of us here that couldn't make out what your counting rates were. Would you tell us what the background counting rate was and what the signal counting rate was?

L. A. Fergason: The signal counting rate was 4.7 thousand counts/min.

J. Merritt: That is much higher than I expected. The background is 80 counts/sec, is that right?

L. A. Fergason: Yes, it's less than 100 counts/sec.

H. Yakowitz (National Bureau of Standards): In the Cambridge probe, the focal circle is not true-focusing through this business of rolling the source along the crystal, and it's really not on a Rowland circle. Have you made a statistical analysis of what this might do to raise or lower your curves as a whole, or will it do this?

L. A. Fergason: I'm not sure.

H. Yakowitz: It might be that if you're off-focus you get something analogous; maybe there is a chemical shift in light elements where you're slightly displaced. Therefore, you're not measuring true peak positions as you swing back and forth across this. Have you analyzed this situation? Because, it occurs to me, that it might shift the entire curve up and down.

L. A. Fergason: No, the only analysis that we have done is that we have run intensity profiles for various elements.

Chairman S. H. Moll: I would like to ask you one question. That is, can you give us any numbers for signal-to-noise ratios, let's say, for silicon, aluminum, and iron which would be measured as the pure element aluminum over, say, the background from uranium. This would give us a feeling for its sensitivity.

L. A. Fergason: If we compare the intensity of pure iron to the background we get from the uranium, it's about a 250 : 1 ratio. I don't know exactly what it is for the aluminum or silicon right now, only that it's a little bit better than that.

P. A. Romans (U.S. Bureau of Mines): I would like to know what you think about the possibility of the electrons which are scattered back from the pole piece exciting the silicon or aluminum in these adjacent standards.

L. A. Fergason: I think this has definitely happened to us, for example, with the nickel. We had a strip of pure nickel element in this mount. I think that was making a contribution to our background.

P. W. Wright (Associated Electrical Industries, Manchester, England): Could I ask you about probe size, beam voltage, and beam current?

L. A. Fergason: In these analyses as reported here, we actually used a line on our probe. We used about a 10-μ line, more for the carbon contamination problem than we did for the averaging effect. The beam voltage for iron and nickel was 25 keV and for silicon and aluminum was 12 keV. The beam current was 4×10^{-7}A.

P. W. Wright: Do I take it you were counting on the same line for up to 15 hr?

L. A. Fergason: Right.

P. W. Wright: I would like to make just one comment on this paper and the previous paper. We normally focus our beam to something less than a micron in our work, but we have counted on scanned areas from 10 μ^2 up to about 300 μ^2 when we do trace analysis.

K. F. J. Heinrich: Did you state how big these jumps are? How far removed is your standard from the sample?

L. A. Fergason: We try to keep that as small as possible, naturally. In most of our work we have been able to keep it down to a 100-μ jump, or less. We align our crystal on the basis of a 400-μ jump, but we always keep them as close as possible on the analysis.

THE APPLICATION OF THE ZIEBOLD
CORRECTION PROCEDURE FOR PROBE DATA
TO THREE TERNARY COPPER-BASE ALLOYS

Richard M. Ingersoll, Jack E. Taylor, and Donald H. Derouin

Anaconda American Brass Company
Waterbury, Connecticut

ABSTRACT

For the quantitative analysis of a 65 Cu–30 Ni–5 Fe alloy, a 96 Cu–3 Si–1 Mn alloy, and a 78 Cu–20 Zn–2 Al alloy, the Ziebold empirical method of correcting electron-microbeam-probe data was used. Four binary standards, of single-phase Cu–Ni, Ni–Fe, Cu–Mn, and Cu–Zn alloys, were cast and the α correction factor found for each element in each binary by Ziebold's relationship $(1 - K)/K = \alpha (1 - C)/C$, where $K = I/I_0$ found in the probe and C is the weight fraction found by wet chemistry. The ARL EMX probe was used at 30 kV with a 25-μ beam diameter to negate inhomogeneities. Experience with these binaries indicated that in the presence of secondary fluorescence, the experimental α values agreed poorly with theoretically calculated α values; however, where secondary fluorescence was negligible, agreement between the experimental and theoretical α values was good. The α values for Cu–Si, Cu–Al, Al–Zn, and Mn–Si alloys, were therefore calculated from the theoretical equations. The α values for Cu–Fe alloys were also calculated from theoretical considerations because single-phase binaries over the composition range of interest could not be made for this system. All these α values were used in Ziebold's ternary equations to correct probe data (again using a 25-μ beam) from specimens of Cu–Ni–Fe, Cu–Si–Mn, and Cu–Zn–Al. These results were compared to wet-chemistry analyses for the same specimens with quite good correlation between the two sets of data. Calibration curves for the binary systems Cu–Ni, Cu–Fe, Ni–Fe, Cu–Mn, Cu–Si, Mn–Si, Cu–Al, Cu–Zn, and Al–Zn were made and are reproduced.

INTRODUCTION

The electron-microbeam-probe analysis of copper-base ternary alloys by conventional methods can be a tedious process, subject to numerous errors. With an increasing demand for analyses of copper with (1) Ni + Fe, (2) Mn + Si, and (3) Al + Zn, a better technique was required which would give accuracy and dependability comparable to wet chemistry results, could be performed with reasonable speed, and was not fraught with error-making possibilities. Such a procedure seemed available in the empirical approach of Ziebold and Ogilvie[1,2] and this technique was explored. An inherent by-product of this procedure is that, since the method is based upon data from standard alloys, it tends to develop greater confidence in probe data among people unfamiliar with probes.

---

[1] References are at the end of the paper.

273

EXPERIMENTAL PROCEDURE

Copper–Nickel–Iron

The Ziebold procedure in a system ABD requires making three sets of binary standard alloys of AB, AD, and BD. In applying this procedure to the Cu–Ni–Fe system, four binary standard alloys of Cu–Ni and four of Ni–Fe were cast. Cu–Fe standards were not made because it is not possible to make single-phase material over the range of composition desired. The Cu–Ni alloys were cast under charcoal cover as 1-in. bars. The Ni–Fe alloy standards were cast in the vacuum furnace and deoxidized with calcium; no residual calcium was found.

Each of the Cu–Ni and Ni–Fe alloy standards were analyzed by wet chemistry to an accuracy of 1% of the amount present, with the results shown in Table I.

Metallographic sections 1 in. in diameter and $\frac{3}{16}$-in. thick were cut from each of the alloys and polished through $\frac{1}{4}$ μ diamond. The specimens prepared in this way were put into the ARL EMX probe. Analysis was done using a 25 μ beam to make the specimens appear homogeneous to the electron beam.[3] 30 kV accelerating potential was used in all analyses. Pure copper, electroplated nickel, and Armco iron were used as standards. The Cu K_α line, the Ni K_α line, and the Fe K_α line were used in the analysis.

The analysis was done for each element by first taking 20 readings from the standard, 20 from the binary alloy, and then 20 from the standard again. From the standard data, an I_0 uncorrected (I_{0u}) for each element was calculated, and from the alloy standard, an I uncorrected (I_u) value found for each element. Background was found on the standard by taking ten readings two Bragg-angle degrees higher and the same lower than the peak both before and after the analysis. Backgrounds for the standard were the average of these 40 readings and the background for the alloy was calculated from the following:

$$\text{B.G.}_{\text{alloy}} = \frac{I_u}{(I_{0u})_{\text{Ni}}} \text{B.G.}_{\text{Ni}} + \frac{I_u}{(I_{0u})_{\text{Fe}}} \text{B.G.}_{\text{Fe}} + \frac{I_u}{(I_{0u})_{\text{Cu}}} \text{B.G.}_{\text{Cu}} \qquad (1)$$

where B.G.$_{\text{alloy}}$ is the background of the alloy, $(I/I_0)_A$ is the intensity ratio of A uncorrected for background, and B.G.$_A$ is the background of pure element A. The background for each standard was subtracted from I_{0u} to give I_0 and the alloy background was subtracted from I_u to give I. The ratio I/I_0 was then found and set equal to K. In all this work, counting techniques were used, and accuracy in all was about 1%.

Table I. Analysis of Cu–Ni and Ni–Fe Alloy Standards

Alloy number	Copper (wt. fraction)	Nickel (wt. fraction)	Iron (wt. fraction)
10785	0.1927	0.8073	
10786	0.3940	0.6060	
10787	0.5969	0.4030	
10788	0.7995	0.2005	
11016		0.0509	0.9491
11015		0.2514	0.7486
10828		0.5047	0.4951
10885		0.8991	0.1004

Ziebold's equation for binaries is expressed as

$$\frac{(1-K)}{K} = \frac{\alpha(1-C)}{C} \qquad (2)$$

where $K = I/I_0$, C is chemical composition (wt. fraction), and α is the proportionality constant. The constant α can be found in either of two ways. In the first method, knowing C and K, equation (2) can be solved for α. The α's found for the four binary standards are averaged to give a usable α value. The greater the number of times this is done, the more accurate α becomes. The second method involves (1) plotting C/K versus C for each determination for each alloy, (2) finding the least-squares fit of a straight line through these points, and (3) noting the value of C/K at $C = 0$. This is the value of α. Since the data for the iron in the Ni–Fe binary standards showed a large amount of scatter (see

Table II. Iron Contents in Ni–Fe Binary Standards

	10885	11015	10828	11016
$C/K(1)$	0.614	0.723	0.772	0.953
$C/K(2)$	0.695	0.764	0.988*	1.030
$C/K(3)$	0.674	0.784	0.838	not done
$C/K(4)$	0.677	0.764	0.865	0.985
C/K average	0.665	0.759	0.825	0.989
C (wt. fraction)	0.1004	0.2514	0.4951	0.9491

* This value not used in computing average.

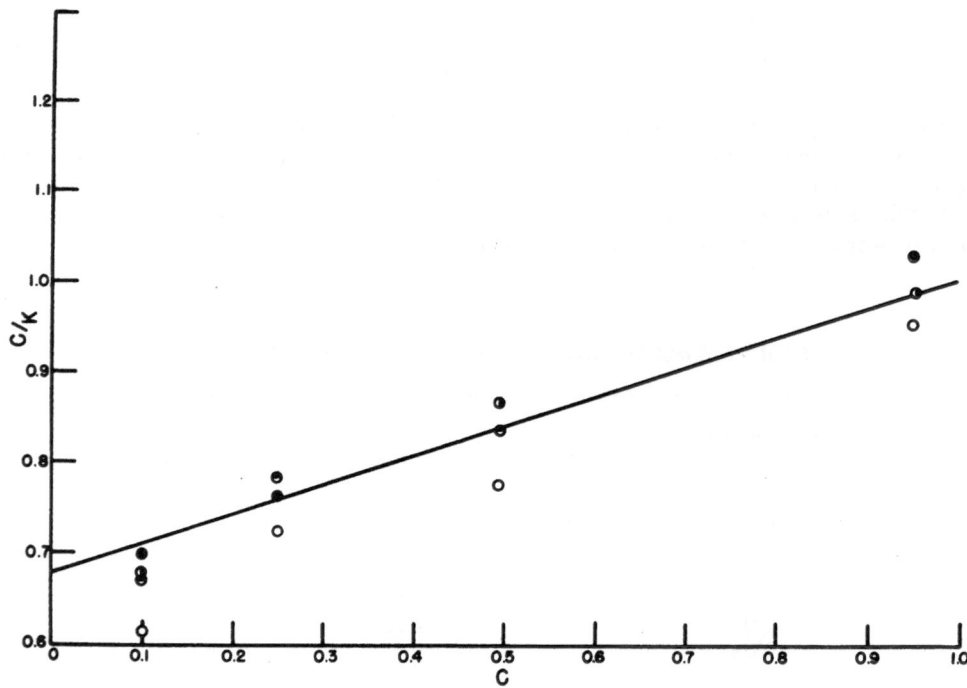

Figure 1. C/K versus C for iron in four Ni–Fe standards.

Table II), the second method was used; the plot is shown in Figure 1. Actually, the line shown is not a least-squares fit but is a straight line drawn visually through the points. This approximates the least-squares fit quite well and gives good results. In Figure 1, scatter is noted around the straight line and α is seen to be 0.67. It was found that a somewhat better fit could be obtained by averaging the values for each binary. This plot is shown in Figure 2 and α is noted to be 0.65. This value does not agree with a theoretical calculated α of 0.74 (using any combination of any of the available absorption and fluorescent corrections) nor with Ziebold's value[4] of 0.80. Nevertheless, $\alpha = 0.65$ is the correction factor found with this probe for iron in Fe–Ni alloys.

For the nickel in the Ni–Fe alloys, the C/K and C values found from the probe data and chemistry results are shown in Table III. The α value for the nickel in Ni–Fe alloys was found from the data in Table III in the same way as the iron and was found to be 1.18. The theoretical corrections indicate that α should be 1.21 and Ziebold in this procedure found 1.20.[4] In constructing the Fe–Ni calibration curves (Figure 5), C values for nickel and iron were assumed and the corresponding K values found from equation

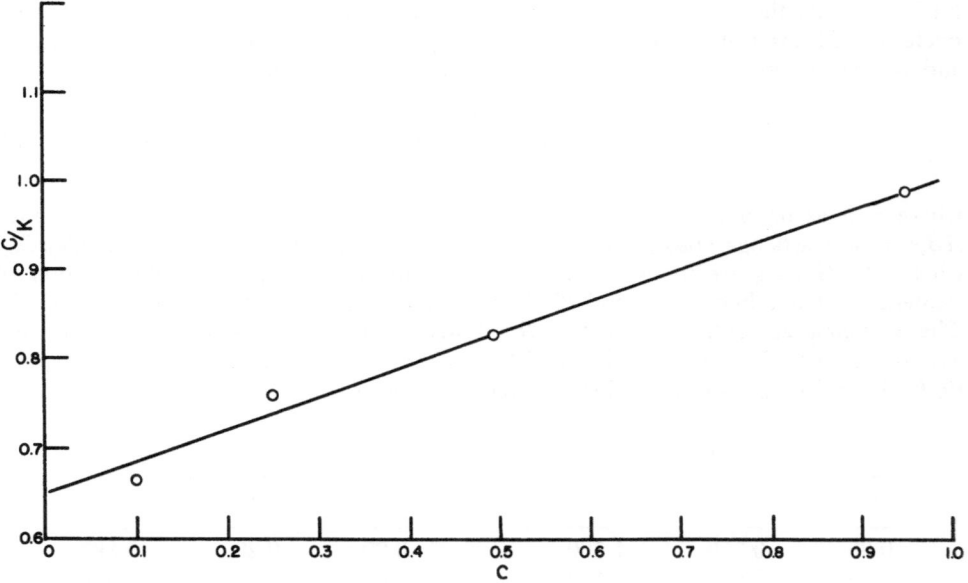

Figure 2. C/K versus C using averaged C/K values for four Ni–Fe standards.

Table III. Nickel Contents in Ni–Fe Binary Standards

	10885	11015	10828	11016
C/K(1)	1.020	1.060	1.120	1.190
C/K(2)	0.900	0.965	1.090	1.110
C/K(3)	1.003	1.050	1.160	not done
C/K(4)	0.988	0.998	1.160	1.210
C/K average	0.978	1.018	1.133	1.170
C (wt. fraction)	0.8991	0.7486	0.5047	0.0509

(2) using the proper experimental α for nickel or iron. For the Cu–Ni alloys, equation (2) was solved for α and the values found for the four alloys for each element averaged as shown below.

	Cu			Ni	
C	K	α	C	K	α
0.1927	0.190	1.02	0.8073	0.844	0.77
0.3940	0.396	0.99	0.6060	0.655	0.81
0.5969	0.607	0.96	0.4030	0.458	0.83
0.7995	0.805	0.97	0.2005	0.236	0.81
	Average	0.99		Average	0.80

One would expect both the α for copper and nickel to approximate 1; however, fluorescence of the nickel is observed since α is < 1. This is caused by fluorescence of the Ni K_α peak by the Cu K_β. As mentioned previously, no Cu–Fe binary standards were made up. Theoretical α values were, therefore, calculated using Castaing's absorption and Wittry's fluorescence correction[5,6] using the following equation:

$$\frac{I}{(I_0)_A} = K_A = \frac{f}{f_0}\left(1 + \frac{G}{f}\right)C_A \qquad (3)$$

where K is as previously, f is Castaing's absorption parameter for element A in alloy AB,[5] f_0 is Castaings absorption parameter for pure element A,[5] $G = (G_1)(G_2)(G_3)$, where G_1, G_2, G_3 are found from Wittry's[6] graphs, and C_A is weight fraction of element A. Then, knowing C and K, the corresponding α was found from equation (2). The Castaing absorption curve for 29 kV was used and the fluorescence correction was found for 52.5° take-off angle and 30 kV. K values were found for the Cu K_α and Fe K_α lines. For Cu–Fe alloys these are tabulated below.

	Cu			Fe	
C	K	α	C	K	α
0.80	0.770	1.200	0.20	0.220	0.887
0.60	0.562	1.170	0.40	0.423	0.928
0.40	0.362	1.175	0.60	0.620	0.928
0.20	0.180	1.139	0.80	0.814	0.913
	Average	1.171		Average	0.911

At this point in the procedure, α values are available for Cu–Ni, Cu–Fe, and Ni–Fe binary systems and are summarized below. It should be noted that absorption produces an $\alpha > 1$ and fluorescence an $\alpha < 1$.

Cu–Ni	Cu–Fe	Ni–Fe
$\alpha_{Cu-Ni}^{Cu} = 0.99$	$\alpha_{Cu-Fe}^{Cu} = 1.171$	$\alpha_{Ni-Fe}^{Ni} = 1.18$
$\alpha_{Cu-Ni}^{Ni} = 0.80$	$\alpha_{Cu-Fe}^{Fe} = 0.91$	$\alpha_{Ni-Fe}^{Fe} = 0.65$

A sample of a Cu–Ni–Fe alloy designated as No. 10622-3 was analyzed by wet chemistry to be as shown below.

Element	Wt. fraction	Element	Wt. fraction
Cu	0.6448	Ni	0.2976
Zn	< 0.0010	P	0.00012
Pb	0.00005	Mn	0.0047
Fe	0.0527	C	0.00007

This specimen was polished through the $\frac{1}{4}$-μ diamond and put into the probe. The probe beam size was set at about 25 μ to make the specimen appear homogeneous to the electron beam. Again, 30-kV acceleration potential was used. I, I_0, and backgrounds were found in the same fashion as those found for the binary standard alloys. The K values were calculated and the appropriate values of K and α substituted in Ziebold's ternary equation, which is

$$\frac{1 - K_A}{K_A} = \left(\frac{\alpha_{AB}{}^A C_B + \alpha_{AD}{}^A C_D}{C_B + C_D}\right)\left(\frac{1 - C_A}{C_A}\right) \tag{4}$$

This equation was set up for copper, nickel, and iron as A, B, and D, respectively, and the three simultaneous equations solved for copper, nickel, and iron fractions. The K values used and the C values found are shown below, in comparison with the chemical analysis.

Element	K	C (wt. fraction)	Wet chemistry (wt. fraction)
Cu	0.644	0.6530	0.6448
Ni	0.324	0.2878	0.2976
Fe	0.0776	0.0615	0.0527
		1.0023	0.9951

For comparison with the Ziebold method, weight fractions were then calculated from these same K values using Birks' MEATAXE approach.[7] Results are shown below.

Element	K	C (wt. fraction)	Wet chemistry (wt. fraction)
Cu	0.644	0.6535	0.6448
Ni	0.324	0.3158	0.2976
Fe	0.0776	0.0734	0.0527
		1.0427	

Since the agreement between the Ziebold results and the wet chemistry appeared quite good and certainly better than Birks' technique, (1) calibration curves for Cu–Ni, Cu–Fe, and Ni–Fe alloys were drawn and are shown in Figures 3, 4, and 5, and (2) the method was extended to Cu–Mn–Si and Cu–Zn–Al alloys.

Copper-Manganese-Silicon

Binary standards for Cu–Mn alloys can be made single-phase up to 35% manganese; but single-phase standards cannot be made for Cu–Si or Mn–Si alloys over the range of composition desired. In the work with the Cu–Ni–Fe material, wherever fluorescence was negligible, the theoretical α factor was found to agree very closely with the experimental α factor. Therefore, the α values for the Cu–Si and Mn–Si alloys were calculated similarly to the way those for the Cu–Fe alloys were found. Equation (3) was used to find K for the copper in Cu–Si alloys and manganese in Mn–Si alloys, and was modified to contain the Poole and Thomas[8] atomic number correction for the silicon as follows:

$$\frac{I}{I_0} = K = \left(\frac{f}{f_0}\right)\left(1 + \frac{G}{f}\right)\left(\frac{C_A \kappa}{\Sigma_i \kappa_i C_i}\right) \tag{5}$$

where I, I_0, K, f, f_0, G, and C are as previously and $\kappa =$ is the Poole and Thomas atomic number correction factor.[8,9] The K values and C values thus found were substituted into equation (2) and the α values found (Table IV). The κ values used were derived from a graph published by Ziebold and Ogilvie[9] and were 1.55 for silicon, 1.41 for manganese, and 1.35 for copper. The absorption correction was made using Castaing's curve for 29 kV, and Wittry's fluorescence correction for 30 kV and 52.5° take-off angle. The Cu K_α, Mn K_α, and Si K_α lines were used.

The α factors for Cu–Mn alloys were determined by first casting four Cu–Mn alloys under charcoal, analyzing these alloys by wet chemistry, finding K, and then solving for α with equation (2), in the same way as was done for Cu–Ni alloys. A 25-μ beam was used to make the specimen appear homogeneous to the electron beam. Pure copper and pure manganese were used as standards; the Cu K_α and Mn K_α lines were used for the analysis, which was done at 30 kV. Table V lists the data. Theory predicts an α of 1.163 for copper and 0.929 for manganese.

To construct the Cu–Mn calibration curve (Figure 6), the above C and K values were used, along with three additional points found by assuming C, knowing α, and

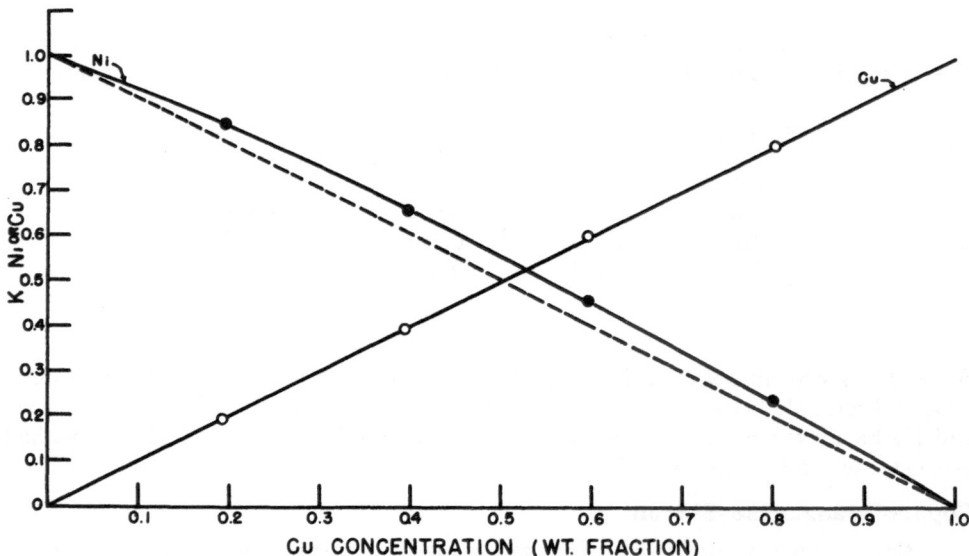

Figure 3. Calibration curves for Cu–Ni alloys.

finding K from equation (2). These numbers were as shown below.

Cu			Mn	
C	K		C	K
0.50	0.461		0.5	0.545
0.30	0.2678		0.7	0.737
0.10	0.0867		0.9	0.917

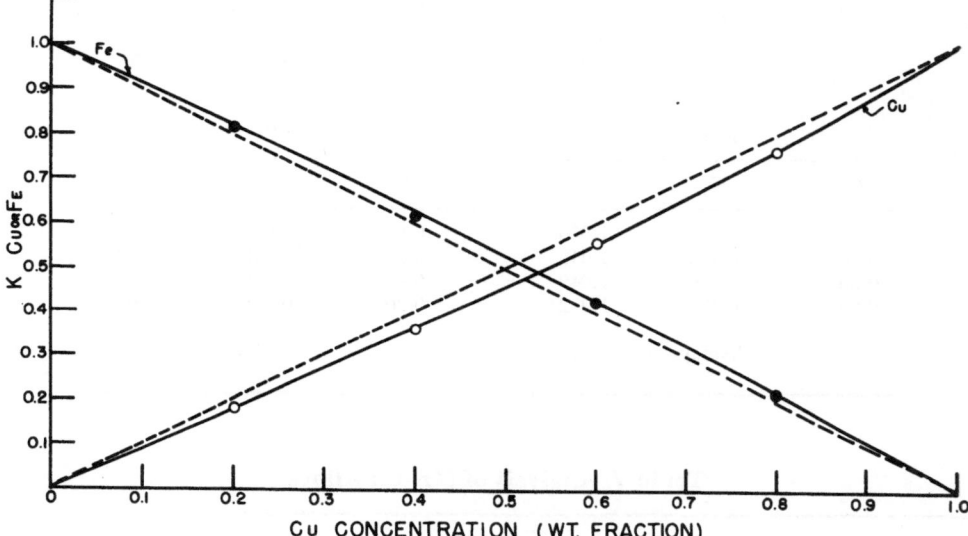

Figure 4. Calibration curves for Cu–Fe alloys.

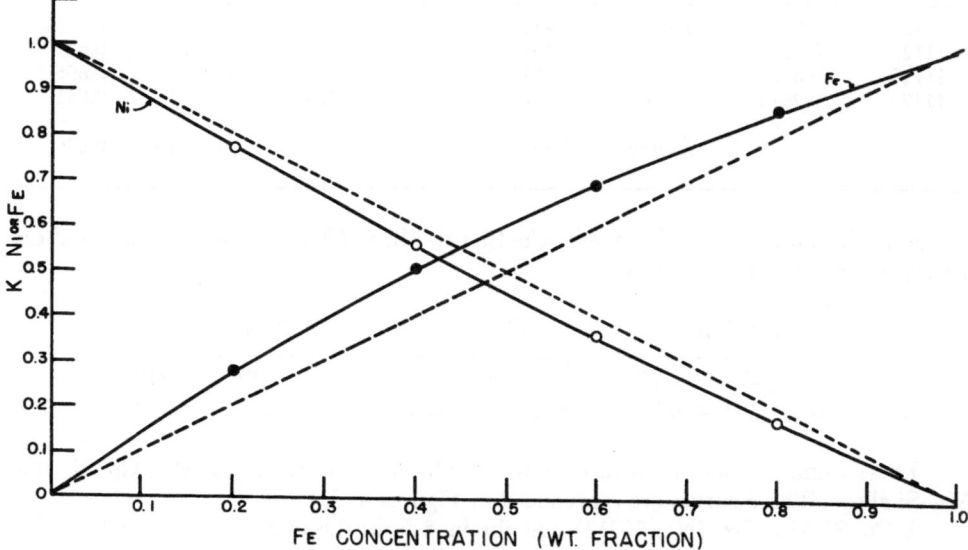

Figure 5. Calibration curves for Ni–Fe alloys.

Table IV. Analysis of Cu–Si and Mn–Si Alloys

			Cu–Si			
	Cu				Si	
C	K	α		C	K	α
0.20	0.198	1.012		0.80	0.648	2.18
0.40	0.398	1.012		0.60	0.405	2.21
0.60	0.597	1.012		0.40	0.228	2.26
0.80	0.798	1.012		0.20	0.096	2.36
	Average	1.012			Average	2.23

			Mn–Si			
	Mn				Si	
C	K	α		C	K	α
0.20	0.1928	1.050		0.80	0.676	1.920
0.40	0.3892	1.048		0.60	0.438	1.922
0.60	0.5892	1.049		0.40	0.258	1.918
0.80	0.7936	1.042		0.20	0.115	1.917
	Average	1.048			Average	1.919

Table V. Analysis of Mn–Cu Alloys

Alloy no.	Cu			Mn		
	C	K	α	C	K	α
11121	0.9596	0.952	1.210	0.0404	0.0444	0.9084
11122	0.8844	0.872	1.194	0.1156	0.1373	0.823
11123	0.8316	0.815	1.121	0.1684	0.2013	0.805
11124	0.7619	0.734	1.166	0.2381	0.2993	0.851
		Average	1.173		Average	0.836

Available now are the α factors for Cu–Si, Cu–Mn and Mn–Si alloys. These numbers are reproduced in summary below.

Cu–Mn	Cu–Si	Mn–Si
$\alpha_{Cu-Mn}^{Cu} = 1.173$	$\alpha_{Cu-Si}^{Cu} = 1.012$	$\alpha_{Mn-Si}^{Mn} = 1.048$
$\alpha_{Cu-Mn}^{Mn} = 0.836$	$\alpha_{Cu-Si}^{Si} = 2.23$	$\alpha_{Mn-Si}^{Si} = 1.919$

These numbers were used to construct calibration curves of Cu–Mn, Cu–Si, and Mn–Si alloys. These curves are reproduced in Figures 6, 7, 8.

A Cu–Si–Mn alloy (No. 10711) was obtained from the mill, a section analyzed by wet chemistry, and an adjacent section analyzed in the probe. Similarly to the technique used for the Cu–Ni–Fe specimen, pure copper, pure manganese, and pure silicon were

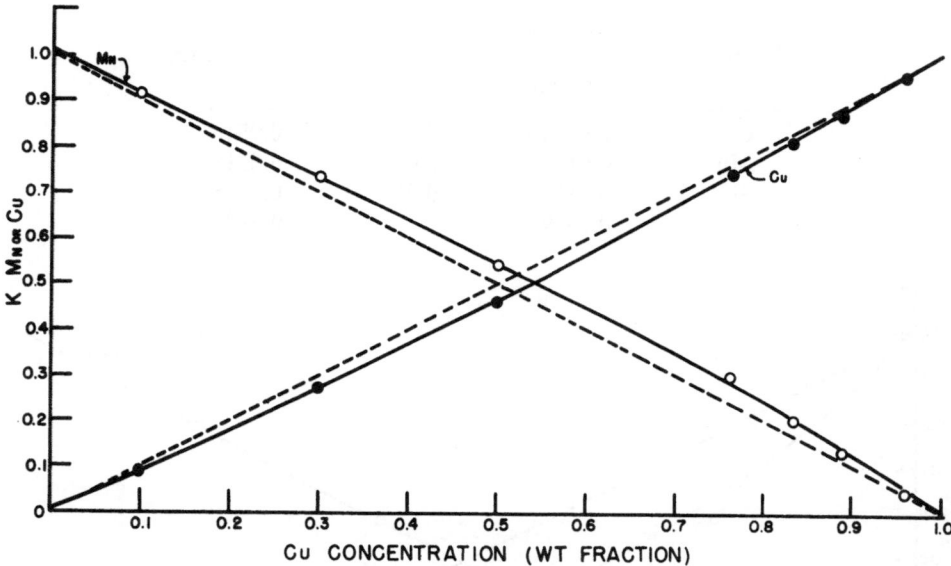

Figure 6. Calibration curves for Mn–Cu alloys.

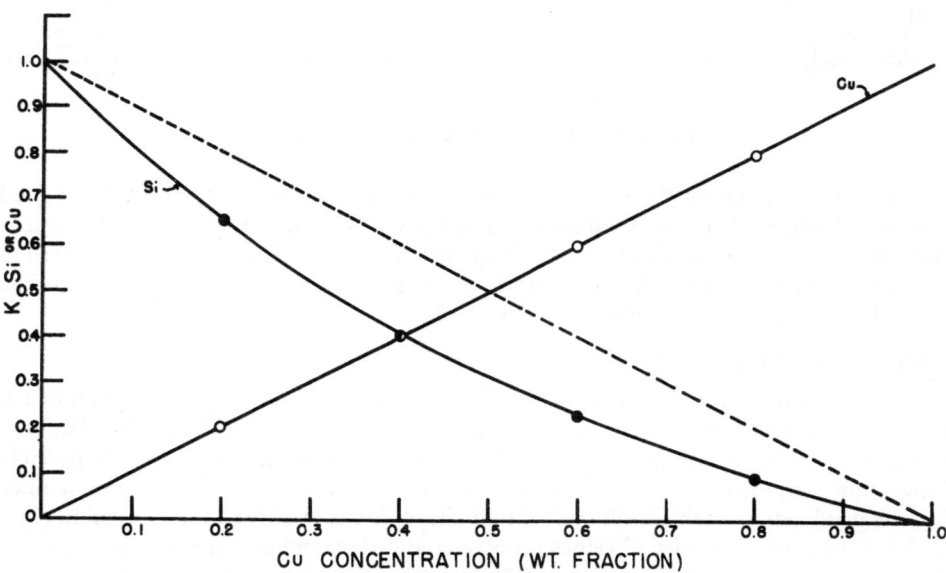

Figure 7. Calibration curves for Cu–Si alloys.

used as standards, and the analysis was done at 30 kV. I and I_0 were found as was done previously. From I and I_0, K was calculated and, along with the proper α values, substituted into equation (4) to give C. The results are listed below.

| Element | Probe | | Wet chemistry |
	C (wt. fraction)	K	
Cu	0.9468	0.949	0.9548
Mn	0.0118	0.0199	0.0118
Si	0.0363	0.019	0.0325
	0.9949		0.9991

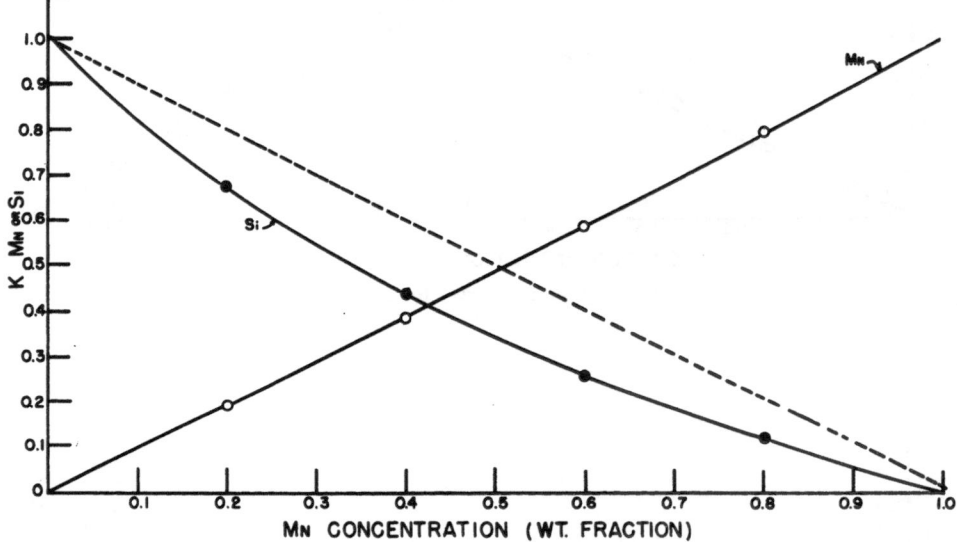

Figure 8. Calibration curves for Mn–Si alloys.

The correlation between the wet-chemistry and the probe results are again quite good. This technique has been used numerous times for Cu–Mn–Si analyses on weld problems and the results reported elsewhere by Bray and Lozano.[10] The work with Cu–Mn–Si alloys has developed considerable confidence in this technique and in probe work in general among people unfamiliar with this work.

Copper–Zinc–Aluminum

Concurrently with the work on Cu–Si–Mn alloys, the technique was extended to Cu–Zn–Al alloys. This system required Cu–Zn, Cu–Al, and Zn–Al binaries. Unfortunately, the making of satisfactory single-phase Cu–Al and Zn–Al binaries is not possible. Since, however, the major corrections are absorption and atomic number, and remembering the good results with Cu–Si and Mn–Si alloys, it was decided to use equation (3) to find K values from assumed C values for copper and zinc, and equation (5) for aluminum and then find α values from equation (2). The procedure followed was identical to that used for Cu–Si and Mn–Si alloys. The Poole and Thomas atomic number correction factor used for aluminum was 1.59 and the corrections were done for the Cu K_α, the Zn K_α and the Al K_α lines. Table VI lists the data for Cu–Al and Zn–Al alloys.

For the Cu–Zn system, four alloys were cast under charcoal and analyzed by wet chemistry. I and I_0 were determined using a 25-μ beam similar to the method used on the

Table VI. Analysis of Cu–Al and Zn–Al Alloys

| | Cu–Al | | | | | |
| | Cu | | | | Al | |
C	K	α		C	K	α
0.20	0.2008	0.996		0.80	0.589	2.792
0.40	0.4012	0.994		0.60	0.3375	2.945
0.60	0.6012	0.994		0.40	0.1838	2.980
0.80	0.8008	0.9975		0.20	0.0755	3.055
	Average	0.9954			Average	2.943
	Zn–Al					
	Zn				Al	
C	K	α		C	K	α
0.20	0.2008	0.996		0.80	0.579	2.908
0.40	0.4012	0.994		0.60	0.3303	3.040
0.60	0.6012	0.994		0.40	0.1779	3.095
0.80	0.8008	0.9975		0.20	0.0735	3.155
	Average	0.9954			Average	3.05

Table VII. Analysis of Cu–Zn Alloys

| | Cu | | | Zn | | |
Alloy	C	K	α	C	K	α
1	0.9001	0.904	0.989	0.0985	0.0911	1.089
2	0.810	0.813	1.015	0.1868	0.1914	0.971
3	0.6996	0.697	1.007	0.3004	0.3037	0.984
4	0.5209	0.531	0.962	0.4791	0.4874	0.967
		Average	0.9931		Average	1.003

Cu–Ni–Fe alloys, and α was found using equation (1). The analysis was done at 30 kV on the Cu K_α and Zn K_α lines. From I and I_0, K was calculated and equation (2) solved for α. Table VII lists the data. The theory predicts α values close to 1 for both copper and zinc. Available now are the α factors for Cu–Zn, Cu–Al, and Zn–Al alloys. These numbers are reproduced in summary below.

Cu–Zn	Cu–Al	Zn–Al
$\alpha_{Cu}{}^{Cu-Zn} = 0.9931$	$\alpha_{Cu}{}^{Cu-Al} = 0.9954$	$\alpha_{Cu}{}^{Zn-Al} = 0.9954$
$\alpha_{Zn}{}^{Cu-Zn} = 1.003$	$\alpha_{Al}{}^{Cu-Al} = 2.943$	$\alpha_{Al}{}^{Zn-Al} = 3.05$

These data were used to draw the calibration curves for Cu–Zn, Cu–Al, and Zn–Al alloys shown in Figures 9, 10, and 11.

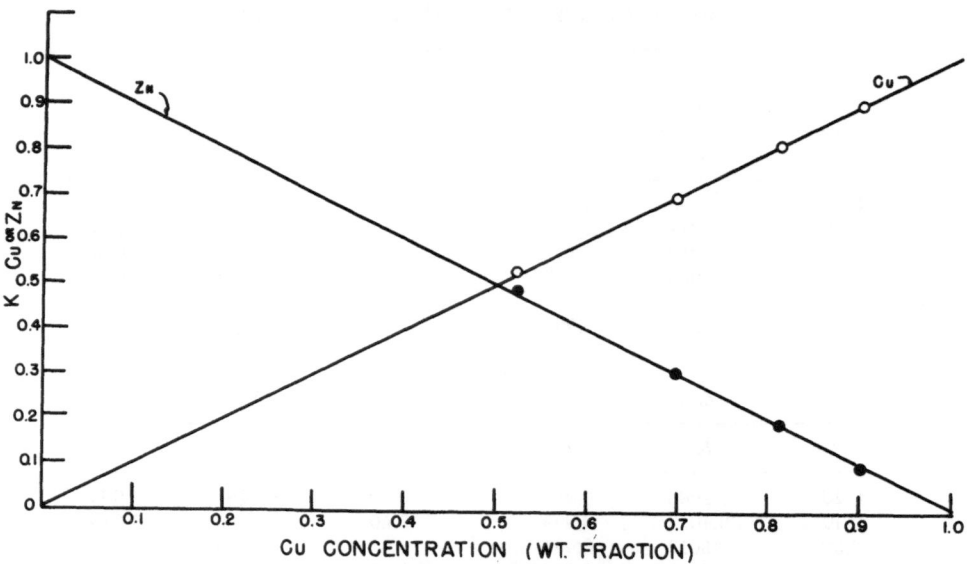

Figure 9. Calibration curves for Cu–Zn alloys.

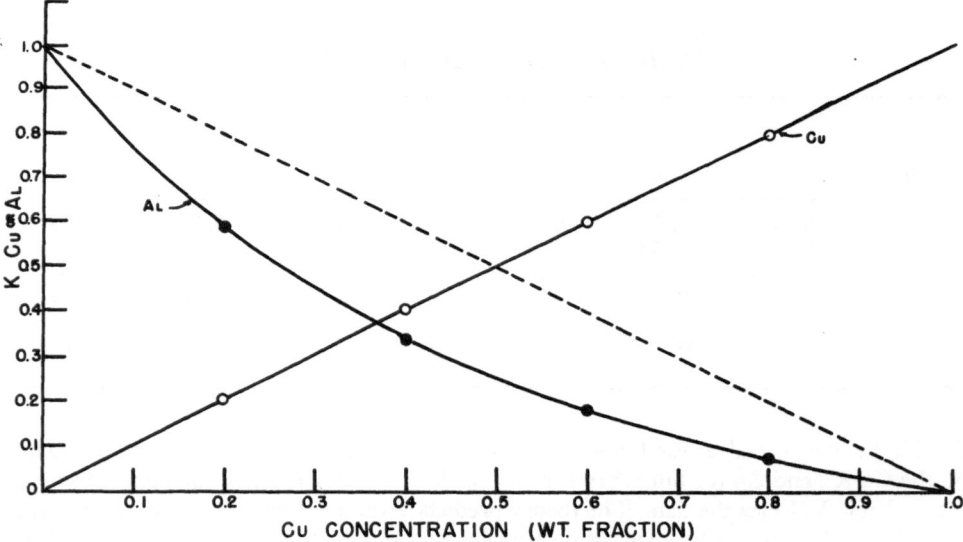

Figure 10. Calibration curves for Cu–Al alloys.

Using these data, a sample of a Cu–Zn–Al alloy (Anaconda American Brass Alloy No. 687) obtained from the mill was analyzed by wet chemistry and by probe as was done previously for the Cu–Ni–Fe and Cu–Mn–Si alloys. Equation (4) was applied with the results shown below. Again, the correlation of probe data with wet chemistry is quite good.

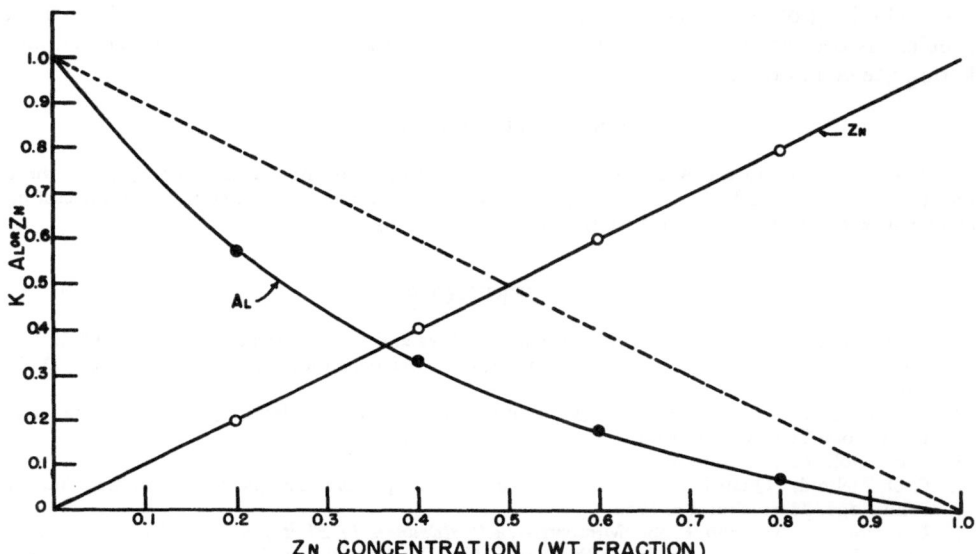

Figure 11. Calibration curves for Zn–Al alloys.

Element	C (wt. fraction)	K	Wet chemistry
Cu	0.7694	0.768	0.7746
Zn	0.2190	0.214	0.2045
Al	0.0228	0.00398	0.0208
As			0.0001
	1.0112		1.0000

SUMMARY AND CONCLUSIONS

The Ziebold correction procedure for electron-microbeam-probe data was applied to ternary alloys of Cu–Ni–Fe, Cu–Mn–Si, and Cu–Zn–Al. Correlations of probe results with wet chemical analyses of the same pieces of metal were good, and it is concluded that, at least in the three systems studied, the Ziebold technique is sufficiently reliable to be highly useful. The fact that this technique is based on alloy standards (wherever possible) seems to inspire confidence in the results among people not too familiar with probe methods.

While this work was done primarily to evaluate the method with respect to ternary alloys, a useful by-product is the set of nine binary calibration curves necessarily developed during the work. These will be useful in handling binary system data, as well as data on other ternary systems involving some of the same elements.

Three problems were encountered in this work, namely: (1) getting good binary alloy standards, (2) getting good wet chemical results on binary standards, some of which are in composition ranges not commonly analyzed, and (3) solving Ziebold's ternary equations. The first one requires careful consideration of the phase diagram of the system, and metallurgical examination of standards for suitability. The second requires checking and rechecking of analytical results, by more than one technique if possible. The third problem is best handled by writing a computer program for solving the equations, and letting the computer do the work.

ACKNOWLEDGMENTS

Thanks are due to the Anaconda American Brass Company for permission to publish these results, and to Thomas M. Lucas and Yngve Dahlin of that organization for supervising the casting of standards and supervising the analytical work, respectively.

REFERENCES

1. T. O. Ziebold, "Analysis of Multicomponent Alloys and an Empirical Correlation of Calibration Data," presented at the 126th Meeting of the Electrochemical Society, Washington, D.C., 1964.
2. T. O. Ziebold and R. E. Ogilvie, "Quantitative Electron Microanalysis," presented at the MIT Microprobe School, Cambridge, Massachusetts, 1963.
3. S. H. Moll, private communication.
4. T. O. Ziebold, "Quantitative Electron Microanalysis," presented at the MIT Summer Course, Cambridge, Massachusetts, 1963.
5. R. Castaing, "Electron Probe Microanalysis," *Advances in Electronics and Electron Physics, Vol. 5*, L. Martin (ed.), Academic Press Inc., New York (1960), p. 317.
6. D. B. Wittry, Technical Report, USCEC Report 84–204, University of Southern California, Los Angeles (1962).
7. L. S. Birks, *Electron Probe Microanalysis*, Interscience Publishers, Inc., New York, p. 125.
8. D. M. Poole and P. M. Thomas, "Quantitative Electron-Probe Mircoanalysis," *J. Inst. Metals* **90**: 228 (1962).
9. T. O. Ziebold and R. E. Ogilvie, "Quantitative Analysis with the Electron Microanalyzer," *Anal. Chem.* **35**: 622 (1963).
10. R. S. Bray and L. Lozano, "Controlling Weld Segregation to Avoid Cracking in a Cu-Si-Mn Alloy," *Welding J.*, (September, 1965).

DISCUSSION

L. S. Birks (U.S. Naval Research Laboratory): I am not arising to defend the meat-axe approach which I coined several years ago as a faster means of getting the wrong answer than some of the other means, but rather to criticize the idea of defocusing the electron beam to try to compensate for inhomogeneities in the sample. I think this is potentially wrong; the reason I think so is that if you consider the sample to be made up of two components completely separated, you must necessarily get for a 50–50 composition exactly 50 I/I_0 for each of them. What you are trying to do is find what you would get for a 50–50 mixture of the two on a homogeneous basis. You are directly contradicting the matrix effects which you are trying to compensate for when you defocus the beam on an inhomogeneous sample.

R. M. Ingersoll: We felt that the reason that we could use the large beam was that we were working with cast alloys, particularly with the nickel–iron alloys, which show many inhomogeneities. Rather than going to much trouble of breaking them down, we felt that this was an easy way to do it. I feel that it gave quite good results.

Chairman S. H. Moll: I would personally like to make a comment on that. The idea of using a defocused beam to average out inhomogeneities in standards may have—I won't state that for a fact—originated in our laboratory. At the time that we presented our first discussions or papers of this approach we indicated that by no means could the theory explain why this method seemed to work so well, for the same reason that Dr. Birks gave. If you were to extrapolate again in your mind the samples completely divided into two pure components representing an average chemistry, all that we can say is that if the defocused beam averages, let's say in a two-phase alloy over many small areas of the individual phases, we find that the average intensity ratio obtained for this multi-phase alloy fits the calibration curves from the same system obtained with single-phase alloys just as well as data from single-phase alloys which exist in the same binary system or ternary system. This is experimental evidence, we don't mean to defend it on any theoretical basis.

L. S. Birks: Don't you think that it is far better to use an averaging by making analyses at several points rather than defocusing, which is not sufficient to give you a proper error from the contributing errors in your analysis.

Chairman S. H. Moll: Of course, one would have to make enough analyses to be sure you approached a representative sampling of the volume percentage of the individual areas. This is the approach that the chemists have used for many years. If you want to measure an inhomogeneous sample, you make several analyses at representative points and you average those results.

H. Yakowitz (U.S. Bureau of Standards): Just a comment on the iron–nickel system. There is a set of alloys—in fact the one that Ziebold determined is on the iron–nickel system—that are homogeneous at the micron level. Did you run any of those to check the α value using the 1-μ beam rather than the 25-μ beam?

R. M. Ingersoll: No, we used only our own alloys which we made up ourselves.

H. Yakowitz: I make the comment—I don't want to commit Dr. Goldstein to anything—he has the alloys and I think that probably he could loan you two or three of them. You could check them with the micron spot and see if there would be a difference in the α value. You can homogenize yours following the technique which Goldstein used when he was at MIT. You only need to anneal them for about 24 hr at 1300°C. This will homogenize them according to the diffusion coefficients which Goldstein tried.

R. M. Ingersoll: Actually, the as-cast nickel–iron alloys are not particularly inhomogeneous. As-cast they are quite homogeneous. You run into a much bigger inhomogeneity problem in the copper–nickel alloys and it is almost impossible to get them homogeneous.

H. Yakowitz: Yes, you have a partition factor in that case.

K. F. J. Heinrich (U.S. Bureau of Standards): I think the point should be made that some of us do object to this blowing up of the beam, not on the basis that we cannot explain why it works, but that from all we know we must suppose it will not work when serious inhomogeneity is present. If my task would be to gain the confidence of people, I would rather gain the confidence of the people who know something about this method than those who don't know anything about it.

R. M. Ingersoll: This is very true, but I have to work with the people I am doing the work for.

A. Taylor (Westinghouse Electric Corporation): A few years ago while I was at the Denver Conference, I described a graphical method of solving ternary systems which were being analyzed by fluorescence analysis. It strikes me that the method could equally well be applied to this micro-probe method by using a suitable ternary plot. I think it would actually solve all those equations by just placing your pencil on the point where you get the correct ratio of counts. I am wondering whether you tried this method out and whether it proved satisfactory in practice.

R. M. Ingersoll: I am vaguely familiar with the technique that you are describing, but I haven't tried it. It might be very worthwhile to try it.

A. Taylor: It seems that on the binary side the curves are the same, and so within the ternary plot all the problems of the fluorescence emission plus the absorption will be automatically taken into account. I don't think you'll need to program your work for a computer at all, but just do it graphically.

W. J. Wittig (Union Carbide Corporation): I have a much easier question for you. Have you done any studies on chromium with these elements, or in particular with aluminum?

R. M. Ingersoll: No. These are the only three alloys I have studied with this system.

HOMOGENEITY CHARACTERIZATION OF NBS SPECTROMETRIC STANDARDS II: CARTRIDGE BRASS AND LOW-ALLOY STEEL*

H. Yakowitz, D. L. Vieth, K. F. J. Heinrich, and R. E. Michaelis

National Bureau of Standards
Washington, D.C.

ABSTRACT

Most modern instrumental methods of analysis depend on the use of known standards of composition for calibration. Newer analytical techniques, such as the solids mass spectrometer, laser probe and, especially, the electron-probe microanalyzer have reduced the amount of a sample which can be analyzed quantitatively to a range of about 0.1 to as small as 0.00005 µg. As a corollary to these microanalytical advances, homogeneity requirements have become severe to meet analytical standards. This paper describes a continuation of the National Bureau of Standards' effort to characterize more fully existing standards as to suitability for the new microanalytical techniques.[1] An NBS cartridge brass sample in both the wrought (NBS-1102) and chill cast forms (NBS-C1102), as well as a low-alloy steel sample (NBS-463), have been investigated by means of electron-probe microanalysis and optical metallography. Some 17 elements are contained in the brass, while 25 elements are found in the steel. Results for 10 elements in the steel and 6 elements in the brass are presented. In the steel, iron, nickel, copper, and silicon are essentially distributed homogeneously at micron levels, while manganese, tantalum, niobium, zirconium, sulfur, and chromium are not. In the brass, copper and zinc are distributed homogeneously at micron levels while lead, sulfur, aluminum, and silicon are not. Electron-probe microanalyzer results indicate that both NBS-1102 and NBS-C1102 brass are suitable for use as a calibration standard for electron-probe microanalysis as well as other microanalytical techniques, such as the solids mass spectrometer. The results for brass have been corroborated by a number of laboratories using the electron-probe analyzer.

INTRODUCTION

The modern-day analyst is confronted with many difficult problems in investigating materials to serve our rapidly advancing technology. He is increasingly asked either to (1) make determinations of chemical elements present in a material at extremely low levels of concentration—parts per million level and lower—or (2) provide quantitative analysis of minute sample volumes, a few cubic microns in size. Use of the solids mass spectrometer and the electron-probe microanalyzer are examples of modern instrumental methods of analysis used by the analyst to aid in solving these new problems. The full potential of these powerful analytical tools normally cannot be realized, however, unless they are calibrated by the use of standard reference materials.

Analytical standards serve the analyst in several important ways: (1) calibration of analytical equipment, (2) checking methods of analysis and analytical techniques, (3)

developing new or improved methods of analysis, and (4) evaluating the accuracy of analytical techniques. The new microanalytical techniques place severe requirements on the analytical standards, particularly with respect to homogeneity.

Although considerable research effort is being devoted to the preparation of analytical standards designed especially for the microchemical techniques, as yet none are available. The purpose of this report is to describe a continuation of the NBS effort to more fully characterize existing standards as to their suitability for the new microanalytical techniques.[1] Once characterized, these standards will serve during the interim period until more suitable standards designed especially for microanalytical use can be prepared and certified.

MATERIALS INVESTIGATED

In a previous study, low-alloy steel NBS-461 was fully characterized and found to be sufficiently homogeneous so that any present microanalytical technique can be carried out with little chance of inaccuracy due to inhomogeneity.[1] In the light of these encouraging results, it was decided to provide characterization of another low-alloy steel in this series.

On the basis of metallographic examination, it appeared that low-alloy steel NBS-463 was the best of the remaining seven standards insofar as microstructural homogeneity was concerned. Furthermore, NBS-463 steel gave evidence of having even fewer inclusions than NBS-461. For these reasons, NBS-463 steel was chosen for study. The composition of this steel is given in Table I.

To extend this work into the realm of non-ferrous materials, a decision also was made to investigate samples of recently prepared NBS brass standards. Cartridge brass samples designated 1102 (wrought) and C1102 (chill-cast) were chosen. These samples have the same analysis, but different metallurgical structures, and certification has been made for some 11 trace elements, in addition to those normally specified. Both the wrought and chill-cast forms were characterized in order to observe possible differences or difficulties attendant on the use of either form.

Table I. NBS Spectrographic Low-Alloy Steel Standard No. 463.
Provisional Certified Analysis (Revision May 5, 1965)

Element	Wt.%	Element	Wt.%
Fe	(96.02)*	B	0.0012
C	0.19	As	0.10
Mn	1.15	W	0.10_5
P	0.031	Zr	0.20
S	(0.02)	Nb	0.19_5
Si	0.41	Ta	0.15
Cu	0.47	Al	0.02_5
Ni	0.39	Co	0.01_3
Cr	0.26	Pb	0.012
V	0.10	Ag	(< 0.0002)
Mo	0.12	Ge	(0.002_5)
Sn	0.013	O	(0.007)
Ti	0.010	N	(0.00_6)

* Values in parentheses are not certified, but are given for information on the composition. Iron percent is by difference.

Table II. NBS Spectrographic Cartridge Brass Standards 1102 and C1102. Provisional Certified Analysis (Revision of August 20, 1962)

Element	Wt.%	Element	Wt.%
Cu	72.8$_5$	As	0.004
Zn	17.1$_0$	Be	0.0000$_3$
Pb	0.020	Bi	0.000$_5$
Fe	0.011	Cd	0.004$_5$
Sn	0.006	Mn	0.004$_5$
Ni	0.005	P	0.004$_8$
Al	0.0007	Si	(0.002)*
Sb	0.005	Ag	0.0010
		Te	0.000$_3$

* Values in parentheses are not certified, but are given for information on the composition.

The metallurgical procedure followed in the preparation of the standard has been described in detail elsewhere.[2] The composition of the brass is shown in Table II. The entire slab from which the brass standards were taken was checked for macro-homogeneity by several methods and found to be entirely satisfactory.[2]

PROCEDURE AND RESULTS FOR BRASS

Specimen Preparation

It is a generally accepted precept that a polished, unetched sample, free from surface defects, is the most desirable one for electron-probe examination. Since brass is a "soft" metal, it was feared that smearing might occur on polishing. The procedure commonly adopted is to polish, etch to remove the smeared layer and also to reveal small underlying scratches, repolish, re-etch, and repeat until the desired surface is achieved. The final surface should be free from scratches under examination at several magnifications, when the sample undergoes a full rotation while being viewed in polarized light.

As a first step, the hardness values of the wrought and cast brasses were compared with those of pure copper. The Knoop hardness number (200-g load) of the wrought material was 135, while that of the cast materials was 117. Pure copper registered a Knoop value of 75. Based on these results, it was felt that the two brass forms could be polished with the same technique, and that the smearing problem would be roughly comparable to that encountered with copper.

The samples were polished on water-lubricated SiC papers through 600 grit. They were then finished on 6-μ diamond on rayon followed by $\frac{1}{4}$-μ diamond on microcloth. The surfaces were scratch-free, but a large number of what appeared to be grey inclusions were observed in the wrought and cast brasses, the number being larger in the cast form. No such inclusions were observed in pure copper carried through the same procedure.

Under microprobe examination, these "inclusions" were found to be essentially silicon in both forms of the brass. They were cathodoluminescent under the beam, glowing a dull blue. Since silicon is present only at a 20-ppm level in the brass, it was suspected that the inclusions were really artifacts due to polishing. Therefore, the samples were repolished on Al_2O_3 papers and finished on the diamond as before. Under these conditions, far fewer inclusions were observed. However, microprobe examination indicated some aluminum in the inclusions but no silicon. It was concluded that the abrasive was driving itself into the brass during rough polishing.

Figure 1. NBS-1102 brass (wrought); 100 × (reduced for reproduction 35%).

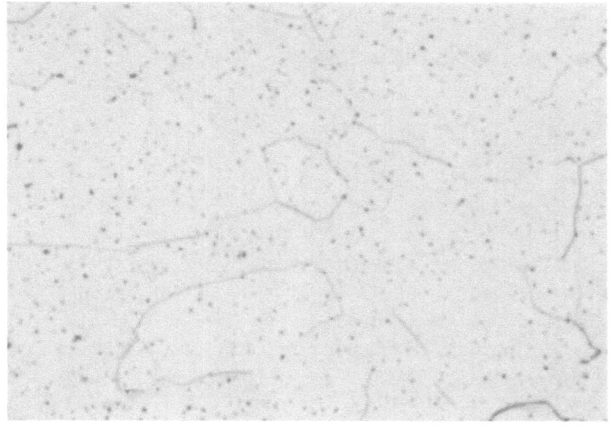

Figure 2. NBS-C1102 brass (cast); near chill-cast face; 50 × (reduced for reproduction 35%).

As a check to determine whether this disturbing phenomenon was peculiar to only the NBS standards, some ordinary commercial brass was obtained. After receiving the same preparation as the standards, similar inclusions were observed. Under microprobe examination, silicon and no aluminum was found after SiC papers were used, while the reverse was observed after Al_2O_3 papers were used. Thus, the phenomenon does not appear to be confined to the NBS standards, but appears to be general.

In view of these results, a different sample preparation procedure was required. At first, microtomy followed by electropolishing was tried. Finally, the as-received brass was given a deep electropolish directly. The bath was 40% H_3PO_4 in water, running at 2–3 V and relatively low current densities. The result was a "hilly" surface suitable only for inclusion examination. Samples prepared in this fashion were used only for inclusion identification.

For structural examination and microprobe examination for matrix homogeneity, the procedure using Al_2O_3 papers was adopted, since only a small amount of abrasive pickup was noted. To minimize time on the Al_2O_3 papers, a Norbide (BN) abrasive wheel can be used; this may, however, pit the sample.

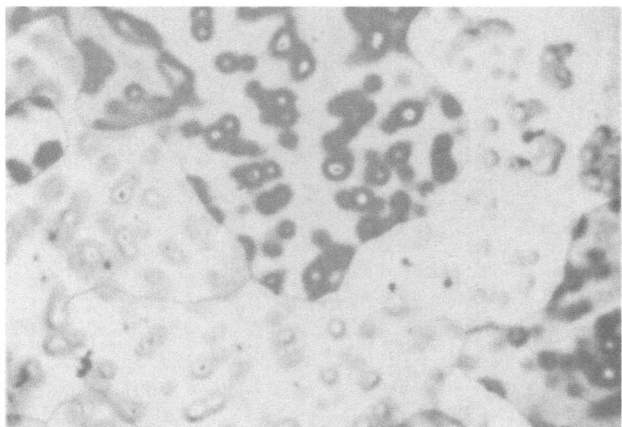

Figure 3. NBS-C1102 brass (cast), about ¾ in. above chill-cast face. Note subdendritic structure; 50× (reduced for reproduction 35%).

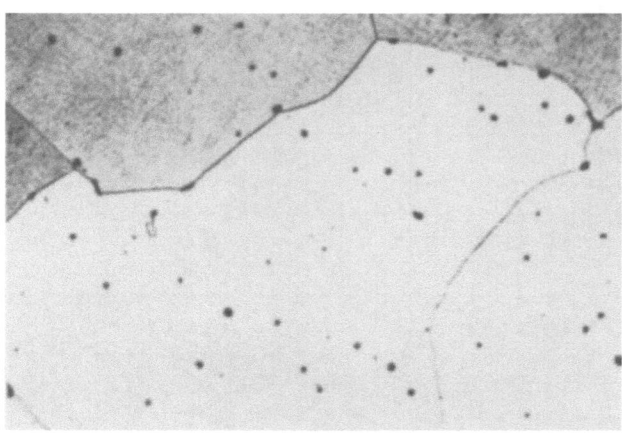

Figure 4. NBS-C1102 brass (cast). Typical blowholes in cast-copper base alloys are shown; 200× (reduced for reproduction 35%).

The structure of the wrought brass is shown in Figure 1. The grains are seen to be twinned, reasonably equiaxed, and variable in size. The ASTM grain size is 0.035–0.045 mm, based on comparison with the standard charts.[3]

The structure of the cast brass near the chill-cast face is shown in Figure 2, while the structure ¾ in. above the chill-cast face, i.e., at the top of the issued standard, is shown in Figure 3. As expected, the dendritic cross section at the top is larger than that at the chill-cast face. For the wrought structure, 75 to 100 grains occupy the same area as the average dendrite cross-section at the chill-cast face. For the wrought structure, 300 to 400 grains occupy the same area as the average dendrite cross section at the top of the chill-cast standard.

The cast form exhibits a porosity which is typical of contracted gas blowholes in cast-copper alloys.[4] Figure 4 shows several small gas holes near the chill-cast face. These irregularly shaped holes range in size from 1 to about 10 μ and are located primarily at the dendrite boundaries.

Some true inclusions were observed in both the wrought and cast forms after the sample preparation problems were overcome. These are extremely small and appear to be essentially round, indicating that they are spheres.

The size of the inclusions ranges from submicron to about $2\,\mu$ in diameter. They are a dull grey in color. In the cast form, they are randomly distributed and are, on the average, slightly larger at the top of the sample than near the chill-cast face. In the wrought samples they are usually found at the grain boundaries; several specimens cut from different samples of the standard confirmed this.

By counting inclusions in the polished and etched section, it was determined that about 200 are present lying in the entire cast face of the standard as issued, i.e., $1\frac{1}{4} \times 1\frac{1}{4} \times \frac{3}{4}$-in. samples. If these are assumed to be spheres of 2-μ diameter, there are then about two such inclusions per cubic millimeter of metal. The ratio of brass volume to inclusion volume is over one hundred million to one. If it is assumed that the sample for the usual microanalytical technique is a sphere $50\,\mu$ in diameter, the probability of obtaining one inclusion in such a sample is negligible. The probability for a random 1-μ probe to strike such an inclusion is less than one in two million.

Inclusion Identification

The true inclusions were subjected to qualitative electron-probe analysis. In both the cast and wrought forms, the results were essentially identical. The inclusions are cathodo-luminescent; they glow with an orange-gold color under electron excitation at 30 keV. Wavelength scans were taken of several inclusions in both forms of the brass.

The presence of reasonably large amounts of lead and sulfur was noted. Some variable amounts of zinc, silicon, and aluminum were also found. The concentration of these elements varied considerably from inclusion to inclusion. No other elements were indicated to be present. Approximately 20 such inclusions were examined in each form of the brass. A set of typical scanning display pictures (Figure 5) illustrates the presence of lead, sulfur, silicon, aluminum, and zinc, and the absence of copper.

These results indicate that lead and silicon cannot be uniformly distributed within the brass. Although not appearing on the certificate of analysis, chemical determinations had been made for sulfur with an averaged result of 0.002%. The sulfur probably was introduced during the melting from the charcoal cover.[2] The aluminum may have been present as an impurity in one of the constituents of the brass. Figure 5 shows that the aluminum and silicon appear to be at the same location within the inclusions. Unfortunately, the presence or absence of other impurities, such as carbon, which might have originated from the same source, could not be checked with the electron-probe micro-analyzer available.

Homogeneity of the Matrix

The homogeneity of the sample in copper and zinc was investigated by means of the electron-probe microanalyzer. Two separate channels using LiF analyzing crystals and sealed proportional detectors were employed. The instrument was operated at 30 keV with probe currents of 100 nA. The probe diameter was approximately $1\,\mu$. The count rate for copper was on the order of 13,500 counts/sec, whereas the count rate for zinc was on the order of 4000 counts/sec. Signal-to-background ratios for zinc were greater than 100 : 1.

A set of random mechanical line scans was run on each brass form as a preliminary step. These were designed to show up fairly gross inhomogeneities of copper and zinc, and also to see whether or not random striking of a blowhole or inclusion would occur. Some three scans of about 5 mm total length were run on both forms of the brass. No effect of striking an inclusion or hole was observed, since there was never any drastic reduction in intensity of the copper and zinc signals simultaneously.

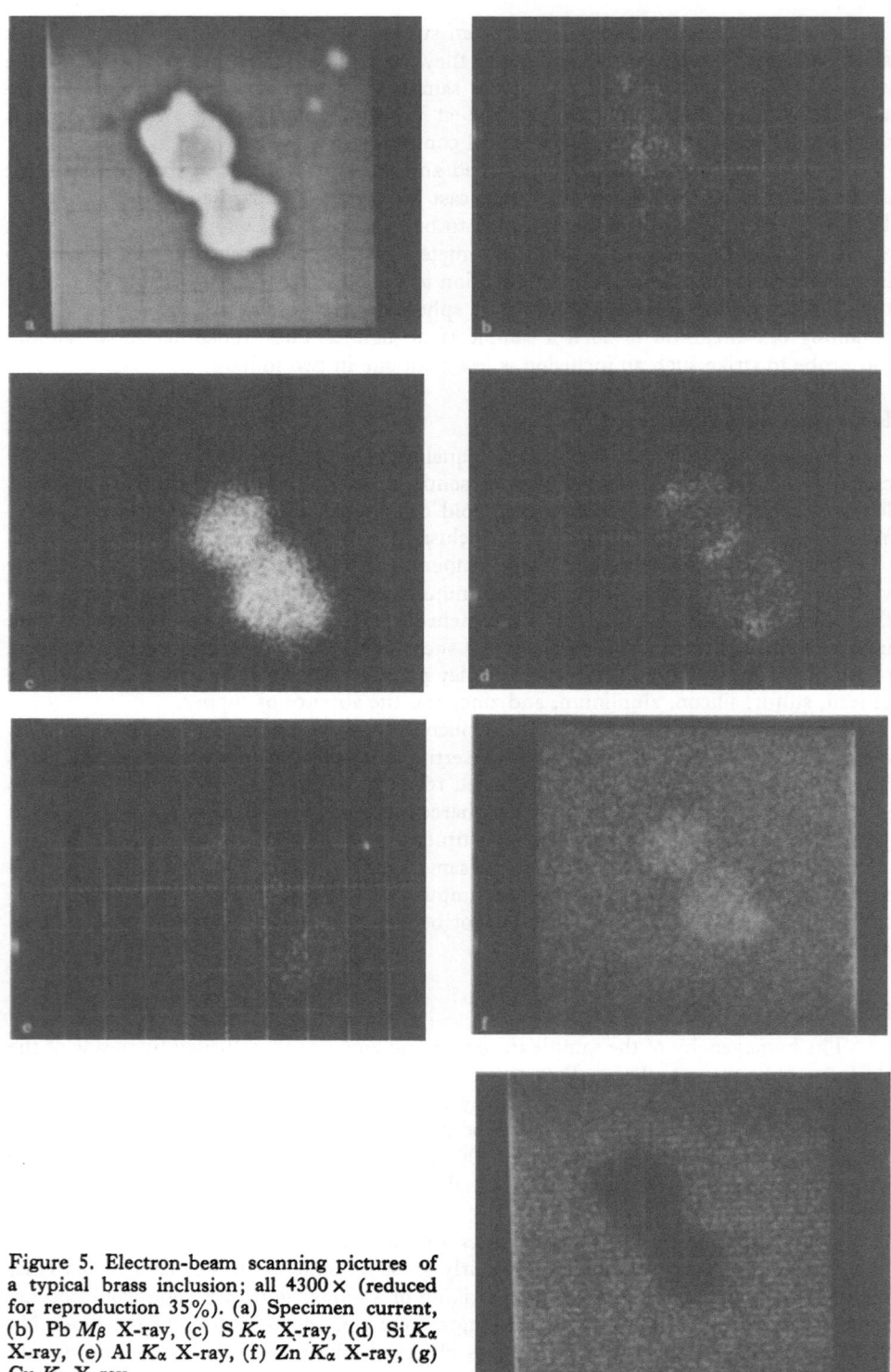

Figure 5. Electron-beam scanning pictures of a typical brass inclusion; all 4300 × (reduced for reproduction 35%). (a) Specimen current, (b) Pb M_β X-ray, (c) S K_α X-ray, (d) Si K_α X-ray, (e) Al K_α X-ray, (f) Zn K_α X-ray, (g) Cu K_α X-ray.

The scan rate was 1.6 μ/sec; hence, about 8000 counts/μ were obtained for copper and about 2500 counts/μ were obtained for zinc. Therefore the statistical fluctuation (three standard deviations) was approximately 3% for copper and 6% for zinc. No fluctuation greater than this was observed for either copper or zinc in any of the line scans, which represented 15,000 one-micron samplings for each brass form.

Other sources of fluctuation are probe instability and non-flatness of the sample. With the former, the copper and zinc count rates would vary in the same fashion, i.e., both would decrease gradually if the probe current decreased and vice versa. Since copper plus zinc comprises 99.95% of the sample, no real double effect should be observable.

Sample flatness problems can be solved by properly aligning the spectrometers. If the spectrometer is properly aligned, a raising or lowering of the sample produces a decrease in count rate. Hence, a double effect, similar to that for probe instability, would result if the sample height with respect to the spectrometer were changed.

After the line scans were completed, point counts in each form of the brass were taken. 24 points were taken at random in the wrought sample, while 42 points were taken in the cast sample, distributed more or less evenly along the 12 line segments of a "tic-tac-toe" grid extending across the entire specimen. Each point was counted twice for both copper and zinc as an internal check on probe stability. The two counts were always statistically in agreement. The sample was refocused in the light optics at each point. Two separate operators took points on each sample.

The results for the cast form of the brass show that the point giving the highest zinc count rate is +1.3% higher in apparent zinc concentration than the mean. The results for the lowest zinc point give a value 2.2% below the mean. The standard deviation for counting error is 0.37% (mean total number of zinc counts accumulated, N, was 79,224). The standard deviation computed from the 42 determinations of zinc content is 0.79%.

If the mean zinc count is considered to correspond to the certified value of 27.10%, the analytical range corresponding to two standard deviation limits is 26.7 to 27.5%. The average sampling was about 10 μ^3 of the brass per point; this corresponds to about 10^{-11} g of brass. Results for both brass forms are given, using the foregoing notation, in Table III.

Table III. Analytical Results for Copper and Zinc in NBS–C1102 and 1102 Cartridge Brass

Element	Form	Mean counts accumulated[a]	Line-to-background ratio	Coefficient of variation for counting	Standard deviation (%)[b]	Analytical range of composition[c]
Zn	Cast	79,224	30 : 1	0.0036	0.79	26.7–27.5
Cu	Cast	131,715	150 : 1	0.0027	0.72	71.8–73.8
Zn	Wrought	79,263	30 : 1	0.0036	2.2	25.9–28.3
Cu	Wrought	283,687	150 : 1	0.0019	1.0	71.1–74.6

[a] Background-corrected.
[b] Computed from experiment.
[c] Two standard deviation limits, at 3-μ level of special resolution.

Table IV. Electron Probe Microanalysis Results for NBS–C1102 and 1102 Brasses

Element	Form	Apparent/(%)	Actual/(%)
Cu	Cast	73.4	72.5
Zn	Cast	27.8	27.8
Cu	Wrought	73.4	72.5
Zn	Wrought	27.7	27.7

Electron Probe Microanalysis Results

As the final proof, it was deemed necessary to perform a complete electron-probe microanalysis of each brass standard for both copper and zinc. This was done as a separate experiment approximately 10 days after all other data had been taken.

The voltage used was 30 kV, while a constant incident beam current of 50 nA was employed. This resulted in a count rate of about 10,000 counts/sec in pure copper and 5500 counts/sec in pure zinc, using LiF crystals in conjunction with sealed proportional counters. Some five points each of copper and zinc were recorded on both brasses. Background was counted $\pm$ 0.05 Å off-peak for copper and $\pm$ 0.1 Å off-peak for zinc. It is interesting to note that the background count for both copper and zinc was about the same in the pure element as in the brass. The X-ray emergence angle was 52.5°.

The lines examined were Zn K_α and Cu K_α. The Duncumb-corrected Philibert equation[5] was used for absorption correction in zinc; the correction required was found to be less than 0.1%. The Castaing fluorescence correction for Cu K_α being fluoresced by Zn K_β was used.[6] Mass absorption coefficients for both K_α and K_β lines were taken from Heinrich.[7] The results are shown in Table IV.

As can be seen, the results in Table IV agree very well with the certified values of 72.85 for copper and 27.10 for zinc. As predicted, the agreement between the cast and wrought forms is excellent. These results have been corroborated by six other electron-probe laboratories who received wrought and cast pieces cut at random from different samples of the brass.

It is suggested that because of the very small difference between apparent and true concentrations, these brasses can be used to calibrate electron-probe microanalyzers. Thus, an operator could check to see that the instrument is giving proper relative intensity data at various times during an experiment by merely taking five minutes to check the copper and zinc values in the brass.

SUMMARY AND CONCLUSIONS FOR BRASS

The cast form of the cartridge brass designated NBS-C1102 shows no evidence of inhomogeneity greater than about 1.6% of the amount present for either copper or zinc for samples of 10^{-11} g. Therefore, this is the maximum degree of inhomogeneity to be expected for any technique sampling a larger amount of the brass. The wrought form shows an apparently larger degree of inhomogeneity, which is 2.0% for copper and 4.4% for zinc. Nevertheless, neither of these figures is prohibitive when micron-sized analytical samples are to be investigated.

Both forms contain small inclusions, which are comprised of variable amounts of lead, sulfur, aluminum, zinc, and silicon, but no copper. These inclusions are easily avoided in the electron-probe microanalyzer; the probability of striking an inclusion

Figure 6. NBS-463 steel, as-polished; 200× (reduced for reproduction 35%).

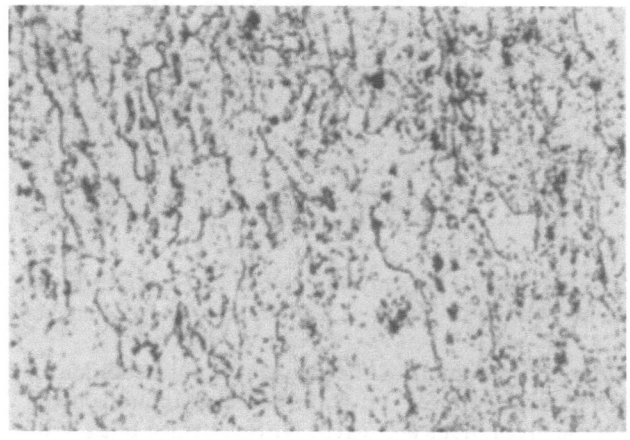

Figure 7. NBS-463 steel, etched structure in longitudinal section; 1000× (reduced for reproduction 35%).

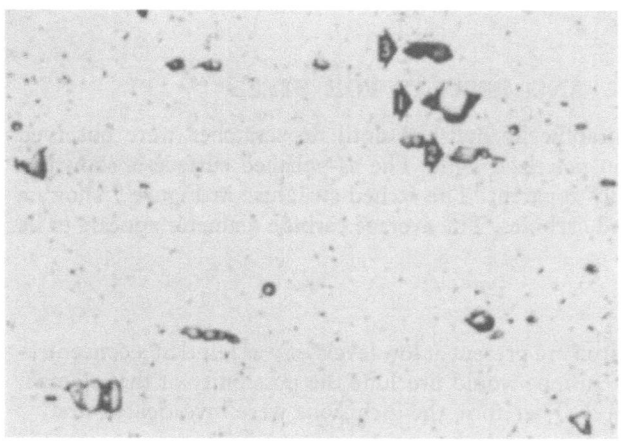

Figure 8. NBS-463 steel; three distinct types of inclusions; 1000× (reduced for reproduction 35%).

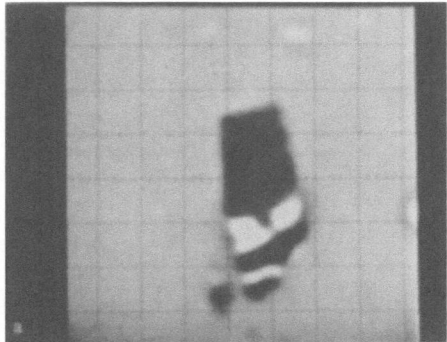
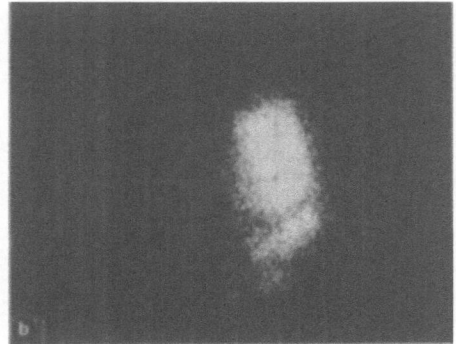

Figure 9. Electron-beam scanning pictures of steel inclusion I; all 430× (reduced for reproduction 35%). (a) Specimen current, (b) Zr $L\alpha$ X-ray, (c) Cr $K\alpha$ X-ray.

lying beneath the polished surface is negligible for any take-off angle. Likewise, the probability of the inclusions affecting any analysis by 1% is negligible for those elements not present in the inclusions.

Care must be taken in the surface preparation of the brass to avoid occlusion of materials. Care also should be taken to make sure than an undisturbed structure is being analyzed. It is asserted that when these precautions are taken, NBS cartridge brass 1102 or C1102 can be used for any present day microanalytical technique, with little chance of inaccuracy due to inhomogeneity. Electron-probe microanalysis results confirm this assertion.

PROCEDURE AND RESULTS FOR STEEL

Steel samples were metallographically polished until no scratches were observed when the specimen was viewed in polarized light. The as-polished surface is shown in Figure 6; the inclusions are readily apparent. The etched structure in Figure 7 shows a large number of small spheroidized carbides. The average carbide diameter appears to be 1–2 μ.

Inclusion Identification

Since all constituents except iron are present at low levels, it was felt that a concentration of any element within the inclusions would preclude the possibility of that element being homogeneous within the steel. Therefore, the inclusions were investigated first.

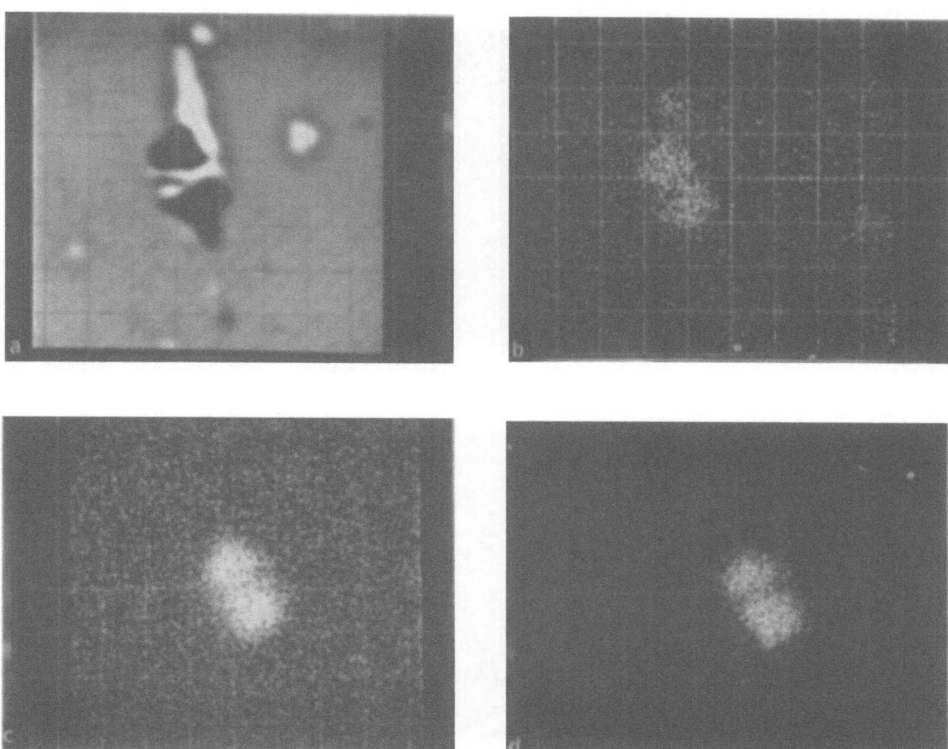

Figure 10. Electron-beam scanning pictures of steel inclusion II; all 4300 × (reduced for reproduction 35%). (a) Specimen current, (b) Zr $L\alpha$ X-ray, (c) Ta $L\alpha$ X-ray, (d) Nb $L\alpha$ X-ray.

Surprisingly, three distinct types of inclusions were observed. These are shown marked in Figure 8. The first is a plain straight-edged small inclusion which is yellow and is fairly plentiful. Such inclusions were found to contain a large proportion of zirconium and some chromium. Scanning electron-probe pictures illustrate this (Figure 9). No other elements of atomic number greater than 11 could be found.

The second type of inclusion resembles a streamer. It is a composite and gray, and contains zirconium, tantalum, niobium, and a trace of sulfur. No other detectable elements were found. Scanning pictures are shown in Figure 10 for all elements except sulfur, which gave too weak a signal for photography.

The third type of inclusion is not plentiful. It is a small grayish composite which is cathodoluminescent, glowing a light beryl-blue under the electron beam. These inclusions contain only zirconium and sulfur among the detectable elements. It is surmised that they consist largely of ZrO_2. The scanning photographs are shown in Figure 11. It should be mentioned that iron is depleted in all three types of inclusions.

Tests for Other Elements

A mechanical line scan of manganese concentration was run at a rate of 1.6 μ/sec. The line-to-background ratio was 6 : 1. A portion of the recording of this scan is shown in Figure 12. The manganese concentration is seen to fluctuate as much as 20% about the mean over relatively short distances. Area scans confirmed the observation that manganese is not uniformly distributed.

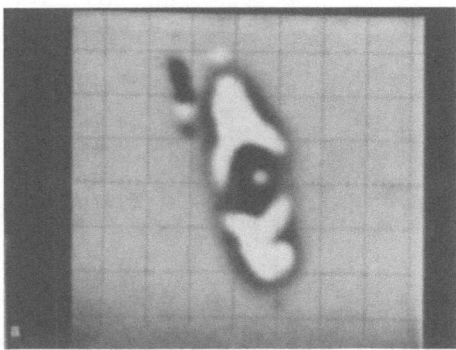
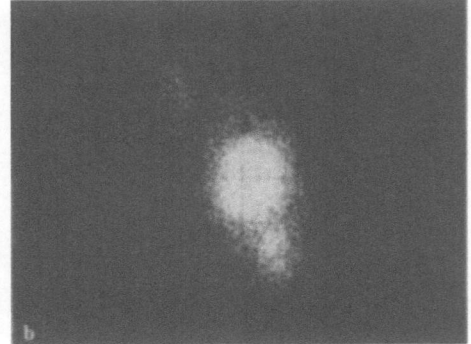
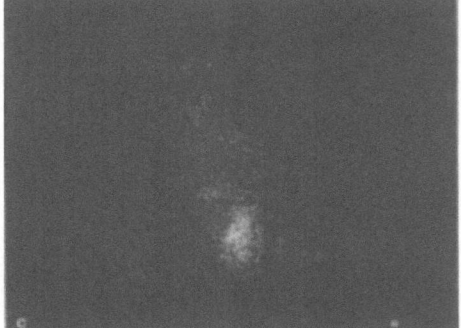

Figure 11. Electron-beam scanning pictures of steel inclusion III; all 4300 × (reduced for reproduction 35%). (a) Specimen current, (b) Zr L_α X-ray, (c) S K_α X-ray.

Point counts away from inclusions were made for iron copper, nickel, and silicon. Two different operators alternated in taking the data. Each point was randomly picked and counted twice in the same fashion as the brass; the sample was refocused in the light optics at each point. The results for each of these elements are presented in Table V.

SUMMARY AND CONCLUSIONS FOR STEEL

Of the ten elements in the steel investigated, five were found to be concentrated in one or more of three distinct types of inclusions. These were tantalum, niobium, sulfur, chromium, and zirconium. It is therefore concluded that the steel is unsuitable as a microanalytical standard for these elements.

Additionally, manganese was found to be nonuniformly distributed. This element is not concentrated in inclusions. However, four elements, namely iron, copper, nickel, and silicon, were found to exhibit uniform distribution within the matrix of the steel. Each is depleted within the inclusions.

Assuming roughly the same number of inclusions in NBS-463 steel as was determined present in NBS-461 steel (the actual number in NBS-463 is less), approximately the same probabilities concerning the inclusions should apply.[1] Thus, the probability of a random 1-μ probe striking an inclusion is about 1 in 350 and about one inclusion should occur per 50-μ sphere of steel sample. For a sampling procedure, which scans a large area of the surface, a number of elements may show compositional fluctuations from one test to another.

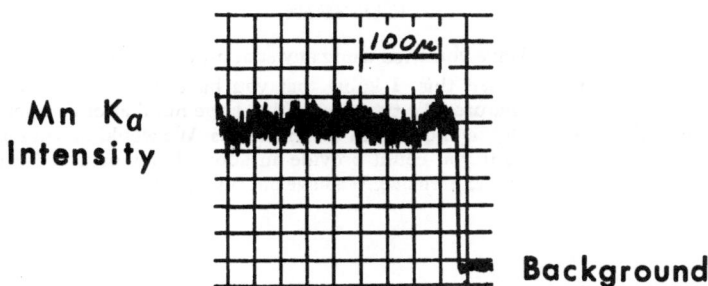

Figure 12. Portion of line scan for manganese in NBS-463 steel.

It is concluded that copper, nickel, and silicon in NBS-463 steel can be analyzed by any microanalytical technique with little chance of error due to inhomogeneity. Iron is a borderline case if one is seeking absolute accuracy. The elements tantalum, niobium, sulfur, chromium, zirconium, and manganese should not be used in the steel for microanalytical standardization.

Table V. Analytical Results for Iron, Copper, Nickel, and Silicon in NBS-463 Steel

Element	Total number of points examined	Mean counts accumulated[a]	Line-to-background ratio	Coefficient of variation for counting	Standard deviation (%)[b]	Analytical range of composition[c]
Fe	32	105,114	21.1 : 1	0.0032	1.6	92.9–99.2
Cu	34	10,236	2 : 1	0.0154	4.4	0.43–0.51
Ni	34	7,474	1 : 1	0.0195	6.0	0.35–0.43
Si	25	24,374	4 : 1	0.0078	2.9	0.39–0.43

[a] Background-corrected.
[b] Computed from experiment.
[c] Two standard deviation limits, at 3-μ level of spatial resolution.

REFERENCES

1. R. E. Michaelis, H. Yakowitz, and G. A. Moore, "Metallographic Characterization of an NBS Spectrometric Low-Alloy Steel Standard," *J. Res. Natl. Bur. Std. A* **68**: 343 (1964). Issued in expanded form as *Natl. Bur. Std. (U.S.) Misc. Publ. 260-3* (1964) 17 pp.
2. R. E. Michaelis, L. L. Wyman, and R. Flitsch, "Preparation of NBS Copper-Base Spectrochemical Standards," *Natl. Bur. Std. (U.S.), Misc. Publ. 260-2* (1964), 36 pp.
3. "Standard Methods for Estimating the Average Grain Size of Metals," *1964 Book of ASTM Standards*, Part 31 : 225 (1964) Plate III.
4. E. P. Polushkin, *Defects and Failures of Metals*, American Elsevier Publishing Co., Inc., New York (1956), p. 47 *ff*.
5. P. Duncumb and P. K. Shields, "Effect of Critical Excitation Potential on the Absorption Correction in X-ray Microanalysis." Tube Investments Research Laboratories Technical Report No. 181 (1964), 9 pp., 7 figures.
6. R. Castaing, "Application of Electron Probes to Local Chemical and Crystallographic Analysis," thesis, University of Paris (1951), pp. 83 *ff*.
7. K. F. J. Heinrich, X-ray Absorption Uncertainty, *The Electron Microprobe*, John Wiley & Sons, Inc., New York (1966), p. 296.

DISCUSSION

Chairman Sheldon H. Moll: Have you tried using any spot other than a focused electron beam?

H. Yakowitz: We have not tried this. I know that you have done it for eight-component systems and other systems of this nature. We try, even with a large number of components, to keep them low in composition and usually tie them up in the inclusions. We would be willing to entertain a study program with large beams if you could provide an alloy of this nature. I know that you feel that it works; we have just had no experience. We have not run at anything higher than a focused-beam size.

ELECTRON PROBE MICROANALYSIS OF ZINC-BEARING HEXAGONAL FERRITES

J. R. Shappirio

U.S. Army Electronics Command
Fort Monmouth, New Jersey

ABSTRACT

The electron probe is shown to be an effective tool for the analysis of the series of ferrimagnetic oxides referred to as the hexagonal ferrites. This series of compounds, containing barium, Fe^{3+}, and a divalent metal cation, is formed by an ordered stacking of basic structural units in varying ratios. The ideal, complex stoichiometry of these polytype-like mixed-layer structures can be computed from X-ray unit cell data; the various structures and their predicted stoichiometry are reviewed. Results of electron probe analysis of zinc-bearing single-crystal hexagonal ferrites are compared with theoretical values, the various correction procedures applied to the probe data are presented, and the limitations of the method in the analysis of hexagonal ferrites are discussed. The information obtained from this study has laid the groundwork for the determination of chemistry in substituted members of the hexagonal ferrite group, and will contribute significantly to the interpretation of the magnetic properties exhibited by these compounds.

INTRODUCTION

The hexagonal ferrites are a chemically complex but structurally interrelated group of ferrimagnetic oxides containing barium, Fe^{3+}, and a divalent metal cation consisting commonly of one or more of the group zinc, manganese, copper, magnesium, Fe^{2+}, cobalt, nickel, etc. Considerable interest has been attached to research on the hexagonal ferrites, because of their large anisotropy fields and high permeabilities at gigacycle (10^9) frequencies. The interpretation of the magnetic properties of these compounds, which are in part a function of the concentration and lattice position of the substituent cation, is critically dependent on the ability to determine both gross chemical stoichiometry as well as local chemical variation within a given crystal. Furthermore, it becomes highly desirable to perform the analysis on the same single crystal from which the magnetic property measurements are obtained. The electron probe thus becomes the best and perhaps the only means presently available for obtaining such a microvolume analysis. The present paper describes the results obtained to date on single-crystal flux-grown zinc-bearing members of the hexagonal ferrite group. Such studies have shown that the various phases within the group are amenable to probe investigation, and that this approach will be of extreme value in characterizing the role of substituent metal cations in the modification of the magnetic properties of the hexagonal ferrites.

STRUCTURE AND CHEMISTRY

Five original members of the hexagonal ferrite group were first described by Braun[1] and were all shown to have crystal structures closely related to one of the five, $BaFe_{12}O_{19}$

---

[1] References are at the end of the paper.

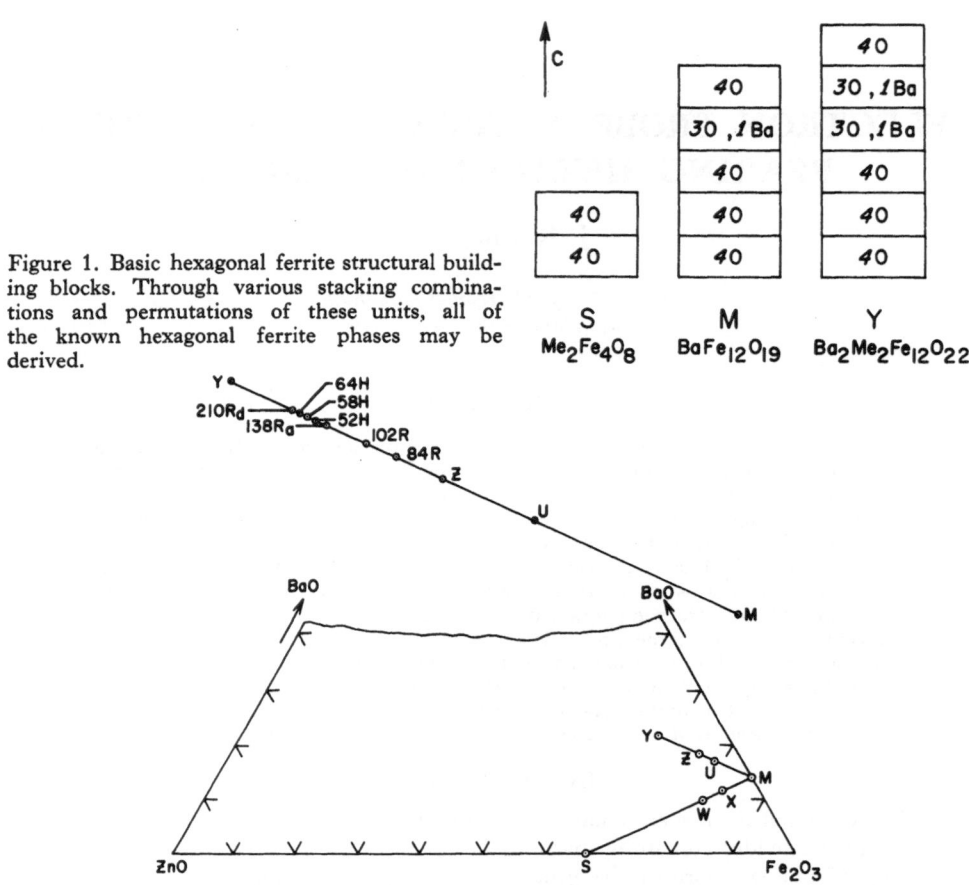

Figure 1. Basic hexagonal ferrite structural building blocks. Through various stacking combinations and permutations of these units, all of the known hexagonal ferrite phases may be derived.

Figure 2. BaO–ZnO–Fe$_2$O$_3$ composition diagram, showing chemical relationships among representative phases in the hexagonal ferrite group.

(the magnetoplumbite structure). A sixth member of the group was described two years later by Smit and Wijn.[2] Subsequent to 1957, synthesis experiments in the Ba–Fe–O–Me^{2+} systems were initiated[3] in this laboratory and concurrent X-ray diffraction studies have disclosed the existence of a large number of additional phases.[4,5] At the present time, 50 or more phases have been characterized structurally. In order to understand their chemical relationships, a description of their structures is imperative.

The hexagonal ferrites have been shown to consist of an anion framework of hexagonally arranged oxygen atoms in layers stacked along the c axis. Two basic anion framework layers occur, one of all oxygen atoms, the other having every fourth oxygen replaced by one barium. These two layer types combine in several arrangements to produce three basic structural building blocks which are designated M, Y, and S; these are illustrated in Figure 1. The S-block consists of two 4-oxygen layers, and leads to the well-known spinel stoichiometry Me$_2^{2+}$Fe$_4$O$_8$. The M-block, consisting of four 4-oxygen layers and one barium–oxygen layer, has the stoichiometry BaFe$_{12}$O$_{19}$, barium ferrite. The last block, designated Y, a six-layer unit, consists of four 4-oxygen layers and two barium–oxygen layers, and has the stoichiometry Ba$_2$Me$_2^{2+}$Fe$_{12}$O$_{22}$. All of the known hexagonal ferrite phases can be derived by variation in combinations and permutations of these structures. It should be noted that X-ray diffraction studies in this laboratory have been concerned primarily with the elucidation of such stacking sequences, although studies are now in progress leading to the determination of their detailed structures and the positions of the divalent cation in the lattice. At present it is sufficient to note that such cations are located in general between the anion layers.

Table I lists the stacking sequences and the corresponding stoichiometry as predicted from the X-ray diffraction studies for a representative number of phases in the system. The chemical interrelations are readily seen with reference to a composition diagram such as that reproduced in Figure 2. Here it may be seen that all phases lie on one of two compositional joins, between either the M and Y phases, or the M and S (S = spinel) positions. In addition to the lettered phases, the six original ferrites, the positions of representative more recently described phases are also indicated. These latter phases are represented by number symbols which refer to the number of anion layers necessary to complete the hexagonal unit cell and by a letter suffix denoting that the primitive cell is hexagonal or rhombohedral.

The twelve phases plotted on the composition diagram are chemically representative of the more than fifty phases now known in the hexagonal ferrite group. Single crystals of these phases are typically flat hexagonal plates averaging 1–2 mm in thickness and less than 5 mm in width. The structure of each crystal utilized for this study was confirmed by X-ray diffraction techniques prior to probe investigation, and all are the result of synthesis from $NaFeO_2$ flux.

EXPERIMENTAL RESULTS AND DISCUSSION

All of the probe data were obtained on an ARL instrument operated at 20 kV, 0.04 μA sample current and with a 1–2 μ electron spot size. Analysis has been accomplished using empirical, working curve, techniques as well as through the use of theoretically calculated intensity corrections, and for both approaches, the same collection of X-ray intensities are used. All of the data have been corrected for deadtime and background by conventional techniques; for the calculated intensities the data have been corrected for absorption by the methods proposed by Birks[6] and by Philibert[7] as modified by Colby and Niedermeyer,[8] and for fluorescence by both the Birks[6] method and by the modified

Table I. Stacking Sequences and Corresponding Stoichiometry for Representative Phases

Phase	Stacking	Stoichiometry
M	M	$BaFe_{12}O_{19}$
W	MS	$BaZn_2Fe_{16}O_{27}$
X	M_2S	$Ba_2Zn_2Fe_{28}O_{46}$
Y	Y	$Ba_2Zn_2Fe_{12}O_{22}$
Z	MY	$Ba_3Zn_2Fe_{24}O_{41}$
U	M_2Y	$Ba_4Zn_2Fe_{36}O_{60}$
$84R$	M_2Y_3	$Ba_8Zn_6Fe_{60}O_{104}$
$102R$	M_2Y_4	$Ba_{10}Zn_8Fe_{72}O_{126}$
$138R_a$	M_2Y_6	$Ba_{14}Zn_{12}Fe_{96}O_{170}$
$52H$	M_2Y_7	$Ba_{16}Zn_{14}Fe_{108}O_{192}$
$58H$	M_2Y_8	$Ba_{18}Zn_{16}Fe_{120}O_{214}$
$64H$	M_2Y_9	$Ba_{20}Zn_{18}Fe_{132}O_{236}$
$210R_d$	M_2Y_{10}	$Ba_{22}Zn_{20}Fe_{144}O_{258}$

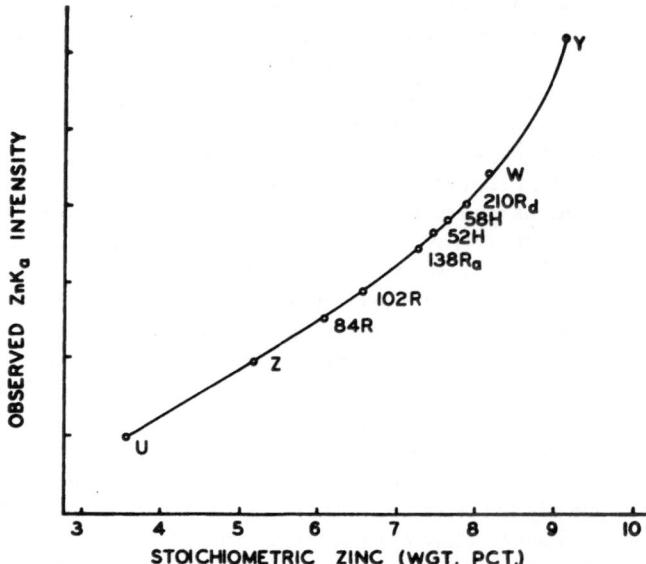

Figure 3. Empirical curve for Zn K_α relative X-ray intensity plotted against zinc concentration predicted by stoichiometry.

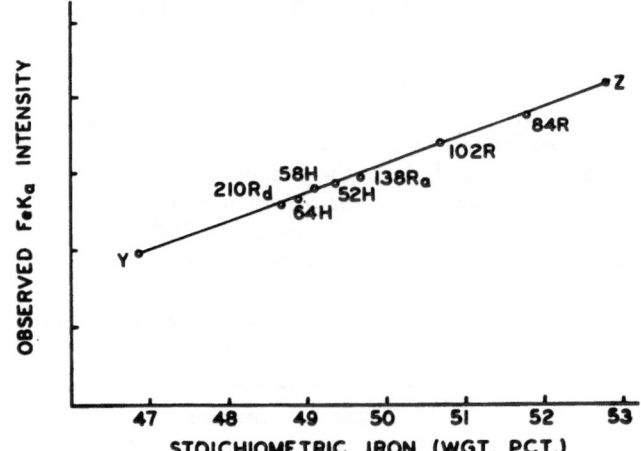

Figure 4. Empirical curve for Fe K_α relative X-ray intensity plotted against iron concentration predicted by stoichiometry.

Wittry[9] technique as tabulated by Colby.[10] The mass absorption coefficients of Heinrich[11] were utilized throughout. For the calculations, the weight fractions of each element were those predicted by the ideal stoichiometry.

The data represent the arithmetic average of from 20 to 30 different analytical points on a given crystal surface, which consisted either of prominent as-grown basal faces or of metallographically polished cross sections cut parallel to the c axis. Although no carbon contamination spot was visually apparent after the 50-sec counting time employed for each element, and although the total X-ray intensity was essentially constant after several repeated measurements for the same element on the same point on the specimen,

Table II. Stoichiometry and Calculated and Measured Relative Intensities

Phase	Stoichiometry (wt. %)			Zinc, I/I_0				Iron, I/I_0			
	Zn	Fe	Ba	Calculated	Recalculated†	Measured	Error*	Calculated (B)	Calculated (W)	Measured	Error*
W	8.2	56.1	8.6	7.8	6.0	5.7	− 5.1	60.2	56.8	57.4	+ 1.0
X	4.8	57.8	10.2	4.6	3.5	3.2	− 8.7	59.5	58.3	57.5	− 1.4
Y	9.2	46.9	19.2	8.8	6.8	7.4	+ 8.8	49.3	46.9	46.1	− 1.7
Z	5.2	52.8	16.2	4.9	3.8	3.6	− 5.3	53.8	52.8	53.3	+ 0.9
U	3.6	55.1	15.0	3.4	2.6	2.5	− 3.8	55.4	55.2	55.4	+ 0.4
$102R$	6.6	50.7	17.3	6.3	4.8	4.8	0.0	52.2	50.7	51.7	+ 2.0
$84R$	6.1	51.8	17.0	5.8	4.5	4.5	0.0	53.1	51.8	53.2	+ 2.9
$138R_a$	7.3	49.7	17.8	6.9	5.3	5.2	− 1.9	51.5	49.8	50.2	+ 0.8
$52H$	7.5	49.4	18.0	7.1	5.5	5.5	0.0	51.3	49.5	50.6	+ 2.2
$58H$	7.7	49.1	18.1	7.3	5.6	5.6	0.0	51.1	49.1	50.2	+ 2.2
$64H$	7.8	48.9	18.2	7.4	5.7	5.7	0.0	50.9	49.0	49.3	+ 0.6
$210R_a$	7.9	48.7	18.3	7.5	5.8	6.0	+ 3.4	50.7	48.8	49.5	+ 1.4

*Error = $\dfrac{\text{measured wt. %} - \text{calculated wt. %}}{\text{calculated wt. %}} \times 100$

† Assumed 5% zinc deficiency and 3% (volume) flux inclusions.

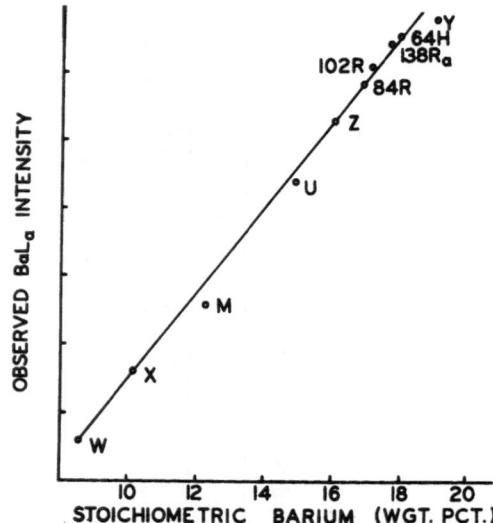

Figure 5. Empirical curve for Ba L_α relative X-ray intensity plotted against barium concentration predicted by stoichiometry.

the sample was mechanically translated perpendicular to the c axis a distance equal to approximately 2–3 times the effective spot size between successive measurements.

The results obtained by the empirical approach are discussed first. For this work, X-ray intensities corrected only for background are plotted against the concentrations predicted by the stoichiometry, several phases being selected arbitrarily as standards. Representative curves obtained in this manner and illustrated in Figures 3, 4 and 5, indicate that the observed X-ray intensity varies sufficiently, even from phases between which relatively small compositional variation exists, so that compositions relative to the phases selected as standards may be determined. Instrumental fluctuations from day to day in absolute intensities measured for any element produce no change in curve shape, but only in distance from any arbitrary base level. Such positional variations may be eliminated by normalizing the intensity values to an arbitrary base determined by any preselected "standard" phase.

The results obtained from such working curves, while confirming that different phases separated even by small chemical differences may be distinguished by probe investigation, do not serve as a check on the stoichiometric values predicted by X-ray diffraction. For this reason, calculations of expected relative X-ray intensities have been made using correction procedures which have been established in the literature as yielding results in reasonably close agreement with determinations by wet chemical techniques.

Strong absorption of Zn K_α radiation by both iron and barium leads to measured intensities, relative to the 100% zinc standard, lower than those predicted stoichiometrically by X-ray diffraction techniques, and by the Castaing first approximation. Correction for absorption by the Birks technique and by the Colby-modified Philibert technique predict the same values for expected relative intensities, below the stoichiometric values by from 0.2 to 0.4 wt.% zinc. The measured relative intensities for all phases lie below the calculated zinc intensities by from 1.3 to 1.7 wt.%, with the exception of the Y and U phases for which the deviation is 2.1 and 0.9 wt.%, respectively, as shown in Table II. Van Hook[12] has suggested a deviation from the stoichiometry for the Zn Y phase, deduced from both wet-chemical and X-ray fluorescence techniques, amounting to 1.92–1.93 rather than 2 atoms zinc per formula unit, and 2.05–2.11 rather than 2 atoms barium. This departure from the stoichiometry, however, is still not sufficient to explain

Figure 6. Electron micrograph of etched surface of $138R_a$ phase showing $NaFeO_2$ flux inclusions. Direct replication; 11,400 × (reduced for reproduction 45%).

the low measured intensities. A possible explanation may be provided by consideration of Figure 6, an electron micrograph of a typical single crystal of the $138R_a$ phase. The high relief materials are $NaFeO_2$ flux inclusions preferentially resistant to the hot HCl employed to strip the direct replica. Preliminary estimates obtained from such micrographs suggest an average flux density of from 2–3 vol.%, comparable to the range of 1–5 % which Van Hook[13] obtained from density determination on $NaFeO_2$ flux-grown ZnY synthesized in his laboratory. Finally, it has been suggested that the inability to reduce the ferrimagnetic resonance linewidth of $NaFeO_2$ flux grown hexagonal ferrites may be due to the incorporation of such flux in the crystals during growth.[14] Calculations for zinc, assuming the Van Hook values rather than the stoichiometric, and a volume of 2% $NaFeO_2$ flux, brings agreement between recalculated and measured zinc to 5% for all phases. The presence of the flux reduces the zinc concentration in the excited volume while simultaneously increasing the absorption correction because of a relative iron increase with respect to zinc in the same volume. To explain the low measured zinc intensities entirely on the basis of flux inclusions necessitates assuming a higher flux concentration than we believe to be consistent with observed X-ray diffraction and electron microscopy studies. A small increase in either the volume figure selected for the flux, or in the departure from stoichiometric values, leads to agreement of better than 2 wt.% zinc (Figure 7). Until such time as direct measurement of the flux volume is possible, we believe the best conclusion to be that zinc in the hexagonal ferrites deviates

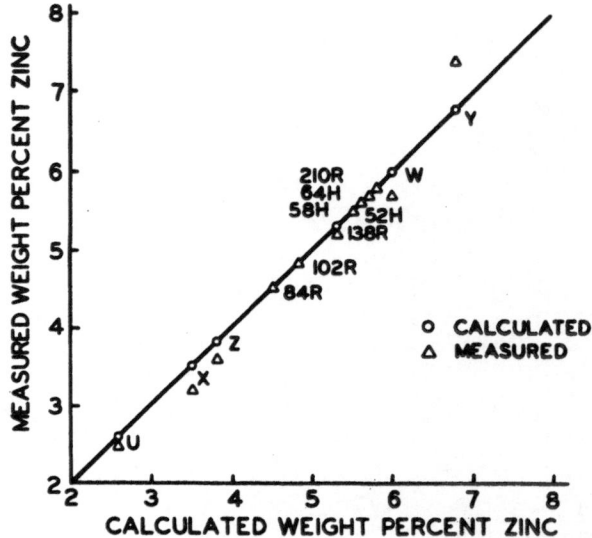

Figure 7. Observed versus calculated zinc concentration for hexagonal ferrite phases. Calculated values obtained assuming 5% stoichiometric zinc deficiency and 2% flux volume.

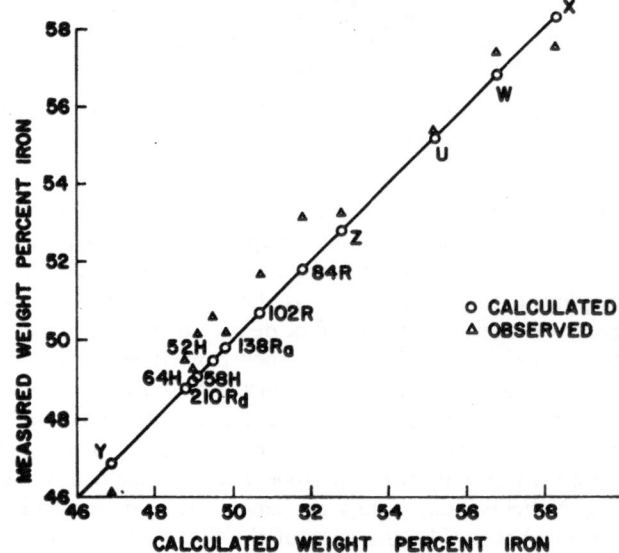

Figure 8. Observed versus calculated (Philibert–Wittry) iron concentration for hexagonal ferrite phases.

from stoichiometry by approximately 3–5 wt.%, and that the presence of flux inclusions combined with this deviation satisfactorily explains the observed zinc intensities.

In the case of the Fe K_α line, absorption and fluorescence corrections by the Birks method and by the Philibert–Wittry methods do not yield the same result for expected relative intensities. Calculated and measured relative intensity ratios are listed in Table II, and it is seen that a somewhat better agreement is provided by the Philibert–Wittry calculation. In Figure 8, measured iron concentrations are plotted against the expected concentration calculated by the Philibert–Wittry methods which yield values with agreement to better than 2 wt.% for iron. If the conclusion stated previously with regard to flux concentrations incorporated in the crystals is correct, we could expect little discrepancy between the measured and calculated values since the wt.% iron in $NaFeO_2$ flux and in hexagonal ferrite deviates by less than $\pm$ 5 wt.% for all phases.

At the present time we have made no determination for barium concentrations by theoretical techniques but are in the process of preparing a series of G calculations which we hope will enable calculation of the excitation of this line by the Wittry methods. At present we are relying on (1) the apparently consistent measurements obtained by empirical techniques, (2) the values obtained for zinc and iron, and (3) considerations of charge neutrality, to suggest that the deviation from stoichiometry exhibited by the zinc, amounting to approximately 1.90–1.93 atoms, is balanced by a small excess of barium of the same order of magnitude.

CONCLUSION

We believe that in spite of the relative chemical complexity of the hexagonal ferrite group, and the close chemical similarity between the phases comprising the group, the probe offers a unique solution to the problems of phase determination and chemical variation studies within the various phases. The results of the present study, although complicated by local inhomogeneities due principally to variation in flux concentration, suggest that hexagonal ferrite single crystals deviate from the stoichiometry predicted by X-ray diffraction. Having shown that the capability for this analytical approach exists, we expect that probe investigation of the substituted ferrite phases is feasible, and should lead to meaningful results in studies of the modification of the magnetic properties of these materials by controlled substitution.

ACKNOWLEDGMENTS

The author wishes to thank Dr. J. A. Kohn and D. W. Eckart for confirmation of the various phases by X-ray diffraction, A. Tauber and R. O. Savage for providing the crystals for study, and C. F. Cook, Jr. for the electron micrograph used in this paper.

REFERENCES

1. P. B. Braun, "The Crystal Structures of a New Group of Ferromagnetic Compounds," *Philips Res. Rept.* **12**: 491–548 (1957).
2. J. Smit and H. P. J. Wijn, *Ferrites*, John Wiley & Sons, Inc., New York (1959), Chapter IX.
3. R. O. Savage and A. Tauber, "Growth and Properties of Single Crystals of Hexagonal Ferrites," *J. Am. Ceram. Soc.* **47**: 13 (1964).
4. J. A. Kohn and D. W. Eckart, Stacking Relations in the Hexagonal Ferrites and a New Series of Mixed-Layer Structures," *Z. Krist.* **119**: 454 (1964).
5. J. A. Kohn and D. W. Eckart, "Mixed-Layer Polytypes related to Magnetoplumbite," *Am. Mineralogist* **50**: 1371 (1965).
6. L. S. Birks, *Electron Probe Microanalysis*, Interscience Publishers, New York (1963).
7. J. Philibert, "A Method for Calculating the Absorption Corrections in Electron Probe Microanalysis," *X-Ray Optics and X-Ray Microanalysis*, Academic Press, Inc., New York (1963), p. 379.
8. J. W. Colby and J. F. Niedermeyer, "Absorption Correction Tables for Microprobe Analysis," National Lead Co., Ohio Report No. NLCO-914, Cincinnati (1964).
9. D. B. Wittry, "Fluorescence by Characteristic Radiation in Electron Probe Microanalysis," University Southern California, Engineering Center Report No. 84-204 (1962).
10. J. W. Colby, "The Correction for Fluorescence by Characteristic Radiation in Microprobe Analysis," National Lead Co., Ohio Report No. NLCO-917, Cincinnati (1964).

11. K. F. J. Heinrich, "Tables of Mass Absorption Coefficients," in: W. M. Meuller, G. R. Mallett, and M. J. Fay (eds.) *Advances in X-Ray Analysis, Vol. 8,* Plenum Press, New York, 1964.
12. J. Van Hook and S. Cvikevich, "Single Crystal Hexagonal Ferrites," Final Report, Contract No. DA-36-039-AMC-00096(E), DA Project No. 1P6 22001 A 057 (1965).
13. J. Van Hook, "Single Crystal Hexagonal Ferrites," Report No. 1, Contract DA-36-039-AMC-00096(E), 1st Quart. Progress Rept. 15 Apr.—14 July 1963.
14. A. Tauber, personal communication.

DISCUSSION

P. Lublin (General Telephone and Electronics Laboratories): Have you experienced any difficulty due to the magnetic nature of the sample?

J. R. Shappirio: I don't believe we've seen any.

D. H. Speidel (Pennsylvania State University): Have you done any phase equilibrium work; for example, you're going on the basis that your compositional variation is part true variation and partly caused by flux. If you don't know what the phase relations as a function of temperature are, could this not be a function of the mechanism of growing your crystal? If you used a completely different technique of crystal growth would you get a completely different result? My point is that, without the phase diagram, you can't say what is actually real and what is actually due to the technique you are using.

J. R. Shappirio: In the case of the phases that we're looking at; the unsubstituted phases, the straight zinc ferrites; they appear to be relatively homogeneous except for the flux inclusions. From different fluxes; barium oxide, boron oxide fluxes; these become much more consistent. Electron micrographs show no inclusions and the data become considerably more consistent. We believe that these things actually exist, although we have done no detailed phase-equilibria studies. I accept the fact that what you say may be true. Considered in the light of the complexity of the system—you have 50 phases on a very small part of this ternary system—one would expect considerable more variation than we actually observe. But these phases do appear to exist.

D. H. Speidel: Since you have such an excellent high-pressure laboratory at Fort Monmouth, it would be very interesting to see what the interrelations of the stabilities of these phases are as a function of pressure, and whether or not the magnetic properties that you are searching for would also be a function of pressure.

J. R. Shappirio: Yes, that point is certainly well taken, and I expect that in the future that will be undertaken. I am sure we are going to get back into it one of these days and this will certainly be one of the problems that will be studied. I also wanted to make a statement that I neglected before, and that is the fact that we do contemplate doing phase-equilibrium studies in the system.

T. Barber (White Sands Missile Range): Have you considered getting any information from infrared absorption on your crystal work? I know we have studied metal oxides in the past in this manner, and it looks like it has some definite promise.

J. R. Shappirio: We have not as yet considered doing this. What we're really trying to establish is whether we have a technique here, and I think we do, that will establish the basic parameters and the chemical composition in the hexagonal ferrites. Thus, we have a handle with which we can determine what the variation is when we substitute a given cation and later we hope to be able to tell not only how much we're putting in, but where it's going.

P. W. Wright (Associated Electrical Industries, Manchester, England): Would you like to say something about specimen preparation of ferrites for microprobe analysis?

J. R. Shappirio: Yes, it's difficult. Of course, when you're dealing with the as-grown basal face, these are generally pretty good. They have been process etched to some extent, but they're of excellent optical quality. This is a far different matter from a cross-section crystal, because ferrites tend to be exceedingly brittle, and we found, much to our horror, that the materials tend to smear. We believe we can eliminate this by light etching. A number of other labs have done this and the micrographs look as if you're able to remove the surface smear. In general, then, what we do is to very carefully polish on successively decreasing diamond grit—not on a lap but on a piece of glass or a piece of cloth—and then follow this with a light etch.

P. W. Wright: What etch do you use?

J. R. Shappirio: There are a number of good etches for the ferrites. Phosphoric acid can be used. We use HCl. It seems to give a little bit more control.

ELECTRON MICROPROBE ANALYSIS OF HIGHLY RADIOACTIVE SAMPLES

V. G. Scotti, J. M. Johnson, and R. T. Cunningham

Pratt and Whitney Aircraft (CANEL)
Middletown, Connecticut

ABSTRACT

An electron microprobe analyzer shielded and modified to accept for analysis radioactive samples reading up to 100 R/hr is described. The following features will be illustrated and discussed: (1) biological shielding, (2) detector shielding, (3) sample transfer system, (4) remote sample loading and focusing system, (5) laboratory layout, and (6) experimental results.

INTRODUCTION

The value of the electron microprobe analyzer for qualitative and quantitive chemical analysis in the fields of metallurgy, mineralogy, solid-state physics, chemistry, and biology has been aptly demonstrated by other workers. In recent years the need for electron microprobe analysis as an aid in evaluating irradiation performance of nuclear fuels and cladding materials has become apparent. Bradbury et al.,[1] have shown, by deliberate sample preparation, that they were able to study the behavior of some solid fission product elements in irradiated UO_2. A study of the distribution of fission products, along with gas analysis and X-ray diffraction, will aid investigators in the study of the dimensional stability of fuels when subjected to neutron bombardment. The EMP is also of great value in the study of fuel-to-cladding compatibility. We have been able to identify intergranular phases, which in the past could only be done by less accurate indirect methods such as light metallography and stain etching techniques.

One of the major obstacles in applying the EMP analysis of irradiated nuclear fuels is the fluorescent background, caused by the fission-product gamma rays (~ 1 MeV) impinging on the spectrometer housing and the structural members of the detector. Although a pulse-height analyzer can be used to discriminate against the high-energy gamma rays, the fluorescent X-ray continuum produced by these gamma rays falls in the same energy band (1–20 keV) as the characteristic X-rays being sought, thus resulting in a lower signal-to-background ratio.

To reduce the gamma activity, Bradbury and his co-workers[1] prepared 0.020-in.-thick wafers of irradiated UO_2. Typical activity levels of these specimens were 12 mR γ at 1 ft and 220 mR $\beta + \gamma$ at 1 ft. This activity permitted the use of the CAMECA probe without modification. The fission product elements molybdenum, ruthenium, and barium were successfully detected under these conditions. In conducting studies of irradiated fuels, deliberate sample preparation as described above is the exception and not the rule. Usually samples selected for microprobe analysis are based upon observations made during light metallographic examination. Further preparation of the sample to reduce

---

[1] References are at the end of the paper.

Figure 1. AMR/3 electron probe microanalyzer before modification.

Figure 2. AMR/3 electron probe microanalyzer after modification.

Figure 3. Modified spectrometer housing and connector tube.

the radioactive background may destroy the very evidence that is to be probed. Thus, the radioactive level of the sample cannot be controlled. The radiation intensity of samples encountered in our work is in the neighborhood of 100,000 mR/hr γ.

To provide the capability of examining highly radioactive samples, it was decided to equip an EMP with lead shielding. The Norelco AMR/3 instrument was selected because it was amenable to modifications, such as displacing the spectrometer housing from the tower, without disturbing the take-off angle or the focusing circle. The bottom loading of the sample made the concept of remote sample loading very attractive. Figure 1 illustrates the Norelco AMR/3 instrument prior to modification and Figure 2 shows the shielding and remote loading sample transfer dolly in place. The following features are described below: (1) shielding, biological and detector, (2) sample transfer system, including remote sample loading and focusing, (3) laboratory layout, and (4) experimental results.

SHIELDING

The lead shielding around the tower provides both biological (personnel) and detector shielding. In order to install shielding between the sample and the detector, the spectrometer housing had to be displaced 3 in. from the tower (Figure 3). The spectrometer housing was raised at the same time to maintain the 15° take-off angle. To maintain the focusing circle, the goniometer arm was extended and the detector-to-crystal distance adjusted. The original connector tube was replaced with a longer connector tube which

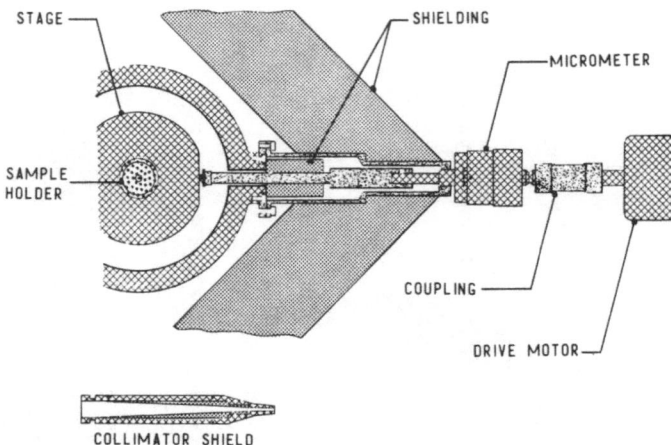

Figure 4. Section through micrometer stage drive and collimator shield.

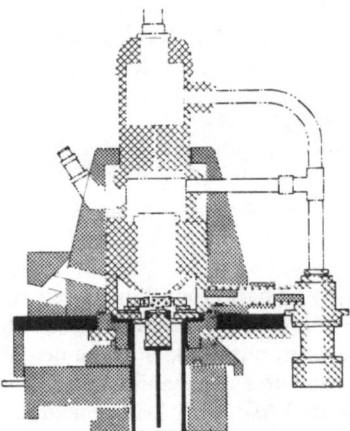

Figure 5. Assembly microprobe.

will be described later. The micrometer stage drives were also modified to extend through the lead shielding (Figure 4). Note the lead shielding incorporated into the extended micrometers to minimize radiation streaming.

The 500 lb of lead shielding was supported by a 1-in.-thick T-shaped boiler plate supported by a versibar frame which is bolted to the concrete floor (Figure 2). This arrangement was necessary to avoid buckling of the table top, which could cause misalignment between the spectrometer and sample chamber. The shielding was constructed of interlocking blocks 3 in. thick at the base of the tower and tapering off to $1\frac{1}{2}$-in. at the top to reduce the total weight of lead. The reduction in shielding was permissible because the condenser, focusing lenses, and other structures in the tower afforded some shielding. We felt that the thickness of the shielding could be reduced in this manner if the specimens were considered a point source. A line drawn from the sample to any point on the outside surface of the shielding shows that at least three inches of lead must be traversed by the gamma rays. The small rectangular lead housing (Figure 2) contains a prism arrangement enabling the operator to look into the specimen chamber.

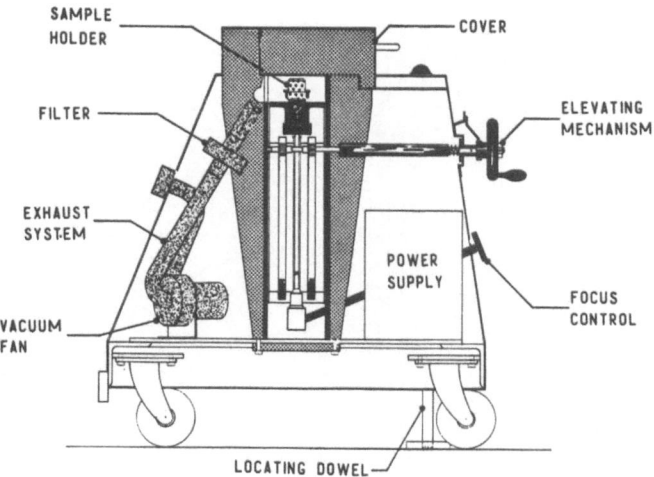

Figure 6. Shielded transport dolly.

The shielding was extended below the sample chamber (Figure 5) by means of a stepped-lead collar; a stainless steel cylindrical insert was provided to maintain precise alignment of the tower and the sample-loading mechanism.

An additional lead shroud was attached to the boiler plate to provide continuity of shielding between the shielded transfer container and the fixed shielding. To eliminate radiation streaming through the tower-pump outport (Figure 5) lead inserts were placed in and around pumpout line.

The collimator previously mentioned was fabricated from W-5%Ni alloy and was used to reduce to an absolute minimum the direct passage of gamma radiation from the sample into the spectrometer housing, while allowing passage of the chracteristic X-rays being measured. Experiments have demonstrated that there is no loss of resolution or line intensity. This alloy was selected because of its high density and reasonable machining properties. The diameter of the opening at the receiving end was $\frac{1}{16}$-in. (in contrast to the $\frac{3}{8}$-in. opening in the regularly supplied connector tube) and graduated to a $\frac{1}{2}$-in. opening at the exit end, thus providing a 3° solid angle.

SAMPLE TRANSFER SYSTEM

The equipment for the transfer of the prepared sample from the hot cell to the microprobe and the mechanics of loading the sample and focusing it will now be described. A lead-shielded container mounted on a dolly was used to transport the sample (Figure 6). Prior to loading the sample, the entire transport dolly was draped with plastic or brown paper to avoid contaminating the exterior surfaces with radioactivity. The dolly was then pushed into the doorway of the hot cell, where manipulators were used to pull open the lid and to crank up the elevating mechanism to raise the sample holder into view. The sample was then placed in the holder, lowered into the shield by means of the elevating mechanism, and the cover pushed closed with the manipulators. The protective drapery was then carefully removed, and the transport dolly was monitored for radioactive contamination.

To load the sample in the probe, the transport dolly was wheeled into the kneehole under the console and precisely indexed by the locating dowels. This required precise

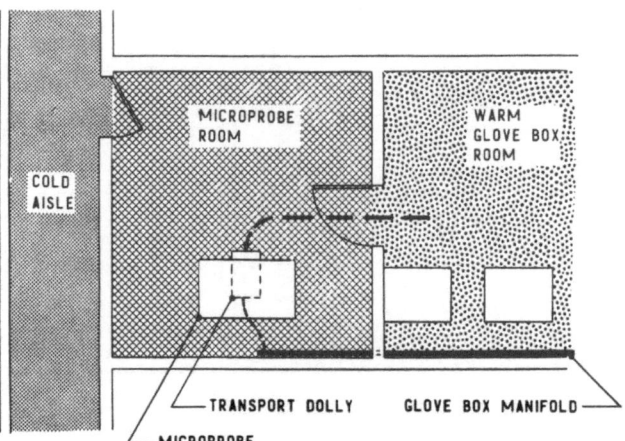

Figure 7. Microprobe laboratory layout.

positioning to permit the sample plate to make an O-ring vacuum seal with the bottom of the tower. The sample stage was also precisely indexed by means of the micrometer stage controls in order that it may accept the sample holder. The cover was then pulled open and the sample plate raised into place by means of the elevating mechanism. When the pointer indicated that the O-ring seal had been made, the column was vacuum-pumped. The mechanical pumping speed gave a rapid indication of the condition of the O-ring vacuum seal. The sample was focused by turning a knob on the face of the transport dolly. The knob drove a gear train which remotely actuated the normal focusing knob.

The transport dolly was equipped with an exhaust system, which kept the shielded container under a negative pressure at all times. The exhaust was discharged through an absolute filter, thus all aerosols and particulate matter were kept under control. A dual power supply was used to power the exhaust system. A 12-V battery was used when the dolly was detached from the probe. When it was in a stationary position under the probe, 125-V house current is used. During this period, a trickle charger was used to recharge the battery. A gage was used to indicate the negative pressure.

Figure 5 illustrates in cross-section how a complete lead-shield envelope is formed around the sample when the transport dolly is in place and the sample is in the instrument. The cross-hatched areas indicate the original structure of the instrument, the dotted areas are the added lead shielding, and the solid black lines are the added steel structural support. The solid black horizontal line represents the 1-in. thick boiler plate which supports the weight of the lead.

LABORATORY ARRANGEMENT

The electron microprobe laboratory was located between the warm and the cold side of the building (Figure 7), so that it could be made a "warm area" or "cold area" as the situation dictated. When the transport dolly containing a radioactive sample was rolled from the hot cell to the probe, the door to the cold aisle was locked and the room entered from the warm side. After the sample was loaded into the instrument and the room was monitored, the warm side door was locked and the room was entered from the cold side. The operator then proceeded with the analysis of the sample in the normal manner. A glove box manifold was used to remove vacuum pump and transport dolly exhausts. It also

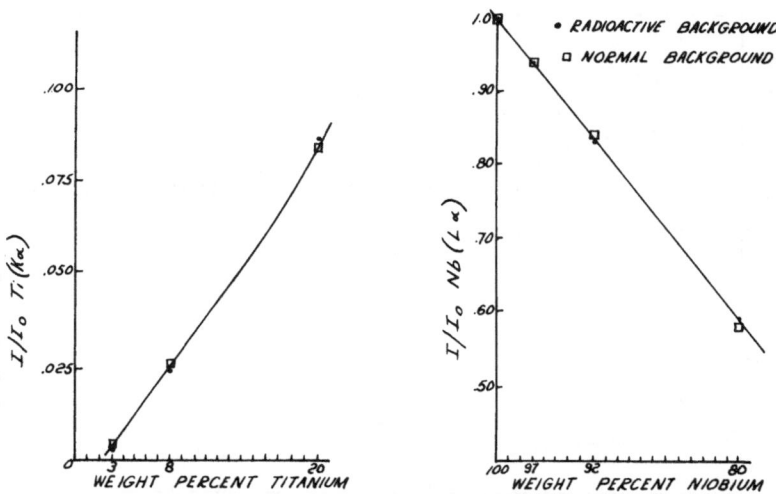

Figure 8. Titanium–niobium calibration curves.

provided an exhaust system for future alpha glove-box installation for the preparation of alpha-emitting materials.

The room had sufficient depth that in the event additional biological shielding was required a concrete or lead wall could be constructed in front of the instrument. Extended electrical connections were provided so that the control panels could be mounted on the proposed shield. Optical relays and stage micrometer drive extensions could also be provided.

EXPERIMENTAL

The modifications and minor changes in instrumentation just described had no apparent effect on measured line intensities. This is supported by data generated in a recent interlaboratory calibration round-robin.

The calibration standard was a specially prepared $NbSi_2$ sample, mounted in bakelite with the pure standards, and carefully polished $\frac{1}{4}$-μ diamond paste. Intensity ratios for the Si K_α, and Nb L_α lines, 0.22 and 0.31 respectively, were in good agreement with calibration curves for the niobium–silicon system previously published by Fornwalt et al.,[2] using an unmodified Norelco AMR/3 probe. The integral line intensity observed for the Si K_α X-ray was in excess of 7000 cps for pure silicon, also in good agreement with the unmodified instrument. In this and all subsequent cases, measurements were made using an electron accelerating potential of 30 kV with a beam size approaching a 1-μ diameter. Specimen current was adjusted to 0.20 μA on niobium.

As part of a series of experiments to establish the capability of the modified instrument, three niobium–titanium wires were mounted in a hole in the center of an irradiated fuel specimen. The sample was then prepared in the normal manner for light metallography and inserted into the instrument as previously described. The radioactivity of the specimen measured 12,000 mR/hr γ at 1 ft. The activity level at the surface of the shielding, however, was 10 mR/hr γ, well within the accepted biological requirements.

Specimen current and back-scatter images were of good quality. It might be noted that the back-scatter detector, a p–n junction type, was situated approximately 1-in.

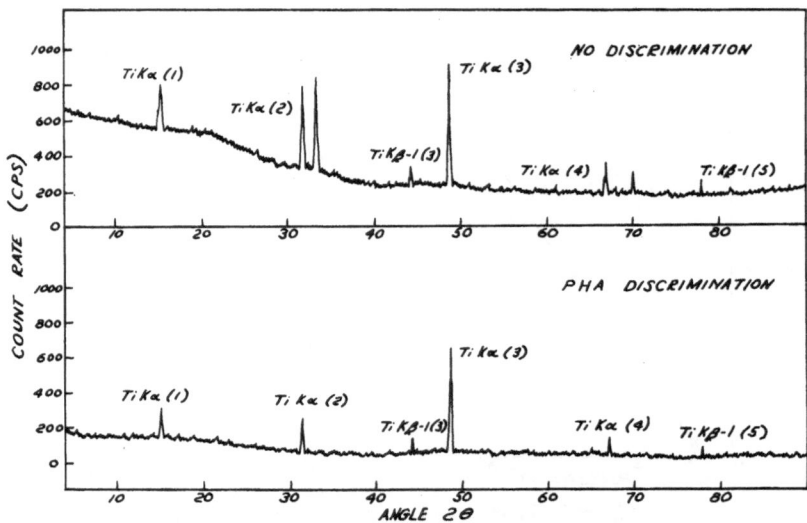

Figure 9. Strip-chart recordings of Nb–3% Ti specimen.

above the specimen, in this case, in a radiation field of 250,000 mR/hr $\beta + \gamma$. Figure 8 correlates intensity ratios with weight percent for the Ti K_α and Nb L_α lines, with and without a high-level radioactive background. The measured line intensities, and hence the intensity ratios, are in terms of net count rates. The curves illustrate the effectiveness of using "net-peak" count rates for analyses, provided the detection and counting systems are sufficiently fast to prevent pulse pileup and spectrum distortion.

Strip chart recordings of the Nb–3%Ti wire in a 12,000 mR/hr radiation field are shown in Figure 9. The general shape of the upper tracing is typical of radioactive samples, showing a general rise in background beginning at about 40° 2θ. More highly radioactive samples show another background maximum at about 60°. The lower tracing illustrates the well-known value of the pulse-height analyzer in reducing the radioactive background.

There is much literature on the theoretical considerations of quantitative microprobe analysis. Investigations of the various theoretical calculation techniques and differences in published mass absorption data reveal a wide spread in calculated intensities. For this reason, it was the practice of this laboratory to use the technique of Ziebold and Ogilvie[3] whenever suitable standards were available.

In general, the performance of the instrument exceeded expectations. Further tests of the instrument's capability were planned as more highly radioactive samples became available, but this was halted by the termination of the technical program.

REFERENCES

1. B. T. Bradbury, J. T. Demant, P. M. Martin, and D. M. Poole, "Electron Probe Microanalysis of Irradiated UO₂," AERE-R-4845, 1965.
2. D. E. Fornwalt, B. R. Gourley, and A. V. Manzione, "A study of the Compatibility of Selected Refractory Metals with Various Ceramic Insulation Materials," CNLM-5942, presented at Electron Microprobe Symposium of Electrochemical Society Meeting, Washington, D.C., October, 1964.
3. T. O. Ziebold and R. E. Ogilvie, *Anal. Chem.* **36**: 322 (1964).

DISCUSSION

H. Yakowitz (National Bureau of Standards): Do you notice any effect on the bending micro-crystal from the background of the hot stuff? Does it make it harder?

V. G. Scotti: We don't have enough working experience to know the effects of radiation on the crystal or any other component in the system.

THE EFFECT OF CHEMICAL COMBINATION ON THE K X-RAY SPECTRA OF SILICON

Donald M. Koffman and Sheldon H. Moll

Advanced Metals Research Corporation
Burlington, Massachusetts

ABSTRACT

Intensities and wavelengths of the K-series lines of silicon were measured using primary excitation in the electron probe. Specifically, intensity and line position data for the Si K spectral lines α_1, α_2, β, α_3, and α_4 were recorded using a high-resolution continuously curved mica-crystal vacuum spectrometer. Spectra were obtained from silicon metal and a wide range of mineral samples, e.g., SiO_2, $Mg_2Si_2O_6$, $CaSiO_3$, $KAlSi_2O_6$, $CaMgSi_2O_6$, $KAlSi_3O_8$, $NaAlSi_3O_8$, $CaAl_2Si_2O_8$, Fe_2SiO_4, and $(Na, K)(Al, Si)_2O_4$. Only slight differences were found to exist among the mineral spectra, but these differed markedly from that obtained from silicon metal.

INTRODUCTION

Generally, the electron-beam analyzer is associated with the determination of chemical composition employing the technique of X-ray fluorescence. Several advances have led to the development of focusing vacuum spectrometers utilizing high-resolution curved mica crystals. Initially, these spectrometers were developed for the purpose of increasing absolute-signal intensities and signal-to-noise ratios in order to improve the precision of chemical analyses. The employment of these spectrometers in the measurement of the spectra of light elements such as silicon, aluminum and magnesium has suggested the feasibility of their use as soft X-ray spectrographs.

An electron-beam microanalyzer has many advantages in this application. A large number of specimens can be surveyed rapidly and with ease. Specimen currents are of the order of 10^{-6} A, reducing the possibility of sample heating, oxidation, or contamination.

Translation of the sample beneath the beam during analysis permits fresh areas to be continuously provided. Visual observation at high magnification ($300\times$) enables one to determine rapidly the stability of the samples. In many cases, decomposition can be eliminated by using an enlarged electron beam ($50\ \mu$) or moving the sample during the spectral analysis, without deleterious effects on the X-ray optical focusing conditions.

With these considerations in mind, investigation of the K spectra of silicon metal and a number of silicates was undertaken.

EXPERIMENTAL

The electron-beam microanalyzer used in this work was designed and built by Advanced Metals Research personnel. The vacuum spectrometer was of the type supplied with the Norelco AMR/3 microanalyzer, including a thin-window flow proportional counter operating with an argon–methane (P–10) flow gas. A curved mica crystal was employed, the radius of curvature being automatically adjusted over the whole spectral range to satisfy the X-ray optical focusing conditions. Recording electronics were standard Norelco equipment except for a low-noise Hamner preamplifier and pulse-height analyzer. The usual vacuum was 5×10^{-5} torr. The silicon was high-purity semiconductor-grade single-crystal material. Silicates were supplied by Dr. Randall Van Schmus of the Air Force Cambridge Research Laboratories.

The α_4/α_3 ratio of the silicon was found not to change during the course of measurement, indicating that sample oxidation was not a problem. The position of the unresolved α_1,α_2 doublet in the minerals was measured relative to silicon metal several times for the determination of the wavelength shift. For a given specimen the α_3, α_4, and β peaks were then measured relative to the α_1,α_2 doublet of that material. In most cases an average was taken over three runs, and the wavelengths were found to agree within ± 0.0006 Å.

Peak intensities were measured and calculated relative to the unresolved α_1,α_2 doublet, which was arbitrarily assigned a value of 1000. For these measurements, the electron accelerating potential was adjusted to insure that the maximum counting rate did not exceed 8,000 cps to eliminate the need for detector dead-time correction. All intensity and wavelength shifts were determined at electron accelerating potentials between 14 and 30 kV.

RESULTS

The complete spectrum for the silicon compounds examined was not recorded in this work. The K lines selected for study were α_1, α_2, α_3, α_4, and β.

K_{α_1,α_2} Doublet

The K_{α_1,α_2} doublet for silicon metal and SiO_2 are shown normalized in Figure 1. As a result, these curves do not show absolute changes in intensity between metal and

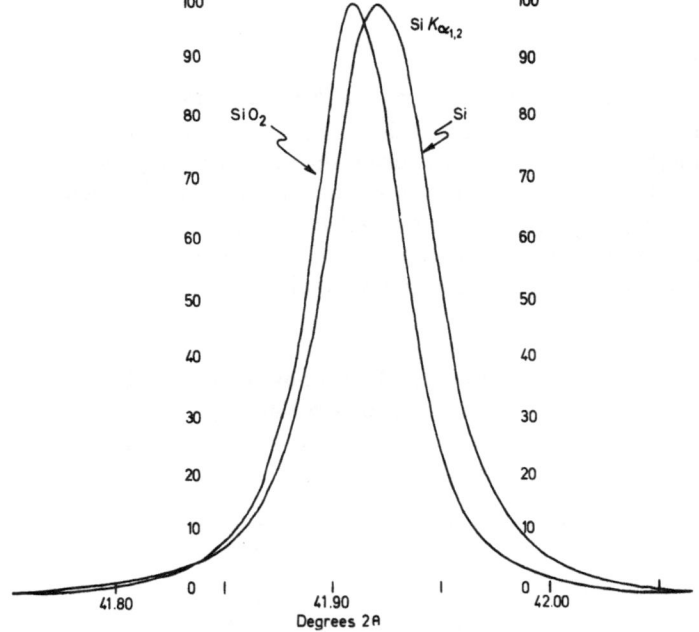

Figure 1. Normalized Si K_{α_1,α_2} for silicon and SiO_2.

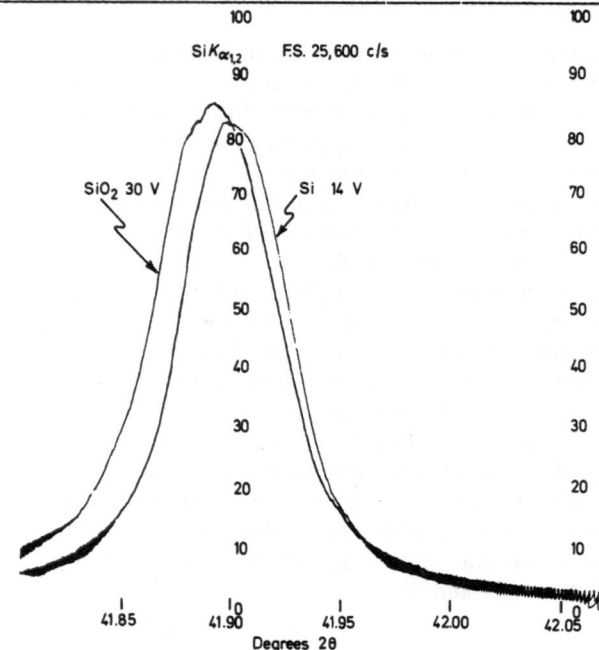

Figure 2. As-recorded Si K_{α_1,α_2} for silicon and SiO_2.

oxide, but merely line-shape and position relationships in the same trace. The line position can be seen to shift to slightly shorter wavelengths when going from metal to oxide. The K_{α_1,α_2} doublet for the silicates showed no observable differences in shape or position compared to that for SiO_2. The unresolved doublet for silicon metal and SiO_2 as recorded and superimposed are shown in Figure 2.

K_α Satellites

The K_{α_3} and K_{α_4} satellites for silicon and SiO_2 are shown as recorded in Figure 3. Although quantitative measurements of the $K_{\alpha'}$ line were not undertaken, it is seen to be somewhat stronger and sharper in the oxide than in the metal, and shifts in the same direction as the K_{α_3} and K_{α_4} lines, i.e., to shorter wavelengths when going from the metal to oxide. Since the intensities of the K_{α_3} and K_{α_4} relative to K_{α_1,α_2} have been shown to be voltage-dependent at low electron acceleration potentials, the α_4/α_3 ratio for silicon was measured as a function of voltage. These results are listed in Table I.

Table I. Ratio of α_4/α_3 for Silicon as a Function of Electron Beam Voltage

Voltage (kV)	α_4/α_3
7.5	0.58
14	0.63
22	0.63
30	0.63

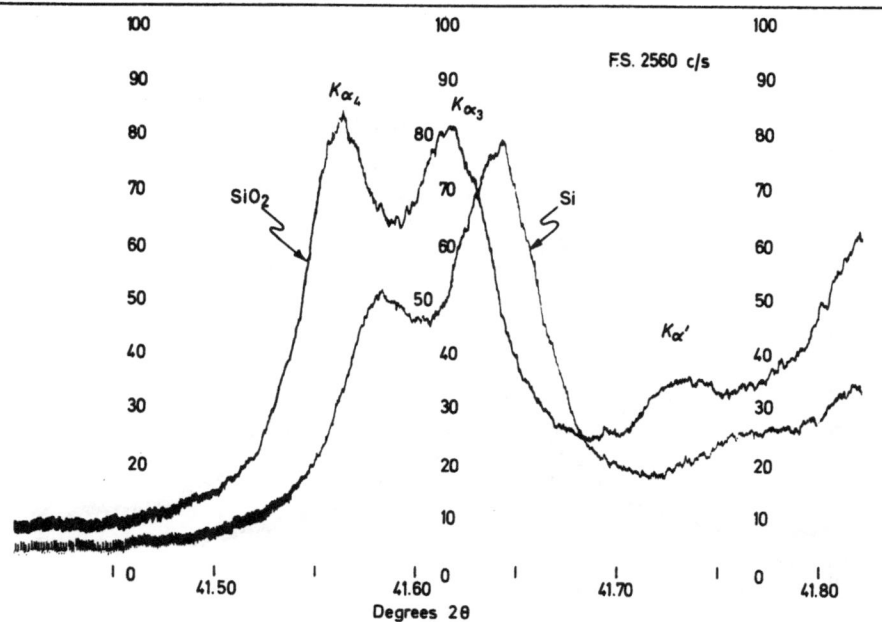

Figure 3. Si K_{α_3} and K_{α_4} satellite lines for silicon and SiO_2 (as-recorded).

It can be seen that the ratio at 7.5 kV differed from that measured at 14 kV. Above 14 kV, the ratio was found to be constant. Therefore, all measurements were conducted in this range. The lower ratio at 7.5 kV also lends evidence to the fact that no oxidation of the sample occurred during analysis. Since the penetration of the electron beam is an order of magnitude less at 7.5 kV than at 14 kV, any surface oxide would contribute greatly to the K spectra. From Figure 3, it can be seen that the K_{α_3} and K_{α_4} intensities are nearly equal in SiO_2. Therefore, if oxide were present on the surface, a lower electron-accelerating potential would result in a higher α_4/α_3 ratio. Since a lower ratio was observed, the presence of oxide is negated.

The spectra of these satellite lines obtained from the silicates was not found to differ in shape or position from that of SiO_2.

The K_β Line

Since in silicon the K_β band is involved in the electron transition from the conduction band, this line is most affected by the state of chemical combination. The K_β bands for silicon and SiO_2 as-recorded are shown in Figure 4. The peak shift was the greatest of all the lines measured. The band shape is also quite different: the symmetrical band shape in SiO_2 shows that in the valence band, all levels are occupied and the material is an insulator, while the more metallic nature of silicon is evidenced by the band asymmetry and approach towards a sharp short wavelength limit. This short wavelength limit would be even sharper in the case of aluminum and magnesium. The band in going from the metal to the oxide is seen to shift to longer wavelengths.

The K_β line shapes observed in the silicates examined were not found to differ significantly from that of SiO_2. Slight differences in peak position were found to exist in the K_β lines of the silicates, compared to SiO_2, but no obvious correlation with chemical combination was found.

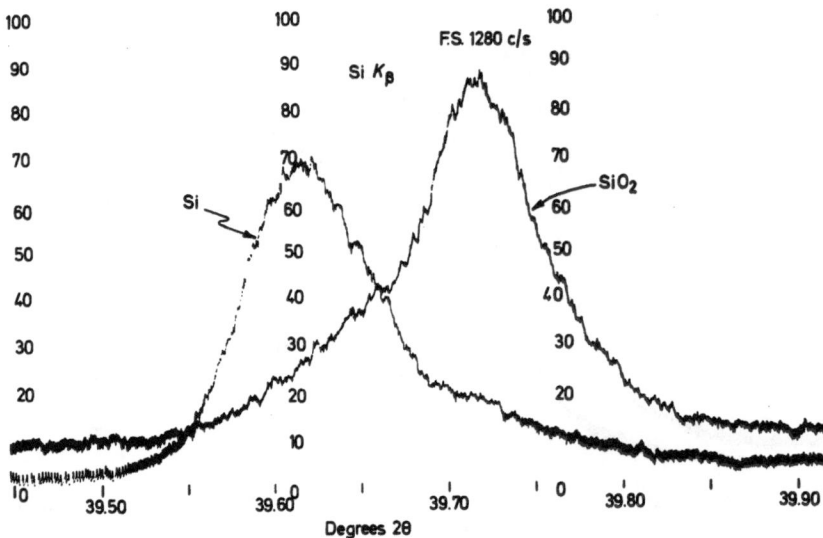

Figure 4. Si K_β line for silicon and SiO_2 (as-recorded).

SUMMARY

A compilation of the line positions and relative intensities for the materials studied are shown in Table II. No significant differences in intensity were observed in comparing the K spectral lines from the silicates with those from SiO_2. The α_4/α_3 ratio for SiO_2 and the silicates was found to be greater than unity in all cases examined.

Table II. Si K Lines from Silicon Metal and Silicates

Material	K_{α_1}		K_{α_3}		K_{α_4}		K_β	
	λ	I	λ	I	λ	I	λ	I
Si	7.1262	1000	7.0803	66	7.0713	41	6.7506	28
SiO_2	7.1244	1000	7.0765	52	7.0673	54	6.7666	30
$Mg_2Si_2O_6$	7.1244	1000	7.0768	50	7.0680	53	6.7617	25
$CaSiO_3$	7.1244	1000	7.0770	51	7.0680	53	6.7594	23
$KAlSi_2O_6$	7.1244	1000	7.0772	52	7.0680	56	6.7649	20
$CaMgSi_2O_6$	7.1244	1000	7.0770	50	7.0682	53	6.7611	25
$KAlSi_3O_8$	7.1244	1000	7.0765	50	7.0678	54	6.7644	22
$NaAlSi_3O_8$	7.1244	1000	7.0766	51	7.0677	53	6.7640	24
$CaAl_2Si_2O_8$	7.1244	1000	7.0763	49	7.0677	51	6.7614	25
Fe_2SiO_4	7.1244	1000	7.0773	50	7.0680	53	6.7630	23
$(Na, K)(Al, Si)_2O_4$	7.1244	1000	7.0763	52	7.0675	56	6.7627	25

CONCLUSIONS

From the results presented above, it can be seen that the K spectra obtained from the silicates do not differ appreciably from that obtained from SiO_2. They do, however, differ from the spectrum obtained from silicon. As a result, the quantitative analysis of silicon in these minerals, which involves the measurement of the intensity of the unresolved K_{α_1,α_2} doublet, would best be performed by using SiO_2 as a standard rather than pure silicon. This procedure would eliminate the effect produced by the observed line shift, i.e., a lower intensity arising from the fact that aligning the spectrometer on silicon metal would correspond to an "off-peak" position for these compounds.

The results present a strong case for the applicability of the electron-beam microanalyzer to the study of the soft X-ray spectra of the elements, since the curved mica-crystal spectrometer has been demonstrated to have the required resolution.

DISCUSSION

J. Holliday (U.S. Steel Research Laboratories): I would like to just make a comment on your statement about the symmetric peaks being indicative of an insulator, while the asymmetrical peak is more representative of a metal. We find, for example, that in the case of titanium nitride, the K emission band from titanium nitride is quite symmetrical while from boron nitride it is asymmetrical. Titanium nitride is a very good conductor, in fact better than titanium metal, while boron nitride is an insulator.

D. M. Koffman: The remarks made were not meant as a general statement. I think that for silicon, aluminum, and magnesium, the statement is valid. The statement has been made by a number of people as a generalization and has been proved a number of times to be not incorrect, but not exactly right.

J. Merritt (Shell Development): I want to ask you a question on the correlation of the bond length and the shift of the K_β line. Would you care to speculate further as to whether you could correlate this further with charge withdrawal?

D. M. Koffman: Although a shift of the K_β line was observed in the silicates studied, an attempt to correlate this shift with the bond length was not made.

E. White: I just wanted to make a comment regarding the Si K_β peak position and the mean SiO bond distance in the silicon tetrahedron. The mean SiO bond length in silicates is a direct indication of the tetrahedral linkage. In other words, as you go from the independent tetrahedral structures to the framework structures you have a progressive change in the mean SiO bond length, and the peak position of Si K_β correlates with this mean bond length. So what chemical parameter you also want to tie in with that doesn't really matter.

THE EFFECT OF CHEMICAL COMBINATION ON SOME SOFT X-RAY *K* AND *L* EMISSION SPECTRA*

David W. Fischer and William L. Baun

Air Force Materials Laboratory
Wright-Patterson Air Force Base, Ohio

ABSTRACT

X-ray spectrochemical analysis is a versatile technique that can be used to determine considerably more than just the elemental composition of a sample. X-ray lines and bands are, in many cases, influenced by the state of chemical combination of the element whose spectrum is being investigated. Significant effects due to changes in bonding are seen in the *K* and *L* spectra of the low atomic number elements which fall in the 20 to 90 Å region. Results using primary excitation, a stearate crystal, and a flow proportional counter are shown for the *K* spectra of boron, carbon, and nitrogen. Chlorine and sulfur *L* spectra from some simple metal compounds of these elements are also shown.

INTRODUCTION

In the last few years, interest in soft X-ray spectroscopy has greatly increased. A considerable amount of this interest is focused on better methods for detecting very light elements, such as oxygen, nitrogen, carbon, and boron, by their X-ray emission spectra.[1] Using these spectra to determine the presence and quantity of a certain element in a chemical compound is not the only use to which they can be put, however. In many cases one can make definite conclusions about the chemical state of an element by studying the peak positions, band shapes, and relative intensities of its emission lines. Chemical combination can cause a very marked change in these lines. Valence, coordination, electronegativity, electrical conductivity and crystal structure may all have some kind of effect on emission spectra, especially those involving transitions from the valence or conduction band.[2-7]

Virtually all of the experimental data in the literature which concerns band shapes, wavelengths, and intensities of X-ray emission lines in the 40 to 100 Å region have been obtained using a ruled grating as the dispersing device.[8-13] There is good reason for this, because until very recently, these gratings were the only effective soft X-ray dispersers available. In the last few years, however, artificial crystals built up of successive mono-layers of long-chain fatty-acid salts have attracted much attention.[1,14,15] These soap-film crystals have been made with $2d$ spacings as large as 130 Å and give high count rates and peak-to-background ratios for a wide range of wavelengths. For many investigators these crystals have a large advantage over the ruled gratings, in that they can be used with the normal commercially available spectrometers.

---

[1] References are at the end of the paper.

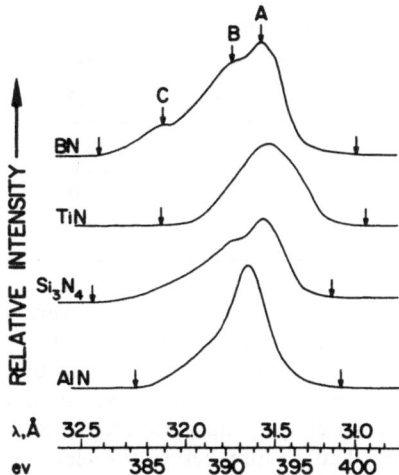

Figure 1. Nitrogen K emission bands from some nitrides.

Most X-ray and microprobe equipment manufacturers are now offering some sort of soft X-ray kit for commercial use. Usually these kits include a soap-film analyzer for working in the very long wavelength region.

Literature data for these soap films, however, are concerned mostly with methods of obtaining high count rates and peak-to-background ratios for the K spectra of the very light elements, especially oxygen and carbon.[1] Little emphasis has been placed on the study of band shapes. This may, in part, be due to the fact that many investigators are under the impression that soap-film crystals cannot compare favorably with ruled gratings where resolution is concerned. As we will show, this is not necessarily true in most cases for the 40 to 100 Å region.

The purpose of this paper, then, is twofold: to show what effects chemical combination has on some soft X-ray emission spectra in the 40 to 100 Å range, and to demonstrate that these average soap films, even when used with a flat-crystal spectrometer, give resolutions comparable to those obtained from a ruled grating instrument for this wavelength region.

EXPERIMENTAL

The soft X-ray apparatus used in this work has been explained in detail in previous publications.[5,16] Basically, it consists of a horizontal flat-crystal Bragg spectrometer mounted in a vacuum chamber. Instrument resolution is no better than that expected from a normal commercially available flat-crystal spectrometer with 0.4° soller-slit collimation. X-rays are produced by direct electron-beam bombardment of the sample. For most of the data reported here, a beam voltage of 5 kV and currents of 2 to 10 mA were used. Dispersion is accomplished with a lead stearate soap film ($2d = 100.79$ Å) consisting of 160 layers and made by the usual Blodgett–Langmuir technique.[1,15–17] The detector is a flow proportional counter with a very thin cellulose nitrate window. Argon–methane (P-10) flow gas is used at about 90 mm Hg pressure.

Normal operating vacuum was 10^{-6} torr. This is not a high vacuum by modern standards and can, in some cases, present serious contamination problems if certain precautions are not taken. To study the C K band it was of utmost importance to insure that no detectable carbon contamination occurred on the target surface during the run. For non-carbon-containing target materials, it was very easy to check for the presence of carbon contamination by rotating the lead stearate crystal and detector to the proper

Table I. Wavelengths and Bandwidths for Nitrogen K Bands

Target	A(λ, Å)	B(λ, Å)	C(λ, Å)	$W_{1/2}$(eV)	W(eV)
BN	31.57	31.71	32.10	6.1	17.7
Si_3N_4	31.56	31.72	—	5.6	15.9
AlN	31.66	—	—	3.4	15.0
TiN	31.53	—	—	5.6	15.0

angle, and checking for the presence of the carbon K emission band. This is a very sensitive method of carbon determination, especially when it is present as a surface contaminant. In many cases, even though the contamination was not visible to the eye, the carbon line was present after only a few minutes of operation when using the cooled brass anode. This problem was eliminated by using a high-temperature strip-furnace tantalum anode assembly. The target was heated to about 200°C for 10 to 15 min before the electron beam was turned on and maintained at least at this temperature during the run. Using this technique, the carbon K band was not detectable; we consider this reasonable evidence that carbon contamination was eliminated, or at least reduced to the point where it was not affecting the spectra of interest.

Wavelengths were determined by using multiple-order reflections of shorter wavelength lines as standards. One of the best standards was found to be the Al $K_{\alpha_{1,2}}$ and K_{α_3} lines from aluminum metal and Al_2O_3.[6,18] No great accuracy is claimed for the resultant wavelengths, but we believe them to be within ± 0.05 Å. The curves shown in the figures are taken from ratemeter scans and have a mean deviation of $\pm 2\%$.

RESULTS

Nitrogen K

Nitrogen K emission bands from a few simple nitrides are shown in Figure 1. The band from boron nitride is in good agreement with those shown by O'Bryan and Skinner[13] and Holliday,[10] though we apparently show a little better resolution. There are no nitrogen K bands from any other compounds shown in the literature so that the BN band is our only point of comparison.

Unfortunately, there are a limited number of nitrogen-containing compounds which lend themselves well to our experimental arrangement. Many potentially interesting compounds, especially the nitrates, are too unstable under electron bombardment to give reliable and reproducible results. It is evident from Figure 1, however, that chemical combination can markedly affect the shape of the nitrogen K emission band.

Boron nitrides gives a band with three distinct maxima labeled A, B and C. A and B are separated by 1.7 eV and are at 31.57 and 31.71 Å, respectively. The weak low-energy band C is at 32.10 Å and is separated from A by 6.5 eV. Table I lists the wavelength values along with the experimental half bandwidths and full bandwidths which have not been corrected for any instrumental or other type broadening effects. The arrows at the long and short wavelength tails of the curves locate the points from which the bandwidths were measured. As has been mentioned many times in the literature, these bands tail off rather slowly, so that the limiting points are somewhat arbitrary.

Unlike BN, the nitrogen K band from Si_3N_4 has only two maxima, A and B. They are at the same wavelength positions as A and B in BN and have the same relative intensities. Because of the absence of band C, the curve for Si_3N_4 is slightly narrower than for BN.

TiN and AlN give only one distinct peak although AlN has what appears to be some weak structure on the low-energy tail. The peak position for AlN falls midway between parts A and B of the bands from BN and Si_3N_4.

It would appear from the positions of the extra structure on the low-energy side of the bands in BN and Si_3N_4 that it is all part of the band and not due to satellites. Because of the small number of compounds from which nitrogen bands could be obtained, it would be rather meaningless to try to correlate changes in shape or position with any one property of the nitrides. Perhaps, as in the oxygen spectrum from oxides,[7] several properties are affecting the spectra at once.

The curves shown in Figure 1 are third order reflections from the lead stearate crystal and cover an angular range of about 135 to 150° 2θ. For these spectra, the peak count rates were 250 to 300 cps at 5 kV and 5 mA with a background of about 30 cps. First-order reflections were much stronger, but poorly resolved. BN gave a first-order count rate for nitrogen K of 12,000 cps at 5 kV and 3 mA.

Certainly, it would be interesting to observe the nitrogen band from nitrates and other nitrogen compounds. Perhaps a system using fluorescence excitation instead of electrons would prove to be the answer, if one could obtain enough intensity.

Carbon K

The carbon K emission band, especially that obtained from graphite and diamond, has been studied by many investigators.[10,13,19] All these spectra were obtained using a ruled grating and all except those shown by Holliday[10] were recorded on photographic film. Most of them are in disagreement concerning the shapes and intensities of various components of the bands. The spectra which we have obtained (Figures 2 and 3) were recorded with a flow proportional counter using the second-order reflection from a lead stearate crystal.

Figure 2 shows the bands obtained from three forms of pure carbon, namely graphite, diamond, and lampblack. The graphite curve agrees fairly well with a few of those found in literature and not as well with others. The main point of disagreement appears to be in the relative intensities of parts A and B of the curve. Holliday, for instance, shows part A to be about half the intensity of part B.[10] In all of the many runs we have made, we get a slightly higher intensity for part A than for part B. Several forms of graphite have been examined, and all give exactly the same band shape. These different forms of graphite have included finely powdered spectrographically pure graphite, thin sections

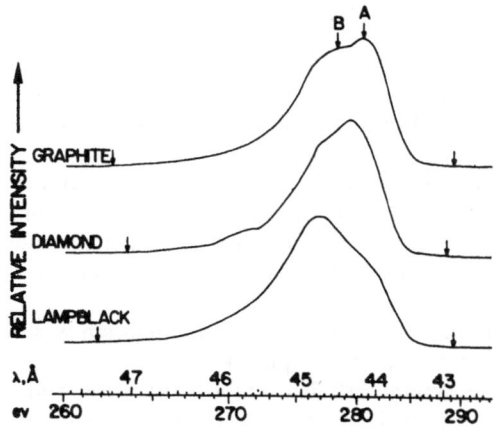

Figure 2. Carbon K emission bands from graphite, diamond, and lampblack.

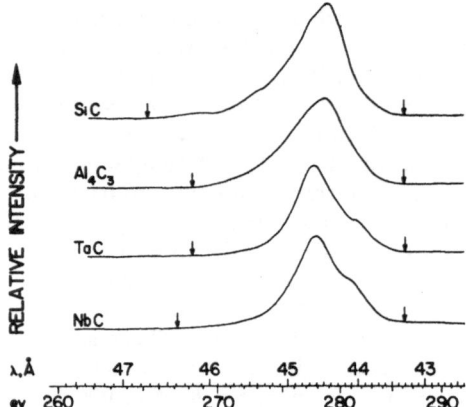

Figure 3. Carbon K emission bands from some carbides.

Table II. Wavelengths and Bandwidths for Carbon K Bands

Target	A(λ, Å)	B(λ, Å)	$W_{1/2}$(eV)	W(eV)
Graphite	44.18	44.62	8.4	27
Diamond	44.36	44.80	7.4	25
Lampblack	43.97	44.77	9.0	28
SiC	—	44.45	5.2	21
Al$_4$C$_3$	—	44.49	5.9	18
TaC	44.08	44.64	4.5	18
NbC	44.10	44.62	4.6	19

of large chunks, and colloidal suspensions of graphite sprayed in a thin film on the anode surface. Temperature seems to affect these intensities slightly. As the temperature is raised, the A to B intensity ratio decreases and is about 1:1 at 1000°C. The curve shown in Figure 2 was obtained at about 300°C. Wavelength values for the peak components agree very well with Skinner and others.[13] Table II lists these values along with the uncorrected half bandwidths and full bandwidths. The arrows on the band tails indicate the points at which the full bandwidths were measured.

The emission band from diamond is in excellent agreement with those shown by Chalklin and Broili et al.[13] This band is narrower than that from graphite and shows some extra weak structure on the low-energy tail.

Several of the literature references show the emission band from lampblack to be the same as that from graphite. Our lampblack band looks much like the graphite band shown by most investigators, but differs considerably from our own graphite band, as can be seen in Figure 2.

It has been suggested that perhaps the different band shape we obtain for graphite has something to do with the large amount of carbon in the analyzing crystal. This hypothesis appears to be inconsistent with our results, however, since our diamond and lampblack curves agree very well with some of those obtained using the ruled gratings which appear in the literature.

It was also found that the carbon spot arising from the carbon contamination when using the cooled brass anode gives the same band shape as lampblack.

Figure 3 shows the carbon K band from a few metal carbides. These bands are all considerably narrower than those obtained from any of the forms of pure carbon and the peak maxima are much sharper. The SiC band shape agrees well with those shown by Broili *et al.* and Chalklin.[13] The NbC curve, on the other hand, is slightly different than that shown by Holliday. As in the case of graphite the difference is in the intensity of the high-energy portion of the curve.

Perhaps such disagreement should not be too surprising. As is mentioned by Kern,[19] these differences may in part be due to the method of recording and to the excitation conditions. Erroneous shapes can be obtained if proper care is not taken to convert photographic blackening to true intensity relationships. It has been shown that the length of time of the exposure can also affect the shapes. Electronic recording techniques, to a large extent, eliminate these sources of error. Kern[19] also states that differing excitation conditions can cause variations in the emission edge. The various intensity maxima seem to be characteristic of the materials, however, and have little or no dependence on the excitation conditions.

Kern shows that the carbon K band is very sensitive to the type of chemical bonding present. He concludes that SiC, diamond, and silicon have, to a large degree, the same type of chemical bonding, i.e., covalent, because of the obvious correspondence of the shape of the C K and Si K_β bands. It appears, however, that such a correlation is questionable. The present authors[20] and others have shown that Kern's Si K_β bands from SiC and elemental silicon are in error. We believe that his spectra are distorted from oxide contamination. The same thing happened to Farineau,[13] who published the Al K_β band from the metal which was contaminated by oxide and unfortunately caused the development of some erroneous theory. There are several other cases of reported band-shape correspondence which have later been shown to be incorrect.

We have obtained the metal-ion emission spectra from the materials shown in Figures 1 and 3, and in no case does there appear to be any obvious correspondence of their shapes with the nitrogen and carbon spectra.

Peak count rates and peak-to-background ratios for the carbon K band depend, of course, on the target material. Graphite, for instance, gave a peak count of 3700 cps at 5 kV and 1 mA for the second-order reflections. NbC gives a count rate of about 300 cps for the same conditions. The first-order reflection is about eight times these values, but affords much poorer resolution. There has been some concern about the high background obtained from the stearate crystals, especially at low angles of reflection. While it is true that at angles below about 20° 2θ the background starts climbing rapidly, it is no problem at higher angles. The second order C K band covers approximately the 115 to 135° 2θ region, and background counts of 25 to 30 cps are normal.

Boron K

The K emission spectra from elemental boron and some boron compounds are shown in Figures 4 and 5. The boron spectrum has been studied previously by several investigators,[10,13,21] but the curves were all obtained using a ruled grating for dispersion and most were recorded using photographic film.

Figure 4 demonstrates the large changes observed when going from elemental boron to boron nitride and boron oxide. These curves are ratemeter scans with the electron beam set at 5 kV and 3 mA. As can be seen, the spectrum from elemental boron consists of only one band, labeled B in the figure. This band is rather metallic-like in shape, being asymmetrical with a fairly sharp short wavelength edge. For B_2O_3, the boron spectrum has an entirely different appearance. The main band B is now quite symmetrical in shape and has shifted by more than 1.0 Å to a longer wavelength. Two new bands, A and C, also appear. Band A is on the short wavelength or high-energy

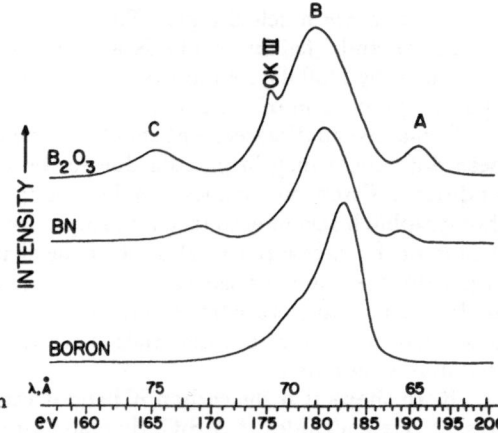

Figure 4. Boron K emission spectra from boron, BN, and B_2O_3.

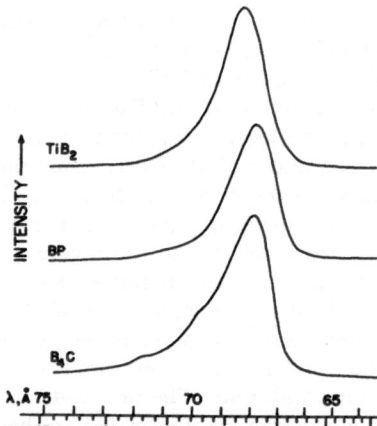

Figure 5. Boron K emission spectra from TiB_2, BP, and B_4C.

Table III. Wavelengths and Half Bandwidths for Boron K Bands

Target	Band	λ, Å	$W_{1/2}$(eV)	Relative intensity	Peak intensity, 5 kV, 3 mA
Boron	B	67.90	5.0	—	12,000 cps
BN	A	65.56	3.0	12	—
	B	68.65	5.9	100	2,700
	C	73.35	3.7	9	—
B_2O_3	A	64.95	4.1	18	—
	B	69.01	7.9	100	2.200
	C	74.94	4.2	20	—
B_4C	B	67.94	5.8	—	7,500
BP	B	67.93	5.2	—	900
TiB_2	B	68.16	4.9	—	3.500
ZrB_2	B	68.03	4.8	—	1,700
NbB_2	B	68.12	5.0	—	1,300
W_2B_5	B	67.99	5.1	—	1,400
SiB_6	B	67.92	4.9	—	1,800
AlB_{12}	B	67.94	5.0	—	3,600

side of the main band, separated by 11.3 eV, and is 18% as intense as B. Band C is on the long wavelength or low-energy side of B, separated by 14.3 eV, and is 20% as strong as B. The wavelengths, half bandwidths, and relative intensities of these bands are listed in Table III. All of the compounds that we have studied which contain a boron–oxygen bond give a boron spectrum exactly the same as that obtained from B_2O_3.

BN gives a boron spectrum which falls somewhere between that of elemental boron and B_2O_3. Band B is symmetrical as in the oxide but it has not shifted as much. Bands A and C also appear but are weaker and closer to the main band than in the oxide. Our spectra from boron, BN, and B_2O_3 are very similar in appearance to those shown by Gwinner and Kiessig,[13] but we detect much larger wavelength shifts in the main band when going from boron to BN to B_2O_3. We also obtain a larger energy difference between extra bands A and C. In the boron interstitial compounds, such as in Figure 5, the bands A and C do not appear. Band B has, in general, the same appearance as that from elemental boron, but has shifted to slightly longer wavelengths. This effect was also noted by Gwinner and Kiessig.

Table III lists the peak intensities obtained from the main boron band for each of the compounds at 5 kV and 3 mA. These intensities have not been corrected for detector window absorption or self-absorption effects. Holliday[10] reports a count rate of about 2000 cps from BN at 3.5 kV and 1 mA, which is slightly higher than we obtain at these conditions. The background count, in general, is less than 50 cps.

The origin of the extra bands A and C which appear in the oxide and nitride seems open to question. O'Bryan and Skinner[13] considered the main band B to be the $2s$ ($2s \rightarrow 1s$ transition) band and A to be the $2p$ ($2p \rightarrow 1s$ transition) band. Holliday[10] disagreed, mostly on the basis of dipole and quadrupole selection rules. He decided that band B was really the $2p$ band but did not explain A and C. We agree with Holliday that the main band should be designated $2p$. Band A is very possibly a satellite band since high-energy satellites are quite common. The origin of band C is not very clear but it provides a very interesting comparison with the Al K_β spectrum[6] as seen in Figure 6. In aluminum metal, the K_β emission band is quite asymmetrical, having a very sharp short wavelength edge somewhat like that observed from elemental boron. For Al_2O_3, the band becomes very symmetrical and shifts considerably, much the same as band B does when going from elemental boron to B_2O_3. Also, in Al_2O_3 an extra band appears on the low-energy side of Al K_β which is not present in the metal. This band is separated from K_β by the same amount as band C is separated from B in B_2O_3. Conventionally, this extra band in Al_2O_3 is called $K_{\beta'}$; its origin is not well understood. This extra band

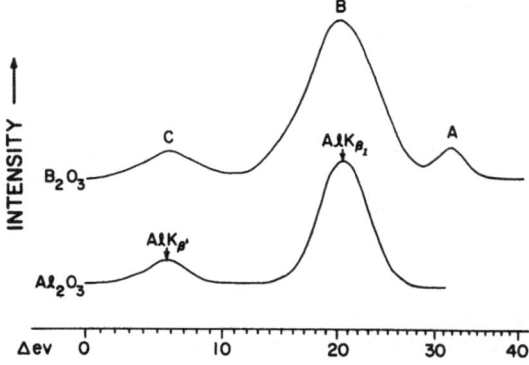

Figure 6. Comparison of boron K and aluminum K_β emission spectra from B_2O_3 and Al_2O_3.

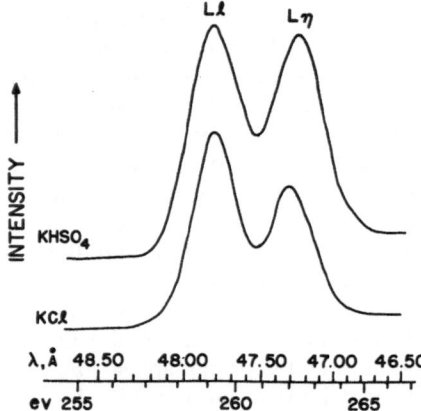

Figure 7. Potassium L_l and L_η lines from KCl and KHSO$_4$.

also appears in the oxides of other elements, such as magnesium and silicon, and oxygen-containing compounds of sulfur and chlorine. A further comparison for band C is evident in the nitrides. In aluminum nitride, for example, $K_{\beta'}$ is present but is weaker and closer to K_β than in Al$_2$O$_3$. The same effect is seen for band C in BN and B$_2$O$_3$. Nevertheless, it is still not clear whether band C is a low-energy satellite band or even possibly the $2s \rightarrow 1s$ transition.

Potassium L

The potassium L emission spectrum is shown in Figure 7. Only the results from two compounds, KCl and KHSO$_4$, are shown, because they represent the largest difference in the spectra that we have observed. It is an unfortunate fact that a large number of potassium compounds decompose when bombarded with electrons under vacuum. As a result, these compounds usually give the spectrum of metallic potassium instead of that from the chemically combined form with which we started. This appears to happen with KCl, for instance, but not for KHSO$_4$. In both instances, the spectrum consists of two lines, L_l and L_η, at 47.83 and 47.27 Å respectively. The only major difference we have observed in the spectrum is a change in the intensity ratio of the L_l and L_η, lines as can be seen in Figure 7.

Very little data appear in the literature concerning the potassium L emission spectrum. Crisp,[22] Siegbahn and Magnusson,[13] and Tyren[13] have reported wavelength values for L_l and L_η, but only Crisp shows the experimental curve. This curve does, however, serve as a very nice comparison of the results obtained from ruled gratings and soap-film crystals for this wavelength region (~ 47 Å). Crisp, using a grating, was barely able to resolve the L_l and L_η lines in the fourth-order reflection. Our curve (Figure 7) is a second-order reflection from a lead stearate soap film and shows very good resolution of the L_l and L_η lines.

For the potassium L spectrum, the electron beam was set at 5 kV and 5 mA. The potassium L_l line from both KCl and KHSO$_4$ had a peak intensity of about 400 counts/sec with a peak-to-background ratio (P/B) of about 10 for the second-order reflection. First order gave about three times these values but with a poorer L_l, L_η resolution. No corrections were made for detector-window absorption or self-absorption in the sample.

Chlorine L

The chlorine L emission spectrum shows much more of a change due to chemical combination than does the potassium L spectrum. A few of the experimental curves are shown in Figure 8. The main band is at about 68 Å. There is not much literature data with

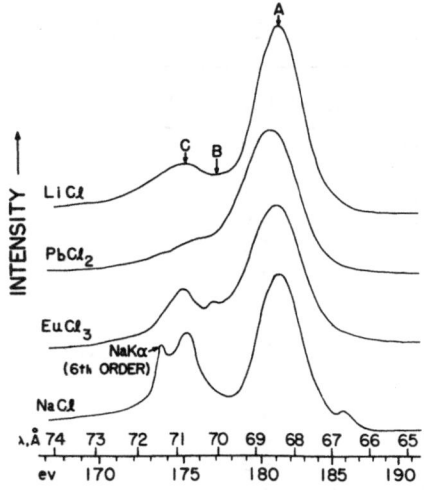

Figure 8. Chlorine L emission spectra from LiCl,
PbCl₂, EuCl₃, and NaCl.

which to compare our results. Siegbahn and Magnusson[13] seem to be the only ones to have published data for chlorine L, although O'Bryan and Skinner[13] added one curve of their own.

Although many of the chlorides decompose when bombarded by electrons, the resulting chlorine L spectrum is, nevertheless, characteristic of the target material since the decomposition consists of evolution of the chlorine. The spectrum of the metal ion left behind is, on the other hand, not characteristic of the compound as stated previously for potassium.

The spectra shown in Figure 8 are the experimental curves and have not been corrected in any way. In general, the chlorine L spectrum consists of a main band, labeled A, and one or two weaker low-energy bands, labeled B and C. Band A is undoubtedly the $L_{2,3}$ band but the origin of B and C is not so obvious. O'Bryan and Skinner were unable to explain them, but suggested, as a possibility, that they are due to double transitions. We wish to add nothing to this suggestion except for a rather minor point. Band C falls exactly where one would expect to see the third-order oxygen K band. The presence of some water of hydration, or maybe some oxychloride or chlorate in the chlorides, could explain this band very simply. We have shown that this band is not due to oxygen, however, since neither the first- nor the second-order reflection is present.

The $L_{2,3}$ band (band A) is much more interesting, because chemical combination has a very definite effect on it. Table IV lists the wavelength values for this band along with band C for several simple chlorides. When one examines the wavelengths listed for the $L_{2,3}$ band a very interesting trend becomes evident. The wavelength increases as the electronegativity value of the metal ion increases. Figure 9 is a graphical plot of this relationship. Pauling's electronegativity values[23] are used for the abscissa. According to Pauling, the difference in electronegativity between the metal ion and the chlorine ion is a rough indication of the amount of ionic character of the chemical bond. If such is the case, our results tend to show that the more ionic the bond in the simple chlorides, the shorter the wavelength of the Cl $L_{2,3}$ band.

The Cl $L_{2,3}$ band appears to be quite symmetrical, its width depending on the metal ion present. Bandwidths at half-maximum intensity for the $L_{2,3}$ band are listed in Table IV for each of the chlorides. These experimental values have not been corrected for any broadening effects. It is difficult to measure the base width of the main band due to the overlap of the low-energy bands, but it appears to be on the order of 12 eV.

Table IV. Wavelengths and Half Bandwidths for Chlorine *L* Bands

Target	Band A (λ, Å)	Band C (λ, Å)	Band A ($W_{1/2}$, eV)
KCl	68.30	70.76	3.6
NaCl	68.37	70.75	3.7
BaCl$_2$	68.38	70.64	3.9
CaCl$_2$	68.42	70.69	3.8
LiCl	68.44	70.78	3.3
SrCl$_2$	68.48	—	3.6
YCl$_3$	68.51	70.74	4.4
EuCl$_3$	68.51	70.82	4.3
ZrCl$_2$	68.55	70.80	4.3
MnCl$_2$	68.58	70.76	3.9
MoCl$_3$	68.60	70.72	4.1
PbCl$_2$	68.61	—	4.5
CuCl$_2$	68.61	70.86	4.0
AgCl	68.63	—	4.0
HgCl	68.63	—	4.3
AuCl$_3$	68.67	70.77	3.8

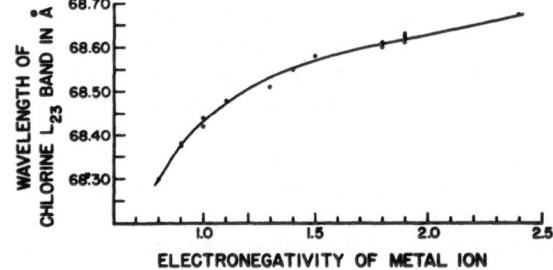

Figure 9. Relationship of wavelength of chlorine $L_{2,3}$ band to electronegativity of metal ion for simple metal chlorides.

There is one band which does not appear in the Cl *L* spectrum, although we fully expected to see it. This is the band arising from neutral chlorine atoms. Fluorine, for instance, in its *K* emission spectrum, shows two fairly strong closely spaced lines on the high-energy side of the main band, which are interpreted as being due to neutral fluorine atoms.[7] The main band is due to fluorine ions. Evidently, then, the chlorides are not ionized in exactly the same manner as the fluorides.

Simple chlorates and oxychlorides give a Cl *L* spectrum which is not very much different from the chlorides. The third-order oxygen *K* spectrum interferes with band C in these compounds but not with the main band. This, incidentally, points up a sometimes very annoying disadvantage of the soap-film crystals. Multiple-order reflections of shorter wavelength lines are quite strong and can seriously interfere with and obscure important parts of the spectrum under study.

The chlorides produce a wide variety of peak count rates and P/B ratios for the Cl $L_{2,3}$ band, depending on the metal ion present and its valence state. With an electron beam of 5 kV and 5 mA, the average values were about 1000 cps peak intensity with a P/B ratio of about 20.

Sulfur L

We have never encountered any emission spectrum which shows as marked a change with chemical combination as does the sulfur *L* spectrum. Some of the curves obtained are shown in Figures 10, 11, and 12, and their wavelength values are listed in Table V. The peak designations 3s, 3p, and 3d are after O'Bryan and Skinner[13] and Tomboulian and Cady.[13]

Sulfur L spectra from elemental sulfur and from various sulfides have appeared in the literature, but nothing is to be found concerning the sulfates.

The shape of the L spectrum from yellow crystalline sulfur (Figure 10) agrees very well with that shown by Tomboulian and Cady, although our wavelength value is a little higher, agreeing better with the value given by Prins and Takens.[13] The low-energy tails of the bands in Figures 10, 11, and 12 are not shown in their entirety because of a slight interference from the second-order carbon K band, which we are unable to completely get rid of when the anode is cold. Separation between the 3s and 3p bands from elemental sulfur is 6.6 eV and the peak intensities are in the ratio of about 2.5:1.

Sulfides give a somewhat different sulfur L spectrum than does elemental sulfur. There are still only two bands present but they are separated by a larger amount. The high-energy band is labeled 3d by O'Bryan and Skinner and is 15 to 20% as intense as the

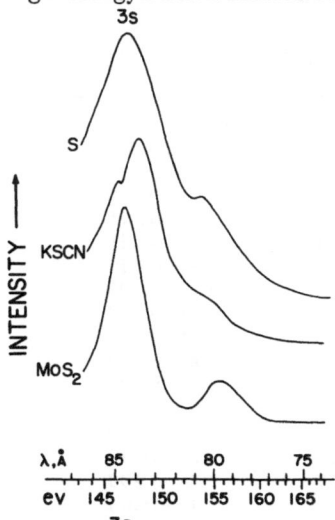

Figure 10. Sulfur L emission spectra from sulfur, KSCN, and MoS₂.

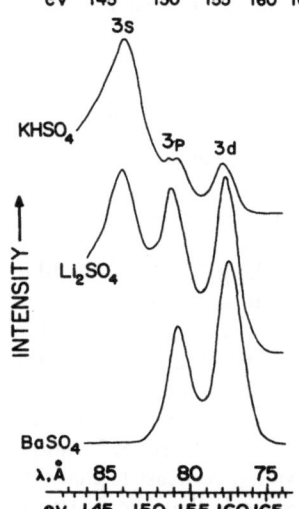

Figure 11. Sulfur L emission spectra from KHSO₄, Li₂SO₄, and BaSO₄.

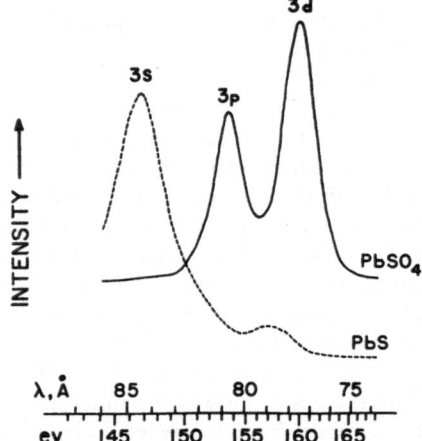

Figure 12. Sulfur L emission spectra from PbS and PbSO$_4$.

Table V. Wavelengths and Intensity Relationships for Sulfur L Bands

Target	$3s$ (Å)	$3s$ (Å)	$3p$ (Å)	Relative intensity $3s:3p:3d$
S	84.10	80.42	—	100:40: 0
ZnS	84.55	—	80.14	100: 0: 15
Cu$_2$S	84.65	—	79.21	100: 0: 15
FeS	84.65	—	79.53	100: 0: 15
CdS	84.72	—	80.07	100: 0: 20
PbS	84.49	—	78.88	100: 0: 10
MoS$_2$	84.41	—	79.55	100: 0: 15
Bi$_2$S$_3$	84.16	—	79.84	100: 0: 20
Li$_2$SO$_4$	83.93	81.00	77.58	100:85: 95
Na$_2$SO$_4$	83.68	80.84	77.75	100:45: 80
K$_2$SO$_4$	83.66	80.81	77.65	100:30: 40
KHSO$_4$	83.72	80.82	77.65	100:35: 40
FeSO$_4$	—	80.95	77.67	0:60:100
BaSO$_4$	—	80.68	77.63	0:45:100
Ag$_2$SO$_4$	—	80.76	77.69	0:50:100
PbSO$_4$	—	80.77	77.52	0:60:100
HgSO$_4$	—	80.90	77.46	0:60:100
KSCN	83.68 84.64 } doublet	80.76	—	100:35: 0
NH$_4$SCN	83.76 84.41 } doublet	80.49	—	100:35: 0

main $3s$ band. In all the sulfides which we have examined, the wavelength of the $3s$ band is greater than from elemental sulfur. There is no apparent correlation between the position of the $3s$ band and any one chemical property of the sulfides. This may very well be due to the fact that several properties, such as electronegativity, crystal structure, and conductivity, are all influencing the spectrum in some manner. The energy separation between the $3s$ and $3d$ bands varies by a rather large amount in different sulfides. In some cases the $3d$ band is very broad and may actually consist of two closely spaced bands.

The sulfur L spectrum from the sulfates (Figure 11) is in striking contrast to that obtained from the sulfides and elemental sulfur. The main difference is that the $3s$ band, which is the strongest band in the sulfides and sulfur, is completely absent in most of the sulfates. This can be seen very clearly in Figure 12, which shows the difference between the sulfur spectrum from PbS and $PbSO_4$. Sulfates of the alkali metals show the $3s$ band in addition to the $3p$ and $3d$ bands. In these compounds, the $3s$ band has shifted substantially to shorter wavelengths, compared to the sulfides. Between CdS and K_2SO_4, for example, the $3s$ band differs in wavelength by more than 1.0 Å. In the heavier metal sulfates, the $3p$ and $3d$ bands are the only ones present and the $3d$ band is the stronger of the two. The energy separation between these two bands varies with the metal ion, being anywhere from 6.1 to 6.8 eV. The $3d$ band is somewhat broader than the $3p$ and both appear to be narrower than the $3s$.

A major difference between sulfides and sulfates is also observed in the sulfur K spectrum[24] and in the sulfur electron spectrum.[25] An additional band, $K_{\beta'}$, is present in the sulfate spectrum but not in the sulfide. The difference is not as dramatic as that observed in the L spectrum, however.

Relative peak intensities for the $3s$, $3p$, and $3d$ sulfur bands from the sulfides and sulfates are listed in Table V.

Thiocyanide compounds of the alkali metals also give a rather interesting spectrum. The $3s$ sulfur band is doubled showing the probable presence of both the S^{2-} and S^{6+} valence states. The curve obtained from KSCN is shown in Figure 10.

CONCLUSIONS

The large changes which can be caused by chemical combination in soft X-ray emission spectra are potentially valuable not only for the theory of electronic band structure of solids but for X-ray spectroscopic analysis as well. Instead of the usual qualitative and quantitative elemental analysis, one can, in many cases, make definite conclusions about the chemical state of the element in question from the shape and position of the spectral lines and bands.

REFERENCES

1. B. L. Henke, "X-Ray Fluorescence Analysis for Sodium, Fluorine, Oxygen, Nitrogen, Carbon, and Boron," in: W. M. Mueller, G. R. Mallett, and M. J. Fay (eds.), *Advances in X-Ray Analysis, Vol. 7*, Plenum Press, New York, 1964, p. 460.
2. W. L. Baun and D. W. Fischer, "The Effect of Valence and Coordination on K Series Diagram and Non-Diagram Lines of Mg, Al, and Si," in: W. M. Mueller, G. R. Mallett, and M. J. Fay (eds.), *Advances in X-Ray Analysis, Vol. 8*, Plenum Press, New York, 1965, p. 371.
3. W. L. Baun and D. W. Fischer, "The Effect of Chemical Combination on K X-Ray Emission Spectra from Magnesium, Aluminum, and Silicon" AFML-TR-64-350, 1964.
4. W. L. Baun and D. W. Fischer, "The Influence of Chemical Combination on Aluminum K Diagram and Non-Diagram Lines," *Nature* 204: 642, 1964.
5. D. W. Fischer and W. L. Baun, "Diagram and Non-Diagram Lines in K Spectra of Magnesium and Oxygen from Metallic and Anodized Magnesium," *Spectrochim. Acta* 21: 443, 1965.
6. D. W. Fischer and W. L. Baun, "Diagram and Non-Diagram Lines in K Spectra of Aluminum and Oxygen from Metallic and Anodized Aluminum," *J. Appl. Phys.* 36: 534, 1965.
7. D. W. Fischer, "The Effect of Chemical Combination on the X-Ray K Emission Spectra of Oxygen and Fluorine," *J. Chem. Phys.* 42: 3814, 1965.
8. F. Fisher, R. S. Crisp, and S. F. Williams, "A Photon Counting Spectrometer for the 50–100 Å Range," *Opt. Acta* 5: 31, 1958.
9. J. E. Holliday, "Soft X-Ray Emission Spectroscopy in the 13 Å to 44 Å Region," *J. Appl. Phys.* 33: 3259, 1962.
10. J. E. Holliday, "Soft X-Ray Emission Spectroscopy in the 10 Å to 150 Å Region," in: E. F. Kaelbe, *Handbook of X-Rays*, McGraw-Hill Book Company, New York, 1965, Chapter 38 (in press).

11. E. D. Piori, G. G. Harvey, E. M. Gyurgy, and R. H. Kingston, "The Soft X-Ray Spectroscopy of Solids," *Rev. Sci. Instr.* **23**: 8, 1952.

12. J. L. Rogers and F. C. Chalklin, "A Geiger Counter Vacuum Spectrometer and its Use for the Study of Soft X-Ray Lines," *Proc. Roy. Soc. (London) Ser. B* **67**: 384, 1954.

13. *Landolt-Bornstein Tables, Vol. 1*, 6th ed., Springer-Verlag, Berlin, 1955, part 4. This reference contains a review by Faessler which includes a compilation of soft X-ray spectra and literature references up to 1955.

14. W. L. Baun and D. W. Fischer, "Preparation and Use of Organic Compounds as Dispersing Devices for Long Wavelength X-Rays," ASD-TDR-63-310, 1963.

15. J. B. Nicholson and D. B. Wittry, "A Comparison of the Performance of Gratings and Crystals in the 20–115 Å Region," W. M. Mueller, G. R. Mallett, and M. J. Fay (eds.), *Advances in X-Ray Analysis, Vol. 8*, Plenum Press, New York, 1965, p. 497.

16. T. C. Furnas and E. W. White, "A Program of Basic Research to Study X-Ray Spectra in the Region 15 to 50 Å," WADD-TR-61-168, 1961.

17. K. Blodgett, "Films Built by Depositing Successive Unimolecular Layers on a Solid Surface," *J. Am. Chem. Soc.* **57**: 1007, 1935.

18. B. Edlén and L. A. Svensson, "The Wavelengths of the Lyman Lines and Redetermination of the X-Unit from Tyrén's Spectrograms," *Arkiv. Fysik.* **28**: 427, 1965.

19. B. Kern, "Si K_β-Banden der Röntgenemissionsspektren von elementarem Silicium, Siliciumcarbid und Siliciumdioxyd," *Z. Physik.* **159**: 178, 1960.

20. W. L. Baun and D. W. Fischer, "K X-Ray Emission Spectra from Silicon and Silicon Compounds," *Spectrochim. Acta.*, 1965 (in press).

21. R. S. Crisp and S. F. Williams, "The Soft X-Ray Emission Spectra of Na, Be, B, Si and Li," *Phil. Mag.* **6**: 365, 1961.

22. R. S. Crisp, "Soft X-Ray Emission from Potassium Metal in the 40 to 1000 Å Range," *Phil. Mag.* **5**: 1161, 1960.

23. L. Pauling, *The Nature of the Chemical Bond*, 3rd ed., Cornell University Press, Ithaca, N.Y., 1960.

24. E. Schnell, "Zur Röntgenfluoreszenzanalyse, 2. Mitt.: Die Relativen Intensitäten der K_β Strahlung Einiger Elemente Der 2 Periode," *Monatsh. Chem.* **94**: 703, 1963.

25. K. Siegbahn, "Electron Spectroscopy for Chemical Analysis," Contract Report No. 2(AF61-052-795), Institute of Physics, Uppsala University, September 1964.

DISCUSSION

N. Spielberg (Philips Laboratories): I wonder if you could give us some information on your operating conditions, X-ray tube vacuum, and voltage?

D. W. Fischer: The vacuum most of the time is somewhere between 10^{-5} and 10^{-6} torr. For almost all of the spectra I showed, our operating conditions are about 5 kV, and somewhere around 1 to 2 mA.

J. Merritt (Shell Development Company): What is the significance of your labels 3s, 3p, and 3d?

D. W. Fischer: Supposedly those are the electron orbits from which these transitions arise.

N. Spielberg: Would you expect to see an effect of voltage on the line shapes of some of these bands?

D. W. Fischer: From what we have done, we have looked at a lot of these from 5 to 10 kV, and in this range we haven't really seen any effect due to voltage.

C. F. Hendee (Bendix Corporation): Is there any correspondence between the band shifts in emission and the electron energy shifts due to composition changes, as has been recently reported out of Siegbahn's laboratory?

D. W. Fischer: You mean electron spectra?

C. F. Hendee: Yes. Have you found any correspondence there at all?

D. W. Fischer: We haven't been able to find any correspondence. We have been very interested, of course, in that work, but we haven't really been able to correlate what he gets with the electron spectra with what we get with the X-ray spectra.

L AND *M* X-RAY SPECTRA IN THE REGION 2–85 Å

Franklin D. Davidson and Ralph W. G. Wyckoff

University of Arizona
Tucson, Arizona

ABSTRACT

A detailed study has been made of the intensities of the *L*, *M*, and *N* X-ray spectra for wavelengths in the region from 2–85 Å. Some of the lines of these spectra involve energy levels which are gradually being depopulated with decrease in atomic number, and this leads to a progressive simplification of the spectra as the wavelength increases. Spectra have also been obtained from several compounds involving these elements in different valence states. Differences in relative intensities have been observed for these compounds, but because of a lack of accurate information about absorption, it is impossible as yet to give such results an adequate interpretation.

INTRODUCTION

As we all know, much of the interest now being shown in the long-wavelength X-ray region is due not only to the desire to extend spectrographic analysis to the lightest elements, but also to the realization that such X-rays can often be made to give useful information about chemical bonding. The state of chemical combination of an element influences its X-ray spectrum both by shifting the wavelengths of emitted spectral lines and by altering the fine structure of the characteristic absorption edges. These effects have long been known and a generation ago were repeatedly studied for the harder radiations. They are, however, so small that very precise measurements are needed to record them; this fact, combined with numerous technical difficulties, discouraged for many years an extension to the longer X-rays where valency differences should be most conspicuous. The mere existence of this meeting demonstrates how our steadily improving techniques for the study of soft X-rays are changing this situation.

Characteristic X-rays arise through the transition of the external electrons of an atom to fill vacancies created by the removal of inner electrons through some process of ionization. The intensity of a line depends, among other things, on the number of electrons of a given energy available to make such a transition, and, as a consequence, we need to know spectral intensities, as well as wavelengths, in order to learn as much as possible about the electron population of an atom from the study of its X-ray spectra. From a review of the periodic table from atom to lighter atom, it is evident that a point will be reached where certain lines must weaken and then disappear through a progressive depopulation of the energy levels involved in producing them. The electron distributions in an atom in different valence states differ in somewhat the same way as do those in atoms of neighboring atomic numbers, and, in favorable cases, these should be reflected in different spectral intensities. An examination of the existing soft X-ray data, however, indicates that they are too fragmentary to throw much light on this point.

Figure 1. Vacuum spectrometer. A signifies the chamber.

As a matter of fact, though wavelengths were accurately determined a generation ago for important lines in the K, L, and M spectra of most elements, the data on many L and M spectra are still incomplete, and accurate intensity observations are restricted to K spectra and the L lines of a few of the heavier elements. As part of our program of research, we are seeking to remedy these deficiencies, giving special emphasis to measurements of the intensities in the longer-wavelength region. Consequently, we have had under way an intensity survey of the L and M spectra of as many elements as possible. The present paper records some of our preliminary results covering the region from approximately 2–85 Å. These results obviously must be followed with more accurate measurements, but they already indicate how these spectra, which are comparatively rich in lines for the heavier elements, become simpler with decrease in atomic number, and, in a few instances, they show how line intensities in the spectrum of a single element depend on the emitting compound. They also emphasize the factors that must be taken into account and the additional information that is needed when seeking to relate an X-ray spectrum to the chemical state of the atom producing it.

METHODS

The vacuum spectrometer used thus far in this survey is shown in Figure 1. It consists of a slightly modified quadrant spectrometer of the type employed in the Philips electron probe, together with a gas-type X-ray tube and secondary sample chamber arranged for windowless operation.[1] The entire outfit is maintained at a mechanical pump vacuum of $\sim 20\ \mu$. Many of the elements in a survey of this sort can be examined only as compounds; this type of quasi-fluorescent excitation is very efficient

[1] References are at the end of the paper.

and has the great advantage that it does not result in the massive chemical de-composition and consequent instability in tube operation consequent on making a non-conductor the anode of an X-ray tube. For the measurements recorded here, the tube anode was of copper and the power consumption was of the order of 150 W, but as our previous work has shown, the efficiency is not greatly altered by change in anode. The radiating samples were sheets or compressed pellets inserted in the chamber (A) to have a position $\sim 1\frac{1}{2}$ in. from the anode. The gratings were either a crystal of mica (for the region ~ 2–13 Å) or a pseudo-grating consisting of a stack of 120 monolayers of lead stearate on leucite (for the region ~ 13–70 Å). Their efficiency in the long-wavelength region is much higher than that of mica and, hence, the data in units of counts/sec-W from the two gratings are not directly comparable. These gratings have been used both flat and bent. The secondary source is too large to be ideally suited to curved-crystal geometry, but trial has shown that it does not result in a notable loss in resolution, but very significantly increases the spectral intensities. The data recorded here have been obtained with a bent crystal or pseudo-grating; they have not been corrected for geometrical factors. The counter was of our own construction, but of conventional design equipped with a stretched polypropylene window of such a thickness that it passed $\sim 80\%$ of the carbon radiation (45 Å) incident upon it. For the shorter-wavelength region (up to ~ 8 Å), the counter gas used was P–10; for longer wavelengths, cleaner and more intense spectra have been obtained with a 50% helium–50% methane mixture that operates ~ 300 V higher than P–10. The signal output of the counter was passed through a Tennelec 100B preamplifier into a conventional ratemeter circuit and recorded in chart form. A few fixed-time measurements were made of intensities, and they are, of course, required for greatest accuracy; however, they are too time-consuming to be justifiable for the survey described here. Intensities were determined as counts per second from height measurements above background on appropriately calibrated charts; they have been expressed either as counts per second per watt of input to the tube (counts/sec-W) for comparisons of the spectra from different elements or as relative intensities (I) referred to a chosen spectral line.

RESULTS

Intensities measured in this fashion[2] for a variety of L and M spectra are collected in Tables I–IV, the line identifications and responsible electron transitions being for the most part those given in the compilation of Blokhin.[3] They bring out clearly the extreme simplicity of the L and M spectra of the lighter elements and the way this simplification proceeds with decrease in atomic number.

Some of the factors that must be taken into consideration when seeking to interpret the intensities shown in these tables are instrumental; others are imposed by the sample which is the source of the radiation. Aside from those that are purely geometric, the chief factors imposed by the instrument are absorption in: (1) the counter gas, (2) the counter window, and (3) the gratings. Since the counter is small enough so that absorption may be incomplete, this absorption and the efficiency of the counter will be drastically different for X-rays of wavelengths on either side of the critical absorption limit of the counter gas. This effect is very apparent when we compare with one another the intensities of the L_α lines from indium and cadmium, or the β_1 lines from silver and cadmium (Table I); their wavelengths lie on either side of the argon K absorption limit at 3.86 Å. The ab-normally high apparent intensity of the M_β line of uranium (Table III) has the same explanation. Correction can, of course, be made using the known or calculated absorption coefficients of argon for these X-rays, though, because certain other essential corrections

Table I. L Spectra Measured with Mica Crystal*

Element	N	$3d \to 2p$ $\alpha_{1,2}(M_{IV,V}-L_{III})$ λ	counts/ sec-W	$\beta_1(M_{IV}-L_{III})$ λ	I	$3p \to 2s$ $\beta_3(M_{III}-L_I)$ λ	I	$3s \to 2p$ $l(M_I-L_{III})$ λ	I	$\eta(M_I-L_{II})$ λ	I	$3p \to 2s$ $\beta_4(M_{III}-L_I)$ λ	I	$4p \to 2s$ $\gamma_3(N_{III}-L_I)$ λ	I	$4d \to 2p$ $\gamma_1(N_{IV}-L_{II})$ λ	I	$\beta_2(N_V-L_{III})$ λ	I
Group A:																			
Cu	29	13.33	3.20	13.05	18	12.07	0	15.30	—	14.87	—	—	—	—	—	—	—	—	—
Zn	30	12.26	9.20	11.99	18	11.16	0	14.05	—	13.69	—	—	—	—	—	—	—	—	—
Ge	32	10.44	12.5	10.17	40	9.63	1.1	11.92	4.0	11.59	2.1	—	—	—	—	—	—	—	—
As	33	9.65	7.4	9.39	51	8.91	1.3	11.05	5.1	10.71	2.2	—	—	—	—	—	—	—	—
Se	34	8.97	5.6	8.72	48	8.30	1.8	10.27	4.9	9.94	2.2	—	—	—	—	—	—	—	—
Group B:																			
Br(NaBr)	35	8.36	11.1	8.11	40	7.75	2.4	9.56	3.0	9.23	2.1	—	—	7.00	0.3	—	—	—	—
Rb(RbCl)	37	7.31	9.5	7.06	59	6.77	8.6	8.35	2.9	8.02	3.8	—	—	—	0	—	—	—	—
Sr(SrCO₃)	38	6.85	15.6	6.61	24	6.35	2.6	7.82	1.9	7.50	1.5	—	—	5.63	0.5	—	—	—	—
Y[Y(NO₃)₃]	39	6.44	3.1	6.20	50	5.97	6.9	7.34	5.9	7.03	4.2	—	—	5.27	1.3	—	—	—	—
Zr	40	6.06	14.3	5.82	44	5.62	7.1	6.90	6.3	6.59	1.6	5.04	nr	4.94	0.7	5.37	1.1	5.57	0
Mo	42	5.40	19.8	5.17	40	5.00	8.6	6.14	2.9	5.83	1.6	4.51	nr	4.37	0.8	4.72	1.5	4.91	3.5
Ru	44	4.84	11.7	4.61	41	4.48	9.2	5.49	4.4	5.19	1.5	4.28	nr	3.89	<0.1	4.17	1.3	4.36	4.8
Rh	45	4.59	7.2	4.36	41	4.24	7.2	5.21	4.4	4.91	1.9	—	—	3.68	1.9	3.93	1.1	4.12	5.0
Ag	47	4.15	9.8	3.93	42	3.82	11	4.70	—	4.41	1.1	3.86	5.4	3.30	2.8	3.51	9.1	3.69	21
Cd	48	3.95	7.6	3.73	78	3.64	11	4.47	3.1	4.18	1.7	3.67	7.8	3.13	3.4	3.33	8.1	3.51	26
In	49	3.77	16.0	3.55	51	3.46	8.6	4.26	1.3	3.97	0.8	3.50	4.9	2.97	2.0	3.15	5.9	3.33	14
Sn	50	3.60	16.3	3.38	37	3.30	6.2	4.06	1.8	3.78	1.4	3.34	4.3	2.83	1.9	2.99	5.2	3.17	13
Sb	51	3.44	12.1	3.22	34	3.15	7.1	3.88	1.4	3.60	1.6	3.18	4.3	2.69	2.4	2.84	4.3	3.02	13

* Measurements for Group A were made with He—CH₄ counter gas, and those for Group B with P-10. In this and the following tables, "nr" signifies an unresolved line. Intensities I are relative to those of α_1, α_2 taken at 100. Wavelength λ is given in angstroms.

Table II. L Spectra Measured with Lead Stearate Pseudo-Grating*

| Atom | N | $3d \to 2p$ | | | | $3s \to 2p$ | | | | |
| | | $\alpha_{1,2}(M_{IV,V}\text{–}L_{III})$ | | $\beta_1(M_{IV}\text{–}L_{II})$ | | $l(M_I\text{–}L_{III})$ | | | $\eta(M_I\text{–}L_{II})$ | |
		λ	counts/sec-W	λ	I	λ	I	counts/sec-W	λ	I
Cl(NaCl)	17	—	0	—	0	67.84	—	1.10	67.25	nr
K(KF)	19	—	0	—	0	47.74	—	4.30	47.23	nr
Ti	22	27.39	0.94	27.02	nr	31.36	36	0.34	30.88	nr
V	23	24.26	1.95	23.85	nr	27.77	36	0.70	27.32	nr
Cr	24	21.67	5.55	21.28	nr	24.79	18	1.00	24.29	nr
Mn	25	19.45	(10)	19.12	17	22.27	17	1.70	21.28	nr
Fe	26	17.57	(17)	17.26	15	20.15	14	2.38	19.73	3.5
Co	27	15.97	(47)	15.66	14	18.30	13	6.11	18.0	3.5
Ni	28	14.57	(54)	14.28	21	16.71	13	7.00	16.28	1.4
Cu	29	13.33	(76)	13.05	20	15.30	7.0	5.32	14.87	2.5
Zn	30	12.26	(157)	11.99	20	14.05	5.3	8.33	13.69	2.4

* The intensities I refer to $\alpha_{1,2}$ as 100.

cannot now be made, we have preferred to present here the crude data. Absorption in the counter window is important for X-rays having wavelengths on either side of the critical limit for carbon (44 Å). Thus, the strong absorption of X-rays immediately shorter than this accounts for the feebleness of the measured intensities for nitrogen K and calcium L spectra and for the M spectra of the elements heavier than rhodium (Table IV). It also explains the apparent increase in intensity of the l,η pair of L lines from potassium compared with titanium (Table II). With an accuracy set by our still approximate knowledge of the absorption coefficients of carbon, correction can also be made for this effect. It could have been reduced by the use of thinner and much more fragile counter windows of collodion instead of polypropylene. The effects of absorption in the crystal, if one is used as grating, are harder to deal with, especially when the radiations lie close to the absorption limits of the elements it contains. They are particularly troublesome when it is, as mica, a natural product the chemical composition of which may be somewhat indefinite and may vary from specimen to specimen; this represents an important argument for preferring an organic crystal, such as EDDT or pentaerythritol, for measurements of the highest accuracy.

Absorption of excited radiation is more complicated in the emitting specimen, especially when it is a compound, and we will not be able to correct for it adequately until many more data are available concerning absorptions in the L and M regions. In contrast to the K spectrum with its single absorption edge, the L region has three and the M five such edges. In the case of many elements, even the positions of these edges are not known with precision, and very little of a quantitative nature is yet known of the changes in absorption that occur across each of these limits. Their role in determining the self-absorption in an emitting sample is strikingly illustrated by the low ratio of the β_1/α lines for copper and zinc compared with those for arsenic and germanium (Table I). This is due to the fact that, for the first two, but not for the others, the L_{III} absorption limit lies between these two lines and thus reduces by $\sim 50\%$ the emerging intensity of the shorter wavelength β_1.

Table III. M Spectra Measured with Mica Crystal*

Element	N	4f → 3d $\alpha_{1,2}(N_{VI,VII}\text{–}M_V)$ λ	counts/sec-W	4f → 3d $\beta(N_{VI,VII}\text{–}M_{IV})$ λ	I	4d → 3p $\gamma(N_V\text{–}M_{III})$ λ	I	4d → 3p $(N_{IV}\text{–}M_{II})$ λ	I	4p → 3d $\zeta_{1,2}(N_{II,III}\text{–}M_V)$ λ	I	4p → 3d $\delta(N_{III}\text{–}M_{IV})$ λ	I	4s → 3p $(N_I\text{–}M_{III})$ λ	I
Gd$_2$O$_3$	64	10.39	0.70	10.23	153	8.83	23	—	—	13.54	34	—	—	—	—
Tb$_2$O$_3$	65	9.92	1.00	9.77	153	8.47	19	—	—	12.95	31	—	—	—	—
Dy$_2$O$_3$	66	9.52	1.22	9.34	125	8.13	12	—	—	12.40	26	—	—	—	—
Ho$_2$O$_3$	67	9.14	1.60	8.95	117	7.85	9	—	—	11.84	20	—	—	—	—
Er$_2$O$_3$	68	8.78	2.12	8.58	86	7.53	6	—	—	11.35	21	—	—	—	—
Yb$_2$O$_3$	70	8.12	6.38	7.89	50	7.01	7	—	—	10.46	10	—	—	—	0.8
Ta	73	7.24	22.5	7.01	40	6.30	2.0	5.56	0.3	9.30	5.0	8.56	0.5	7.60	—
W	74	6.98	17.9	6.74	52	6.08	1.5	5.34	0.8	8.94	3.8	7.36	2.4	7.35	1.0
Pt	78	6.04	18.5	5.82	63	5.31	3.2	4.59	0.6	7.72	7.5	7.09	3.2	6.44	1.1
Au	79	5.84	18.7	5.61	61	5.13	2.9	4.42	0.8	7.45	5.3	6.87	1.4	6.24	1.4
HgO	80	5.62	7.3	5.42	63	4.98	2.8	4.25	0.6	7.25	5.8	6.58	0.5	6.09	1.5
Tl	81	5.46	20.0	5.24	60	4.81	2.5	4.11	0.7	6.96	7.5	6.37	0.4	5.87	1.5
Pb	82	5.29	11.6	5.06	62	4.66	3.9	3.96	0.7	6.73	8.6	6.15	0.9	5.69	10.9
Bi	83	5.12	13.0	4.90	57	4.52	3.1	3.83	0.7	6.51	4.2	—	—	5.53	1.1
Th	90	4.14	6.6	3.93	58	3.67	9.8	3.01	2.3	5.23	4.8	4.90	2.3	4.55	1.5
U	92	3.91	3.4	3.71	138	3.47	21	2.81	3.0	4.94	7.4	4.61	—	4.32	—

* Counter gas was P-10. Intensities I refer to I_α as 100.

Table IV. M Spectra Measured with a Stearate Pseudo-Grating*

Atom	N	$\zeta_{1,2}(N_{II,III}-M_V)$ λ	counts/ sec-W	$\alpha+\beta$ λ	I	N_I-M_{III} λ	I	? λ	I
Mo	42	64.36	7.25	—	—	—	—	55.6	10
Ru	44	52.34	4.31	—	—	—	—	44.8	26
Rh	45	47.67	3.81	—	—	—	—	—	—
Ag	47	39.71	—	—	—	—	—	—	—
Cd	48	36.75	0.15	—	—	—	—	30.9	108
In	49	33.6	0.41	—	—	—	—	26.4	64
Sn	50	31.23	0.17	—	—	—	—	25.2	64
Sb	51	28.76	0.23	—	—	—	—	—	—
Te	52	26.9	0.38	—	—	—	—	21.6	67
I(KI)	53	25.3	0.45	—	—	—	—	20.0	100
Cs(CsCl)	55	22.1	0.98	—	—	16.9	62	—	—
Ba(BaF$_2$)	56	20.59	1.63	—	—	15.9	69	—	—
La(La$_2$O$_3$)	57	19.3	1.88	—	0	14.6	140	—	—
Ce(CeO$_2$)	58	18.3	1.31	13.9	116	—	—	—	—
Pr(Pr$_2$O$_3$)	59	17.4	2.56	13.2	146	—	—	—	—
Nd(Nd$_2$O$_3$)	60	16.5	3.75	12.5	147	—	—	—	—
Sm(Sm$_2$O$_3$)	62	14.9	2.25	11.3	167	—	—	—	—
Gd(Gd$_2$O$_3$)	64	13.54	4.00	10.3	169	—	—	—	—

* I refers to the intensity of $\zeta_{1,2}$ as 100.

Table V. Electron Distribution and L-Series Line Intensities

Atom	N	Electron distribution $M_{II,III}(3p)$	$M_{IV,V}(3d)$	$N_I(4s)$	Intensities (counts/sec-W) $\alpha_{1,2}(M_{IV,V}-L_{III})$	$\beta_1(M_{IV}-L_{II})$
K	19	6	0	1	0	0
Ti	22	6	2	2	0.94	nr
V	23	6	3	2	1.95	nr
Cr	24	6	5	1	5.55	nr
Mn	25	6	5	2	10	2
Fe	26	6	6	2	17	3
Co	27	6	7	2	47	7
Ni	28	6	8	2	54	11
Cu	29	6	10	1	76	15
Zn	30	6	10	2	157	31

The L and M spectra we have been discussing contain several lines whose intensities for a series of elements should vary with the number of electrons in the responsible energy level. One example is furnished by the $\alpha_{1,2}$ and β_1 L lines of the elements from about potassium ($N = 19$) to about zinc ($N = 30$). Both these lines are due to a transition of a $3d$ electron to a vacant $2p$ state and, as can be seen from Table V, it is in this region that the $3d$ shell is being filled. As would be expected, no α and β lines are observed for potassium and lighter elements, and their intensities steadily increase as the $3d$ level becomes filled. A rapidly mounting absorption of the longer wavelengths in the counter window is responsible for some of the loss in intensity with atomic number, but it is not enough to obscure the underlying dependence of intensity on the number of $3d$ electrons.

Table VI. Electron Distribution and L-Series Line Intensities

Atom	N	Electron distribution			Intensities (counts/sec-W)		
		$N_{II,III}(4p)$	$N_{IV,V}(4d)$	$O_I(5s)$	$\gamma_3(N_{II,III}-L_I)$	$\gamma_1(N_{IV}-L_{II})$	$\beta_2(N_V-L_{III})$
Zn	30	0	0	0	0	0	0
Ge	32	2	0	0	0	0	0
As	33	3	0	0	0	0	0
Se	34	4	0	0	0	0	0
Br(NaBr)	35	5	0	0	0.03	0	0
Sr(SrCO$_3$)	38	6	0	2	0.08	0	0
Y[Y(NO$_3$)$_3$]	39	6	1	2	0.04	0	0
Zr	40	6	2	2	0.10	0.16	0
Mo	42	6	5	1	0.16	0.30	0.69
Ru	44	6	7	1	<0.01	0.15	0.56
Rh	45	6	8	1	0.14	0.08	0.36
Ag	47	6	10	1	0.27	0.89	2.06
Cd	48	6	10	2	0.26	0.61	1.98
In	49	6	10	2	0.32	0.94	2.24

Table VII. Electron Distribution and M-Series Line Intensities

Atom	N	Electron distribution	Intensities (counts/sec-W)	
		$N_{VI,VII}(4f)$	$\alpha_{1,2}(N_{VI,VII}-M_V)$	$\beta(N_{VI,VII}-M_{IV})$
Ba(BaF$_2$)	56	0	—	?
La(La$_2$O$_3$)	57	0	—	?
Ce(CeO$_2$)	58	1	—	?*
Pr(Pr$_2$O$_3$)	59	2	—	0.59
Nd(Nd$_2$O$_3$)	60	3	—	0.87
Sm	62	5	—	0.59
Gd	64	7	0.70	1.07
Tb	65	8	1.00	1.53
Dy	66	9	1.22	1.52
Ho	67	10	1.60	1.87
Er	68	11	2.12	1.82
Yb(Yb$_2$O$_3$)	70	13	6.38	3.18
Ta(Ta$_2$O$_5$)	73	14	13.3	5.45
W(WO$_3$)	74	14	14.0	8.96

* For cerium and lighter elements, the β line becomes confused in position with the N_I-M_{III} line.

Another example of the same dependence, seen in the variation in the intensities of the γ_1 and β_2 L lines, due to the transition $4d \rightarrow 2p$ of the elements from about strontium ($N = 38$) to about cadmium ($N = 48$). As Table VI indicates, the intensities of these lines are less and the parallelism with the electron number is less obvious. This is partly because these weak intensities have been less accurately measured and partly because of the different sensitivity of the counter on either side of the argon K absorption limit. Nevertheless, it is seen that both lines are much more intense when the $4d$ shell has been filled.

The M spectra of the rare earths offer a favorable group of elements for study, because the $4f$ level involved in the production of their α and β lines becomes filled in

Table VIII. *L* **Spectral Intensities from
Various Compounds of an Element**

Compound	α	β_1	l
Co	100	14	13
CoO	100	28	13
Ni	100	21	13
NiO	100	31	8
Cu	100	20	7.0
Cu_2O	100	23	5.0
CuO	100	30	6.0
Zn	100	18	—
ZnO	100	28	—
$ZnCO_3$	100	35	—
Ge	100	40	4.0
GeO_2	100	49	3.2
As	100	51	5.1
As_2O_5	100	51	3.5

passing from cerium ($N = 58$) to ytterbium ($N = 70$). As can be seen from Table VII, an unexpected feature of these spectra is the fact that though both α and β are due to $4f \rightarrow 3d$ transitions, their relative intensities are not constant; instead, as the number of $4f$ electrons is reduced, β becomes more intense than α. In doing this, β behaves as do the γ and ζ lines, which have their origins in completed $4d$ and $4p$ levels; their increases in relative intensities (referred to α) are expressions of the decline in intensity of α.

CONCLUSIONS

In the course of our measurements, we have already accumulated a certain amount of data on the relative intensities of lines of an element in several of its compounds. The interpretation of such data remains uncertain because of inadequacies in our knowledge of absorption, but we are including in Table VIII a few observations which will give an idea of the variations that occur. The α and β but not the l lines are due to electrons that could be involved in valency changes of the elements of this table. There are differences in intensity that are not adequately explained by self-absorption in the sample, but we do not yet consider ourselves in a position to interpret them in detail.

Particularly in the case of the M spectra, the results given here are to be considered as only preliminary. In spite of previous photographic work, the spectra of many elements have been so incompletely measured that more work must be done before many lines observed for the lighter elements can even be identified with certainty. The last two columns of Table IV, one tentatively identified as N_I–M_{III} and the other designated by a question mark, are examples. Our data are gradually reducing these gaps and giving increasingly accurate information about those spectral lines in the L as well as in the M spectra, which, similar to those discussed above, arise from incompletely filled shells or ones that contain chemically active electrons.

ACKNOWLEDGMENT

This work was carried out under Grant No. NsG–120 from the National Aeronautics and Space Administration.

REFERENCES

1. R. W. G. Wyckoff and F. D. Davidson, "Windlowless Tubes for X-Ray Spectroscopy," *Rev. Sci. Instr.* **34**: 572, 1963; "Windlowless X-Ray Tube Spectrometer for Light-Element Analysis," *Rev. Sci. Instr.* **35**: 381, 1964.
2. R. W. G. Wyckoff and F. D. Davidson, *J. Appl. Phys.* **36**: 1883, 1965.
3. M. A. Blokhin, *Physik der Röntgenstrahlen*, VEB Verlag Technik, Berlin, 1957.

DISCUSSION

J. Holliday (U.S. Steel Research Laboratories): I was wondering if you have made any corrections for your analyzer?

F. D. Davidson: No, we haven't. There are just the crude data. This is the preliminary survey, and we are going to go back in the regions that we were interested in and study those in more detail and make corrections then.

Chairman W. L. Baun: Historically, I think we all recognize the difficulty in maintaining stability in a gas tube. How do you do it?

F. D. Davidson: It hasn't been too much of a problem. We use an air in-leak with a valve for control and a buffer tank usually for the best stability. We pump it down with a forepump and then we leak air into our tube. It's held pretty good. We've made a few checks, such as looking at an oxygen peak, check for carbon, and let it run for two or three hours at a time, and it hasn't built up any contamination. The oxygen peak hasn't dropped and the carbon hasn't built up on it. It's held steady current for that time.

Chairman W. L. Baun: What kind of total power do you run?

F. D. Davidson: Most of these were from 80 to 150 W. We found with the primary target 8–10 kV and about 10 mA were sufficient.

H. W. Pickett (General Electric X-Ray): More on the gas tube. Do you make any effort to control what the gas is in there or is it just residual air at 20 μ?

F. D. Davidson: It's just residual air, and we do leak air from a little buffer tank that's been pumped down.

D. W. Fischer (Wright-Patterson Air Force Base): In some of the L spectra lines that you showed, especially for the transition metals, we find a very definite dependence of the excitation potential on the relative intensities of these lines. Have you looked at such an effect?

F. D. Davidson: No, we haven't yet. That's one thing we want to do. We want to make a more thorough study of those in particular. Most of these were from the oxides, and we've got a lot of the metals now, and we want to compare the two.

H. K. Herglotz: Regarding the gas-filled tube, what problem is the cathode sputtering in your case?

F. D. Davidson: Apparently we haven't noticed any effect.

H. K. Herglotz: Is it an aluminium cathode?

F. D. Davidson: Yes. We used the copper anode on all of these, and the copper looks very clean after it has been run for a long time.

H. K. Herglotz: Aluminium is the one that sputters least, but still it is noticeable.

CHEMICAL BONDING AND THE SULFUR
K X-RAY SPECTRUM

D. W. Wilbur and J. W. Gofman

Lawrence Radiation Laboratory
University of California
Livermore, California

ABSTRACT

An investigation has been made of the relative K_β intensities in different chemical states of the sulfur atom using the K_α lines, with appropriate corrections, to provide the intensity standards. Both inorganic and organic compounds were included in the study. The data for each compound appear to be reliable to about $\pm 0.5\%$, while the whole series of compounds shows a variation greater than 20% in the corrected K_β/K_α ratios. Energies were also measured, particularly the K_α energies, and their shifts were studied relative to the K_β intensity shifts. The work was done with a plane, single-crystal, helium-path spectrometer with proportional counter and pulse-height analysis for detection. The results are indicative of the usefulness of the method both in clarifying an uncertain chemical state and in studying the electronic structure of the bonded atom.

INTRODUCTION

In 1924 Lindh and Lundquist[1] first showed that the sulfur K X-ray spectrum could be affected by bonding. Since then, many studies have been made showing chemical effects on both the sulfur K_α and K_β lines. Faessler[2] has collected (with references) the most reliable of these measurements up to the end of 1952. The main changes studied have been shifts in K_α and K_β energies and changes in K_β structure. Of particular interest are the high-resolution studies made by Faessler and co-workers of both the K_α and K_β spectra of sulfur.[3,4] More recently, Shuvaev[5] has presented theoretical arguments and experimental evidence (from six sulfur compounds) for a change in intensity of K_β lines with chemical bonding when these lines are caused by transitions of valence electrons. Schnell[6] studied the effect of bonding on the K_β intensities for the third-period elements aluminum through chlorine, and he showed that readily measurable changes occur in all of them. Neither of these authors gave K_α energies with the intensities from their compounds, and the intensity uncertainties were at least $\pm 5\%$ for Shuvaev and probably ± 1-2% for Schnell. Of the twenty compounds that these authors studied, only two were organic.

The object of the work here reported was a more thorough study of K_β intensities in conjunction with K_α energy studies for the same compounds, thus allowing comparison of the two ways of studying chemical states. The sulfur atom was chosen because it was known to show significant changes in these parameters and because of its importance in biology. As in the previous studies, the K_α lines were used as internal intensity standards for the K_β lines.

---

[1] References are at the end of the paper.

THEORY

Shuvaev's[5] treatment of intensities gives for third-period elements a linear relationship between K_β intensity and density of the valence p-character electrons in the region of the $1s$ shell of the atom. This linear relationship was obtained by using the one-electron approximation, by assuming that the shape of the valence-electron wave function in the interior of the atom was independent of chemical bonding, and by assuming that the transition moment integral was significant only in the principal region of $1s$ shell density, where the previous approximation was valid. Calculations showing the effects of these approximations were not included, and it would seem that the latter one particularly should have been investigated. Accepting his relationship, we still have two problems in relating intensity to the net charge on an atom such as sulfur, or in relating $3p$ electron density in the interior of the atom to the charge on the atom. The $3s$ electrons of sulfur may be involved in bonding, and unbonded valence electrons may have their wave functions distorted by neighbor atoms, possibly changing their density in the interior of the atom.[7] Thus, K_β intensity and valence-electron density in the interior of the atom may not always show a linear dependence on the total number of valence electrons, but we can hope the relationship will not be seriously distorted.

Coulson and Zauli[8] and Bendazzoli et al.[9] have made calculations for the sulfur atom to find a relationship between the K_α energy and the number of valence electrons. They used Slater orbitals and considered an isolated atom with 0–8 valence electrons. The results show a small dependence on the steric arrangement of ligands, but, when averaged, the relationship of charge to K_α energy was found to be approximately as follows:

$$\Delta E = \frac{n(n+3)}{8}$$

where ΔE is the K_α energy shift (eV) from neutral sulfur and n is the net charge in units of the electron's charge. Shuvaev[10] has also developed a theory which, to a certain approximation, gives $\Delta E = kn$ for third-period elements, where all valence electrons are involved in bonds. For sulfur, he reports $k = 1.32$ as the value in best agreement with dipole-moment data. Both of these relationships might be seriously affected by distortions of the wave functions of unbonded valence electrons by neighbors, and this is why Shuvaev specifies that his is valid only when all valence electrons are in bonds. Shuvaev used a single shielding constant k for both $3s$ and $3p$ electrons, although his own calculations indicate that these may differ by as much as 14%. He has assumed constant shape for the valence-electron orbitals in the interior of the atom. Lastly, it should be mentioned that Blokhin et al.[7] have compiled a table of approximately additive contributions of different bonding partners to the sulfur K_α shifts and have used the table in studying some compounds of uncertain chemical structure.

EXPERIMENTAL

The spectrometer used was a General Electric XRD-6 with helium path attachment, quartz crystal ($2d = 6.686$ Å), gas-flow proportional counter using a $\frac{1}{4}$-mil Mylar window and a 90% argon and 10% methane gas mixture, EA-75 chromium target tube, and the standard G.E. counting console and 75-kV power supply. Extra collimation and background reduction were provided by replacing the standard sample-side collimator of 0.08-in. blade spacing and $1\frac{5}{8}$-in. length with one of 0.01-in. blade spacing which was constructed in our shop. The counter-side collimator was 3.5 in. long and had a blade

spacing of 0.005 in. The X-ray tube was run at 50-kV constant potential and 30 or 38 mA.*
To supplement the scanning system provided with the unit, an incremental stepping
system (Humphrey Electronics, Morrisville, North Carolina) was used for semiautoma-
tion of the point-by-point data collection. Pulse-height analysis was used.

Samples were of the highest commercial grade available (usually reagent grade)
and were used as fine powders for sample preparation. Each sample was made by taking
this powder in an amount several times that calculated as necessary for effective infinite
sample thickness and pressing it into a $1\frac{1}{4}$-in. ID steel ring using a sample preparation
kit (General Electric). Pressing force ranged as high as six tons, but depended upon the
sample material.

Data for K_α intensities were taken over a 2θ range of 105.50° to 108.00°, either by
scanning and summing the counts during the scan, or, more often, by stepping between
these angles with intervals of 0.02–0.10° 2θ chosen according to the intensities of different
regions, with a total of about 80 points per run. This range included the high-energy K_α
satellites. For K_β intensities, the range was 96.40–100.00° 2θ with the data taken in the
same manner. Both stepping and scanning were done from low-angle to high-angle. The K_β
integration was extended on the high-angle side, because the peaks all seemed to have a
high tail extending far to that side; in fact, it was still declining beyond 100° 2θ. In so far
as we could determine, this tail was a part of the K_β emission of sulfur. There also appeared
to be some sulfur emission in the 95–96° 2θ range, but it was very diffuse and quite low in
intensity and was considered negligible for this study. Stepping was usually the preferred
procedure for data-taking, because it gave information on the structure of the line as well
as on its energy and intensity. However, scanning was an easier way of obtaining intensity,
since it gave an integrated count; the stepping data had to be averaged at each point and
then integrated by a simple rectangular method. A second advantage of scanning was that
less time was required per run, and this was important for samples exhibiting deterioration.

For each compound, two or more independent determinations were made, with
enough data for each to give a calculated standard deviation close to 0.3%; for each
determination, this usually required the average from five or ten sample subgroups of
the following types:

1. Measurement of K_β at 40 sec/step, followed or preceded by measurement of
   K_α, usually on the same sample, at 10 sec/step.
2. Two measurements of K_β at 20 sec/step, separated by a measurement of K_α at
   10 sec/step, using a new sample for each, or, alternatively, one sample for a K_β and
   the K_α and a new sample for the other K_β.
3. Two scan measurements of K_β with one of K_α between them, using a new sample
   for each.

The choice of subgroups depended upon the sample's resistance to deterioration in the
beam. Subgroups were run together so that machine conditions would be about the same
for the whole subgroup. The only compounds showing definitely measurable deterioration
during the irradiation times used here were ethionine, cysteine, and methionine and its
derivatives, in all of which the sulfur emission intensities increased with time of irradiation.
Exploratory studies on methionine, which had the second-fastest deterioration, showed
that, during a 2-hr irradiation when the intensities changed by about 7%, the K_β/K_α
ratio changed less than 1%. The observations seem best explained by a breakup of the
molecule, possibly through decarboxylation, with preferential sublimation of the portions

---

* Some pure sulphur data were taken at 40-kV constant potential and 36 mA, and Na_2SO_4 was run
  at 50-kV full-wave rectified and 56 mA.

not containing the relatively heavy sulfur atom. To minimize the deterioration problem, the irradiation time was limited, usually to about $\frac{1}{2}$ hr or less; in addition, a new sample was used for each K_β and K_α run, in the hope that the deterioration would, as in methionine, similarly affect both intensities.

Several corrections were applied to the measured intensities. A background correction based on measurements at 115.00° 2θ for K_α and at 92.00°* 2θ for K_β was subtracted. The printout from the scalar did not give the units digit, so the intensities had to be corrected by effectively adding 4.5 to all numbers. A small coincidence correction was applied to the K_α lines by determining the approximate dead time of the system ($\Delta t = 1.25\ \mu$ sec) and adding $N_0^2\ (\Delta t/t)$ to N_0, where N_0 is the observed total count in time t. For pure sulfur, this was less than 0.4%, and for all other samples it was less than 0.2%. To eliminate the effects on intensity of different absorptions in different samples, a matrix correction was applied to the observed K_β/K_α ratio. This correction was made for monochromatic chromium K_α excitation; calculations, based partially on the data of Burke and Pettit,[11] showed this approximation to be adequate, especially if one is concerned mainly with comparisons within a series. Shuvaev[5] made no matrix corrections. Schnell used a chromium tube and chromium K_α corrections, but did not specify the conditions of excitation, which varied, so there is additional uncertainty in making comparisons. Also, Schnell used compounds with significant enhancement without correcting for it. The equations we used for the matrix effect were the same as those of Sherman,[12] except that he is lacking a factor of $\frac{1}{2}$ in his equation for enhancement involving a pair of elements and a factor of $\frac{1}{4}$ in his equation for an enhancement chain of three elements. One of us found the error of $\frac{1}{2}$, and we later found that Renaud[13] had pointed out both errors. The mass absorption coefficients except for copper and arsenic were taken from a compilation published as an insert in the *Norelco Reporter*.[14]

For each compound, the K_α position was determined as a shift from that of pure sulfur by running the following series of determinations: pure sulfur sample, two or three samples of compounds, a pure sulfur sample, two or three samples of compounds, *etc.* At least three independent determinations of the K_α shifts were made in this way for each compound. A run consisted of 22–24 points taken across the center of the K_α peak at 40 or 100 sec per point depending upon the intensity. The data for each run were graphed, and the average position was determined by reading both sides of the peak at its half-maximum and three-quarters-maximum. Shifts were then found by comparing the results from the determination on the compound with its preceding sulfur standard runs and succeeding sulfur standard runs.

RESULTS

The variations in K_β structure with bonding are very marked, but the K_α line showed no noticeable structure changes. The K_α full width at half-maximum was 0.320° ± 0.002° 2θ or about 4.77 ± 0.03 eV. Figure 1 shows a typical K_α spectrum over the range of integration with the high-energy satellites making the low hump on the left side, which is about the same height as the maximum K_β intensity. In contrast, the very marked changes in the K_β structure may be seen by comparison of the selected spectra of Figure 2. A prominent low-energy line appears in some compounds and not in others, while changes of several electron volts occur in the widths of the lines. Also, Na_2SO_3 showed a high-energy hump which was unique in this series of compounds. Since our

---

* Methylmethionine sulphonium chloride required a special estimation procedure for its K_β background, since the Cl K_α was at 90° 2θ.

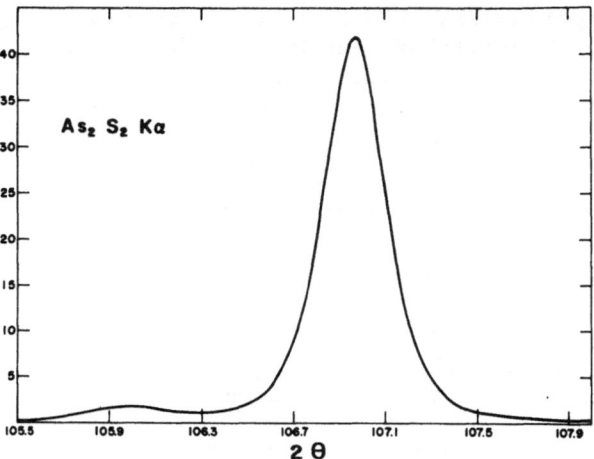

Figure 1. The sulfur K_α from As_2S_2.

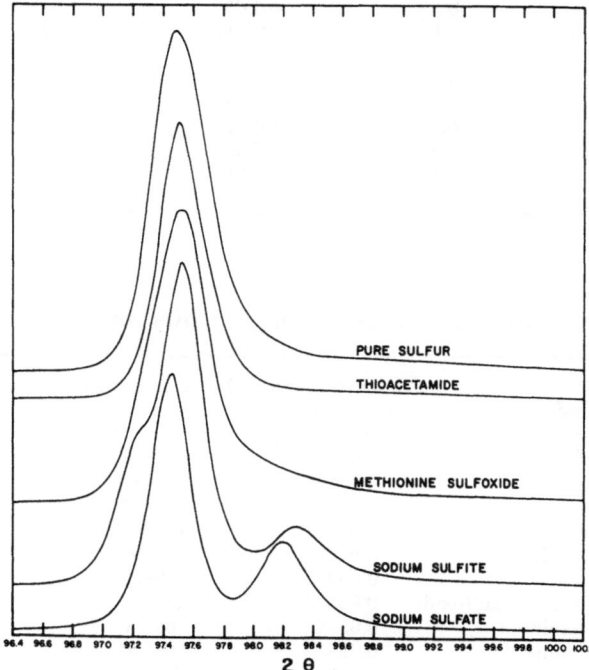

Figure 2. The sulfur K_β for pure sulfur and selected sulfur compounds.

spectrometer was a low-resolution instrument chosen with the main intention of doing integrated intensities, we could not resolve the components of these features as well as might be desired, but certainly Figure 2 shows that large changes were detected. The main K_β peak full width and the shift in position relative to pure sulfur all measured at half-maximum are given for each compound in Table I. Results from individual determinations indicate that the widths should be reliable to about ± 0.03 eV and the positions to about ± 0.02 eV.

Table I. Energy Shift and Width Data for the K_β Line and Energy Data for the K_α Line for Different Chemical States of Sulfur

Compound	K_β Width at $\frac{1}{2}$ max. (eV)	K_β Shift at $\frac{1}{2}$ max. relative to pure sulfur (eV)	K_α Shift relative to pure sulfur (eV)	K_α Experimental standard deviation (eV)
Pure sulfur	8.32	0	0	
Na$_2$SO$_4$	5.95	+1.440	+1.200	±0.001
Na$_2$S$_2$O$_8$	6.70	+1.341	+1.142	±0.003
Sulfamide	7.37	+0.607	+1.043	±0.012
Sulfanilamide	7.86	+0.346	+0.954	±0.002
Na$_2$SO$_3$	7.72	+1.077	+0.806	±0.003
Methionine sulfoxide	8.59	+0.817	+0.391	±0.006
Methylmethionine sulfonium chloride	8.08	+0.321	+0.078	±0.008
L-Cystine	8.11	+0.459	+0.005	±0.002
NaSCN	7.51	+0.569	−0.013	±0.006
L-Cysteine	7.68	+0.576	−0.022	±0.004
L-Methionine	8.52	+0.059	−0.025	±0.003
2-Mercaptoacetanilide carbamate	8.01	+0.511	−0.029	±0.003
CuS	8.32	−0.343	−0.043	±0.005
DL-Ethionine	8.23	−0.095	−0.049	±0.002
Thiourea	6.92	+0.360	−0.076	±0.002
Thioacetamide	6.57	+0.461	−0.080	±0.001
As$_2$S$_2$	7.15	+0.115	−0.119	±0.006

Table II. K_β/K_α Ratios for Sulfur in Different Chemical States

Compound	Measured ratio of net areas	Matrix correction factor	Final corrected ratios
Pure sulfur	0.05357	0.8970	1.000
Na$_2$SO$_4$	0.05618	0.8459	0.989
Na$_2$S$_2$O$_8$	0.05656	0.8487	0.999
Sulfamide	0.05638	0.8599	1.009
Sulfanilamide	0.05869	0.8605	1.051
Na$_2$SO$_3$	0.06068	0.8468	1.069
DL-Methionine sulfoxide	0.06425	0.8582	1.147
Thioacetamide	0.06342	0.8796	1.161
DL-Ethionine	0.06512	0.8620	1.168
L-Methionine	0.06557	0.8620	1.176
L-Cystine	0.06613	0.8620	1.186
Methylmethionine sulfonium chloride	0.06567	0.8738	1.192
Thiourea	0.06651	0.8748	1.211
CuS	0.06658	0.8744	1.212
NaSCN	0.06818	0.8575	1.217
2-Mercaptoacetanilide carbamate	0.06872	0.8606	1.231
L-Cysteine	0.06942	0.8620	1.245
As$_2$S$_2$	0.06856	0.8782	1.253

The K_α positions in electron volts relative to those of pure sulfur are also given in Table I. Experimental standard deviations for each are included with the realization that these measure only reproducibility by our measuring techniques. Except in the case of Na_2SO_3, where the discrepancy is 0.11 eV, our measurements on identical compounds agree adequately with those of Faessler and Goehring, who stated that their uncertainty should be less than 0.1 X or 0.04 eV. In all cases, the conversion to electron volts was made using the relation $E = 12372.42/\lambda$ keV, with λ in ux[15] and $2d = 6673.065$ xu (the average of the measurements of Bearden et al.[16] on two quartz crystals).

The K_β/K_α ratios are given in Table II, which gives the measured ratios, the matrix correction factor, and a set of final corrected ratios relative to pure sulfur at 1.000, obtained by dividing each ratio by the corrected sulfur ratio. The estimated uncertainty for these is about $\pm 0.5\%$, since for the whole series the average separation of two independent measurements was less than 0.5%.* The final ratios should be independent of the spectrometer and therefore appropriate for comparisons.

ANALYSIS OF RESULTS

The almost constant width of the K_α lines in different compounds served mainly to indicate that they were not contaminated by significant amounts of sulfur in a different oxidation state. The range of $0.004°$ 2θ in our K_α widths at half-maximum is considerably less than the maximum range of 0.5 xu in the K_{α_1}, K_{α_2} separations of Faessler and Goehring[3] (which corresponds to a range of $0.014°$ 2θ on our spectrometer). They studied a larger series of compounds which may account in part for the difference.

The sulfur K_α energy data can be converted to charge data if the relations of Coulson and Zauli[8] and Bendazzoli et al.[9] or of Shuvaev[10] are used, but their relationships differ not only in form, but also in the magnitude of the results, with Shuvaev's results indicating much smaller changes in charge for a given K_α shift. For instance, Coulson and Zauli would report a charge of $+2.08$ for the sulfur in Na_2SO_4, whereas application of Shuvaev's relationship with his constant $k = 1.32$ gives a charge of $+0.91$. In connection with this problem, it is worth noting that Coulson and Zauli, as well as Bendazzoli et al., arrived at their results purely by calculation, while Shuvaev derived a certain form for the dependence and then selected his constant to agree with dipole-moment data. Despite these problems, both equations assert that the K_α energies are monotonically increasing functions of the net charge on the atom; hence, one should at least be able to order the compounds by sulphur atom charge. Among the sulfides, even this relationship may fail to hold, however, since the orbital of the unbonded pair of $3p$ electrons may be distorted. Finally, it must be noted that the charges measured using the K emission spectrum are actually those of a sulfur atom about 10^{-15} sec after being singly ionized in the K shell, and the implications of this fact are not well understood.

Turning to the K_β lines, we will first consider their structure. Although the resolution of the spectrometer was low, the large changes in width of the K_β lines sould be correlated with changes in the distribution of valence-electron energies. For instance, the range in valence-electron energies would appear to be more than 2 eV greater for methionine than for Na_2SO_4 at the half-maximum of valence-electron density. This width is probably attributable to several component emissions in which intensity as well as energy varies from compound to compound.[4,17] No simple variation of width with chemical bonding is apparent, but perhaps the widths could be useful relative to quantum-mechanical

---

* An exception is methylmethionine sulfonium chloride, for which the uncertainty is closer to $\pm 1\%$.

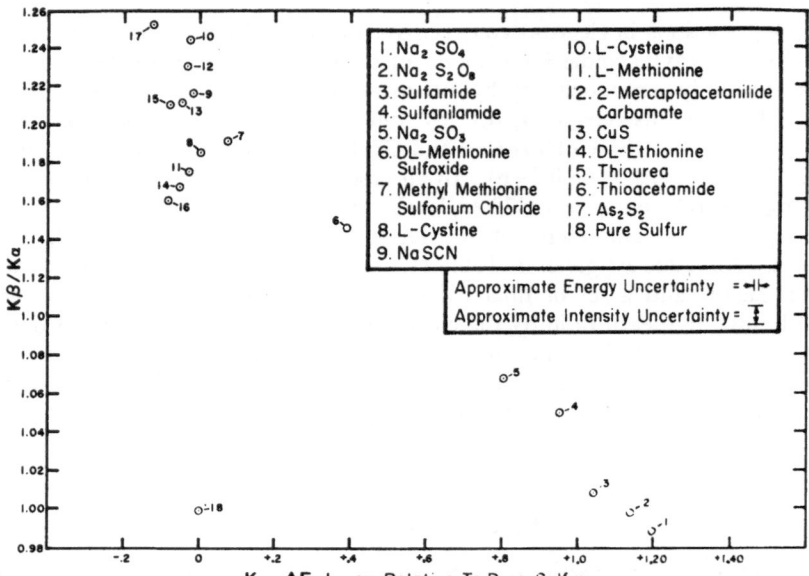

Figure 3. Energy versus K_β/K_α ratio for a series of sulfur compounds.

treatments. The prominent K_β satellite at about 13 eV lower energy than the main K_β emission shows an intensity variation with bonding in the opposite direction from that of the main line, ranging from about one-third the main K_β peak in Na_2SO_4 to about one-seventh the main peak in Na_2SO_3. Also, in this study, it was of measurable intensity only when the sulfur atom was bonded to at least two oxygen atoms. On the basis of energies, Faessler and Schmid[4] attributed this line to the normally forbidden $3s \rightarrow 1s$ transition; Kern[18] thought a similar silicon emission to result from excitons; and Ivanov's discussion[19] indicates he would attribute it to $3p$ character in the $3s$ band. The explanation that it comes from p character introduced into the $3s$ band by hybridization and bonding would seem to agree fairly well with the observation that its intensity increases in a regular manner with decreased electron density on the sulfur atom and, therefore, with stronger sulfur-bonding involvement.

In interpreting K_β intensity variations, the problem of unbonded valence-electron distortion and the fact that the source of the emission is a sulfur atom ionized in the K shell for about 10^{-15} sec are both applicable. In addition, the $3s$ electrons are involved in bonding to different degrees in different sulfur compounds, while the $3s \rightarrow 1s$ transition is presumably forbidden. Despite these problems, we can derive an approximate constant B to use in relating charge and intensity by assuming (after Shuvaev[5]) the following:

$$\Delta(K_\beta/K_\alpha) = -Bn$$

and by assigning the whole K_β intensity of pure sulfur to 4 electrons. This gives $B = 0.25$. Use of this value for B results in about the same relative range of charges as Shuvaev's K_α treatment for the $+4$ and $+6$ valence states of sulfur, but gives much smaller magnitudes for these if pure sulfur is taken as the zero of the scale. In the lower sulfur valences, the intensities give much larger charge changes than the energies, besides a different ordering of the compounds by sulfur atom charge.

Figure 3 is a plot of K_α energy versus K_β intensity. An approximately linear relationship between the two is apparent in the right-hand portion of the graph. However, in the left-hand portion where the sulfides are, there is only a scatter of points falling about the line that can be extrapolated from the points on the right, and, peculiarly, pure sulfur on the lower left is well isolated from any of the other points. The explanation of this seeming agreement in one area and lack of it in another is not obvious, but it appears that the effects of the unbonded $3p$ electrons may be quite different for the two parameters, thus causing the scatter. Since in sulfur compounds of valence -2 or 0 the range is much greater for intensity than for energy, relative to the total ranges observed, one might guess that the effects of unbonded electron changes are much greater on intensity than on energy, but real understanding awaits further theoretical work. Also, there is a good possibility that the approximations used to show a linear relationship between valence-electron density and K_β intensity are only very grossly valid. The approximately linear relation between K_α energy and K_β intensity for sulfur in the high valence states might be explained by partial or complete absence of unbonded electron effects and by involvement of the $3s$ electrons in the bonds to an extent proportional to that of the $3p$ electrons. At the least, Figure 3 provides an empirical, two-dimensional means of relating an uncertain chemical state of a sulfur atom to states that are better characterized.

SUMMARY

The sulfur K_β intensities and K_α energies have been studied in a series of sulfur compounds in an attempt to better understand their relationship to each other and to the chemical state of the sulfur atom, particularly its charge.

The two parameters display an approximately linear relationship for compounds involving sulfur in three or more bonds, thus showing at least some agreement concerning the charges on these atoms. For lower valence states, no obvious relation was found.

A plot of the relationship between K_α energy and K_β intensity has been prepared. This relationship should be independent of the spectrometer employed and hopefully can be used to empirically relate uncertain chemical states to known ones.

ACKNOWLEDGMENT

This work was performed under the auspices of the U.S. Atomic Energy Commission. The complete study from which this report is derived will be published as an AEC document.

REFERENCES

1. A. E. Lindh and O. Lundquist, "Die Struktur der K_{β_1}-Linie des Schwefels," *Ark. Mat. Astr. Fysi.* **18**(14): 3–11, 1924.
2. A. Faessler, "Roentgenspektrum und Bindungszustand," in: *Landolt-Börnstein Zahlenwerte und Funktionen, Sixth edition; Vol. 1, Atom- und Molekularphysik, Pt. 4, Kristalle,* 1955, pp. 769–868.
3. A. Faessler and M. Goehring, "Roentgenspektrum und Bindungszustand—Die K_α-Fluoreszenstrahlung des Schewfels," *Naturwissenschaften* **19**: 169–177, 1952.
4. A. Faessler, and E. D. Schmid, "Uber die Struktur des Roentgen-K_β-Spektrums von Schwefel," *Z. Physik* **138**: 71–79, 1954.
5. A. T. Shuvaev, "Influence of the Chemical Bond on the Energy and Intensity of the X-Ray Lines of Atoms in Compounds," *Izv. Akad. Nauk SSSR Ser. Fiz.* **25**: 986–991, 1961 (Columbia Technical Translations).
6. E. Schnell, "Zur Roentgenfluoreszenzanalyse," 1. Mitt: "Die Intensitätsverhaeltnisse der K-Linien dee Roentgenspektrums von Chlor in Abhängigkeit von der chemischen Bindung," *Monatsh.* **93**: 1383–1387, 1962; 3. Mitt.: "Die Aenderung der Relativen Intensitaeten der K_β-Strahlung der Elemente Schwefel, Phosphor, Silicium, und Aluminium bei Verbindungsbildung," *Monatsh.* **94**: 703–713, 1963.

7. M. A. Blokhin, A. T. Shuvaev, and V. V. Gorskii, "X-Ray Study of the Chemical Bonds in Sulfur Compounds," *Izv. Akad. Nauk SSSR Ser. Fiz.* **28**: 801–804, 1964 (transl. J. A. S. Bradley).
8. C. A. Coulson and C. Zauli, "The K_α Transitions in Compounds of Sulfur," *Mol. Phys.* **6**: 525–533, 1963.
9. G. L. Bendazzoli, P. Palmieri, and C. Zauli, "X-Ray Transitions in Compounds of Sulfur: Frequency and Intensity Shift of K_α transitions," *Boll. Sci. Fac. Chim. Ind. Bologna* **22** (3–4): 97–101, 1964.
10. A. T. Shuraev, "Determination of Ionic Charges in Compounds of the Third-Period Elements by Means of X-Ray Emission Spectra," *Izv. Akad. Nauk SSSR Ser. Fiz.* **28**: 758–764, 1964 (transl. J. A. S. Bradley).
11. E. A. Burke and R. M. Pettit, "Absorption Analysis of X-Ray Spectra Produced by Beryllium Window Tubes Operated at 20 to 50 Kvp," *Radiation Res.* **13**: 271–285, 1960.
12. J. Sherman, "The Theoretical Derivation of Fluorescent X-Ray Intensities from Mixtures," *Spectrochim. Acta* **7**: 283–306, 1955.
13. M. Renaud, "Le Calcul du Transfert de Rayonnement en Fluorescence X., L'effet de Matrice; L'equation de Transfert," *Compt. Rend.* **256**: 3086–3089, 1963.
14. Anonymous, "Table of X-Ray Mass Absorption Coefficients," *Norelco Rept.* **9** (3): 1962.
15. J. A. Bearden, "X-Ray Wavelengths," *U.S. At. Energy Comm. Rept.* NYO-10586, 1964.
16. J. A. Bearden, A. Henins, J. G. Marzolf, W. C. Sauder, and J. S. Thomsen, "Precision Redetermination of Standard Reference Wavelengths for X-Ray Spectroscopy," *Phys. Rev.* **135A**: 899–910, 1964.
17. J. Valasek, "Effects of Chemical Combination on the X-Ray Emission Spectrum of Sulfur," *Phys. Rev.* **43**: 612–614, 1933; "X-Ray Emission Spectra of Sulfides and Sulfates," *Phys. Rev.* **51**: 832–834, 1937.
18. B. Kern, "Die Si K_β-Banden der Roentgenemissionsspektren von elementarem Silicium, Siliciumcarbid, und Siliciumdioxyd," *Z. Physik* **159**: 178–193, 1960.
19. A. V. Ivanov, "L_{23} X-Ray Emission Spectra of Sulfur in Sulfides," *Izv. Akad. Nauk SSSR Ser. Fiz.* **26**: 405–408, 1962 (Columbia Technical Translations).

DISCUSSION

N. Spielberg (Philips Laboratories): Did you notice any shift of the K_β line, the main peak, with compounding?

D. W. Wilbur: Yes.

N. Spielberg: Would that shift be considerably smaller than the K_α shift?

D. W. Wilbur: It's about the same magnitude as the K_α shift, though the shift in width of the line is greater than the shift in position. The shift in position was about $1\frac{1}{2}$ eV total range, and the shift in width was about $2\frac{1}{2}$ eV total range.

N. Spielberg: I seem to recall that Faessler's data seemed to show a greater shift in the K_α lines than the K_β lines. Is that correct?

D. W. Wilbur: No, I think he just decided the K_β lines were not useful, because their structure changes were larger than their shifts.

N. Spielberg: I was wondering if you would comment also on, in the case of silicon as reported earlier, the big shifts occurring with the K_β lines and hardly any with the K_α lines.

D. W. Wilbur: I believe there are significant shifts in the K_α lines, because Faessler has also studied these and reported them at the Maryland Conference on Applied Spectroscopy in 1962.

Chairman W. L. Baun: The shift in the silicon K_α lines is less than 1 eV in going from the metal to the oxide, while the β shift is about $4\frac{1}{2}$ eV, so the β shifts considerably more than the α in the particular system that you are talking about. I was wondering about the $K_{\beta'}$. How far was it separated from the main β band?

D. W. Wilbur: About 13 eV.

Chairman W. L. Baun: Another interesting thing is that, in almost all of the materials in which the atom is coordinated with oxygen, this band shows up between 13 and 15 eV, not just in a given period of elements, but in several periods. This is one thing that led some people to think it was some exciton.

D. W. Wilbur: This may very well be.

A. Vaciago (Catholic University Medical School, Rome): Are you planning any measurement on sulphur satellites, of course, with a different spectrometer?

D. W. Wilbur: Are you talking about K_α satellites?

A. Vaciago: Yes.

D. W. Wilbur: I can't do useful satellite work with this spectrometer. It doesn't have the resolution. I am planning to start work with some high-resolution equipment; then I may look at the K_α satellites.

A. E. Austin: (Battelle Memorial Institute): Do any of your compounds have two or more sulfurs or a chain where one sulfur will be bonded to a group and also to another sulfur, a chain of sulfurs where you have two or more types of sulfur in the same compound?

D. W. Wilbur: No. I always tried to get a unique chemical state because the resolution of the intensities required this.

A. E. Austin: That's one point in the basic. But then the ultimate point in application of this is to take it and look at the more complex ones where they're already mixed up.

D. W. Wilbur: This was just an exploratory study.

DETERMINATION OF ELECTRON DISTRIBUTION AND BONDING FROM SOFT X-RAY EMISSION SPECTROSCOPY

J. E. Holliday

United States Steel Corporation
Monroeville, Pennsylvania

ABSTRACT

From measurements of the shift in wavelength with chemical combination, it appears there is very little transfer of charge between the metal and nonmetal atoms for the Group-IV transition-metal carbides and diborides. For Group-VI transition-metal carbides and beyond, there appears to be a definite transfer of charge from the carbon to the metal atom. The bonding in the Group-IV transition-metal carbides is largely covalent, and metallic for the Group-IV transition-metal diborides. A comparison of shapes of emission bands with crystal structure indicates that, in general, NaCl-type structures have symmetrical peaks with narrow bandwidths compared to hexagonal structure, which have asymmetrical peaks, broad bandwidths, and long tails to the bottom of the bands.

INTRODUCTION

The emission bands of the metal and the nonmetal from transition-metal interstitial compounds of titanium, zirconium, niobium and molybdenum were investigated in order to better understand the bonding and electron distribution in these compounds. It was also desired to find what relation, if any, existed between crystal structure and shape of the emission bands. Much work has been done on the metal K emission[1-4] bands of the transition-metal borides, carbides, nitrides, and oxides, but little work has been done on the metal L and M emission bands. Recently, Fischer[5] has reported wavelength shifts for the L_{III} emission band from transition-metal oxides and Bonnelle[6] has done a large amount of work on iron, cobalt, and nickel oxides. The fact that the K emission band from the nonmetals can be measured on the same samples and the same instrument as the metal emission band aids in understanding the type of bonding and electron distribution in these compounds.

INSTRUMENTATION

The spectrometer for measuring the soft X-ray emission bands has been described elsewhere.[7,8] The targets were bulk material and were metallographically polished before being put in the X-ray chamber, which was then pumped to approximately 5×10^{-8} torr. The surface of the target was cleaned by ion bombardment in the chamber before the spectra were measured. The detector was a flow-proportional counter. For the region being investigated (20–70 Å), the resolution was approximately 0.08 Å. Two sample targets were in the vacuum at the same time. The pure transition metal and graphite were used as standards for wavelength shift and intensity measurements on the compounds. The target potential was 4000 V and the beam current was 1.4 mA.

---

[1] References are at the end of the paper.

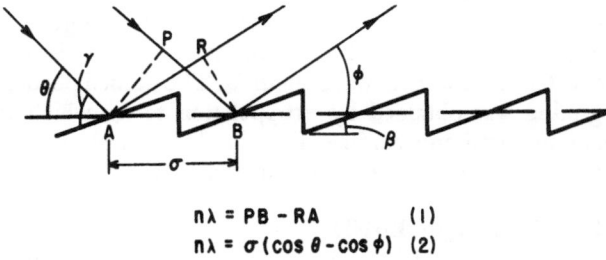

$$n\lambda = PB - RA \qquad (1)$$
$$n\lambda = \sigma(\cos\theta - \cos\phi) \qquad (2)$$

Figure 1. Schematic of blazed grating. Dashed line is the plane of the grating, σ is the grating constant, θ is the glancing angle, γ is the angle made to the reflecting surface, β is the blaze angle, and ϕ is the diffracting angle [reprinted by permission, J. E. Holliday, "Change in Shape of the K Emission Spectra of Light Elements with Chemical Combination," in: T. D. McKinley, K. F. J. Heinrich, D. B. Wittry (eds.), John Wiley & Sons, Inc., New York, 1966].

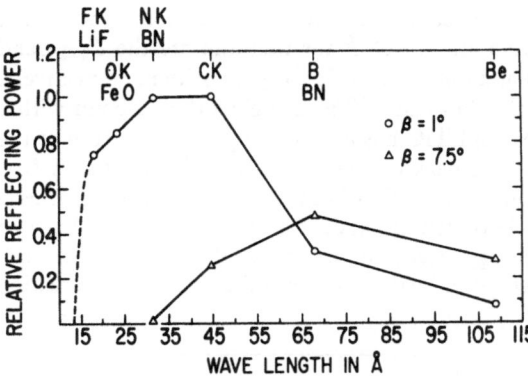

Figure 2. Relative reflecting power referred to the intensity of the C K band at 44.85 Å for two aluminized gratings having 600 lines/mm, concave radius of 1 m, and a $\beta = 1°$ (O) and 7.5° ($\triangle$). For both gratings, $\theta = 2°$, giving a γ of 3° (O) and 9.5° ($\triangle$) [reprinted by permission, J. E. Holliday, "Change in Shape of the K Emission Spectra of Light Elements with Chemical Combination," in: T. D. McKinley, K. F. J. Heinrich, D. B. Wittry (eds.), John Wiley & Sons, Inc., New York, 1966].

Unless otherwise indicated, the analyzer is a 1° blazed replica concave grating with a radius of curvature of 1 m and an aluminized surface ruled with 3600 lines/mm. There are two main advantages to the blazed replica grating.[8,9] On this type of grating it is possible to eliminate piling up at the top of the groove, because the bottom of the groove in the original ruling, which is very sharp, becomes the top of the groove in the first-generation replica. The other advantage to the blazed grating is that the total reflected beam can be reflected into orders other than the zero. To appreciate this, one must understand that the diffraction angle ϕ (Figure 1) at which the wavelength λ of order n occurs is independent of the angle γ that the incident X-rays make with the reflecting surface of the grating. However, the order in which the maximum reflected intensity occurs is dependent on γ. Thus, by keeping the angle θ fixed (the angle that incident X-rays make with the plane of the grating) and changing the blaze angle β (Figure 1), a selective effect can be obtained. The maximum reflected intensity for a blazed grating will occur in the general vicinity of a diffraction angle equal to $(\gamma + \beta)$ or $(\theta + 2\beta)$ for the incident X-rays striking the blaze in the direction indicated in Figure 1. The selective effect can be seen in Figure 2, where β has values of 1 and 7.5° and θ is kept constant at 2°. Since a known source of continuous radiation was not available, the peak intensities of the radiation from the second-period elements were used to measure the relative efficiency of the gratings as a function of wavelength. The intensities were corrected for counter window and $Z^{3/2}$ dependence.[10]

For the 1° blazed grating and $\theta = 2°$, the $(\theta + 2\beta)$ relation predicts that the maximum reflected intensity will occur in the vicinity of 31 Å, which is close to that indicated in Figure 2. For the 7.5° blaze grating, the maximum reflected intensity should occur at approximately 200 Å. From Figure 2, it will be seen that it occurs near 65 Å. The $(\theta + 2\beta)$ relation for determining the maximum reflected intensity is probably over-simplified. However, the general trend that with increasing blaze angle the maximum reflected intensity occurs at longer wavelengths is established from the results shown in Figure 2. The drop in relative reflecting power for wavelengths shorter than the maximum is due to a combination of the selective effect and the critical wavelength λ_c. The dashed portion of the curve for the 1° blazed grating was obtained by extrapolation to λ_c, which is approximately 13 Å for $\gamma = 3°$ and an aluminum surface. For the 7.5° blazed grating, λ_c had increased to approximately 30 Å, because γ has increased from 3 to 7.5°.

TITANIUM $L_{II,III}$ BAND

The Ti $L_{II,III}$ emission band ($3d + 4s \rightarrow 2p$ transition) for titanium, TiO, TiN, TiC, and TiB_2 is shown in Figures 3–5. All the curves were recorded using the point-by-point method, and at least three curves were run on each spectrum to ensure that all details of the spectrum are reproducible. In comparing these curves, some interesting facts become apparent. The peaks at 27.45 and 27.08 Å for titanium are the L_{III} band ($3d + 4s \rightarrow 2p\ J_{3/2}$) transition and L_{II} band ($3d + 4s \rightarrow 2p\ J_{1/2}$) transition, respectively. Ern and Switendick[11] have shown that the $2p$ band of TiO is below the $3d$ band with a gap between them; no gap exists in TiN; and TiC has nearly complete admixture of the $2p$ and $3d$ bands. It does not appear that the peak on the low-energy side of the Ti L_{III} band for TiO, TiN, and TiC is the $2p$ band. A transition from the $2p$ band of the nonmetal to the $2p$ level of titanium is a quadrupole transition, and, if it existed, it would have an intensity much weaker than the low-energy peaks indicated for TiO and TiN. Also, this peak was not present for TiB_2, which gives further evidence that it is not a transition

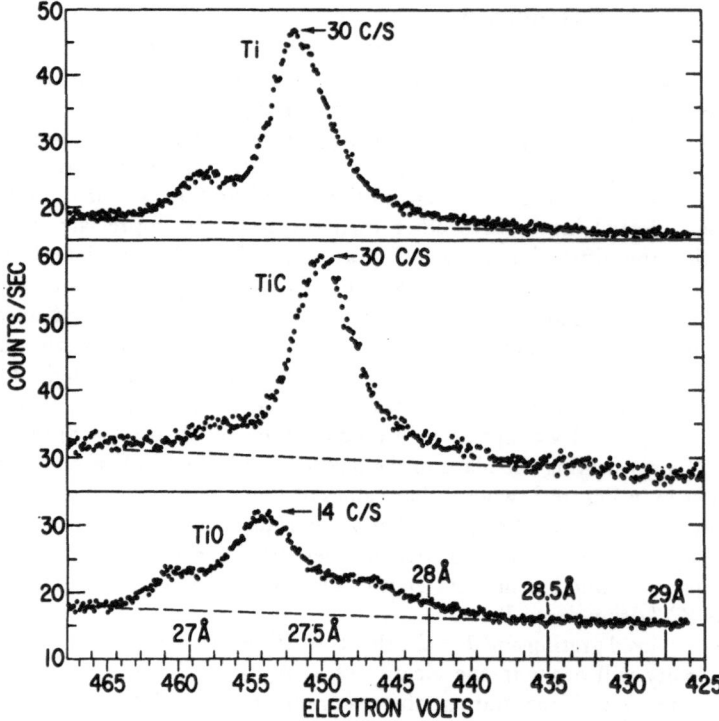

Figure 3. The Ti $L_{II,III}$ emission bands ($3d + 4s \rightarrow 2p$ transition) from titanium, TiC, and TiO. Intensities are the peak intensities above background. The deviation is $\pm 1.5\%$.

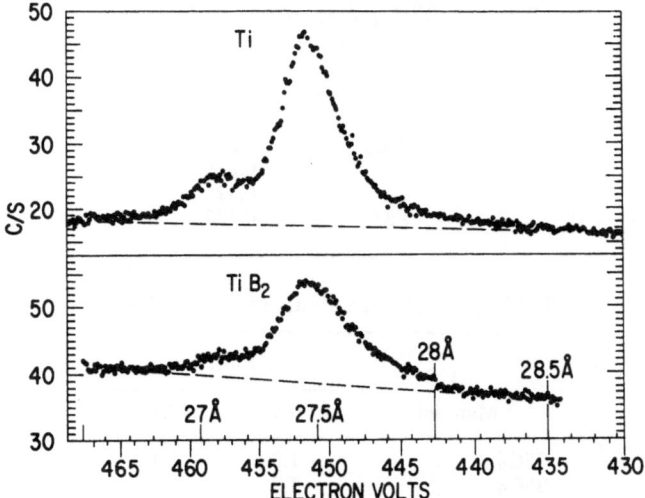

Figure 4. The Ti $L_{II,III}$ emission band ($3d + 4s \rightarrow 2p$ transition) for titanium and TiB$_2$. The deviation is $\pm 1.5\%$.

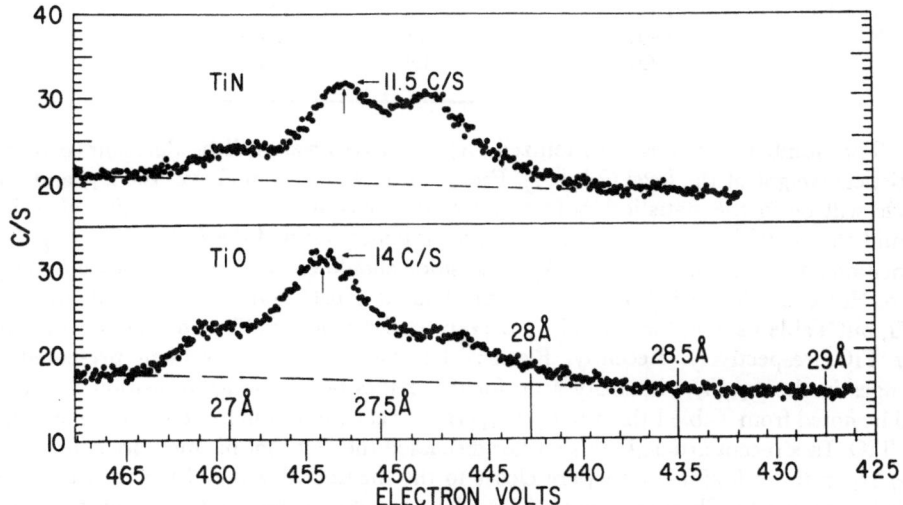

Figure 5. The Ti $L_{II,III}$ emission band ($3d + 4s \rightarrow 2p$ transition) for TiN and TiO. Arrows indicate peak position and deviation of $\pm 1.5\%$.

from the $2p$ band of the nonmetal to the $2p$ band of titanium. On this basis, it would be more reasonable to consider the peaks on the low-energy side of the Ti L_{III} band for TiO, TiN, and TiC as a low-energy satellite.

It is of interest to note that this peak was present for TiO and TiN, which have the NaCl-type structure, but absent for titanium and TiB$_2$, which have the hexagonal structure. This indicates the possibility of a relation between this low-energy peak and crystal structure. Because of the L_{II} and L_{III} band overlap and the low-energy satellite, it is not possible to obtain an experimental bandwidth that can be compared to the theoretical bandwidth of TiO (8.95 eV), TiN (9.25 eV), and TiC (8.75 eV) calculated by Ern and Switendick.[11]

Table I. Ti L_{II}/L_{III} Intensity Ratios

Material	L_{II}/L_{III}
TiC	0.15
TiB$_2$	0.2
Ti	0.275
TiN	0.325
TiO	0.425

Table II. Shifts in Peak of Ti L_{III} Band

Material	Shift	
	ΔE (eV)	$\Delta\lambda$ (Å)
TiC$_{0.96}$	−1.8	+0.11
TiC$_{0.93}$	−1.2	+0.075
TiC$_{0.83}$	−0.6	+0.035
TiB$_2$	—	—
Ti	—	—
TiN	+1.1	−0.07
TiO$_{0.9}$	+1.8	−0.115
TiO	+2.3	−0.14
TiO$_{1.17}$	+3.0	−0.18
TiO$_{1.97}$	+4.4	−0.27

The number of atoms with ionized L_{II} and L_{III} levels will be determined by the statistical weight of the level $(2J + 1)$. From the J values given above, the L_{II} and L_{III} levels will be in the statistical ratio of 1 : 2. If, for some reason, the statistical weight should change with chemical combination, the intensity ratio of the L_η line $(3s \rightarrow 2p\ J_{1/2})$ transition and the L_l line $(3s \rightarrow 2p\ J_{3/2})$ transition should change the same as the L_{II}/L_{III} ratio. However, the Ti L_η/L_l intensity ratio remains constant at ~ 0.117 for titanium and TiO, but Table I shows that the Ti L_{II}/L_{III} intensity ratio is 0.275 and 0.425 for titanium and TiO, respectively. Recently, Fischer[5] has mentioned the problem presented by changes in the L_{II}/L_{III} intensity ratio for the oxides of first-series transition metals. It will be noted from Table I that the L_{II}/L_{III} ratio is at a minimum for TiC and a maximum for TiO. In a recent article,[12] this author discussed the fact that his measurements of the L_{II}/L_{III} ratio of 0.25 for iron were closer to the theoretical value of 0.5 than the value of 0.1 obtained by Skinner et al.[13,14] It will be noted that the L_{II}/L_{III} intensity ratio of TiC is close to the value obtained by Skinner. It is quite possible that his target surfaces were converted to carbides by the carbon in the vacuum system. Some investigators have suggested that the L_{II}/L_{III} ratio would be 0.5 and would remain constant for different chemical combinations if all L_{II} or L_{III} emission band satellite intensities were included. If the peak on the low-energy side of the L_{III} band is considered to be a L_{III} band satellite, then the L_{II}/L_{III} ratio for TiO will be closer to that of titanium, but it would be further from the theoretical value. Thus, consideration of satellites does not seem to solve the L_{II}/L_{III} intensity ratio anomaly.

WAVELENGTH SHIFT AND ELECTRON DISTRIBUTION

Measurements of the wavelength shift in Table II show that relative to titanium the Ti L_{III} band has shifted toward higher energies for TiO, while for TiC the Ti L_{III} band

Table III. Shifts and Halfwidths of Emission Bands

Material	Shift ΔE(eV)	$W_{1/2}\Delta E$(eV)
C K Band:		
Diamond	—	8.1
NbC	−0.25	2.6
ZrC	−0.3	2.5
TiC	−0.3	3.1
Mo$_2$C	−0.9	4.7
Graphite	−2.1	6.4
B K Band:		
Boron	—	4.3
TiB$_2$	−0.4	3.8
ZrB$_2$	−0.4	3.35

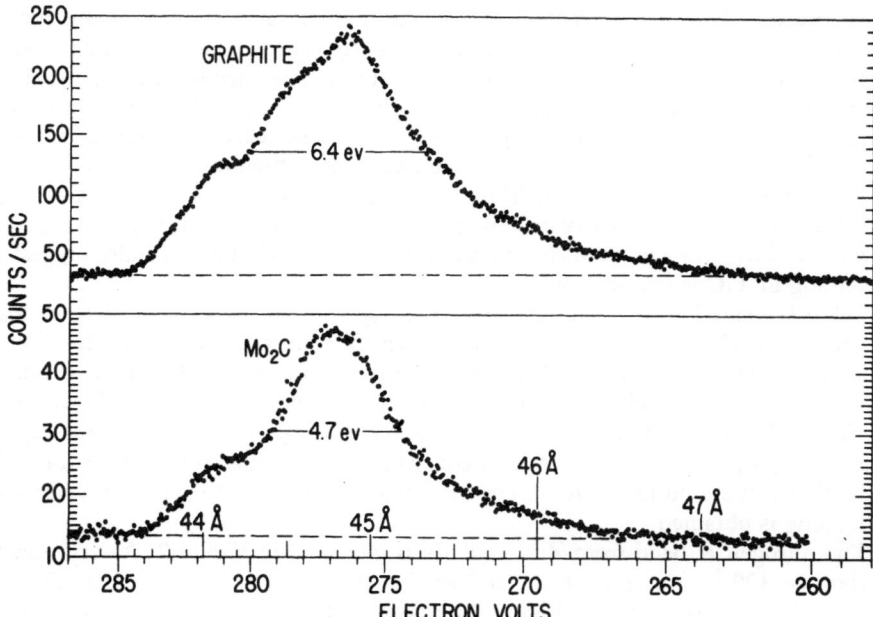

Figure 6. Carbon K emission band from graphite and Mo$_2$C. The deviation is ±1%.

has shifted toward lower energies relative to titanium. For nonstoichiometric compounds of TiO$_x$ ($x = 0.9$ to $x = 1.97$), the Ti L_{III} band shifts toward higher energies with increasing x, but, for TiC$_x$ ($x = 0.83$ to $x = 0.96$), the Ti L_{III} band shifts toward lower energies.

If it is assumed that the titanium atom is neutral in titanium and there is a transfer of charge from titanium to oxygen in TiO, then it would appear from Table II that there is a transfer of charge from the carbon atom to the titanium atom. The theoretical band calculations of Williams and Lye[15] on TiC predict that there is a transfer of 1.33 electrons from the carbon to the titanium atom. In comparing the peak of the C K band of diamond with the peak of the C K band from TiC, ZrC, and graphite in Table III it is found that these peaks have shifted 0.3, 0.3, and 2.1 eV, respectively, in reference to the C K peak from diamond. The reason for the shift in wavelength between graphite and diamond is that in diamond there are four bonding electrons per atom, while in graphite there are three bonding electrons per atom and the fourth electron is mobile. Thus, the X-rays would reflect a more positive carbon atom from graphite than diamond. The carbon atom in diamond is essentially neutral and has been considered to be the best example of a covalent band. As a consequence in relating wavelength shift to ionization of the carbon atom, the shift in wavelength should be compared to diamond rather than graphite. The 0.3-eV shift for the C K band from TiC does not reflect the charge transfer to the titanium atom indicated in Table II or that predicted by Williams and Lye. The shift in the peak of the B K band from TiB_2 and ZrB_2 relative to B K band from boron is 0.4 eV and is close to the shift of the C K band from TiC and ZrC. However, there is a large difference in the shift of the Ti L_{III} band from TiB_2 and TiC relative to titanium (Table II). This discrepancy can be explained if it is assumed that the X-rays from the titanium atom in titanium reflect a positive atom rather than a neutral atom. The fact that an atom can be positive in a pure element was indicated above by the wavelength shift of the C K band from graphite. Thus, the increase in negativity of the titanium atom indicated in Table II for TiC could be largely due to the conduction electrons becoming more closely bound to the titanium atom. The increase in the number of electrons bound to the titanium atom will result in additional chemical bonds which could account for the high melting temperature of TiC. The lack of a change in the Ti L_{III} wavelength from TiB_2 indicates that the free electrons in titanium do not become more associated with the titanium in the TiB_2 atom, but remain relatively free. This would account for the high conductivity of TiB_2 (TiB_2 is a better conductor than titanium) and the fact that TiB_2 has a lower melting temperature than TiC. These results suggest that the bonding in TiC is covalent, while it is metallic in TiB_2.

Table III and Figure 6 indicate there is a big shift in the C K band from Mo_2C compared to the C K band from the Group-IV and V transition-metal carbides. Thus, it is quite possible that, for the carbide of the transition metals beyond Group V, the carbon atom has a definite ionic character. This was shown by the work of Sidhu,[16] who found from X-ray diffraction measurements that carbon is a positive ion in the compound $Mn_{1.75} Co_{2.25} C_{0.93}$. However, measurements will have to be made on the other transition-metal carbides and diborides before a complete understanding of the electron distribution and bonding is obtained.

The nitrogen K band from ZrN and BN is indicated in Figure 7. A very thin film of carbon (~ 100 Å) was put on the BN so that the target was conducting. The N K band from TiN is not shown in Figure 7 because of the presence of the Ti L_l line only 0.25 Å from the N K peak of TiN. Within experimental error, there was no observed shift between the peaks of the N K bands from BN, TiN, and ZrN. The peak of the B K band from BN has shifted 2.9 eV relative to the B K band from boron. This amount of shift in λ indicates that the boron in BN has a large positive charge compared to the boron atom in TiB_2. Since BN is a good insulator, the $2p$ electron and some of the $2s$ electrons are not mobile, but are transferred to the nitrogen atom. Because the N K peaks for BN, TiN, and ZrN all have the same wavelength, nitrogen is also a negative ion in TiN and ZrN. Thus, the increase in positive charge in the titanium atom in TiN relative to the titanium atom in titanium is the result of electrons being transferred from the metal to the nitrogen atom. From the above analysis, nitrogen is an electron acceptor in TiN and ZrN, and there appears to be an ionic character to the bond between the metal and the nitrogen.

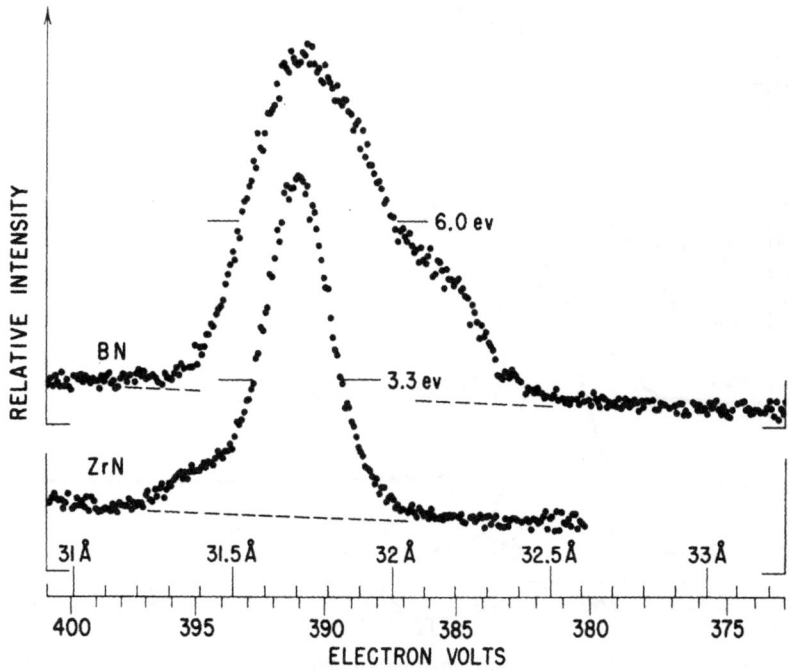

Figure 7. Nitrogen K emission band from BN and ZrN. The peak height has been normalized. The deviation is $\pm 2\%$.

BAND SHAPE, CRYSTAL STRUCTURE, AND BONDING

The possibility of a relation between crystal structure and emission band shape was indicated earlier. By comparing the above emission band shapes with crystal structure, it was found that,. in general, emission bands from hexagonal-type structures have asymmetrical peaks, broad bandwidths, and long tails to the bottom of the band. For NaCl-type structures, the emission bands have symmetrical peaks, narrower widths, and less tailing to the bottom of the band. This fact is clearly shown in Figure 8, where the C K band for the NaCl-type carbides is symmetrical with less tailing to the bottom of the band and has a halfwidth of approximately a third that of the C K band from hexagonal graphite. In addition, the C K band from hexagonal Mo_2C (Figure 6) has a halfwidth $W_{1/2}$ about twice that of the C K bands from the NaCl. Also, the Ti L_{III} band is slightly more asymmetrical for titanium and TiB_2 than for TiC. The asymmetry index for the Ti L_{III} band from titanium and TiB_2 is 1.3 compared to 1.0 for TiC.

That the shape of the band is related to the crystal structure is evident for the N K bands in Figure 7. The N K band from ZrN is symmetrical and has a halfwidth approximately one-half that of the N K band from BN, which has a hexagonal structure. Metallic bonding does not always produce emission bands that have shapes generally associated with metals, that is, a narrow emission edge and an asymmetrical shape. Even though ZrN is a better conductor than zirconium, the N K band from ZrN has a symmetrical peak with no emission edge. On the other hand, the N K band from BN has an appearance closer to that of a metal, yet it is a very good insulator.

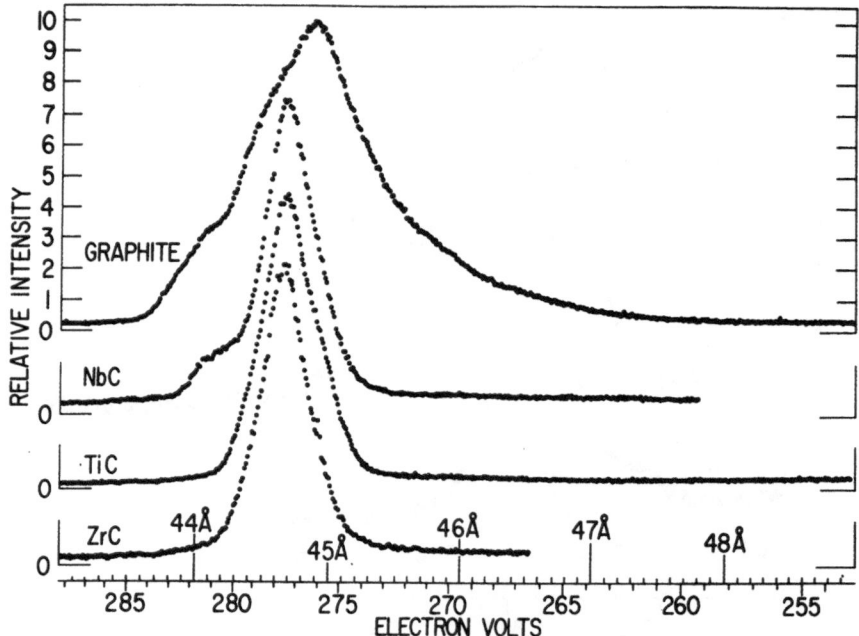

Figure 8. Carbon K emission band from graphite, NbC, TiC, and ZrC using a 1° blazed replica grating with 2160 lines/mm. All peak heights have been normalized. The deviation is ±1%.

The asymmetrical peaks and long tails to the bottom of the band characteristic of hexagonal structures are also seen for hexagonal TiB_2 and ZrB_2 (Figure 9). The B K band from TiB_2 shows more structure than Fischer and Baun[17] obtained from a lead stearate crystal. This is probably because the 0.08-Å resolution obtained from the 3600-lines/mm grating and 20-μ slit is greater than that obtained from a soft film in the 60–70-Å region. The B K band from ZrB_2 shows even more structure. There is a hump on the low-energy side ~1.3 eV from the main peak. The significance of the presence of this hump in ZrB_2 and its absence in TiB_2 is not known at present. Inner-level smearing cannot be the answer, since the inner level for boron should be the same for both TiB_2 and ZrB_2.

As indicated earlier, the bond between the metal and the carbon atoms in the Group IV and V transition-metal carbides appears to be largely covalent. Because of this, the shape of the metal and carbon emission bands should be the same. In order for this to be observed experimentally, it is necessary that both emission bands reflect the same symmetry and that the inner level does not smear out the details of one of the bands. The Zr M_V band ($5p \rightarrow 3d$ transition) from ZrC is shown in Figure 10. It will be seen that in comparing the C K band in Figure 8 ($2p \rightarrow 1s$ transition) and the Zr M_V band from ZrC that the shape of the bands is not the same. Inner-level smearing cannot be the reason for not seeing a peak 2.1 eV on the high-energy side of the main peak from the C K band from ZrC. The C K band is sufficiently narrow that a peak 2.1 eV from the main peak would be evident in the shape of the band. Before this anomaly can be explained, a better understanding of the relation between bonding and band shape will have to be obtained.

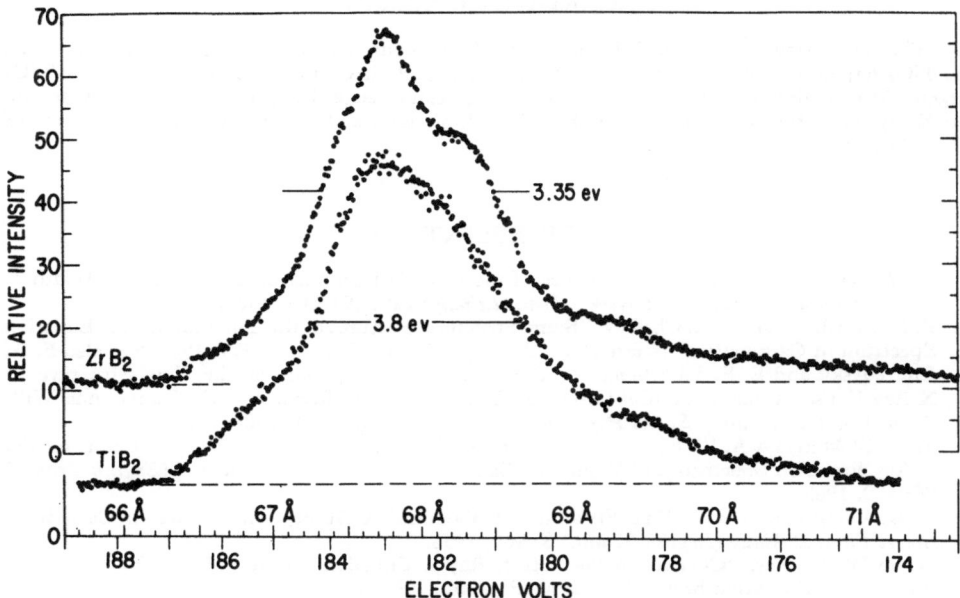

Figure 9. Boron K emission band from ZrB$_2$ and TiB$_2$. The deviation is $\pm2\%$. The peak heights have been normalized.

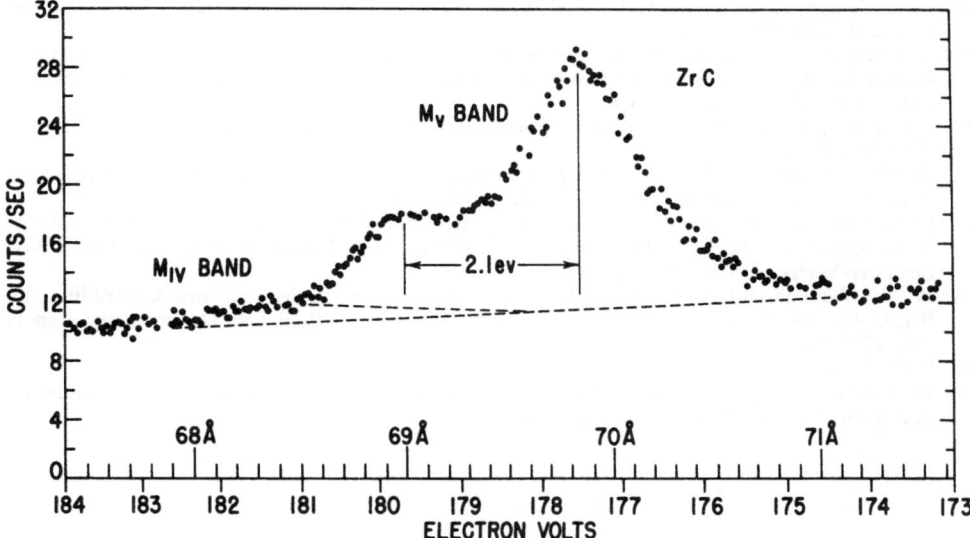

Figure 10. The Zr M_{IV} and M_V emission bands ($5p \rightarrow 3d$ transition) for ZrC. The background is shown by the dashed line; the grating is a 1° blazed replica grating with 2160 lines/mm. The deviation is $\pm2\%$.

ACKNOWLEDGMENTS

The author would like to thank A. Panson of Westinghouse Research Laboratories for supplying the TiO_x targets and also Robert G. Lye of the Union Carbide Corporation for supplying the TiC_x targets. At this laboratory, the author would like to acknowledge the helpful discussions with and the X-ray diffraction work of Leo Zwell and E. J. Fasiska, and also the technical assistance of W. A. Hester.

REFERENCES

1. A. Meisel and W. Nefedow, "Influence of the Chemical Binding on the Shape and Width of X-Ray Emission Lines," *Z. Physik. Chem. (Leipzig)* **219**: 194–204, 1962.
2. E. E. Vainshtein and V. I. Chrikov, "Some Structural Features of the Titanium X-Ray Emission Spectrum in Carbonates," *Soviet Phys. Doklady (English Transl.)* **7**: 724, 1963. See also E. A. Zhurkovskii and E. E. Vainshtein, "A Comparative Investigation of the Fine Structure of X-Ray Emission Bands for K-β Group of Titanium in the Metal and its Compounds With Some Light Elements," *Soviet Phys. Doklady (English Transl.)* **4**: 1308, 1960.
3. M. A. Blokhin and A. T. Shuvaev, "Concerning the Influence of the Chemical Bonds on the X-Ray Emission Spectrum of Titanium," *Bull. Acad. Sci. USSR Phys. Ser. (English Transl.)* **26**: 429, 1962.
4. E. E. Vainshtein et al., "Fine Structure of Titanium K-Absorption Spectra in Carbides," *Soviet Phys. Doklady (English Transl.)* **3**: 960, 1958.
5. David W. Fischer, "Changes in the Soft X-Ray L Emission Spectra with Oxidation of the First Series Transition Metals," *J. Appl. Phys.* **36**: 2048, 1965.
6. Christiane Bonnelle, "Contribution à L'Étude des Métaux de Transition du Premier Groupe, du Cuivre, et de Leurs Oxydes par Spectroscopie X dans le Domaine de 13–22 Å," Theses de Docteur Es-Sciences Physiques, L'Université de Paris, 1964.
7. J. E. Holliday, "A Soft X-Ray Spectrometer Using a Flow Proportional Counter," *Rev. Sci. Instr.* **31**: 891, 1960.
8. J. E. Holliday, "Soft X-Ray Spectroscopy in the 10 Å to 150 Å Region," *Handbook of X-Rays*, McGraw-Hill Book Co., New York, 1966, Chapter 38.
9. J. E. Holliday, "Blazed Replica Gratings for Soft X-Ray Spectroscopy," *J. Opt. Soc. Am.* **52 (1312)**: WB17, 1962.
10. A. J. Campbell, "K X-Ray Yields from Elements of Low Atomic Number," *Proc. Roy. Soc. (London)* **A274**: 319, 1963.
11. V. Ern and A. C. Switendick, "Electronic Band Structure of TiC, TiN, and TiO," Technical Report No. 192, Laboratory for Insulation Research, Massachusetts Institute of Technology, Cambridge, Massachusetts, 1964.
12. J. E. Holliday, "Soft X-Ray Emission Spectroscopy in the 13 Å to 44 Å Region," *J. Appl. Phys.* **33**: 3259, 1962.
13. H. W. B. Skinner, T. Bollen, and J. E. Johnston, "Notes on Soft X-Ray Spectra Particularly of the Fe Group Elements," *Phil. Mag.* **45**: 1070, 1954.
14. D. H. Tomboulian, "The Experimental Methods of Soft X-Ray Spectroscopy and the Valence Band Spectra of the Light Elements," in: S. Flugge (ed.), *Handbuch der Physik*, Vol. *XXX*, Springer-Verlag, Berlin, 1957, p. 244.
15. W. S. Williams and R. G. Lye, "Research to Determine the Mechanisms Controlling the Brittle–Ductile Behavior of Refractory Cubic Carbides," Technical Documentary Report ML-TDR-64-25, Part II, 1965.
16. S. S. Sidhu, private communication.
17. D. W. Fischer and W. L. Baun, "The Effect of Chemical Combination on Soft X-Ray Emission Spectrum of Boron," Technical Report AFML-TR-65-360, 1965.

CHEMICAL EFFECT ON X-RAY
ABSORPTION-EDGE FINE STRUCTURE

E. W. White and H. A. McKinstry

Materials Research Laboratory, Pennsylvania State University
University Park, Pennsylvania

ABSTRACT

Existing theories of X-ray absorption-edge fine structure do not adequately explain details of spectra observed for solids. However, the possibility that X-ray absorption spectra might eventually be used as tools for the characterization of new insulator materials has prompted the study of an extensive selection of simple oxides of known chemistry and structure. The complete *K* absorption fine structure has been measured for some forty simple oxides of six elements of the fourth period. The *L* absorption-edge spectra have been measured for metallic lead and several lead oxides. The several isostructural and polymorphic sets included among these oxides, as well as the Magnéli phases for three of the elements, have made it possible to study the effects of valence, coordination number, electron configuration, and crystal structure. The applicability of current theories of the fine structures are discussed in the light of these findings. An automated single-crystal spectrometer is described.

INTRODUCTION

X-ray absorption edges and extended fine structure are affected by the chemical state of the absorbing element. Considerable evidence, both theoretical and experimental, has shown that absorption structures may be affected by valence, bond type, coordination number, interatomic distance, crystal structure, and temperature. However, long-range order appears to have little or no effect. The precise nature of the chemical effect has not been resolved and existing theories do not explain the details of experimental results.

This study was undertaken in an attempt to correlate *empirically* X-ray absorption-edge features with known crystal chemical parameters for a large number of well-characterized materials. An experimental technique has been developed which facilitates the collection of complete absorption-edge curves and allows up to seven materials to be analyzed simultaneously under identical conditions. Several pure metals and their oxides were selected for study because of the interest in oxides at this laboratory. The extensive collection of some 35 well-characterized oxides of transition-metal elements available in this laboratory is used in this study. By confining the study to compounds having a common anion, it is hoped that the effect of valence, coordination number, and electron configuration can be more completely evaluated. Several isostructural series and polymorphic sets are included among these samples. Several Magnéli phases in the titanium–oxygen and vanadium–oxygen systems are also included.

The theories that have been advanced to explain the undulating absorption coefficients on the high-energy side of the absorption edge have been summarized by Parratt[9] and by Azároff.[1] The fine structure is usually divided into two regions. The Kossel structure

[1] References are at the end of the paper.

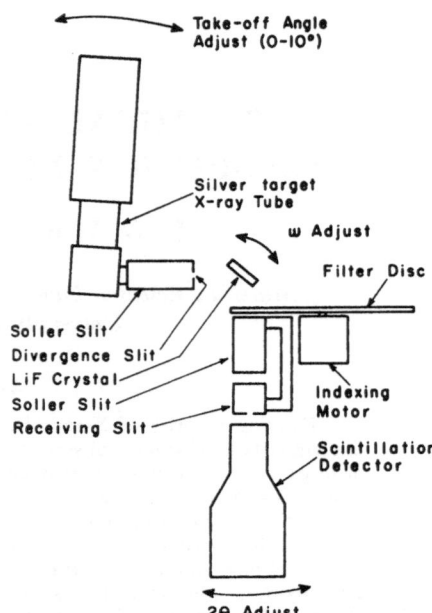

Figure 1. Schematic of automated single-crystal
              spectrometer.

includes features of the edge and extends several electron volts beyond. It is thought to
result from electronic transitions from filled levels in the atom to unoccupied "optical
levels." The Kronig structure begins somewhere in the region where the Kossel structure
ends and continues out for several hundred electron volts. It is referred to as the extended
X-ray absorption fine structure. The explanations that are used to describe the origins of
the Kronig structure must somehow consider the environment of the atoms in the solid
material. However, since glasses and amorphous materials show the same Kronig structure as the corresponding crystals, and since two theories by Kozlenkov[5] and Lytle[6]
use infinite potentials outside a small region and obtain remarkably good agreement with
the experimental positions of successive maxima, the final explanations are not yet
available. The data presented here emphasize the need for a theory that will predict the
variation of amplitudes and the shifts in position of the absorption features as functions
of crystal chemical parameters.

EXPERIMENTAL PROCEDURE

A Picker X-ray diffraction system as described by Furnas and White[3], was modified
to operate as an automated single-crystal spectrometer. A schematic of the spectrometer
is shown in Figure 1. A silver-target X-ray diffraction tube having a focal spot about
1×15 mm was operated at 15 keV and 25 mA for all the experiments. It was used at a
1.5° take-off angle for lead, germanium, iron, and manganese and at 2.5° for the remaining
elements. Silver was chosen as the target because it has no characteristic lines in the
wavelength range of interest. The observed continuum yield at low take-off angles is no
greater from silver than from lighter targets, such as copper, owing to the higher self-
absorption by silver. Operation at higher accelerating voltages did not result in higher
observed continuum yields for most of the edges.

**Table I. Adopted Wavelengths of K Absorption Edges
and Calculated Dispersion for LiF Crystal**

Element	Wavelength (Å)	Dispersion (eV/0.01° 2θ)
Titanium	2.4973	0.56
Vanadium	2.2690	0.71
Chromium	2.0701	0.89
Manganese	1.8964	1.09
Iron	1.7433	1.32
Germanium	1.1165	3.46

The slit system consisted of two 4° soller slits, a divergence slit, and a receiving slit. The $\frac{1}{12}$° divergence slit and 0.02° receiving slit were used for the lead, germanium, iron, and manganese edges. The $\frac{1}{12}$° divergence and 0.05° receiving slits were used for the remaining elements. The diffractometer radius was 5.73 in. A lithium fluoride crystal $(2d = 4.0272 \text{ Å})$ was used for all edges. The crystal was rocked (ω adjust) to obtain optimum alignment for each edge. Table I lists the K absorption edge wavelengths and calculated dispersions for this crystal.

Specimens were prepared as films of uniform thickness. Each specimen was ground under alcohol to -325 mesh. A stock solution of 2:1 ethylene dichloride–formvar was added to the powder in an agate mortar using a ratio of about one part solution to one part powder by volume. Ethylene dichloride was added while the contents were being mixed with the pestle to a thin slurry. The slurry was painted onto teflon strips $1 \times 1 \times \frac{1}{8}$ in., one layer at a time, using a $\frac{1}{2}$-in. brush. Each coat was allowed to evaporate to dryness before the next layer was brushed on. The orientation of brush strokes was changed from coat to coat. Typical films consisted of 10–15 coats. The composite weight was checked until the desired thickness was accumulated. This varied from about 3 mg/cm² of titanium to about 8 mg/cm² of iron, as the element. The formvar served as a binder, which allowed one to remove and handle the films easily. Teflon was chosen as substrate since formvar does not adhere tightly to its surface. An X-ray powder diffractometer scan was made of a portion of each film as a final check of sample identification and to be certain that the crystallinity of the specimen had not been affected by grinding.

Samples were mounted on an eight-position filter disk. The disk was driven by an eight-position Ledex Digimotor. Position number one was left open so that I_0 values could be recorded. The data-collection procedure was to measure I_0 and the transmitted intensity for seven specimens at each angular increment. X-ray intensity data were recorded on the basis of preset time (40 sec for lead; 100 sec for germanium, iron, manganese, and chromium, and 200 sec for vanadium and titanium). Count data were automatically punched into IBM cards, each card containing the I_0 and transmitted X-ray intensity for the seven samples, at a given 2θ value. The spectrometer was automatically stepped at increments of 0.01° 2θ for lead, germanium, iron, and manganese and 0.02° 2θ for chromium, vanadium, and titanium. A helium tunnel, fashioned from 0.001-in. mylar sheet, was used for the vanadium and titanium runs.

A scintillation detector was used for all the experiments. The signal was processed by a pulse-height analyzer to eliminate harmonics and low-energy background noise. All data were processed by computer to generate plots of normalized μX versus 2θ. Additional experimental details can be found in the thesis by White.[12]

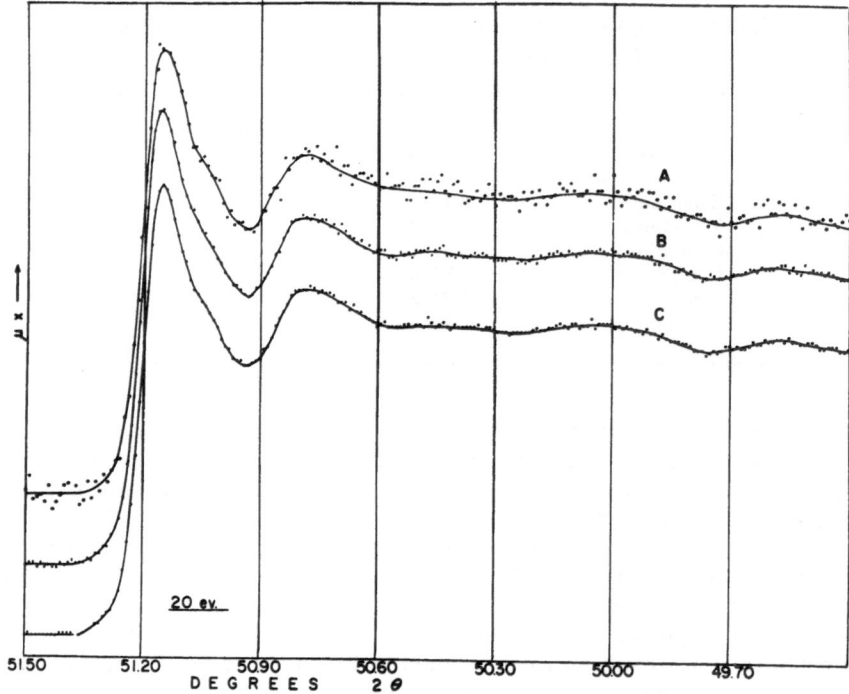

Figure 2. Absorption curves for different thicknesses of α-Fe$_2$O$_3$.

RESULTS

Thickness Effects

Curves A, B and C of Figure 2 for three thicknesses of α-Fe$_2$O$_3$ films are used to illustrate the validity of comparing normalized curves. The maximum μX value for curve A is 0.44; for curve B, 1.29; and for curve C, 2.04. Thus, it is shown that, when normalized, curves appear to be comparable and free of thickness effects at this level of resolution. The only difference noticed among these curves is a result of variations in precision of measurement of the individual μX values.

Schematic Absorption Curve

The absorption spectra for all the oxides can be discussed in terms of the schematic absorption curve shown in Figure 3.

Point A is the energy of the initial increase in absorption measured at the point of intersection of a line through the background and a line tangent to the slope of the edge. Point B corresponds to the energy of a small absorption maximum that is often absent.

Point C indicates the position of the edge at one-half height. It is often used as the reference energy of the edge in the literature. Point D usually corresponds to the energy of the first major absorption peak. It is often assumed to correspond to the energy of the transition $1s \rightarrow 4p$. Point E, when present, is usually the second resolved maximum. Point F is generally the third resolved maximum, possibly corresponding to the energy at which a $1s$ electron escapes the energy-level system of the "ionized atom." Maxima beyond F are sequentially labelled with capital letters as indicated in Figure 3.

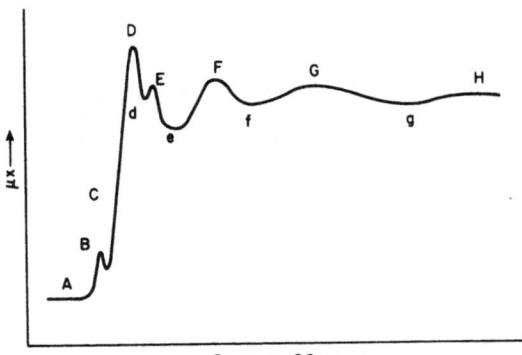

Figure 3. Schematic absorption curve.

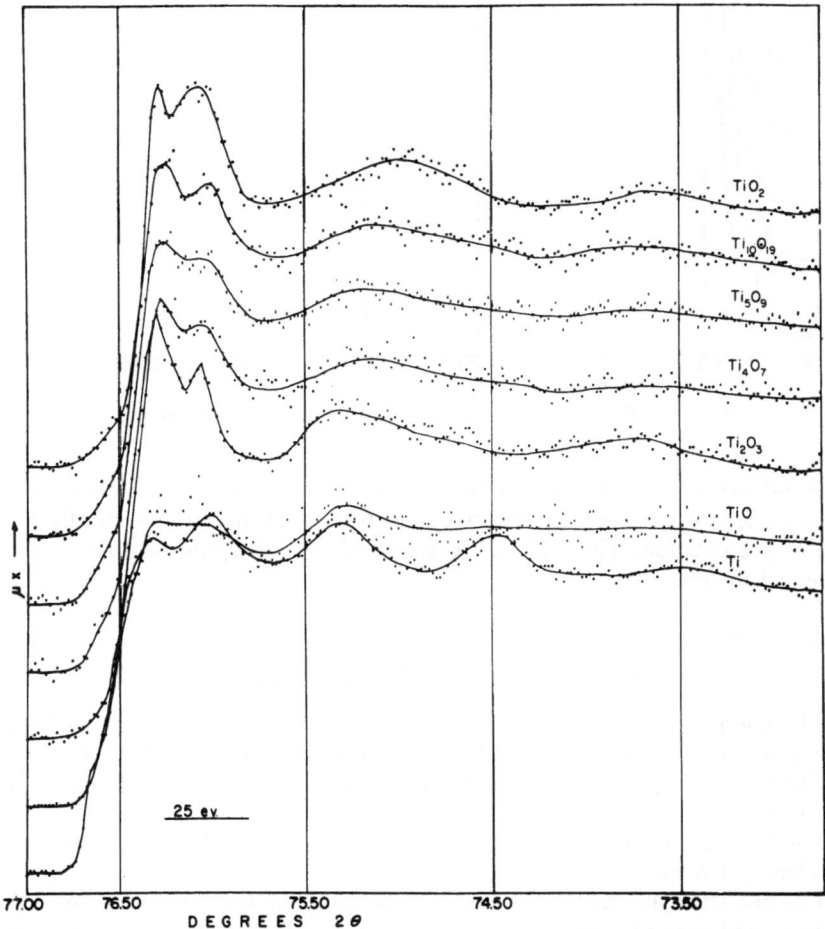

Figure 4. *K* absorption spectra for titanium metal and oxides.

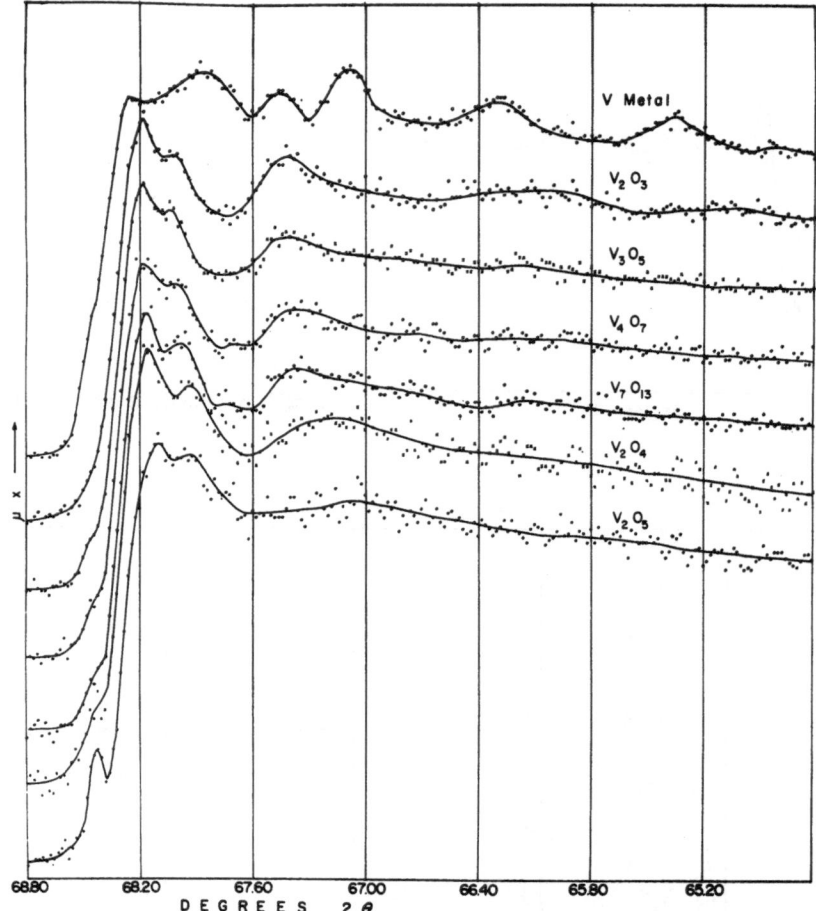

Figure 5. *K* absorption spectra for vanadium metal and oxides.

Figures 4–9 show the normalized absorption curves for most of the materials included in this study. In these figures, successive curves have been displaced vertically for convenience of presentation. Table II lists energies of the absorption features having set Point A equal to zero.

General Trends

The following qualitative observations follow from these data:

1. The energy of the leading edge (A), the point at which absorption begins to increase, is essentially constant for all compounds of a given element.
2. The first absorption peak (B) is generally resolved for cations of charge four or greater.
3. The position of the main edge (C) tends to shift toward higher energies as the valence increases.
4. The fine structure for the cations in oxide compounds tends to be most strongly developed near the edge, then smooths out rapidly toward higher energies. On the other hand, metals tend to have low-amplitude structure near the edge, while

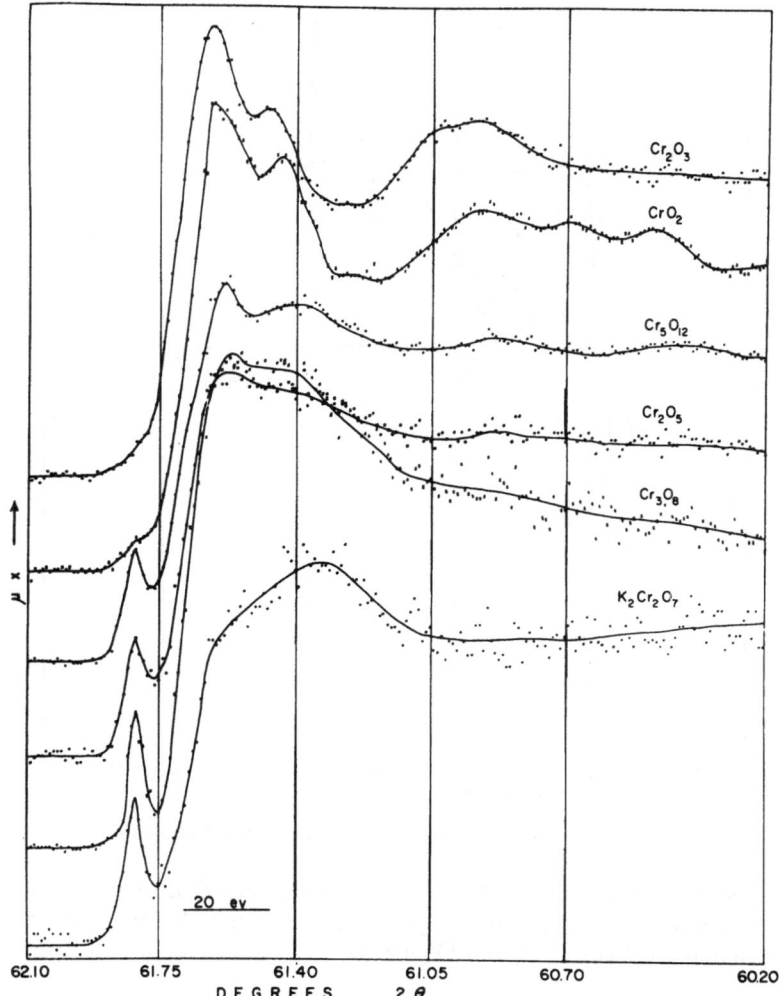

Figure 6. *K* absorption spectra for chromium oxides.

their fine structures are well-resolved for several hundred electron volts from the edge.

5. The D, E, and F maxima tend to shift toward higher energies, the higher the valence.

6. For a given element, the intensity ratio of the D to E maxima is highest for low cation charges and gradually diminishes as the cation charge increases (chromium and titanium oxides, especially). The E maximum is occasionally unresolved (GeO_2 and α-Fe_2O_3).

7. The energy difference between points D and F varies from about 45 eV (MnO) to about 75 eV (GeO_2).

Van Nordstrand,[11] on the basis of an extensive experimental survey, has grouped these spectral features into four types. His Type I spectra would include the Cr_2O_3 spectrum of Figure 6. The Type II spectra which he assigned to octahedral complexes with linear ligands CN and CO is similar to the anatase spectrum in Figure 13. He

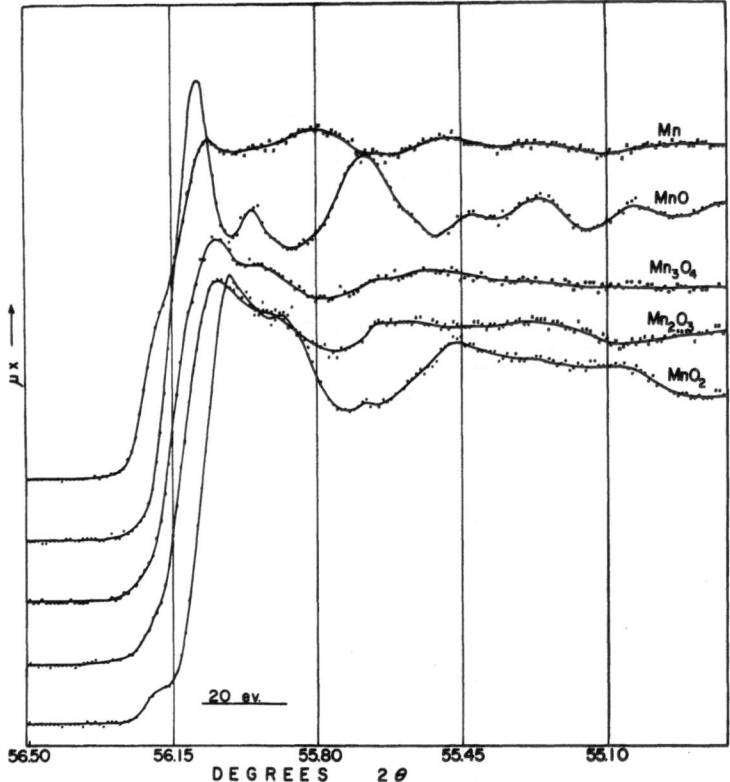

Figure 7. *K* absorption spectra for manganese metal and oxides.

classified metals as Type III spectra. His Type IV spectra include the high-valence cations, such as Cr^{+6} in $K_2Cr_2O_7$ in Figure 6. It appears from this study that there are progressive gradations among the Type I, II, and IV spectra.

Parameters of Crystal Chemistry

Effects of Valence. Several changes occur in the absorption structure which can be qualitatively correlated with changes in valence of a particular element. Unfortunately, no single feature or parameter has been isolated which gives a strong correlation with the valence state for all the elements studied. One of the most consistent correlations is observed for the increased resolution and amplitude of peak B for the oxides of vanadium and chromium. It appears, for example, that one could use this feature to determine the oxygen content for unknown samples in the chromium–oxygen system. There is also a consistent shift in the D peak position with valence for the chromium oxides. Also, for chromium, the D/E peak intensity ratios shift consistently with valence.

Dodd and Kaup[2] have proposed utilization of the shift in the D and F peaks of iron to measure the ferric/ferrous ratio in clay minerals. This is probably a valid technique so long as the iron is all in six-fold coordination. It is noted in Figure 8 that the D and F peaks positions for Fe_3O_4 are in essentially the same position as for α-Fe_2O_3, instead of being two-thirds of the distance from FeO toward α-Fe_2O_3. This may be due to the fact that iron in Fe_3O_4 is in both tetrahedral and octahedral coordination. The same sort of break in trend is noted in the case of Mn_3O_4.

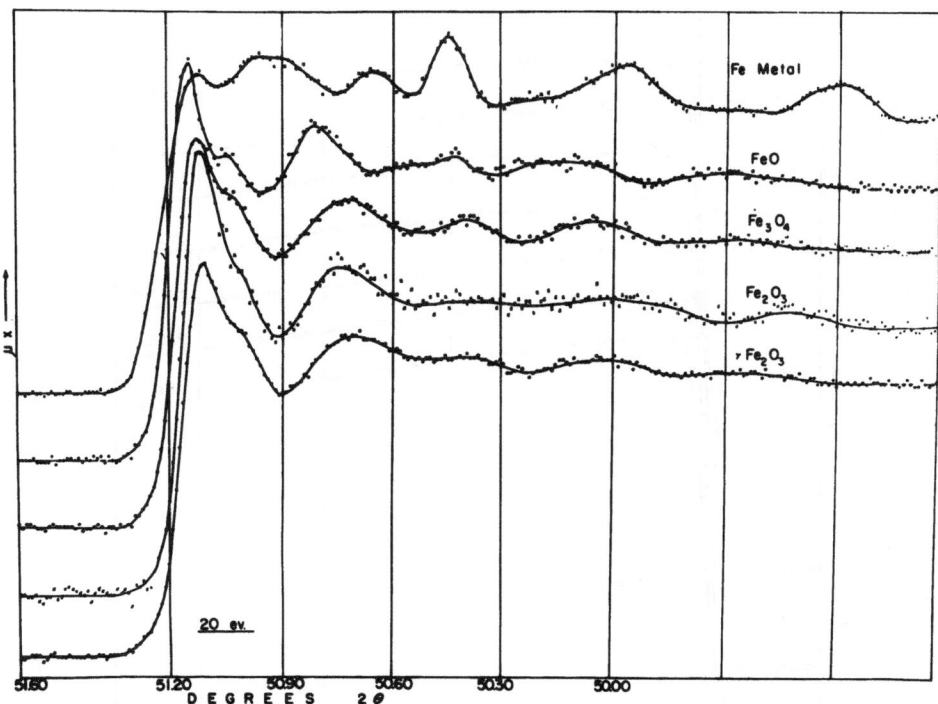

Figure 8. K absorption spectra for iron metal and oxides.

The oxides of lead were run primarily for the purpose of establishing whether or not one could hope to study heavy-metal compounds by this technique. Figure 9 summarizes the results. It is seen that fine structures can be resolved on the L_{III} edge and that the amplitude of the fine structure tends to increase with increasing valence of the lead.

Effect of Bond Type. Absorption spectra for metals are distinctly different from those for oxide compounds. The initial sharp absorption features common to the lower-valence cations are not present in the metal spectra. On the other hand, the extended fine structure is consistently well-resolved (except for lead) for several hundred electron volts beyond the edge.

Effect of Electron Configuration. No correlation has been observed between d electron configuration and absorption fine structure. For example, Ti^{+4} in rutile and Cr^{+6} in $K_2Cr_2O_7$ are isoelectronic, but their absorption structures are entirely different.

Isostructural Series. Absorption spectra for three oxides with the sodium chloride structure have been redrawn to a linear energy scale in Figure 10. These curves are quite dissimilar indicating that simple crystal structure and valence considerations are not adequate to explain absorption structure. Apparently, the metallic character of the bonding in TiO contributes to the gross difference between it and the other compounds. The MnO and FeO curves are somewhat similar. The main difference is that the extended fine structure of MnO is better resolved than for FeO. The equivalent maxima in the FeO structure are shifted slightly to higher energies, as the Kronig theory would predict, for the cubic structure having the smaller cell dimension.

Figure 11 gives the spectra for the corundum-structure oxides. The spectra for Ti_2O_3, V_2O_3, and Cr_2O_3 have similar structures, especially with respect to the D, E,

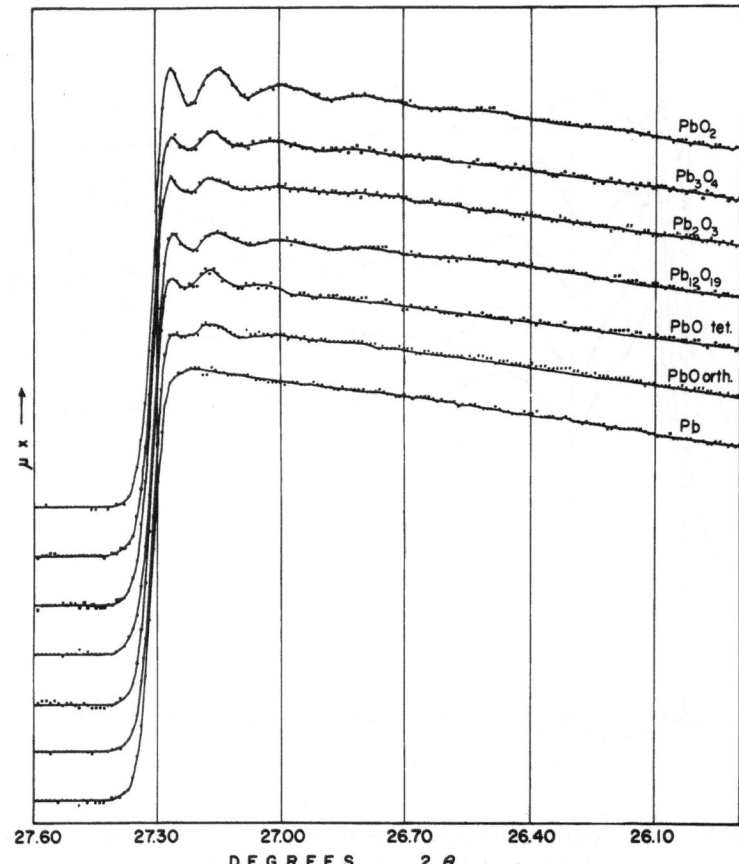

Figure 9. L_{III} absorption spectra for lead metal and oxides.

and F maxima. However, the spectrum for α-Fe_2O_3 is markedly different. This difference may arise from variations in cation–cation interactions across shared octahedral faces as proposed by Goodenough.[4] In the refinement of these corundum structures, Newnham and de Haan[8] have determined that α-Fe_2O_3 has an anomalously large cation-to-cation distance across the shared octahedral face.

The absorption spectra in Figure 12 for the three transition-metal rutile structures (TiO_2, MnO_2, and CrO_2) are very similar, while the spectrum for GeO_2 is quite different.

Effects of Polymorphism. Several polymorphic sets included among the results serve to illustrate the effect of coordination and polyhedral linkage. In Figure 13, spectra for the two polymorphs of TiO_2 show that major changes can occur as a result of subtle differences in crystal structure. In rutile each octahedron shares two edges, but in anatase each TiO_6 octahedron shares four edges. The radical dissimilarity in the D and E maxima between these polymorphs would suggest that these maxima are not simply caused by electron transition to "optical-like" orbitals, but rather that they must be strongly affected by next nearest neighbors.

From Figure 8, it is seen that the extended fine structure for γ-Fe_2O_3 is similar to that of Fe_3O_4. The γ-Fe_2O_3 sample contains 10% SiO_2 for purposes of phase stabilization.[10]

Table II. Energies of Absorption Features (in eV) with Point A Set to Zero

Sample	Å (2θ)	A	B	C	D	d	E	e	F	f	G	g	H	h
Ti (metal)	76.72	0	—	6.5	22.0	28.7	39.2	57.8	79.2	108.0	127.8	160.8	193.0	
TiO	76.68	0	—	8.2	22.0	26.5	33.2	55.5	76.0					
Ti₂O₃	76.60	0	—	6.2	16.5	25.5	30.0	51.2	80.8	130.5	161.2			
Ti₄O₇	76.70	0	—	12.2	23.0	29.8	36.0	57.8	86.2					
Ti₅O₉	76.71	0	—	13.0	24.8	32.0	38.0	58.2	85.5	148.2	178.0			
Ti₁₀O₁₉	76.71	0	—	14.2	24.8	32.0	38.0	54.2	89.0	146.0	169.0			
TiO₂ (rutile)	76.63	0	—	13.8	24.8	31.5	38.2	49.5	55.0	64.0	82.2	106.0s	117.0s	144.5s
TiO₂ (anatase)	76.70	0	—	16.0	23.0	26.5	35.2	56.8	99.5	147.8	177.8			
V (metal)	68.57	0	—	10.0	21.8	28.2	51.0	69.0	79.8s	90.5	106.8	139.2	166.2	215.0
V₂O₃	68.58	0	—	14.0	28.0	36.0	40.0	61.0	83.2s	138.5	177.8	226.2	265.2	
V₃O₅	68.55	0	—	14.0	26.0	32.5	36.0	58.2	81.2s	106.8				
V₄O₇	68.56	0	—	14.0	26.8	36.8	39.5	59.5	84.0s					
V₇O₁₃	68.55	0	—	14.8	26.8	33.2	40.2	60.5	86.5s					
V₂O₄	68.58	0	—	16.8	28.0	40.2	45.2	71.0	100.0s					
V₂O₅	68.56	0	10.0	17.5	32.5	38.0	45.2	70.0	104.5s					
Cr₂O₃	61.89	0	—	11.0	22.0	31.5	36.0	53.0	88.5					
CrO₂	61.90	0	—	14.5	23.5	33.5	40.0	62.0	85.0	99.5	107.0	119.5	127.0	144.0
Cr₅O₁₂	61.89	0	5.0	13.0	25.5	31.5	43.0	72.5	92.0	111.0	132.5			
Cr₂O₅	61.89	0	5.0	14.0	26.0	35.0	42.0	71.5	88.5					
Cr₃O₈	61.89	0	5.0	14.0	28.0	32.5	39.0							
K₂Cr₂O₇	61.89	0	5.0	16.5	30.0	34.0	47.5	81.5						
Mn (metal)	56.26	0	—	7.5	21.5	29.0	60.5	67.9	83.3	97.6	105.3	126.4	147.6	
MnO	56.215	0	—	5.0	13.5	22.0	28.5	38.5	75.5	86.5	93.0	104.0	116.0	
Mn₃O₄	56.225	0	—	8.5	21.0	49.0	65.0	68.5	80.5					
Mn₂O₃	56.21	0	—	7.5	19.0	29.0	33.5	49.5	63.5	82.5	104.5			
MnO₂	56.24	0	—	15.0	25.0	35.5	40.0	55.0	60.5	64.5	84.5	99.0	106.5	117.5
Fe (metal)	51.362	0	—	9	23	30	48	72	85	97	112	129	140	150
FeO	51.32	0	—	6	16	26	30	39	61	81	106	122	151	182
Fe₃O₄	51.30	0	—	8	18	—	—	47	72	97	115	136	160	189
α-Fe₂O₃	51.29	0	—	6	14	—	—	44	64	93	113	129	170	203
γ-Fe₂O₃	51.30	0	—	6	17	—	—	47	71	96	113	131	158	194
FeS	51.35	0	—	7	23	48								
SrFeO₃	51.34	0	—	13	25	—	—	55	85	104	114	128	135	140
GeO₂ (rutile)	32.24	0	—	8	20	47	76	132	156	184	212	233	268	314
GeO₂ (quartz)	32.25	0	—	14	23	56	91	146	184					
Glass 1	32.25	0	—	10	23	54	91	146	187					
Glass 7	32.25	0	—	10	27	52	91	142	187					
Glass 8	32.245	0	—	12	25	49	89	137	182					
Glass 18	32.25	0	—	12	24	54	91	142	191					
Ge-III	32.25	0	—	13	27	63	94	131	162	191	234	279	332	382
Ge-I	32.25	0	—	14	30	47	94	128	158	196	223	272	325	
Ge-70°	32.25	0	—	13	30	50	63	80	97	128	158	199	234	279
Ge-300°	32.25	0	—	13	30	50	63	80	97	134	162	196	230	282

The largest difference observed between two polymorphs has been for the case of GeO₂ (rutile) and GeO₂ (quartz). It can be seen from Figure 14 that these two spectra are quite different, perhaps as a result of the germanium coordination change from six in the rutile polymorph to four in the quartz form.

Two polymorphs of elemental germanium have been studied (Figure 15). Ge-I is the diamond-structure, semiconducting polymorph. Ge-III, formed under high pressure, is body-centered-tetragonal. The only differences in the two absorption spectra appear near the edge (between the D and F maxima). The extended fine structure is the same in both cases.

The absorption spectra for the tetragonal and orthorhombic polymorphs of PbO (Figure 9) do not appear to have significantly different structures at this level of spectral resolution.

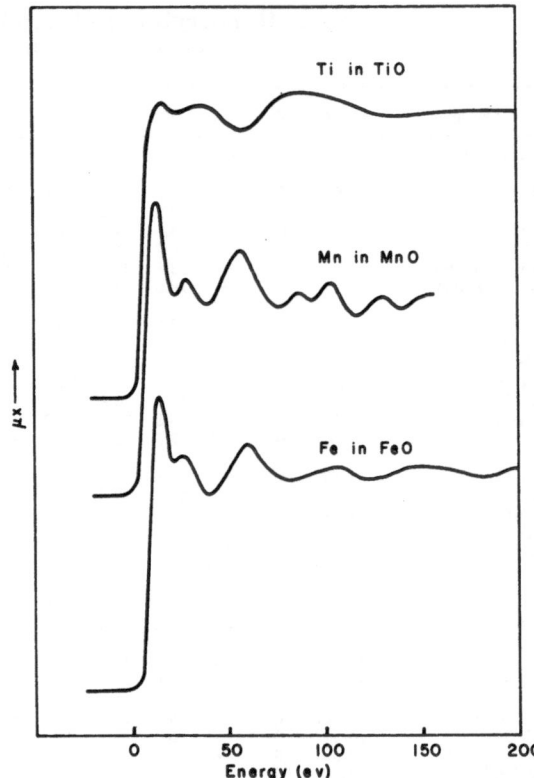

Figure 10. K absorption spectra for sodium-chloride-structure oxides.

Effects of Crystallinity. The K absorption spectra for the two polymorphs of GeO_2 and the four germanium-containing glasses have been reproduced in Figure 14. Glass number 1 is GeO_2 with 1 mol.% SiO_2. This glass was cloudy, showing evidence of phase separation. Its absorption spectrum is distinctly equivalent to the crystallized quartz polymorph, suggesting that the glass also has Ge^{+4} in four-fold coordination and that the basic structure must be essentially the same as for the crystalline form. Furthermore, it appears that the apparent phase separation cannot be caused by some of the Ge^{+4} going into octahedral sites unless the quantity of octahedrally coordinated germanium is very small (certainly less than 10%).

Glasses 7, 8, and 18 are compositions from the system CaO–Al_2O_3–GeO_2–SiO_2. Glass 7 has the molar composition: 3.5 Al_2O_3–19.3 CaO–72.38 GeO_2–4.82 SiO_2. Glass 8 has the composition: 3.5 Al_2O_3–19.3 CaO–57.9 GeO_2–19.3 SiO_2. Glass 18 has the composition: 3.5 Al_2O_3–14.48 CaO–77.2 GeO_2–4.82 SiO_2.

All the absorption structures for these glasses are identical to crystalline GeO_2 (quartz) in spite of the wide compositional range and in spite of the fact that the cation neighbors for germanium are not always germanium atoms. Likewise, the lack of long-range order appears not to influence the spectra.

A second example of the lack of a long-range-order effect is shown for the case of X-ray amorphous and crystalline thin films of germanium metal. Sample Ge-70° in Figure 15 is for an amorphous thin film of germanium that was deposited in vacuum on an aluminum substrate maintained at 70°C. The Ge-300° sample is crystalline as a result of having been deposited on a heated substrate. Both films have identical absorption structure, thus

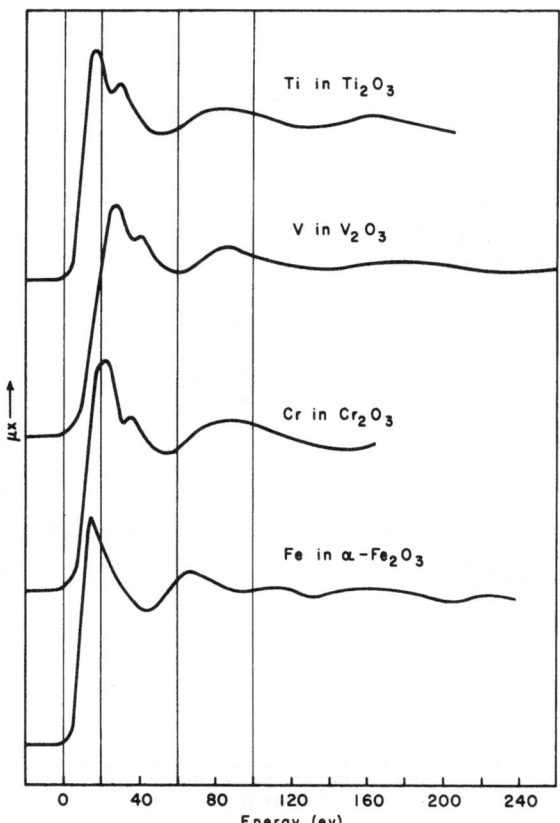

Figure 11. K absorption spectra for corundum-structure oxides.

supporting the thesis that extended fine structure is independent of the degree of long-range order. These results for germanium are in general agreement with the recent work of Lytle.[6]

SUMMARY

An X-ray diffraction system was modified to function as an automated single-crystal absorption spectrometer, capable of step-by-step measurement of direct beam and absorbed X-ray intensities for up to seven materials at one time. The raw intensity data were processed by computer to obtain normalized μX versus 2θ curves. Point-by-point comparison of absorption curves for several compounds of a given material was greatly facilitated by this technique, because effects of long-term electronic and mechanical drift were eliminated.

This study was primarily confined to the first transition-series elements. Oxide compounds were used almost exclusively in order to hold constant any role played by the coordinating anion. Thirty-four simple oxides (synthesized under controlled conditions of oxygen pressure, temperature, composition), eleven miscellaneous compounds and glasses, and nine metal specimens were studied. These samples included several sets showing effects of valence, bond type, structure (including change of first and second coordination), and d electron configuration. The factors found to be the most significant

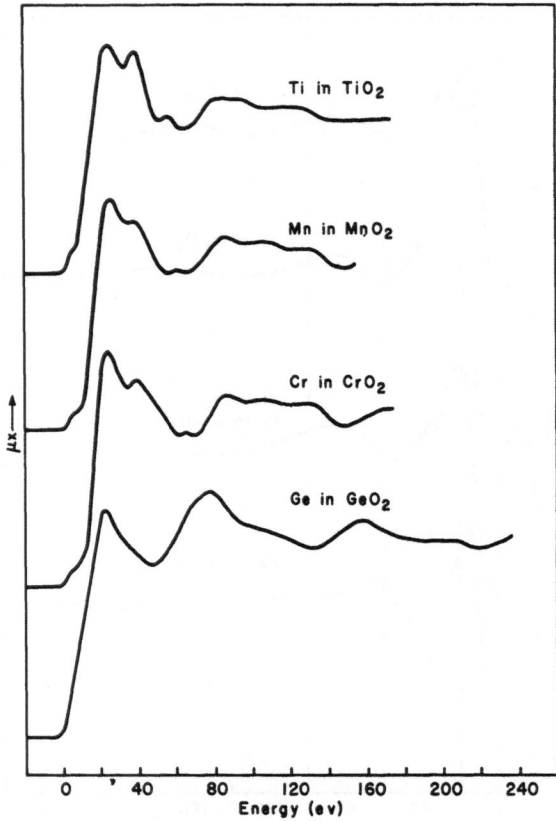

Figure 12. K absorption spectra for rutile-structure oxides.

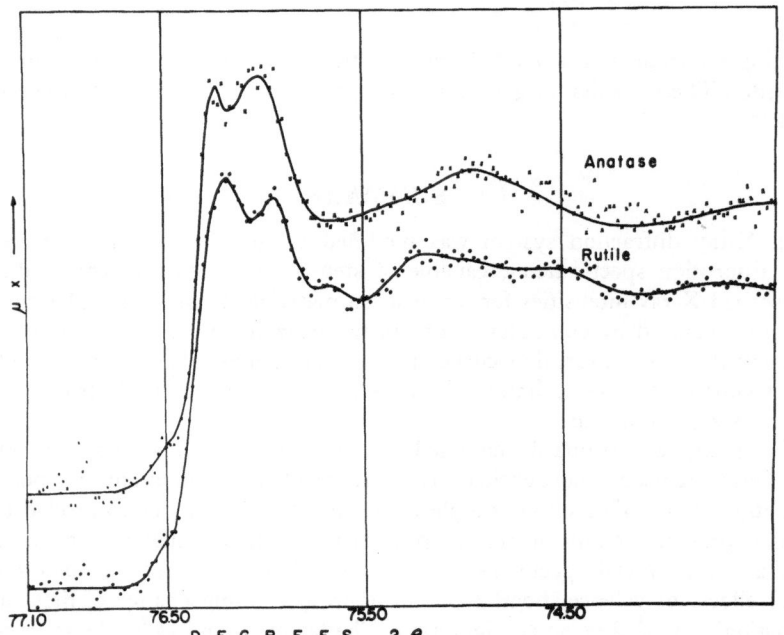

Figure 13. K absorption spectra for rutile and anatase.

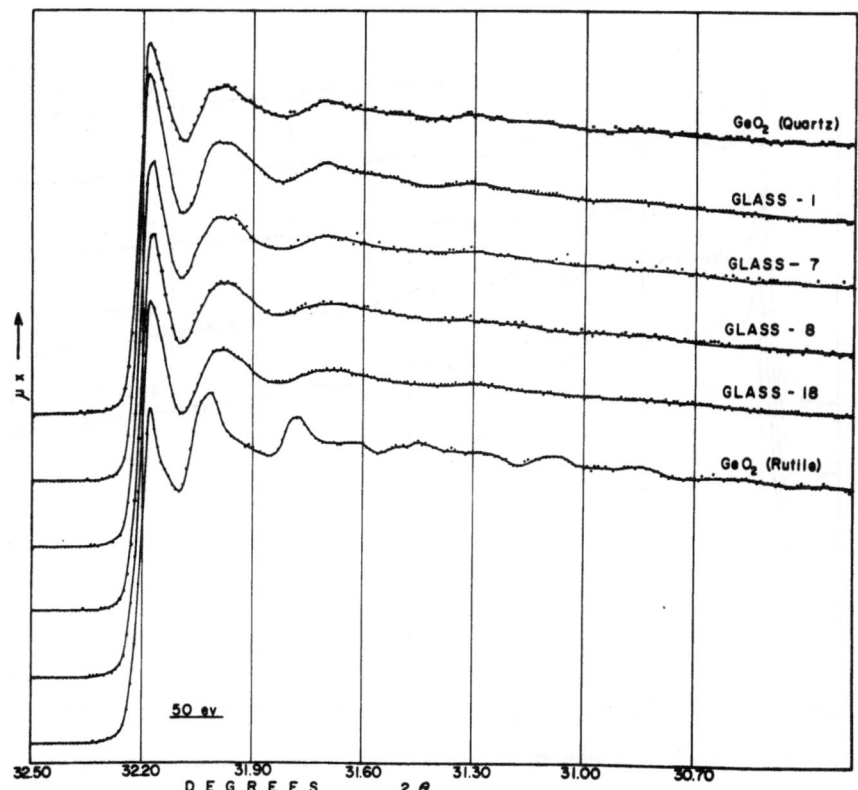

Figure 14. K absorption spectra for GeO_2 polymorphs and selected glasses.

in controlling the absorption fine structure are bond character, nearest neighbor environment (primary coordination number), and valence. Longe-range order (lro) appears to be of little or no consequence, as found earlier by Van Nordstrand,[11] Nelson *et al.*,[7] and Lytle.[6] Furthermore, the next-neighbor cation appears to have no effect, at least in the case of the germanium-containing glasses. The d electron configuration also has no apparent effect on the absorption structure.

From these findings, it is seen that any theory which does not adequately incorporate effects of bonding, cation charge, and crystal structure (including primary and secondary coordination) is not complete.

ACKNOWLEDGMENTS

The writers wish to thank the many members of the faculty of the Materials Research Laboratory for having supplied most of the specimens used in this study. We especially acknowledge the contributions of Dr. W. B. White. The Penn State Computation Center is acknowledged for performing the computer calculations. This work was supported by the Advanced Research Projects Agency, Contract Number SD-132 on Crystal Preparation Research.

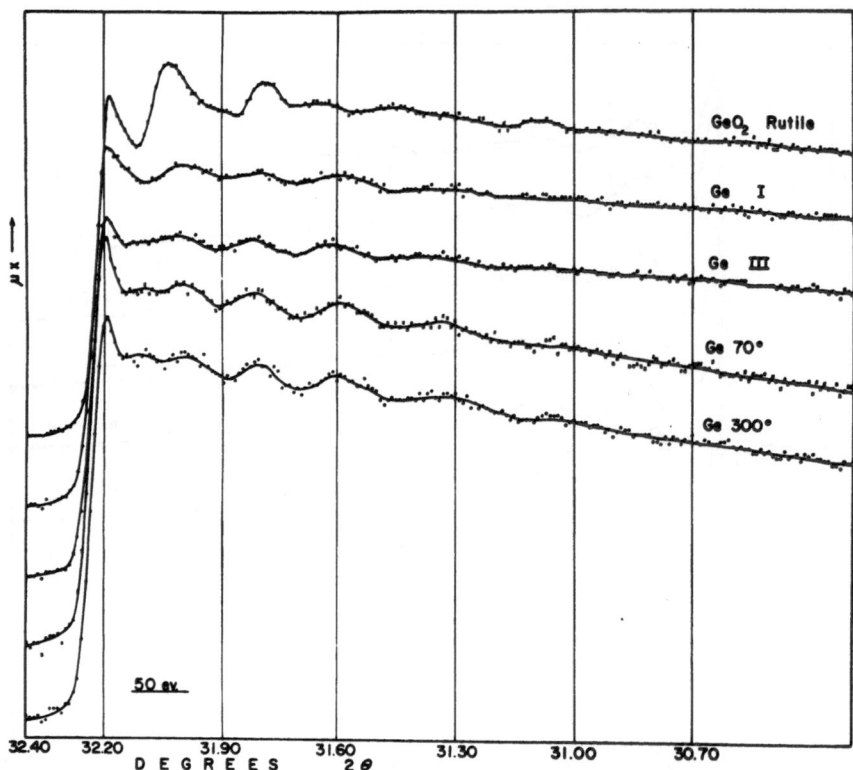

Figure 15. K absorption spectra for germanium metals and GeO₂ (rutile).

REFERENCES

1. L. V. Azároff, "Theory of Extended Fine Structure of X-Ray Absorption Edges," *Rev. Mod. Phys.* **35**(4): 1012, 1963.
2. C. G. Dodd and D. J. Kaup, "Determination of Iron Oxidation States in Clay Minerals by X-Ray Absorption Edge Fine-Structure Spectrometry," *Clay Minerals Bull.* **5**(30): 290, 1963.
3. T. C. Furnas, Jr., and E. W. White, "New Instruments for X-Ray Analysis," in: W. M. Mueller (ed.), *Advances in X-Ray Analysis*, Vol. 4, Plenum Press, New York, 1961, pp. 521–537.
4. J. B. Goodenough, "Direct Cation–Cation Interaction in Several Oxides," *Phys. Rev.* **117**: 1442–1451, 1960.
5. A. I. Kozlenkov, "Theory of the Fine Structure of X-Ray Absorption Spectra," *Bull. Acad. Sci. USSR Phys. Ser. (English Transl.)* **25**(8): 968–987, 1961.
6. F. W. Lytle, "X-Ray Absorption Fine Structure in Crystalline and Noncrystalline Materials," presented at the International Conference on Physics of Non-Crystalline Solids, Delft, Netherlands; Boeing Scientific Research Laboratories Report D1-82-0361, Seattle, Washington, 1964.
7. W. F. Nelson, I. Siegel, and R. W. Wagner, "*K* X-Ray Absorption Spectra of Germanium in Crystalline and in Amorphous GeO₂," *Phys. Rev.* **127**: 2025, 1962.
8. R. E. Newnham and Y. M. de Haan, "Refinement of the α-Al₂O₃, Ti₂O₃, V₂O₃, and Cr₂O₃ Structure," *Z. Krist.* **117**: 235, 1962.
9. L. G. Parratt, "Electronic Band Structures of Solids by X-Ray Spectroscopy," *Rev. Mod. Phys.* **31**(3): 616–645, 1959.
10. D. W. Strickler, "The 1: 5 and Defect Spinels in the System Li₂O · Fe₂O₃ · Al₂O₃," M.S. Thesis, The Pennsylvania State University, University Park, Pennsylvania, 1959.

11. R. A. Van Nordstrand, "X-Ray Absorption Edge Spectroscopy of Compounds of Chromium, Manganese, and Cobalt in Crystalline and Non-Crystalline Systems," in: V. D. Frechette (ed.), *Non-Crystalline Solids*, John Wiley & Sons, New York, 1960, pp. 168–198.

12. E. W. White, "Chemical Characterization of Materials by X-Ray Spectroscopy," Ph.D. Thesis, The Pennsylvania State University, University Park, Pennsylvania, 1965.

DISCUSSION

A. Taylor (Westinghouse Research Laboratories): You showed curves of the absorption spectra of F_3O_4 and Fe_2O_3. Have you done anything with the cubic phase Fe_3O_4 when it's non-stoichiometric and goes all the way over to Fe_2O_3? If not, would you expect a change towards the Fe_2O_3 type, or would you expect the Fe_3O_4 type of spectrum that you showed?

E. W. White: The extended fine structure of the cubic Fe_2O_3 and that of Fe_3O_4 are almost identical. However, the leading edge structure of α-Fe_2O_3 and γ-Fe_2O_3 are quite similar. So, as you go from γ-Fe_2O_3 to Fe_3O_4, the extended fine structure stays essentially the same.

A. Taylor: May I ask a second question and then I will sit down and then you can expound. As you know, in the theory of alloys, when you get the Hume–Rothery phases, you get, say, alloys which are body-centered-cubic when you get an electron–atom ratio of 3 : 2. Now, you get alloys, such as FeAl, NiAl, and CoAl, which have the transition element, and for that you have to give the transition element zero valence to fit in with the Hume–Rothery rule. Does that mean that if you took emission or absorption spectra there of the transition element it would be the same as if you were dealing with the free element rather than the element in the alloy? Or would you expect differences on the basis of what you know of these spectra?

E. W. White: My experience with metals and alloys is almost nil. I would really like to refer this second question to Dr. Azároff or to Dr. Holliday.

THE EFFECT OF CHEMICAL
COMBINATION ON X-RAY SPECTRA: OPEN
DISCUSSION

J. Merritt (Shell Development): This is quite a minor point, but it may get some discussion started. There has been reference several times during this afternoon to the self-absorption effect. We were intrigued by the work that Baun and Fischer did on the L_2/L_3 ratio, and their study showed that this varied for the iron transition elements with accelerating voltage. We said this is the L_3 edge in between these two lines, and so Chet Muller of our laboratories prepared a very thin sample (he estimates it was about 1000 ± 500 Å) and found that the L_2/L_3 ratio did not vary with accelerating voltage, which I think confirms the idea that it is an absorption effect, and I presume it was due to the L_3 edge.

D. W. Fischer (Wright-Patterson Air Force Base): The only comment I have is you did, as you said, check the bulk samples that were larger than that in thickness and did come up with the dependence of accelerating potential on the intensity ratios.

J. Merritt: They were very close to your dependence.

G. A. Walker (Minnesota Mining and Manufacturing Company): I would like to refer back to the discussion on resonance and ionic bonding. I would like to say first of all I am a physicist, not a chemist, and I am not fully conversant with chemical bonding. However, I can understand that, in an ionic bond where an electron goes from one state to another, one would have a distinct wavelength shift. If you have a resonance bond, I'd like to ask Dr. Holliday or Dr. Azároff would this not give you a line broadening instead of a line shift? Would it not also explain some of these bumps, maybe, where your electron is resonating between two energy states and the intensity of the main peak to the bump tells you how long it spends in one state and how long in another. Could you not, if this were the case, be able to tell something about the resonating bond, for instance, the shape of the resonating bond? I don't know of any evidence of this; I am merely speculating to see if this would be the case. If it is, there is a lot of work that could be done on the shape of chemical bonds.

J. E. Holliday: All I can say is that it's an interesting thought, and I think it could be pursued, but that's about as much as I can say at this time. While I'm up here, I would like to say one thing. The increase of negative charge on the titanium atom does not depend upon any of the measurements on wavelength shift of the carbon K band. It depends strictly on the titanium metal atom itself and the wavelength shift of the Ti L_{III} band. I want to make this clear. It is based upon the fact that, for nonstoichiometric compounds of TiO_x, the Ti L_{III} band shifts towards higher energies, but for TiC_x the Ti L_{III} band shifts toward lower energies relative to that for titanium. However, that the titanium atom is more negative in TiC than in pure titanium has been found both from our experiments and the theoretical band calculations of Williams and Lye on TiC_x. Thus, there is both experimental evidence and theoretical calculations, so we feel that this is enough to make this idea worthy of consideration.

N. Spielberg (Philips Laboratories): To come back to this question of the dependence on voltage of the ratio L_2/L_3 line intensities. Liefeld at New Mexico State University looked at the $L\alpha$ line of copper, nickel, and zinc, as a function of voltage.* He found that the shape of this line changes, and he, of course, was able to correlate this to the absorption structure. However, when he got down to a low enough voltage so that this shape did no longer change and there was no longer this particular self-absorption effect, he then found another effect, which was due to satellite excitation. When he dropped his voltage far enough so as not to excite the L_2 states, but still enough to excite the L_3 states, the shape of the line changed considerably. So one would have to worry about making some of these comparisons; either one must include the satellites always or one must exclude them always.

---

* R. J. Liefeld, *Bull. Am. Phys. Soc.* II, **10**, 549, 1965; D. Chopra and R. J. Liefeld, *Bull. Am. Phys. Soc.* II, **9**, 404, 1965.

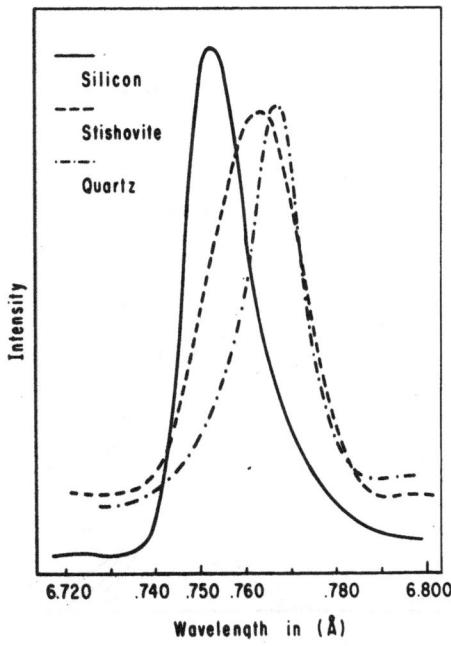

Figure 1. Si K_β peaks from silicon, stishovite, and quartz.

W. L. BAUN (Wright-Patterson Air Force Base): As we showed last year on the aluminum, up to three to four times the excitation potential of the parent line, there is a very marked effect on the satellite line ratio and on the actual intensity of it.

T. C. YAO (Shell Development Company): I was going to offer something on the chemical bonding, but while we are on the form or shape of lines, I would like to offer some suggestion as to the caution about taking too much interpretation of the shape of line inasmuch, as everyone knows, the shape of lines has a lot to do with the composition at the time you measure it, and also the geometry of the sample. I offer the suggestion that as we examine the shape of lines we look at them. Now to go back to the chemical state of bonding, actually we have a lot of evidence in front of us telling us that there is really no definite line we can draw to determine if it is ionic or covalent one way or another. On the evidence that if you examine the line position of various M lines to a metal ion, the line does shift, as someone has shown, according to the electrical negativity of the element, as Pauling illustrated earlier. Therefore, we see that the valence shell must be moving in and out. There is no definite line we can draw and definitely say this is ionic bonding one way or another.

J. E. HOLLIDAY: Faessler has shown that the wavelength shift is related to the charge on the atom, as Yao has stated above. If, for example, there is a wavelength shift which indicates an increase in charge on atom A in going from the pure element to the compound and decrease in charge on atom B, then one interpretation is that there has been a transfer from atom A to atom B and there is an ionic character to the bond between A and B. On the other hand, if there was no change in wavelength in going from the element to the compound for both elements A and B, then it can be assumed that there is no transfer of charge and the bonding between A and B is covalent.

E. W. WHITE: Figure 1 illustrates the sort of results we have obtained using an ARL microprobe with ADP spectrometer to record the silicon K_β emission band. It has been found that the peak shifts about 0.02 Å when going from elemental silicon to alpha quartz. There is also a peak shift of about 0.005 Å between the alpha quartz and the stishovite (rutile structure) polymorphs of SiO_2. The Si K_β peak positions from silicon in some 30 silicates have been found to vary in the range between that from quartz to that from stishovite. This shift has been correlated with the mean silicon–oxygen distance of the silicates. There is also a correlation between peak position and the kind of silicate tetrahedral linkage present in the silicates. The framework silicate structures give Si K_β peaks closest to that for quartz, while the isolated tetrahedra silicate structures give peaks close to that from stishovite. We have also used the Si K_β line shift to study the valence of silicon in

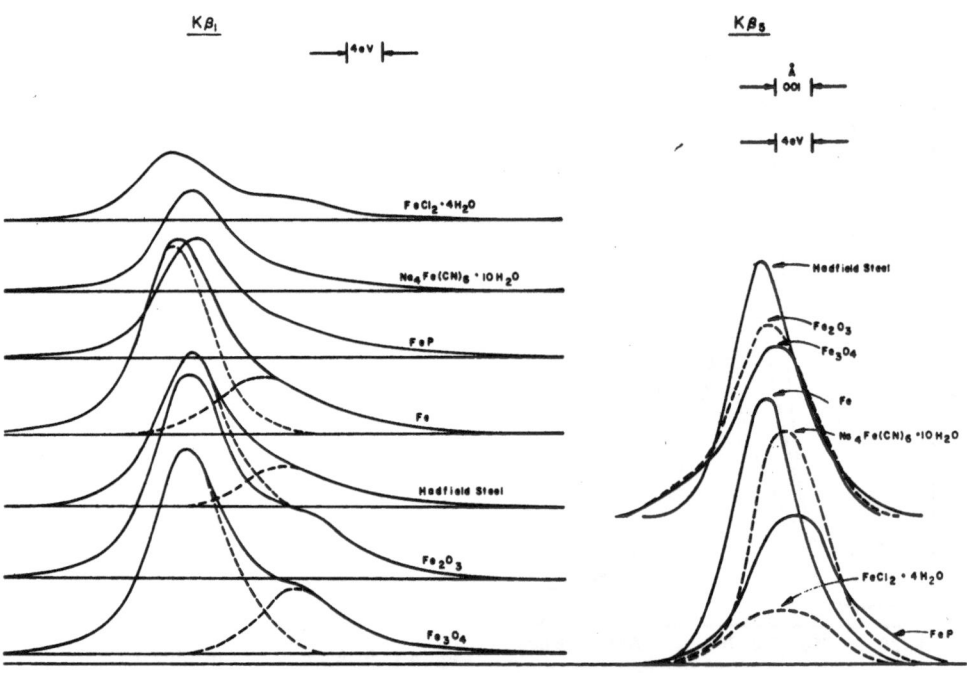

Figure 2. Fe K_{β_1} and K_{β_5} spectral profiles.

SiO thin films. Our results show that the so-called SiO materials actually consist of an intimate mixture of elemental silicon and SiO_2 instead of divalent silicon.

W. L. BAUN: I think the last comment by Dr. White was one of the most powerful uses of the emission and absorption—especially emission—as he has shown. The material need not be crystalline and a great deal can be found out about amorphous films.

J. MERRITT (Shell Development): This comment is made in order to get Bill Baun to say something about some of these data he has on photoelectron spectroscopy. The few things that have appeared in the journals indicate that, for example, with copper in copper oxide, if you compare how the inner levels shift, the interesting thing is that everything inside seems to shift about the same amount, of the order of about 4 or 5 eV. This is a general pattern. So the X-ray shifts end up with only 0.5 or 1 eV difference. However, if we could use photoelectron spectroscopy, we may have a much more powerful tool.

CHAIRMAN W. L. BAUN: I'm sure we have a much more powerful tool, and you pretty well summarized it. This is a general thing. We haven't been able to correlate any of the changes in the photoelectron spectra with the shifts that we've seen, but, on the other hand, the photoelectron work is still pretty new.

G. GORDON (Lawrence Radiation Laboratory): I would like to comment on this X-ray intensity versus the pulse-height analysis. Just last week we did some work on copper K radiation, which is around 8 kV, and found that by increasing the intensity and plotting it out on a 400-channel pulse-height analyzer, at a high intensity, roughly 50 times operating at 1 and 50 mA, the range went from a peak plotted at 8 kV down to what would be equivalent to 5 kV, everything else being the same using a number 6 xenon proportion counter.

G. A. WALKER (Minnesota Mining and Manufacturing Company): All these samples that we have been talking about today I presume are polycrystalline samples and I assume the beam intercepts few grains, at least, at a time. I wondered if any of you have taken a sample with a large grain and considered if there is any difference between the wavelengths of the radiation coming from the center of the grain and the grain boundary. If you are averaging over a lot of grains, you are bound to get grain boundary effects where the bonding may be different, and I was wondering if this may explain some of these sidelines. Has anyone looked at large grains to see if there is any difference between the center of the grain and the grain boundary as far as wavelength is concerned?

W. L. BAUN: I don't know of anybody who used the probe for looking at each area.

E. W. WHITE: To my knowledge, no one has observed a "chemical effect on X-ray spectra" at grain boundaries. One would not expect to see such an effect unless there were differences in valence or coordination number between the center of the grain and the grain boundary. Also, the electron probe excites a rather large volume of sample (typically a hemisphere of 1–5 μ in diameter) so that one might not resolve the grain boundary adequately.

CHAIRMAN W. L. BAUN: I think the range of 6-eV electrons in aluminum is roughly 6000 Å.

G. A. WALKER: Work presently being done involves looking at grains in a sample and the interior of a sample. This could be a little bit different. I know it's not much, but I just wondered if it would have any effect here.

C. P. GAZZARA (Watertown Arsenal): I have one slide which I would like to show. First of all, there are a couple of qualifying statements. I am not a spectroscopist. Figure 2 was prepared by Dr. Richard Weiss from the same agency. He made some comments which I might reiterate. First of all, he thought he would show this exploratory slide, or representative of exploratory work, to provoke some discussion and some useful comments, but I think if he were here he would find that this group doesn't need any provocation. Figure 2 shows the K_{β_1} and the K_β of iron emission lines in various forms. The compounds are shown. The iron, itself, is shown, I think it's the fourth line down. They are all on the same energy scale. We can see shifts and we can also see (which to him was quite important) the splitting of the line, i.e., the development of the second line. It is especially strong with Fe_3O_4 at the bottom. First of all, Hadfield Steel, which you see there, is 12% manganese and 1% carbon in iron. It is an austenite stabilizer. Comments that he made follow. What is the splitting due to? Iron might complicate the issue a little bit. He feels that the splitting of these two lines is more pronounced with the increase in magnetic moment in the iron atom. Secondly, the question that comes up is: Is this second line a satellite or is it an actual splitting of the K_{β_1} line? For example, he suggests that the split could be due to spin orbit coupling, but he can't imagine the magnitude of the second peak to be so large from this effect. The energy scale, incidentally, increases to the right. If there are any comments on this, I'd like to receive them and convey them back to Dr. Weiss. This work was done on the standard Norelco fluorescence unit using a lithium fluoride analyzing crystal and using the 400 reflection to try to get better resolution.

CHAIRMAN W. L. BAUN: There is some detailed Russian work on the K_{β_5} and K_{β_1} in which they show something similar to this.

A. VACIAGO (Catholic University Medical School, Rome): I fully agree with all that has been said here and I just want to add something. In my department, we have started to make some calculation in connection with Coulson. The experimental arrangement is as usual. I have a research student named Giatiturco, who has done all the work and I am going around talking about it. Just to gain some confidence about this calculation, I just want to add that the Coulsen and Dowley calculations are a very nice internal check. The K_α transition as an experimental value, as you all know, is 2306.85 eV for sulfur and the calculated value with the wave function in the same approximation that was used for working out the chemical shift gave for the same transition the value of 2310.42 eV. I wanted to make this remark in order to give confidence for the other calculations we have made on sulfur satellites. For the satellites, the only experimental values available, as far as we know, are old values given in 1934 by Parratt on some sulfites. They are there for the α', α_3, α_3', α_4. We have tried to explain these values making the appropriate wave function calculation in the standard approximation of j–j intermediate coupling and using the integrals calculated for the mono-ionized states and making some perturbation on the coefficients in order to take into account the double ionized state we have here. We have the three different states—sulfates, sulfur, and sulfites. These values must be compared with those values. The comparison is quite good. These are the differences. Shown in Table I are the differences between the computed values and the experimental values which I remember are referred to this valence state. Do not take into account the second line given for α'. There was a possibility of two different assignments, but we figured this is the value for the proper functions for α'. What I would like to point out is there is quite a big chemical shift, and, therefore, experimental measurements would be welcome to test this chemical shift in the sulfur satellites. If we work out these differences, I mean the distance between the satellite line and the α_2 line, these are the experimental values having been calculated from the differences between these values and the α_2 value for the same state, which is given in Faessler's work. The same differences for computed values are given for the three different states and, again in these cases, both terms of the differences are theoretical, both are computed for the sulfur state. Also, α_2 is computed for sulfates, sulfur, and sulfites. As you can see, the two lines move together and the moment of magnitude of the computed separation is of the same order of the experimental separation. I would also like to say (in answer to some comment by Dr. Azároff) we have made computation on electrons emitted by sulfur in different states, and indeed electron spectroscopy, as you would expect, is much more sensitive to chemical state than this because you really have the shift of the energy level. Our computation has been very well oriented, but they give shifts up to 10% going from sulfates to sulfites. We are now improving this computation.

Table I

Transition	Experimental line value for sulfur (eV)	Calculated line value for three sulfur states			
		S^{-2}		S^0 (eV)	S^{+6} (eV)
		(eV)	($\alpha_{calc} - \alpha_{exp}$)		
α' $^1S[1s2s] \rightarrow {}^1P[2s2p]$	2315.83	2317.84	2.01	2324.44	2338.00
α_3 $^3P[1s2p] \rightarrow {}^3P[2p^2]$	2321.28	2321.83	0.55	2328.37	2341.99
α_3' $^3S[1s2s] \rightarrow {}^3P[2s2p]$	2322.03	2325.22	3.19	2331.79	2345.26
α_4 $^1P[1s2p] \rightarrow {}^1D[2p^2]$	2324.14	2328.27	4.13	2334.80	2349.09
α_2	2306.53	2299.26		2305.80	2319.40
$\alpha' - \alpha_2$	9.30	18.6		18.6	18.6
$\alpha_3 - \alpha_2$	14.75	22.5		22.5	22.5
$\alpha_3' - \alpha_2$	15.50	25.9		26.0	25.9
$\alpha_4 - \alpha_2$	17.61	29.0		29.0	29.6

DETERMINATION OF INTERATOMIC DISTANCES FROM X-RAY ABSORPTION FINE STRUCTURE

F. W. Lytle

Boeing Scientific Research Laboratories
Seattle, Washington

ABSTRACT

A pragmatic, easily applied theory of the extended X-ray absorption fine structure found on the high-energy side of absorption edges is developed. This simple model can be successfully applied to determinations of interatomic distances, and it accounts for differences in Kronig structure on different absorption edges in the same material. Quantitative agreement with experimental data is good.

INTRODUCTION

In spite of a number of rigorous attempts, the extended X-ray absorption fine structure found on the high-energy side of absorption edges has not been adequately explained. (The recent review by Azároff[1] contains an evaluation and the original references.) Evaluation of the theories by comparison with experimental data is difficult because either the calculations are very complex or the experimental data contain insufficient information. While the more exact theories may be interesting from the standpoint of the physics involved, a simple, easily applied treatment would be more useful to those who have only a pragmatic interest in the phenomenon. This paper is an attempt to formulate such a theory and illustrate its capability for determining interatomic distances. Portions of this work were presented and are contained in the proceedings[2] of the 1964 Delft Conference on Physics of Non-Crystalline Solids.

During the X-ray absorption event, the X-ray is absorbed by the atom with conservation of energy and momentum. The energy balance is as follows:

$$h\nu - E_k = E \tag{1}$$

where $h\nu$ is the initial energy of the X-ray photon, E_k is the initial binding energy of the electron, and E is the kinetic energy of the ejected electron. The ejected electron carries away the excess energy, while the momentum of the photon can be absorbed by the massive atomic nucleus. If I_0 is the X-ray intensity without an absorber and I is that with an absorber in the X-ray beam, the familiar absorption equation is as follows:

$$I = I_0 \exp\left(- \mu x\right) \tag{2}$$

or

$$\mu x = \ln I_0/I \tag{3}$$

[1] References are at the end of the paper.

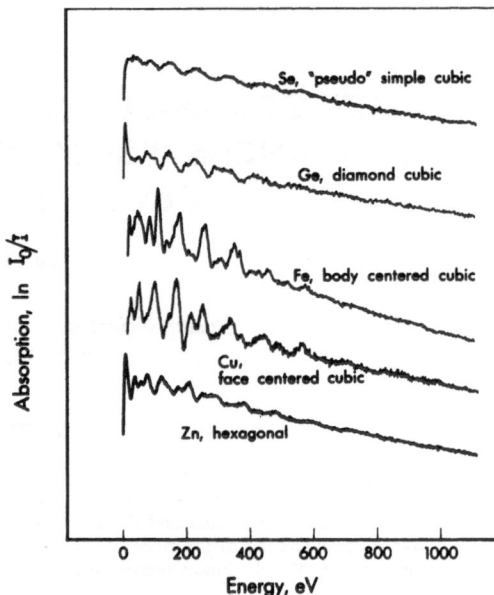

Figure 1. Kronig structure in various lattices.

where μ is the linear absorption coefficient and x is the absorber thickness. The experiment is performed by measuring I and I_0 at suitably close wavelength intervals and plotting $\ln I_0/I$ as a function of wavelength (converted to the energy of the ejected electron).

Historically and physically, there is good reason for noting two separate areas in the fine structure—that within approximately 50 eV of the absorption edge and that at greater energy, which have been named (after early investigators) the Kossel and Kronig structures, respectively. The close-in structure contains much information significant to and dependent on the chemical bonding and symmetry in which the atom is involved; however, our primary interest is the Kronig structure, which is due to the interaction of the ejected electron with surrounding atoms and is dependent upon the interatomic distance and symmetry. Examples of Kronig structure in various lattices are shown in Figure 1.

EXPERIMENTAL TECHNIQUE

A block diagram of the apparatus is shown in Figure 2. This consists of a conventional X-ray generator with tungsten- or silver-target X-ray tube and single-crystal spectrometer (Siemens) utilizing narrow slits (10–100 μ), a special crystal holder and positioning stage with LiF or quartz crystal, scintillation counter, pulse-height analyzer, and electronic scaler timer. The numerical resolving power was approximately 2000–3000. A convenient graphic comparison of resolution is the full width at half height of standard reference lines; the width at half maximum intensity of the $W L_{\gamma_1}$ line was 16 eV compared to 10–13 eV for a double-crystal spectrometer. By alternately recording X-ray intensity with the absorber in (I) and out (I_0) of the X-ray beam, many instrumental drift problems were minimized. This was achieved within the cryostat by the bellows-hinged absorber changer illustrated in Figure 3(b). The device is actuated by a stepping motor on top of the cryostat. Although the cryostat may be used with liquid helium, most of the thermal smearing[3] is removed at liquid-nitrogen temperature (77.4°K), and the data reported

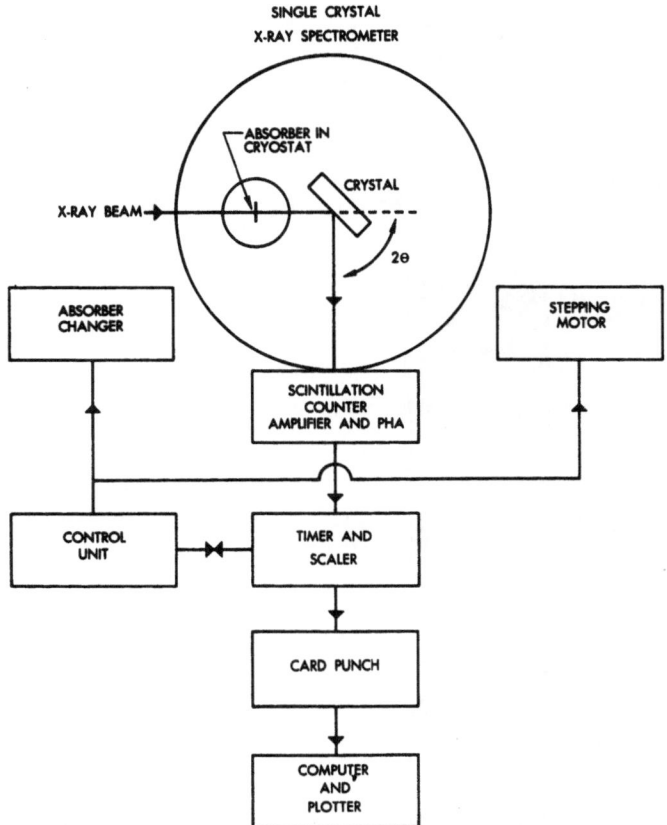

Figure 2. Block diagram of X-ray equipment.

here were obtained at this temperature. Absorbers that did not have high thermal conductivity were sandwiched between thin aluminum foils. Absorbers prepared by evaporation were evaporated on and supported by this same aluminum foil, which does not influence the fine structure in this wavelength region.

An interconnecting control unit keyed by the scaler-timer changed the absorber position and actuated the spectrometer stepping motor appropriately. The X-ray counting statistical accuracy was always better than 0.2%. Data were taken in 1–2 eV intervals. The output from the timer scaler was punched and (printed) on IBM cards and the absorption curves calculated and automatically plotted on the IBM 7090 computer and Orthomat plotter. The computer program accurately located the true energy position of the absorption curve by comparison with reference lines always included in the data. A subroutine also corrected the data for coincidence losses in the counting circuits which became appreciable on strong emission lines. The finished absorption curves are plotted as the product of linear absorption coefficient and absorber thickness ($\mu x = \ln I_0/I$) as a function of energy in electron volts. The zero of energy corresponds to the first part of the absorption edge of the appropriate metal. These values and the wavelengths of the reference lines are those tabulated by Bearden.[4]

All metallic absorbers were thin metal foils, well annealed and 1–5 μ thick. The selenium in Figure 1 was ground to a powder, cast in plastic, and supported on aluminum

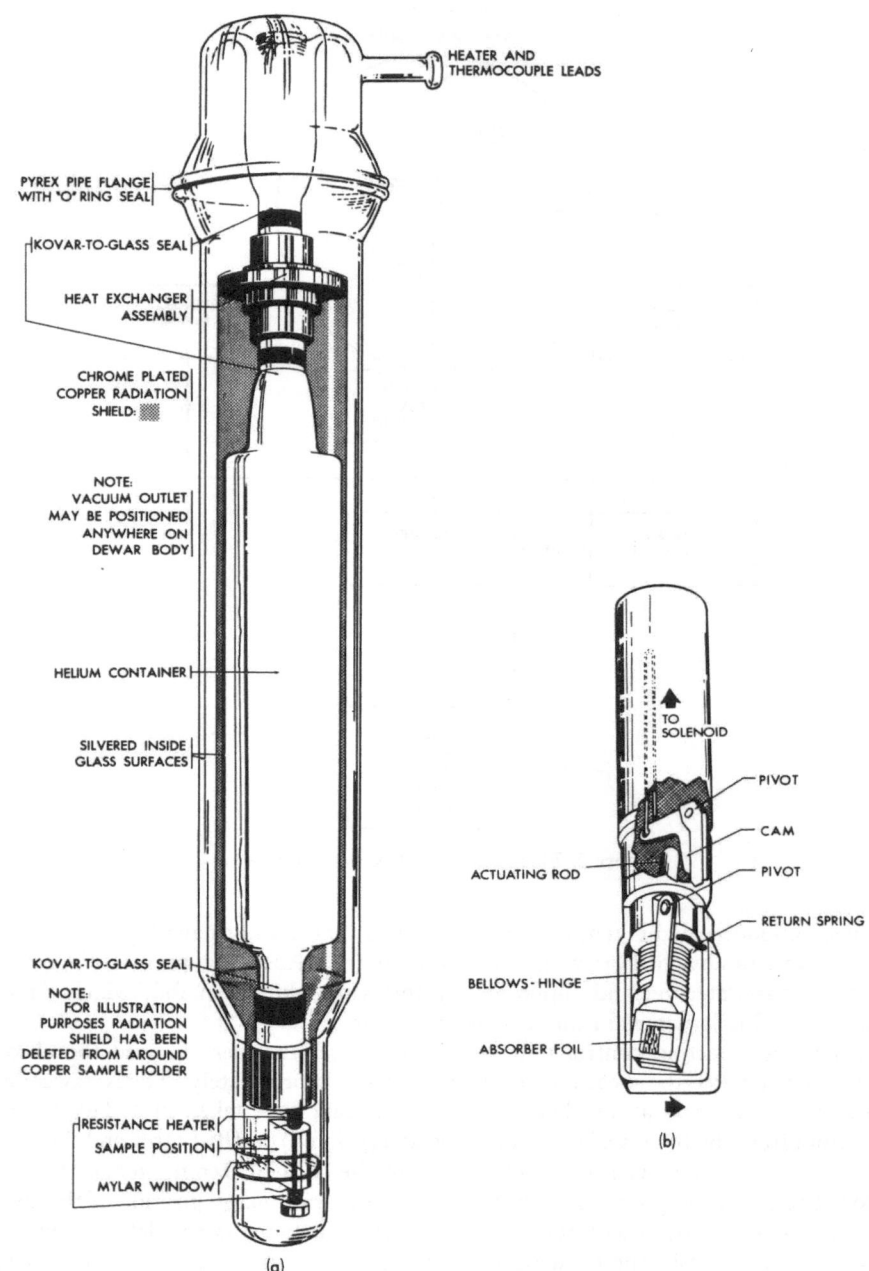

Figure 3. X-ray cryostat for (a) X-ray diffraction and (b) X-ray absorption.

foil. The crystal structure was actually monoclinic (red variety), but is roughly a distorted simple cubic and is used merely to illustrate the Kronig structure from that type of lattice. Germanium was prepared in two ways—by the procedure discussed above for selenium and by vapor deposition onto aluminum foil with subsequent annealing to

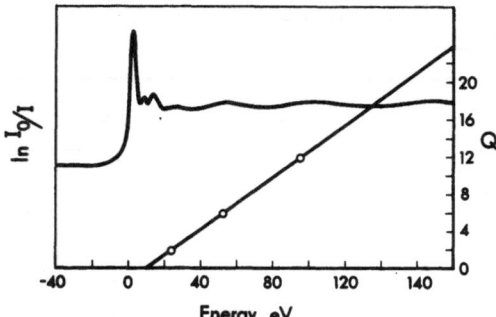

Figure 4. Kronig structure and plot of E vs. Q for gaseous $GeBr_4$.

develop the crystal structure. There was no difference in the Kronig structure or X-ray diffraction patterns of the two varieties. The iron-containing amorphous carbon was prepared in the same way as the selenium.

THEORY

The problem consists of electron waves interacting with a surrounding potential array and is similar to band-structure calculations. In the method of Wigner and Seitz, the wave equation is solved in one atomic polyhedron (unit cell), while choosing boundary conditions so that the wave functions join smoothly to waves in adjoining cells. The method depends on the reduction of the problem to spherical symmetry which makes the solution much easier.

Similar concepts and methods are used here. A nearly spherical polyhedron is constructed in the lattice and approximated by a sphere of equivalent volume with radius r_s. The wave equation then can be solved in the sphere in the usual way for any state $\psi(r)$ as follows:

$$\nabla^2\psi(r) + \frac{2m}{\hbar^2}[E - V(r)]\psi(r) = 0 \tag{4}$$

where E is the kinetic energy and $V(r)$ is the potential seen by the electron. Accurate treatment of $V(r)$ is difficult and involves considerable complication of the solution. We make the following gross simplification reasoning as follows:

1. The absorption spectrum of a monatomic gas, such as argon,[5] has an optical series-limit character and has been explained by atomic dipole transitions to levels belonging to the absorbing atom. Electrons with much higher energy are also influenced by the initial symmetry of the electron as is evident in the gold L-absorption spectra in Figure 7. We assume that symmetry is conserved for all electrons during the absorption process, and the most favorable transitions will be those conforming to the well-known dipole selection rules where the final states involved in Kronig structure are defined by the surrounding atoms.

2. The experimental data for amorphous materials[2] and polyatomic gases ($GeBr_4$ is discussed later) suggest that near-neighbor atoms are most important in the effect. The construction of the atomic polyhedron is dependent upon and includes the effect of the near neighbors on the absorbing atom.

3. Whenever the expanding electron wave front has a node in the neighborhood of the surrounding atoms, the electron wave will fulfill conditions for resonance in a way analogous to resonance in any cavity. This condition will increase the amount of energy absorbed by the system (an allowed transition) and is identified with absorption maxima.

Table I. Energy Levels Q for a Particle in a Spherical Cavity in Units of $h^2/8m\, r_s{}^2$

n	$l = 0$		$l = 1$		$l = 2$		$l = 3$		$l = 4$	
1	1.0	(1s)	2.04	(2p)	3.37	(3d)	4.95	(4f)	8.80	(5g)
2	4.0	(2s)	6.04	(3p)	8.40	(4d)	11.0	(5f)	13.9	(6g)
3	9.0	(3s)	12.0	(4p)	15.4	(5d)	19.0	(6f)	22.8	(7g)
4	16.0	(4s)	20.0	(5p)	24.4	(6d)	29.0	(7f)	34.0	(8g)
5	25.0	(5s)	30.0	(6p)	35.3	(7d)	41.0	(8f)	47.0	(9g)
6	36.0	(6s)	42.0	(7p)	48.6	(8d)	55.2	(9f)	61.8	(10g)
7	49.0	(7s)	56.0	(8p)	63.8	(9d)	71.1	(10f)	78.8	(11g)
8	64.0	(8s)	72.0	(9p)	80.5	(10d)	89.0	(11f)	97.9	(12g)
9	81.0	(9s)	90.0	(10p)	99.4	(11d)	109	(12f)	119	(13g)
10	100	(10s)	110	(11p)	120	(12d)	131	(13f)	142	(14g)
11	121	(11s)	132	(12p)	143	(13d)	155	(14f)	167	(15g)
12	144	(12s)	156	(13p)	168	(14d)	181	(15f)	195	(16g)
13	169	(13s)	182	(14p)	195	(15d)	208	(16f)	223	(17g)
14	196	(14s)	210	(15p)	224	(16d)	238	(17f)	254	(18g)
15	225	(15s)	240	(16p)	255	(17d)	270	(18f)	287	(19g)
16	256	(16s)	272	(17p)	288	(18d)	304	(19f)	322	(20g)

4. Therefore, we let $V(r) = \infty$ at r_s and $V(r) = 0$ within the sphere. This reduces the problem to a particle-in-a-box and is easily solved. An absorption maximum will occur whenever E satisfies the following equation:

$$E = \frac{h^2}{8mr_s{}^2}\, Q \tag{5}$$

where h is Planck's constant, m is the electron mass, and Q are the zero roots of the half-order Bessel function which appears in the radial part of the solution of the wave equation. Table I lists the Q values for a particle in a spherical cavity in units of $h^2/8m\, r_s{}^2$. The wave functions for this model have an arrangement of radial and angular nodes similar to atomic wave functions and, hence, are similarly named, i.e., 1s, 2p, etc. Furthermore, the familiar dipole selection rules ($\Delta l = \pm 1$) are inherent in the model and restrict the transitions from any initial symmetry to a specific set of Q levels. For example, the K-absorption will be to levels of p symmetry where $Q = 2.04$, 6.04, 12.0, 20.0, etc. For L shell absorption, the L_{II} and L_{III} transitions will be to s and d levels and the L_I transitions to p levels. Also analogous to the atomic scheme, the principal quantum number is n, the number of radial nodes is $n-1$, the angular quantum number l is the number of angular nodes, and the degeneracy of each level is $(2l + 1)$. A simple data analysis scheme is obtained when E is plotted against Q where the slope of the resulting straight line contains r_s. If M is the slope, the following relation (in angstrom units) holds:

$$r_s{}^2 = \frac{37.60}{M} \tag{6}$$

To apply this technique to actual absorption data a number of factors must be considered. First, the Kossel structure must be separated from the Kronig structure in order that the first absorption maximum may be assigned to the first Q, etc. This is most easily illustrated with an actual example where the difference is apparent. Figure 4

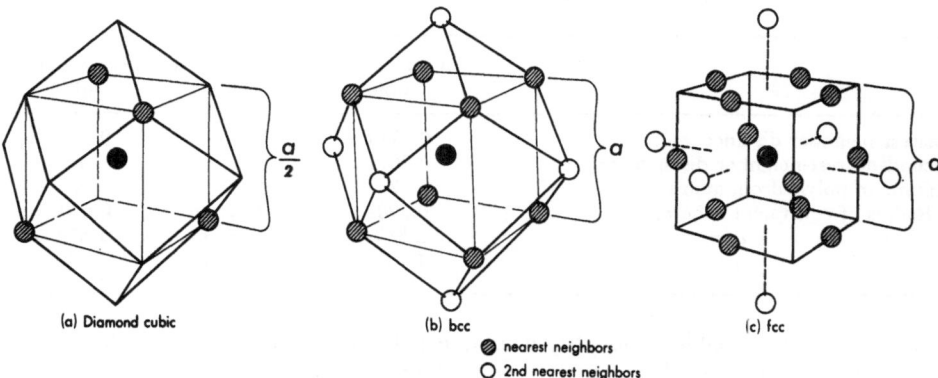

Figure 5. Atomic polyhedra for (a) diamond-cubic, (b) body-centered-cubic, and (c) face-centered-cubic lattices.

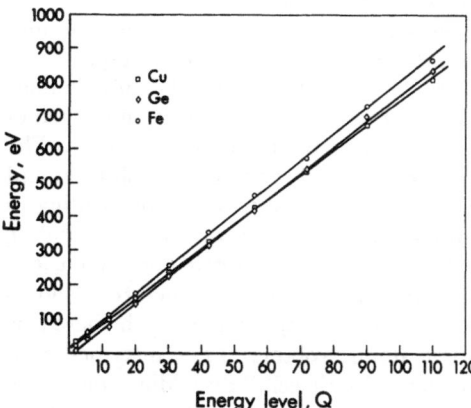

Figure 6. Plot of E vs. Q for germanium, iron, and copper.

shows the data of Glaser[6] for gaseous $GeBr_4$. The first three, close-spaced absorption maxima have the form of an optical series-limit and are clearly of the Kossel structure, while the broader, widely-spaced maxima beginning at about 21 eV are of the Kronig structure. Figure 4 also contains a plot of E vs. Q where the ordered pairs (E, Q) are (21.3, 2.04) ,(49.7, 6.04), and (92.5, 12.0). It is apparent that the straight line is a good fit to the points and, from the slope, $r_s = 2.29$ Å. In the $GeBr_4$ molecule, the germanium atom is in the center with tetrahedrally located bromine atoms at a distance Ge–Br $= 2.32$ Å. In an isolated molecule, the appropriate r_s for our spherical box model will be the interatomic distance. Note that the values are in fair agreement.

In solids, the discrimination between Kossel and Kronig structure is not always easy; the effect of neighboring atoms broadens the levels and smears the structure; however, the ground state of the model depends on $r_s{}^2$ and Q where Q is either 1.0, 2.04, or 3.27, *etc.* Since the magnitude of r_s can be estimated, the first Kronig maximum can be identified. In the event the (E, Q) pairs are improperly matched, the plot will not be linear and reassignment of the (E, Q) pairs and replotting will be necessary.

The derivation of a unit cell in solid materials was determined in a more or less empirical manner simply by plotting the absorption data E vs. Q and evaluating r_s. When this was done for a number of materials having the same (or nearly the same) crystal

Table II. Description of Atomic Polyhedra

	Diamond-cubic	Body-centered-cubic	Face-centered-cubic
Nearest-neighbor distance, r_1	$0.4330\,a$	$0.8660\,a$	$0.7071\,a$
Second-nearest-neighbor distance, r_2	$0.7071\,a$	a	a
Volume of polyhedron, atoms	2	4	4
r_s Radius of equivalent sphere,	$0.3908\,a$	$0.7814\,a$	$0.6203\,a$
$a =$	$2.559\,r_s$	$1.280\,r_s$	$1.612\,r_s$
$r_1 =$	$1.108\,r_s$	$1.108\,r_s$	$1.140\,r_s$

structure, it was possible to identify a unique polyhedron for each particular type of lattice. This experimental approach produced the polyhedra shown in Figure 5. A description of each is given in Table II. Note that the r_s determined from the E vs. Q plot is the radius of a sphere with volume equal to the volume of the respective polyhedron. The conversion of this unit to the usual interatomic distance is also indicated. All dimensions in Table II and Figure 5 are in units of the usual orthogonal unit cells for each lattice with unit cell dimension a, which contain eight, two, and four atomic volumes for the diamond-cubic, body-centred cubic, and face-centered-cubic lattices, respectively.

The atomic polyhedra were constructed after the general method originally demonstrated by Wigner and Seitz, as discussed by Slater.[7] Draw lines from an atom to each of its neighboring atoms and then construct planes perpendicularly bisecting these lines. The smallest closed cell will contain one atomic volume, the next, two atomic volumes, and so on. (Brillouin[8] discusses and illustrates this kind of cell in lattices.) By proceeding in this way, we construct nearly spherical polyhedra of known volume with the surrounding atoms located at the boundaries of the polyhedra and which still retain the dimensions and symmetry of the lattice. The polyhedron for the diamond-cubic structure is shown in Figure 5(a). The four nearest neighbors are located at $\sqrt{3}a/4$ and are each shared by four cells; therefore, the polyhedron contains two atomic volumes. (The inscribed cube is included to show the relationship to the orthogonal unit cell.) By adding four other atoms [Figure 5(b)], the body-centered-cubic polyhedron is obtained. The distance from center to nearest neighbor is $\sqrt{3}a/2$, and the cell clearly contains four atomic volumes. In the face-centered-cubic lattice, an alternate representation for the orthogonal unit cell is shown in Figure 5(c), where a central atom is surrounded by twelve nearest neighbors located at $a/\sqrt{2}$ on the edges of a surrounding cube. By constructing the closed figure involving the fourth and fifth neighbors, a nearly spherical polyhedron enclosing four atomic volumes will be obtained. (A drawing of the figure was not attempted.)

This scheme of construction creates nearly spherical, identically shaped polyhedra, which contain the symmetry and dimensions of each lattice and could be used for a more exact calculation which does not involve a spherical approximation. The approximation to a sphere of equivalent volume still identifies the lattice because of the different relative dimensions of each. Furthermore, the boundaries of each polyhedron are located on atom positions where intuitively we expect the ejected electron–crystal field interaction to occur. It is even more significant that, when Kronig structures for materials of each lattice type are analyzed in terms of these models, good agreement exists.

RESULTS

The Kronig structure for germanium, iron, and copper shown in Figure 1 is plotted as E vs. Q in Figure 6. The energies of absorption maxima and the Q values assigned to

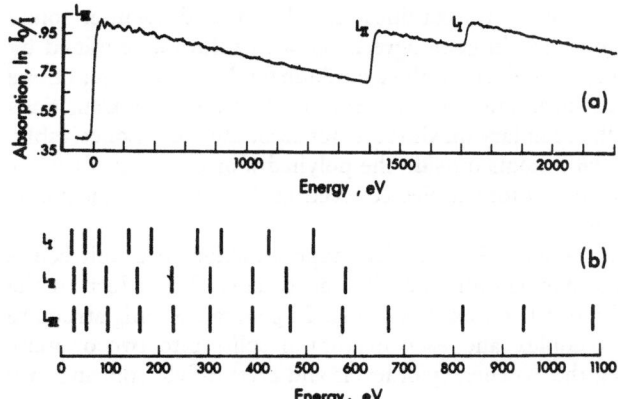

Figure 7. (a) Gold L-absorption Kronig structure. (b) Energies of L_I, L_{II}, and L_{III} Kronig structure.

Table III. Energy of X-Ray Absorption Maxima (in eV)

Q	Ge	Fe	Cu	Q	Au L_{III}	Au L_{II}	Q	Au L_I
—	—	—	15	1	18	20	2	22
2	10	20	25	4	28	30	6	51
6	53	48	49	9	52	52	12	77
—	—	84*	66*	16	98	97	20	140
12	80	110	95	25	165	160	30	185
—	108*	138*	135*	36	232.	230	42	279
20	148	174	157	49	305	308	56	327
—	202*	206*	202*	64	392	396	72	425
—	—	219*	—	81	470	463	90	515
30	230	256	237	100	574	583		
—	—	304*	302*	121	668			
42	320	352	322	144	820			
—	—	408*	356*	169	946			
—	—	429*	390*	196	1087			
56	420	459	417	225	1245			
—	—	518*	467*	256	1427			
72	545	574	537					
—	—	644*	594*					
90	695	724	670					
110	—	862	802					

* Asterisk identifies secondary absorption maxima.

Table IV. Comparison of Interatomic Distances

Material	Distance	Value (Å)	Value determined by r_s (Å)	Percent difference
GeBr₄	Ge–Br	2.32	2.29	−1.3
Ge	Ge–Ge	2.45	2.45	—
Fe	Fe–Fe	2.48	2.43	−2.0
Cu	Cu–Cu	2.56	2.58	+0.8
Au	Au–Au	2.88	2.97	+3.1

each are listed in Table III. The points fit straight lines very closely, and the interatomic distances determined from the slopes are in good agreement with values determined by X-ray diffraction (Table IV). These data illustrate the technique for K-absorption in three types of lattice. Note that all of the prominent maxima can be identified with transitions to p symmetry; however, the small, secondary maxima are not accounted for. Conceivably, they could be due to interaction with atoms outside the polyhedra or due to inadequacy of the approximation of pure p symmetry for the ejected electron. In any event, the model as presented cannot account for them.

In materials with noncubic symmetry, the spherical approximation to a unit cell is no longer valid even if the atomic polyhedron could be constructed. The effect on the Kronig structure is apparent in Figure 1 in hexagonal zinc. In general, Kronig structure in noncubic materials is more complex and each maximum splits into two or more components. Arguments involving the degeneracy of levels and crystal-field splitting may make sense in some cases.[2]

The L-absorption Kronig structure of gold is shown in Figure 7(a). In Figure 7(b), energies of absorption maxima for each edge are compared. Note that the L_{II} and L_{III} structures are very nearly the same, and the L_I structure is obviously different. Actually, the L_{III} structure extends under the L_{II} edge and the L_{II} structure under the L_I edge, but they have been separated into their respective parts in Figure 7(b) and Table III. This was accomplished by recording the Kronig structure at two temperatures (77.4 and 295°K) and using the change in amplitude of Kronig structure with temperature to separate them; the temperature smearing effect is largest on high-energy maxima.[3] Finally, in Figure 8 the energies of absorption maxima are plotted as a function of Q, where the L_{II} and L_{III} structures are identified with s states and the L_I structure with p states of Table I. All the structure fits one straight line (as it should) and the interatomic distance calculated from the slope agrees reasonably well with the known value (Table IV). Similar data obtained in this laboratory for the L-absorption structure of platinum, tantalum, and tungsten also give good agreement in both the selection rules and interatomic spacing.

Perhaps the greatest potential use of Kronig structure is in the investigation of noncrystalline materials—gases, liquids, glasses, and amorphous solids. Data for a variety of solid materials were treated elsewhere,[2] and gaseous $GeBr_4$ has been discussed in this paper. A further example of the potential of the technique is given in Figure 9 where the

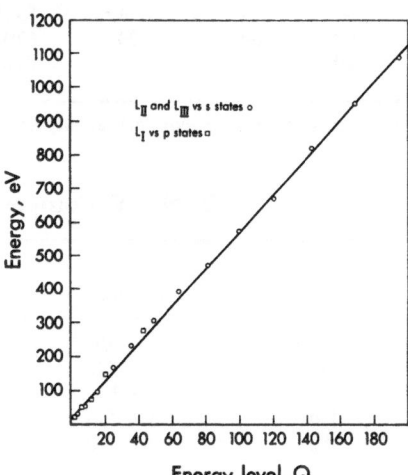

Figure 8. Gold L-absorption E vs. Q plot.

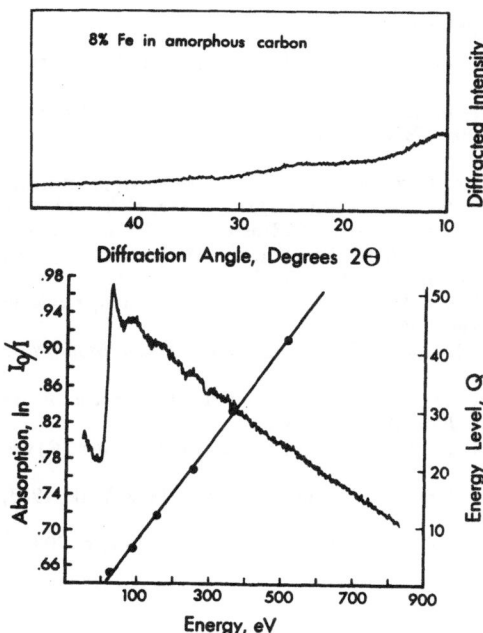

Figure 9. X-ray diffraction and Kronig-structure data from amorphous carbon containing 8 wt.% Fe.

X-ray diffraction powder pattern (Cu K_α radiation), the iron Kronig structure, and a plot of E vs. Q for amorphous carbon containing 8 wt.% Fe are shown. The diffraction pattern gives no useful information as far as structure is concerned; however, the Kronig structure is present and the E vs. Q plot determines $r_s = 1.71$ Å. It is clear from the diffraction pattern that the iron is not present as a separate phase, but is incorporated in the carbon structure. While it is not possible to easily define a unit cell, we know that r_s is approximately equal to an interatomic distance, and in this case, $2r_s = 3.42$ Å, which uniquely establishes the position of the iron atom as lying between the carbon layers, since Wells[9] has reported that the interlayer separation of carbon rings in poorly crystalline carbons is 3.44 Å.

SUMMARY

The simple model presented here is successful in determining interatomic distances and accounting for the differences in Kronig structure on different absorption edges in the same material. The quantitative agreement with experimental data is certainly more than would be expected; however, we've only asked the theory to account for the two qualities specifically built-in, i.e., interatomic distance and spherical symmetry (which produces the dipole selection rules). Nothing can be said about transition probabilities, line shape, Kossel structure, or the smaller irregularities present in the Kronig structure. The physical concept of electrons trapped in a box is certainly not an adequate picture of the phenomenon and is merely a result of the approximations used to make the calculations simple. A theory of Kronig structure which incorporates all that is known in the physics of the interaction of electrons and matter has yet to be developed, but, for those who are interested in understanding materials by using the extended X-ray absorption fine structure, the model presented here works and can be easily understood and applied.

REFERENCES

1. L. V. Azároff, "Theory of Extended Fine Structure of X-Ray Absorption Edges," *Rev. Mod. Phys.* **35**: 1012, 1963.
2. F. W. Lytle, "X-Ray Absorption Fine Structure in Crystalline and Non-Crystalline Materials," in: J. A. Prins (ed.), *Physics of Non-Crystalline Solids*, North-Holland, Amsterdam, 1965, p. 12.
3. F. W. Lytle, "X-Ray Absorption Fine Structure Investigations at Cryogenic Temperatures," in: J. R. Ferraro and J. S. Ziomek (eds.), *Developments in Applied Spectroscopy, Vol. 2*, Plenum Press, New York, 1963, p. 285. See also F. W. Lytle, "X-Ray Diffractometric Examination of Low Temperature Phase Changes in $SrTiO_3$," in: W. M. Mueller, G. Mallett, and M. Fay (eds.), *Advances in X-Ray Analysis, Vol. 7*, Plenum Press, New York, 1964, p. 136.
4. J. A. Bearden, *X-Ray Wavelengths NYO-10586*, U.S. Atomic Energy Commission, Division of Technical Information, Oak Ridge, Tennessee, 1964.
5. L. G. Parratt, "Electronic Band Structure of Solids by X-Ray Spectroscopy," *Rev. Mod. Phys.* **31**: 616, 1959.
6. H. Glaser, "The Absolute Absorption Coefficient of Germanium and the Fine Structure in the *K* Edge of Some of its Compounds," *Phys. Rev.* **82**: 616, 1951.
7. J. C. Slater, "The Electronic Structure of Solids," in: S. Flugge (ed.), *Handbuch der Physik, Vol. 19*, Springer-Verlag, Berlin, 1956, p. 1.
8. L. Brillouin, *Wave Propagation in Periodic Structures*, second edition, Dover, New York, 1953, p. 152.
9. A. F. Wells, *Structural Inorganic Chemistry*, third edition, Clarendon, Oxford, 1962, p. 709.

X-RAY TECHNIQUES IN THE 1 TO 400 Å RANGE

Andrew A. Sterk

American Machine & Foundry Company
Alexandria, Virginia

ABSTRACT

X-ray work in the 10 to 400 Å range usually requires, for maximum sensitivity, new methods of X-ray generation, analysis, and detection. A comparative survey includes X-ray generation by fluorescence, electron bombardment, and proton bombardment. Dispersive analysis with crystals and gratings is described. Finally, windowless photoelectric multipliers are evaluated as X-ray detectors.

Results show that typical efficiences range from 10^{-4} to 10^{-2} photons/electron or photons/proton, with the latter value as a practical upper limit. An efficiency of 8.5% has been measured for a KAP crystal covering the wavelength range up to 25 Å, while an original gold-plated grating has a maximum measured efficiency of 20%. Counter efficiency may range from 1 to 50%, depending on wavelength. Total efficiency for a grating-type spectrometer in the 20 to 400 Å range has been measured to 2×10^5 photons/cm²-count.

INTRODUCTION

As is well known, X-ray techniques for wavelengths above 10 Å are different, and sometimes substantially so, from methods used for the shorter wavelengths. In the following, we will describe some of the experimental methods, results, and instrumentation hardware which was used for the generation, analysis, and detection of soft X-rays with the emphasis on wavelengths above 10 Å.

X-RAY GENERATION

There are three practical methods of X-ray excitation above 1 Å, namely, excitation by protons, electrons, and primary X-rays. A comparison of the three approaches for wavelengths below 2.7 Å is given by L. S. Birks.[1] The general conclusion from that comparison shows that the excitation efficiency decreases with wavelength using primary X-rays and increases with wavelength with the use of both electrons and protons. Data for electron excitation efficiency for longer wavelengths up to 44 Å is given by M. Green.[2] The author reported on efficiency of excitation with proton bombardment for wavelengths up to 44 Å[3] which may exceed 10^{-3} quanta/proton and approaches a value of 10^{-2} for longer wavelengths. It was also pointed out that the proton excitation method has the unique characteristic of not producing a continuous spectrum but only pure line radiation. The proton-excited method of X-ray production has been used extensively during most of the experimental work described later on and appears to provide the only high-intensity, monochromatic source for wavelengths > 10 Å.

---

[1] References are at the end of the paper.

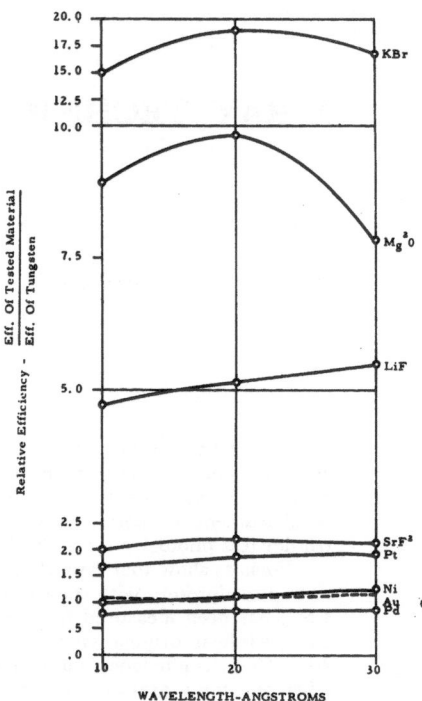

Figure 1. Photoefficiencies of several materials in the wavelength range 10–30 Å.

DETECTORS

Scintillation detectors, which have efficiencies approaching 100% for wavelengths below 2.5 Å, are not suitable for soft X-ray detection. Photomultiplier noise amplitudes exceed the signal in the soft X-ray region and the requirement of hermetically sealing the scintillating crystal means an intolerable intensity loss due to X-ray attenuation in the sealing material. The ionization type of X-ray detector, which, depending on its inherent gain, falls in the proportional or Geiger counter category, is usable in principle for very soft X-rays. The major practical difficulty is, however, the requirement of a gas-tight window. In practice, it is impossible at the present state of the art, to make completely leak-free thin windows. Therefore, counters of the gas-flow type have been used, where the lost gas leaking through the thin X-ray window is continuously replenished. The thin films used for windows must have low X-ray absorption and are therefore made from metals like aluminum or compounds like aluminum oxide or plastics. In any case, the absorption characteristics of these films are highly nonuniform, consisting of one or a few transmission bands covering a narrow selected wavelength range. The main application is, therefore, for special cases where detection of only one wavelength or a small group of wavelengths is required.

A third group of detectors useful for soft X-rays is based on the photoelectric effect. Briefly, an incident X-ray quantum will, upon bombardment of a suitable photocathode, generate a loose electron which is collected, amplified, and registered by electronic means. The photoelectric effect is strongly dependent on the energy of incident radiation, photocathode surface composition, and angle of incidence between X-rays and the surface of the photocathode. Generally speaking, the efficiency increases with increase in wavelength, a decrease in angle of incidence, and is highest for certain alkali halides. Typical values may vary from a few percent to 100%, with 20% as an average. Figure 1 shows the relative efficiency of a few photocathode materials, with tungsten as a standard, for normal incidence and for wavelengths between 10–30 Å.

With an absolute efficiency for tungsten of about 2%, we see that a KBr photocathode will yield an efficiency of at least 30%. For small angles of incidence of the order of 5–8°, the efficiency can be doubled and tripled and will approach the 100% limit. Some of the more sensitive photocathodes are somewhat hygroscopic and care has to be exercised in their use. Work presently in progress is aimed at arriving at a non-hygroscopic material with high sensitivity. Photoelectrons emitted by the cathode have to be multiplied before entering a regular electronic amplifier. This is most conveniently done with a continuous strip multiplier where electrons, upon repeatedly striking the strip surfaces, are multiplied and finally collected by an anode.

Detector output can be measured in terms of integrated DC anode current, or for best sensitivity, as individual pulses each corresponding to a primary event. In this case, the short output pulse width (nanoseconds) requires a high-speed pulse amplifier or some pulse integration before amplification.

While the detector gain is subject to considerable statistical fluctuations, it is possible to achieve a counting plateau of 5%/100 V, eliminating the need for a highly regulated HV power supply. Detectors of this type have the general advantages of reliability, ruggedness, and simplicity. They can be exposed to air (non-operating) and are extremely noiseless. They are not sensitive to light, but have to be protected against charged particles, like an open-window X-ray tube. A recent report[4] described a design without the usually required magnetic field. Examples of application to a flat-crystal spectrometer and a grazing-incidence grating spectrometer will be shown in the next section.

MONOCHROMATORS

The two methods to be described are design extensions of X-ray monochromators using diffracting crystals and optical monochromators using diffraction gratings.

The mechanical design for both systems is based on a total instrument exposure to vacuum, including drive motor, detectors, and electronics. The use of materials with low outgassing rates, dry lubrication of all moving parts, and the avoidance of sealing problems with mechanical feedthroughs allow for rapid pump-down and ultimate pressures in the 10^{-6} torr range. Each of the two methods is being pushed toward the limit of its wavelength range. In the present stage of development, the wavelength spectrum can be divided into the region of crystal monochromators usable to approximately 100 Å, a second region extending from 15 to 400 Å where diffraction gratings in grazing incidence are used, and finally the region above 400 Å, where gratings at normal incidence are applicable. We will describe our work with crystal monochromators first.

The device which was used in our work is shown in Figure 2. It is a conventional Bragg X-ray spectrometer consisting of an X-ray collimator, a flat diffracting crystal turning through an angle θ, and an X-ray detector covering an angular range 2θ. The main feature which makes this particular device useful for soft X-ray work is the small size and the possibility of operating the whole device in a hard vacuum. This has been achieved by using an X-ray detector of the windowless photomultiplier type, which will operate reliably in a hard vacuum. The mechanical motion of the spectrometer is accomplished in the vacuum with a stepping motor without the need of mechanical feedthroughs; the electronics for the X-ray detector is transistorized and therefore also operable in vacuum. Various types of crystals, including LiF, EDDT, KAP, and soap films, have been used with this arrangement. The wavelength range up to approximately 100 Å

Figure 2. Soft X-ray crystal spectrometer.

Table I. KAP Crystal Spectrometer

λ (Å)	Line	2θ	$\Delta\lambda$ (Å)	Efficiency (counts/photon-cm^2)
7.12	Si K_α	31°08′	0.223	5 × 10^{-4}
8.34	Al K_α	36°41′	0.220	5 × 10^{-4}
9.89	Mg K_α	43°50′	0.215	9 × 10^{-4}
13.31	Cu L_α	60°40′	0.200	1 × 10^{-3}
17.5	Fe L_α	82°39′	0.174	2 × 10^{-4}
23.6	O K_α	125°50′	0.106	1 × 10^{-4}

using a soap film crystal is presently being explored and we will report here on the performance with the KAP crystal covering a wavelength range from 6 to 25 Å. Detector performance was already known, therefore, the two main parameters of interest are crystal efficiency and resolution. For this purpose, a source of monochromatic X-rays generating appropriate K and L lines in this wavelength range is used. The source is calibrated directly using the same detector and electronics as those being used in the spectrometer, and crystal efficiency is calculated directly from the ratio of the two intensity measurements, with and without the crystal. Resolution is measured by scanning through selected spectral lines and measuring their width.

The experimental results show that the resolution is somewhat poorer than expected for the KAP crystal. The measured value is 0.425° (2θ), whereas the expected value determined by the collimator divergence is 0.35°. It is believed that this result is due to slight non-uniformity of the crystal structure. The measured resolution is constant for the wavelength range from 8.3 to 23.6 Å. Crystal efficiency measured on three crystals gives an average value of 8.3%, which is again independent of the wavelength within the operating range of the spectrometer. The efficiency and wavelength resolution of the

Figure 3. Grazing incidence grating spectrometer.

Figure 4. Grating mount.

complete monochromator including the detector are shown in Table I. The detector used has a platinum cathode; therefore, a potential increase of one order of magnitude in efficiency is possible with a more sensitive cathode. The mechanical drive consists of the crystal shaft and 2:1 gear train, and is energized by a stepping motor. This type of motor which operates in vacuum can be remotely programmed for speed and direction of rotation by electronic means, which obviates the need for mechanical gear switching devices. The advent of soap-film crystals with d spacings of 100 Å and higher will make it feasible to cover most of the range of practical interest with this type of monochromator.

We will consider now the grating techniques and finally make a comparison between the two methods. The grazing incidence grating spectrometer which was used in our work is shown in Figure 3. Figure 4 is a close up of the grating mount and zero-order trap. The basic geometry is determined by a small source of radiation which is realized by an entrance slit, and a curved grating lying on the focusing Rowland circle, seen in upper left hand corner of Figure 3. Radiation diffracted by the grating is focused along the periphery of the Rowland circle. An exit slit moving along the circle and followed by a photoelectric detector serve as the sensing element, with the wavelength being determined by the instantaneous position of the exit slit. The basic parameters, such as spectrometer wavelength range, efficiency, resolution, and angle of view (source size), are determined by the grating constants, as well as the spectrometer geometry. A highly restrictive factor in the choice of spectrometer layout is the condition that the incident (and therefore also the diffracted) radiation has to strike the grating surface at a small angle, i.e. in grazing incidence. The basic mathematics of such a system has been treated extensively by Lukirskii;[5] we will summarize here only the main features that describe the performance in the soft X-ray region, most of which is not completely amenable to theoretical treatment.

The wavelength range is given by the grating equation:

$$m\lambda = \sigma(\sin \alpha - \sin \beta) \tag{1}$$

where m is the order of diffraction, λ is the wavelength of radiation, σ is the grating spacing, α is the angle of incidence measured from grating blank normal, and β is the angle of diffraction measured from grating blank normal. Resolution is determined by the following equation derived by Holliday:[6]

$$\Delta\lambda = \frac{\sigma S}{\rho m}\left(1 + \frac{\cos \beta}{5}\right) \tag{2}$$

where S is the slit width, ρ is the grating radius, and $\Delta\lambda$ is the resolution (minimum separation that can be resolved).

High resolution is always a desirable feature, which can be achieved according to (2) by minimizing σ and maximizing ρ. The slit opening S should not be made too small, because this reduces the effective entrance area. Practical limitations like overall size and grating ruling techniques will then determine the final values for ρ and σ. The angle of incidence α is already fixed to a fairly large value by the condition of grazing incidence.

In order to cover the design range from 25 to 400 Å we chose $\alpha = 88°$ and selected a grating with the following constants: $\sigma = 17,361$ Å (576 l/mm) and $\rho = 998.8$ mm. This determined all parameters of equation (1) and therefore the basic spectrometer layout. The slit width S in equation (2) was selected to be 25 μm and the resultant theoretical value of $\Delta\lambda = 0.5$ Å checked closely with the measured value, as shown in Table II.

Maximum efficiency is achieved by optimizing the reflectivity of the grating. It is advantageous, for this purpose, to give the grating a so-called blaze (see Figure 5), which means that the grooves will be inclined by an angle x, also called blaze angle, with respect to the grating blank surface. In order to derive a value for x, we start with the following condition:

$$(\alpha - x) = (\beta + x)$$

or

$$x = \frac{\alpha - \beta}{2}$$

and introducing this into equation (1) for $m = 1$,

$$\lambda_x = 2\sigma \sin x \cos(\alpha - x) \tag{3}$$

where λ_x is the wavelength of maximum intensity for the blaze angle x. This relation was confirmed experimentally, although the agreement was only approximate, which is

Table II. Grating Spectrometer

λ (Å)	Line	β	Δλ (Å)		Efficiency (counts/photon-cm²)
			calculated	measured	
13.31	Cu L_α	86°59′30″	0.44	0.47	7.6 × 10⁻⁷
44.50	C K_α	85°26′	0.44		2.8 × 10⁻⁶
113.0	Be K_α	83°09′30″	0.44		5.2 × 10⁻⁶
304.0	He II	79°04′45″	0.45		4 × 10⁻⁶

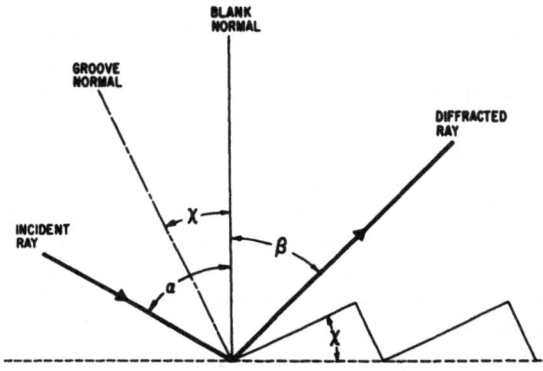

Figure 5. Blazed grating geometry.

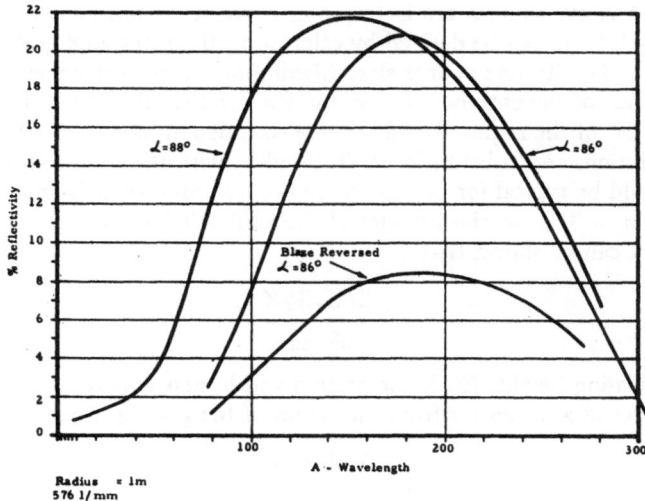

Figure 6. Blazed grating reflectivity.

not too surprising considering the fact that a number of other contributing factors were not included.

The experimental data in Figure 6 show the difference in reflectivity for the same blazed grating, with the incident radiation going with and against the blaze angle. The curve corresponding to the correct position, shows the expected maximum for two

Table III. Relative Higher-Order Reflectivities

Crystal or grating	2nd order 1st order	3rd order 1st order	4th order 1st order	5th order 1st order	6th order 1st order
LiF crystal	0.111	0.01235			
EDDT crystal	0.0378	0.00143			
KAP crystal	0.02	0.0004			
Blazed grating	0.1	0.04	0.016	0.0064	0.0025

values of the angle of incidence α. As expected for larger values of α, the optimum response falls at shorter wavelengths. Finally, the absolute reflectivity of the grating may exceed 20%, as indicated in the measured curves. Similar measurements on a replica grating showed identical results, except that the reflectivity was down by a factor of two as compared to the original grating.

Below 100 Å, blaze angles become very small and the contribution of the blaze to increased reflectivity becomes doubtful. The blaze has, however, another significant advantage in reducing higher-order reflections. This was determined by comparing measurements at higher orders for a blazed grating and a grating without blaze. For the plain grating, higher-order reflections were quite pronounced, in some cases exceeding the intensity of the first order. For the blazed grating, results are summarized in Table III, which also lists comparative values for crystals.

The geometry of the grazing incidence grating spectrometer imposes limitations on the effective source size seen by the grating, which has to be considered with extended sources. As most X-ray sources are isotropic in character, emitting uniformly in most direction, the useful flux can be derived by calculating the solid angle subtended by the spectrometer and the effective source size. Maximum intensity is achieved by locating the source, in place of the entrance slit, on the Rowland circle. The main condition for this approach is use of the required small source size, determined by equation (2), which defines the resolution, and substituting S (slit width) with the source width S_s. Values of 25–50 μm would be typical for average resolution requirements. Defining a geometric efficiency factor $\eta_g = \Omega A_e$ as the product of the subtended solid angle Ω and effective source area A_e, it can be shown that

$$\eta_g = \frac{2LS_sW_gZ}{\rho^2 \cos\alpha} \tag{4}$$

where L is the grating height, W_g is the grating width, and Z is the source height. For an extended source at a distance r from the center of the grating it can be shown that

$$\eta_g = \frac{2LS_sW_gZ}{r\rho} \tag{5}$$

A typical value of η_g calculated by substituting actual numbers into equation (4) is about 10^{-4}; for a value of $r = 30$ cm and using equation (5), $\eta_g = 10^{-5}$ is the value which results.

Figure 7 shows a typical wavelength scan for a vanadium X-ray target under electron bombardment. The scan spectrum extends from 11 to 375 Å. Both the vanadium L line at 24.5 Å and the vanadium M emission band at 340 Å are recorded. The scale factor has been changed during the run and drops from a maximum value of 3×10^5 for the

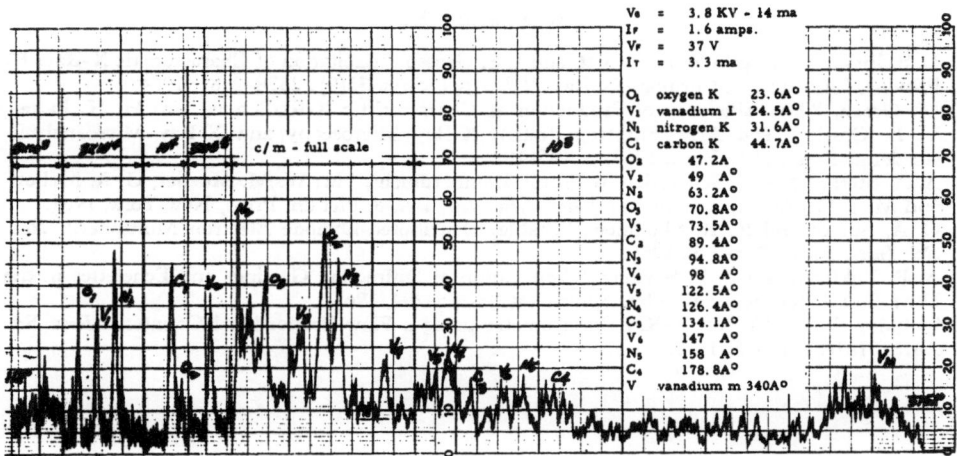

Figure 7. Vanadium target.

low wavelength end of the spectrum to 10^3 for the longer wavelengths. Besides the two previously mentioned lines, we notice also the higher-order reflections of the vanadium L line and first higher-order reflections of contaminants, including oxygen, nitrogen, and carbon. The background level is due mainly to first- and higher-order reflections of the X-ray continuum created at the target.

SUMMARY

Excitation of X-ray wavelengths above 10 Å by proton bombardment is very efficient with proton energies up to 100 kV. The generated radiation is highly monochromatic, consisting of characteristic lines only. This obviates the need for a source monochromator in many uses.

The windowless photoelectric detector with electron multiplier makes a simple and reliable detector for the very soft X-ray range. Operation at a small angle of incidence and alkali halide photocathodes results in detection efficiences of better than 50% for the wavelength region above 25 Å. An efficiency of better than 10% has been realized in the 10–25 Å region. The detector output is not proportional to the incident energy, but adequate plateaus of 5%/100 V are feasible.

Both flat-crystal and curved-grating spectrometers have been operated successfully in the soft X-ray region above 10 Å. The crystal spectrometer is presently limited to the region below 100 Å by the lack of efficient crystals with d spacings greater than 100 Å. The grating spectrometer is a more complex device, requiring precise alignment. A good original grating has a reflectivity of more than 20% above 100 Å, which assures a good overall spectrometer efficiency, provided the X-ray source is small and close enough so as not to exceed the limited viewing angle of the grating spectrometer (about 1°). A mechanical and electronic design which permits complete spectrometer immersion in vacuum allows for rapid operation and the use of windowless X-ray sources.

ACKNOWLEDGMENTS

The author wishes to acknowledge the design contributions and skillful experimental work of William P. Saylor and Charles L. Marks.

REFERENCES

1. L. S. Birks, R. E. Seebold, A. P. Batt, and J. S. Grosso, "Excitation of Characteristic X-Rays by Protons, Electrons, and Primary X-Rays," *J. Appl. Phys.* **35**(9): 2578–2581, 1964.
2. M. Green, "The Efficiency of Production of Characteristic X-Radiation," in: H. H. Pattee, V. E. Cosslett, and Arne Engstrom (eds.), X-Ray Microscopy and X-Ray Microanalysis, Academic Press, New York, 1963, p. 185–192.
3. A. A. Sterk, "X-Ray Generation by Proton Bombardment," in: W. M. Mueller, G. R. Mallett, and M. J. Fay (eds.), *Advances in X-Ray Analysis, Vol. 8*, Plenum Press, New York, 1964.
4. C. A. Spindt and K. R. Shoulders, "Stable, Distributed-Dynode Electron Multiplier," *Rev. Sci. Instr.* **36**(6): 775–779, 1965.
5. A. P. Lukirskii and E. P. Savinov, "Application of Diffraction Gratings and Echelettes in the Ultra Soft X-Ray Region," *Opt. Spectr.* **14**(2): 285–294, 1963.
6. Holliday, J. E., "A Soft X-Ray Spectrometer using a Flow Proportional Counter," *Rev. Sci. Instr.* **31**(8): 891–895, 1960.

A BLAZED-GRATING SCANNING SPECTROM-
ETER FOR ULTRASOFT X-RAYS SUITABLE
FOR USE IN AN ELECTRON MICROPROBE

J. B. Nicholson and M. F. Hasler

Hasler Research Center
Applied Research Laboratories, Inc.
Goleta, California

ABSTRACT

The importance of operating within the blaze angle is emphasized by a series of measurements of ultrasoft X-rays obtained with a new grazing-incidence, grating spectrometer. A formula to maximize output intensity has been derived which predicts a variation of input angle with changes of wavelength. The spectrometer is designed to utilize this function. The grating spectrometer was installed into an Applied Research Laboratories Analyst's Microprobe. Since the AMX furnishes an ultrasoft spectrometer with a fatty acid pseudocrystal as standard equipment, it is possible to compare the two systems simultaneously on the same sample at the same wavelength. The grating and crystal systems are compared for peak intensity and line-to-background ratio.

INTRODUCTION

Experimental data have been presented previously which indicated that blazed gratings can be used effectively in the ultrasoft X-ray region.[1-3] The information gained in evaluating the potentialities of X-ray gratings has led to the formulation of several equations which form a hypothesis for grating selection and utilization. A grating scanning spectrometer has been designed to confirm these basic concepts.

With the current interest in microanalysis, it was deemed most worthwhile to place the spectrometer into an electron microprobe and compare the efficiency of the grating system with the curved crystal system. Since both dispersive systems are symmetrical about the emitting sample surface, this provides a unique opportunity to determine how the two systems complement each other.

This investigation then will outline the parameters for operating a blazed grating at maximum efficiency and will compare the grating with a curved lead stearate decanoate pseudocrystal for peak intensity and line-to-background ratio.

THEORETICAL CONSIDERATIONS

Blazed-Grating Equation

The effect of the grating blaze angle on the diffraction efficiency of ultrasoft X-rays has been previously demonstrated.[3] If a blazed grating is to operate at maximum efficiency, the incoming radiation must be specularly reflected from the ruling groove to the

---

[1] References are at the end of the paper.

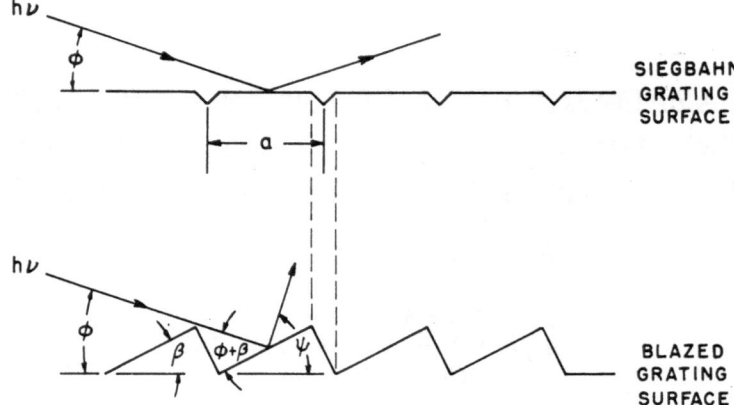

Figure 1. A comparison of grating ruling techniques.

same angular position on the focal circle at which the diffracted line appears. To ensure this effect, constraints are placed on the diffracted output angle, which are a function of the input angle and the grating blaze angle. The grating equation is as follows (Figure 1):

$$n\lambda = a(\cos\phi - \cos\psi) \tag{1}$$

where λ is the wavelength, n is the order, a is the grating constant, ϕ is the grazing angle of incidence, and ψ is the angle of the emergent ray. For small angles, a series approximation of the trigonometric terms yields the following:

$$n\lambda = \tfrac{1}{2}a(\psi^2 - \phi^2) \tag{2}$$

The incoming radiation which is specularly reflected from the facet surface will appear at an angle $\phi + 2\beta$, where β is the blaze angle of the grating. This is the restriction placed on ψ. The grating equation is rewritten in the following form:

$$n\lambda = 2a\beta(\phi + \beta) \tag{3}$$

Equation (3) uniquely defines an angle ϕ for any $n\lambda$ and stipulates how the grating must be rotated in conjunction with $n\lambda$.

Critical Angle

The blazed-grating equation must be modified in an additional way. If the incoming radiation is to be specularly reflected from the ruling land, the input angle must not exceed the critical angle; in fact, for maximum intensity, it should be somewhat less. Reflectivity curves for ultrasoft X-rays indicate a gradual drop-off in intensity as the critical angle is approached. This effect has been illustrated experimentally[4,5] and explained theoretically.[6]

The angle of reflection from the ruling land can be related to the critical angle θ_c in the following way:

$$(\phi + \beta) = f\theta_c \tag{4}$$

where f is a fraction less than unity (usually 0.50) and where[7]

$$\theta_c = \sqrt{2\delta} = \left(\frac{e^2\lambda^2 N}{\pi mc^2}\right)^{\tfrac{1}{2}} \tag{5}$$

where e is the mass of the electron, λ is the incoming wavelength, N is the number of electrons per cubic centimeter in substrate, m is the mass of an electron, and c is the velocity of light. Evaluation of constants and substitution into the grating equation yields:

$$n = 2\beta a\, fk\sqrt{N} \qquad (6)$$

where $k = 2.992 \times 10^{-7}\,\text{cm}^{-1}$. Notice that wavelength cancels out and that the only important parameters are the grating constant a, the blaze angle β, and the number of electrons per cubic centimeter in the substrate, N.

These arguments and their experimental justification are outlined more completely in another paper[8] and will not be elaborated upon any further here.

Equations (3) and (6) then can be used in the following way: If gratings are to be selected on the basis of their ruling constant, blaze angle, and coating, equation (6) can be evaluated for f. If this value exceeds unity, the grating is not suitable for X-ray diffraction. If f is approximately 0.50, it will probably work at maximum efficiency. If f is about 0.75, intensity will suffer, but physical discrimination against short wavelengths should improve.

GRATING MONOCHROMATOR

A blazed grating with an excellent groove profile and surface finish[3] was obtained from C. F. Mooney of the Grating Research Department at Bausch & Lomb, Rochester, New York. It was a 1-m, 600-grooves/mm, aluminized replica. If it is assumed that the surface is principally Al_2O_3, an evaluation of f [from equation (6)] using the nominal blaze angle of 1° yields a value of 0.565.

The first project model of the scanning grating monochromator was designed for installation into the Analyst's Electron Microprobe (AMX). It operates in either of two modes. First, the input angle to the grating can be varied from 3° to 13° by moving the grating along the $52\frac{1}{2}$° takeoff-angle center line and rotating it an amount proportional to its linear travel. This is accomplished by a lead screw, and, thus, any angle within this range can be obtained. In this mode, the input angle is set to some value and the secondary slit is scanned along the focal circle. Secondly, the grating motion and secondary-slit motion are connected so that the input angle ϕ and wavelength λ vary according to equation (3). Thus, the system is always operating at maximum efficiency (patent is pending on this motion).

Figure 2 shows the essential features of the spectrometer. The grating is attached to a holder with two screws (low center of picture). The grating is 0.50 × 0.75 in. and ruled over the entire surface with the rulings 0.50 in. in length. The two feedthroughs adjust grating deflection and an edge which can be moved toward the center of the grating. This serves as a baffle to keep unwanted radiation away from the detector area. The edge limits the effective area of the grating surface, because, by bringing it close to the surface, only the center portion of the grating is used. This allows some interesting experiments pertaining to effective blaze angle.

The grating pole and the secondary slit are connected to a movable plate into which a slot of $\frac{1}{8}$-in. width has been machined. This facilitates alignment and also ensures that both grating and secondary slit are on the focal circle at all times. A threaded shaft passes through the intersection of the secondary slit and the focal circle so that, as the slit moves, it always faces the grating pole. The secondary slit is mounted directly in front of the detector body and has two permanent magnets acting as charged-particle deflectors mounted on it facing the grating.

Figure 2. Grating scanning spec-
        trometer.

EXPERIMENTAL PROCEDURE

The grating scanner was pre-aligned mechanically in the following way. The mono-chromator plate was located on a flat surface plate so that the two surfaces were parallel. A microscope the body of which was located parallel to the surface plate was focused onto a machinist's square and the cross hair adjusted perpendicular to the surface plate. The grating was rotated until the rulings were perpendicular to the surface plate and parallel to the secondary slit. The microscope was then sighted through the secondary slit to the edge of the grating. The grating tilt was then adjusted by shims between the grating and its holder.

After the preliminary mechanical alignment, the grating scanner was inserted into the AMX (Figure 3). Stops are provided so that the entire system locates properly. The system's alignment was checked optically by installing a grain-of-wheat lamp at the inter-section of the electron beam and sample surface. The monochromator has a plate ad-justment similar to that used in the crystal spectrometer which, in effect, rotates the focal circle through the point source (primary slit). The center of rotation is a point 10.7 in. along the take-off angle center line.

At low grazing angles, images are very astigmatic. The line image of the point source produced by the grating is used at the zero-order position, and an optimum focus is achieved by alternately adjusting the grating deflection and spectrometer plate ad-justment. If the system is properly pre-aligned, the focused line appears at the secondary slit position.

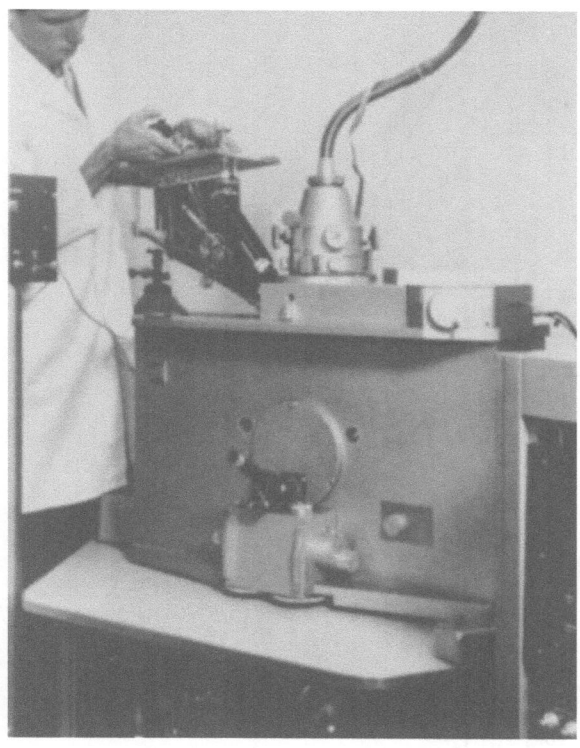

Figure 3. Installation of the grating spectrometer into the Analyst's Electron Microprobe (AMX).

A standard crystal spectrometer utilizing a lead stearate decanoate pseudocrystal was also installed with the AMX. The system was pumped down to approximately 10^{-5} torr. The accelerating potential was set to 5 kV with a sample current of 0.1 μA. The final alignment procedure on the spectral line is essentially the same for both systems. At a suitable low wavelength, the secondary slit is scanned across the line for each incremental change in plate adjustment. The position giving maximum intensity is noted on the scanner dial. A high wavelength is chosen, and the same procedure is repeated with the deflection adjustment. For the crystal spectrometer, Mg $K_\alpha \times$ 3 (29.667 Å) was used for the plate adjustment and Mg $K_\alpha \times$ 8 (79.112 Å) was used for the deflection adjustment. For the grating monochromator, O $K_\alpha \times$ 1 (23.7 Å) was used for the plate adjustment and C K_α up to the fourth order was used for the deflection adjustment.

The readout system consisted of a flow-proportional counter with a 5000-Å, aluminized formvar window. The flow gas was 88% A–12% CO_2 at several centimeters above atmospheric pressure. A high-gain pre-amplifier ($\sim$ 80 $\times$) was connected directly to the counter anode wire. The signal lead was connected to a Hamner non-overload amplifier and pulse-height analyzer, the output of which was connected to a Hamner scalar timer and a Moseley X–Y recorder through an electronic integrator.

Effective Blaze Measurement

The input angle to the grating was varied from 3 to 8° in 0.25° increments. The peak intensity of the line was recorded and plotted as a function of angle. Spectral lines investigated were F K_α, O K_α, N K_α, C K_α, B K_α, and Be K_α.

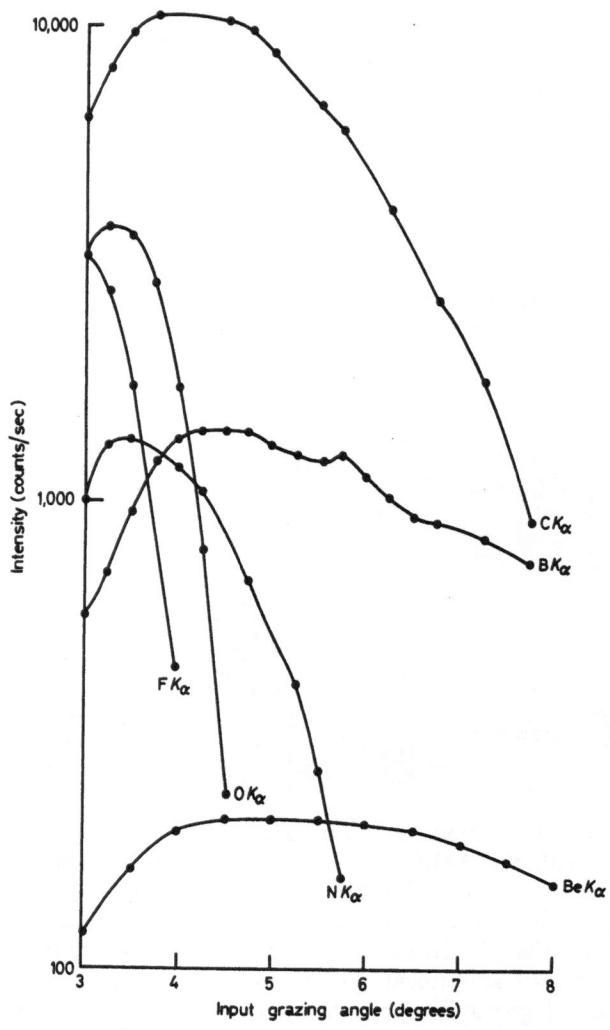

Figure 4. Grating efficiency versus input grating angle for a 600-grooves/mm, 1-m, aluminized grating.

Grating and Crystal Comparison

The two dispersive systems were compared in the following way: C K_α × 1 from graphite and N K_α and B K_α from BN were the spectral lines used. Ratemeter traces of those lines were obtained. Peak intensity and line-to-background ratio measurements were made. The detector had a 0.020-in. slit mounted directly on it. For O K_α and shorter wavelengths, a KAP crystal was used in place of the lead stearate decanoate pseudocrystal.

The detector and associated electronics were installed into the grating scanner and the above measurements made. The detector and electronics were then replaced into the crystal scanner to ensure that no changes had taken place.

An adjustable secondary slit was installed onto the grating scanner. The above measurements were repeated at slit widths of 100 and 50 μ.

RESULTS

The results of the grating-efficiency measurements are presented in Figure 4. The grating used for this investigation had a regular groove profile and reasonable surface smoothness.

Since the maximum intensity does change with wavelength, the possibility of a scanning monochromator that always operates at maximum efficiency now seems feasible. Not only is the advantage of speed readily apparent, but physical discrimination is also possible. At 4.5°, for example, O $K_\alpha \times 2$ (47.24 Å) can be effectively discriminated against in the presence of C K_α (44 Å). The ratio of count rates of O $K_\alpha \times 1$ is approximately 50, when measured at 3.25°, as compared to 4.5°. This effect would be most advantageous where analysis of very low concentrations is necessary. The same argument applies for N $K_\alpha \times 2$ (63.2 Å) near the B K_α (67.2 Å) peak. The ratio of count rates is approximately 10 when N $K_\alpha \times 1$ is measured at 3.50°, as compared to 5.75°. With the pseudocrystal, the input and output angles are equal, and, therefore, background can occur through specular reflection as well as crystal diffraction. This rapidly increases background as the input angle is decreased. This effect is eliminated with gratings, since the zero order is not superimposed on the diffracted line.

Grating and Pseudo-Crystal Comparison

The experimental results of the grating and crystal comparison are tabulated in Table I. All data were taken with an accelerating potential of 5 kV, a sample current of 0.1 μA, and a beam current of approximately 0.3 μA. This corresponds to about 0.5 mW of power at the sample surface.

Table I. Comparison of Peak Intensities and Line-to-Background Ratios for a Blazed Grating and a Lead Stearate Decanoate Pseudocrystal in the Electron Microprobe

| Spectral line | Sample | λ (Å) | Peak intensity (counts/sec)/line-to-background ratio | | |
| | | | Grating | | Crystal 0.020-in. Slit |
			100-μ Slit	50-μ Slit	
O K_α	Al$_2$O$_3$	23.62	3619 —— 30	2123 —— 38	47* — 70
N K_α	BN	31.60	1600 —— 13	1262 —— 23	922 —— 10
C K_α	Graphite	44.7	16,000 —— 86	9277 —— 110	4910 —— 51
B K_α	BN	67.2	1495 —— 19	778 —— 22	598 —— 22
Be K_α	Beryllium foil	114	230 —— 45	115 —— 49	—

* KAP crystal.

In all cases, peak intensity is superior to that found with the diffracting crystal. The effect is most startling at O K_α, where the count rate is at least 30 times greater. This is due to the use of KAP crystal, which must be utilized at very large input angles. N K_α is nearly a factor of 2 better; C K_α, a factor of 3; and B K_α, over a factor of 2. Be K_α was not obtainable with the pseudocrystal.

Peak-to-background ratios are comparable. The intensities obtained at two slit widths are presented to illustrate that the slit widths are still not narrow enough to maximize line-to-background ratio. In fact, as the slit width is narrowed, until it approaches the natural line width, line-to-background ratio should improve, but the line intensity should not change appreciably. The data indicate a line intensity drop of $\cong 40\%$ for a 50% change in slit width. This is due to imprecise alignment. A grating monochromator is far more difficult to align than a single-crystal spectrometer. The slit width must be very narrow (25 μ or less) if resolutions and line-to-background ratios are to be improved. With more precise alignment, a considerable improvement of these factors should occur.

At the slit widths presented, the resolution of the grating was approximately that of the pseudocrystal for all wavelengths. This will most certainly improve as is evidenced by Holliday's work,[1,9] in which slit widths of 25 μ were used.

Since the absolute efficiency of the grating used in this study is known for C K_α,[3] some conclusions as to the speed of the crystal can be obtained. The ratio of solid angle accepted by the crystal to that accepted by the grating is about 6.7. However, the count rate at C K_α is 3.26 higher for the grating than for the pseudocrystal. This gives a factor of 22 for the grating speed as compared to that of the pseudocrystal. The absolute grating efficiency for this grating[3] at C K_α is 6%; therefore, the crystal efficiency is approximately 0.27%. There are undoubtedly possibilities for improvement in future gratings. If the surface can be smoothed so that irregularities are small compared to the wavelength of the radiation being measured, then perhaps up to 25% efficiencies could result. This combined with good groove profiles should produce a very efficient dispersion device for X-rays.

SUMMARY AND CONCLUSIONS

It has been shown that the blazed grating is an efficient dispersion medium for ultra-soft X-rays. Not only is its inherent good resolving power available, but also it can provide intensities greater than any crystal yet obtainable for wavelengths over 20 Å. Therefore, for microsource systems where low intensities are commonplace, the grating scanner should be a valuable instrumental addition. Also, in fluorescence analysis, plane gratings with Soller slits could be used to take advantage of the large source areas available and provide spectrometers of high efficiency.

REFERENCES

1. J. Holliday, "Soft X-Ray Emission Spectroscopy in the 13 Å to 44 Å Region," *J. Appl. Phys.* **33**(11): 3259, 1962.
2. J. B. Nicholson and D. B. Wittry, "A Comparison of the Performance of Gratings and Crystals in the 20–115 Å Region," in: W. M. Mueller, G. R. Mallett, and M. J. Fay (eds.), *Advances in X-Ray Analysis, Vol. 7*, Plenum Press, New York, 1964, pp. 497–511.
3. J. B. Nicholson, C. F. Mooney, and G. L. Griffin, "The Effect of Grating Blaze Angle on the Diffraction Efficiency of Ultra-Soft X-Ray Radiation," in: W. M. Mueller, G. R. Mallett, and M. J. Fay (eds.), *Advances in X-Ray Analysis, Vol. 8*, Plenum Press, New York, 1965, pp. 301–314.
4. B. L. Henke and J. C. Miller, "Ultra-Soft X-Ray Interaction Coefficients," AFOSR TN-59-895, Physics Division Air Force Office of Scientific Research, Washington, D.C., 1959.

5. A. P. Lukirskii, E. P. Savinov, O. A. Ershov, and U. F. Shepelev, "Reflection Coefficients of Radiation in the Wavelength Range from 23.7 Å to 113 Å for a Number of Elements and Substances, and The Determination of the Refractive Index and Absorption Coefficient," *Opt. Spectry. USSR (English Transl.)* **16**(2): 314, 1964.
6. R. Wuerker, "Spectral Reflectance by Solids of Carbon *K* Radiation," Ph.D. Thesis, Stanford University, Palo Alto, California, 1960.
7. A. H. Compton and S. Allison, *X-Rays in Theory and Experiment*, second edition, D. Van Nostrand, New York, 1935.
8. M. F. Hasler and J. B. Nicholson (to be published).
9. J. Holliday, *Handbook of X-Rays*, McGraw-Hill, New York (to be published).

DISCUSSION

J. E. Holliday (U.S. Steel): You mentioned that you did experiments with different blaze angles. Did I understand you correctly?

J. B. Nicholson: Not for this particular work, but I have worked with blaze angles ranging from 1° up to 7°55′.

J. E. Holliday: Would you care to comment on those results?

J. B. Nicholson: I found that the 7°55′ blaze angle performed in much the same way that yours did. It was very poor in intensity. The carbon K_α line was less than 1% efficient. We feel that the low blaze is almost essential for optimizing in the ultrasoft X-ray regions.

J. E. Holliday: I noticed that your carbon K_α band had a maximum intensity for a 1° blaze and a glancing angle of about 4°. Is that right?

J. B. Nicholson: Yes.

J. E. Holliday: Did you ever try using a 2° glancing angle and a 2° blaze?

J. B. Nicholson: Yes, we have.

J. E. Holliday: Did you find the results the same?

J. B. Nicholson: No, I didn't. It turned out to be less, by a factor of about 2. However, the grating was rough and irregular.

A. A. Sterk (American Machine and Foundry Co.): I assume you used the curved grating. Is this correct?

J. B. Nicholson: Yes, that's right.

A. A. Sterk: What was the radius of curvature?

J. B. Nicholson: One meter.

A. A. Sterk: Could you say a few words about your alignment procedure?

J. B. Nicholson: Yes. The difficult part is to make sure the primary slit appears in the same position that the spot does. We place a light source at the secondary slit and shine it backwards toward the grating with the system in air and look at the focused line at the position where the point source would be. It turns out to be curved and fairly sharp, and you can optimize parameters by getting an optimum focus there.

A. A. Sterk: Thank you.

H. K. Herglotz (E. I. du Pont): Are you familiar with the work of Franks and Sayce?

J. B. Nicholson: Yes.

H. K. Herglotz: Would you like to comment on that?

J. B. Nicholson: Yes. I think that the work that they did was interesting. They worked mostly with the Siegbahn type of grating and got very smooth lands, and this is the thing that we're interested in. It's difficult to do with blazed gratings, because you must deform the surface when you rule. If we could achieve the perfection of surface that they have, I think that we would have a very-high-speed system.

H. K. Herglotz: I don't have exact figures, but it seems to me that without any blazed angle you can't achieve your efficiencies.

J. B. Nicholson: The problem with the Siegbahn grating as opposed to the blazed grating is that you cannot deflect some of the zero-order reflection into the positions where the diffracted line appears. With the Siegbahn gratings, you have a very strong zero order and not necessarily strong diffracted lines. This has been attributed partly to the surface and also the to fact that, as you rule, you turn a chip up and it scatters radically. The blazed grating is replicated so that the chip appears in the trough. We are currently studying improvements for surface finish.

A. A. Sterk: I would like to make another comment on the Siegbahn grating and the blazed grating. From some data that were measured at Cornell by Tomboulian and from some of our measurements, it looks like the Siegbahn grating is definitely inferior as far as the higher-order reflection goes, which was already pronounced as the blazed grating suppresses the higher-order reflections. Furthermore, the latest Bausch and Lomb gratings are superior in efficiency to the best Siegbahn grating that we have ever seen.

A. K. Baird (Pomona College): What limits, if any, are placed on beam scanning using a grating in a probe?

J. B. Nicholson: As long as the beam scans parallel to the grating ruling in the plane of a primary slit, there should be no trouble. We are investigating the effect of scanning the beam in a direction perpendicular to the rulings.

APPLICATION OF MULTILAYER ANALYZERS TO 15–150 Å FLUORESCENCE SPECTROSCOPY FOR CHEMICAL AND VALENCE BAND ANALYSIS

Burton L. Henke

Pomona College
Claremont, California

ABSTRACT

Methods and instruments have been developed for efficient spectroscopic analysis in the 15 to 150 Å ultrasoft X-ray region using specially designed Langmuir–Blodgett type multilayer analyzers with a flow proportional counter detection system. Simple adaptations of the basic Philips vacuum spectrograph are employed with a high-intensity, demountable, ultrasoft X-ray source to excite fluorescence in the lighter elements for chemical and valence band analysis.

INTRODUCTION

Ultrasoft X-ray spectroscopy is of considerable value in the following applications to materials analysis.

Quantitative Analysis for the Light Elements

The shortest wavelengths which are emitted by the light elements are F K (18.3 Å), O K (23.6 Å), N K (31.6 Å), C K (44.6 Å), B K (67.8 Å), and Be K (114 Å). Ultrasoft X-ray spectroscopy has been applied to extend the versatile and efficient method of fluorescence chemical analysis into this light element region.

Valence Electron Energy Band Analysis

The valence energy-level structure is reflected in the spectral distribution of the radiation which is emitted when valence electrons undergo transitions to "holes" in the relatively sharp inner energy states. Most of the useful spectral bands for this type of solid-state analysis lie in the ultrasoft X-ray region.

Until recently, the techniques of ultrasoft X-ray spectroscopy have been considered relatively difficult. Nevertheless, the great importance of this spectroscopy is clearly evident. For this reason the author has joined others during the past five years in the development of simplified, efficient methods for low-energy X-ray spectroscopy which can be applied directly to the above-mentioned research areas. The current results and application of the ultrasoft X-ray program of this laboratory are presented here with the hope of encouraging others to advantageously apply these new approaches to material analysis.

SPECTROSCOPY WITH MULTILAYER ANALYZERS

The approach used here is based upon relatively simple adaptations of a standard vacuum spectrograph (Philips). The design of this instrument has proven to be versatile and to have the necessary close optical coupling between the source and the sample. With this instrument, efficient ultrasoft X-ray spectroscopy has been gained through an optimization of the source, analyzer, and detection techniques.[1-5]

Until recently, essentially all of the X-ray spectroscopy in the wavelength region beyond 25 Å has been with grazing incidence diffraction spectrographs. Modern techniques for generating small-angle blazing in the replicated gratings have greatly increased the efficiency of this approach by directing more of the energy into the diffracted orders and out of the zero-order reflection.[6] Nevertheless, multilayer analyzers made by the Langmuir–Blodgett technique of depositing monolayers from fatty acids have been developed and chosen for this work because of the following considerations:

1. The total solid angle of acceptance for a grating system is relatively small, and only for a point or line source is the amount of energy reflected, using focusing optics, comparable to that for a multilayer system in the 15 to 150 Å region. For large projected source areas, such as those mostly involved in this work, the crystal analyzer is much more efficient simply because collimated flux from the large source can be utilized by the crystal set at a Bragg angle, which is usually very large compared with the grazing angle of grating operation.

2. The large $2d$ spacing multilayer analyzers can be employed precisely as, and interchangeably with, conventional crystals and with a conventional vacuum X-ray spectrograph.

3. The grating analyzer can yield 2 to 5 times higher resolution in the 15 to 150 Å region—but with a special grazing incidence spectrograph and with considerably more difficulty in alignment and operating procedures. The higher resolution capability is dependent upon correspondingly narrow localization of a small source, which is difficult to achieve, for example, if this source is the focal spot of an electron microprobe.

4. Probably the greatest inherent disadvantage of the multilayer analyzer is its lower resolving power capability as compared to gratings. However, as will be demonstrated below, most of the work of current interest in ultrasoft X-ray spectroscopy requires an analyzer resolution well within that presented by the multilayer systems.

The most successful multilayer analyzers which have been developed in this laboratory are 50 to 150 d layer systems of lead myristate, lead stearate, lead lignocerate, and lead melissate of $2d$ spacings equal to 80, 100, 130, and 160 Å, respectively. It has been found possible to generate these "crystals" with sufficiently high order so as to yield resolving powers approximately equal to the number of layers involved. The inherent resolution limit is set essentially by the penetration depth of the ultrasoft radiations. For the 15 to 150 Å region, resolving powers in the range of 50 to 150 are obtainable with these analyzers.

A detailed description of the preparation of these multilayer analyzers has been presented.[3,4] The lead stearate crystals are now available commercially with "light-element analysis kits" from several U.S. companies.

[1] References are at the end of the paper.

Table I. Light-Element X-Ray Fluorescence Analysis

Element	Line	X-ray tube anode	Crystal analyzer	Proportional counter voltage	Fluorescent sample		Analysis sensitivity	
					Element	Sample	P/B (counts/sec)	3σ MDL (wt. %)
O	O K (23.6 Å)	Cu	PbS	2325	53.3 % O	SiO₂ (silicon dioxide)	$\dfrac{13297}{759}$	0.03
N	N K (31.6 Å)	Cu	PbS	2370	56.4 % N	BN (boron nitride)	$\dfrac{3840}{462}$	0.11
C	C K (44.6 Å)	Cu	PbS	2410	100 % C	Graphite	$\dfrac{21681}{696}$	0.04
B	B K (67.8 Å)	C	PbS	2490	43.6 % B	BN (boron nitride)	$\dfrac{12444}{108}$	0.01
Be	Be K (114 Å)	C	PbL	2600	100 % Be	Beryllium foil	$\dfrac{1855}{132}$	0.2

Conditions: A demountable X-ray tube is used in a Norelco (Philips) vacuum spectrograph. The carbon source is obtained by painting a copper anode with colloidal graphite. Crystal $2d$ spacings are 100.6 Å for lead stearate multilayer, and 130 Å for lead lignocerate multilayer. The proportional counter has a side window of 2 cm in diameter and 7 cm long, with a 40 μ tungsten anode wire. The counter gas is methane at one atmosphere pressure and with a flow rate of 0.5 CFH. X-ray tube and counter windows: two double layers of Formvar, 30 μg/cm² each, mounted upon 70 % transmission, 200-mesh nickel grid. Collimation: 3.7° from sample and 2.7° from crystal. Angles are measured by soller blade spacing/soller blade length. Analysis Sensitivity: peak (counts/sec)/background (counts/sec) data for excitation source at 6 kV and 330 mA. The analysis sensitivity is indicated by the 3σ minimum detectable limit (MDL) as given by the relation

$$\text{MDL (wt. \%)} = \frac{0.3\ W\sqrt{B}}{P - B}$$

for 100 sec counting time Peak, P and background B are in counts/sec and correspond to a measured sample of known weight percent W.

APPLICATION

Light-Element Fluorescence Analysis

Some examples of peak and background counting rates for the fluorescence K radiations are summarized in Table I for the light elements oxygen through beryllium, within some standard sample materials. These data were taken with excitation, collimation, choice of multilayer analyzer, and pulse-height discrimination selected so as to optimize the analysis sensitivity as defined here by the 3σ minimum detectable limit criterion. The analysis sensitivities which are listed are intended only as indications of the present capabilities of the methods as described here. In previous reports on fluorescence analysis,[3,4] the author has listed results gained from measurements of low percentage concentrations of the light elements from a series of National Bureau of Standards samples. (Since the publication of those results the corresponding analysis sensitivities have been improved by factors of two or three, mostly due to improvements in the multilayer analyzers.)

It should be noted that, very much as in conventional fluorescence analysis, the final accuracy of these measurements depends strongly upon the particular analysis problem, as dependent upon such factors as matrix effects, sample preparation, and availability of good calibration standards.

The actual recorder tracing for these preliminary runs on analysis sensitivity are shown in Figure 1. These are presented in order to illustrate the wavelength "spread" of the measurements as resulting from the broad collimation chosen for a particular optimized sensitivity. Adjacent elements among the oxygen-through-beryllium group are generally well resolved by their K radiations even with this "open" collimation. However, more collimation may well be important, with some nitrogen–oxygen combinations, for example. The $L_{2,3}$ bands of chlorine, sulphur, and phosphorous are also presented here to illustrate the possibilities of matrix interference in the ultrasoft X-ray region. These radiations were measured under the same conditions as obtained for those of the lighter elements.

As is described below, the light elements oxygen through beryllium emit K bands that may shift in wavelength, change in shape, and even split into multiple bands according to the energy-level structure of the valence state. However, for these K bands in particular these changes are small and often unnoticed within the low-resolution spectroscopy associated with fluorescence analysis with broad collimation. Nevertheless, for precise quantitative analysis of the lightest elements it may often be required to use standards which are similar chemically to the system being measured. In order to illustrate the magnitude and character of these shifts, measurements are presented in Figure 2 of the band spectra differences for boron, beryllium, and silicon as elements and as oxides. These were measured using 0.3° collimation from the sample rather than the 4° collimation as used for the measurements illustrated in Figure 1.

Valence Electron Band Analysis

The physical and chemical properties of solids (electrical, optical, crystalline, etc.) can be correlated with the energy-level structure of the loosely bound valence electrons.[7–15] Because of the growing interest in solid-state physics, a considerable amount of theoretical work has been presented in recent years on these correlations. This effort has made clear a considerable need for more associated experimental data than is currently available.

Perhaps the most promising of all methods for the measurement of the density of

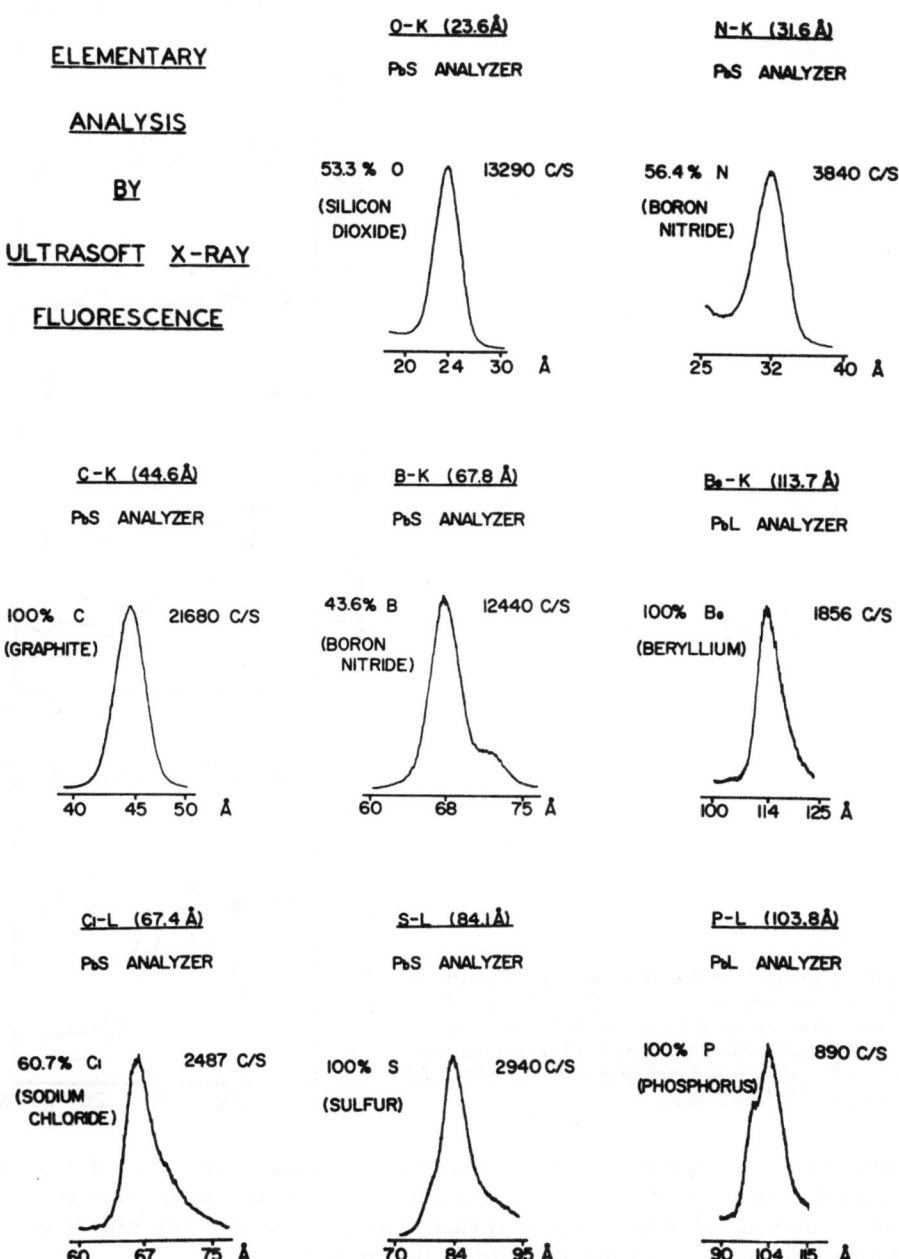

ELEMENTARY

ANALYSIS

BY

ULTRASOFT X-RAY

FLUORESCENCE

O-K (23.6Å)
PbS ANALYZER

53.3 % O   13290 C/S
(SILICON
DIOXIDE)

N-K (31.6 Å)
PbS ANALYZER

56.4 % N   3840 C/S
(BORON
NITRIDE)

20 24 30 Å

25 32 40 Å

C-K (44.6Å)
PbS ANALYZER

100% C   21680 C/S
(GRAPHITE)

B-K (67.8 Å)
PbS ANALYZER

43.6% B   12440 C/S
(BORON
NITRIDE)

Be-K (113.7 Å)
PbL ANALYZER

100% Be   1856 C/S
(BERYLLIUM)

40 45 50 Å

60 68 75 Å

100 114 125 Å

Cl-L (67.4 Å)
PbS ANALYZER

60.7% Cl   2487 C/S
(SODIUM
CHLORIDE)

S-L (84.1Å)
PbS ANALYZER

100% S   2940 C/S
(SULFUR)

P-L (103.8Å)
PbL ANALYZER

100% P   890 C/S
(PHOSPHORUS)

60 67 75 Å

70 84 95 Å

90 104 115 Å

Figure 1. Recorder tracings and peak counting rates on K and L emission bands of light elements, illustrating the wavelength separation of the elements with essentially open collimation and operating parameters as listed in Table I.

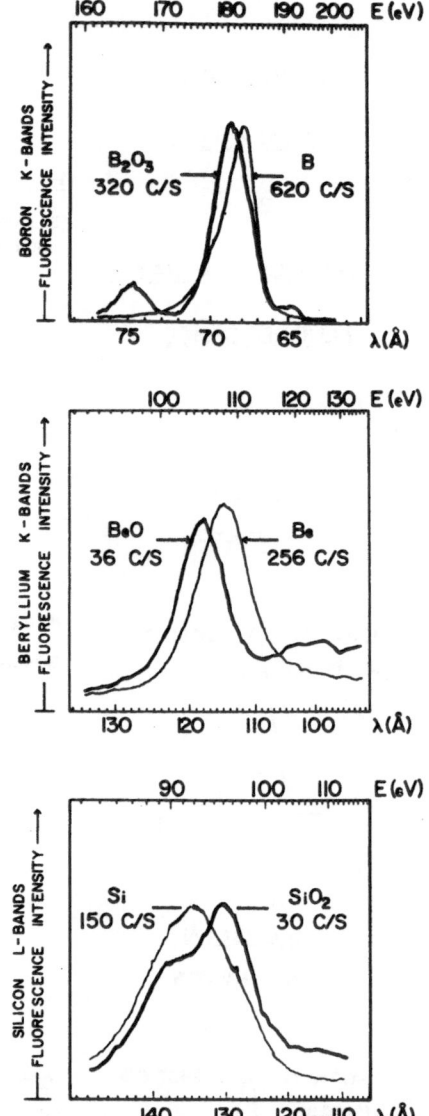

Figure 2. Effect of oxidation as evidenced in the K band of boron and beryllium and the $L_{2,3}$ bands of silicon. Spectra excited by C K (45 Å) radiation and analyzed by a lead stearate crystal of 100-Å $2d$ spacing for boron, and by a lead melissate crystal of 160-Å $2d$ spacing for Be K and Si $L_{2,3}$.

valence-electron energy levels is that of ultrasoft X-ray spectroscopy. To a first approximation, the emission spectral bands resulting from the transitions of valence-state electrons into "holes" within a relatively sharp energy state of a nearby inner level of the atom have an intensity distribution given by the relation

$$I(E) \sim N(E)[\nu^3 f(E)] \tag{1}$$

where $I(E)$ is the spectral line distribution reduced as a function of energy E (E is linearly related to photon energy $h\nu$; it is often convenient to measure E from the lowest energy state in the valence band), $N(E)$ is the density of energy levels at energy E, and the factor $[\nu^3 f(E)]$ is the corresponding transition probability from energy level E. In certain simple

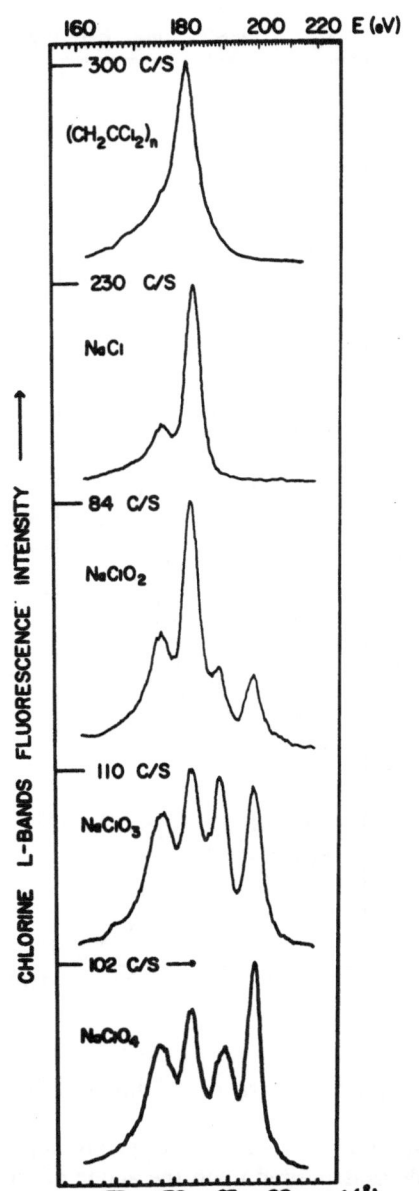

Figure 3. Fluorescence spectroscopy of the $L_{2,3}$ bands of chlorine as emitted from the polymer Saran, sodium chloride, sodium chlorite, sodium chlorate, and sodium perchlorate. Spectra excited by C K (45 Å) radiation and analyzed with a lead stearate crystal of 100-Å $2d$ spacing.

systems, $f(E)$ may be approximated as a constant, and the experimentally derived function $I(E)/\nu^3$ is interpreted as directly representing the form of $N(E)$.

The inner levels adjacent to the valence band are the sharpest available for final states and therefore involve transitions which yield emission spectra of maximum information. Such emission spectra lie within the ultrasoft X-ray region. For chlorine, for example, the L_3 level is about 0.1 eV in width, and the L spectra lie in the ultrasoft 60 to 75 Å region. The K level for this element is about 0.5 eV in width and the K band spectra lie in the 4 Å region and are relatively weak. Transitions between states within

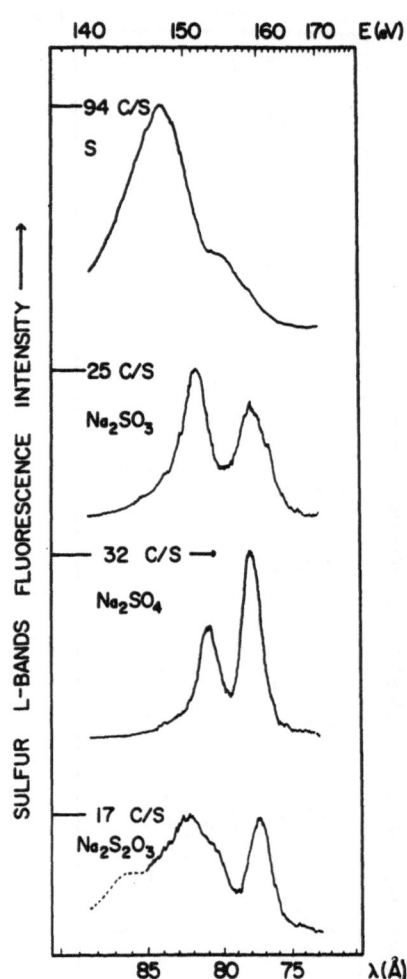

Figure 4. Fluorescence spectroscopy of the $L_{2,3}$ bands of sulphur as emitted from elementary sulphur, sodium sulphite, sodium sulphate and sodium thiosulphate. Spectra excited by C K (45 Å) radiation and analyzed with a lead stearate crystal of 100-Å $2d$ spacing.

the valence band yield broadened, structureless, optical spectra of minimum $N(E)$ information.

All of the reported work on valence-band analysis by ultrasoft X-ray spectroscopy has been by direct electron excitation of the sample as deposited upon the anode of an X-ray tube. All of the previous work, with but one exception,[15] has been with spectrographs using diffraction gratings at grazing incidence for analyzers. This relatively difficult experimental approach has been appropriate in order to gain sufficiently high intensity with high resolution. It has been most successful for samples which are conducting. Non-conducting samples, as deposited upon an X-ray tube anode, must be coated with some conducting film, such as evaporated carbon. Many sample materials decompose or chemically shift under the action of electron-beam excitation.

The high-efficiency techniques which have been developed and described[1-5] for secondary excitation of ultrasoft X-ray spectra have made it possible to demonstrate that a considerable amount of the valence-electron band analysis can be most effectively

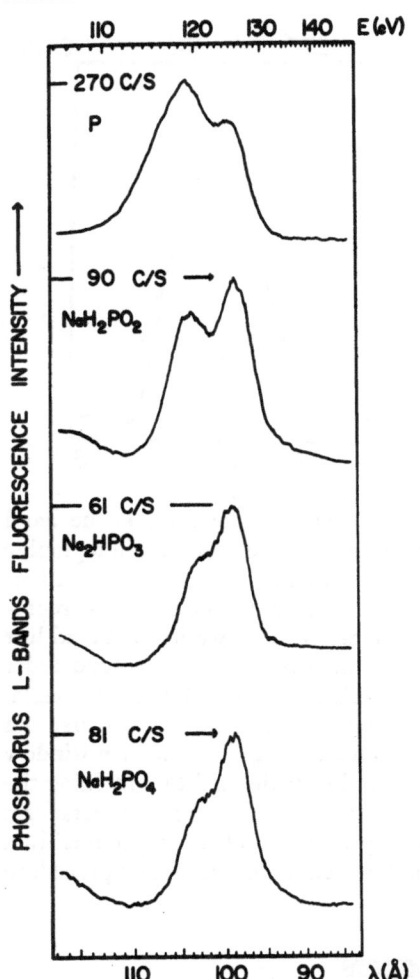

Figure 5. Fluorescence spectroscopy of the $L_{2,3}$ bands of phosphorus as emitted by elementary phosphorus, sodium hypophosphite, sodium orthophosphite and sodium orthophosphate. Spectra excited by C K(45 Å) radiation and analyzed with a lead lignocerate crystal of 130-Å $2d$ spacing.

accomplished by this approach. The advantages gained are the relative simplicity of this type of spectroscopy, along with its ability to maintain the sample free of contamination and decomposition effects.

A few examples of the variation of the position of the fluorescence ultrasoft X-ray spectral bands in Figures 3, 4, and 5 are shown. The spectra, as presented, are taken directly from recorder tracings on which are inserted the appropriate wavelength and energy axes. These spectra therefore have not been corrected for the instrumental window. Figure 3 presents the Cl $L_{2,3}$ spectral bands from the polymer Saran, sodium chloride, sodium chlorite, sodium chlorate, and sodium perchlorate. This remarkable variation in the band spectra is equally dramatic as evidenced in a similar series of S $L_{2,3}$ spectra from several valence states of sulfur (Figure 4), and of P $L_{2,3}$ spectra of several valence states of phosphorous (Figure 5).

The typical working resolution of the grating spectrograph method is about one to five times than it is for this type of crystal spectroscopy in the 15–150 Å region. Nevertheless, upon comparing many of our results with those derived by grating spectrographs, the difference often seems very slight. This is in part due to the fact that many band

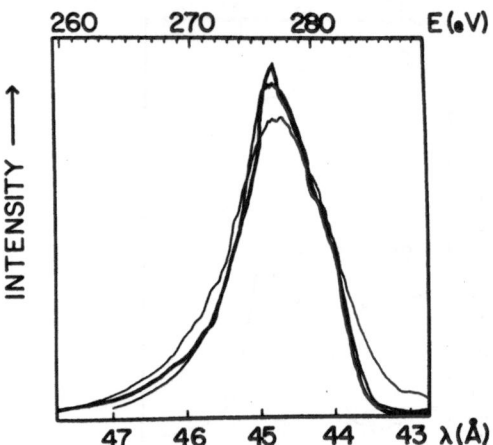

Figure 6. Comparison of the C K band from graphite as measured by a typical grating spectrograph with a photon counter (broad line) and with a lead stearate analyzer of 50 d layers in first- and in second-order reflections.

spectra do not involve detailed structure. A known exception, however, is the band spectra of graphite, which has been demonstrated by high-resolution photographic grating spectroscopy to exhibit a considerable amount of fine structure. A grating-spectrometer measurement of this band as measured with a photon counter and representative of many such which have appeared in the literature is shown in the broad line profile of Figure 6. Superimposed upon this profile is the first-order and second-order spectral profiles taken with a lead stearate crystal in this laboratory. The instrumental window with this stearate multilayer analyzer was 6 and 3 eV in width for the first and second order, respectively, as measured at half-height on an essentially Gaussian window shape. By step-scanning such measurements to gain good statistics and by unfolding the effect of the instrumental window from the observed profile (using Fourier transform techniques and a high-speed computer) it has been possible to obtain reproducibly a considerable amount of the fine structure of this graphite band. This unfolding procedure will be presented in a later report.

ACKNOWLEDGMENTS

The author gratefully acknowledges the invaluable assistance of Robert Lent and Stig Johansson of the research and technical staff of the Pomona College Physics Department, and of Robert Elgin, Lawrence Evans, Jonathan Jerome, and Eric Smith. This work was supported by the U.S. Air Force through the Air Force Office of Scientific Research under Grant No. 689-64. A part of the student support was by the Undergraduate Research Program of the National Science Foundation.

REFERENCES

1. B. L. Henke, "Production, Detection and Application of Ultrasoft X-Rays," *X-Ray Optics and X-Ray Microanalysis*, Academic Press, Inc., New York, 1964, p. 157.
2. B. L. Henke, "Sodium and Magnesium Analysis—Method," *Advances in X-Ray Analysis*, *Vol. 6*, Plenum Press, New York, 1963, p. 361.
3. B. L. Henke, "X-Ray Fluorescence Analysis for Sodium, Fluorine, Oxygen, Nitrogen, Carbon and Boron," *Advances in X-Ray Analysis, Vol. 7*, Plenum Press, New York, 1964, p. 469.
4. B. L. Henke, "Some Notes on Ultrasoft X-Ray Analysis—10 to 100 A Region," *Advances in X-Ray Analysis, Vol. 8*, Plenum Press, New York, 1965, p. 269.
5. A. K. Baird and B. L. Henke, "Oxygen Determination in Silicates and Major Elemental Analysis of Rocks by Soft X-Ray Spectrometry," *Anal. Chem.* 37: 727, 1965.
6. J. Nicholson, "The Effect of Grating Blaze Angle on the Reflection of Ultrasoft X-Ray Radiation," *Advances in X-ray Analysis, Vol. 8*, Plenum Press, New York, 1965, p. 301.

7. H. W. B. Skinner, "The Soft X-ray Spectroscopy of the Solid State," *Rept. Progr. Phys.* **5**: 257, 1939.

8. H. W. B. Skinner, "The Soft X-Ray Spectroscopy of Solids. I. K- and L-Emission Spectra from Elements of the First Two Groups," *Trans. Roy. Soc. A* **239**: 95, 1940.

9. H. M. O'Bryan and H. W. B. Skinner, "The Soft X-Ray Spectroscopy of Solids. II. Emission Spectra from Simple Chemical Compounds," *Proc. Roy. Soc. (London), Ser. A* **176**: 229, 1940.

10. D. H. Tomboulian, "The Experimental Methods of Soft X-Ray Spectroscopy and the Valence Band Spectra of the Light Elements," *Handbuch der Physik* **30**: 246, 1957.

11. L. G. Parratt, "Electronic Band Structure of Solids by X-Ray Spectroscopy," *Rev. Mod. Phys.* **31**: 616, 1959.

12. J. E. Holliday, "Soft X-Ray Emission Spectroscopy in the 13- to 44-A Region," *J. Appl. Phys.* **33**: 3259, 1962.

13. J. E. Holliday, "Soft X-Ray Emission Spectroscopy in the 10 to 150 A Region," *Handbook of X-Rays*, McGraw-Hill Book Company, New York, 1965, Chapter 38.

14. W. L. Baun and D. W. Fischer, "The Effect of Chemical Combination on Some Soft X-Ray K and L Emission Spectra," *Advances in X-Ray Analysis*, this volume, p. 329.

15. D. W. Fischer and W. L. Baun, "The Effect of Chemical Combination on Long Wavelength K and L X-Ray Spectra," *Anal. Chem.* **37**: 902, 1965.

MATRIX AND PARTICLE SIZE EFFECTS IN ANALYSES OF LIGHT ELEMENTS, ZINC THROUGH OXYGEN, BY SOFT X-RAY SPECTROMETRY

K. W. Madlem

Pomona College
Claremont, California

ABSTRACT

With the development of soft X-ray sources by Henke, it is now possible to extend the range of quantitative analyses by X-ray spectrometry into the very light element range (magnesium through boron). Techniques of specimen preparation, recognized as one of the most critical aspects of quantitative X-ray analysis in the heavier elemental ranges, present new problems, especially with the physically and chemically heterogeneous materials encountered in mineral and rock analysis. This paper presents the results of new tests extended into the light and very light element range. It is shown that most of the problems previously found are intensified as lower atomic number elements are considered; conditions of the specimen surface are critical and require precise control of particle size. In addition, significant absorption by relatively heavier elemental constituents, either in a separate phase or within the same phase, causes poor precision and is difficult to predict. Thus, the combination of matrix absorption within and between discrete phases, combined with the difficulty of grinding different phases to a uniform particle size, suggests that fusion techniques are required for best precision. Without fusion, both scatter and bias are introduced through non-linear calibrations, with each calibration dependent upon specimen composition and particle size. It is concluded that only with nearly uniform specimens, where gross deviations are to be checked (e.g., process control) can ground rock powders be used with success in light element analyses.

INTRODUCTION

When the Geology Department of Pomona College first acquired an X-ray spectrograph in 1959, the ability to analyze for common chemical constituents of silicate rocks (oxygen, sodium, magnesium, aluminum, silicon, phosphorus, sulfur, potassium, calcium, titanium, manganese, and iron) was limited by the useful range of tungsten excitation and associated X-ray optics, to aluminum and heavier elements. Aluminum, the third most abundant element of interest, has poor fluorescent yield, with low peak-to-background ratio, precluding high precision results within reasonable counting times; phosphorus and sulfur, as minor constituents, could not be measured quantitatively. Because silicate rocks are composed essentially of silicon, oxygen and aluminum, with minor amounts of sodium, magnesium, potassium, calcium and iron, complete major element analyses by X-ray spectrometry were not then possible with the equipment available.[1]

[1] References at the end of the paper.

In 1961, Burton L. Henke, of the Physics Department, Pomona College, suggested the use of demountable X-ray sources of his design as a means of reaching the very light elements, sodium and magnesium, in silicate analysis. In the following three years, soft and very soft fluorescence analysis techniques were further extended in Henke's laboratory, to include fluorine, oxygen, nitrogen, carbon, boron, and beryllium, as well as trace amounts of phosphorus and sulfur. The improvements included development of organic analyzing crystals with large d-spacings, very thin tube and detector windows, and optimum selection of target materials.[2-5] These techniques were adapted by us for the complete quantitative analyses of geological samples.[6-8]

As a result of this extension into the very soft X-ray region, new problems arose in evaluating analytical precision and obtaining the best quantitative analyses. These problems were primarily those related to specimen preparation and standard calibration. Studies were made of each phase of specimen preparation and led to the present method of preparation for silicates, using lithium tetraborate flux for analyses of all major and minor elements of atomic number 11 and above.[9]

In some applications of soft X-ray analysis, however, either the samples are not appropriate for fluxing techniques (e.g., carbonates and hydrous minerals, where fusion losses would be difficult to evaluate), or the speed of specimen preparation is more important than obtaining the best possible precision and accuracy. It is therefore necessary when considering fluorescent yield to evaluate the effects of variance of particle size and shape in an infused sample and differing chemical properties of co-existing mineral matrices. Any analytical work on mineral mixtures, using X-ray spectrometry, is complicated by these variables, whose effects on precision are interrelated in such a way that recognition of their separate contributions is difficult. Mineralogical differences contribute to the density, shape, and size of ground particles, causing some particles to remain elongate or platy and some to yield differentially in size, even through extensive grinding. The resulting particle size and shape heterogeneity produces matrix problems in which the resulting X-ray signal and signal-to-noise ratios for a given element may be more directly related to the mineralogy of the sample than to the chemistry of the sample. For analyses performed in the long wavelength region (where excitation is largely a surface phenomenon), control over these variables is essential for high-precision results.

PREVIOUS WORK

The following assumptions, based on previous work,[10-12] can be made for elemental analyses reaching to the soft X-ray regions, but not necessarily within:

1. Transition zones can be found for each element followed by a small particle-size range where change in size brings no further increase in intensity.
2. This zone differs for each element and for each concentration of a particle size within a specimen.
3. The transition zone moves toward the finer sizes for the lighter elements.
4. Particles in the large size ranges are found to give lower intensities with poorer precision.
5. Working within a transition zone brings highly variable results.
6. Constant-sized matrices gave the most reproducible results.
7. For lighter elements, fine grinding is apt to place the analysis in the center of a transition zone.
8. Fusions give better precision, but in general require more elaborate techniques and introduce more chance of gross error.

9. Internal standards can be used with some success.
10. Heavy absorbers have been used successfully for elements heavier than manganese, but appear to offer no advantage in the lighter elements.
11. Extremely long grinding times tend to improve the precision of some analyses.

For elements in the soft and ultra-soft X-ray regions, no experimental work has been possible until the development of the Henke tube. This paper attempts to provide some guides on particle-size and matrix effects in light and very light elemental analysis by X-ray spectrometry. For this study, tests from previous experiments cited above were extended into the light and very light element range using various compounds of oxygen. The following areas were explored: (1) the effects of particle size and varying matrices using soft X-ray sources and thin detector windows; (2) the location of the transition zones for different particle sizes and concentrations for the various elements studied; (3) the importance of the surface of the specimen: and (4) the effects of heavier elements on the lighter elements in the same specimen.

PROCEDURES

Materials Analyzed

The materials used were: carbonates—Na_2CO_3 (technical grade powder), $MgCO_3$ (magnesite), $CaCO_3$ [limestone (99% calcite)], $FeCO_3$ (siderite), $ZnCO_3$ (smithsonite); oxides—SiO_2 (optically pure quartz), MnO_2 (technical grade powder); silicate rock—biotite quartz monzonite (standard no. 5, Pomona College). Weight percentage of elements were: O (48.5), Na (3.59), Mg (0.51), Al (8.15), Si (31.75), P (0.06), K (2.81), Ca (1.47), Ti (0.30), Fe (2.12), total (99.26). Mineral percentages by volume were: oligoclase (45), microcline perthite (26), quartz (24,) biotite (4); total (99).

The bulk minerals and rock were first reduced in a jaw crusher to about $\frac{1}{8}$ in. chips, then ground in a Bico rotating plate mill. The resulting particles were sifted for 20 min. in a Ro-Tap shaker, using sieves of 1100, 991, 701, 495, 351, 246, 175, 124, 88, and 61 μ. A portion of the size fraction received from the pan was saved. Further size reduction of Bico-milled grains was done in a Pitchford Pica mill using three steel balls in a standard tool steel vial, for 1, 3, and 5 min. Still further reduction was attempted by grinding in a mullite mortar with distilled water as a lubricant. Size analysis of particles is discussed below.

Fusions used in some of the tests had 65% $Li_2B_4O_7$ as a flux. The glass obtained was ground using the same procedures as for the unfused material. All specimens were pressed into briquettes for X-ray analysis at approximately 24,000 psi. Bakelite powder was used as a backing and rim; no internal binder was necessary.[13]

X-Ray Instrumentation

Philips Electronic Instruments vacuum-path spectrographs, modified to accept the Henke demountable soft X-ray sources, were used for all determinations. Conditions of excitation, collimation, window materials, analyzing crystals, and detection are summarized in Table I. The reader is referred to earlier papers by Henke,[3–5] Baird and Henke,[7] and Baird et al.,[6] for details of the instrumentation and operation. All results reported in this paper are net signal-over-background in counts per second, or wt.% element by reference to established standards.

Particles Sizes and Precision of Specimen Preparation

A test of the reproducibility of sieving was done using one set of mixed particle sizes of quartz monzonite rock as received from the Bico Mill. These particles were sieved

Table I. X-Ray Instrumentation and Conditions

Element	O	Na	Mg	Si	Ca	Mn	Fe	Zn
Excitation	Henke Cu L	Henke Al K		Henke Ag L	Philips FA–60		Tungsten	
Tube kV–mA	6–330	12–150		12–150	30–25		20–10	16–10
Tube window	Formvar	6μ Al foil		25μ Be	1500μ Be			
Crystal	Lead stearate decanoate	Gypsum		EDDT				LiF
2θ Angle	27.40°	103.50°	81.95°	108.71°	44.89°	27.66°	25.35°	41.80°
Primary collimation	$4 \times 1 \times 0.5$ in.			$4 \times 0.6 \times 0.035$ in.	$4 \times 0.6 \times 0.015$ in.			
Flowcounter collimation	$0.5 \times 1 \times 0.022$ in.	$0.5 \times 1 \times 0.011$ in.						
CTR window	Formvar	6μ Al foil		6μ Aluminized mylar				
CTR gas	Methane	P–10						
CTR voltage	2340	1640		1658	1506	1475	1450	1405
PHA E	10.0V	6.0V	6.6V	8.7V	10.2V	8.0V	4.2V	3.6V
PHA ΔE	50V	19.8V	21.0V	24.0V	16.8V	15.0V	19.8V	16.8V
Spec. vac.	$125\ \mu$	$150\ \mu$			None			

Table II. Precision of Sieving

Sieve No.	Total weight (g)	Size ranges (μ)*										
		1040–991	701	495	350	246	175	124	88	61	Pan	Loss
1.	82.9	18.6	14.7	10.4	7.9	6.6	5.0	4.5	3.3	2.5	9.0	—
2.	82.1	18.3	14.1	10.6	7.9	6.8	5.0	4.5	3.3	2.4	8.8	0.83
3.	81.6	17.4	14.5	10.7	7.9	6.9	5.0	4.5	3.4	2.4	8.6	1.34
4.	81.1	16.5	14.6	11.0	8.0	6.8	5.1	4.7	3.4	2.4	8.3	1.86
5.	80.9	16.2	14.7	11.1	8.0	7.3	5.1	4.4	3.3	2.3	8.3	2.06
6.	80.5	15.5	14.9	11.2	8.1	7.3	5.0	4.6	3.3	2.3	8.1	2.38
Mean weight*		17.1 ±1.3	14.6 ±0.3	10.8 ±0.3	7.9 ±0.07	6.9 ±0.28	5.0 ±0.04	4.5 ±0.08	3.3 ±0.05	2.4 ±0.07	8.5 ±0.3	
C Percent		7.2	1.8	3.0	1.0	3.7	0.9	1.9	1.5	3.2	4.0	

* The results of six sievings, weighed each time for all sieve sizes used.

six times, 10 min. each time, using a Ro-Tap shaker. Each of the resulting six sets of size fractions was weighed. The results for nine separate sizes, ranging from 1040 μ to less than 61 μ, are shown in Table II. Evaluation of the grinding for the finest particles, below pan size, is not easily done directly, but can be estimated from the reproducibility of counting rates from replicate preparations, each ground the same length of time in the same equipment. (Phase contrast microscopy at 1200× of 300 grains from each of the mineral powders ground for 1, 3, and 5 min. in the Pica mill give an average overall size range of 5.7–2 μ.)

In order to make an estimate of the reproducibility of ground products, it is first necessary to know the variances associated with instrument counting statistics, the placement of samples in the spectrograph (e.g., orientation effects, etc.) and the briquetting processes used to form coherent samples from the powders. To determine these, a ground fusion was split and four briquettes prepared; each briquette was read (for Na K_α) four times, each in a different orientation. A fusion was chosen for the test because the glass gives the closest possible approach to uniform physical properties, minimizing variations which could affect the grinding. Analysis of variance of the results showed no significant contribution to the total error over the measured counting statistics for either briquetting or orientation. It was concluded, therefore, that variations in fluorescent yield from mineral powders, due to possible grain size or shape variations, could be isolated (if they existed) in the tests to follow.

Several unfused ground compounds were then studied to check the reproducibility of characteristic X-ray emission from grindings, sievings, and Pica millings. In these tests, duplicate preparations were made of each of the nine sieve sizes, plus powders from three different grinding times in the Pica mill, and duplicate spectrograph readings. The study thus included 12 different sizes, two separate sievings, two briquettes of each size in each sieving, and two duplicate readings of each briquette. All compounds tested gave similar results: (1) a small but significant (α 0.05) component of variance between pressings of briquettes compared to the counting error, (2) a small (in some cases not significant) component of variance due to sievings within one size, and (3) a large component due to different sizes. From this it was concluded that the specimen preparation techniques used were adequate to reproduce a given size of particle, and that the effects on X-ray emission intensities due to differing sizes could be detected in the tests to follow.

PARTICLE SIZE EFFECTS IN SIMPLE COMPOUNDS

So as to isolate the many variables occurring in multicomponent rock analyses, which tend to mask particle size effects, and confound matrix effects, experiments were designed in an ordered plan using samples from simple through to complex in composition and physical properties. The first tests were performed using simple compounds (oxides and carbonates) alone. Changes in intensity of characteristic X-rays from light elements with varying particle sizes between samples were noted (Figure 1). In the sieved size ranges, a great deal of scatter can be seen. Despite this, there is a significant linear trend for increasing counting rate with decreasing particle size. In the Pica-milled sized ranges, there is no significant change in intensity with increased grinding within the times used. It appears that all Pica-milled specimens lie on a line of the type $y = k$, where counting rate is independent of particle size. Two independent intensity-versus-size curves result for each element, one for sieved sizes, and one for Pica-milled sizes. If one attempted to join the two curves (as shown in Figure 1), the single curve would appear to define a zone of abrupt change in X-ray intensity with particle size, within which it would be impractical

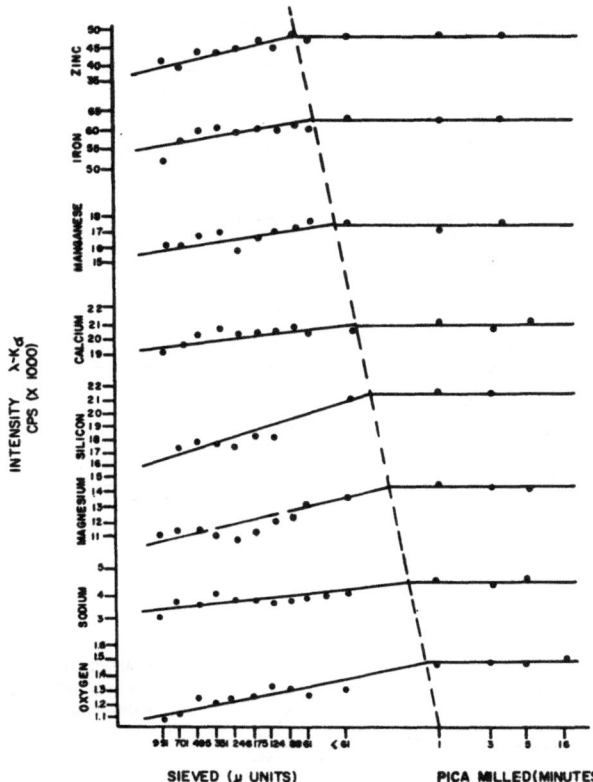

Figure 1. X-ray intensity changes with sieved and Pica-milled ground compound powders.

to work. The start of this zone is indicated by a broken diagonal line. However, assumption of a single curve is not necessarily correct, as these are indeed two separate curves, resulting from two separate techniques of sizing. We have shown statistically that these intensity values are reproducible within each size range and, therefore, each curve can be considered valid.

The effect of inhomogeneity of particle size within samples of one compound was studied by mixing large and small particles, 50% by weight. This was done for each compound previously used. The smallest sizes, shown to yield highest intensities in previous tests, were used as the constant size per sample, to each of which the larger size particles were added. Three different sizes were used to represent the larger size range.

For each element analyzed, unmixed samples of each grain size were run as references to check intensity losses. As seen in Figure 2, these reference samples (solid dots) gave an approximately linear decrease in intensity with increasing grain size, plus a large scatter. When the larger particles were mixed in with the smaller, constant-sized particles, there was very little change in intensity (circles). Thus, these results might suggest that it would be possible to eliminate the particle size effect by mixing large grains of interest with small. However, the findings are from specimens composed of a single compound with no other elements contributing to possible matrix effects.

MATRIX AND PARTICLE SIZE EFFECTS IN MIXTURES OF TWO COMPOUNDS

In order to isolate possible matrix problems, mixed particle size effects with two compounds per sample were studied. Most combinations of Na_2CO_3, $MgCO_3$, $CaCO_3$,

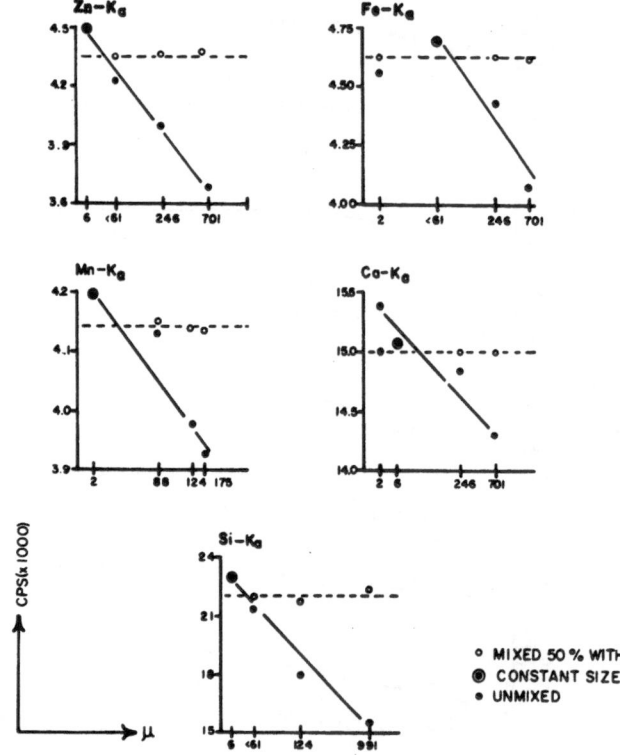

Figure 2. X-ray intensity changes with sieved and Pica-milled ground powders mixed 50% by weight within each compound.

○ MIXED 50 % WITH
◉ CONSTANT SIZE
● UNMIXED

$FeCO_3$, $ZnCO_3$, SiO_2 and MnO_2 were used. This was done in three groups. The first group involved 50% mixtures of one small particle size (Pica-milled) diluted with a larger (701 μ) size. The light element measured was in the compound of small particle sizes. The second group was composed of specimens prepared from 50% mixtures of two compounds, each composed of small-sized particles (Pica-milled). Samples for the third group were a matrix of small particles (Pica-milled) in large particles (701 μ). Characteristic fluorescence was measured from elements in the larger-sized particles. (The samples used in the first and third groups were identical except for the elemental line measured.) Concentration of compounds in all of these samples was determined by volume, in an attempt to overcome the effects of varying densities. X-ray intensity changes recorded in these samples can be correlated with both particle size and matrix changes. Two runs for each elemental determination were made in the spectrograph, one in forward and one in reverse order, to eliminate possible counting-rate drift with time. Four readings were made on each sample and their net mean counting rates used. An undiluted sample containing the element whose characteristic line was measured was used as a standard in each case. The percent intensity losses recorded from the mixtures are shown in Table III. With a 50% reduction of compound, on would expect, ideally, a like reduction of intensity. This was generally true for mixtures of the same particle size. When it did not apply for those specimens containing mixed particle sizes, a microscopic examination of each sample surface was made. The 50% mixtures of two compounds of uniform particle size were each found to have a surface distribution of 50% by area. Mixtures containing 50% large and 50% small particles showed a surface distribution of about 90% small to about 10% large particles (Figure 3). Using infinite thickness values, the fluorescence from a sphere of 701 μ in diameter was calculated in a 50% mixture of Pica-milled

Table III. Percent Intensity Loss in Mixed Particle Sizes and Compounds

		Compound B							
λ-K_α	Compounds mixed 50% by volume	Na_2CO_3	$MgCO_3$	SiO_2	$CaCO_3$	MnO_2	$FeCO_3$	$ZnCO_3$	Size index*
Na	Na_2CO_3	0	8	11	7	—	4	10	1
		—	60	63	49	60	56	54	2
Mg	$MgCO_3$	3	3	20	3	1	1	1	1
		45	0	52	43	—	49	54	2
Si	SiO_2	6	7	—	6	19	4	14	1
		52	47	—	50	45	49	52	2
Ca	$CaCO_3$	15	7	12	—	15	11	24	1
		50	50	69	—	51	69	59	2
		91	83	95	—	—	—	—	3
Compound A: Mn	MnO_2†	35	—	50	55	—	50	—	2
		55	77	76	85	—	—	—	3
Fe	$FeCO_3$	32	30	45	48	—	—	44	2
		88	69	93	88	—	—	—	3
Zn	$ZnCO_3$	43	—	39	48	—	—	—	2
		77	—	75	69	62	65	—	3

Mean percent signal loss for all compounds	Observed	Predicted	Theoretical
1. Small with large	9.5	10.0	10.96
2. Small with small	48.7	50.0	—
3. Large with small	83.0	90.0	89.50

* Size index (relative particle size of compound A and compound B):
1. Small with large (pica mill with 701 μ).
2. Small with small (pica mill with pica mill).
3. Large with small (701 μ with pica mill).

† MnO_2 large size–264 μ.

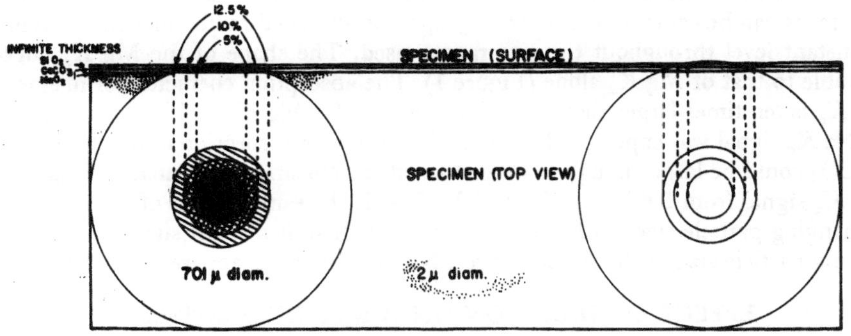

Figure 3. Surface area exposed to fluorescence in a 50% mixture by weight, of 701 μ and 3 min Pica-milled particles.

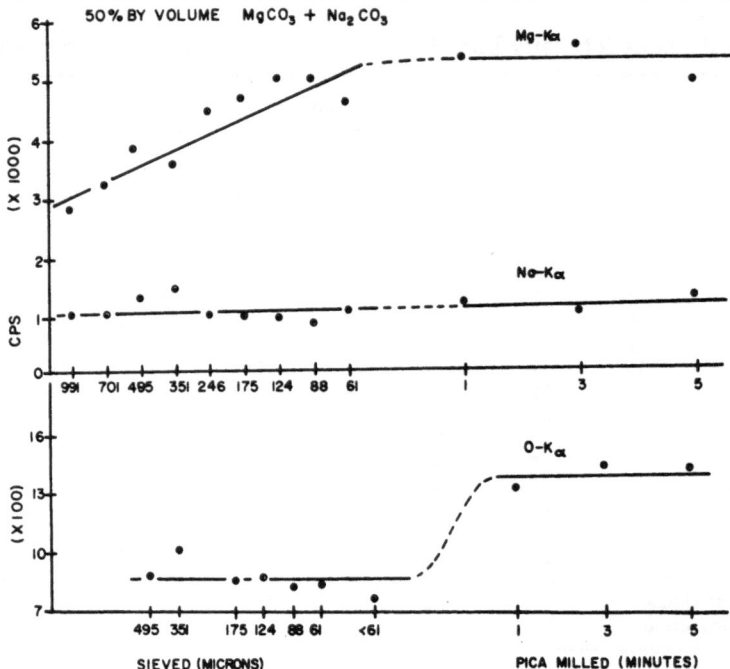

Figure 4. X-ray intensities of Na K_α, Mg K_α, and O K_α fluorescence in 50% mixtures of Na$_2$CO$_3$ and MgCO$_3$, with various sieved sizes and Pica-milled times.

particles of about 2 μ in diameter. The mean values for loss of signal over that of 100% compound was found to be about 90% for the 701 μ particles in a small matrix, and about 10% for small particles in a large matrix (Table III). Thus, the resulting intensity losses found in mixed compounds of differing particle size can be predicted from sample distribution at or near the specimen surface. Gross deviations from this predicted intensity ratio can be assumed due to additional matrix effects of absorption or enchancement.

Further observations of these effects were made over a wider range of particle sizes using mixtures of Na$_2$CO$_3$ and MgCO$_3$, 50% of each by volume. The particle size effect was essentially eliminated by using one particle size for both compounds within any one mixture. Figure 4 shows the resulting signal loss or gain through the particle size ranges used for the K_α line intensities for magnesium, sodium, and oxygen. When compared with the results from unmixed powders (Figure 1), the effects of absorption or enhancement can be estimated. The Na K_α signal is absorbed by the matrix and remains at a constant level throughout the size ranges used. The shape of the Mg K_α curve is comparable to that of Mg K_α alone (Figure 1). The absorption coefficient of magnesium for Na K_α is ten times larger than that of magnesium for Mg K_α radiation; enhancement of the Na K_α signal was expected. It is difficult to see how absorption of the Na K_α signal in MgCO$_3$ could be large enough to cause this effect. No simple explanation is apparent. The O K_α signal from 50% Na$_2$CO$_3$ and MgCO$_3$ is the same as that of oxygen in SiO$_2$ with changing particle size, with the exception that there is a slight signal decrease with decreasing particle size in the sieved ranges. No explanation is apparent for this effect.

EFFECT OF DILUTION ON X-RAY INTENSITIES

In order to study the effect of dilution on X-ray intensities, lithium carbonate was used as a dilutant in analyses for sodium, magnesium, and oxygen. Separate analyses

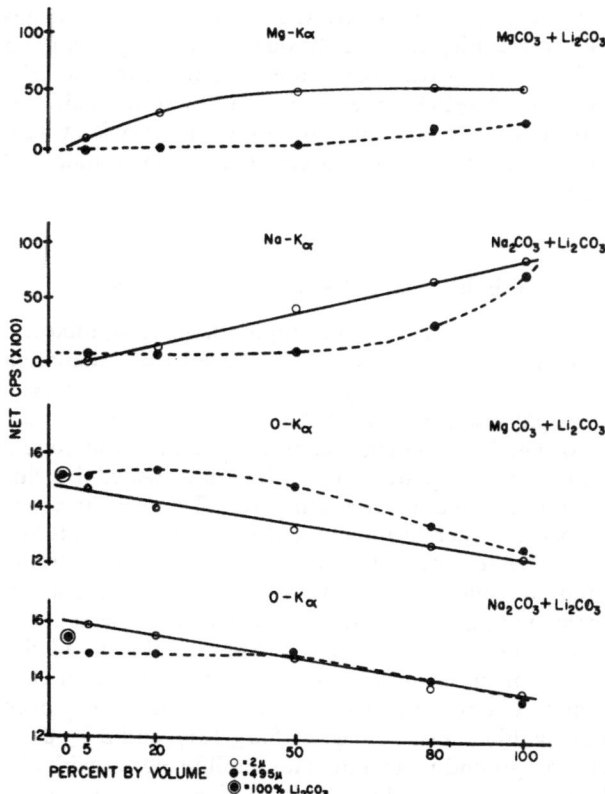

Figure 5. MgCO₃ and Na₂CO₃ in 0%, 5%, 20%, 50%, 80%, and 100% dilutions of Li₂CO₃, in 495 μ sieved particles and 3 min Pica-milled particles.

were made in the 495 μ and 2 μ (Pica-milled) sizes. The dilutant was a very finely powdered reagent-grade chemical, used as received from the producer. Mixtures were by volume percent because of the low density of the Li_2CO_3 powder. Dilutions of 80, 50, 20 and 5% were prepared. The mixed powders were shaken in the Pica mill for 30 sec (in plastic vials to avoid further size reduction). The resulting specimens could be thought of as representing two closed systems, with 100% Li_2CO_3 as one end member of each system and 100% Na_2CO_3 or $MgCO_3$ as the other end member. Assuming a linear relation exists between weight percent element and X-ray counting rate, the effects of particle size on near-surface fluorescence can be studied. As has been shown in Figure 3 and Table III, predictions can be made concerning relative intensities for percent compound and particle sizes. The 50% mixtures of 2 μ (Pica-milled) particles with the finely powdered Li_2CO_3 should have a specimen surface distribution of about 50% for each, and the signal reduction of Na K_α and Mg K_α from 100% $MgCO_3$ samples should be around 50%. The results (Figure 5) show this to be the case only for Na K_α. No explanation for Mg K_α intensities is made. In the mixtures of large (496 μ) particles with the fine dilutant, there should be a surface distribution with a ratio of about 10 : 90 for the 50% dilutions. The expected signal loss of around 90% would yield a non-linear curve. The results shown in Figure 5 show the curves to be quadratic for both Na_2CO_3 and $MgCO_3$ mixtures, and Mg K_α shows a 75% loss and Na K_α an 87% loss. The curves for O K_α from these same mixed samples (Figure 5, lower part) show a general decrease in counting rate as the amount of $MgCO_3$ or Na_2CO_3 increases. (Li_2CO_3 contains 65% oxygen versus 45% Na_2CO_3 and 57% in $MgCO_3$.) For mixtures of small particles the decrease is approximately linear. In mixtures containing large particles of $MgCO_3$ and Na_2CO_3, the decrease in O K_α intensity is markedly quadratic, presumably due to the large exposed surface area of Li_2CO_3 particles, which contain relatively more oxygen. Only when the volume of larger particles of $MgCO_3$ or Na_2CO_3 exceeds 50% is there a significant decrease in the O K_α signal. Thus, the effect of a dilutant upon X-ray intensity is at least approximately related to the differences in particle size between matrix compounds and the compound containing the element of interest.

MATRIX AND PARTICLE SIZE EFFECTS IN ROCK ANALYSIS

All of the tests reported above on particle size effects in simple compounds, mixtures of simple compounds, and varying weight percents within each suggest that attainment of small, uniform grain sizes should be critical when analyzing silicates containing several different minerals, each with different physical properties of hardness, cleavage, and fracture. In order to provide a guide to possible grinding techniques for multicomponent silicates, analytical results from splits of the quartz monzonite rock treated by fine sieving, Pica-milling, and extensive mortar-grinding were compared. This was first performed with fused samples of the rock in order to minimize the influence of differing physical properties of the minerals. The test here was to determine which grinding technique gave the best and most reproducible X-ray intensities from a single substance (a complex lithoborosilicate glass containing at least 13 major and minor elements).

Two separately prepared fusions were crushed and sieved. Splits of the sieved size ranging from 264 to 351 μ were saved for pressing. The remaining material from each fusion was Pica-milled for 1 and 4 min. The resulting particle sizes had previously been determined to be, for the two times, within a mean range of $5.0 \pm 1.0\ \mu$ and $2.7 \pm 0.3\ \mu$ respectively. Splits of this material were ground in an automatic mullite mortar with distilled water as a lubricant, for 10, 30, and 60 min. The resulting size ranges were not determined. The mullite grindings from one fusion were dried under a heat lamp after decanting. A loss of fines was anticipated. The product from the other fusion was dried on an electric hot plate without decanting. All the resulting powders were pressed into briquettes, two for each size range. The resulting intensity changes (expressed as apparent weight percent element) were studied with respect to apparent change in particle size (Figure 6). From the consistent loss of signal in the mortar-ground specimens, it is obvious that the chances are high for a decreased precision with increased handling during ultra-fine grinding. The relatively constant high signals from Pica-milled specimens are consistent with results from previous studies. These curves are representative of three distinct grinding methods and should not be considered as continuous. The conclusion is reached that high-speed impact-milling produces the best results in a homogeneous sample of a complex substance.

A comparison of analytical results from Pica-ground powders of fused and unfused rock was next made, in order to assess the additional complications of dealing with different minerological properties in one sample. Splits of fusions and unfused rock were ground for $\frac{1}{4}$, 1, 4, 16, and 32 min. in the Pica mill. Four briquettes were prepared from the products of each grinding time for both fusions and rock, making a total of 40 briquettes for analysis. The data in Table IV compares the mean weight percents of magnesium, aluminum, silicon, potassium, and iron and their standard deviations, determined for each grinding time. The comparisons show that (1) determined elemental concentration

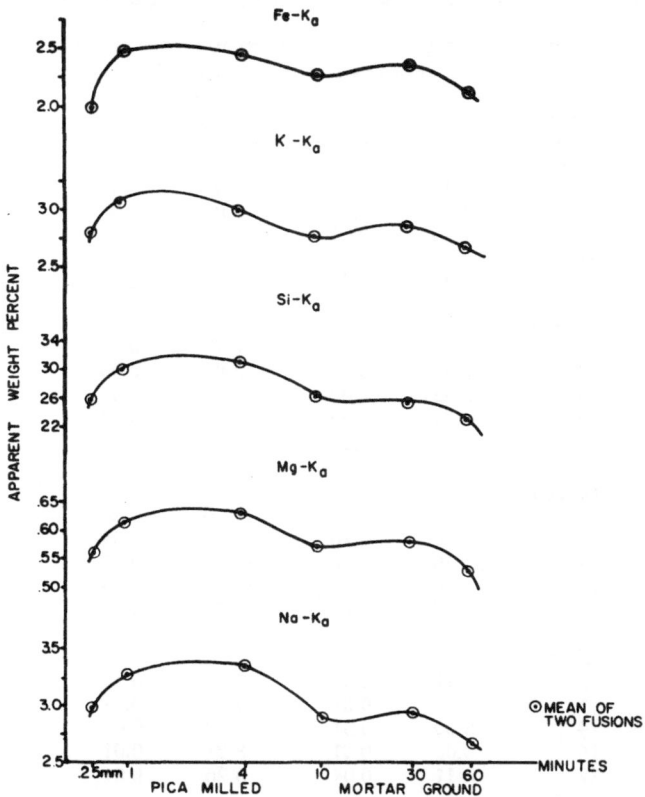

Figure 6. Signal variations resulting from 3 sizing techniques; (1) sieving, (2) Pica-milling, and (3) mullite mortar grinding, for the indicated times, of two fusions.

is independent of the length of grinding time (over $\frac{1}{4}$ min) for fusions, (2) the determined elemental concentrations vary directly with length of grinding time (up to 32 min) for unfused powders, and (3) the standard deviations of determined elemental concentrations are generally larger, more variable, and decrease with increased grinding time for unfused powders. The marked percentage increase in determined elemental concentrations for magnesium, potassium, and iron in unfused rock powders with increased grinding can be related to the biotite mica (relatively rich in these elements) in the rock. The apparent relative volume of this mineral on the surfaces of the final briquettes used for analysis must increase in response to increased grinding, because the platy structure of the mineral precludes uniform grain shape. This effect is undoubtedly enhanced by the compacting pressures used in briquetting, which cause a rotation of thin, platy grains to near parallelism with the surface of the briquette. The rock used in this test contains only 3.5% mica by volume. With samples containing larger amounts of micaceous minerals (e.g., muscovite, biotite, and chlorite, in rocks such as schists, gneisses, shales, and mudstones) we would expect this effect to be more pronounced. For further discussions of these problems, see papers by Volborth[14] and Baird et al.[15]

Table IV. Effect of Grinding Fused and Unfused Rock Powders

Minutes	Fusions		Ground Rock	
	Weight*	SD	Weight*	SD
Iron				
¼	2.43	0.02	2.01	0.09
1	2.45	0.01	2.09	0.02
4	2.46	0.01	2.44	0.03
16	2.43	0.01	2.66	0.01
32	2.44	0.01	2.90	0.01
Potassium				
¼	2.94	0.01	2.95	0.03
1	2.93	0.00	2.86	0.00
4	2.92	0.00	2.96	0.01
16	2.91	0.00	2.87	0.00
32	2.94	0.01	2.85	0.00
Silicon				
¼	30.73	0.25	29.70	0.20
1	31.20	0.10	30.58	0.04
4	31.40	0.06	31.52	0.12
16	31.56	0.05	31.82	0.09
32	32.05	0.12	31.49	0.05
Aluminum				
¼	7.91	0.07	7.13	0.15
1	8.01	0.04	7.75	0.07
4	8.00	0.01	8.10	0.02
16	8.03	0.01	8.21	0.01
32	8.11	0.04	8.26	0.01
Magnesium				
¼	0.64	0.02	0.46	0.03
1	0.64	0.01	0.62	0.01
4	0.66	0.01	0.58	0.01
16	0.74	0.01	0.63	0.01
32	0.67	0.01	0.59	0.01

* Mean of 4 preparations for each grinding time.

SUMMARY

The results reported here on particle size effects serve to confirm earlier studies with heavier elements. In addition, they suggest that there is a direct relationship in light element analyses between atomic number and maximum permissible particle size, which should be used for the best analytical results. At $Z = 30$, particles as large as 100 μ give high intensity and reproducible analyses; at $Z = 8$, it is necessary to use sizes in the range of 5–10 μ to achieve optimum signal and precision. Tests with both simple compounds of light elements and homogenized (fused) complex compounds emphasize that it is the character of the surface of the sample rather than the chemistry of the grains in the surface that controls analytical precision. Thus, many of the effects commonly attributed to the "matrix" are in fact due to grain size and shape heterogeneity only. For this reason, the application of correction factors for presumed absorption or enhancement can be risky in very light element analyses.

The sizing of particles to be used for light element analyses presents serious problems. Even though sieving is highly reproducible, the method can be used only for the coarse sizes. In addition, sieving is selective according to grain shape. Elongate particles will pass sieves of a mesh opening equal to the intermediate grain dimension. These elongate grains tend to orient parallel to compacted specimen surfaces and thus bias the analyses. Platy grains offer similar but more severe problems.

High-speed impact mills, now available commercially in several forms, provide rapid and precise means for achieving the necessary $1-10\mu$ particle sizes which seem to be required for light element analyses. All impact-milled substances tested gave very reproducible results, essentially independent of length of grinding time, *providing* that the material was *physically* (not necessarily chemically) homogeneous to start. Milling is, unfortunately, highly selective to the characteristics of hardness, cleavage, and fracture of grains. Thus, fusion techniques, long advocated as the best means of minimizing "matrix effects", assume new importance as a means for minimizing particle size effects. In analyses of mixtures of grains with differing physical properties, it is evident that close control on milling times should be made. Even with close control on grinding times, there is danger of analytical bias if the sample mineralogy varies. Possible improvements, through more extensive grinding in mortars, seem impractical because of (1) increased specimen handing errors, (2) accentuation of unequal granular character of grains, and (3) increased handling time compared to fusion techniques.

ACKNOWLEDGMENTS

The author is indebted to the following staff members of the Geology Department, Pomona College, for their assistance: Dr. A. K. Baird, for suggesting the subject of research and giving valuable assistance in the writing of this paper, and E. E. Welday, for reviewing the paper and for his professional advice throughout this research. This work was conducted as part of a program on the application of X-ray analysis to elemental distributions in rocks, which research is supported by a grant (GP–1336) from the National Science Foundation.

REFERENCES

1. A. K. Baird, R. S. MacColl, and D. B. McIntyre, "A Test of the Precision and Sources of Error in Quantitative Analysis of Light, Major Elements in Granitic Rocks by X-Ray Spectrography" in: W. M. Mueller, G. R. Mallett, and M. J. Fay (eds.), *Advances in X-Ray Analysis, Vol. 5*, Plenum Press, New York, 1962, pp. 412–422.
2. B. L. Henke, "Microanalysis with Ultrasoft X-Radiations," in: W. M. Mueller, G. R. Mallett, and M. J. Fay (eds.), *Advances in X-Ray Analysis, Vol. 5*, Plenum Press, New York, 1962, pp. 285–305.
3. B. L. Henke, "Sodium and Magnesium Fluorescence Analysis–Part I: Method," in: W. M. Mueller, G. R. Mallett, and M. J. Fay (eds.), *Advances in X-Ray Analysis, Vol. 6*, Plenum Press, New York, pp. 361–376.
4. B. L. Henke, "X-Ray Fluorescence Analysis for Sodium, Fluorine, Oxygen, Nitrogen, Carbon, and Boron," in: W. M. Mueller, G. R. Mallett, and M. J. Fay (eds.), *Advances in X-Ray Analysis, Vol. 7*, Plenum Press, New York, 1964, pp. 460–488.
5. B. L. Henke, "Some Notes on Ultrasoft X-Ray Fluorescence Analysis—10 to 100 Å Region," in: W. M. Mueller, G. R. Mallett, and M. J. Fay (eds.), *Advances in X-Ray Analysis, Vol. 8*, Plenum Press, New York, 1965, pp. 269–284.
6. A. K. Baird, D. B. McIntyre, and E. E. Welday, "Sodium and Magnesium Fluorescence Analysis—Part II. Application to Silicates," in: M. W. Mueller, G. R. Mallett, and M. J. Fay (eds.), *Advances in X-Ray Analysis, Vol. 6*, Plenum Press, New York, 1963, pp. 377–388.
7. A. K. Baird and B. L. Henke, "Oxygen Determinations in Silicates and Total Major Elemental Analysis of Rock by Soft X-Ray Spectrometry," *Anal. Chem.* **37**: 727–729, 1965.

8. A. K. Baird, D. B. McIntyre, and E. E. Welday, "Soft and Very Soft Fluorescence Analysis: Spectrographic and Electronic Modifications for Optimum, Automated Results," *Develop. Appl. Spectr.* **4**: 3, 1965.
9. E. E. Welday, A. K. Baird, D. B. McIntyre, and K. W. Madlem, "Silicate Sample Preparation for Light–Element Analysis by X-Ray Spectrography," *Am. Mineralogist* **49**: 889–903, 1964.
10. F. Bernstein, "Application of X-Ray Fluorescence Analysis to Process Control," in: W. M. Mueller, G. R. Mallett, and M. J. Fay (eds.), *Advances in X-Ray Analysis, Vol. 5*, Plenum Press, New York, 1962, pp. 486–499.
11. F. Claisse and C. Samson, "Heterogeneity Effects in X-ray Analysis," in: W. M. Mueller, G. R. Mallett, and M. J. Fay (eds.), *Advances in X-Ray Analysis, Vol. 5*, Plenum Press, New York, 1962, pp. 325–354.
12. E. L. Gunn, "The Effect of Particles on Surface Irregularities on the X-Ray Fluorescent Intensity of Selected Substances," in: W. M. Mueller, G. R. Mallett, and M. J. Fay (eds.), *Advances in X-Ray Analysis, Vol. 4*, Plenum Press, New York, 1961, 382–400.
13. A. K. Baird, "A Pressed-Specimen Die for the Norelco Vacuum-Path Spectrography," *Norelco Reporter* **8**: 108, 1961.
14. A. Volborth, "Biotite Mica Effect in X-Ray Spectrography Analysis of Pressed Rock Powders," *Am. Mineralogist* **49**: 634, 1964.
15. A. K. Baird, D. A. Copeland, D. B. McIntyre, and E. E. Welday, "Note on 'Biotite Effect in X-Ray Spectrograph Analysis of Pressed Rock Powders', A. Volborth," *Am. Mineralogist* **50**: 792, 1965.

DISCUSSION

D. H. Speidel (Pennsylvania State University): How can you be sure that you have had no change in composition?

K. W. Madlem: You mean through grinding time?

D. H. Speidel: Yes. For example, we have shown at Penn State that in a mortar powdered by a graduate student, it is possible to get up to 25 kilobars of heat. You are bound to lose some carbonate as CO_2. Another example would be oxygen loss in iron. Certainly oxidation would occur unless you were extremely careful. Either one of these would be enough to change the ratios that you were using. Have you tested for this or have you considered it?

K. W. Madlem: As I mentioned earlier, the mullite mortar ground values were worthless. An apparent loss in weight per cent due to signal intensity loss occurred after 30 min grinding. The contamination which occurred could have been due in part to oxidation. No attempt was made to control these changes in this study. None of the carbonates were ground in this manner for just such a reason. A biotite quartz monzonite rock powder was used instead. This test was performed to show that this method of grinding is not good because these changes are likely to occur, resulting in poor precision.

A. K. Baird (Pomona College): I would like to point out, in response to the previous discussion, that the shapes of all the curves for differing grinding times within any one grinding method, were the same for all elements measured, ranging from oxygen to zinc, for intensity changes.

K. W. Madlem: For the carbonate and oxide minerals, all grinding was done on a high-impact mill in timed batches of 4 min each. Between grinding, the steel vials were allowed to cool in order to overcome the adverse effects of too much heat.

E. L. Gunn (Esso Research and Engineering): Did you investigate the crystallographic properties of your various components, that is, the materials you studied? This may be related to the previous question. If so, did you notice any effects that you might attribute to crystallography?

K. W. Madlem: The materials were all studied for their purity optically before being selected for use in these studies. After the 4 min grinding time on the Pica mill, they were again studied. No optical changes were noted at this time. In the mullite mortar grinding we noted some changes after the 10 min grinding time. The effect of fusion which occurred resulted in a somewhat glassy matrix, with altered optical properties.

THE CHARACTERISTIC X-RAYS FROM BORON AND BERYLLIUM

R. C. Ehlert and R. A. Mattson

General Electric Company
Milwaukee, Wisconsin

ABSTRACT

Lead stearate and lead lignocerate multilayer soap-film structures are used to disperse the K emission lines of boron and beryllium respectively. Data are presented showing the dependence of the peak height and half-width on the number of layers in the lignocerate structure. Spectra are presented and compared for the pure element and several compounds of each element. Both electron and X-ray excitation are used. Detection is by a thin-window flow-proportional counter.

INTRODUCTION

The characteristic emission spectra from elemental boron and beryllium and from compounds containing these elements have been reported in the literature.[1-7] In all this previous work excitation was by an electron beam. Here spectra excited by both electron and X-rays are reported and compared.

EXPERIMENTAL PROCEDURE

The soft X-ray vacuum spectrometer built by the X-Ray Department of the General Electric Company was described in detail previously.[8] It consists of a windowless tube, focusing optics of the Johann type, and a flow-proportional counter having a 1500-Å parlodion window. Either electron or X-ray excitation may be employed to excite the characteristic spectra from the sample. For X-ray excitation, shown in Figure 1(A), the anode is made positive with respect to the cathode and the sample. The primary X-rays strike the sample at an angle of 45°, while the secondary X-rays leave the sample at an angle of 90°. An area of the sample $\frac{5}{16} \times \frac{3}{4}$ in. is exposed to the X-rays. Most of this area is viewed by the optics. For electron excitation, shown in Figure 1 (B), the sample is made positive with respect to the cathode. A bias voltage can and usually is applied to the electrode which was the anode in the previous configuration. This bias voltage makes it possible to control the position at which the electron beam strikes the sample. When the beam focal spot is brought in line with the optics, the counting rate is maximized. Other features of the spectrometer include a variable-width source slit, a tunable crystal holder, and an optional detector slit. Both the source-slit width and crystal holder are adjustable from outside the vacuum system. For the data presented in this paper, a multilayer soap-film structure is used as the analyzing crystal.[9,10]

Soft X-ray spectra are often complex, consisting of a main peak and one, two, or

---

[1] References are at the end of the paper.

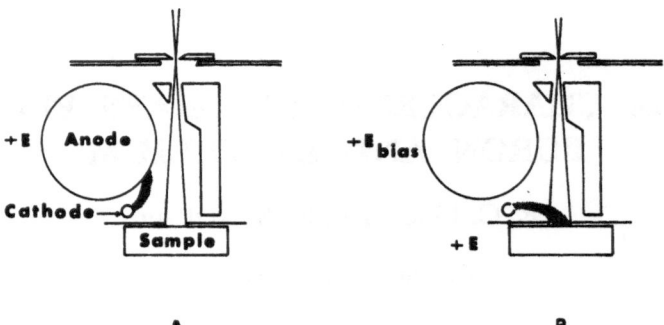

Figure 1. Electrode configuration; (a) X-ray excitation, (b) electron excitation.

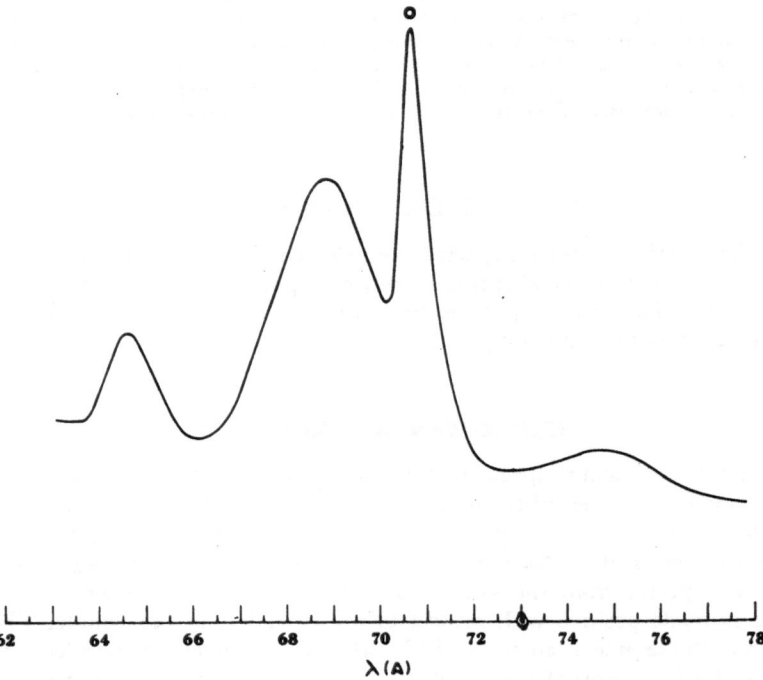

Figure 2. Characteristic radiation from Pyrex excited by 3 kV electrons. No pulse-height selection.

more satellites, the positions and relative intensities of which are dependent on the chemical and physical state of the element being analyzed. In addition, the presence of high orders of more energetic characteristic X-rays from other elements present in the sample or adsorbed on its surface can distort the observed spectrum. An example will illustrate the difficulties that can occur. Figure 2 shows the spectrum from boron present in Pyrex glass. Here the boron makes up 4% of the sample's weight. The sample is excited by 3 kV electrons impinging on a piece of Pyrex coated with a layer of carbon to make it conducting. (All non-conducting samples are treated in this manner.) The analyzing crystal is a 60-layer lead stearate soap-film structure. Four peaks are present, one of which

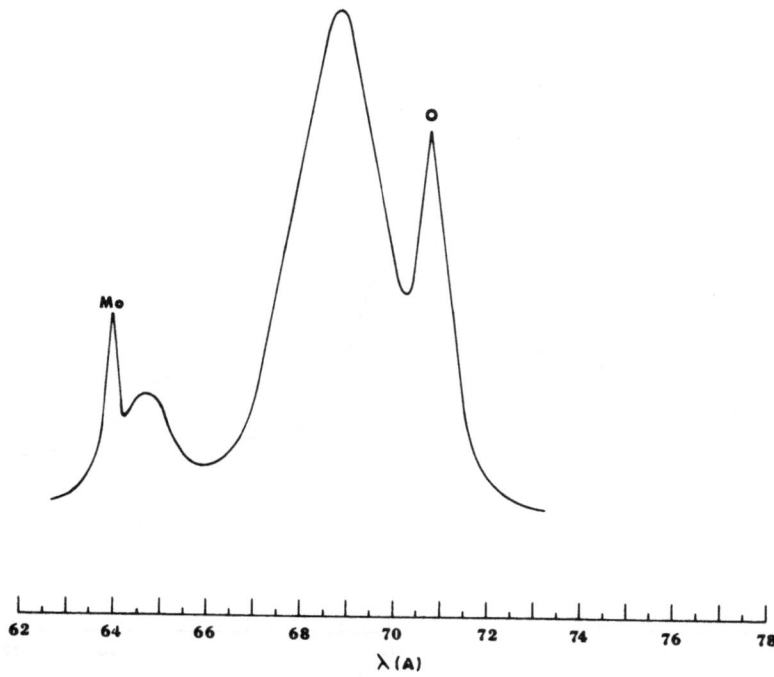

Figure 3. Characteristic radiation from Pyrex excited by the X-rays from a
molybdenum target tube. No pulse height selection.

is sharper than the other three. This sharp peak is the third-order O K_α line. The other
peaks are due to boron. Such a spectrum consisting of a main peak and both a high and
low energy satellite is typical of the borates. The intensity of the oxygen peak can be
reduced relative to the main boron peak by exciting the sample with molybdenum
radiation. Both the continuum and the characteristic M_ζ line at 64.38 Å are effective in
exciting the boron. Figure 3 shows the result. Here the oxygen line is smaller than the
main boron line. However, an additional peak due to the M_ζ molybdenum X-rays
scattered off the sample is observed. The extraneous oxygen peak can be eliminated
from the spectrum by using pulse-height selection as shown in Figure 4. This curve was
obtained using electron excitation. When using pulse-height selection in the soft X-ray
region, some of the desired pulses are lost. However, this sacrifice must be made if the
correct spectra are to be observed. In the data which follows, pulse-height selection is
always used to discriminate against unwanted lines.

BORON

Not all boron-containing compounds exhibit this type of spectrum. A single unsymmetrical peak is observed for elemental boron, boron carbide, aluminum boride
(AlB_{12}), and titanium diboride. These spectra have peaks which occur at nearly the same
wavelength and have nearly the same width. The similarity of the four spectra is shown
in Figures 5 and 6, in which relative intensity is plotted as a function of wavelength.

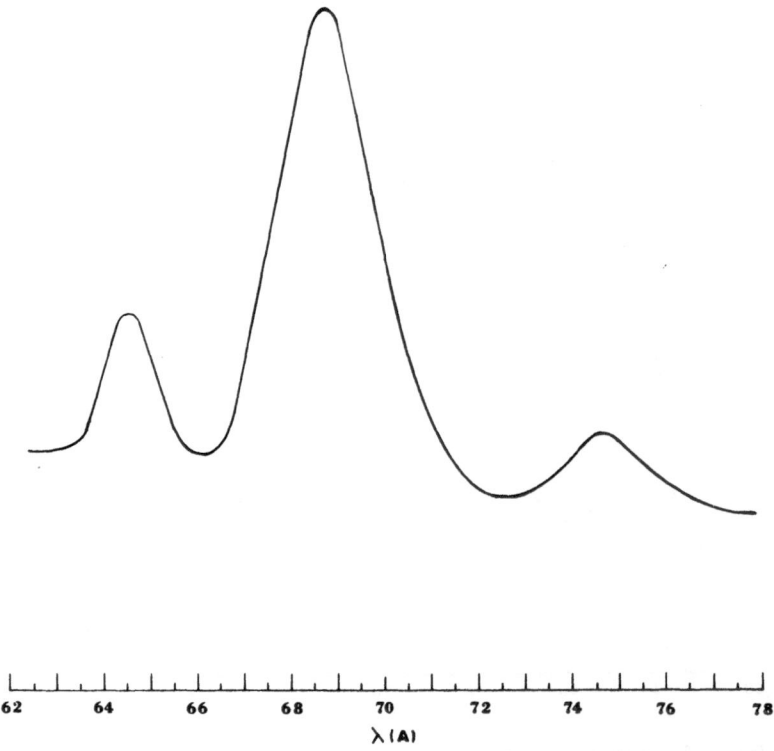

λ (A)

Figure 4. Characteristic radiation from Pyrex excited by 3 kV electrons using
pulse-height selection.

In Figure 5 elemental boron and boron carbide are shown, while in Figure 6, aluminum
boride and titanium diboride are shown. The curves are taken from ratemeter traces
and represent the average of two measurements normalized to give the same peak height.
The errors present depend on the counting rate, which is usually quite large for electron-
excited spectra. For example, for pure boron an input power of 0.125 W will give a peak
counting rate of 3500 cps. This result is obtained with a 15-mil source slit and a 36-mil
counter tube slit. By removing the counter tube slit, the counting rate can be increased
by a factor of 3.5 without impairing the resolution. For boron in Pyrex, an input power
of 0.125 W would give a peak counting rate of only 90 cps. When the Pyrex sample was
run the input power was increased to 0.6 W, which gave a peak counting rate of 430 cps.
The energy of the electron beam is 3 kV in all cases and current regulation is employed.
The effect of using an electron beam of a different energy will be discussed later.

Spectra from the borates are more complex, as was illustrated by the spectrum
from boron in Pyrex. For H_3BO_3, B_2O_3, and $Na_2B_4O_7$, similar spectra are obtained.
Figure 7 compares the spectra from H_3BO_3 and B_2O_3 and Figure 8 compares the spectra
from $Na_2B_4O_7$ and B_2O_3. The peak height occurs at nearly the same wavelength in each
case, although small differences are present. Likewise, small differences also occur in the
position and relative intensity of the satellite lines. The relative intensities of the satellite
lines depend on the energy of the incident electrons. The intensity of the low-energy

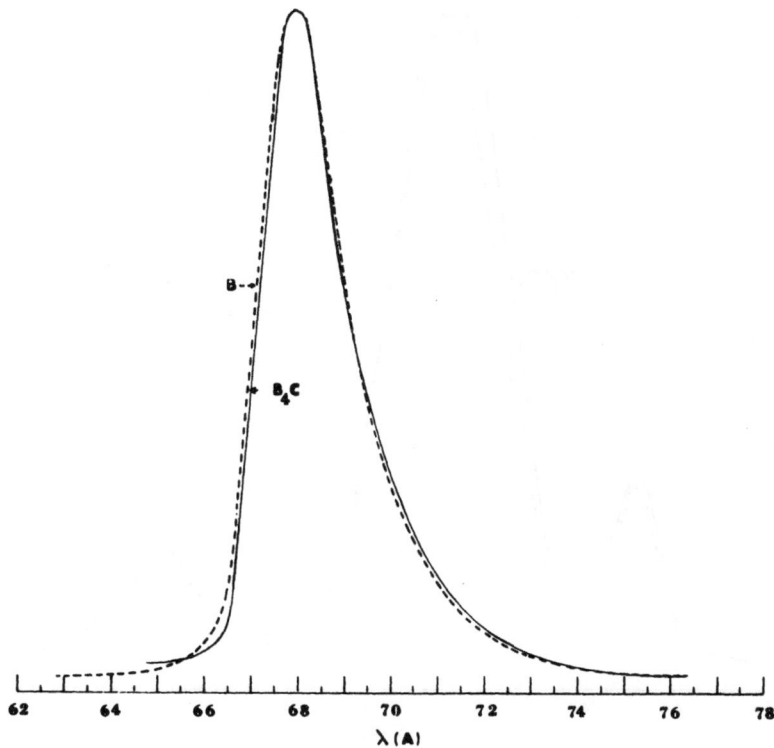

Figure 5. B K emission spectrum from boron and B_4C excited by 3 kV electrons.

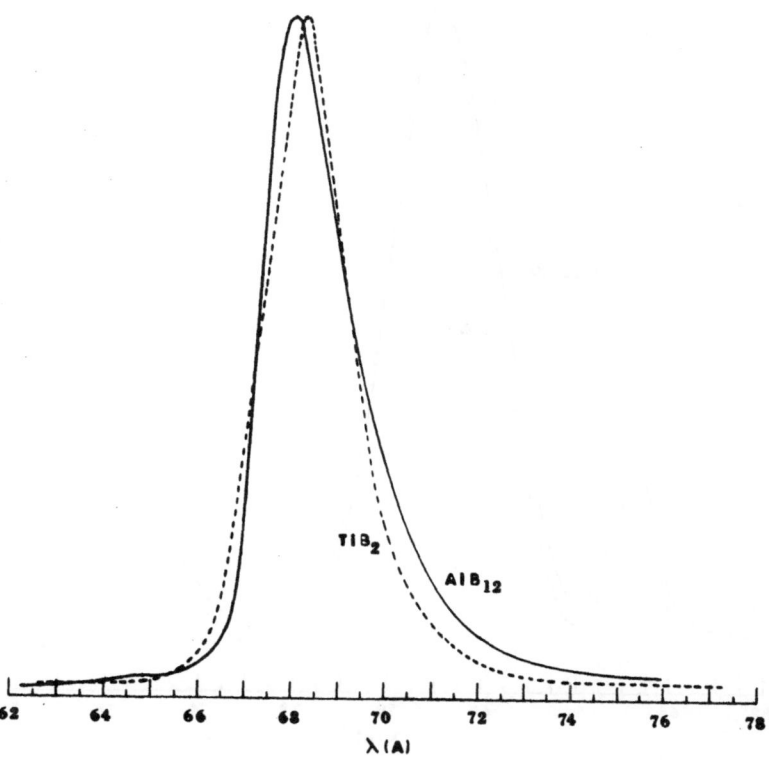

Figure 6. B K emission spectrum from AlB_{12} and TiB_2 excited by 3 kV electrons.

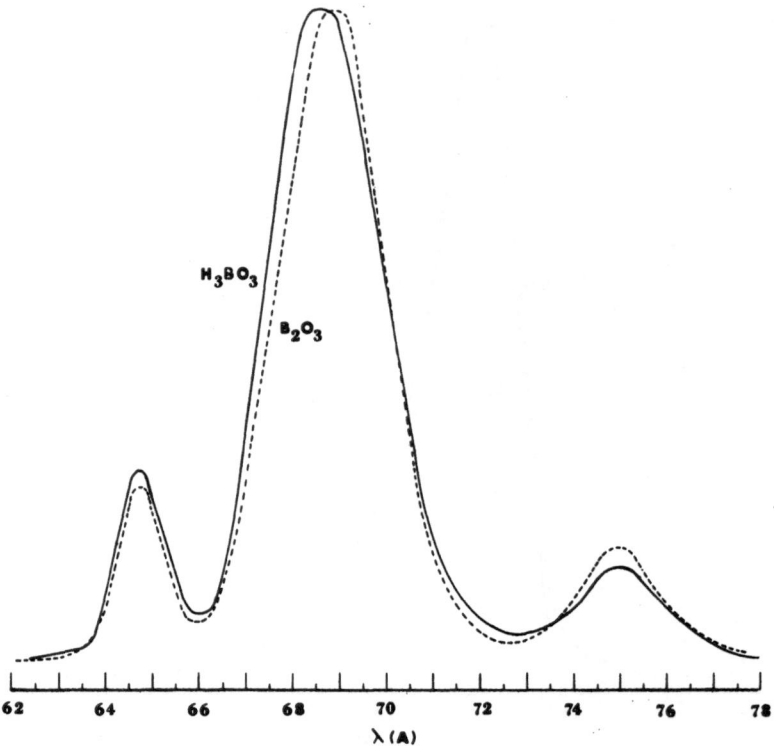

Figure 7. B K emission spectrum from H_3BO_3 and B_2O_3 excited by 3 kV electrons.

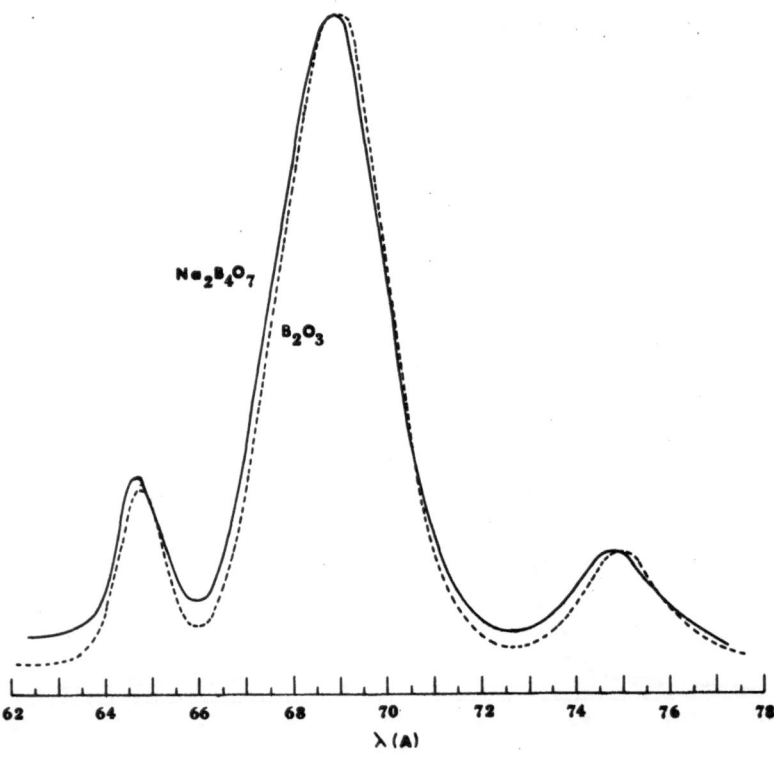

Figure 8. B K emission spectrum from $Na_2B_4O_7$ compared with that from B_2O_3. Both spectra excited by 3 kV electrons.

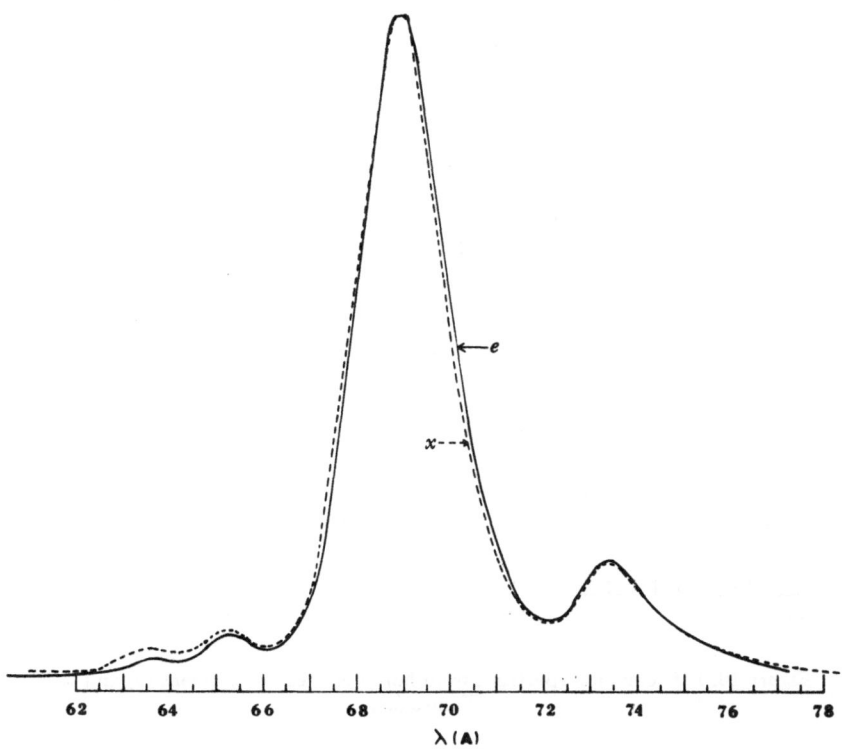

Figure 9. B K emission spectrum from BN excited by electrons (e) and C K X-rays (x).

satellite does not change much as the energy is increased from 1.5 to 6.4 keV. However, the high-energy satellite intensity falls from 35 to 25% of the main peak intensity as the voltage is increased over this range.

All of the spectra with the exception of boron in Pyrex were also excited by C K X-rays. For most of the samples, the spectra obtained with X-ray excitation are nearly identical to those obtained with electron excitation. Examples are shown in Figures 9 and 10, where boron nitride and elemental boron respectively are compared for the two modes of excitation. These figures show that the half-width is slightly smaller and the peak-to-background ratio slightly larger for X-ray excited spectra. Operating conditions for the windowless carbon anode tube were 6 kV and 50 to 100 mA, which gives typical counting rates of 200–500 cps. Again, the 36-mil slit was present in front of the counter tube. There was no need for these samples to be conducting, hence no carbon film was applied. The similarity of these spectra with those obtained by electron bombardment indicates that the carbon film has little or no effect on the shape of the observed spectra. On the basis of this data, there would seem to be little, if any, advantage to using X-ray excitation for a qualitative analytical tool unless, of course, a particular sample could be damaged or decomposed by electron bombardment. However, for quantitative work, the presence of surface contamination may be prohibitive, making X-ray excitation necessary.

Some samples, sodium borate in particular, give somewhat different spectra for the two modes of excitation. The two spectra shown in Figure 11 agree well except for the satellite line located at 64.75 Å. This line is weak for the X-ray-excited case.

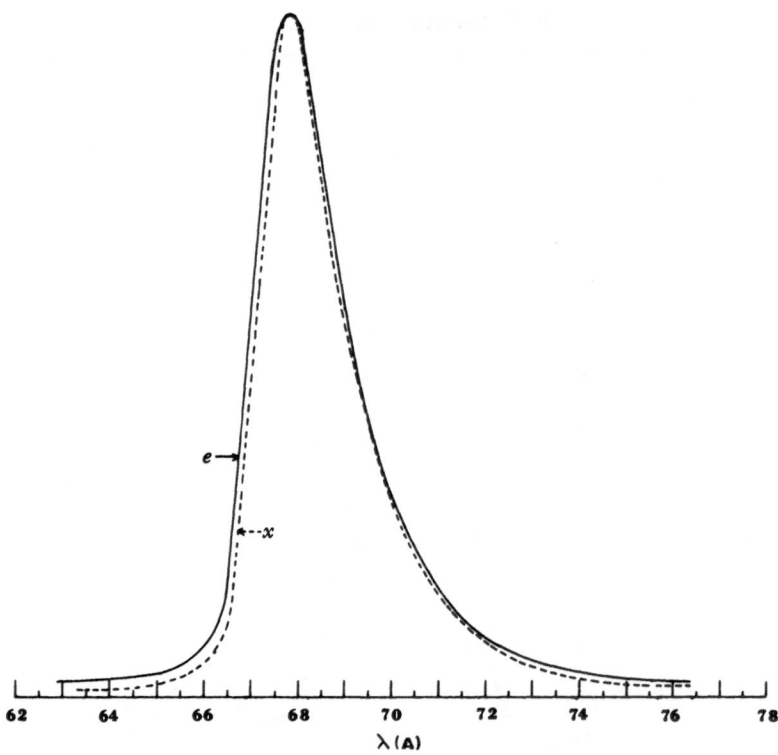

Figure 10. B K emission spectrum from boron excited by electrons (e) and C K X-rays (x).

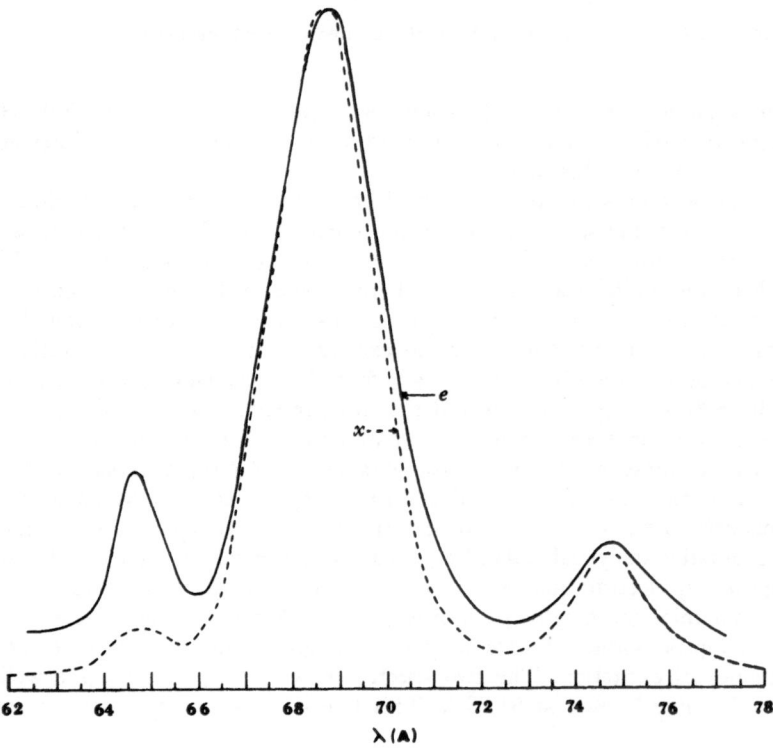

Figure 11. B K emission spectrum from $Na_2B_4O_7$ excited by electrons (e) and C K X-rays (x).

Table I. Position and Half-Width of Various Peaks of
Boron and Boron Compounds*

Sample	Form	Band	Excitation	$\lambda(\text{Å})$	$W_{\frac{1}{2}}(\text{eV})$
B	Powder	Main	$e\dagger$	67.94	6.1
			$x\ddagger$	67.93	5.6
B_4C	Powder	Main	e	67.89	5.8
			x	67.96	5.5
BN	Solid	Satellite		65.25	3.7
	Disk	Main	e	68.86	6.0
		Satellite		73.36	6.3
		Satellite		65.21	4.0
		Main	x	68.79	6.1
		Satellite		73.36	6.9
B_2O_3	Pellet	Satellite		64.78	3.9
	Compacted	Main	e	68.96	7.3
	With Graphite	Satellite		74.96	5.6
		Satellite		64.68	4.6
		Main	x	68.89	7.4
		Satellite		74.86	5.7
H_3BO_3	Compacted	Satellite		64.71	4.4
	Pellet	Main	e	68.68	8.0
		Satellite		74.94	6.6
		Satellite		64.68	4.6
		Main	x	68.94	7.8
		Satellite		74.82	6.7
Pyrex	Solid	Satellite		64.43	3.9
	Disk	Main	e	68.82	6.4
		Satellite		74.64	4.1
AlB_{12}	Solid	Main	e	67.94	5.9
	Disk		x	67.89	5.5
TiB_2	Solid	Main	e	68.18	5.7
	Disk		x	68.18	4.1
$Na_2B_4O_7$	Single	Satellite		64.68	9.1
	Crystal	Main	e	68.86	7.5
		Satellite		74.79	7.4
		Satellite		64.68	4.6
		Main	x	68.94	7.8
		Satellite		74.82	6.7

* The eighth- and ninth-order Al K_α line used for calibration.

$\dagger$ e = electron excitation.

$\ddagger$ x = X-ray excitation.

For B_2O_3, the two spectra also differ only in the intensity of the high-energy satellite. The intensity of this peak is 0.3 that of the main peak for electron excitation and 0.25 that of the main peak for X-ray excitation. For H_3BO_3, the satellite peaks have the same relative intensity with respect to the main peak for the two forms of excitation. However, the main peak is slightly narrower for the X-ray excited case.

The position and half-width of the various peaks are tabulated in Table I using the eighth- and ninth-order Al K_α line for calibration. For each sample the position of a peak obtained by electron excitation differs by at most 0.1 Å (2.7 eV) from the position of the peak obtained by X-ray excitation. Fischer and Baun[3] have reported similar data for boron and for boron present in BN, B_2O_3, B_4C, TiB_2, and AlB_{12}. The positions of the peaks reported here differ by at most 0.1 Å from those reported by them. The half-widths for electron excitation are also in good agreement. Earlier work by investigators using a photographic emulsion as the detector and a grating as the analyzer are not in good agreement with current data with regard to the position of the lines.

BERYLLIUM

To detect the characteristic X-rays from beryllium a crystal having a $2d$ spacing on the order of 125–135 Å is needed. Several lead lignocerate multilayer structures with a $2d$ of 131 Å were built up. Since beryllium X-rays are not very penetrating, the number

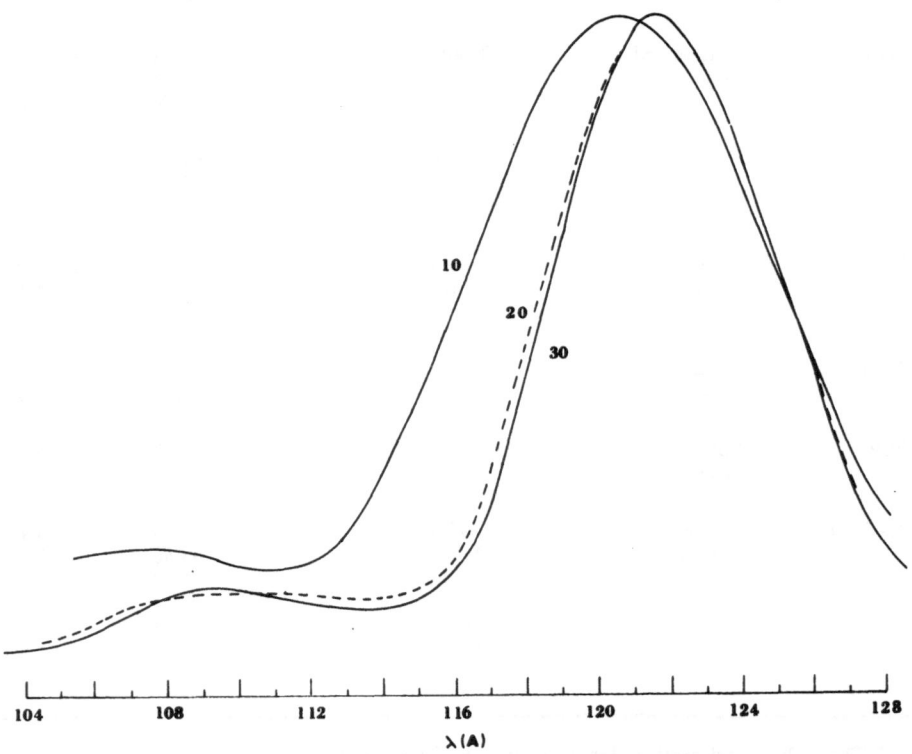

Figure 12. Be K emission spectra from BeO shown for 10-, 20-, and 30-layer lead lignocerate soap-film structures.

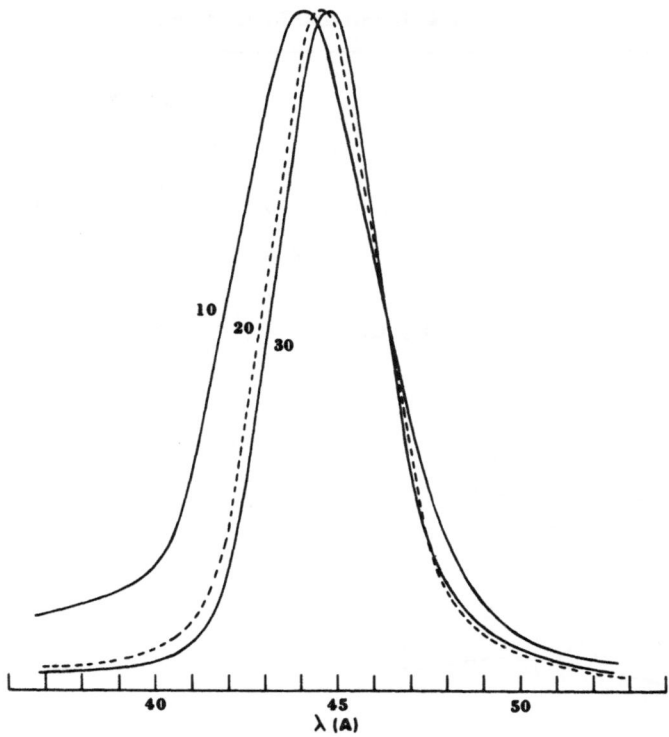

Figure 13. C *K* emission spectra shown for 10-, 20-, and 30-layer lead
lignocerate soap-film structures.

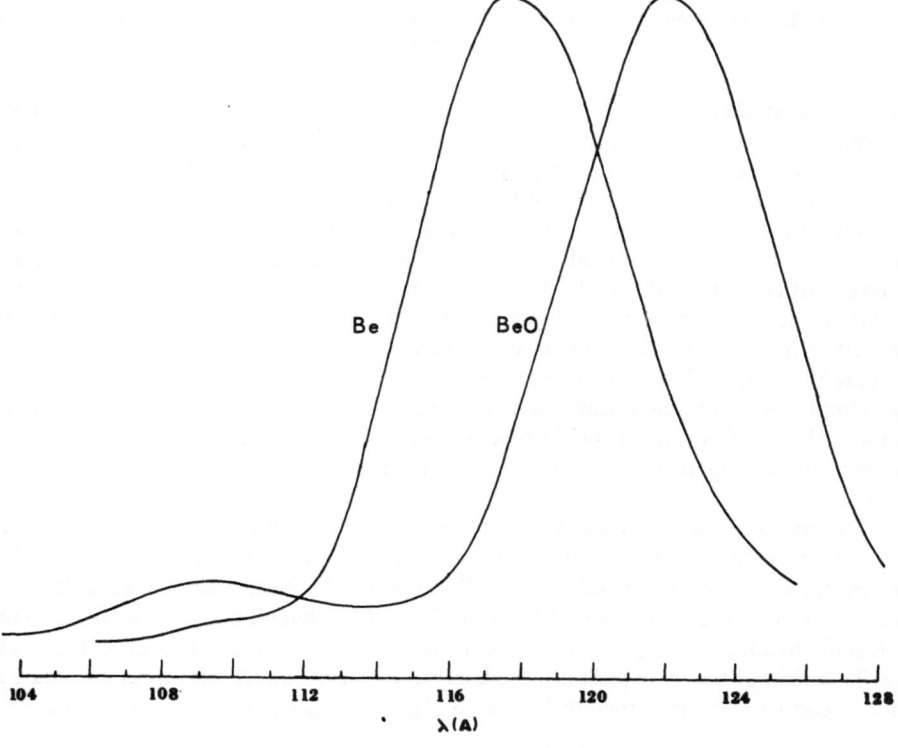

Figure 14. Be *K* emission spectra from beryllium metal and BeO excited by 3 kV electrons.

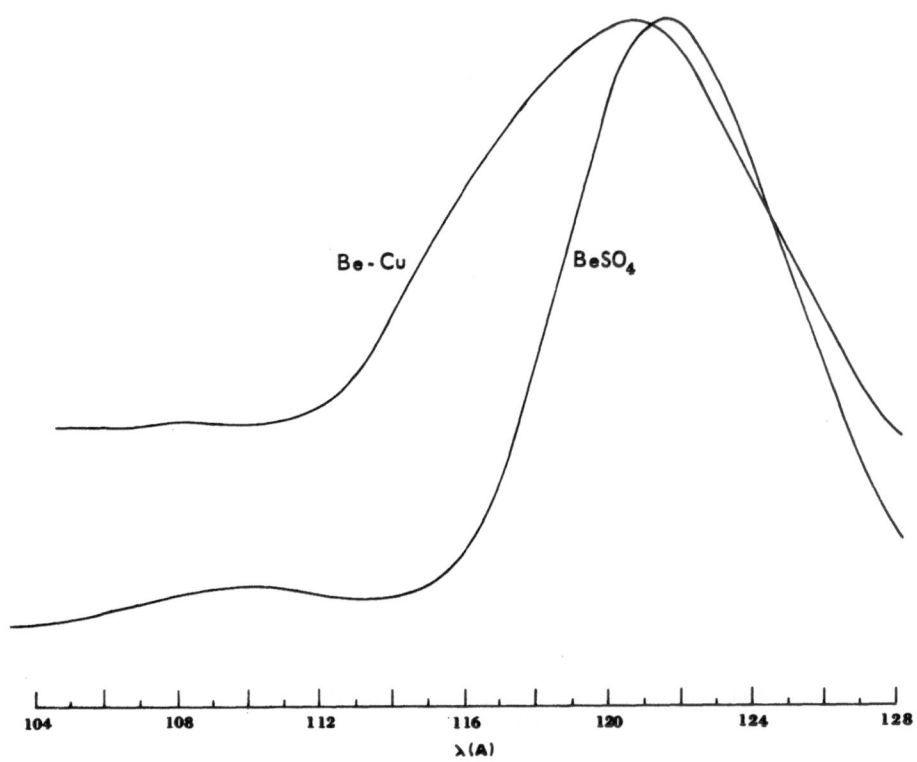

Figure 15. Be K emission spectra from $BeSO_4$ and Be–Cu alloy containing 2% beryllium excited electrons.

of layers required for the structure to be infinitely thick is small. The effect of the number of layers on the dispersing power of the structure is shown in Figure 12, in which the emission spectrum from beryllium present as an oxide is shown. The numbers near the curves indicate the number of double layers present in the structure, the number 20 referring to the dashing curve. For this figure, the ordinates are normalized so that the peak heights are all the same, although they actually increase from 1300 cps/W for the 10-layer structure to 2900 cps/W for the 20-layer structure and to 5100 cps/W for the 30-layer structure. Some increase in resolution is obtained in using 30 double layers rather than 20, but the half-width and peak-to-background ratio do not change much. To compare this data with that for more penetrating radiation, Figure 13, is presented. Here the second-order diffracted beam for C K X-rays is shown for the same three structures. Again, a shift is observed in the $2d$ spacing. The half-width decreases and the peak-to-background ratio increases. For the measurements which follow, the 30-layer structure is used.

The characteristic X-rays have been observed for beryllium metal and for beryllium in beryllium oxide, in beryllium sulfate, and in an alloy with copper. Figure 14 shows the spectra for beryllium metal and beryllium oxide. In both cases the excitation is by 3 keV electrons. A satellite line exists for both forms, although it is very weak. The main peak shifts to a lower energy in the oxide while little or no shift is observed in the satellite line. This shift of the main peak is also observed for beryllium X-rays from its sulfate and from its copper alloy, as shown in Figure 15. Again the satellite line does not shift. As the

Table II. Characteristic X-Rays for Beryllium, BeO, BeSO₄, and Be–Cu

Sample	Form	Excitation	λ (Å)	$W_{\frac{1}{2}}$(eV)
Be	Metal	e^*	117.43	6.48
		x†	117.24	6.02
BeO	Powder	e	121.84	6.12
		x	121.70	6.63
BeSO₄	Powder	e	121.79	6.65
Be–Cu	Metal	e	120.82	7.80

* e = electron excitation.
† x = X-Ray excitation.

goniometer is now set up, it is impossible to swing to a larger angle to determine if a low-energy satellite exists.

The beryllium present in the copper alloy makes up only 2% of the weight of the material. Counting rates of 900 cps are obtained for an input power of 0.78 W (3 kV and 260 μA). In the sulfate, it makes up 18% of the weight and in the oxide 36% of the weight. For the latter, counting rates on the order of 4000 cps are obtainable for an input power of 0.45 W (3 kV and 150 μA). For these measurements, the source slit was opened to a width of 80 mils.

X-ray excitation with carbon radiation was used for beryllium metal and beryllium oxide. The spectra are essentially the same as those obtained with electron excitation. The position and half-width of the various lines for both modes of excitation are given in Table II.

CONCLUSIONS

On the basis of the data reported in this paper it is evident that the emission spectrum from the light elements is essentially the same for electron and X-ray excitation. In some cases the strength of a satellite line will be stronger if the spectrum is excited by electrons rather than by X-rays. These differences when they exist may be valuable in the identification of a compound or class of compounds and may also some day aid in the development of a satisfactory theory for this phenomena.

ACKNOWLEDGMENT

The authors wish to thank Robert Agenten, who assisted in taking the data.

REFERENCES

1. D. Coster and S. Hof, "On the Emission Spectra of Some Oxides and Pure Elements in the Soft X-Ray Region," *Physica* 7: 655, 1940.
2. R. S. Crisp and S. E. Williams, "Soft X-ray Emission Spectra of Sodium, Beryllium, Silicon, and Lithium," *Phil. Mag.* 6: 365, 1961.
3. D. W. Fischer and W. L. Baun, "The Effect of Chemical Combination on Long Wavelength K and L X-Ray Spectra," presented at 16th Pittsburgh Conference on Analytical Chemistry and Applied Spectroscopy, March, 1965.
4. E. Gwinner and H. Kiessig, "The Influence of Binding Energy on the Boron *K* Lines," *Z. Physik.* 107: 449, 1937.
5. A. Hautot and J. Serpe, "On the *K* Emission Line of Boron," *J. Phys. radium* 8: 175, 1937.

6. H. W. B. Skinner, "The Soft X-Ray Spectroscopy of Solids. I *K*- and *L*-Emission Spectra from Elements of the First Two Groups," *Phil. Trans. Roy. Soc. London Ser. A* **239**: 95, **1940**.
7. H. W. B. Skinner, and J. E. Johnston, "Soft X-Ray Bands from Dilute Alloys," *Proc. Cambridge Phil. Soc.* **34**: 109, 1938.
8. R. A. Mattson, "Some Measurements of Carbon *K* Excitation in a New Ultrasoft X-Ray Spectrometer," *Advances in X-Ray Analysis, Vol. 8*, p. 333, University of Denver, Plenum Press, New York, 1965.
9. R. C. Ehlert, "The Diffraction of X-Rays by Multilayer Stearate Soap Films," in: W. M. Mueller, G. R. Mallett, and M. J. Fay (eds.), *Advances in X-Ray Analysis, Vol. 8*, Plenum Press, New York ,1965, p. 325.
10. R. C. Ehlert, "Overturning of Monolayers," *J. Colloid Sci.* **20**: (4) 387, 1965.

DISCUSSION

N. Spielberg (Philips Laboratories): In your comparison of the photon and the electron excitation of one of the boron compounds, could you indicate what the kV was in the electron excitation and what your carbon excitation was?

R. C. Ehlert: The carbon excitation in the electron beam voltage was 3 kV.

N. Spielberg: It would be interesting if you could get your electron-beam voltage down to the same energy as the carbon excitation and see whether this difference still persists.

A. A. Sterk (American Machine and Foundry Company): We have plotted beryllium spectra with a grating spectrometer and think we show much more detailed structure than your spectra. The resolution of that spectrometer was $\frac{1}{2}$Å. Could you tell us what the resolution of your crystal spectrometer was?

R. C. Ehlert: I'm not sure what that was.

A. A. Sterk: There are no data on wavelength resolution.

R. C. Ehlert: No.

J. E. Holliday (U.S. Steel): On your beryllium spectrum, did you check this with other authors? I don't believe many have shown a satellite on the low energy.

R. C. Ehlert: I did check this with Skinner's data. They do not show a satellite line. Also the position of the peaks that they observed was different from ours. I might say that there was some disagreement between their data and ours for the boron spectra, also. But as sort of added proof that our data might be correct, we are in good agreement with the boron data which was obtained by Fischer and Baun at Wright Field. At most our positions differ by about 0.1 Å, and I think this is about as good as our data are.

J. E. Holliday: But theirs was done with a grating.

R. C. Ehlert: A grating, and they had quite good resolution.

J. E. Holliday: Lead stearate, not a grating.

R. C. Ehlert: Fischer and Baun's work was done with lead stearate, yes. Ours was done with lead stearate for the boron and lead lignostearate for the beryllium.

J. E. Holliday: But there is a disagreement between results obtained from lead stearate and from gratings. I don't think there were any satellites ever shown for the beryllium low-energy side from gratings. So this shows that there is some discrepancy between the lead stearate and the gratings which will have to be resolved.

R. C. Ehlert: Yes.

J. B. Nicholson (Applied Research Laboratories): Did you measure the width at half-maximum for the beryllium line on pure beryllium for both the direct excitation and the fluorescence?

R. C. Ehlert: Yes; for pure beryllium with electron excitation the half-width was about $6\frac{1}{2}$ eV and it was about 6.0 eV for the X-ray-excited case.

S. R. Colberg (U.S. Naval Ordnance Test Station): I was wondering if you had any explanation for the very large peak shift between beryllium and beryllium oxide.

R. C. Ehlert: No, I do not.

B. L. Henke (Pomona College): Rechecking our beryllium with the lead stearate we do not get the satellite structure on the low-energy side.

R. C. Ehlert: This satellite was on the high-energy side. We were unable to swing to a larger angle to see if the low-energy satellite existed. With the lead lignostearate structure that we used we started scanning at about 150° and scanned downward; this just picked up the edge of the peak. So we have no data as to whether a satellite line would be there or not. In all of the figures which I have shown, energy is increasing to the left.

Chairman S. P. Ong: Maybe I should insert a remark in here. It is possible that the satellite is due to another phenomenon because of the interference that has occurred similar to optical interference in thin film. This occurs when the stearate layer is too thin. We have seen that with 10 to 20 layers of stearate that you get reflections similar to optical reflections. I don't know whether this is the case here or not.

THE APPLICATION OF A SOFT X-RAY SPECTROMETER TO STUDY THE OXYGEN AND FLUORINE EMISSION LINES FROM OXIDES AND FLUORIDES

R. A. Mattson and R. C. Ehlert

General Electric Company
Milwaukee, Wisconsin

ABSTRACT

Results are shown for O and F K emission lines from several oxide and fluoride compounds. The spectra are examined for both structure and wavelength. Excitation of the spectra is by both electrons and X-rays. The emission spectra from several oxygen- and fluorine-bearing gases and the influence of the composition of the diffracting crystal on the shape of the observed oxygen spectra are also discussed. Curved crystals of potassium acid phthalate and lead stearate, and a thin-window flow proportional counter, are used to obtain the results.

INTRODUCTION

The intent of this paper is to indicate what role a versatile soft X-ray spectrometer might play in the study of X-ray spectra between 5 and 120 Å. Here we shall confine our remarks only to the results obtained for various oxide and fluoride samples. Elsewhere in these proceedings[1] we discuss the results obtained for samples containing boron and beryllium. Others [2-7] have measured the X-ray spectrum of fluoride and oxide samples which have been excited by electrons; here we shall compare some spectra excited by X-rays with those excited by electrons. The spectra from several oxygen- and fluorine-bearing gases and the influence of the composition of the diffracting crystal on the observed oxygen spectrum are also discussed. No attempt will be made to interpret the spectra.

EXPERIMENTAL

The fully focusing curved crystal X-ray spectrometer used in these measurements has been described elsewhere,[8] so only the X-ray source configuration need be described here. Four variations of the X-ray source were used and are depicted in Figure 1. The entire source assembly is housed in an aluminum enclosure which can be pumped separately by an ion pump.

In the X-ray excitation mode [Figure 1(a)] the sample and cathode are maintained at ground potential while the anode assumes any positive potential up to 16 kV. For electron excitation, the sample is maintained positive with respect to the filament [Figure 1(b)] while the conventional X-ray anode acts as a bias electrode to deflect the electron

---

[1] References are at the end of the paper.

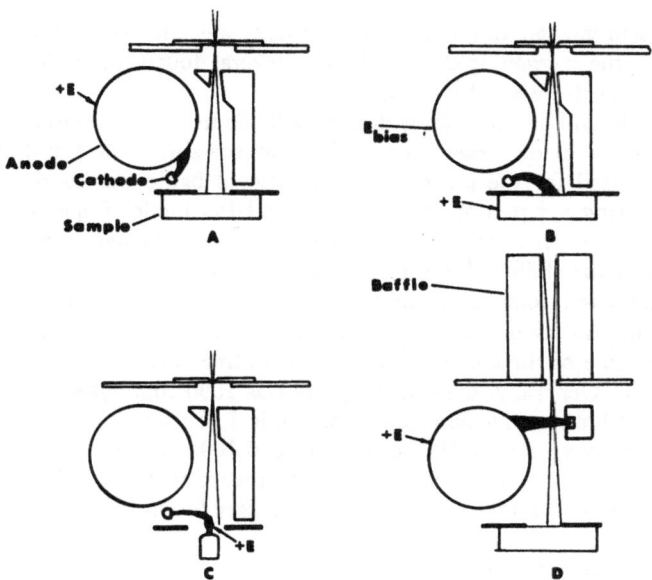

Figure 1. Schematic representation of X-ray source. (a) X-ray excitation, (b) electron excitation, (c) electron excitation of heated wire, (d) electron excitation of gaseous sample.

beam. The sample and the sample mask are the only positive high-voltage elements in the source volume. The X-ray optics are constrained to look only at the sample so that no contribution is made to the background by the sample mask. A third configuration [Figure 1(c)], which is a special case of the electron excited mode, was also used. In this mode, a wire, ribbon, or helix was suspended across two insulators such that it could be held at a positive voltage. All other surfaces including those beneath the wire were at ground potential. Electrons were then accelerated to the wire. The temperature of the wire could be varied by ohmic heating such that surface contamination could be evaporated away. Measurements of interactions of gases with materials at various temperatures could be observed in this way and "erased" when the measurement was over.

To excite a gas, a slightly different configuration was used [Figure 1(d)]. An electron beam was caused to pass through a portion of the volume viewed by the X-ray optics. Some of the electrons interact with the gas molecules, causing emission of X-rays. The energy of the electron in this arrangement is dependent on the electric field potential between the cathode and anode. An electron gun would be preferable to this arrangement so that electron energies would be definitely known.

Gas is admitted into the box in the area normally used for ion cleaning of the anode. The system pressure is measured in the bell jar so that actual "sample" pressures are higher than the measured pressures. Nominal system pressures for gas excitation were 3×10^{-4} torr and electron currents up to 12 mA were used. Fluorescence excitation of any surface viewed by the optics was geometrically prevented; hence, only doubly scattered photons could enter the X-ray optics.

EXPERIMENTAL

A 1-in. diameter counter tube with a 1-μ polypropylene window was used to detect both the fluorine and oxygen radiation. A bent mica crystal ($2d = 19.937$) was used for the fluoride measurements, and KAP ($2d = 23.632$) was the crystal which was mainly used to measure the oxygen spectra, although several fourth-order lead stearate and first-order RbAP spectra were also taken.

The source-slit and detector-slit widths were 0.020 in. for the fluoride measurements and 0.030 in. for the oxide measurements. The source slit was set at 0.080 in. and no detector slit was used for the gas data. An additional baffle was used to prevent the detector from viewing any part of the source in this latter mode of operation.

Normal operating pressures were 8×10^{-6} torr during excitation of the solids, but were raised by a factor of 50 to 100 for excitation of gases.

All data were taken from a strip chart recorder, the time constant of which was varied from 2 to 8 sec, depending on the counting rate. This factor may have a subtle influence on the shape of some of the curves. Maximum counting rates varied depending on the mode of excitation. Generally, the counting rate was 1000 counts/sec or greater for primary excitation of both oxides and fluorides. Typical of counting rates for secondarily excited fluoride and oxygen were 1000 counts/sec and 100 to 200 counts/sec, respectively. All measurements were repeated at least twice to insure reproducibility of the data. A constant potential of 8 kV across a copper anode was used for all X-ray excitation measurement.

Samples were made conductive by coating with a thin carbon film. X-ray excited data were taken of some materials with and without the carbon film to verify that there was no spectral change. Some materials would not withstand electron bombardment; hence,

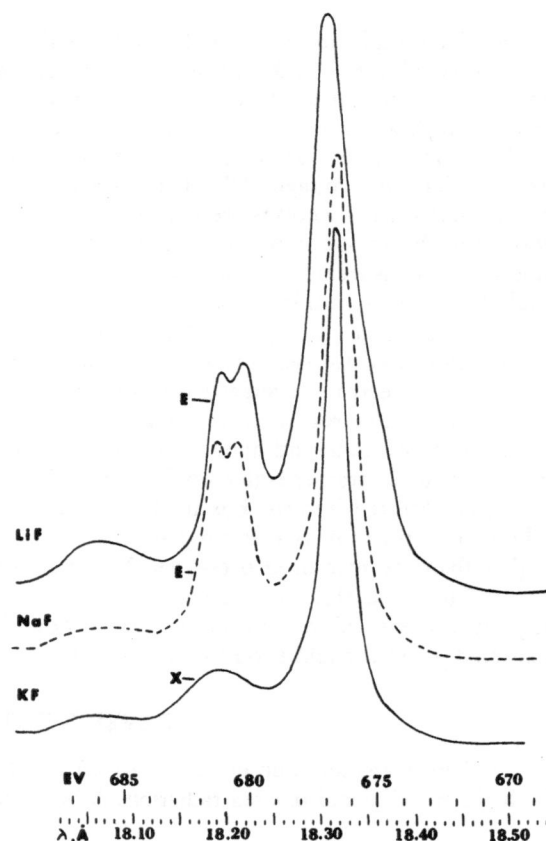

Figure 2. Fluorine emission lines from cubic fluorides (NaCl type).

only X-ray-excited spectra were obtained. Most of the fluorides were single-crystal specimens while the oxides were pelletized reagent grade materials.

FLUORIDE RESULTS

Fluorides of four different crystal types were examined with both X-ray and electron excitation. The crystal types were (1) cubic (NaCl type), (2) cubic (CaF_2 type), (3) tetragonal (SnO_2 type), and (4) cubic ($CaTiO_2$ type). O'Bryan and Skinner,[2] Fischer,[5] and Sawada et al.[7] have shown the same spectra for the first crystal type except that a high-energy band at 685 eV seems more predominant in all our data shown in Figure 2. This same band is present in all of the fluorides, but varies in amplitude and energy. There is evidence of still another higher energy band (not shown), the amplitude of which is about that of background for some fluorides. The KF spectra shows high-energy satellite structure not reported by Fischer. Prior to analysis, the polycrystalline KF sample was kept in a vacuum oven for four days to dry any moisture. KF, NaF, PbF_2, and CaF_2 were the only polycrystalline samples which were used. The same results were obtained for a single crystal of CaF_2 as for the polycrystalline form.

Two materials of the second group, CdF_2 and PbF_2 (Figure 3), show a low-energy satellite as well as the high-energy satellites. Significant differences are evident in the asymmetrical main peaks as well as in the first high-energy satellite band. Doublet structure is found only in the CaF_2.

A dissimilarity amongst spectra of the third type, shown in Figure 4, is again found. MgF_2 has no evident low-energy line, but has a very pronounced high-energy satellite

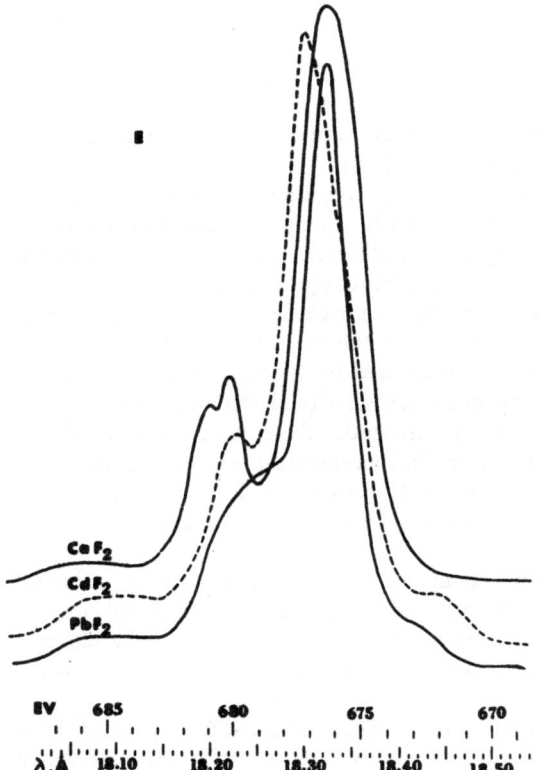

eV 685   680   675   670

λ,Å   18.10   18.20   18.30   18.40   18.50

CaF₂
CdF₂
PbF₂

Figure 3. Fluorine emission lines from cubic fluorides (CdF_2 type).

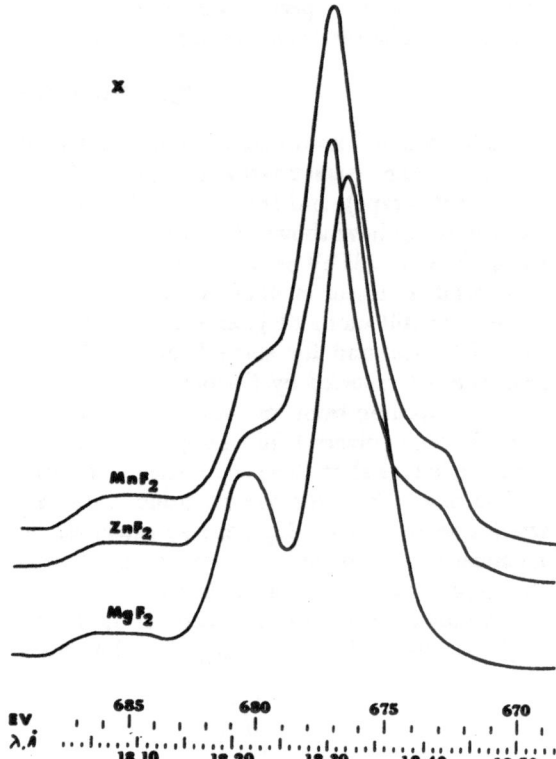

Figure 4. Fluorine emission lines from
tetragonal fluorides (SnO$_2$ type).

compared to the MnF$_2$ and ZnF$_2$. These later two spectra are quite similar and are unlike any other spectra of those measured except, perhaps, Teflon.

Correspondence among the spectra of the fourth type, Figure 5, is the highest for any crystal type which was measured. An apparently unresolved low-energy satellite provides the main difference between each spectrum. Differences in the spectra observed between X-ray and electron excitation are not great. Figures 6(a) and (b) show a superposition of spectra obtained for the two modes of excitation for CaF$_2$ and LiF spectra. High-energy satellites are reduced in amplitude relative to the main peak and the intensity ratio of the doublet changes when the sample is excited by X-rays. Also, a low-energy satellite is more in evidence with secondary excitation than with electron excitation. There do not appear to be further differences in the spectra under the experimental conditions used here. It is felt that by scanning the spectra in incremental steps and accumulating sufficient counts at each interval the statistical precision would be improved such that a more meaningful comparison could be made. Subtle differences in the spectra would then become more evident.

The effect of the energy of the exciting X-rays or electrons may strongly influence the spectra which is emitted. All primary excitation measurements were taken with electron energies considerably in excess of the K absorption edge. When the electron energy was increased, the intensity of the high-energy satellite would fall relative to the peak.

Figure 7 depicts several miscellaneous spectra taken under differing conditions. SF$_6$ has a narrow peak with both high- and low-energy satellites with energies which

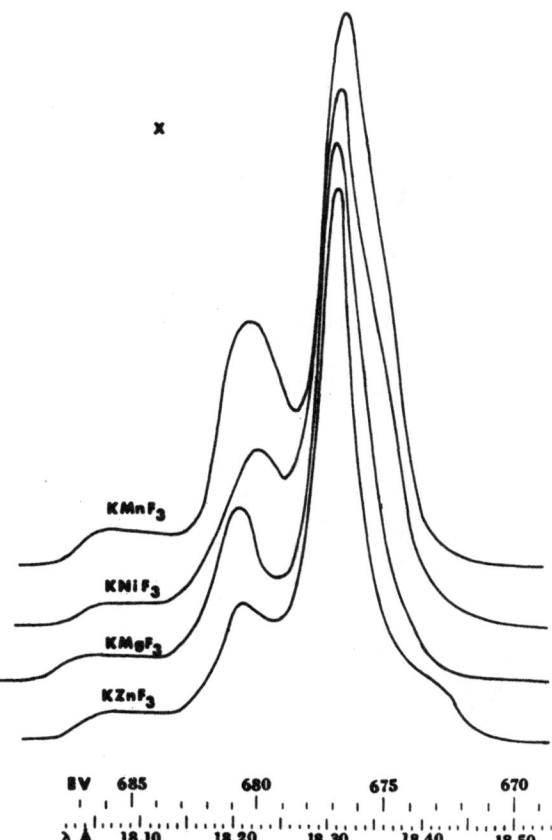

Figure 5. Fluorine emission lines from cubic fluorides (CaTiO₂ type).

differ little for that of the main peak. X-ray excited Teflon appears to have the same structure except that the satellites are separated more from the peak. If SF_6 gas is introduced when a sample such as copper or aluminum is bombarded by electrons, a fluoride is formed. Two examples of what might be CuF_2 and AlF_3 are also shown in Figure 7. The deposit formed on magnesium metal yielded a spectrum identical to that of the single-crystal MgF_2. A deposit on a nickel surface sublimed, whereas the deposits on aluminum, copper, and magnesium were stable.

To illustrate what role a soap-film "crystal" might play in spectral analysis, consider the lead stearate curves in Figure 8. Here the increase in resolution is quite evident as the order increases. The fifth-order spectra for LiF is equivalent to that obtained with first-order KAP. There is no influence of the crystal composition on the relative heights of the fluorine peaks. Crystal composition does, however, affect the spectra of various oxides obtained in fashions similar to those described above.

OXIDE RESULTS

Nearly all the data in the literature on oxide spectra were obtained using concave diffraction gratings to disperse the spectra. The results for beryllium, boron, and fluorine compounds obtained with the single-crystal spectrometer used here agree with much

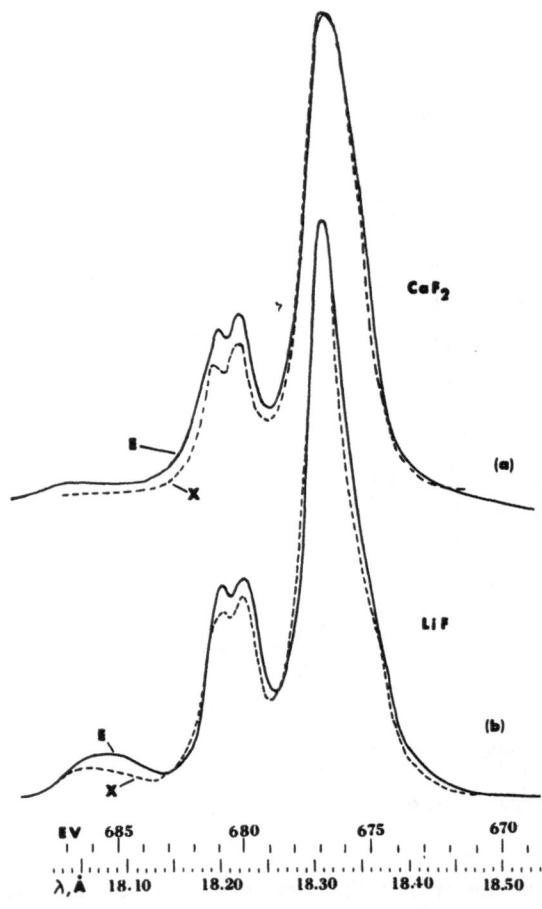

Figure 6. (a) Fluorine emission lines from CaF₂, (b) fluorine emission lines from LiF.

of the reported data. Also carbon and nitrogen not reported in these proceedings do not greatly differ with reported grating data. All measurements of the oxygen spectra with a KAP crystal, however, showed the existence of a strong line with an energy of about 6eV greater than the main reported peaks. At best, inflections in the curves were found with those instruments with sufficient resolving power to distinguish between the energies. RbAP and lead stearate crystals showed a less intense high-energy peak than did the KAP. Flat crystals produce the same results. Electron-excited spectra from Al₂O₃ and MgO appear as shown in Figure 9 when viewed with lead stearate (fourth-order) and with KAP. There is more detail in the KAP curves as might be expected. The same features are found in both lower-energy peaks. Lukirskii et al.[4] and O'Bryan and Skinner[2] show an inflection where the peaks exist for MgO. Fischer[9] has corroborated these data and has provided data showing that other phthalate crystals give still other ratios. The ratios of the low-energy peak intensity to the high-energy peak intensity are tabulated in Table I.

Table I. Ratios of Low-Energy to High-Energy Peak Intensity

Crystal	Peak ratio
KAP	0.75:1
CsAP	0.62:1
NaAP	0.46:1
NH$_4$AP	0.30:1
RbAP	0.19:1
Lead stearate	0.17:1
Barium stearate	0
Lead lignocerate	0

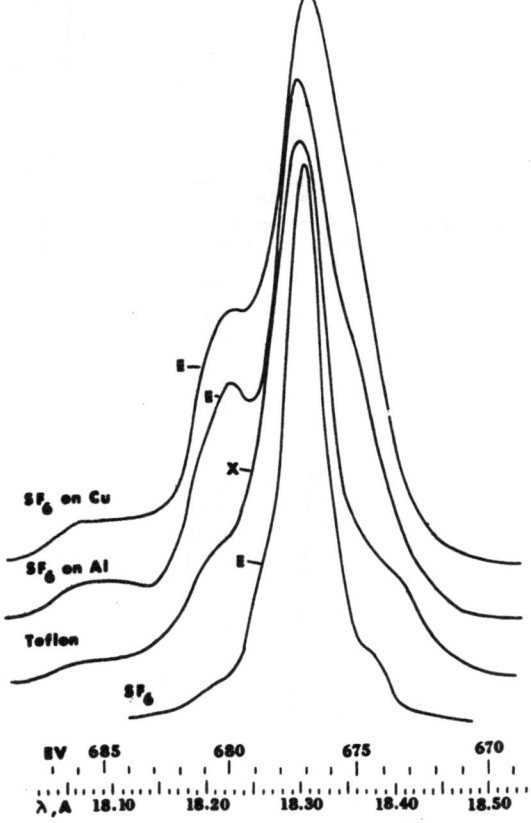

Figure 7. Fluorine emission lines from unclassified fluorides.

MgO and SiO$_2$ spectra are shown for both modes of excitation in Figure 10. X-ray excitation reduces the relative intensity of the higher energy peaks. The loss of detail is probably due to the greater statistical variations present due to lower counting rates. Nominal intensities were 1000 counts/sec with 5 kV and 200 μA for primary excitation and 50 to 500 counts/sec for 9 kV, 70 mA for secondary excitation. To obtain comparable intensities for the geometry used in the spectrometer, 20,000 times more power is required for X-ray excitation than for electron excitation.

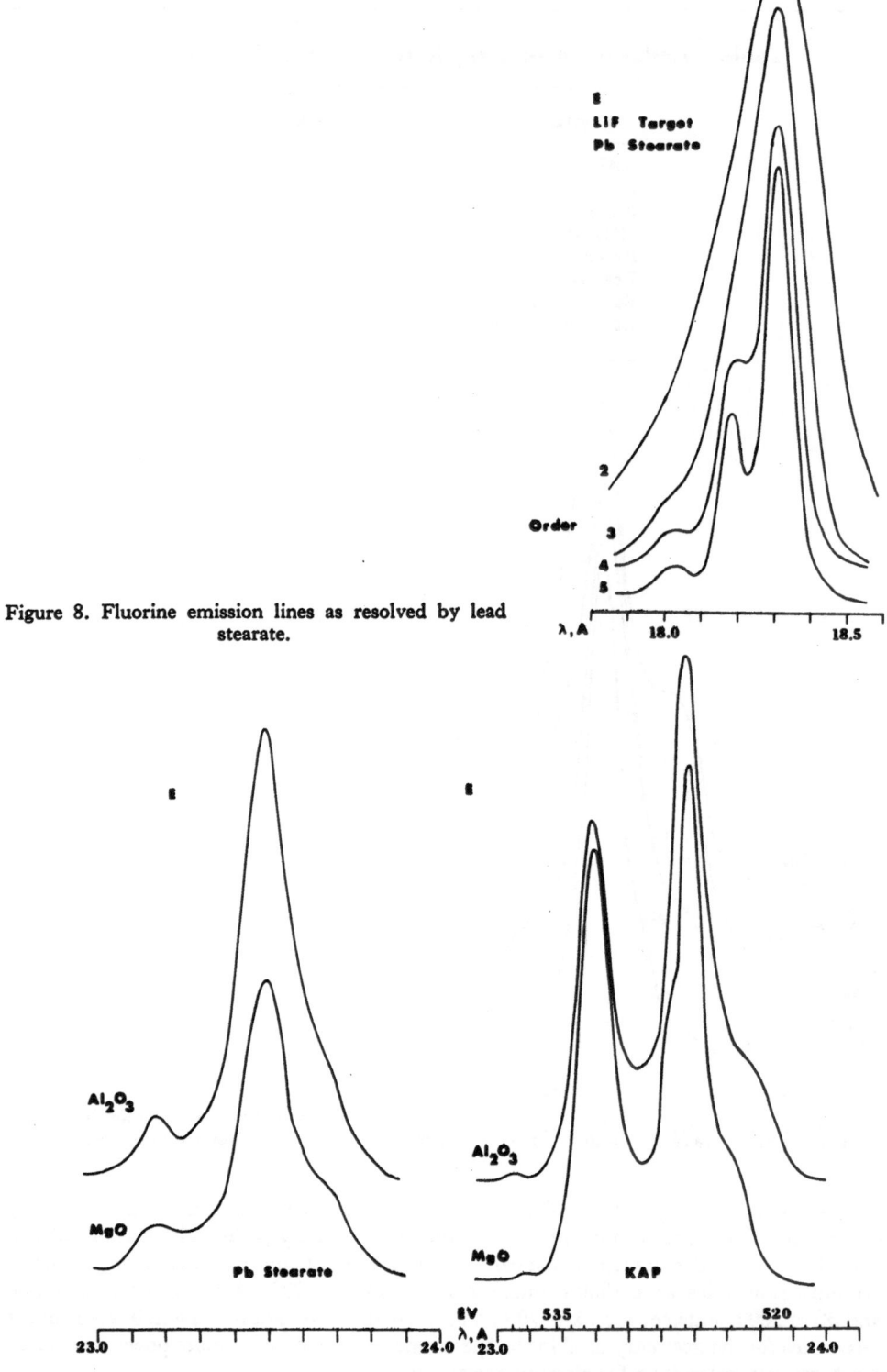

Figure 8. Fluorine emission lines as resolved by lead stearate.

Figure 9. Oxygen emission lines from Al₂O₃ and MgO using KAP and lead stearate crystals.

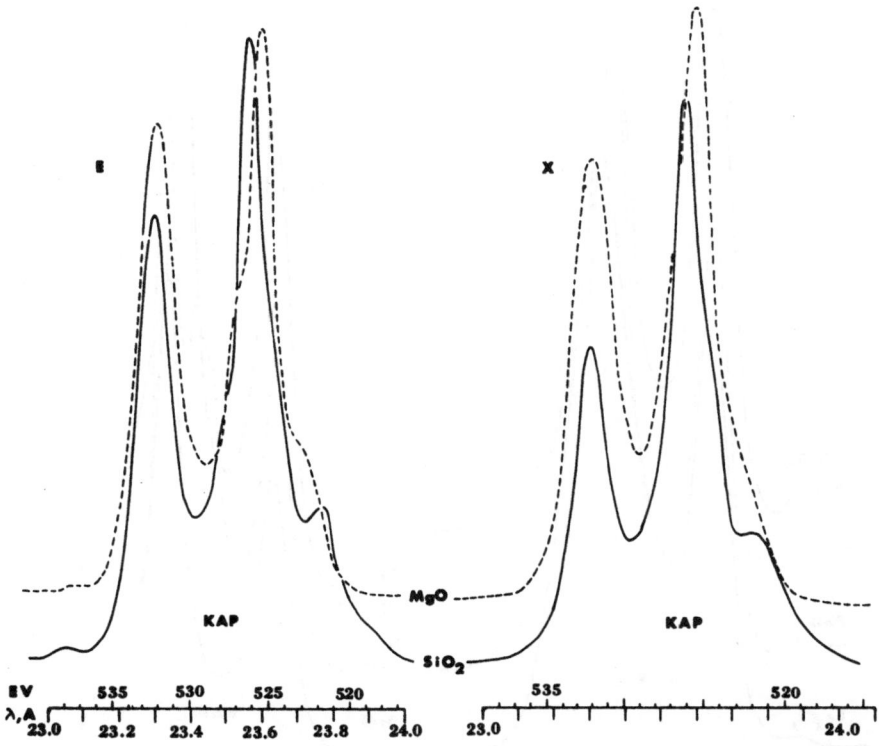

Figure 10. Oxygen emission lines from SiO_2 and MgO with X-ray and electron excitation.

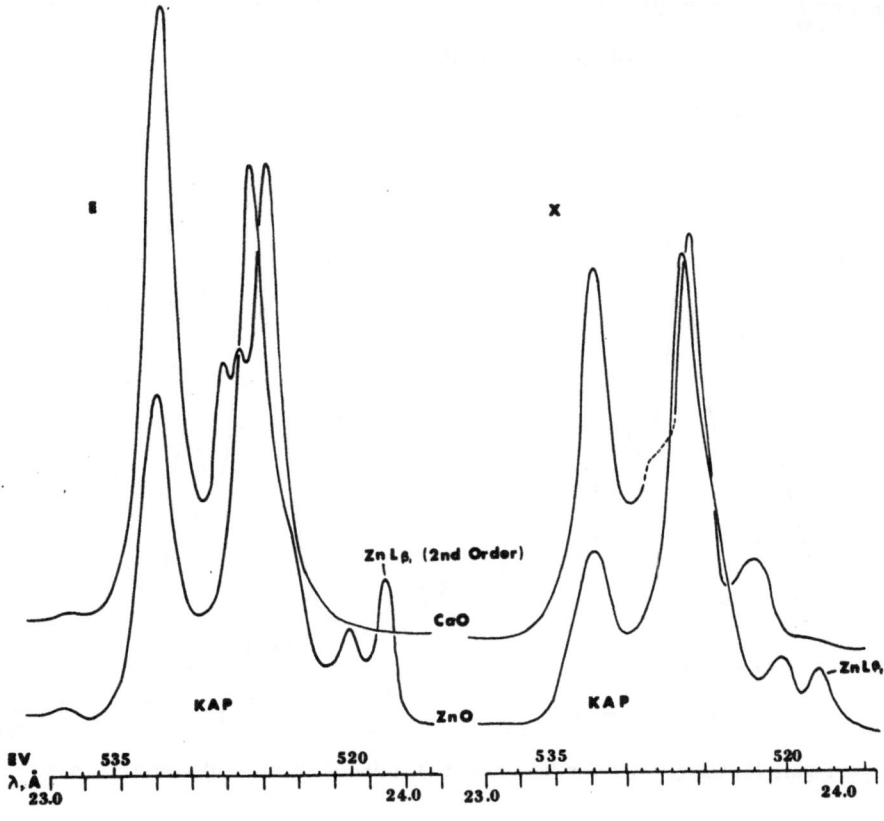

Figure 11. Oxygen emission spectra of CaO and ZnO with X-ray and electron excitation.

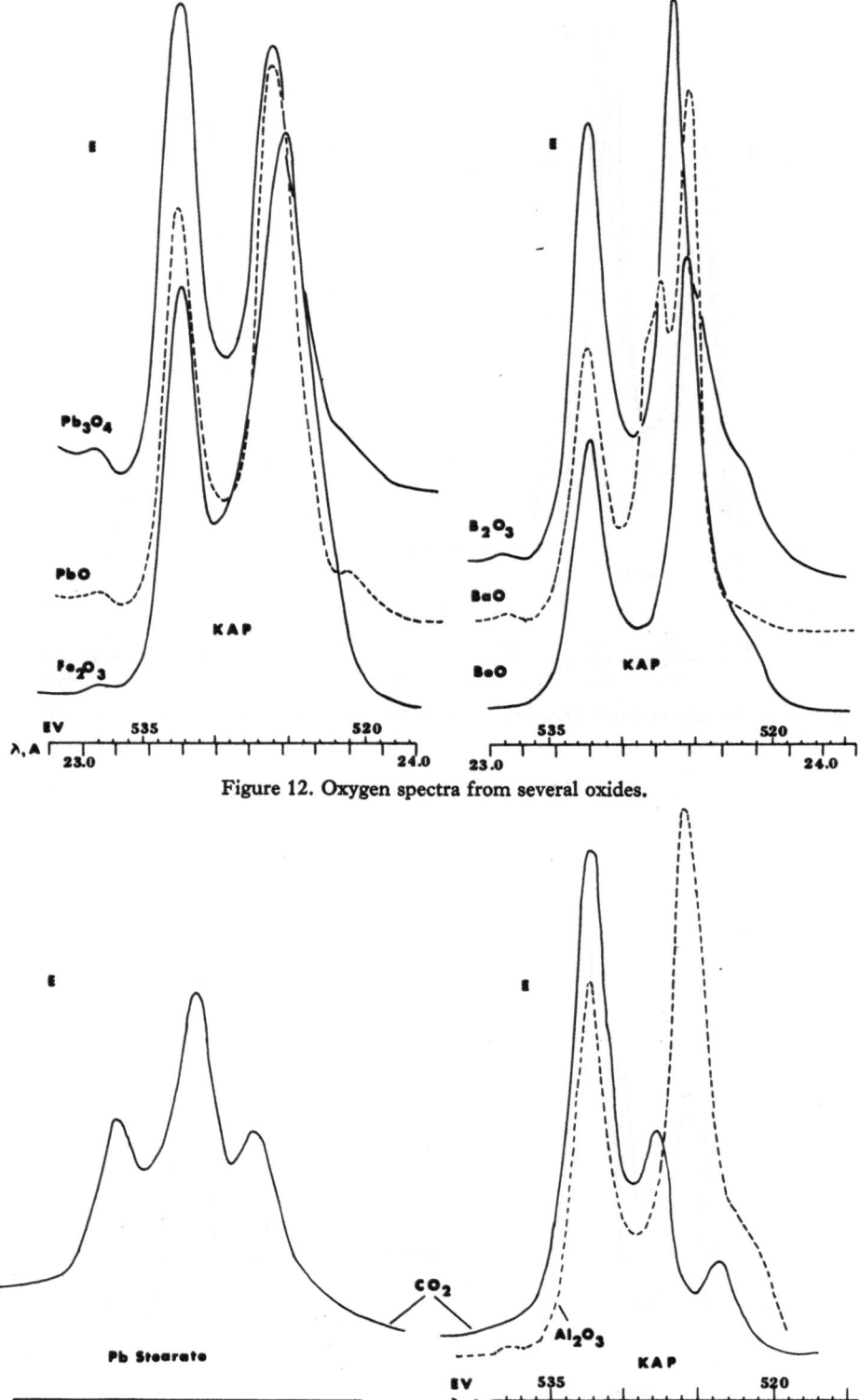

Figure 12. Oxygen spectra from several oxides.

Figure 13. Oxygen emission lines from CO_2 using KAP and lead stearate crystals.

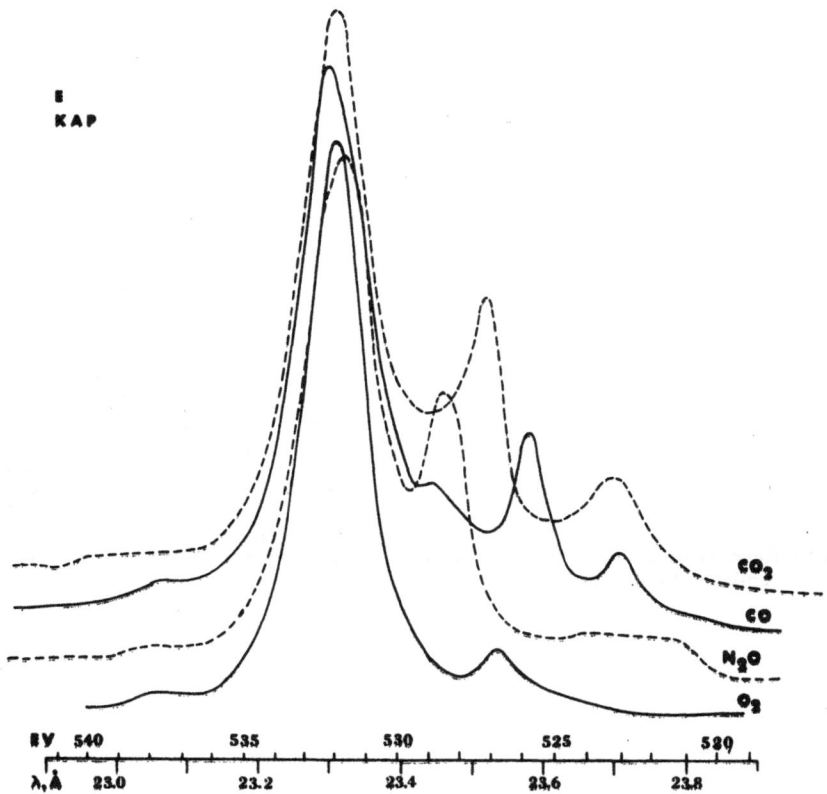

Figure 14. Oxygen emission lines from N_2O, CO_2, CO, and O_2 using a KAP crystal.

Figure 11 shows the more complex structure of CaO and ZnO. With X-ray excitation, a low-energy satellite appeared in the CaO spectra, while the Zn L_β (second-order) line was reduced in amplitude. For CaO, the complex structure between the two peaks was not resolved with X-ray excitation and the relative amplitude of the high-energy peak was reduced. A small high-energy peak at about 534 eV is observed for all electron excited samples. Figure 12 has an assortment of electronically excited curves which show either little structure such as Fe_2O_3, or complex structure, as BaO, with intermediate energy satellites, and B_2O_3, with two unresolved low-energy satellites. It is seen that the high-energy peak is relatively stationary while the low-energy peak shifts, depending on the chemical composition of the oxide.

Interesting spectra can be obtained for gases containing oxygen. The following two curves (Figure 13) were taken with the source operating in the mode shown in Figure 1(D). Again, the spectra appears different depending on the crystal which is used. Figure 13 illustrates the difference observed with KAP and lead stearate. Only the high-energy peak is suppressed. The Al_2O_3 curve is shown for reference only. Figure 14 depicts the oxygen spectra from four gases. O_2 shows the simplest structure, with only one low-energy satellite and a high-energy band. CO has three low-energy peaks, and CO_2 has two. There is apparently some shifting of the main peaks; the half-widths, however, are nearly the same.

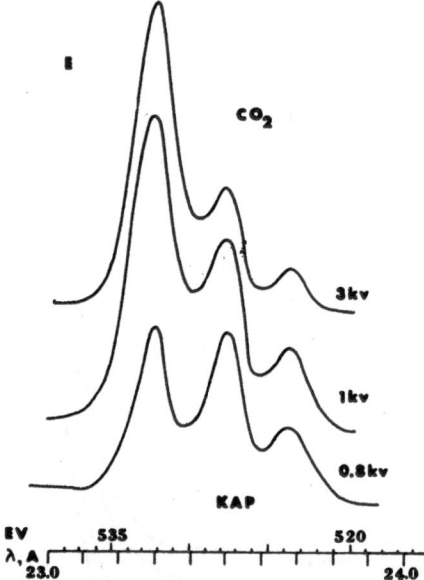

Figure 15. Effect of excitation potential on oxygen emission lines from CO_2.

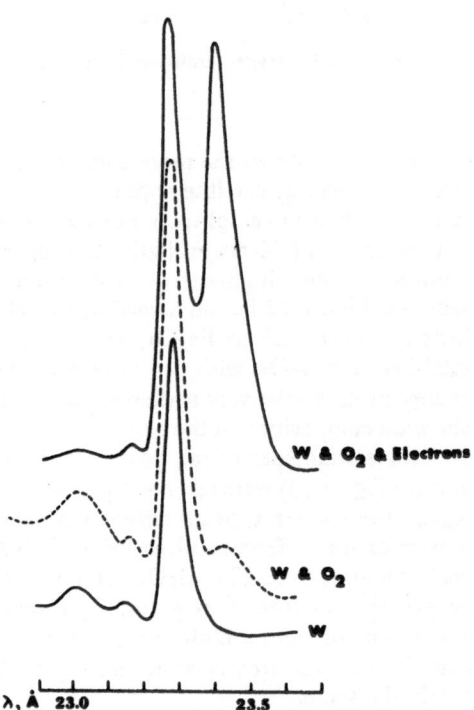

Figure 16. Oxygen emission lines from tungsten helix.

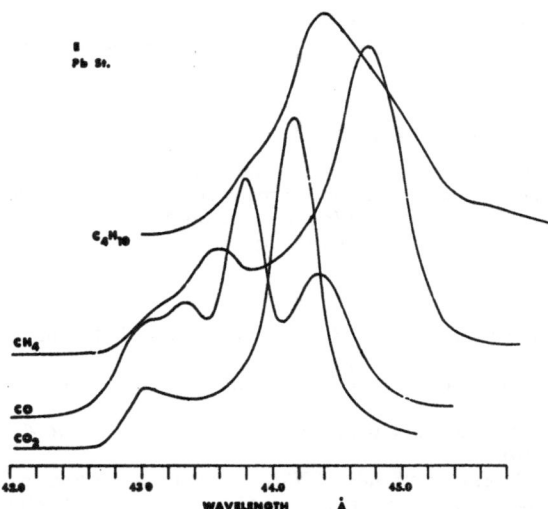

Figure 17. Carbon K emission lines from C_2H_{10}, CH_4, CO, and CO_2 using a lead stearate crystal.

Liefeld[10] has reported that the energy of incident electrons caused distortion of satellite emissions from some transition-group metals. Similarly, Fischer[11] has shown a variation in the L_{II}/L_{III} intensity ratio with excitation potential.

To illustrate the effect of electron energy on the oxygen spectra from gases, Figure 15 shows the decrease in the high-energy peak as the excitation voltage is lowered. The three curves were plotted with different ordinate scales, hence they appear different. Voltages above 3 kV do not affect the intensity ratio greatly, nor does the increased voltage increase the counting rate.

Finally, Figure 16 shows a typical measurement of a reaction of a gas with a solid. The solid is a tungsten helix, which can be heated to 2800°C and issues the spectrum shown in the bottom curve. Even though the helix is at white-heat temperatures, the high-energy peak persists. If oxygen gas, at gage pressure of 4×10^{-4} torr, is allowed to surround the tungsten at room temperature for 30 min, a spectrum like the dashed curve will result. No electrons struck the tungsten for that result. If the pressure were reduced to 1×10^{-4} and an electron beam of 200 μA were allowed to strike the helix, then the upper curve results. (This ordinate is 2.5 times the two lower curves.) By heating the helix, the spectra are reduced to that of the lower curve. The time taken to attain that result depends on the temperature. At about 500°C the reaction proceeds rapidly. The bottom curve remains the same whether gas at 4×10^{-4} torr is present or not, which indicates that a sufficient number of electrons are not present to produce appreciable spectra by excitation of the gas. CO and CO_2 produce a spectrum similar to the center curve even with electron excitation. Apparently, the electrons provide energy to allow the oxygen to react rapidly with the metal to produce an oxide. Again, the excitation voltage strongly influences the spectrum in a manner similar to that shown in Figure 15.

CONCLUSIONS

The foregoing paragraphs indicate that a single-crystal spectrometer can be quite useful in examining the many facets of soft X-ray spectroscopy. The results presented here indicate some measure of performance of an instrument of this type. Improvements can be made in both technique and in instrumentation. Interpretation of the results requires that one be cognizant of the parameters affecting the spectra and the degree of their influence. It is hoped that some better feeling for some of these parameters has been brought out in the above presentation.

ACKNOWLEDGMENTS

The authors wish to acknowledge the participation of Robert Agenten who made many of the measurements, Dr. James Fergeson of Bell Telephone Laboratories for many of the fluoride samples, and to thank David W. Fischer of the Materials Laboratory, Wright-Patterson Air Force Base, for the use of the RbAP crystal and permission to present the phthalate crystal data.

REFERENCES

1. R. C. Ehlert and R. A. Mattson, "The Characteristic X-Rays from Boron and Beryllium," this volume, p. 456.
2. H. M. O'Bryan and H. W. B. Skinner, "The Soft X-Ray Spectroscopy of Solids," *Proc. Roy. Soc. (London) Ser. A* **176**: 229, 1940.
3. F. Tyren, "Precision Measurements of Soft X-Rays with a Concave Grating," *Nova Acta Regiae Soc. Sci. Upsaliensis* **12**: 7–66, 1940.
4. A. P. Lukirskii, T. M. Zimkina, and I. A. Brytov, "Study of X-Ray Spectra in the Region of Wavelengths Greater than 15 Å Using a Spectrometer with a Gold-Covered Diffraction Grating," *Op. Spectr.* **16**: 372–5, April, 1964.
5. D. W. Fischer, "Effect of Chemical Combination on the X-Ray *K* Emission Spectra of Oxygen and Fluorine," *J. Chem. Phys.* **42**: 3814, 1 June, 1965.
6. J. E. Holliday, "Soft X-Ray Emission Spectroscopy in the 13 to 44 Å Region," *J. Appl. Phys.* **33**: 3259, November, 1962.
7. M. Sawada, F. Zembei, and J. Kosugi, "Influence of Lattice Bonding on the *K*-Series of Fluorine," *J. Proc. Physico-Math Soc. Jap, 3rd Ser.* **26** (1–2), January–February, 1944.
8. R. A. Mattson, "Some Measurements of Carbon *K* Excitation in a New Ultra-Soft X-Ray Spectrometer," paper presented at the Fourteenth Annual Conference on the Applications of X-Ray Analysis, Denver, Colorado, August 24–26, 1965.
9. D. W. Fischer, Private Communication.
10. R. J. Liefeld, "*L* Emission from First Transition Metals Near Threshold Excitation," presented at the International Conference on Physics of X-Ray Spectra, Cornell University, Ithaca, New York, 22–24 June, 1965.
11. D. W. Fischer, "The Use of Soft X-Rays for Materials Analysis: The *L* Emission Spectra of the Fe Group Transition Metals and Their Oxides," ML TDR 64–263, September, 1964.

DISCUSSION

N. Spielberg (Philips Laboratories): What was your target in the X-ray excitation cases?

R. A. Mattson: This was copper in all cases.

N. Spielberg: Perhaps something such as atomic number 24 or 23 in the *K* series might give you a line which is a little closer.

R. A. Mattson: Perhaps.

E. Davidson (Applied Research Labs): If I understood you right, you operated some of your experiments at 10^{-4} torr.

R. A. Mattson: For the gaseous excitation, yes.

E. Davidson: How did you prevent contamination of the anode?

R. A. Mattson: The anode certainly was contaminated but we were only looking at the excitation of the gas itself.

E. Davidson: But what happened to your analysis as you went on?

R. A. Mattson: Certainly we changed the conditions. We cleaned the anodes. These were all separate runs.

E. Davidson: How often did you have to clean it?

R. A. Mattson: These were results of two or three repeated sets of data. There were no contamination effects.

C. F. Hendee (Bendix Corporation): We have seen some amazing changes in spectra due to changes in composition or excitation conditions. I haven't heard any comment about the possibility of ghosts in spectra. Certainly in the optical spectra region using gratings one is concerned with ghosts if the gratings are not good. Is this a factor in these soap-film crystal or soft X-ray grating spectra?

R. A. Mattson: I am sure ghosts are possible and they are probably more possible in using curved crystal optics. But in the spectra that we did get, we have used both flat and curved crystals for the data that we have shown for the oxide cases. Certainly this is a consideration.

J. Merritt (Shell Development): Pursuing that ghost argument, I thought ghosts arose from some sort of super periodicity in the ruling of gratings. It is hard to see how this would arise in a soap-film crystal.

R. A. Mattson: I interpreted the remark to mean crystals in general.

N. O. Smith (Fordham University): I notice there are more peaks for carbon monoxide than for carbon dioxide. Is this conceivably connected with the fact there are more contributing structures for carbon monoxide than for carbon dioxide?

R. A. Mattson: Well, as I said at the beginning of the talk, I won't try to interpret the spectra, but if you would ask the question to see how many carbon peaks you have for the same gas, I would be delighted to show the next slide if it is still available [Figure 17]. There are actually the same number of carbon peaks for the CO—I think there are three—for $C K$ and there are very marked differences in the carbon gaseous spectra that you have seen. CO_2 is the bottom curve. The CO has the three curves and there is a big difference between the methane and the butane. We were quite surprised to see this. We had a freon gas which we also used, a liquid, which we evaporated. We got very much the same spectra as the top one. So this may be another interesting area for investigation.

USE OF THE SOFT X-RAY SPECTROGRAPH AND THE ELECTRON-PROBE MICROANALYZER FOR DETERMINATION OF ELEMENTS CARBON THROUGH IRON IN MINERALS AND ROCKS

A. K. Baird and D. H. Zenger

Pomona College
Claremont, California

ABSTRACT

The major elements in common rocks are of low atomic number, but analyses of high precision are possible by soft X-ray spectrography if several grams of rock sample are available. The electron-probe microanalyzer is shown to complement this established method by permitting analyses of particles as small as 1 μ in diameter. This paper describes applications of these methods to the analysis of the major and minor elements of silicate, carbonate, and phosphate minerals and rocks.

Elements of particular interest are as follows: carbon in particles enclosed in carbonate rocks; oxygen, as the major constituent of the specimens; phosphorus in phosphatic nodules and apatites; manganese and iron, as colorations in fossil shells; and the group oxygen, sodium, magnesium, aluminum, silicon, potassium, calcium, and iron as complex segregations and zonations within single crystals of several mineral phases.

If the bulk composition of a rock is known, and also the chemistry of the constituent minerals, it is possible to compute quantitative mineralogic analyses of high precision. Thus, the combined use of soft X-ray spectrography and electron-probe microanalysis can provide quantitative chemical and mineralogical information on the earth's crust on all scales from thousands of square miles (by means of appropriate sampling) down to the scale of 1 μ.

BULK ANALYSIS BY X-RAY SPECTROGRAPHY

For the past five years, X-ray spectrography has been used in our laboratory for the analyses of silicates, carbonates, and phosphates. During this period significant improvements have been made in X-ray sources,[1,2] analyzing crystals,[2] thin-window detection systems,[3] specimen preparation methods,[4] and automated data control.[5,6] These improvements in instrumentation and techniques as applied to the analyses of geological samples have been summarized recently by Baird *et al.*[6,7]

A summary of precisions of various methods of silicate analysis is shown in Table I. Between 1961 and 1965 both the lower atomic number limit and the precision of X-ray analysis have been greatly improved. At present the X-ray method compares favorably with all other techniques for analyses of major and minor light elements.

Included in this comparison are figures for fast neutron activation applied by us to oxygen and silicon in various rocks and minerals.* With the capability of determining

---

[1] References are at the end of the paper.

* Equipment: Technical Measurement Corporation Activation 111, dual pneumatic transfer system, 400 channel analyzer and integrater–resolver. For oxygen the reaction O^{16}—N^{16} (6-7meV, 7.3 sec half-life); for silicon the reaction Si^{28}—Al^{28} (1.8 meV, 2.3 min half-life); both induced by 30 sec irradiation with 14 meV fast neutrons at a flux of 10^{10} neutrons/sec.

Table I. Analytical Precisions of Methods of Silicate Analysis

| Element | Wt.% | Standard deviations (wt.%) | | | | | |
| | | X-ray | | Wet chemistry | Emission | Flame | Neutron activation |
		1961	1965				
O	48.0	—	0.4	0.3	—	—	0.5
Na	2.6	—	0.02	0.15	0.11	0.07	—
Mg	0.5	—	0.01	0.15	0.12	—	—
Al	9.0	0.32	0.04	0.19	0.53	—	—
Si	30.0	0.20	0.10	0.14	1.10	—	0.5
P	0.06	—	0.006	—	—	—	—
K	3.7	0.05	0.03	0.21	0.15	0.07	—
Ca	2.5	0.03	0.01	0.07	0.14	—	—
Ti	0.2	—	0.003	—	—	—	—
Fe	3.0	0.04	0.02	0.21	0.14	—	—

Table II. Accuracy of Silicate Analyses (%)

(1) Cation sum (X-ray)	(2) Oxygen (X-ray)	(3) Oxygen (activation)	Sum (1) + (2)	Sum (1) + (3)
52.63	46.8	45.4	99.43	98.03
55.81	44.6	43.4	100.41	99.21
51.26	48.6	49.2	99.86	100.46
50.90	47.5	47.9	98.40	98.80
50.76	47.9	48.5	98.66	99.26
50.75	47.1	48.0	97.85	98.75
55.16	44.0	44.3	99.16	99.46
54.76	45.9	44.1	100.66	98.86
50.75	47.6	47.8	98.35	98.55
52.64	47.6	46.9	100.24	99.54
52.40	47.6	49.2	100.00	101.60
50.95	48.7	48.6	99.65	99.55
55.23	45.9	44.1	101.13	99.33
54.73	46.2	45.6	100.93	100.33
51.92	47.4	48.2	99.32	100.12
51.46	49.2	48.9	100.66	100.36
		Mean	99.67	99.51
		Standard deviation	0.99	0.89
		Relative standard deviation	1.0	0.9

oxygen, both by X-ray spectrography and by activation analysis, all major and minor elements of silicates can be measured. This enables us to evaluate the *accuracy* of our analyses as well as the *precision* by the degree of approach to a summation of 100% by weight. Table II shows these summations for sixteen of our working standards. Oxygen values determined by both methods can be compared and the close agreement of mean summations (99.67% and 99.51%) is noted.

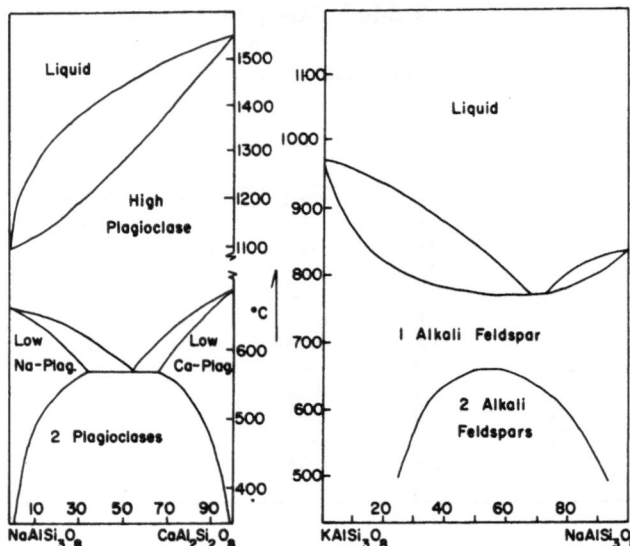

Figure 1. Generalized binary phase diagrams for the feldspar family of minerals.

ROCK AND MINERAL VARIABILITY

Rapid, precise, and accurate bulk analysis of rocks by X-ray spectrography has permitted us to study the variabilities of rocks on scales ranging from single drill core segments to areas over thousands of square miles.[8-10] These studies have shown that chemical variability can be high in small samples, requiring the bulking of many specimens to form a composite sample for analysis. Though the chemical composition of bulked samples is easily determined, the mineralogical composition is more difficult to obtain. With either large grain size or small, optical techniques become inadequate and many thin sections must be examined for statistically valid results.

Unfortunately, even in rocks of medium grain size, individual grains of a given mineral phase can be highly variable: compositional zoning and intergrowths or exsolutions of two or more phases are very common. Furthermore, partial to complete chemical replacements or alterations of minerals, postdating their initial growth, are found almost universally in rocks. Examples of some of these problems are shown in simplified phase diagrams for the feldspar family minerals (Figure 1). In addition to complex solidus–liquidus relationships, subsolidus exsolutions can produce phase mixtures, some of which may not be resolved with conventional light optics (magnifications to about 400×). Correlation of optical properties with chemistry is, therefore, seriously handicapped, and even if the phases are of adequate size for optical study the available chemical information is still inadequate.

We conclude that mineral phase studies by bulk analysis, by transmitted light optics, by physical separation (magnetic or density), or by quantitative X-ray diffraction are not possible with the same precision, resolution, and speed presently attainable for whole rock analyses. It is in this application that the electron-probe microanalyzer complements the other established techniques.

ELECTRON-PROBE MICROANALYSIS

Light Element Capability

In most nonmetallic geologic samples the elements of main interest are of atomic number 26 (iron) and lower. Light-element long-wavelength capability of the probe is,

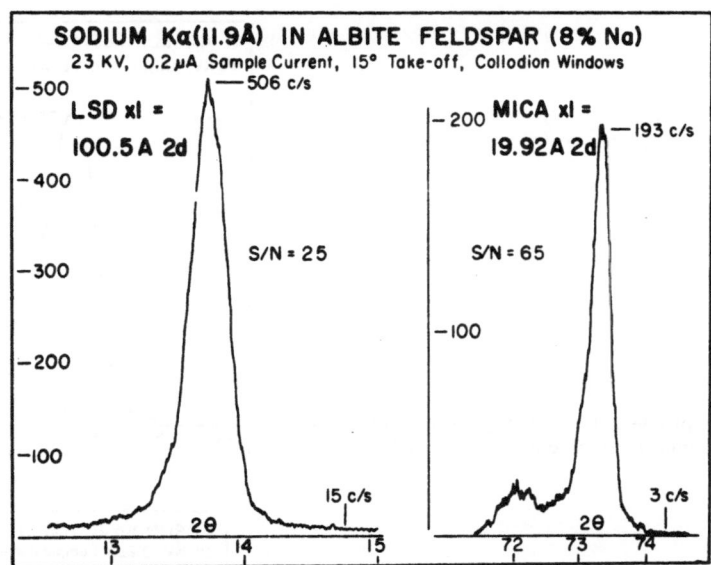

Figure 2. Comparison of Na K_α line profiles with lead stearate decanoate analyzing crystal (LSD) and biotite mica. Sample is albite, 8% sodium. P-10 counter gas and counter voltage of 1600 V.

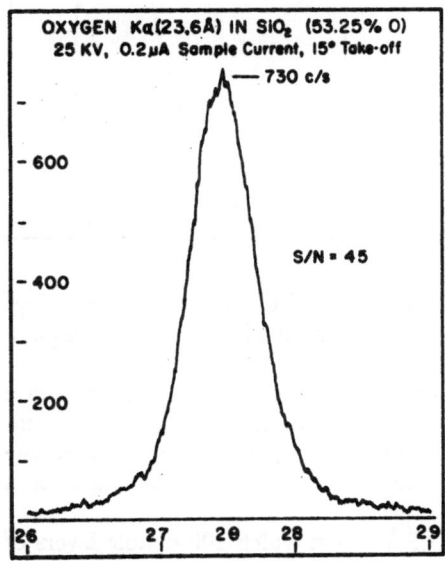

Figure 3. Line profile of O K_α. Sample is quartz, 53.25% oxygen. 100% methane counter gas and counter voltage of 2750 V.

therefore, very important. The Philips Electronic Instruments AMR/3 probe in use in our laboratory is equipped with a rotatable turret of four windows and a blanking-plate which separates the high vacuum of the column from the poorer vacuum of the spectrometer. Windows currently being used are Mylar (for the shorter wavelengths with the spectrometer at atmosphere), polypropylene, and two collodion windows (for the soft and ultrasoft wavelengths with the spectrometer pumped to about 30 μ). The collodion

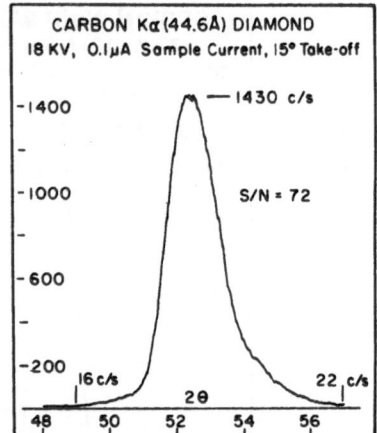

Figure 4. Line profile of C K_α. Sample is diamond. 100% methane counter gas and counter voltage of 2800 V.

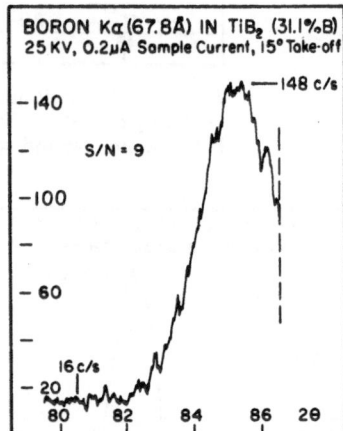

Figure 5. Line profile of B K_α. Sample is TiB₂. 100% methane counter gas and counter voltage of 2850 V.

windows are made by floating three drops of 1% solution on distilled water. After drying, the film is picked up on a loop-frame and cemented to a circular window mount equipped with an O-ring seal for the turret.

The gas flow proportional counter of conventional side-window design also has a collodion window which is supported by a Buckbye–Mear screen of 70% transmission. P-10 counter gas is used in analyses of fluorine and heavier elements; 100% methane for boron, nitrogen, oxygen, and carbon. A Tennelec 100B low-noise preamplifier on a 2.5 ft. cable from the detector is used.

The analyzing crystal is bent mica ($2d = 19.5$ Å) upon which 100 double layers of lead stearate decanoate ($2d = 100.5$ Å) have been deposited by the Henke method of forming multilayer analyzers.[3] In analyzing for boron through sodium, the appropriate Bragg angles for stearate are selected; above sodium, reflections from the underlying mica are used. Thus, the single spectrometer is capable of covering the entire spectral range from B K_α (67.8 Å) upward to the short wavelengths. The only changes required over this range are selection of Bragg angle, selection of turret window, and change of counter gas at about 20 Å.

Examples of the performance of the AMR/3 in the very soft region are shown in

Figures 2 through 5 for Na K_α, O K_α, C K_α, and B K_α. In all line profiles shown, appropriate PHA baselines and windows were selected using an oscilloscope connected to the amplifier output.

Sample Preparation and Optical Viewing

Electron-probe designs have been controlled largely by the needs of metallurgy, a field in which probe analysis has won wide acceptance. Only in recent years have geologists, biologists, and medical researchers applied the instrument to their analytical problems. In many of these later applications, the conventional, 90° incident reflected-light optics, useful in viewing opaque samples, has proven very inadequate.

For about 100 years, geologists have been using transmitted, polarized light microscopy of relatively low magnification (to 400×) to study mineralogical and petrological samples in thin slices (30 μ) mounted between glass. An example, exsolved $NaAlSi_3O_8$ (blue-green) in $KAlSi_3O_8$ (yellow-red) is shown in Figure 6(a). This view was taken through a conventional petrographic microscope at 30× magnification in doubly polarized light.

The AMR/3 probe is equipped with a substage microscope illuminator (in addition to a vertical illuminator) and a rotatable light polarizer. The binocular viewing optics (Zeiss) contains a rotatable polaroid light analyzer which can be moved in and out of the optic path. Thus, the AMR/3 probe has the features of a petrographic microscope combined with electron excitation. The importance of this combination is shown in Figures 6 and 7. In each group and pair of photos, the same areas have been photographed through the probe optics (300×) changing only the mode of illumination: Exsolutions [Figures 6(b), (c), (d)], intergrowths [Figures 7(a), (b)], and compositional zoning [Figures 6(e), (f) and 7(c), (d)] are invisible in reflected light. In reflected light, only surface features can be seen, and the mineral-phase locations and interrelationships, of most importance in geological probe analysis cannot be discerned.

The preparation of thin polished sections is merely an extension of older established techniques used for petrographic microscopy. In our analyses, a modified version of J. V. Smith's procedures are used (personal communication): The procedure is as follows:

(1) A conventional microscope slide (about $1 \times 1\frac{7}{8}$ in.) is lapped to standard thickness (0.047 in.) using a Buehler holder (model 30–8000). (2) The smooth (fine-lapped) surface of the rock or mineral chip is cemented to the frosted surface of the slide using Lakeside 70, epoxy or plastic cement. (3) The slide and chip are diamond-sawed to about 80-μ thickness using a vacuum holder [11] on a sliding-bench saw. (4) A Buehler holder (model 30–8001) is used to lap the section to 50 μ; final lapping (000 abrasive) is done on a stationary glass plate. (5) A Felker diamond drill bit, mounted in a drill press chuck, is used to cut a $\frac{15}{16}$ in. circle of thin section and glass. Thorough cleaning to remove lapping compounds is then performed (an ultrasonic cleaner is recommended). (6) The circular thin section is cemented to a holder (cork or wood dowel) and polished through the usual stages of 6-, 1-, and $\frac{1}{4}$-μ diamond abrasive on a Sampson–Patmore ore polisher. Sections are cleaned carefully between polishing stages. For critical smoothness, the diamond polishing can be followed by alumina down to the 0.05-μ size. (7) After removal from the holder, and careful surface cleaning, standard vacuum-evaporation techniques for a carbon or aluminum electrically conducting surface follow.

These procedures have the advantage that no transfer and recementing of a polished thin slice is necessary, thus eliminating the possibility of wavy (even if polished) surfaces, which can cause variations in X-ray take-off angle. The method also allows one to use

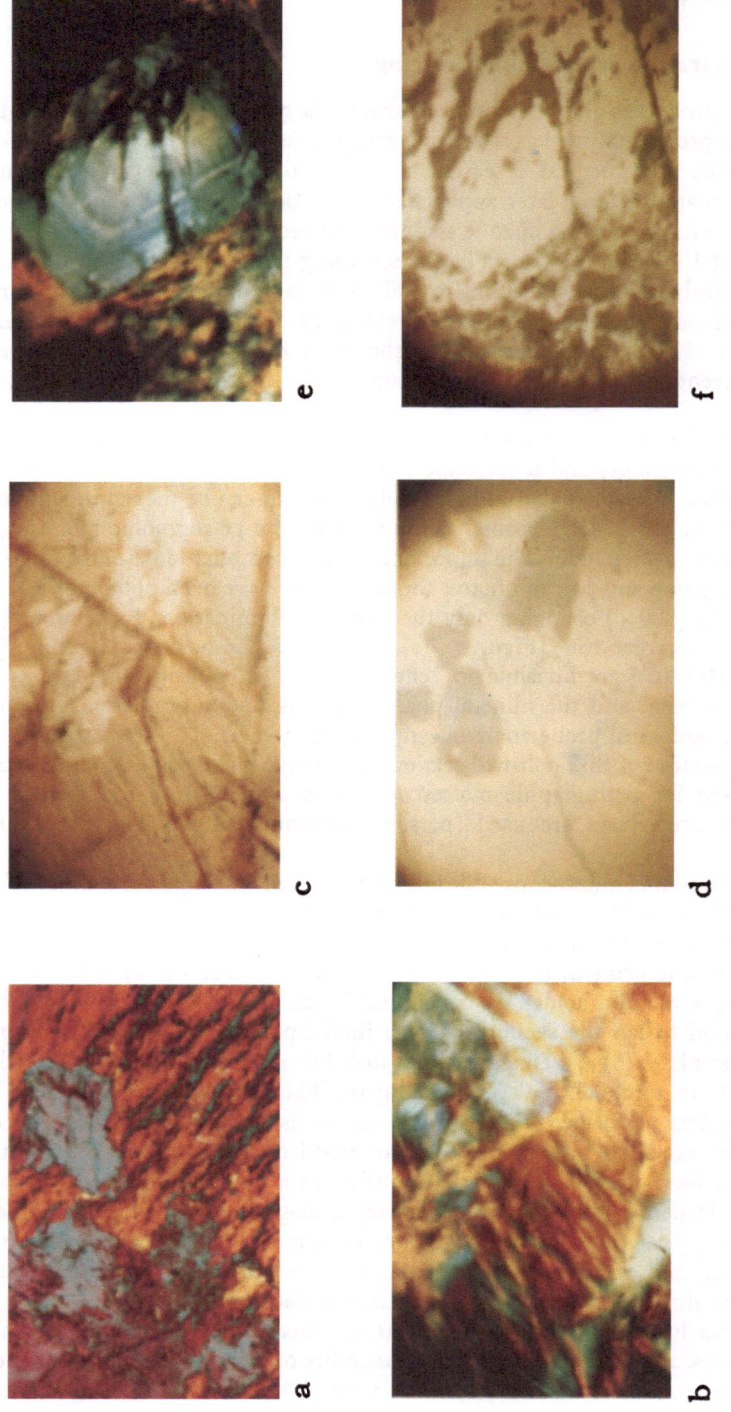

Figure 6. (a) Perthitic feldspar. NaAlSi$_3$O$_8$ blue-green; KAlSi$_3$O$_8$ yellow-red. 30× through petrographic microscope (polarizer and analyzer inclined at 80° to emphasize phase relations). (Reduced for reproduction 30%.) (b), (c), and (d) Perthitic feldspar of 6 (a). 300× photographed through probe optics; (b), doubly polarized transmitted light [colors as in 6 (a)]; (c), plane polarized transmitted light; (d), reflected non-polarized light. Objects visible only in (c) and (d) are surface features. (Reduced for reproduction 30%.) (e) and (f) Compositionally zoned plagioclase feldspar. 300× photographed through probe optics; (e), doubly polarized transmitted light; (f), reflected nonpolarized light. (Reduced for reproduction 30%.)

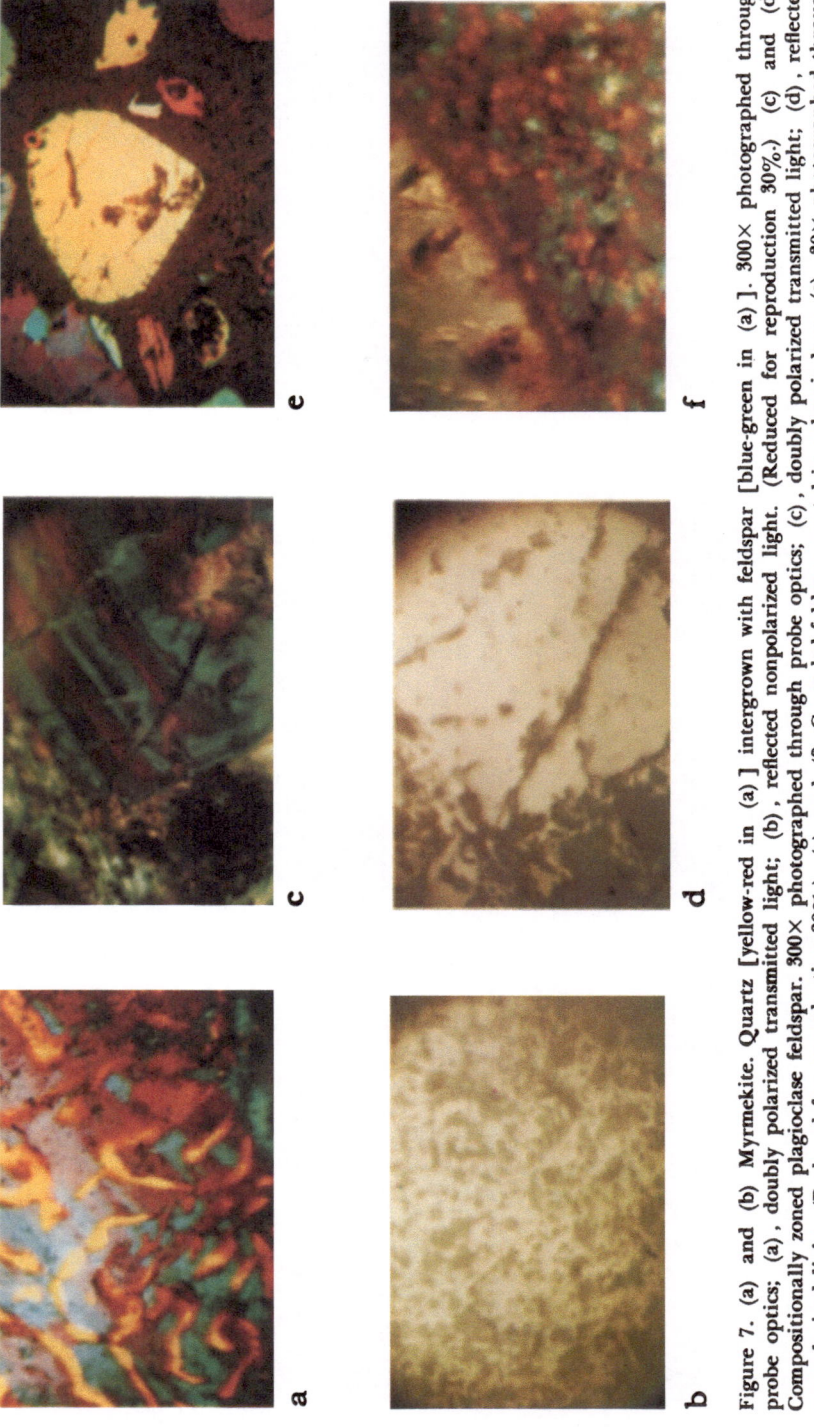

Figure 7. (a) and (b) Myrmekite. Quartz [yellow-red in (a)] intergrown with feldspar [blue-green in (a)]. 300× photographed through probe optics; (a), doubly polarized transmitted light; (b), reflected nonpolarized light. (Reduced for reproduction 30%.) (c) and (d) Compositionally zoned plagioclase feldspar. 300× photographed through probe optics; (c), doubly polarized transmitted light; (d), reflected nonpolarized light. (Reduced for reproduction 30%.) (e) and (f) Corroded feldspar crystal in volcanic lava; (e), 30× photographed through standard petrographic microscope, doubly polarized transmitted light; (f), corroded edge of crystal with brown reaction rim crosses upper left of field. 300× photographed through probe optics in doubly polarized transmitted light. (Reduced for reproduction 30%.)

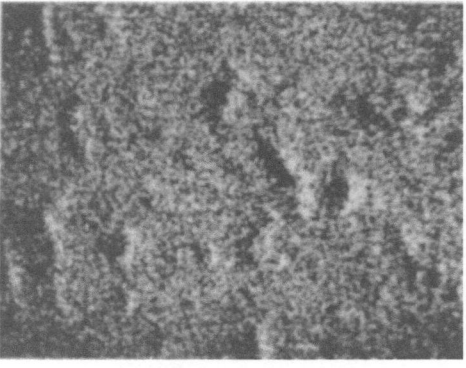

a   b

Figure 8. Boron in TiB₂. Beam scan images at 400×. (a), sample current; (b), B K_α. Conditions of analysis as in Figure 5. (Reduced for reproduction 35%.)

previously prepared petrographic thin sections. With such sections, the coverslip is easily removed by first freezing the slide. After removal of the coverslip cement with a suitable solvent, steps (5) through (7) above are followed.

Qualitative Analysis

The importance and value of qualitative analysis in studying phase relationships by microprobe techniques for elements of higher atomic number than sodium is well established. The light element spectrometer now extends the possibilities of point-to-point examination, stage traversing, and beam scanning down to atomic number 4, in the ultrasoft X-ray region. Examples of these applications are:

1. *Boron* distribution in a nonhomogeneous sample of TiB₂. Figure 8(a) sample current display at 400×, shows a pattern which correlates well with the B K_α signal, Figure 8(b). After rotation of the sample through 90°, the same pattern is obtained showing that the variations in boron are real and that the pattern is unrelated to possible surface irregularities.

2. *Oxygen* distribution in and around an ilmenite [(Fe, Mg, Mn) TiO₃] grain. This example, part of a study of specimen preparation techniques, shows concentrations of oxygen [O K_α, 400×, Figure 9(b)] in cracks between portions of the grains. These cracks are visible in the high-contrast sample current image [Figure 9(a)]. This sample was polished with 0.05-μ Al₂O₃ compound and insufficiently cleaned before analysis.

3. Complete light element analyses of phase relations in intergrowths and ex-solutions of the minerals quartz (SiO₂), orthoclase (KAlSi₃O₈), and albite–oligoclase (90% NaAlSi₃O₈–10% CaAl₂Si₂O₈). Figure 10 shows stage traverses across a complex perthitic–myrmekitic grain similar to that shown in Figure 7(a). The sympathetic behavior of O K_α and Si K_α, and the antipathetic behavior of the alkalies and calcium, is pronounced. This example is typical of many others faced by the petrographer where it is important to know phase relations on a micron scale and optical methods are inadequate.

4. *Carbon* in carbonaceous material enclosed in limestone and dolomite. The distribution of small plates of high carbon content are shown at 800× in Figures 11(a) and (b). This light element is easily resolved even though the matrix of the sample is carbonate [(Ca, Mg)CO₃]. Figure 11(c), Ca K_α distribution, shows the regions of high C K_α signal to be devoid of Ca.

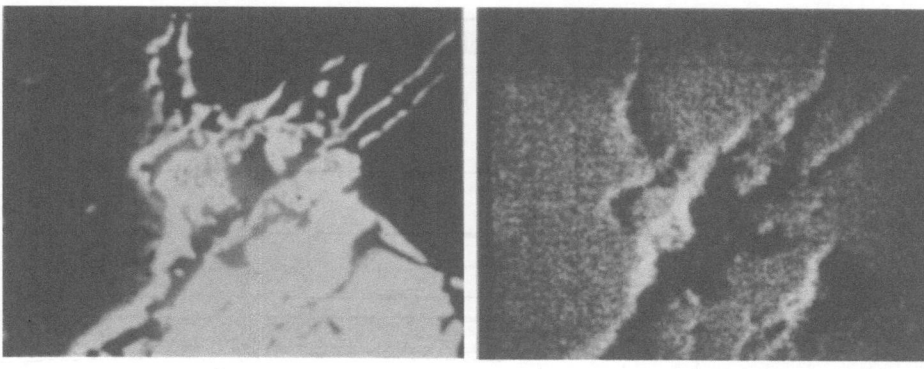

a   b

Figure 9. Ilmenite (Fe, Mg, Mn) TiO₃ in a silicate matrix. Beam scan images at 400 ×. (a) sample current; (b) O K_α. Conditions of analysis as in Figure 3. (Reduced for reproduction 35%.)

5. *Sodium* in alkali feldspars. Figures 7(e) and 7(f) show low and high magnification optical views of a feldspar grain with a pronounced reaction rim against what once was a silicate melt of a volcanic lava of dacitic composition. The optical properties of the grain suggested it was sanidine ($KAlSi_3O_8$). Beam scan images at 400 × for Na K_α and K K_α [Figures 12(a) and 12(b) reveal, however, that the grain contains sodium and is particularly enriched in that element in its reaction rim].

6. *Magnesium* in dolomite rhombs enclosed in limestone. The origin of the crystalline phase dolomite, apparently at the expense of calcite, remains an unresolved problem in sedimentary petrology. Knowledge of the distribution of the magnesium-rich dolomite phase can provide genetic clues. In Figure 13, the sharply bounded crystal faces of dolomite growing in a calcite matrix are clearly defined. The dark oval in the center of the sample current display [Figure 13(a)] was caused by electron beam damage (see discussion below).

7. *Phosphorus* in phosphatic spheroids (possibly oolites). Small, spherical growths of phosphorus-rich material (related to apatite and other amorphous phosphatic materials such as gall or kidney stones) are of importance both to geologists and biologists. In Figure 14 the phosphorus, calcium, and silicon distributions are shown for a spheroid in a chert matrix. It appears as if the spheroid grew around a quartz or chert fragment and included chert around part of its margin.

Quantitative Analysis

Unlike bulk X-ray spectrography, where the amount of care in sample preparation largely controls the precision of quantitative analyses, quantification of probe data is complicated by several additional factors resulting from electron excitation and from inherent sample inhomogeneities. Usually it is not possible or practicable to homogenize probe samples because the inhomogeneities are the features of interest. Restricting this discussion to light and very light element analyses of geological samples, the following problems are apparent:

Correction Parameters. Corrections for X-ray absorption, X-ray enhancement, and atomic number effect are poorly defined for the light elements and for complex systems of up to a dozen elements. In many mineralogical analyses, all effects causing deviations from linear calibrations are grouped into a single "α-correction" which is

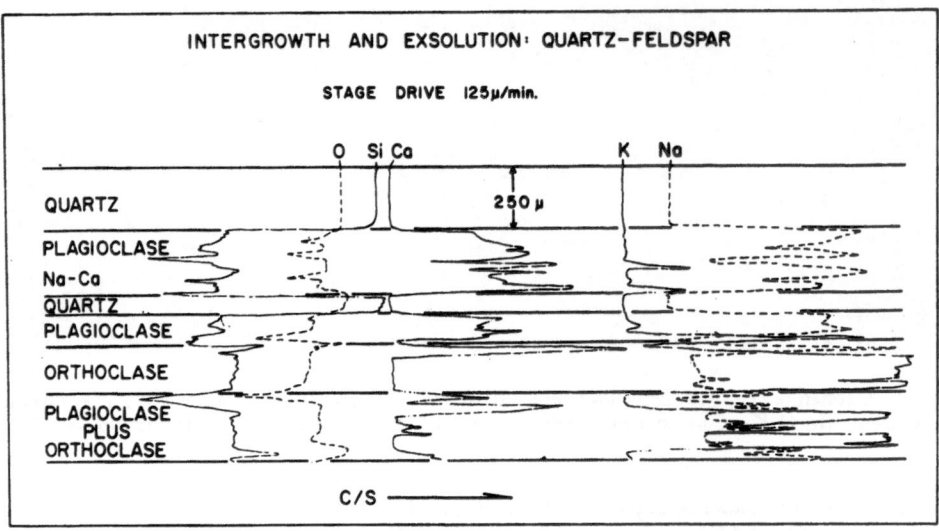

Figure 10. Perthitic myrmekite. Intergrowths and exsolutions of SiO_2, $KAlSi_3O_8$ and $CaAl_2Si_2O_8$. Stationary beam, about 1 μ in diameter; specimen stage traversed at 125 μ/min.

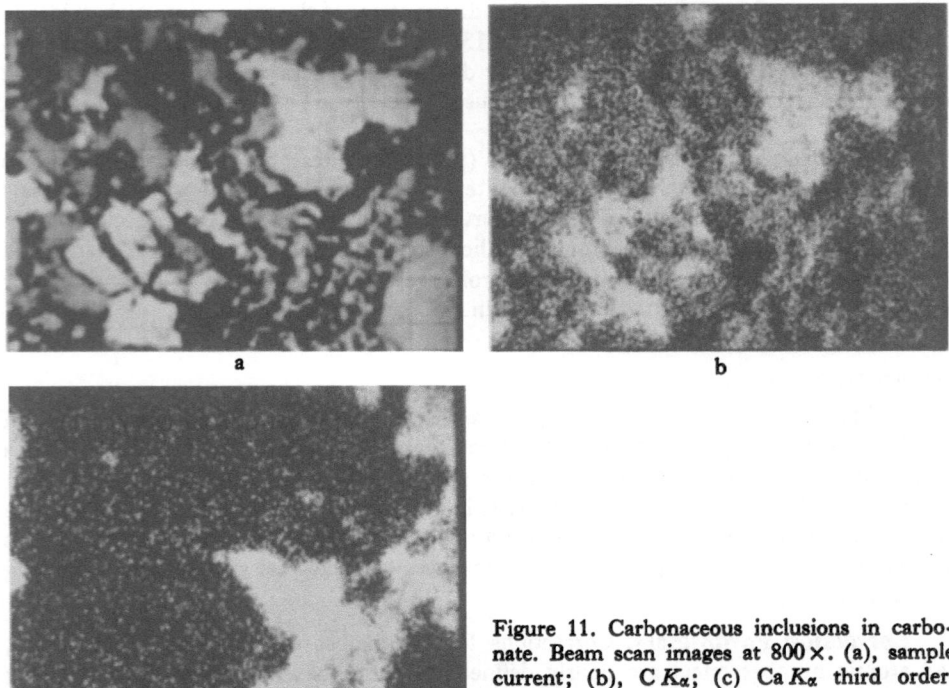

Figure 11. Carbonaceous inclusions in carbonate. Beam scan images at 800 ×. (a), sample current; (b), C $K\alpha$; (c) Ca $K\alpha$ third order. Conditions for carbon analysis as in Figure 4. (Reduced for reproduction 35%.)

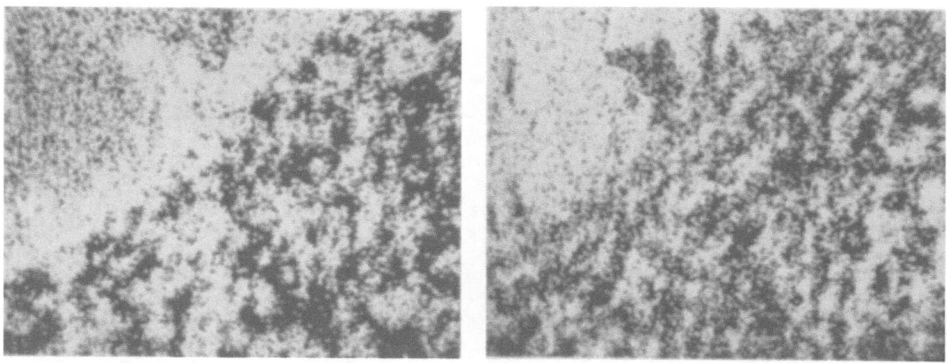

a   b

Figure 12. Beam scan images (400 ×) of alkalis in corroded feldspar of Figure 7(f). (a), Na K_α;
(b), K K_α third order. (Reduced for reproduction 35%.)

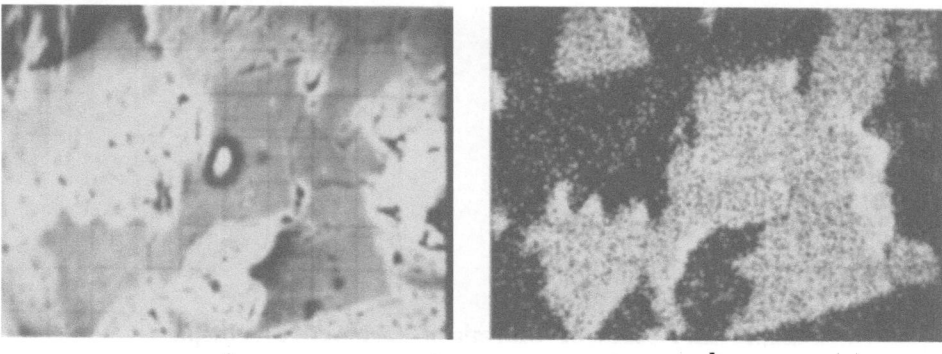

a   b

Figure 13. Dolomite crystals growing in a limestone. Beam scan images at 400 ×. (a), sample current;
(b), Mg K_α. (Reduced for reproduction 35%.)

determined empirically. This approach overcomes the present lack of knowledge of individual corrections, but can lead to errors if applied over too wide a compositional range, or if undetected compositional changes occur in the system.

Mineral Variability. In addition to grain inhomogeneities of interest (diffusion across boundaries, exsolutions, intergrowths, etc.), the question of reproducibility of the measured counting rate from a single spot is of utmost importance. Before being concerned with whether this or that author's corrections give 1% (or 5%) precision, one must determine specimen variability. "Analysis of variance" provides a convenient means for estimating the magnitudes of these errors. As one example, eight different feldspars were selected from a crystallography collection for their optical clarity and apparent homogeneity. Each was analyzed in replicate by bulk X-ray spectrography. Three grains of each feldspar were mounted in plastic for probe analysis, each grain with a different crystallographic orientation. Nine spots per grain were examined for sodium, potassium, calcium, aluminum, and silicon taking precautions to avoid beam damage (see discussion below). The results of this test (Table III) gave measures of counting error, intra-grain variability, and variabilities between grains of the same mineral. (Obvious differences between minerals are discussed below.) It is clear from this test that the variabilities within

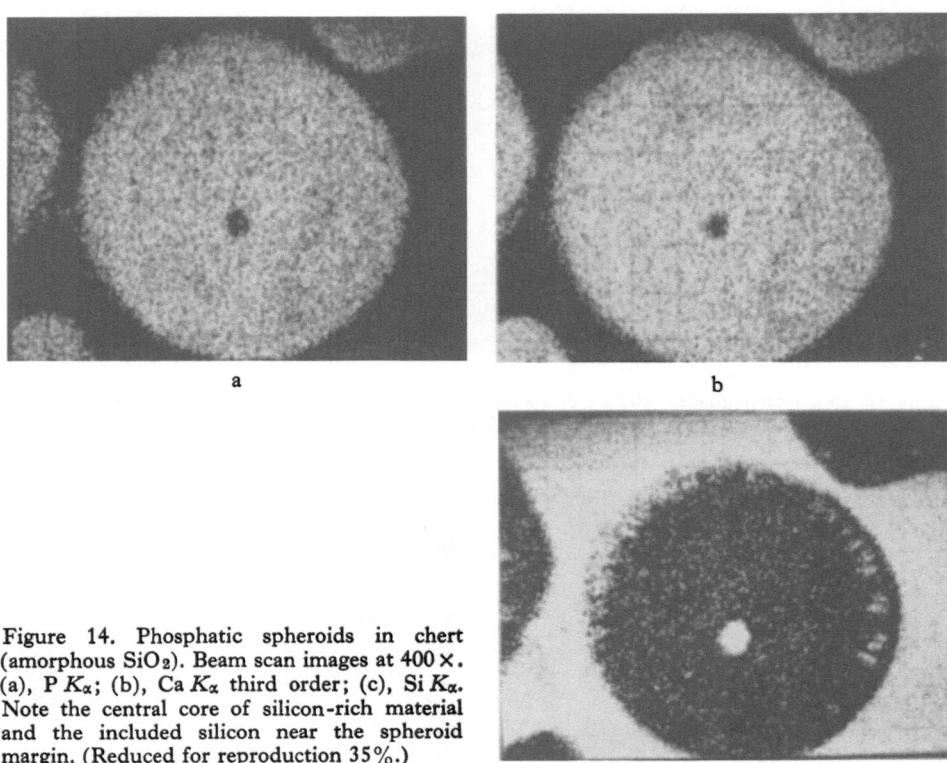

Figure 14. Phosphatic spheroids in chert (amorphous SiO_2). Beam scan images at 400×. (a), P K_α; (b), Ca K_α third order; (c), Si K_α. Note the central core of silicon-rich material and the included silicon near the spheroid margin. (Reduced for reproduction 35%.)

Table III. Results of Analysis of Variance: 8 Feldspars, 3 Grains of Each, 9 Spots Measured per Grain

Element	Ranges of Relative Standard Deviations of Counting Rates			
	Counting Statistics	Within Grains	Between Grains	Total
Na	0.7–0.9	3.0– 8.0	2.0– 7.0	3.0–12.0
Al	0.5–0.7	3.4– 6.9	2.0–19.2	4.0–19.8
Si	0.3–0.5	2.4– 6.7	1.2–17.2	2.6–18.0
K	0.9	4.0–11.5	1.7– 7.2	4.0–13.5
Ca	0.5–0.8	5.1– 7.7	4.7– 7.7	4.5–12.2

individual grains (even when the grains were selected for a high degree of homogeneity) is large, and may be larger than some of the errors possibly introduced when using poorly established correction factors. A finely focused electron beam ($\sim 1\mu$) was used in this test. It should be noted that increasing the spot size is not a legitimate means of "averaging out" grain variabilities. The answer lies in correct *sampling* of the grain at several spots (if possible) in the same fashion that stockpiles are analyzed by random grab-samples, or rocks are field-collected by drilling at pre-established map grid intersections.

Specimen Damage. Carbon contamination of the analyzed spot is only one cause of variable counting rate in mineral analysis. The relatively low thermal conductivity of

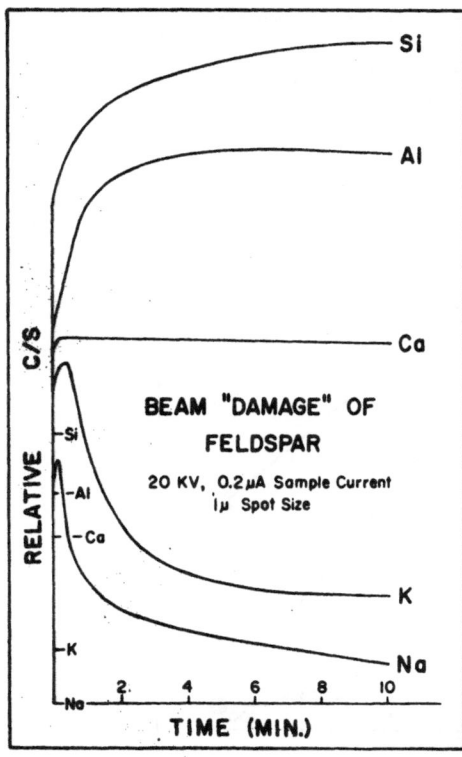

Figure 15. Relative changes in counting rates with time for feldspar analysis. Each curve assumes electron beam turned on simultaneously with recorder. Base levels of 0 counts/sec are indicated. Curves include both real count gains or losses and recorder lag at the short elapsed times.

some minerals, such as feldspars and carbonates, results in temperatures sufficiently high to volatize alkalis and to release CO_2 under normal probe operating conditions. Figure 15 shows the relative increase in counting rate with time for Si K_α and Al K_α, and the corresponding reduction in counting rate for K_α signals from the alkalis. Clearly, conditions of specimen current per unit of specimen area (or volume) versus the allowable counting period should be established for good quantitative results.

An example of this approach, for Na K_α and Si K_α, is shown in Figure 16 relating mW/μ^2 to the time for a 5% change in counting rate (up or down). Electron beam sizes were estimated from beam-scan line images of sample current across a razor blade edge at a magnification of 2400 ×. The two lines shown in Figure 16 should be parallel and at a slope of 45°; experimental error is, therefore, present, but general guidelines can still be established.

For Na K_α it is clear that any sample loading over about 5 mW/μ^2 will result in unreasonably short counting periods (< 10 sec.) Either the beam diameter will have to be increased (with no loss of signal), or the sample current reduced (with loss of signal). These considerations of specimen damage suggest, then, that the use of as large a spot as possible will give best results. Note, however, that the spot size increase is not made in an attempt to average chemical inhomogeneities.

Despite the inadequacies of present correction parameters, and the problems of grain variability and beam damage, useful quantitative analyses can be made by reference to closely related standards. For example, in the tests of grain variability described above, several feldspars were analyzed for their major constituents by X-ray spectrography. In Figure 17 these determined bulk elemental values are plotted against the mean net

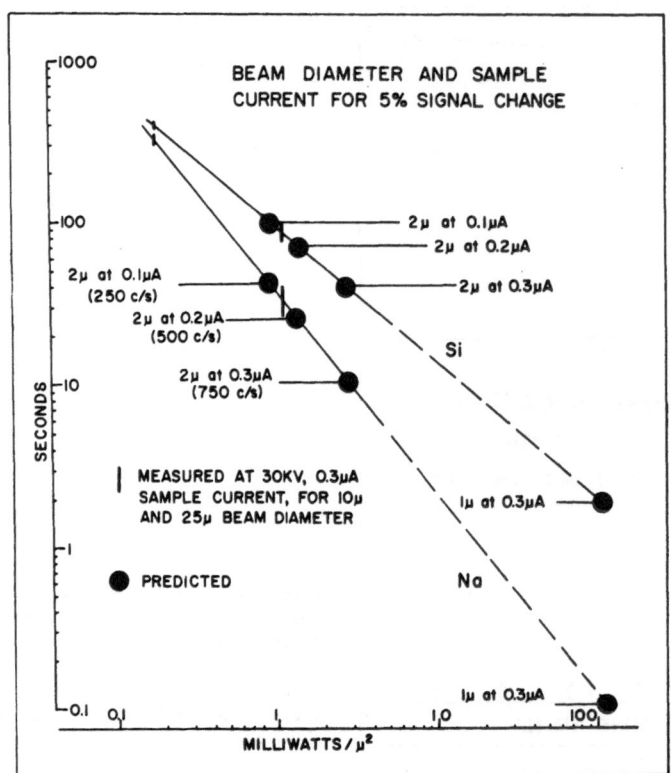

Figure 16. Specimen current loading versus counting time for different beam sizes and sample currents. See text for discussion.

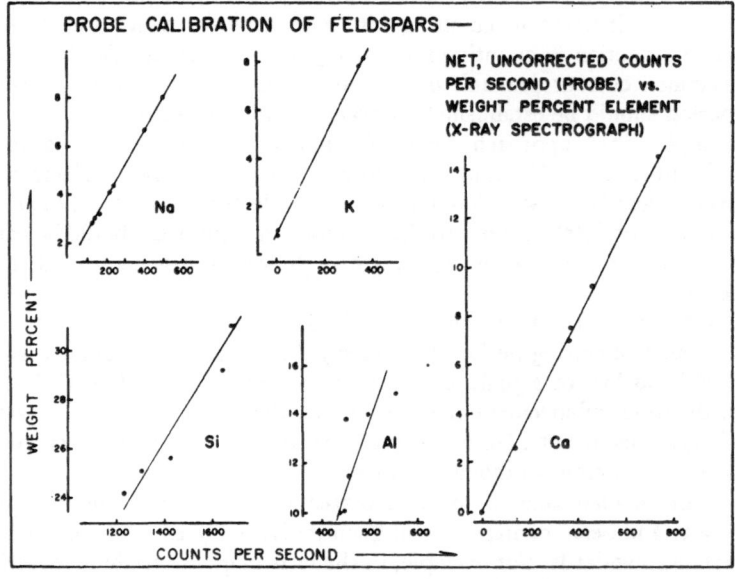

Figure 17. Calibrations for the major and minor elements of feldspars without using correction factors.

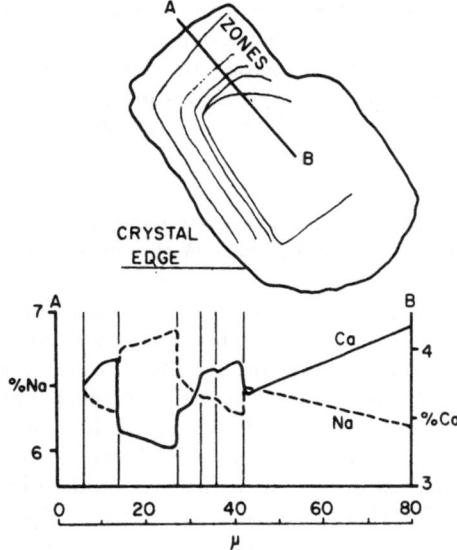

Figure 18. Sodium and calcium variations in compositionally zoned plagioclase. Tracing of grain is from Figure 6(e).

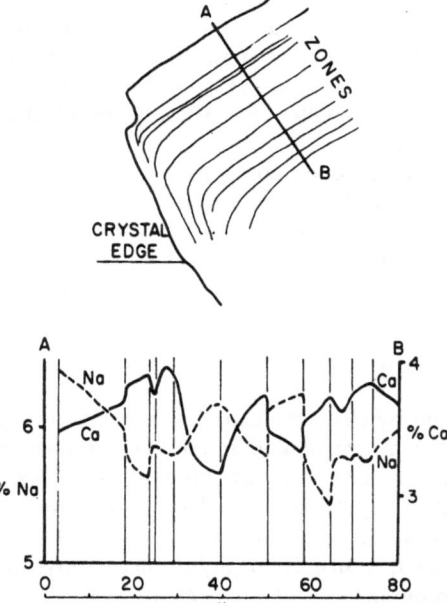

Figure 19. Sodium and calcium variations in compositionally zoned plagioclase. Tracing of grain is from Figure 7(c).

counting rates of the 27 spots per feldspar, which were determined by the electron probe. For sodium, potassium, and calcium (the main variables of the feldspar family of minerals), the correlations are excellent. No correction factors have been applied to the calibrations shown in Figure 17.

In application, the variations of sodium and calcium in the system $NaAlSi_3O_8$–$CaAl_2Si_2O_8$ have been determined for the plagioclase grains shown in Figures 6(e) and 7(c). Tracings of the grain outlines, compositional zones, and positions of traverses are shown in Figures 18 and 19. Correlation of this quantitative data with the known phase chemistry of the plagioclase system (Figure 1) is of great importance in genetic studies of rocks. Other mineral systems of common rocks have been studied by this and other laboratories, and similar calibrations of the probe data by empirical means have been found to be the most successful, providing that adequate standards are available.

Computation of Mineral Composition

It was noted earlier in this paper that petrologists are often severely handicapped in obtaining quantitative estimates of the mineral composition of a rock. With a knowledge of the bulk rock chemistry, however, the elemental weight percentages can be recast to a set of standard minerals (referred to as "normative minerals") with standard compositions. Rules for the calculations are given by Johannsen,[12] and McIntyre[13] has written a Fortran program for the IBM 7094 which will handle 150 rock analyses per minute. If the input data is from X-ray spectrography (rather than conventional wet chemistry) estimates of the Fe^{2+}/Fe^{3+} ratios must be obtained. In the future it is possible that X-ray line-shift techniques[14] may provide this information directly. For our present work, however, the mean values of several thousand wet chemical analyses of igneous rocks were studied by multivariate analysis and it was found that $\log(Si/Fe^{2+})$ has a marked correlation ($r = 0.94$) with total iron. Predictions from the regression equation are used, and even though the ratios are only approximate, a simple test shows that no serious error is introduced for our rocks. For a single rock analysis the normative mineralogy was recomputed with Fe^{2+}/Fe^{3+} ranging from 1/1 to 5/1. Orthoclase, ilmenite, and the anorthite content ($CaAl_2Si_2O_8$) of plagioclase were unchanged; other minerals varied as follows: quartz, 18.5 to 19.6%; plagioclase, 53.1 to 53.2%, hypersthene, 3.1 to 5.3%, and magnetite 0.5 to 1.5%. The ranges for these major constituents are smaller than those introduced by errors in the analytical determinations of the elements.

Several of the standard normative minerals computed are not common to the granitic rocks we are studying; other minerals, such as micas and amphiboles, common to granite, are not computed. These latter minerals are omitted in the standard set because their chemistries have not been well established. The electron probe offers a rapid means of analysis of these minerals and the standard norm can then be modified to provide a closer approach to the real mineralogy.

CONCLUSIONS

Soft X-ray spectrography, both by X-ray and by electron excitation, can be a rapid, precise, and accurate method for the analyses of all major and minor constituents of geological samples of atomic number 5 and higher. The combination of bulk analysis, mineral-phase analysis, and high-speed computation provides complete information on what elements and minerals are present in rocks and in what quantities.

The long-established technique of polarized, transmitted light microscopy has proven to be an important part of electron-probe analysis of non-opaque samples. In most cases, the rapid identification of mineral phases by their optical characteristics permits the probe operator to find immediately the sample areas of critical interest.

ACKNOWLEDGMENTS

The authors are indebted to Edward Welday and Kathleen Madlem for help in developing sample preparation techniques for the electron probe and for their continued help in our X-ray analyses. The methods discussed in this paper were developed with the support of the National Science Foundation (grants G19075 and GP1336) for a study of the granitic rocks of Southern California.

REFERENCES

1. B. L. Henke, "Sodium and Magnesium Fluorescence Analysis—Part I: Method," in: W. M. Mueller, G. R. Mallett, and M. J. Fay (eds.), *Advances in X-Ray Analysis, Vol. 6*, Plenum Press, New York, 1963, p. 361.
2. B. L. Henke, "X-ray Fluorescence Analysis for Sodium, Fluorine, Oxygen, Nitrogen, Carbon, and Boron," in: W. M. Mueller, G. R. Mallett, and M. J. Fay (eds.), *Advances in X-Ray Analysis, Vol. 7*, Plenum Press, New York, 1964, p. 460.
3. B. L. Henke, "Some Notes on Ultrasoft X-Ray Fluorescence Analysis—10 to 100 Å Region," in: W. M. Mueller, G. R. Mallett, and M. J. Fay (eds.) *Advances in X-ray Analysis, Vol. 8*, Plenum Press, New York, 1965, p. 269.
4. E. E. Welday, A. K. Baird, D. B. McIntyre, and K. W. Madlem, "Silicate Sample Preparation for Light Element Analyses by X-ray Spectrography," *Am. Mineralogist* **49**: 809, 1964.
5. D. B. McIntyre, "Fortran II Programs for X-ray Fluorescence," Technical Report 13, Department of Geology, Pomona College, 1964.
6. A. K. Baird, D. B. McIntyre, and E. E. Welday, "Soft and Very Soft Fluorescence Analysis: Spectrographic and Electronic Modifications for Optimum, Automated Results," *Develop. Appl. Spectr.* **4**: 3, 1965.
7. A. K. Baird and B. L. Henke, "Oxygen Determinations in Silicates and Total Major Elemental Analysis of Rocks by Soft X-Ray Spectrometry," *Anal. Chem.* **37**: 727, 1965.
8. A. K. Baird, D. B. McIntyre, E. E. Welday, and K. W. Madlem, "Chemical Variations in a Granitic Pluton and its Surrounding Rocks," *Science* **146**: 258, 1964.
9. A. K. Baird, D. B. McIntyre, and E. E. Welday, "Granitic Rocks of the San Bernardino Mountains, California: Chemical Composition and Variability Within 2000 Square Miles," *Geol. Soc. Am. Spec. Papers* **82**: 6, 1965.
10. A. K. Baird, D. B. McIntyre, E. E. Welday, and D. M. Morton, "A Test of Chemical Variability and Field Sampling Methods, Lakeview Mountain Tonalite, Southern California Batholith," *Calif. Div. Mines and Geol., Shorter Cont.*, in press.
11. D. A. Copeland, "A Simple Device for Trimming Thin Sections," *Am. Mineralogist* **50**: 1128, 1965.
12. A. Johannsen, *A Descriptive Petrography of the Igneous Rocks, Vol. 1*, University of Chicago Press, Chicago, 1931, p. 89.
13. D. B. McIntyre, "Fortran II Program for Norm and Von Wolff Computations," Technical Report 14, Department of Geology, Pomona College.
14. B. L. Henke, "Application of Multilayer Analyzers to 15–150 Å Fluorescence Spectroscopy for Chemical and Valence Band Analysis," this volume, p. 432.

DISCUSSION

D. H. Speidel (Pennsylvania State University): On your last diagram, going across the zoning, did you try to make temperature plots? For example, if this is a true zone crystal, you would expect the composition to be fairly close in the middle and to differ as you go out because of the solvus curve. It didn't seem to do this.

A. K. Baird: No, it doesn't. I searched our mineralogy collections to find the most exaggerated apparent zoning I could for this test, and I don't care to make a geological "case" from these results. I merely wanted to find out if I could get an antipathetic behavior between the sodium and calcium values in stage traverses across plagioclase feldspars, despite the serious problems associated with beam damage on such specimens. Temperature plots were not tried. However, in the binary system albite–anorthite (complete solid solution without maxima or minima), I would expect crystal centers to be relatively high in calcium and low in sodium, not "close" in composition as you state, assuming chemical disequilibrium controlled the zoning; I would expect the outer zones to be the reverse.

K. F. J. Heinrich (National Bureau of Standards): Could you please repeat your argument on the measurement of the beam width? I understood that you said that the extended variation of the current corresponds to twice the beam diameter.

A. K. Baird: The situation is this: As soon as the beam hits the edge of the razor blade we should expect an increase in the sample current. We would expect that increase to continue until the beam passed off of the razor blade edge.

K. F. J. Heinrich: Yes, and this distance corresponds to one times the beam diameter, not twice.

A. K. Baird: Yes.

K. F. J. Heinrich: The center of the beam has moved exactly one time around the beam diameter.

A. K. Baird: What I am trying to say is this, that the sample current line images do not show abrupt intensity changes at high magnification on the scanner (2400 ×).

K. F. J. Heinrich: Yes, I understand that.

A. K. Baird: The sample current appears to increase slowly over the edge. This is the difficulty in establishing where you take the blade-edge position and thus the beam perimeter.

K. F. J. Heinrich: But the point is that in order to move the beam from being completely on the left field to completely on the right field, you move it one diameter, not two diameters.

A. K. Baird: Yes, that's right. My original statement of two diameters was incorrect.

K. F. J. Heinrich: And if you go farther, you wouldn't have any further beam intensity.

A. K. Baird: I didn't word this properly. What I should have said is that the turnover is not sharp and it puts a great uncertainty into deciding just where to draw the limit, and so I say as a conclusion that we are dealing with a beam that is probably 1 to 2 μ. I can't say for certain that it is a micron or less based on this test. I think that it's certainly less than three microns.

A "WINDOWLESS" X-RAY FLUORESCENCE TUBE AND A HIGH-RESOLUTION DIFFRACTOMETER

George Walker

Minnesota Mining and Manufacturing Company
St. Paul, Minnesota

ABSTRACT

A system is described whereby X-ray fluorescence analysis of elements below atomic number 19 (potassium) is undertaken with a "windowless" demountable X-ray tube. Comparison is made with 100-μ- and 30-μ-thick beryllium windows. Deflection plates are required before or behind the sample to eliminate scattered electrons. A high-resolution X-ray diffractometer is also described by making use of a fine focal line on the target. Burn-through due to localized heating on the target is eliminated by using a thermocouple arrangement.

DESCRIPTION

The arguments for and against the demountable X-ray tube are, I am sure, familiar to all in the X-ray analysis field. It is the purpose of this paper to point out some work done by the author while working for Associated Electrical Industries in Great Britain.

The work was carried out on an A.E.I. Raymax 60[1] X-ray diffraction unit, designed by Dr. R. Witty, with whom the author was working. This instrument lends itself to both X-ray diffraction and X-ray fluorescence analysis with the same X-ray tube (different targets). The tube is continuously evacuated with a diffusion pump, although, with a "getter" pump, one can obtain a virtually sealed-off tube with no contamination. The vacuum-path spectrometer is attached to the X-ray tube, as shown in Figure 1, for X-ray fluorescence. For X-ray diffraction, windows A are used.

Window B in Figure 1 provides a serious drawback to the analysis of elements below potassium on the atomic number scale, due to absorption of the exciting X-ray quanta. This absorption is proportional to $\rho\lambda^3 Z^3 X$, where ρ is the density of window, λ is the wavelength of radiation, Z is the atomic weight of window, and X is the thickness of window. For maximum efficiency of excitation, the primary X-ray wavelength should be just shorter than the characteristic wavelength of the element being excited, because of the falling off of the absorption curve on either side of the characteristic absorption edge.[2] For magnesium, this wavelength is about 9.45 Å. If the X-ray tube window filters this radiation out of the primary beam, then one depends on the less efficient short-wavelength continuous or characteristic radiation from the target.

To compensate for absorption in the X-ray tube window, an increase in power can be effected or the thickness of the window reduced (or eliminated entirely). The effect of reducing the window thickness is reported by Spielberg.[3] An increase in power can only be carried so far without exceeding the power rating of the target, after which a reduction

[1] References are at the end of the paper.

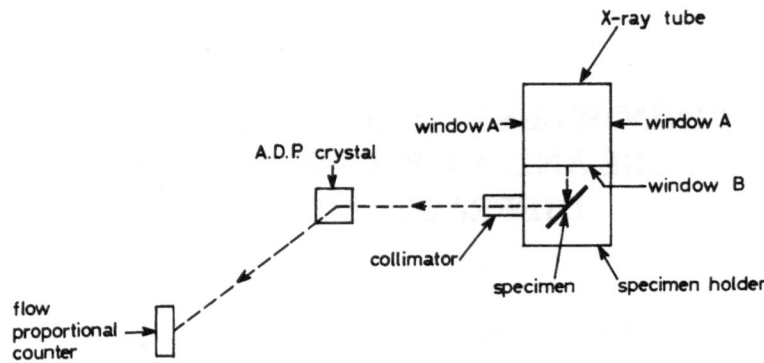

Figure. 1. Relation of specimen holder to X-ray tube and vacuum-path spectrometer.

in window thickness is the only possible course. It is further seen that if one depends on continuous radiation from the target for excitation, then an increase in tube voltage is the best way to get an increase in intensity, using the equation[4]

$$I_{cont} = AiZV^m$$

where I_{cont} is the intensity of continuous radiation from target, A is a constant, i is the tube current, Z is the atomic weight of target, V is the tube voltage, and m is approximately 2. Since this energy is spread over a range of wavelengths, an increase in power is not as efficient as in the characteristic radiation case[5] as follows:

$$I_{char} = Ci(V - V_k)^n$$

where I_{char} is the intensity of characteristic radiation, C is a constant, i is the tube current, V is the tube voltage, V_k is the excitation voltage, and n is approximately 1.5. This intensity is spread over about 0.001 Å (measured at half-maximum intensity) in most lines. The ratio of characteristic to continuous radiation increases rapidly with voltage up to about three times the characteristic excitation voltage, after which it levels off.[6] It is therefore seen that if a target can be choosen with a K, L, or M characteristic line just shorter, in wavelength, than the element wavelength, and which will not be greatly absorbed by the window, then one will obtain a maximum excitation efficiency.

On a continuously evacuated X-ray tube very thin beryllium windows can be tolerated even if they are porous. An 18-mm-diameter, 30-μ-thick beryllium window was used to study the effects of filtration on magnesium and chlorine excitation. Table I shows the effects of (1) going from a 100-μ-thick beryllium window to a 30-μ-thick beryllium window, (2) going from a tungsten target to a molybdenum target, and (3) going to no window in the X-ray tube. The third condition was accomplished with the experimental arrangement shown in Figure 2. When the window was first removed, there was a problem with scattered electrons but this was overcome with the deflector plates shown in Figure 2. The voltage across the plates had to be increased with increasing tube voltage. The scattered electrons still strike the sample, giving rise to an increase in characteristic radiation as well as continuous radiation from the sample. The continuous, or background, radiation thus produced was not too troublesome, but could be reduced by placing the deflection plates in front of the sample. An ADP crystal and flow proportional counter were used for these experiments.

In diffractometry experiments, the basic geometry requires a good focal line on the target for good resolution, together with a high intensity at the counter.[7-9] Because of

Table I. Effects of Changes in Beryllium-Window Thickness, Target Material, and Absence of Window

Specimen*	λ (Å)	Target	λ(Å)		Relative intensity†		
			K_α	L_α	100-μ window	30-μ window	No window
Cl	4.72	Mo	0.71	5.41	0.71	1.22	3.71
		W	0.21	1.47	1	1.7	5
Mg	9.88	Mo			1.9	9.83	63.6
		W			1	4.9	27.8

* The chlorine was present as a low-percentage element in a heavy-element matrix. The magnesium was also present in a heavy-element matrix, containing about 10% magnesium.

† Intensities are given as peak-to-background and are normalized on tungsten with a 100-μ window (after Spielberg[3]).

the ease of target inspection and replacement, more bias can be used on the cathode assembly, giving a much sharper focal line on demountable tubes. Due to heat considerations on the target, sealed-off tubes can only tolerate a focal line of about 0.8 mm, giving the required safety factor without burn-through.

Experiments with a demountable X-ray tube proved that a small platinum–platinum–rhodium thermocouple could be placed behind the target (see Figure 3), and when localized heating occurs, which would eventually cause burn-through, the thermocouple switches the power off, preventing water infiltration. The thermocouple was attached to the target via a thin film of high-temperature resin.

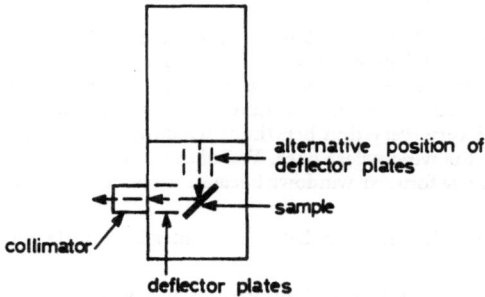

Figure 2. Position of deflector plates.

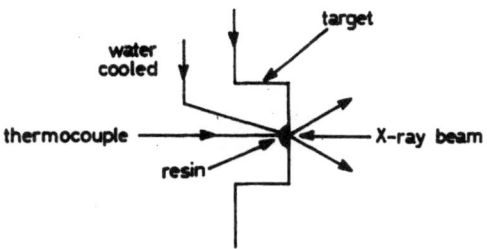

Figure 3. Protection of target against burn-through.

With the Raymax 60, a focal line of about 0.1 mm in width has been obtained. With this focal line, and using a collimating system to an X-ray tube angle of 3°, Cu $K_{\alpha_1,_2}$ separation was obtained down to a Bragg angle of 8.325° from a powdered ADP sample.

CONCLUSION

I would like to say that due to the versatility with which a demountable tube lends itself to research, there can be no reasonable objection to the so-called maintenance problems (which are few with present day vacuum technology) associated with vacuum equipment.

ACKNOWLEDGMENTS

The author wishes to thank Associated Electrical Industries of Great Britain for permission to publish this paper. Thanks are especially due to Dr. R. Witty and Mr. P. Wright, without whom this paper would not have been possible.

REFERENCES

1. H. S. Peiser, in: *X-Ray Diffraction by Poly-Materials*, Institute of Physics, 1960, p. 64.
2. Kurt Togel, "Vacuum X-ray Fluorescence Analysis for Elements with Atomic Number between 12 and 22," *Siemens-Rev.* 28(5): 153, 1961.
3. N. Spielberg, "Tube Target and Inherent Filtration as Factors in Fluorescence Excitation of X-Rays," *J. Appl. Phys.* 33(6): 2033, 1962.
4. B. D. Cullity, *Elements of X-Ray Diffraction*, Addison-Wesley, Reading 1956, p. 6.
5. B. D. Cullity, *ibid.* p. 7.
6. H. S. Peiser, *op. cit.* p. 67.
7. J. B. Nelson, in: *X-Ray Diffraction by Poly-Materials*, Institute of Physics, 1960, p. 88.
8. G. W. Brindley, in: *X-Ray Diffraction by Poly-Materials*, Insitute of Physics, 1960, p. 148.
9. G. W. Brindley, *op. cit.* 165.

DISCUSSION

G. Walker: I would like to bring up something. I was speaking to Dr. Henke about the formvar window he uses in his systems. I know that the feeling is that you get contamination here, especially if you use bad samples. Many of the companies that we were doing this work for worked with refractory materials and we never noticed too much contamination, without use of a formvar window. There were times when we used very, very thin beryllium windows merely as a vacuum separation. This is the reason we went to the windowless ones. We didn't have a window there and it made the tube more useful. We did not use formvar windows because we felt they might introduce problems.

T. C. Martin (Texas Nuclear Corporation): Would you describe your counter in a little bit more detail, please?

G. Walker: This was a flow proportion counter using a 5-μ aluminized Mylar window. We used an argon–methane mixture as a flow gas.

ON THE PRIMARY X̊-RAY ANALYZER

Shizuo Kimoto and Masayuki Sato

Japan Electron Optics Laboratory Co., Ltd.
Tokyo, Japan

and

Hitoshi Kamada and Takuzi Ui

University of Tokyo
Tokyo, Japan

ABSTRACT

The primary X-ray analyzer is used for nondestructive spectrochemical analysis of solid specimens. Accelerated electron beams bombard the specimen surface directly and generate primary X-rays which are measured in a vacuum spectrometer. The method of primary X-ray spectroscopy is superior to the fluorescence X-ray spectroscopy because (1) detectable sensitivity for such light elements as magnesium and aluminum is very high, and (2) the correction of the measured value for self-absorption of X-rays by the specimen itself is low. The performance of the instrument and applications are reported.

INTRODUCTION

In recent years, the analysis of light elements in a solid specimen has been carried out by X-ray spectroscopy. There are two methods available for the soft X-ray excitation. One is a fluorescence method,[1] the other is an electron excitation method.[2,3] In electron excitation, it is easy to obtain suitable excitation conditions for soft X-rays by selection of the accelerating voltage and current density of the electron beam. Moreover, it is much easier to correct the measured values obtained by the electron excitation method than it is to correct values obtained by the fluorescence procedure. However, electron excitation may give rise to difficulties such as heating up of the specimen and charge-up of non-electroconductive materials.

This paper presents the electron excitation method as a useful technique to obtain sufficient X-ray intensity and high peak-to-background ratio for the light elements, using techniques to avoid the above-mentioned difficulties.

INSTRUMENTATION

Figure 1 shows the construction of the primary X-ray analyzer. The electron optical system consists of a self-biased gun and a condenser lens. The electron beam strikes the specimen vertically and the X-rays thus generated have a take-off angle of 45°. The diameter of the electron beam which strikes the specimen is approximately 5 mm and

---

[1] References are at the end of the paper.

508

Figure 1. A general view of the primary X-ray analyzer.

Figure 2. Photograph showing the inside of a vacuum X-ray spectrometer.

the incident power is from 0.5 to 5 W in normal use. The high voltage of the gun ranges from 1 to 50 kV, continuously variable.

Figure 2 shows the inside of the X-ray spectrometer, consisting of a 0.15° soller slit for incident X-rays, two analyzing crystals of 14 × 29 mm, a 0.3° soller slit for diffracted X-rays, and two different types of detectors. A magnetic deflector is located at the window of the spectrometer to prevent scattering of the electrons from the specimen.

The gas-flow proportional counter, using P-10 at atmospheric pressure, is used for soft X-rays. The window is 10 × 10 mm and is constructed of a 0.2-μ collodion film supported by a 250-mesh, 40%-transmission, nickel grid for the wavelengths longer than

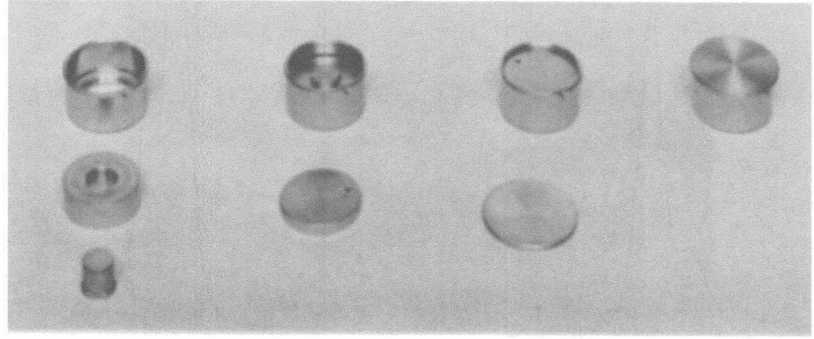

Figure 3. Photograph showing a holder for standard specimen dis-assembled.

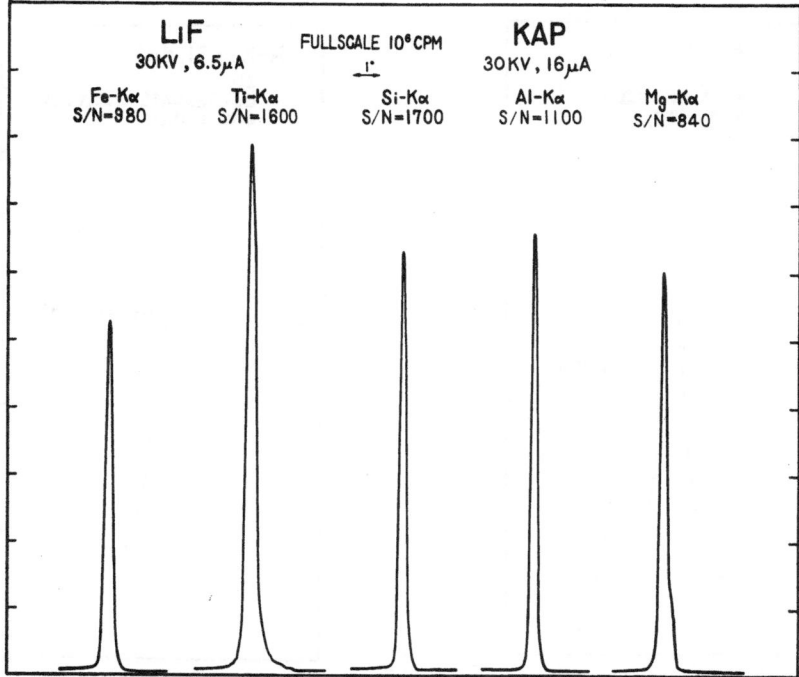

Figure 4. The spectra of the characteristic X-rays of iron, titanium, silicon, aluminum, and magnesium.

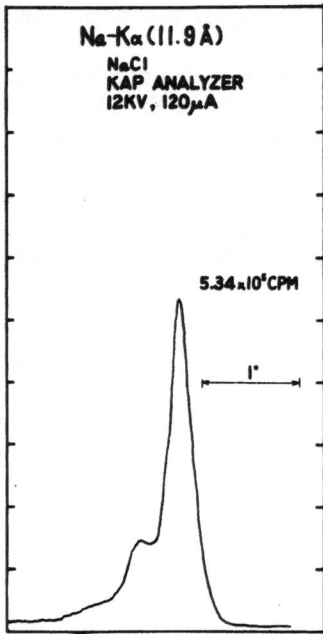

Figure 5. The spectrum of sodium in NaCl crystal coated with evaporated aluminum thin film.

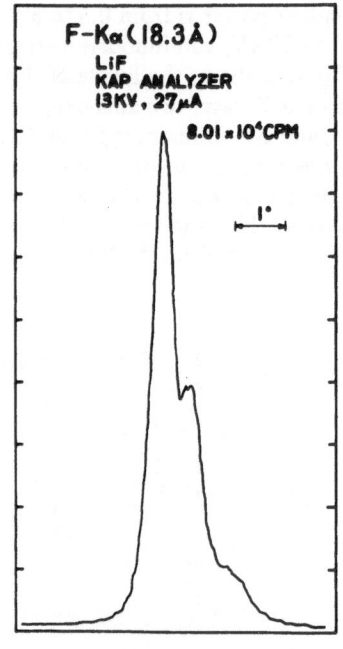

Figure 6. The spectrum of fluorine in LiF crystal coated with evaporated aluminum thin film.

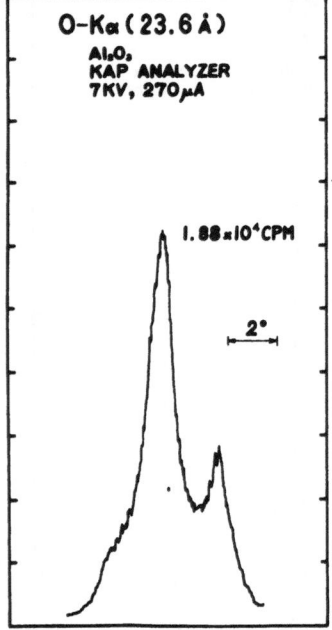

Figure 7. The spectrum of oxygen in Al_2O_3 crystal coated with evaporated carbon thin film.

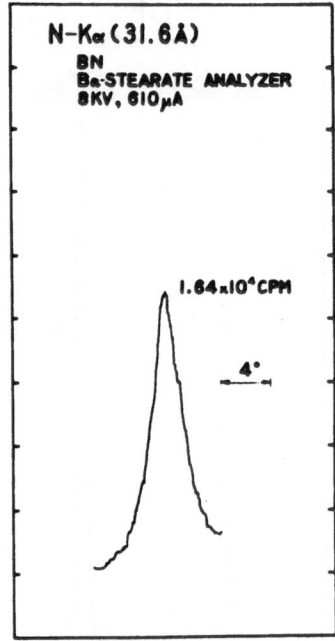

Figure 8. The spectrum of nitrogen in BN coated with evaporated aluminum thin film.

sodium, and a 4-μ Mylar film for the wavelengths shorter than magnesium. A scintillation counter is used for the X-ray detection of heavy elements.

Specimen holders are insulated from the ground by a Teflon bushing, enabling the absorbed electron intensity to be measured by a built-in micro-microammeter.

Absorbed electron current is useful to monitor the intensity of incident electrons or the physical topography and chemical composition of the specimen surface.

In order to measure the average chemical values of a specimen surface, the holder is so arranged that its movement is governed by a heart cam while under electron bombardment. This mechanism is also effective to prevent heat-up and contamination of the specimen. An optical microscope of 8× magnification permits viewing the specimen during the electron bombardment. Figure 3 shows the holder for a standard specimen, a small piece of pure metal. Vacuum during operation is 5×10^{-5} mm Hg or better.

EXPERIMENTS AND RESULTS

Figure 4 shows the typical spectra and also shows the X-ray intensity, peak-to-background ratio, and the required power for magnesium, aluminum, silicon, and iron as pure metals. The spectra of elements lighter than magnesium are shown in Figures 5 to 10.

A specimen which has no electroconductivity must be coated with some conductive material (such as carbon or aluminum) to prevent charge-up. Figure 11 shows an example of qualitative analysis.

Non-corrected calibration curves for silicon in low-alloy steel where exciting voltage is used as a parameter are shown in Figure 12. From these curves, the most suitable excitation voltage was determined as 25 kV and the detectable limit of silicon in low-alloy

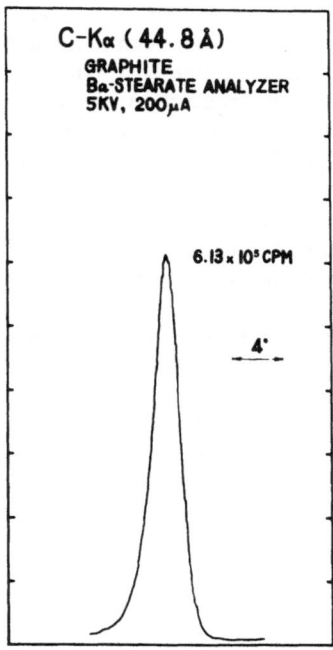

Figure 9. The spectrum of carbon in a graphite crystal.

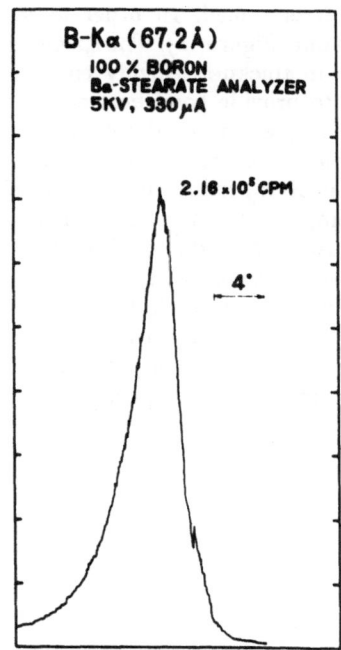

Figure 10. The spectrum of boron in pure boron.

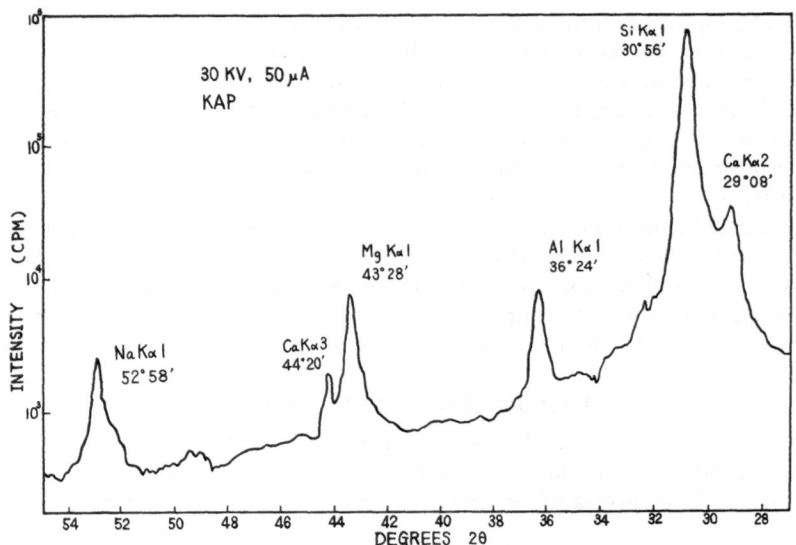

Figure 11. Qualitative analysis of a type of glass coated with evaporated carbon thin film.

steel was estimated at 0.005%, this value being calculated for a 1-min counting time. Moreover, from the fact that every measured value can be plotted in a straight line, it can be inferred that it may be easy to correct the measured values.

As has been said earlier, electron excitation will cause the heating of a specimen unless it is cooled. In order to measure this heating effect, some experiments were carried out (Figure 13). The specimen used was a type of firebrick 10 mm in diameter and 10 mm in thickness, firmly embedded in a brass block in a standard 1-in. specimen holder to provide good heat conductivity and a large heat capacity. A chromel–alumel thermocouple was used for measuring the temperature on the specimen surface. The experiment was conducted by supplying 30 kV and 30 μA to the specimen, which gave sufficient X-ray intensities for light elements such as magnesium, aluminum and silicon. The temperature on the specimen surface rose 60° from room temperature after 30 min of electron bombardment. It may be concluded that the heating of a specimen would not be disadvantageous to measurements without special cooling equipment for the specimen, if the specimen is carefully prepared.

The influence of surface finish is shown in Figure 14. Specimens used in this experiment were three low-alloy steels, A, B and C whose concentrations of silicon were 0.96%, 0.41%, and 0.23%, respectively. Both A and C specimens were used as monitors for

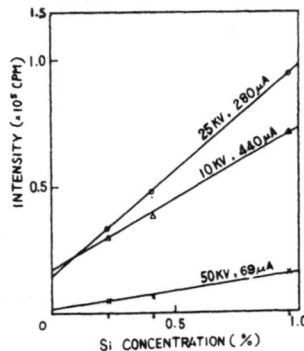

Figure 12. Calibration curves for silicon under various exciting voltages.

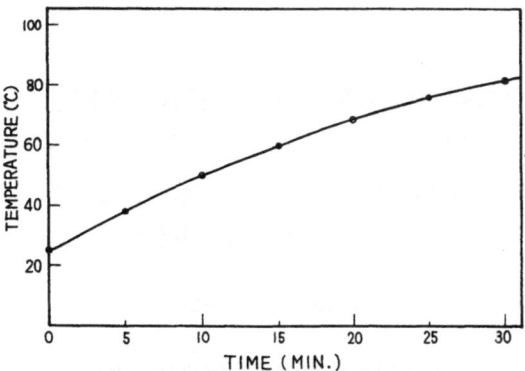

Figure 13. The heat-up of specimen surface as a function of measuring time. Specimen is a type of firebrick. Experimental condition: 30 kV, 30 μA non-cooling.

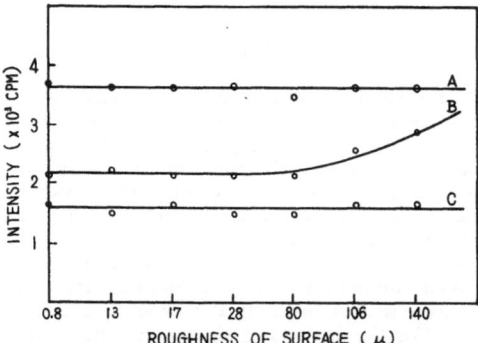

Figure 14. The influence of surface finish.

checking the experimental condition, and A and C curves show that the experimental condition did not change at each measuring period. Specimen B was scratched with emery paper, increasing the surface roughness from 0.8 to 140 μ.

The detection of carbon in stainless steel is now being studied. The preliminary detectable value is 0.013% for a 1-min counting time. On the other hand, the detection of oxygen and nitrogen is rather difficult because of their absorption into the window of the counter.

ACKNOWLEDGMENT

The authors wish to express their appreciation to H. Kohinata of JEOL for his help in performing this experiment.

REFERENCES

1. B. L. Henke, "X-Ray Fluorescence Analysis for Sodium, Fluorine, Oxygen, Nitrogen, Carbon, and Boron," in: W. M. Mueller, G. R. Mallett, and M. J. Fay (eds.), *Advances in X-Ray Analysis*, Vol. 7, Plenum Press, New York, 1964, p. 460.
2. J. G. M. Fox, "X-Ray Spectrometric Determination of Light Elements Using Direct Electron Excitation," *J. Inst. Metals* **91**: 239, 1962–63.
3. N. Spielberg, "Sensitivity of X-ray Spectro-chemical Analysis of Steel by Electron Excitation," in: *X-Ray Optics and X-ray Microanalysis*, Academic Press, New York, 1963.

DISCUSSION

N. Spielberg (Philips Laboratories): Did I understand that you rotate the specimen under the electron beam?

M. Sato: Yes.

SOME OBSERVATIONS ON THE USE OF CERTAIN ANALYZING CRYSTALS FOR THE DETERMINATION OF SILICON AND ALUMINUM*

Frank L. Chan

Aerospace Research Laboratories
Wright-Patterson Air Force Base, Ohio

ABSTRACT

During the past several years, a number of analyzing crystals have been prepared in the U.S. Air Force Aerospace Research Laboratories. Crystals such as alkaline acid malonates, sucrose, pentaerythritol (PET), and several others have been grown with the sole purpose of application in X-ray fluorescence analysis of silicon, aluminum, and other elements of low atomic number. Crystals from natural sources, such as quartz, mica, and gypsum, have also been procured from different parts of the world for this purpose.

Among the analyzing crystals so prepared, pentaerythritol gave the highest count rate for silicon and aluminum. However, since this crystal is organic in nature, great care must be exercised in handling this type of crystal in order to obtain constant count rates. For the analysis of silicon and aluminum, α-quartz crystal gave a somewhat low count rate, but this crystal has certain advantages over the organic crystals, and can be used for analysis of materials having high silicon and aluminum content.

For instance, when using a conventional X-ray emission vacuum spectrograph of standard make, with the latest alterations, one of the PET crystals attained a count rate of 101,870 cps for pure silicon when the instrument was operated at 60 kVP and 50 mA with a negligible background count. With a constant potential attachment and operated at 60 kVP and 34 mA, 115,600 cps with negligible background count was obtained for silicon. The count rate for aluminum was of the same order of magnitude with somewhat higher background count.

Several sets of standards have been procured and small amounts of silicon and aluminum in these standards have been analyzed by the latest modified vacuum spectrograph. These results are under study and the limit of detection calculated. Procedures and results are described and discussed.

INTRODUCTION

X-ray fluorescence analysis has advanced to such a stage that is now includes light elements with atomic numbers smaller than 22. Elements such as phosphorus, silicon, aluminum, and magnesium, which previously were too light to be analyzed by this method, are now readily and successfully analyzed. With a special setup, elements such as carbon, boron, and beryllium have been analyzed by this method. Not only are precision and accuracy by this method comparable to the conventional chemical methods, but also the speed of carrying out the analysis surpasses most available methods. Whereas chemical methods must necessarily sacrifice the samples taken for analysis, the X-ray fluorescence method in most cases does not destroy or damage the samples. Normally, the optimum range of interest for an element is somewhere between 5 to 20%; however, recent advances make it possible to analyze elements that appear in trace amounts.

X-ray fluorescence has the disadvantage of not being an absolute method; that is, unlike the chemical methods, it requires a calibration curve for quantitative analysis of elements which are of interest. Materials used for the construction of a calibration curve should be of the same nature and matrix as that of the unknowns; otherwise, the X-ray fluorescence method may not yield quantitative results.

Analysis of silicon and aluminum by X-ray fluorescence, like analysis of other elements, is confined mostly to the surface, and therefore the nature of the surface, with respect to aggregation as well as absorption and enhancement, must be taken into consideration. For best results the calibration curve must be constructed of samples closely related to the unknown samples.

The determination of elements of low atomic number has not been very successful in the past. The two main reasons are (1) lack of efficient analyzing crystal and (2) no suitable counter for detecting the long-wavelength X-rays. Recently, X-ray vacuum spectrographs with electronic devices and counters capable of determination of these elements have become available.

It is generally agreed that in dealing with wavelengths between 0.5 and 2.5 Å, an analyzing crystal such as lithium fluoride gives the best results as far as intensity and dispersion are concerned. However, for longer wavelengths, there are only certain specific crystals best suited for certain wavelengths. It has been found from this work that PET gives the best results as far as count rate is concerned. Crystals of PET prepared in these laboratories gave in excess of 115,000 cps for pure silicon with negligible background count. This count rate could be increased if the Mylar film on the No. 7 flow proportional counter were replaced by thinner materials, such as polypropylene or other materials. However, such thin materials are not as durable as the one in use. Unlike the Mylar film in use, the thin film made from other materials requires frequent replacement.

INSTRUMENTATION

To carry out the experiments, two different instruments were used. For the study of interplanar spacings of different crystals such as PET, quartz, mica, and malonic acid compounds, two types of X-ray cameras capable of taking both powder patterns and single-crystal rotation photographs were used. For the analysis of silicon and aluminum in different alloy samples, an up-to-date vacuum spectrograph was used.

For X-Ray Single-Crystal Fluorescence Rotation Photographs and Powder Patterns

The cameras used for these purposes were designed by the author and have been or are in the process of being patented.[1-3] Powder patterns as well as single-crystal rotation photographs can readily be taken using a General Electric or Norelco camera with attachments. These cameras and attachments were placed on the Norelco (or GE) generator. For the tracing of reciprocal lattice spots and spectral lines in the powder patterns, a Leeds and Northrup microphotometer was used.[4]

For X-Ray Fluorescence Analysis: the Vacuum Spectrograph

The instrument used was a SPG-3 vacuum spectrometer manufactured by the GE X-ray Department. The characteristic secondary X-ray beam, produced by a specimen

[1] References are at the end of the paper.

excited by the primary X-ray beam from the target, was diffracted by a flat analyzing crystal. This spectrograph can also be operated in a vacuum, in helium, or in air as circumstances demand. For the determination of silicon or aluminum, as in the present case, a vacuum spectrograph is necessary because of excessive attenuation of the long wavelength by air. The spectrograph is the XRD 6 installation, which includes an XRD 6 high-voltage supply, an EA 75 X-ray dual target tube, and an SPG-4 detector.

To reduce the cost of installation and operation, an XRD 6 D/F unit was also installed, which uses the common X-ray generator and the SPG-4 detector. Any one of these instruments can be put in its operating position with a set of connectors. The power to these instruments was provided with a special cable and transformer connected to the main power supply line to avoid electrical disturbances from other instruments and furnaces in the laboratory. These instruments can be operated at 75 kV and 50 mA.

The spectrometer is equipped with a vacuum pump of the Welch Duoseal type. The entire chamber can be evacuated in the course of a few minutes to a vacuum of 10 (or less) μ.

The goniometer in the spectrometer is designed for high precision and accurate angular measurements. The 2θ angles can be scanned manually or automatically in increasing or decreasing angles. It can be scanned at four different speeds, depending on the circumstances. The worm-gear can be disengaged from the goniometer and any desired 2θ angles can quickly be set. The vacuum chamber has an opening at the top. It measures $10\frac{5}{8}$ in. in diameter and is used for changing and adjustment of Soller slits, counter tubes, and analyzing crystals. Care should be taken to keep the vacuum chamber free from moisture, by installation of a mercury trap and a drying train.

The EA-75 X-ray target tube has dual targets, one of chromium and one of tungsten. It is located above the sample changer. A chromium target is used for silicon and aluminum analysis. The samples taken for analysis are positioned at 30° to the horizontal for optimum geometry. A maximum of four samples can be placed in the vacuum spectrograph for each run.

Two analyzing crystals can be placed inside the vacuum chamber side by side, and any one of these crystals can be turned to its reflecting position by a remote control mechanism without breaking the vacuum system. The sample holder is provided with sample masks and has an opening of 1.9 × 2.6 cm. For analysis of aluminum, a brass or nylon mask is used, and for silicon, an aluminum mask is used. The sample is held against the mask by a spring.

The spectrometer is made to accommodate two counters; a No. 7 flow proportional counter is placed in front of the scintillation counter. They can be operated individually or simultaneously. For silicon and aluminum analysis, the No. 7 flow proportional counter is used, and the flow gas is a mixture of 10% methane and 90% argon. The units also provide a constant potential setup.

The SPG No. 4 detector has a scaler–timer combination, an amplifier, pulse-height selector, ratemeter, strip-chart recorder, and a digital computer. The No. 7 flow proportional counter, for normal run, is operated at 1575 V and the pulse-height selector is adjusted to $E = 2$ V and $\Delta E = 4$ V.

EXPERIMENTAL

X-Ray Diffraction Procedure

The procedures previously described have been followed for the taking of single-crystal rotation photographs and powder patterns.[4] The X-ray generator was operated at 50 kVP and 20 mA to decrease exposure time. Normally, with such operating conditions, the duration of exposure is 45 min to 1 hr for single-crystal rotation photographs. For powder patterns, the time is somewhat longer.

X-Ray Fluorescence Procedure

The procedure for fluorescence analysis has likewise been described.[5] A solid sample with a suitable smooth surface, such as alloy samples, needs very little preparation. The operating characteristics of the vacuum spectrograph have been described.

RESULTS

X-Ray Diffraction Study of PET Crystal

Attempts have been made to study a number of compounds for possible use as analyzing crystals for such elements as silicon, aluminum, and other elements of low atomic number. These compounds were first examined for possible reflecting planes by taking X-ray powder patterns and single-crystal rotation photographs. A number of these compounds were under examination.

In the study of PET crystals for use in analysis of silicon and aluminum by X-ray fluorescence, it was noted that the American Society for Testing and Materials gave card number 3-0214 for pentaerythritol, in its 1964 compilation,[6] fifteen spectral lines with corresponding interplanar spacings. These spectral lines have not been indexed in the original ASTM card file. However, practically all these lines have been indexed by Smith.[6]

For the present study, a number of X-ray powder patterns of PET have been prepared from C.P. materials and recrystallized several times. One of the powder patterns using the author's attachment is shown in Figure 1. This pattern was taken with a copper target tube and the Cu K_β radiation removed by nickel filter. The pattern was taken with Wilson's technique and with a pure silicon sample specially prepared by Dow Corning Corporation (Sample X-17-19-5-1) as reference on the same film as shown.

Figure 1. X-ray diffraction powder patterns taken on the same film with the author's camera attachment, Wilson technique. Top, recrystallized PET; bottom, pure silicon.

Figure 2. Single-crystal rotation photograph of PET taken with the author's camera attachment. $a_0 = 8.081$; $c_0 = 8.716$; $c/a = 1.4331$; space group: I4̄.

Table I. Interplanar Spacing with Index and Relative Intensity of Spectral Lines of PET*

d	I/I_0	hkl
4.98	48	011
4.39	100	002
4.30	27	110†
3.05	72	200
2.62	2	103†
2.60	46	211
2.50	26	202
2.15	7	004 (220)
1.99	29	031
1.93	25	222 (150)
1.76	3	312
1.66	11	015
1.53	3	400
1.46	6	006
1.33	3	413
1.247	3	044†
1.204	4	226
1.124	3	053 (433)
1.052	1	253 (406)†

* Cu K_α radiation; nickel filter. The ASTM card[6] as reported by Dow Chemical Company gave essentially the same for d values and relative intensity.

† Not reported previously.

The lattice constants were determined from the single-crystal rotation photographs taken on two axes, namely the c axis and a axis. These constants were calculated from the K_{α_1} and K_{α_2} doublets in the back-reflection region of the cylindrical camera based on the procedure published earlier.[4] The single-crystal rotation photograph with the crystal rotating on its c axis is shown in Figure 2. These lattice constants ($a = 6.082$, $b = 1.433$) were fed into the computer to obtain the interplanar spacings for possible h, k, and l indexes. The print-out values are then compared with d values calculated from the microphotometric tracing of the X-ray powder pattern of PET.

By the use of Wilson's technique for the X-ray powder pattern in conjunction with the microphotometric trace of the pattern, one can precisely determine the two spots in the cylindrical camera 180° apart, one of these spots being the entrance of the incident beam and the other the focal point of the X-ray. The interplanar spacings of any spectral line can be determined with high precision by this method.

The relative intensity of the spectral lines is based on the blackening of the X-ray film with time of exposure. The method is essentially that described by Klug and Alexander.[7] The measured interplanar spacings d and the corresponding relative intensity are shown in Table I.

Lattice constants, number of molecules per unit cell, specific gravity, space group, and structural data have been determined and studied by a number of investigators, notably Shiono et al.[8] A neutron diffraction study of PET has also been made.[9]

Analysis of Silicon and Aluminum with the Vacuum Spectrograph

A number of analyzing crystals of suitable d spacings such as PET, sugar, compounds of malonic acid, and others were prepared. Many of these have very poor reflectivity to the X-ray wavelengths of interest. Others, such as EDDT and ADP, were procured from commercial sources. A few of the results are shown in Table II. PET undoubtedly gives the highest count rate with a surprisingly low background count. A sample of pure quartz having silicon content of 46.73% gave 35,607 cps using a PET crystal prepared in these laboratories. Crystals are shown in Figure 3 and 4. This count rate is about $2\frac{1}{2}$ times the count rate of an EDDT crystal purchased commercially. These count rates

Table II. Various Analyzing Crystals for the Determination of Silicon and Aluminum*

Crystal	Sample taken for analysis	
	Quartz (46.73% Si)	National Bureau of Standards high-temperature Alloy No. 1205 (6.68% Al)
	Si K_α (cps)	Al K_α (cps)
PET	35,607	1,666
EDDT	13,728	474
Quartz (10$\bar{1}$0)	9,222	—
Mica	4,185	159

* Conditions: full wave rectification; 100 sec counts; chromium target, 60 kVP, 50 mA (modified X-ray tube); No. 7 flow proportional counter tube (0.00015-in. aluminized Mylar); Soller: 0.020 × 1$\frac{1}{2}$ in.; vacuum system; $E = 2V$; $\Delta E = 4V$.

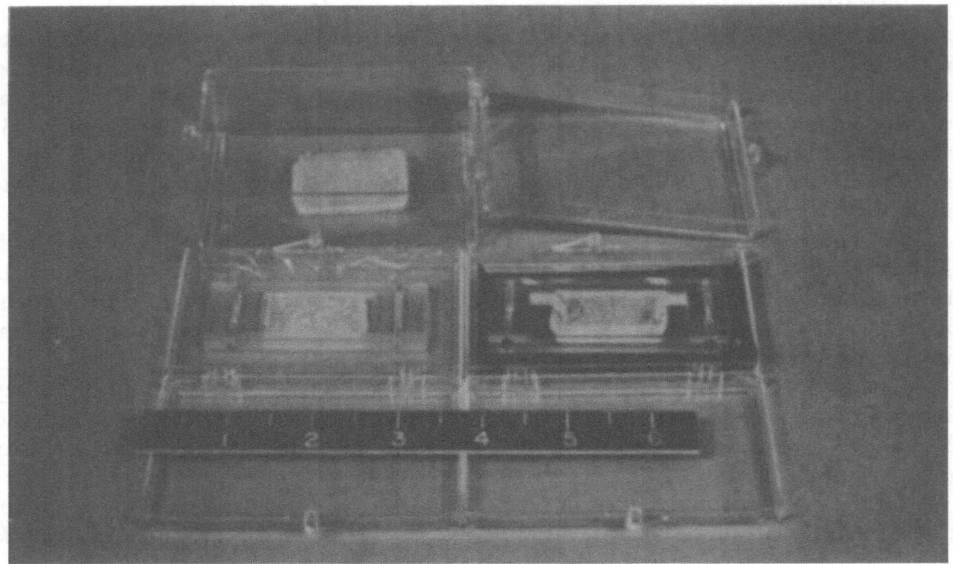

Figure 3. Some PET crystals prepared by the author.

are based on the new modification of the dual target tube by the General Electric X-Ray Department in May, 1965. The modification is essentially the shortening of the gap between the X-ray targets and the sample, and the increase of the number of baffle plates in the soller tunnel so as to increase the resolution. Operating on a full wave rectification and 60 kVP and 50 mA, pure silicon gave 101,870 cps and, on using a constant potential attachment, 115,600 cps.

Figure 4. Some quartz crystals prepared by the author.

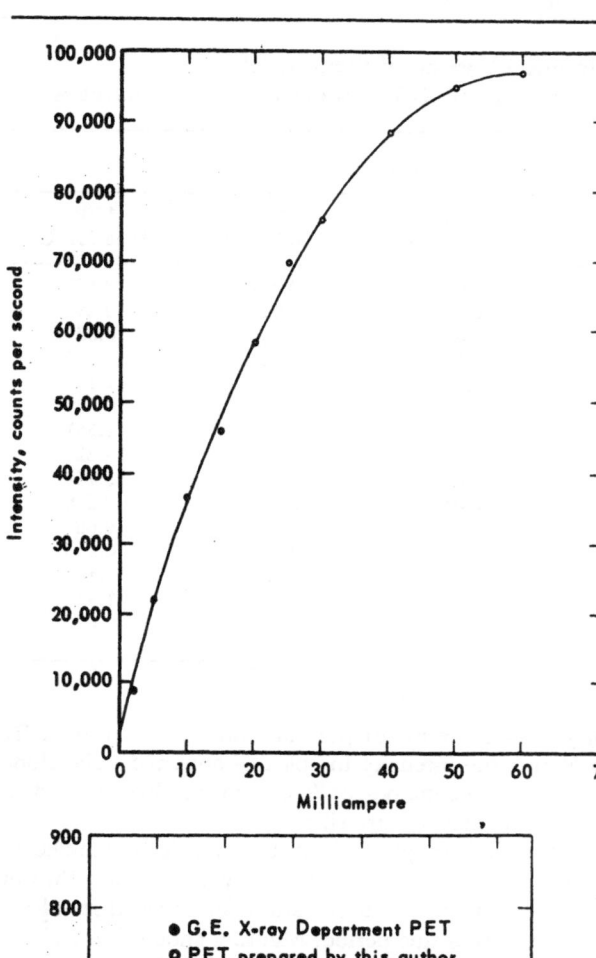

Figure 5. X-ray intensity in relation to applied current. Conditions: PET analyzing crystal (C.P.2); full wave rectification; 10 sec counts; chromium target, 50 kVP (modified X-ray tube); Soller: 0.020 × 1½ in.; No. 7 flow proportional counter tube (0.00015-in. aluminized Mylar); vacuum system; no pulse-height selector.

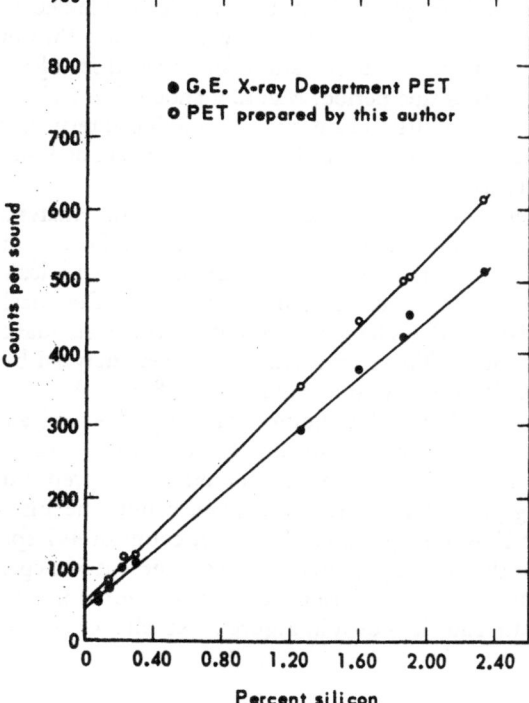

Figure 6. Determination of silicon in Aluminum Corporation of America's aluminum-base alloy samples. Conditions: PET analyzing crystal (C.P.2); full wave rectification; 100 sec counts; chromium target, 50 kVP, 60 mA (modified X-ray tube); Soller: 0.020 × 1½ in.; No. 7 flow proportional counter tube (0.00015-in. aluminized Mylar); vacuum system; $E = 2V$; $\Delta E = 4V$.

Table III. National Bureau of Standards'
High-Temperature Alloy Standards for X-Ray Spectrographic Analyses

Element	Composition of standard (%)			
	1190 Udimet 500	1203 Inco 713-A	1204 Inco 713-B	1205 Inco 713-C
C	(0.10)	(0.01)	(0.03)	(0.19)
Mn	0.61	0.31	0.41	0.29
Si	0.22	0.86	0.56	0.63
Cu	0.093	0.19	0.12	0.056
Ni	51.9	75.5	70.6	67.5
Co	19.1	—	—	—
Fe	(0.6)	(1.4)	(3.1)	(1.55)
Cr	17.00	11.90	12.75	13.82
Mo	3.80	3.01	4.28	5.75
W	0.08	<0.01	0.028	0.019
Al	2.83	4.34	5.60	6.68
Ti	3.57	1.09	0.63	0.36
Zr	0.11	0.055	0.12	0.46
Nb (Cb)	<0.01	1.00	1.31	1.95
Ta	<0.01	0.34	0.46	0.67

In Figure 5 the results of analysis of a sample of pure aluminum are shown. By varying the current from 2 mA to 60 mA the intensity in cps was obtained. The slope decreases as the applied current increases. A count rate of 96,666 cps has been recorded, which is of the same order of magnitude as that of pure silicon.

Since PET is an organic compound, it is fragile and must be carefully handled to avoid damage to the surface. It must be placed in a desiccator when not in use. Two of these crystals have been in constant use and under observation for more than $1\frac{1}{2}$ yr. The count rate has been very consistent during this period. Mention should be made at this time that a quartz crystal, when cut on a 1010 plane gives a suitable interplanar spacing for silicon determination. In spite of its poor reflectivity the crystal is not easily broken or affected by atmospheric conditions.

Three sets of standards have been obtained from different sources for the analysis of silicon and aluminum. These standards are the following: (1) the National Bureau of Standards' Nos. 1190, 1203, 1204, and 1205, (2) the Aluminum Company of America's aluminum-base alloy standards Nos. 96 to 104 inclusive, and (3) the Dow Chemical Company's magnesium-base alloy standards. The elements present in these standard samples are tabulated in Tables III, IV, and V. These standards have been analyzed by the fluorescence method. A few of the results are plotted in Figures 6, 7, 8, and 9.

The results of the count rate from silicon in the Aluminum Company of America's standard sample using two different PET crystals are shown in Figure 6. One of these crystals belongs to the General Electric X-Ray Department and the other was prepared by the author. By subtracting the background counts from the gross count, the PET crystal belonging to the General Electric X-Ray Department has a net count of 447 cps on a 2% silicon content. The net count for the crystal prepared by the author was 530 cps, which amounts to a 20% higher slope, as shown in Figure 6. Analysis of aluminum in Dow Chemical Company's magnesium-base alloy is shown in Figure 7. Results for silicon appear to have similar characteristics.

Table IV. Aluminum Company of America's Standard Samples of Aluminum-Base Alloy

Element	Composition of standard (%)*								
	No. 96, SSB214-A-2	No. 97, SSA214-B-18	No. 98, SS214-C-18	No.99, SAC640-13	No. 100, SAC642-16	No. 101, SAC599-7	No. 102, SAC600-9	No. 103, SAC601-33	No. 104, SAC602-8
Cu	0.082	0.10	0.051	0.040	0.13	0.023	0.055	0.11	0.21
Fe	0.37	0.27	0.26	0.077	0.35	0.12	0.15	0.40	0.58
Si	1.89	0.22	0.15	0.084	0.30	1.26	1.60	1.86	2.32
Mn	0.15	0.14	0.13	0.054	0.24	0.45	0.12	0.26	0.064
Mg	4.01	3.99	3.94	3.08	4.58	2.98	3.54	4.06	4.68
Zn	0.083	1.95	0.080	—	—	0.026	0.045	0.087	0.13
Ni	0.034	—	0.035	—	—	—	—	—	—
Cr	0.036	—	0.033	—	—	—	—	—	—
Ti	0.15	0.048	0.11	0.074	0.17	0.022	0.049	0.11	0.17
Pb	—	—	—	—	—	—	—	—	—
Bi	—	—	—	—	—	—	—	—	—
Sn	—	—	—	—	—	—	—	—	—

* Standard is given by sample number and Alcoa designation.

Table V. Dow Chemical Company's Standard Samples of Magnesium-Base Alloy

Element	Composition of standard (%)*				
	66484	66485	66486	66487	66492
Al	6.83	8.76	9.21	9.94	3.05
Mn	0.15	0.30	0.17	0.10	0.46
Zn	2.59	2.20	1.99	1.76	0.95
Fe	0.034	0.055	—	—	0.023
Cu	0.023	0.013	0.018	0.056	0.012
Ni	0.0027	0.0065	0.0064	0.0047	0.0038
Pb	0.0259	0.0205	0.0028	0.0042	—
Sn	0.042	0.033	0.0012	0.0011	—
Si	0.043	0.097	0.007	0.166	0.002
Ca	—	—	—	—	0.135

* Standard is given by Dowmetal designation.

Analyses of silicon and aluminum in the National Bureau of Standards' samples are shown in Figures 8 and 9, respectively. A study of the count rate statistics reveals that in the NBS standard samples, the count rate is of the same order of magnitude for both silicon and aluminum. For instance, sample No. 1203 has 0.86% silicon and 4.34% aluminum. The count rate for silicon in the sample was 283.7 cps with a background count of 31.0 cps, which amounts to 246 cps for 1% silicon. A similar calculation gave 223.2 cps for 1% aluminum. Other calculations from these samples lead to a similar conclusion.

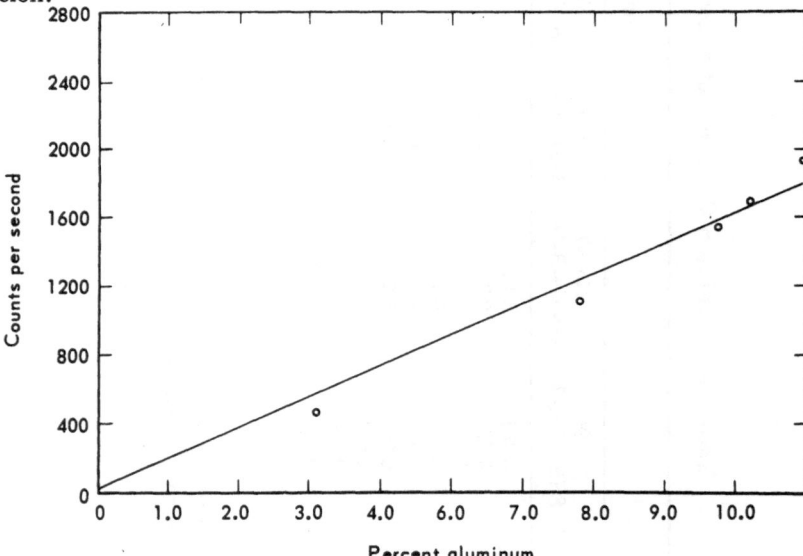

Figure 7. Determination of aluminum in Dow Chemical Company's magnesium-base standard alloy samples. Conditions: PET analyzing crystal (C.P.2); full wave rectification; 100 sec counts; chromium target, 50 kVP, 60 mA (modified X-ray tube); Soller: 0.020 × 1½ in.; No. 7 flow proportional counter tube (0.00015-in. aluminized Mylar); vacuum system; $E = 2V$; $\Delta E = 4V$.

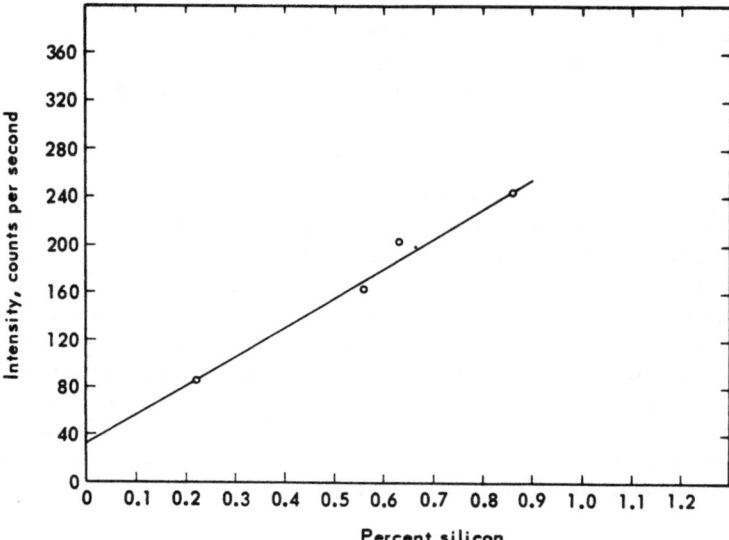

Figure 8. Determination of silicon in National Bureau of Standards high-temperature alloy. Conditions: PET analyzing crystal (C.P.2); full wave rectification; 100 sec counts; chromium target, 50 kVP, 60 mA (modified X-ray tube); Soller: 0.020 × 1¼ in.; No. 7 flow proportional counter tube (0.00015-in. aluminized Mylar); vacuum system; $E = 2V$; $\Delta E = 4V$.

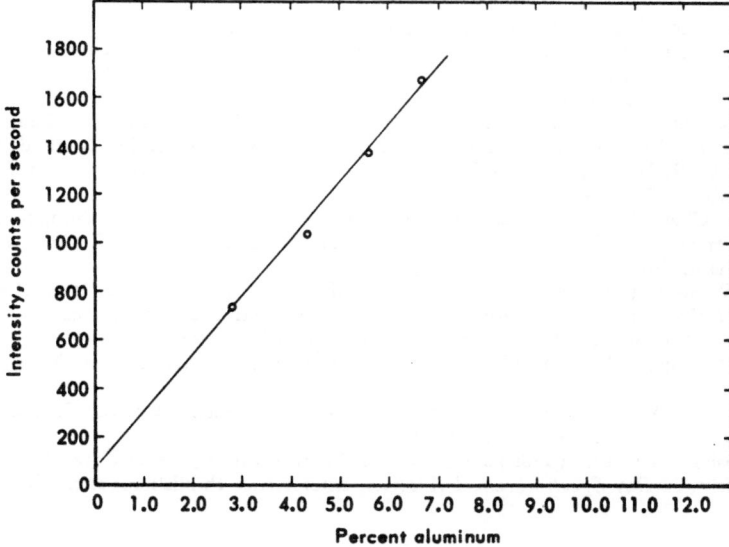

Figure 9. Determination of aluminum in National Bureau of Standards high-temperature alloy. Conditions: PET analyzing crystal (C.P.2); full wave rectification; 100 sec counts; chromium target, 50 kVP, 60 mA (modified X-ray tube); Soller: 0.020 × 1¼ in.; No. 7 flow proportional counter tube (0.00015-in. aluminized Mylar); vacuum system; $E = 2V$; $\Delta E = 4V$.

The silicon content in these samples varies between 0.22% and 0.86% as shown in Table III. The corresponding count rate varies between 85.3 and 244.0 cps. The slope factor $\Delta C/\Delta I$ was found to be 0.0043 and the $\Delta I/\Delta C$ was found to be 24.8 cps for 0.1%. The extrapolated background was 30.74 cps and for 1σ is 0.55 cps. The minimum detection limit is then $3\sigma \times$ the slope factor or $1.65 \times 0.0043 = 0.0066\%$.[10]

SUMMARY

A number of possible analyzing crystals for the determination of silicon and aluminum by the X-ray fluorescence method have been investigated. Among these crystals, PET, to date, appears to give the highest count rate for silicon and aluminum with a very low background count. Single-crystal rotation photographs have been taken with the author's X-ray camera attachments for the PET crystals on two different axes, and the lattice constants were used to calculate the interplanar spacings and for indexing purposes. X-ray powder patterns have likewise been taken for this crystal and the spectral lines indexed. A number of sets of standard samples have been obtained from several sources. The amount of silicon or aluminum varies from set to set. Various aspects of X-ray fluorescence analysis of these elements have been studied and presented, using the analyzing crystals and the different sets of standards.

ACKNOWLEDGMENTS

The author would like to thank Dr. H. W. Pickett, of the General Electric X-Ray Department, Milwaukee, Wisconsin, for his comments on some of the results presented and also Elsie Kling of the same company for taking part of the X-ray data.

REFERENCES

1. Frank L. Chan, "X-Ray Camera Attachment," U.S. Patent No. 3,079,500.
2. Frank L. Chan, "Spherical Goniometer Head and Adapter," U.S. Patent No. 3,160,748.
3. Frank L. Chan, "Some Modifications of the X-ray Instruments and Their Utilization to the Study of Analytical Problems," *Norelco Reptr.* **10**: 133, 1963.
4. Frank L. Chan, "Determination of Zinc Sulfide and Cadmium Sulfide in Solid Solutions of Small Single Crystals Used for Semiconductors by X-ray and Chemical Methods," in: W. M. Mueller, G. R. Mallett, and M. J. Fay (eds.), *Advances in X-Ray Analysis, Vol. 5*, Plenum Press, New York, 1962, p. 142.
5. Frank L. Chan, "A Study of Silicon Determination in Organic-silicon Compounds by X-ray Fluorescence with Vacuum Spectrograph," paper presented at SAC Conference held at Nottingham, England, July 1965.
6. Joseph V. Smith (ed.), Index (Organic) to the X-ray Powder Data File, 1964, *ASTM Special Technical Publication 48-N1*, American Society for Testing and Materials, Philadelphia, Pa. [see also X-ray Powder Data File Set 1–5 (Revised) ASTM, 1960].
7. Harold P. Klug and Leroy E. Alexander, *X-ray Diffraction Procedures*, John Wiley & Sons, Inc., New York, 1954.
8. R. Shiono, D. W. J. Cruickshank, and E. G. Cox, "A Refinement of the Crystal Structure of Pentaerythritol," *Acta Cryst.* **11**: 389, 1958.
9. Ian Hvoslef, "A Neutron Diffraction Study of Pentaerythritol," *Acta Cryst.* **11**: 383, 1958.
10. L. S. Birks, *X-Ray Spectrochemical Analysis*, Interscience Publishers, Inc., New York, 1959.

X-RAY SPECTROGRAPHIC ANALYSIS OF AUTOMOTIVE COMBUSTION DEPOSITS WITHOUT THE USE OF CALIBRATION CURVES

J. C. Wagner and F. R. Bryan

Ford Motor Company
Dearborn, Michigan

ABSTRACT

An X-ray spectrographic technique requiring no calibration curves is used to determine chlorine, bromine, and lead in automotive combustion deposits. Other elements, such as zinc and barium, can probably be determined by the same X-ray technique. The inert dilution method, described by Sherman[1] in 1957, is employed. Improvements in sample preparation have been made to make the procedure more reliable. Results are calculated by the use of a formula, which is shown to be valid even though sample and standard have composition differences which cause interelement-effect errors if calibration-curve techniques are used. Adequate agreement is obtained between the inert dilution X-ray method and chemical procedures.

INTRODUCTION

X-ray spectrographers are usually careful to see that the matrix of a sample is very close in composition to that of the standards used to calibrate the method. Otherwise, large errors may occur as a result of differences in absorption or enhancement. In 1957, Sherman[1] proposed a method of avoiding this problem by employing mathematical equations to take into account absorption and enhancement effects. His work was experimentally incomplete because suitable techniques for the handling of powder samples apparently had not been developed. Sherman's equations have remained as a theoretical explanation of intensities in X-ray spectrography. Very little work has been reported in which an analyst has used these equations to make a quantitative analysis. Subsequent to the publication of Sherman's work, improved methods for dealing with powder samples have been found. For example, the Bleuler rotary mill can crush a sample in 3 min grinding time to such a small size that particle size effects are virtually eliminated.

Recently, we were asked to analyze several automotive combustion deposits. The construction of calibration curves was not practical because the matrices of the samples were expected to vary widely, thereby necessitating several calibration curves for each element. This situation presented a good opportunity to test the equations developed by Sherman.

SHERMAN'S INERT DILUTION METHOD

This procedure requires a standard sample which has been analyzed for the elements of interest and which is somewhat similar in composition to the unknown. Both the

[1] References are at the end of the paper.

standard and the unknown are intimately mixed in known proportions with an arbitrarily chosen diluent not containing the elements for which analyses are to be made. Counting rates are measured, and a mathematical equation is used to calculate results. If the sample and standard are both diluted in the same ratio, the following equation is used:

$$C_2 = C_0 \; \frac{T_1 - T_0}{T_3 - T_2}$$

where C_2 is the concentration of the element in the unknown (S_2), C_0 is the concentration of the element in the standard sample (S_0), T_1 is the measured reciprocal intensity (expressed in seconds for a given number of counts) for the mixture of S_0 and diluent, T_0 is the measured reciprocal intensity for S_0, T_3 is the measured reciprocal intensity for the mixture of S_2 and diluent, and T_2 is the measured reciprocal intensity of S_2.

APPLICATION OF INERT DILUTION METHOD
TO THE ANALYSIS OF COMBUSTION DEPOSITS

The combustion deposits used in these experiments were collected from vehicles subjected to severe operating conditions resulting in abnormally large deposits, thus providing sufficient samples for both chemical and X-ray analysis. Table I shows the chemical analyses of typical materials used in this investigation. The following examples of chlorine, bromine and lead determinations by the inert dilution method represent samples that would cause complications if calibration curve techniques were used.

Chlorine Determination

The mass absorption coefficient for Cl K_{α_1} by bromine is very high (1600). If calibration curves were used to determine chlorine content, separate curves would be necessary for samples having appreciable differences in bromine. As shown below, the inert dilution method yields fairly accurate results for chlorine when sample and standard differ markedly in bromine content (0.8% and 4.5% respectively).

% Bromine		% Chlorine in Sample 409	
Sample 409	Standard 410	By X-ray	By wet chemistry
0.8	4.5	12.8	11.4

Bromine Determination

The mass absorption coefficient of Br K_{α_1} by lead is 345. In spite of this large value, accurate bromine analyses can be made by the inert dilution method when sample and standard differ markedly in lead content (60.8% and 31.1%, respectively). This is shown below.

% Lead—chemical		% Bromine in Sample 412	
Sample 412	Standard 413	By X-ray	By wet chemistry
60.8	31.1	0.6	0.5

Table I. Analysis of Typical Automotive Combustion Deposits Under Study in These Experiments

Sample number	Element percentage present								
	Pb	Cl	Br	C	P	Ba	Zn	S	Al
404	59.6	1.9	6.4	8.0	2.6	0.8	0.4	2.6	0.2
405	60.7	1.6	6.2	7.1	2.6	0.4	0.5	2.5	0.1
406	59.6	1.2	5.8	7.0	2.8	0.2	0.6	2.9	0.1
409	63.4	11.4	0.8	7.8	1.6	1.0	0.8	1.3	0.04
410	61.7	5.2	4.5	5.0	2.7	1.1	0.8	2.1	0.05
412	60.8	11.7	0.5	9.2	1.3	0.1	0.8	1.2	0.04
413	31.1	11.4	0.6	6.9	1.5	0.02	0.5	1.0	0.01
414	60.0	1.0	4.2	5.4	3.3	0.02	0.5	1.8	0.01
415	64.4	1.0	5.0	5.7	3.2	0.02	0.2	2.3	0.01
420	23.7	2.5	3.2	39.2	1.8	1.0	0.2	1.6	0.02

Lead Determination

The Pb K_{α_1} counting times for sample 420 and standard 404 are not far apart (12.1 sec and 15.3 sec, respectively), as the data below indicate. In spite of the proximity of these counting times, the lead content of standard 404 is more than twice as much as that of sample 420. The lead analysis of sample 420 by the inert dilution method is close to the value obtained by wet chemical methods (25.7% and 23.7%, respectively).

% Lead—chemical		Pb K_{α_1} counting times for 3×10^5 counts(sec)		% Lead—X-ray
Sample 420	Standard 404	Sample 420	Standard 404	Sample 420
23.7	59.6	12.1	15.3	25.7

ANALYSES OF FIVE TYPICAL DEPOSITS BY THE INERT DILUTION METHOD

X-Ray Equipment and Technique

X-ray instrument parameters are shown in Table II. The X-ray unit is a Philips vacuum spectrograph equipped with a chromium target thin-window tube.

Table II. X-Ray Instrument Parameters

Element	Analyzing crystal	wavelength	Order	kV	mA	Pulse-height analyzer	Pressure	Detector
Pb	LiF	L_{α_1}	1	35	25	No	Atmospheric	Scintillation counter
Br	LiF	K_{α_1}	2	35	25	Yes[a]	Atmospheric	Flo-counter P-10 gas
Cl	EDT	K_{α_1}	1	35	25	Yes[b]	35–50 μ	Flo-counter P-10 gas

[a] To eliminate interference from Ba L_{γ_4}, first order.
[b] To eliminate interference from Pb L_{α_1}, fourth order.

Sample Preparation

Samples are in the form of pellets which have been pressed from powders under a load of 12,000 lb. Pellet diameter is either $\frac{1}{2}$ in. or 1 in. All powders (except the diluent) are crushed for 3 min in a Bleuler rotary mill to avoid particle size effects. Mixing of samples with diluent is accomplished by shaking for 15 min by means of a small paint mixer. The vial holding the sample contains three polyethylene balls to aid in the mixing. The diluent used in these experiments is −200-mesh nickel powder. Dilution ratio is 5 parts by weight of nickel to 1 part of sample.

Time Required

Typical counting times are shown in Table III. Time consumption, although somewhat greater than when calibration curve techniques are used, is several hundred percent less than required by wet chemical methods. Grinding time in the Bleuler rotary mill is 3 min, and pelletizing requires 2 min per sample. Mixing of the diluent with the sample by means of a paint mixer requires about 15 min. Mixing of several samples can be carried out simultaneously.

Table III. Typical Counting Times

Element sought	Percent of element in sample	Sample	Total counting time (sec)	Sample area exposed to X-ray beam (in. in diameter)
Pb	60.7	405	470	$\frac{1}{2}$
Br	0.6	412	800	$\frac{1}{2}$
Cl	11.7	412	475	$\frac{3}{4}$

Comparison with Wet Chemical Results

Lead, bromine, and chlorine analyses of five typical samples determined by wet chemical techniques and the inert dilution method are shown in Table IV. Maximum deviation between results obtained by the two methods is within 10% of the amount present except for three determinations in the 1% range.

Table IV. Comparison of Analyses of Automotive Combustion Deposits by X-Ray and Wet-Chemical Methods

Sample Number	%Lead		% Bromine		% Chlorine	
	By X-ray	By wet chemistry	By X-ray	By wet chemistry	By X-ray	By wet chemistry
405	60.0	60.7	6.2	6.2	1.5	1.6
406	59.8	59.6	5.9	5.8	1.7	1.2
412	57.5	60.8	0.6	0.5	10.7	11.7
414	56.3	60.0	4.3	4.2	1.5	1.0
415	60.3	64.4	5.2	5.0	1.2	1.0

SUMMARY AND CONCLUSIONS

Chlorine, bromine, and lead in automotive combustion chamber deposits can be determined by an X-ray spectrographic technique which does not require calibration curves. Data are presented which show that usable results are obtained, even though samples and standards have differences in composition that would cause interelement-effect errors if calibration curve techniques were used.

REFERENCE

1. Sherman, Jacob, "A Theoretical Derivation of the Composition of Mixable Specimens from Fluorescent X-Ray Intensities," in: W. M. Mueller, G. R. Mallett, and M. J. Fay (eds.), *Advances in X-Ray Analysis, Vol. 1*, Plenum Press, New York, 1957, p. 231.

DISCUSSION

B. S. Sanderson (National Lead Company): What did you use to dilute with?

J. C. Wagner: For these samples we used nickel powder, −200-mesh nickel powder, but other diluents can be used. We have tried borax, which works, and molybdenum, which also works.

B. S. Sanderson: What was the range of dilution?

J. C. Wagner: We used five parts of diluent by weight to one part of sample.

AUTHOR INDEX

Bold numbers refer to papers in this volume

SUBJECT INDEX